Shop Manual for

Medium/Heavy Duty Truck Electricity and Electronics

Shop Manual for

Medium/Heavy Duty Truck Electricity and Electronics

Sulev Oun

Jack Erjavec
Series Advisor
Professor Emeritus, Columbus State Community College
Columbus, Ohio

Australia Canada Mexico Singapore Spain United Kingdom United States

NOTICE TO THE READER

Delmar Staff
Director: Alar Elken
Executive Editor: Sandy Clark
Developmental Editor: Allyson Powell
Editorial Assistant: Matthew Seeley
Executive Marketing Manager: Maura Theriault
Channel Manager: Mona Caron

Executive Production Manager: Mary Ellen Black
Senior Production Coordinator: Karen Smith
Project Editor: Christopher Chien
Art/Design Coordinator: Cheri Plasse

Printed in the United States of America
4 5 6 7 8 9 XXX 09 08 07 06 05

For more information, contact Delmar at
5 Maxwell Drive, Clifton Park, NY 12065,
or find us on the
World Wide Web at *http://www.delmar.com*

Library of Congress
Cataloging-in-Publication Data

Oun, Sulev.
Shop manual for medium/heavy duty truck electricity and electronics / Sulev Oun.
p. cm. — (Today's technician)
Includes index.
ISBN 0-8273-7006-7
1. Trucks—Electric equipment. 2. Trucks—Electronic equipment. I. Oun, Sulev. II. Series.

TL272. S46 2000
629.25'4'0288—dc21 00-064411

Asia (including India):
Thomson Learning
60 Albert Street, #15-01
Albert Complex
Singapore 189969
Tel 65 336-6411
Fax 65 336 7411

Australia/New Zealand:
Nelson
102 Dodds Street
South Melbourne, Victoria 3205
Australia
Tel 61 (0)3 9685-4111
Fax 61 (0)3 9685-4199

Latin America:
Thomson Learning
Seneca 53
Colonia Polanco
11560 Mexico D. F. Mexico
Tel (525) 281-2906
Fax (525) 281-2656

Canada:
Nelson
1120 Birchmount Road
Toronto, Ontario
Canada M1K 5G4
Tel (416) 752-9100
Fax (416) 752-8102

UK/Europe/Middle East:
Thomson Learning
Berkshire House
168-173 High Holborn
London WC1V 7AA
United Kingdom
Tel 44 (0)171 497-1422
Fax 44 (0)171 497-1426

Business Press
Berkshire House
168-173 High Holborn
London WC1V 7AA
United Kingdom
Tel 44 (0)171 497-1422
Fax 44 (0)171 497-1426

Spain:
Paraninfo
Calle Magallanes 25
28015 Madrid
España
Tel 34 (0)91 446-3350
Fax 34 (0)91 445-6218

Distribution Services:
ITPS
Cheriton House
North Way
Andover,
Hampshire SP10 5BE
United Kingdom
Tel 44 (0)1264 34-2960
Fax 44 (0)1264 34-2759

International Headquarters:
Thomson Learning
International Division
290 Harbor Drive, 2nd Floor
Stamford, CT 06902-7477
USA
Tel (203) 969-8700
Fax (203) 969-8751

CONTENTS

Photo Sequences

Job Sheets

PREFACE

Unlike yesterday's mechanic, the technician of today and for the future must know the underlying theory of all systems and be able to service and maintain those systems. Today's technician must also know how these individual systems interact with each other. Standards and expectations have been set for today's technician, and these must be met in order to keep the world's medium and heavy duty trucks running efficiently and safely.

The *Today's Technician* series, by Delmar Thomson Learning, features textbooks that cover all mechanical and electrical systems of medium and heavy duty trucks. Principal titles correspond with the eight major areas of ASE (National Institute for Automotive Service Excellence) certification.

Each title is divided into two manuals: a Classroom Manual and a Shop Manual. Dividing the material into two manuals provides the reader with the information needed to begin a successful career as a medium and heavy duty truck technician without interrupting the learning process by mixing cognitive and performance-based learning objectives.

Each Classroom Manual contains the principles of operation for each system and subsystem. It also discusses the design variations used by different manufacturers. The Classroom Manual is organized to build upon basic facts and theories. The primary objective of this manual is to allow the reader to gain an understanding of how each system and subsystem operates. This understanding is necessary to diagnose the complex truck systems.

The understanding acquired by using the Classroom Manual is required for competence in the skill areas covered in the Shop Manual. All of the high priority skills, as identified by ASE, are explained in the Shop Manual. The Shop Manual also includes step-by-step instructions for diagnostic and repair procedures. Photo Sequences are used to illustrate many of the common service procedures. Other common procedures are listed and are accompanied with fine-line drawings and photographs that allow the reader to visualize and conceptualize the finest details of the procedure. The Shop Manual also contains the reasons for performing the procedures, as well as when that particular service is appropriate.

The two manuals are designed to be used together and are arranged in corresponding chapters. Not only are the chapters in the manuals linked together, the contents of the chapters are also linked. Both manuals contain clear and thoughtfully selected illustrations. Many of the illustrations are original drawings or photos prepared for inclusion in this series. This means that the art is a vital part of each manual.

The page layout is designed to include information that would otherwise break up the flow of information presented to the reader. The main body of the text includes all of the "need-to-know" information and illustrations. In the side margins are many of the special features of the series. Items such as definitions of new terms, common trade jargon, tools lists, and cross-references are placed in the margin, out of the normal flow of information so as not to interrupt the thought process of the reader.

Jack Erjavec, Series Advisor

Classroom Manual

To stress the importance of safe work habits, the Classroom Manual dedicates one full chapter to safety. Included in this chapter are common safety practices, safety equipment, and safe handling of hazardous materials and wastes. This includes information on MSDS sheets and OSHA regulations. Other features of this manual include:

Cognitive Objectives

These objectives define the contents of the chapter and define what the student should have learned upon completion of the chapter.

Each topic is divided into small units to promote easier understanding and learning.

References to the Shop Manual

Reference to the appropriate topic in the Shop Manual is given whenever necessary. Although the chapters of the two manuals are synchronized, material covered in other chapters of the Shop Manual may be fundamental to the topic discussed in the Classroom Manual.

Marginal Notes

New terms are pulled out and defined. Common trade jargon also appears in the margin and gives some of the common terms used for components. This allows the reader to speak and understand the language of the trade, especially when conversing with an experienced technician.

Cautions and Warnings

Throughout the text, cautions are given to alert the reader to potentially hazardous materials or unsafe conditions. Warnings are also given to advise the student of things that can go wrong if instructions are not followed or if a nonacceptable part or tool is used.

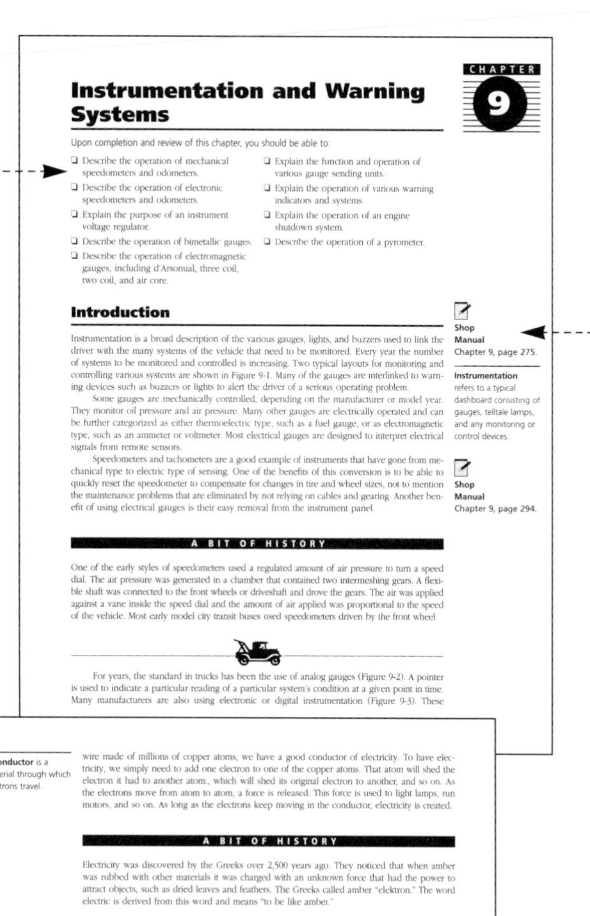

CHAPTER 9

Instrumentation and Warning Systems

Upon completion and review of this chapter, you should be able to:

- ❑ Describe the operation of mechanical speedometers and odometers.
- ❑ Describe the operation of electronic speedometers and odometers.
- ❑ Explain the purpose of an instrument voltage regulator.
- ❑ Describe the operation of bimetallic gauges.
- ❑ Describe the operation of electromagnetic gauges, including d'Arsonual, three coil, two coil, and air core.
- ❑ Explain the function and operation of various gauge sending units.
- ❑ Explain the operation of various warning indicators and systems.
- ❑ Explain the operation of an engine shutdown system.
- ❑ Describe the operation of a pyrometer.

Introduction

Shop Manual Chapter 9, page 275.

Instrumentation is a broad description of the various gauges, lights, and buzzers used to link the driver with the many systems of the vehicle that need to be monitored. Every year the number of systems to be monitored and controlled is increasing. Two typical layouts for monitoring and controlling various systems are shown in Figure 9-1. Many of the gauges are interlinked to warning devices such as buzzers or lights to alert the driver of a serious operating problem.

Instrumentation refers to a typical dashboard consisting of gauges, telltale lamps, and any monitoring or control devices.

Some gauges are mechanically controlled, depending on the manufacturer or model year. They monitor oil pressure and air pressure. Many other gauges are electrically operated and can be further categorized as either thermoelectric type, such as a fuel gauge, or as electromagnetic type, such as an ammeter or voltmeter. Most electrical gauges are designed to interpret electrical signals from remote sensors.

Shop Manual Chapter 9, page 294.

Speedometers and tachometers are a good example of instruments that have gone from mechanical type to electric type of sensing. One of the benefits of this conversion is to be able to quickly reset the speedometer to compensate for changes in tire and wheel sizes, not to mention the maintenance problems that are eliminated by not relying on cables and gearing. Another benefit of using electrical gauges is their easy removal from the instrument panel.

A BIT OF HISTORY

One of the early styles of speedometers used a regulated amount of air pressure to turn a speed dial. The air pressure was generated in a chamber that contained two intermeshing gears. A flexible shaft was connected to the front wheels or driveshaft and drove the gears. The air was applied against a vane inside the speed dial and the amount of air applied was proportional to the speed of the vehicle. Most early model city transit buses used speedometers driven by the front wheel.

For years, the standard in trucks has been the use of analog gauges (Figure 9-2). A pointer is used to indicate a particular reading of a particular system's condition at a given point in time. Many manufacturers are also using electronic or digital instrumentation (Figure 9-3). These

A **conductor** is a material through which electrons travel.

wire made of millions of copper atoms, we have a good conductor of electricity. To have electricity, we simply need to add one electron to one of the copper atoms. That atom will shed the electron it had to another atom., which will shed its original electron to another, and so on. As the electrons move from atom to atom, a force is released. This force is used to light lamps, run motors, and so on. As long as the electrons keep moving in the conductor, electricity is created.

A BIT OF HISTORY

Electricity was discovered by the Greeks over 2,500 years ago. They noticed that when amber was rubbed with other materials it was charged with an unknown force that had the power to attract objects, such as dried leaves and feathers. The Greeks called amber "elektron." The word electric is derived from this word and means "to be like amber."

An **insulator** is a material through which electrons travel very poorly or not at all.

Insulators are materials that don't allow electrons to flow through them easily. Insulators are atoms that have seven or eight electrons in their valence ring. The electrons are held tightly around the atom's nucleus and they can't be moved easily. Insulators are used to prevent electron flow or to contain it within a conductor. Insulating material covers the outside of most conductors to keep the moving electrons within the conductor.

Semiconductors are neither conductors nor insulators.

In summary, the number of electrons in the valence ring determines whether an atom is a good conductor or insulator. Some atoms are not good insulators or conductors; they are called semiconductors. The following describe a conductor, insulator and semiconductor:

1. Conductor: An atom with three or less electrons
2. Insulator: An atom with five or more electrons
3. Semiconductor: An atom with four electrons

The electrons in the atoms of a conductor can be freed from their outer orbits by forces such as heat, friction, light, pressure, chemical reaction, and magnetism. When electrons are moved out of their orbit, they form an electrical current under proper conditions. Insulators are required to control the routing of the flow of electricity.

WARNING: Any broken, frayed, or damaged insulation material requires replacement or repair to the conductor. Exposed conductors can result in a safety hazard and circuit component damage.

CAUTION: The human body is a conductor of electricity. When performing service on electrical systems, be aware that electrical shock is possible. Although the shock is usually harmless, the reaction to the shock can cause injury. Observe all safety rules associated with electricity. Never wear jewelry when servicing or testing the electrical system.

The term **circuit** means a circle consisting of a path for electron flow from the voltage source to load components, and a return path back to the voltage source.

Electricity Defined

In order to use electricity to provide the power necessary to start a truck and operate various electrical components, electricity must be put into motion. Simply stated, electricity is the movement of electrons through a conductor. This movement of electricity in a circuit can easily be compared to compressed air, where the air under pressure is pushed through an air line (conductor) to an actuator or component such as a brake chamber and performs a function.

16

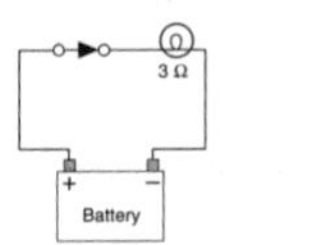

Figure 2-9 Simplified light circuit with 3 ohms of resistance in the lamp.

To show how easily this works, consider the 12-volt circuit in Figure 2-9. This circuit contains a 3-ohm lightbulb. We want to find the current in the circuit. By covering the I in the circle we see the formula we need, I = E/R. Then we plug in the numbers: I = 12/3. Therefore, our circuit current is 4 amps.

A BIT OF HISTORY

Georg S. Ohm was a German scientist in the 1800's who discovered that all electrical quantities are proportional to each other and therefore have a mathematical relationship.

Ohm's Law is important to your understanding of electricity beyond your ability to calculate values. Based on Ohm's Law we can see that current will increase if we decrease the resistance and not change the voltage. Likewise, if we increase the resistance, current will decrease. Ohm's Law explains what happens when we change things in a circuit and when something goes wrong in a circuit. For example, look at Figure 2-10: On the left side is a 12-volt circuit with a 3-ohm lightbulb. This circuit has 4 amps of current flowing through it. If we add a 1-ohm resistor to the same circuit (as shown on the right), we now have a total resistance of 4 ohms. Because of the increased resistance, the current dropped to 3 amps. The lightbulb will be powered by less current and will be less bright than it was before we added the additional resistance.

Another point to consider is voltage drop. Before we added the 1-ohm resistor, the source voltage (12 volts) was dropped by the lightbulb. With the additional resistance, the voltage drop of the light bulb decreased to 9 volts. The remaining 3 volts were dropped by the 1-ohm resistor. We know this from using Ohm's law. When the circuit current was 4 amps, the lightbulb had 3 ohms of resistance. To find the voltage drop we multiply the current by the resistance:

$$E = I \times R \text{ or } E = 4 \times 3 \text{ or } E = 12$$

When we added the resistor to the circuit, the lightbulb still had 3 ohms of re
the current in the circuit decreased to 3 amps. Again we can determine the voltage
tiplying the current by the resistance:

$$E = I \times R \text{ or } E = 3 \times 3 \text{ or } E = 9$$

The voltage drop of the additional resistor is calculated in the same way: E =
3 volts. The total voltage drop of the circuit is the same for both circuits; howeve
drop at the lightbulb changed. This also would cause the light bulb to be dimmer.

22

A Bit of History

This feature gives the student a sense of the evolution of truck systems. This feature not only contains nice-to-know information, but also should spark some interest in the subject matter.

Magnetic sensors are used to measure speeds, such as engine, vehicle, and shaft speeds. These sensors typically use a permanent magnet. Rotational speed is determined by the passing of blades or teeth in and out of the magnetic field. As a tooth moves in and out of the magnetic field, the strength of the magnetic field is changed and a voltage signal is induced. This signal is sent to a control device, where it is interpreted.

EMI Suppression

As manufacturers began to increase the number of electronic components and systems in their vehicles, the problem of EMI had to be controlled. The low-power integrated circuits used on modern vehicles are sensitive to the signals produced as a result of EMI. EMI is produced as current in a conductor is turned on and off. EMI is also caused by static electricity that is created by friction, resulting from tires and their contact with the road, or from fan belts contacting the pulleys.

EMI can disrupt the vehicle's computer systems by inducing false messages to it. The computer requires messages to be sent over circuits in order to communicate with other computers, sensors, and actuators. If any of these signals are disrupted, the engine and/or accessories may turn off.

Electromagnetic Interference (EMI) is an undesirable creation of electromagnetism whenever current is switched on and off.

A choke is an inductor in series with a circuit.

EMI can be suppressed by any one of the following methods:

1. Adding a resistance to the conductors. This is usually done to high-voltage systems such as the secondary circuit of the ignition system.
2. Connecting a capacitor in parallel and a choke coil in series with the circuit.
3. Shielding the conductor or load components with a metal or metal-impregnated plastic.
4. Increasing the number of paths to ground by using designated ground circuits. This provides a clear path to ground that is very low in resistance.
5. Adding a clamping diode in parallel to the component.
6. Adding an isolation diode in series to the component.

Summaries

Each chapter concludes with summary statements that contain the important topics of the chapter. These are designed to help the reader review the contents.

Summary

- ❑ An atom is constructed of a complex arrangement of electrons in orbit around a nucleus. If the number of electrons and protons are equal, the atom is balanced, or neutral.
- ❑ A conductor allows electricity to easily flow through it.
- ❑ An insulator does not allow electricity to easily flow through it.
- ❑ Electricity is the movement of electrons from atom to atom. In order for the electrons to move in the same direction, an electromotive force (EMF) must be applied to the circuit.
- ❑ The electron theory defines electron flow as motion from negative to positive.
- ❑ The conventional theory of current flow states that current flows from a positive point to a less positive point.
- ❑ Voltage is defined as an electrical pressure and is the difference between the positive and negative charges.
- ❑ Current is defined as the rate of electron flow and is measured in amperes. Am
amount of electrons passing any given point in the circuit in one second.
- ❑ Resistance is defined as opposition to current flow and is measured in ohms (Ω
- ❑ Ohm's Law defines the relationship between current, voltage, and resistance. It is
law of electricity and states that the amount of current in an electric circuit is inve
tional to the resistance of the circuit, and is directly proportional to the voltage in

Terms to Know

Capacitance
Circuit
Conductor
Electromagnetism
Electromotive force (EMF)
Equivalent series load
Induction
Insulator
Ohm's Law
Parallel circuit
Semiconductors
Series circuit

Terms to Know

A list of new terms appears next to the Summary. Definitions for these terms can be found in the Glossary at the end of the manual.

- ❑ Wattage represents the measure of power (P) used in a circuit. Wattage is measured by using the power formula, which defines the relationship between amperage, voltage, and wattage.
- ❑ Capacitance is the ability of two conducting surfaces to store voltage.
- ❑ Direct current (DC) results from a constant voltage and a current that flows in one direction.
- ❑ In an alternating current (AC) circuit, voltage and current do not remain constant. AC current changes direction from positive to negative and negative to positive.
- ❑ For current to flow, the electrons must have a complete path from the source voltage to the load component and back to the source.
- ❑ The series circuit provides a single path for current flow from the electrical source through all the circuit's components, and back to the source.
- ❑ A parallel circuit provides two or more paths for current to flow.
- ❑ A series-parallel circuit is a combination of the series and parallel circuits.
- ❑ The equivalent series load is the total resistance of a parallel circuit plus the resistance of the load in series with the voltage source.
- ❑ Voltage drop is caused by a resistance in the circuit that reduces the electrical pressure available after the resistance.
- ❑ Kirchhoff's voltage law states that the total voltage drop in an electrical circuit will always equal the available voltage at the source.

Review Questions

Short answer essay, fill-in-the-blank, and multiple-choice type questions follow each chapter. These questions are designed to accurately assess the student's competence in the stated objectives at the beginning of the chapter.

Review Questions

Short Answer Essays

1. List and define the three elements of electricity.
2. Describe the use of Ohm's Law.
3. List and describe the three types of circuits.
4. Explain the principle of electromagnetism.
5. Describe the principle of induction.
6. Describe the basics of electron flow.
7. Define the two types of electrical current.
8. Describe the difference between insulators, conductors, and semiconductors.
9. Explain the basic concepts of capacitance.
10. What does the measurement of "Watt" represent?

Fill-in-the-Blanks

1. ______________ are negatively charged particles. The nucleus contains positively charged particles called ______________ and particles that have no charge called ______________.
2. A ______________ allows electricity to easily flow through it. An ______________ does not allow electricity to easily flow through it.
3. For electrons to move in the same direction, there must be a ______________ applied.

42

Shop Manual

To stress the importance of safe work habits, the Shop Manual also dedicates one full chapter to safety. Other important features of this manual include:

Performance Objectives

These objectives define the contents of the chapter and define what the student should have learned upon completion of the chapter.

Although this textbook is not designed to simply prepare someone for the certification exams, it is organized around the ASE task list. These tasks are defined generically when the procedure is commonly followed and specifically when the procedure is unique for specific vehicle models. Imported and domestic model trucks are included in the procedures.

CHAPTER 5

Battery Diagnosis and Service

Upon completion and review of this chapter, you should be able to:

- ❑ Demonstrate and follow all safety precautions and rules associated with servicing the battery.
- ❑ Perform a visual inspection of the battery, cables, hold downs, and boxes.
- ❑ Perform a proper maintenance of the complete battery system.
- ❑ Test a conventional battery's specific gravity and interpret the result.
- ❑ Perform an open circuit test and interpret the results.
- ❑ Perform a load or capacity test of any battery.
- ❑ Perform a 3-minute charge test to determine battery sulfation or a possible charging system problem.
- ❑ Properly charge a battery.
- ❑ Perform several battery drain tests and determine the cause of the problem.
- ❑ Perform a voltage drop test to determine the condition of battery connections.
- ❑ Properly jump-start a vehicle whether it is a 12-volt or 24-volt system.

Basic Tools

Basic mechanic's tool set

Appropriate service manual

Introduction

The battery is the heart of the electrical system of the truck. Most of the truck's electrical circuits obtain their power supply from the battery in one way or another. Therefore, a battery with a problem can affect any of the truck's electrical circuits. If a problem is suspected in any of the truck's electrical systems, the battery condition must be checked and corrected if needed, prior to any other electrical system diagnosis.

The battery's function becomes much more difficult as the environmental conditions change from normal to extreme cold or even extreme heat. It is very important that the batteries in trucks are maintained properly. That is why every preventive maintenance (PM) schedule includes some form of battery checking. It makes more sense to find a possible problem before a failure occurs. Of course, the worst-case scenario would be for a truck not to start when it is scheduled to get to or be at a specific destination. Keep in mind that a commercial vehicle plays a role in commerce and it has to be profitable.

Preventive maintenance is a scheduled function performed on vehicles to maintain performance and prevent on-road breakdowns. A good fleet places top priority on this scheduled maintenance.

Most shops have tools for testing that range from simple hydrometers, hand-held draw testers, and multimeters to more sophisticated test equipment such as the VAT-40, manufactured by the Sun Electrical Corporation (Figure 5-1). Some tests in this chapter will require the procedures that are used with a VAT-40. However, there are many other testers that incorporate the same procedures. The trend is towards computer-based testers that conduct tests automatically after a test is selected (Figure 5-2). Test equipment manufacturer's procedures should always be followed very closely.

CUSTOMER CARE: Many computer-based testers determine the health of a battery. A test might indicate that a battery should be replaced, yet the engine starts with no problem. The customer might doubt your diagnosis. However, explain that environmental factors such as temperature can affect the battery's starting ability. Although the battery starts fine during the heat of the day, when the temperature drops overnight, the vehicle might not start in the morning. You will have to let the customer know that this is part of preventive maintenance.

General Precautions

The Consumer Product Safety Commission's National Eye Injury Surveillance System (NEISS) discovered some startling statistics. In a one-year period, they found that there were more than

105

Tools Lists

Each chapter begins with a list of the Basic Tools needed to perform the tasks included in the chapter. Whenever a Special Tool is required to complete a task, it is listed in the margin next to the procedure.

Marginal Notes

Page numbers for cross-referencing appear in the margin. Some of the common terms used for components, and other bits of information, also appear in the margin. This provides an understanding of the language of the trade and helps when conversing with an experienced technician.

Photo Sequences

Many procedures are illustrated in detailed Photo Sequences. These detailed photographs show the students what to expect when they perform particular procedures. They also can provide a student a familiarity with a system or type of equipment, which the school may not have.

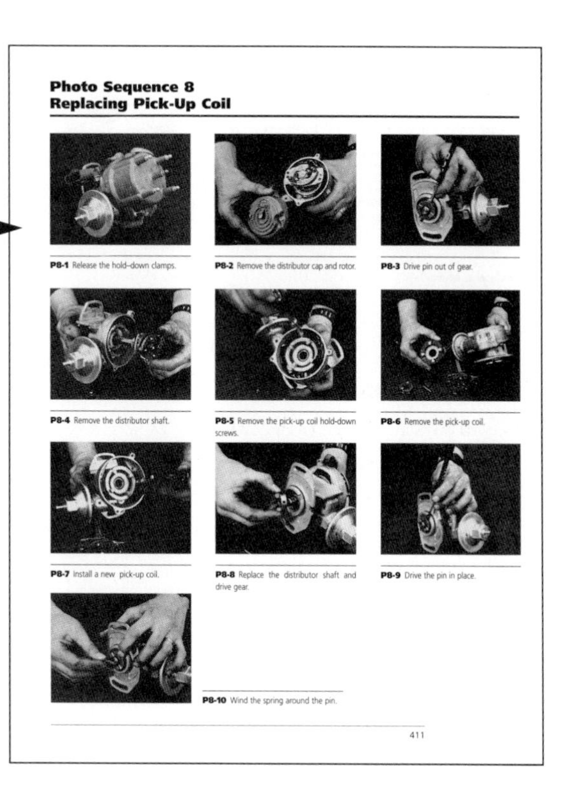

Photo Sequence 8
Replacing Pick-Up Coil

P8-1 Release the hold-down clamps.

P8-2 Remove the distributor cap and rotor.

P8-3 Drive pin out of gear.

P8-4 Remove the distributor shaft.

P8-5 Remove the pick-up coil hold-down screws.

P8-6 Remove the pick-up coil.

P8-7 Install a new pick-up coil.

P8-8 Replace the distributor shaft and drive gear.

P8-9 Drive the pin in place.

P8-10 Wind the spring around the pin.

411

References to the Classroom Manual

Reference to the appropriate topic in the Classroom Manual is given whenever necessary. Although the chapters of the two manuals are synchronized, material covered in other chapters of the Classroom Manual may be fundamental to the topic discussed in the Shop Manual.

Customer Care

This feature highlights those little things a technician can do or say to enhance customer relations.

Service Tips

Whenever a short-cut or special procedure is appropriate, it is described in the text. These tips are generally those things commonly done by experienced technicians.

Cautions and Warnings

Throughout the text, cautions are given to alert the reader to potentially hazardous materials or unsafe conditions. Warnings are also given to advise the student of things that can go wrong if instructions are not followed or if a nonacceptable part or tool is used.

Job Sheets

Located at the end of each chapter, the Job Sheets provide a format for students to perform procedures covered in the chapter. A reference to the ASE and/or NATEF tasks addressed by the procedure is referenced on the Job Sheet.

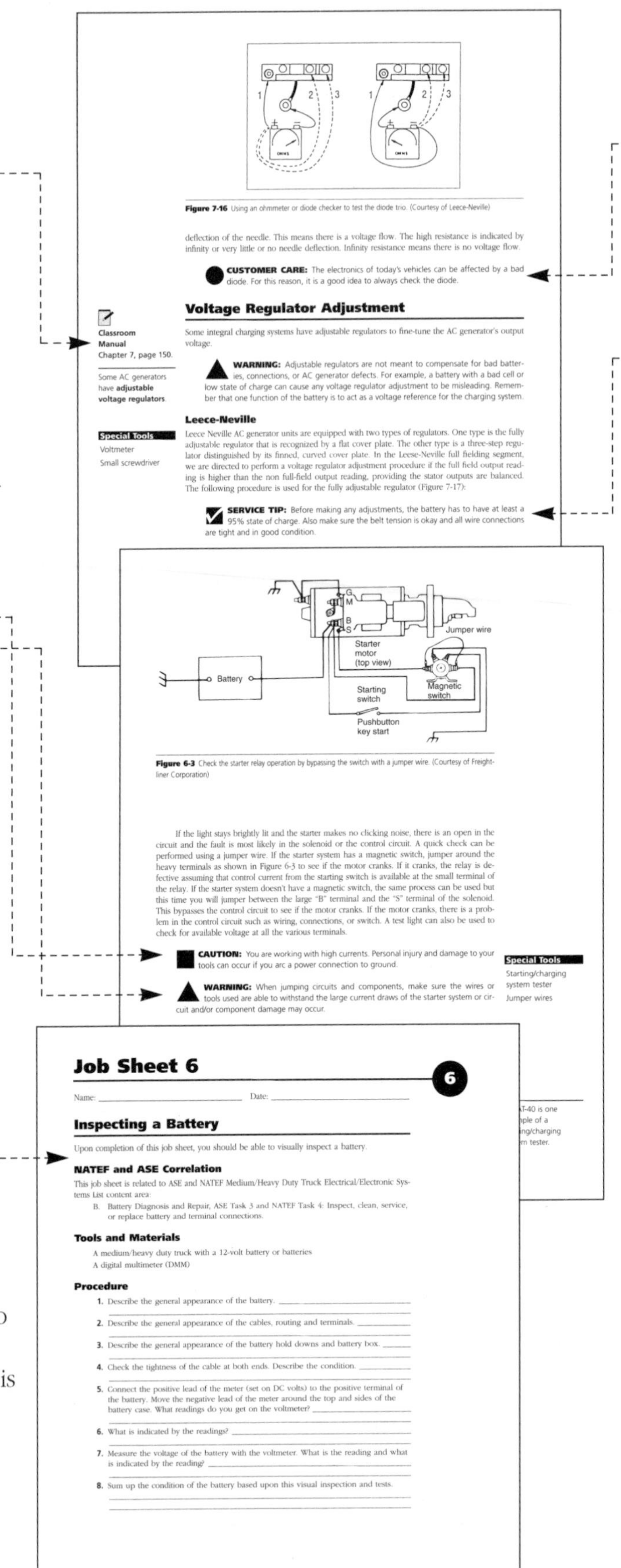

Figure 7-16 Using an ohmmeter or diode checker to test the diode trio. (Courtesy of Leece-Neville)

deflection of the needle. This means there is a voltage flow. The high resistance is indicated by infinity or very little or no needle deflection. Infinity resistance means there is no voltage flow.

CUSTOMER CARE: The electronics of today's vehicles can be affected by a bad diode. For this reason, it is a good idea to always check the diode.

Voltage Regulator Adjustment

Classroom Manual Chapter 7, page 150.

Some AC generators have **adjustable voltage regulators**.

Some integral charging systems have adjustable regulators to fine-tune the AC generator's output voltage.

WARNING: Adjustable regulators are not meant to compensate for bad batteries, connections, or AC generator defects. For example, a battery with a bad cell or low state of charge can cause any voltage regulator adjustment to be misleading. Remember that one function of the battery is to act as a voltage reference for the charging system.

Leece-Neville

Special Tools
Voltmeter
Small screwdriver

Leece Neville AC generator units are equipped with two types of regulators. One type is the fully adjustable regulator that is recognized by a flat cover plate. The other type is a three-step regulator distinguished by its finned, curved cover plate. In the Leese-Neville full fielding segment, we are directed to perform a voltage regulator adjustment procedure if the full field output reading is higher than the non full-field output reading, providing the stator outputs are balanced. The following procedure is used for the fully adjustable regulator (Figure 7-17):

SERVICE TIP: Before making any adjustments, the battery has to have at least a 95% state of charge. Also make sure the belt tension is okay and all wire connections are tight and in good condition.

Figure 6-3 Check the starter relay operation by bypassing the switch with a jumper wire. (Courtesy of Freightliner Corporation)

If the light stays brightly lit and the starter makes no clicking noise, there is an open in the circuit and the fault is most likely in the solenoid or the control circuit. A quick check can be performed using a jumper wire. If the starter system has a magnetic switch, jumper around the heavy terminals as shown in Figure 6-3 to see if the motor cranks. If it cranks, the relay is defective assuming that control current from the starting switch is available at the small terminal of the relay. If the starter system doesn't have a magnetic switch, the same process can be used but this time you will jumper between the large "B" terminal and the "S" terminal of the solenoid. This bypasses the control circuit to see if the motor cranks. If the motor cranks, there is a problem in the control circuit such as wiring, connections, or switch. A test light can also be used to check for available voltage at all the various terminals.

CAUTION: You are working with high currents. Personal injury and damage to your tools can occur if you arc a power connection to ground.

WARNING: When jumping circuits and components, make sure the wires or tools used are able to withstand the large current draws of the starter system or circuit and/or component damage may occur.

Special Tools
Starting/charging system tester
Jumper wires

Job Sheet 6

Name: ______________ Date: ______________

Inspecting a Battery

Upon completion of this job sheet, you should be able to visually inspect a battery.

NATEF and ASE Correlation

This job sheet is related to ASE and NATEF Medium/Heavy Duty Truck Electrical/Electronic Systems List content area:

B. Battery Diagnosis and Repair, ASE Task 3 and NATEF Task 4: Inspect, clean, service, or replace battery and terminal connections.

Tools and Materials

A medium/heavy duty truck with a 12-volt battery or batteries
A digital multimeter (DMM)

Procedure

1. Describe the general appearance of the battery. ______________
2. Describe the general appearance of the cables, routing and terminals. ______________
3. Describe the general appearance of the battery hold downs and battery box. ______________
4. Check the tightness of the cable at both ends. Describe the condition. ______________
5. Connect the positive lead of the meter (set on DC volts) to the positive terminal of the battery. Move the negative lead of the meter around the top and sides of the battery case. What readings do you get on the voltmeter? ______________
6. What is indicated by the readings? ______________
7. Measure the voltage of the battery with the voltmeter. What is the reading and what is indicated by the reading? ______________
8. Sum up the condition of the battery based upon this visual inspection and tests. ______________

Instructor Check ______________

135

Case Studies

Case Studies concentrate on the ability to properly diagnose the systems. Each chapter ends with a case study in which a vehicle has a problem, and the logic used by a technician to solve the problem is explained.

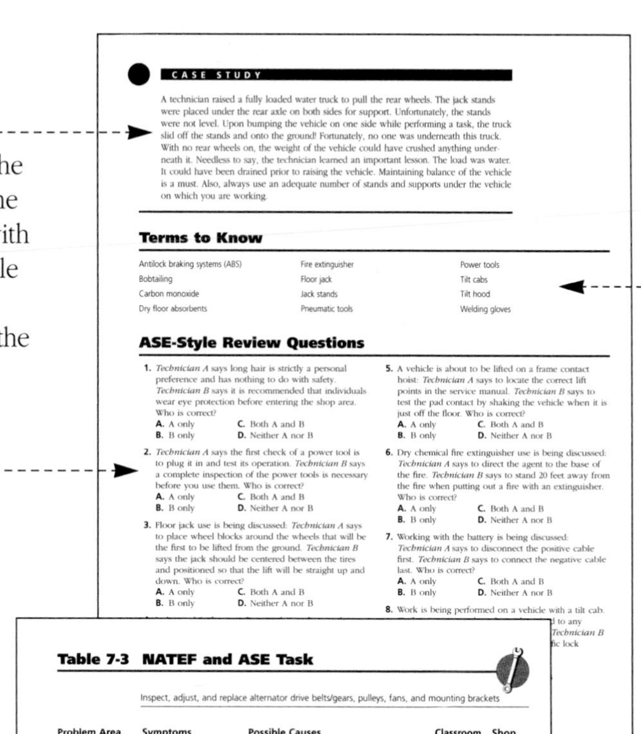

CASE STUDY

A technician raised a fully loaded water truck to pull the rear wheels. The jack stands were placed under the rear axle on both sides for support. Unfortunately, the stands were not level. Upon bumping the vehicle on one side while performing a task, the truck slid off the stands and onto the ground! Fortunately, no one was underneath this truck. With no rear wheels on, the weight of the vehicle could have crushed anything underneath it. Needless to say, the technician learned an important lesson. The load was water. It could have been drained prior to raising the vehicle. Maintaining balance of the vehicle is a must. Also, always use an adequate number of stands and supports under the vehicle on which you are working.

Terms to Know

Antilock braking systems (ABS)
Bobtailing
Carbon monoxide
Dry floor absorbents
Fire extinguisher
Floor jack
Jack stands
Pneumatic tools
Power tools
Tilt cabs
Tilt hood
Welding gloves

ASE-Style Review Questions

1. *Technician A* says long hair is strictly a personal preference and has nothing to do with safety. *Technician B* says it is recommended that individuals wear eye protection before entering the shop area. Who is correct?
   A. A only **C.** Both A and B
   B. B only **D.** Neither A nor B
2. *Technician A* says the first check of a power tool is to plug it in and test its operation. *Technician B* says a complete inspection of the power tools is necessary before you use them. Who is correct?
   A. A only **C.** Both A and B
   B. B only **D.** Neither A nor B
3. Floor jack use is being discussed: *Technician A* says to place wheel blocks around the wheels that will be the first to be lifted from the ground. *Technician B* says the jack should be centered between the tires and positioned so that the lift will be straight up and down. Who is correct?
   A. A only **C.** Both A and B
   B. B only **D.** Neither A nor B
5. A vehicle is about to be lifted on a frame contact hoist: *Technician A* says to locate the correct lift points in the service manual. *Technician B* says to test the pad contact by shaking the vehicle when it is just off the floor. Who is correct?
   A. A only **C.** Both A and B
   B. B only **D.** Neither A nor B
6. Dry chemical fire extinguisher use is being discussed: *Technician A* says to direct the agent to the base of the fire. *Technician B* says to stand 20 feet away from the fire when putting out a fire with an extinguisher. Who is correct?
   A. A only **C.** Both A and B
   B. B only **D.** Neither A nor B
7. Working with the battery is being discussed: *Technician A* says to disconnect the positive cable first. *Technician B* says to connect the negative cable last. Who is correct?
   A. A only **C.** Both A and B
   B. B only **D.** Neither A nor B
8. Work is being performed on a vehicle with a tilt cab. ... d to any ... *Technician B* ... fic lock

Terms to Know

Terms in this list can be found in the Glossary at the end of the manual.

ASE-Style Review Questions

Each chapter contains ASE-style review questions that reflect the performance objectives listed at the beginning of the chapter. These questions can be used to review the chapter as well as to prepare for the ASE certification exam.

Diagnostic Chart

Chapters include detailed diagnostic charts linked with the appropriate ASE and/or NATEF task. These charts list common problems and most probable causes. They also list a page reference in the Classroom Manual for better understanding of the system's operation and a page reference in the Shop Manual for details on the procedure necessary for correcting the problem.

Table 7-3 NATEF and ASE Task

Inspect, adjust, and replace alternator drive belts/gears, pulleys, fans, and mounting brackets

Problem Area	Symptoms	Possible Causes	Classroom Manual	Shop Manual
REPLACE AND ADJUST DRIVE BELT	Charging system output below specifications	1. Slipping or worn drive belt	137	177
REPLACE AC GENERATOR PULLEY	Belts wear prematurely	1. Pulley bent		
REPLACE AC GENERATOR FAN	Noises	1. Bent fan blades	148	198

Table 7-4 NATEF and ASE Task

Perform charging system voltage and amperage output tests; determine needed repairs

Problem Area	Symptoms	Possible Causes	Classroom Manual	Shop Manual
OUTPUT TESTING	Charging system producing low or no output	1. Faulty regulator 2. Open stator 3. Shorted stator 4. Open field coil 5. Shorted field coil 6. Worn or slipping belt 7. Worn brushes	134	187

Table 7-

Problem Area

VOLTAGE DR... TESTING OF ... CHARGING SYSTEM

ASE Practice Examination

A 50-question ASE practice exam, located in the appendix, is included to test students on the content of the complete Shop Manual.

Appendix A

ASE Practice Examination

1. A nonfunctioning right front turn signal light is being diagnosed on a heavy duty truck. *Technician A* says it would be a good idea to check the bulb and socket first, then the signal light switch. *Technician B* says that it may be necessary to check for proper ground connections at the front of the truck. Who is correct?
   A. Technician A **C.** Both A and B
   B. Technician B **D.** Neither A nor B
2. A heater/AC blower circuit is being tested using a DVOM. *Technician A* says to check for a voltage drop across the blower motor; it should be less than 0.04 volts. *Technician B* says to check resistance through the fuse; it should be 50% in ohms of the fuse rating in amps. Who is correct?
   A. Technician A **C.** Both A and B
   B. Technician B **D.** Neither nor B
3. *Technician A* says that when checking amperage, the meter must always be connected in the circuit being tested in parallel. *Technician B* says that checking for amperage with a DVOM only must be limited to circuits with less than 10 amps. Who is correct?
   A. Technician A **C.** Both A and B
   B. Technician B **D.** Neither nor B
4. *Technician A* says that prior to checking a circuit using an ohmmeter you should insure that all power is off or disconnected. *Technician B* says that the reading "OL" on the ohm scale of a DVOM indicates extremely low resistance. Who is correct?
   A. Technician A **C.** Both A and B
   B. Technician B **D.** Neither nor B
5. To determine the voltage drop between the battery and the starter switch, you would do which of the following while cranking the engine?
   A. Connect an ammeter between the battery ground post and the live side of the starter switch
   B. Connect a voltmeter between the positive battery post and the live side of the starter switch
   C. Connect a voltmeter between the start post of the solenoid and the solenoid ground post
   D. None of the above; it is impossible to determine the voltage drop of this section of the starter
6. The fuse on an interior light continues to blow when the light is turned on. *Technician A* says to remove the fuse and bulbs, then check for continuity from any connection to ground with an ohmmeter. Very low resistance will indicate a definite short to ground. *Technician B* says that as long as resistance is very low from the switch to the bulb connections, the problem is not in the wiring. Who is correct?
   A. Technician A **C.** Both A and B
   B. Technician B **D.** Neither nor B
7. *Technician A* says that you can check a capacitor or condenser with an ohmmeter using the same techniques and procedures as used in checking a diode. *Technician B* says that when a capacitor is charged it will show "OL" on a DVOM ohm scale. Who is correct?
   A. Technician A **C.** Both A and B
   B. Technician B **D.** Neither nor B
8. You suspect a broken wire on a gas tank gauge circuit between the gauge and sending unit; however, the suspect wire is very difficult to trace the entire distance. Which of the following scales on a DVOM would be most helpful in proving your suspicion?
   A. AC volts **C.** HTC
   B. Ohms **D.** Milliamp
9. *Technician A* says that the symbol for a common point on most electrical diagrams is the letter "C". *Technician B* says that the symbol for a positive connection or hot point on a diagram is the letter "P". Who is correct?
   A. Technician A **C.** Both A and B
   B. Technician B **D.** Neither nor B
10. *Technician A* claims that electrical components that do a lot of work or give a lot of light require a significant amperage. *Technician B* claims that electrical components that draw significant amperage from a 12-volt system must have low electrical resistance. Who is correct?
    A. Technician A **C.** Both A and B
    B. Technician B **D.** Neither nor B
11. *Technician A* says that when checking electronic components you should be sure to use a DVOM with high internal resistance. *Technician B* says you can check for current flow through electronic components using a 12-volt test light. Who is correct?
    A. Technician A **C.** Both A and B
    B. Technician B **D.** Neither nor B

493

Reviewers

We would like to extend a special thank you to those who saw things we overlooked and for their contributions:

David Biegel
Madison Area Technical College

Douglas Bradley
Utah Valley State College

Dennis Chapin
Rogue Community College

Jeffry N. Curtis
Bellingham Technical College

Tom Hogue
Santa Ana College

Rudy Hrynkiw
Fairview College

Winston Ingraham

Dan Jeffrey
Northland Career Center

Kenneth W. Kephart
Central Texas College

Terryl Lindsey
OSU – Okmulgee

John Murphy
Centennial College

Dennis Vickerman
Southeast Technical Institute

Contributing Companies

We would also like to thank these companies who provided technical information and art for this edition:

Bendix Commercial Vehicle Systems
BET. Inc.
Caterpillar, Inc.
Cecil R. Dodd
Cummins
Delco Remy Division
Detroit Diesel Corporation
Eaton Corporation
Freightliner Corporation
General Motors Corporation
Goodson Shop Supplies
Heavy Duty Trucking
Horton, Inc.
Jacobs Vehicle Systems™
Kysor/Cadillac Corporation
Mac Tools
Mack Trucks, Inc.
Midtronics
MPSI
Navistar International Truck Corp.
OTC/SPX Corporation
Peterbilt Motors Co.
Prestolite Leece Neville
Robert Bosch Corporation
Sun Electric Corporation
The Budd Company
The Maintenance Council (TMC) of the American Trucking Association
Truck-Lite
Volvo Trucks North America, Inc.

Portions of materials contained herein have been reprinted with permission of General Motors Corporation, Service Operations.

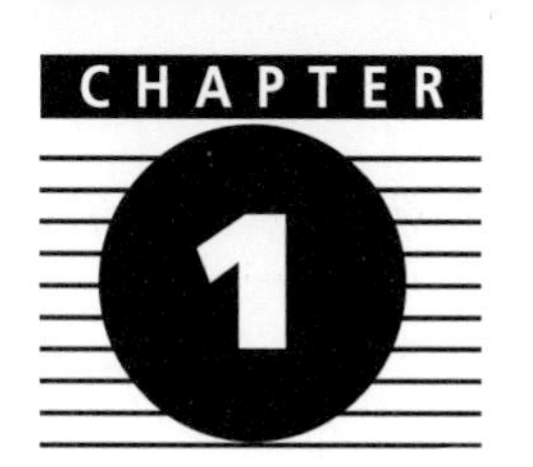

Safety

Upon completion and review of this chapter, you should be able to:

- ❏ Describe safe dress and appearance standards.
- ❏ Describe the qualities of a safe work area.
- ❏ List the procedure for inspecting hand and power tools.
- ❏ Operate a vehicle properly.
- ❏ Raise and support a vehicle safely.
- ❏ Tilt a cab properly.
- ❏ Explain welding and fabricating precautions.
- ❏ Demonstrate proper battery safety.
- ❏ Demonstrate proper use of fire extinguishers.

Introduction

There can never be enough time devoted to practicing shop and personal safety procedures and habits. Too many times we learn the procedures when we first enter the trucking industry, and then we get complacent and form bad habits once we become veterans. When a mishap occurs, only then do we say to ourselves, if only I would have done this differently, perhaps this mistake would not have occurred. For this reason, we need to take this safety chapter seriously and learn proper procedures to safeguard ourselves and our co-workers from any injuries, which could range from minor to life threatening. Most of all, adhere to these procedures throughout your life. Not only is this an obligation, but in most shops it is also a requirement.

Basic Tools

Floor jack
Jack stands
Wheel blocks
Service manual

Proper Attire

Classroom Manual
Chapter 1, page 1.

A **tilt hood** is a complete hood and fender assembly that is tilted or lifted away from the front of the vehicle.

Proper clothing in medium/heavy duty truck shops is very important. Most trucks have tilt hoods and tilt cabs, allowing better access to the engine compartment. The danger with this design is the exposure of the technician to many moving parts. For this reason, clothes need to be properly fitted. Wearing baggy clothes or not wearing clothes properly, such as leaving shirttails out, can cause clothing to get caught in moving parts. The danger is more prevalent when performing electrical work, such as cranking engines to diagnose a starting problem, running engines to check charging, or checking other electrical systems. In addition, trucks have many steps and other obstacles in which it is necessary for the technician to maneuver around them to perform various tasks. Any loose clothing can potentially snag on stationary or moving objects and may cause stumbling or some type of injury. Figure 1-1 shows a properly attired technician.

Also, be aware that rings, bracelets, watches, and other jewelry can get caught in moving parts. When performing electrical work, the metal in jewelry acts as a conductor that can cause shorts when it comes in contact with electrical current. Not only can you harm yourself, but also these shorts can damage the computer components found in today's trucks.

Many components of medium/heavy duty trucks are much heavier than their automotive counterparts. Therefore, proper footwear should be considered in the event that a heavy object, such as a starter, should fall. Tennis and jogging shoes offer very little protection.

Many functions on medium/heavy duty trucks require heating, burning, and welding. Gloves should be worn during those operations to prevent burns, cuts, and scrapes.

Safety Glasses

You should wear some type of Occupational Safety and Health Administration (OSHA) approved eye protection whenever and wherever there is a possibility of dirt or metal particles becoming airborne (Figure 1-2). Dirt, grease, or rust can get into your eyes and cause serious injury. Safety

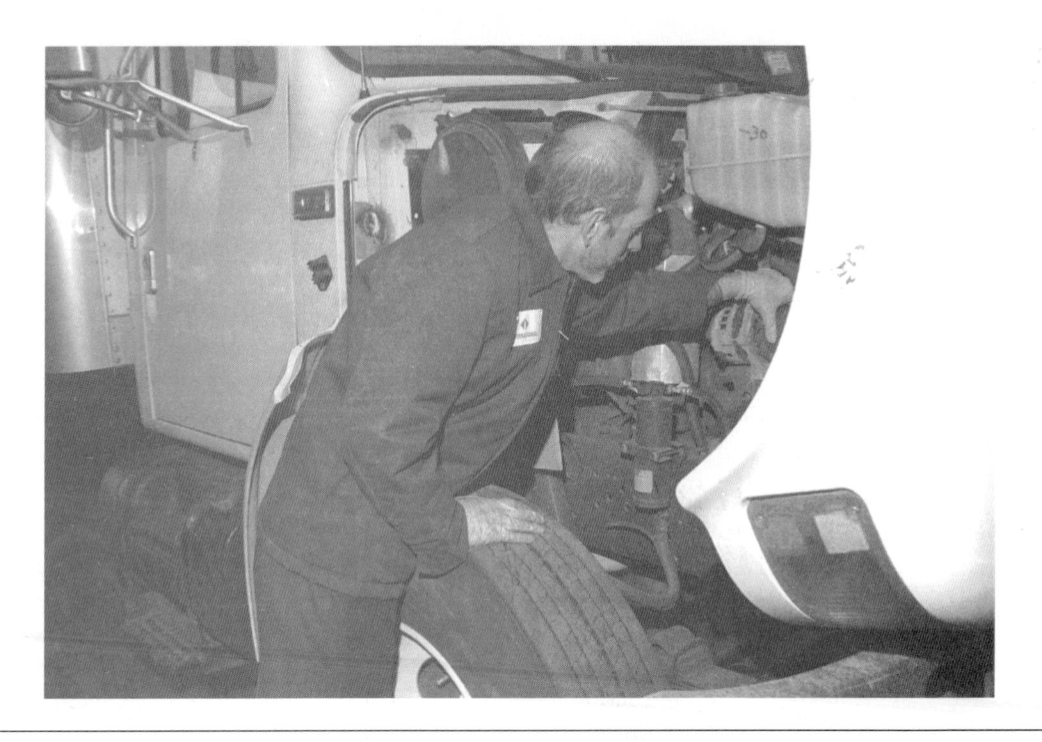

Figure 1-1 Shop coats can be worn to protect clothing.

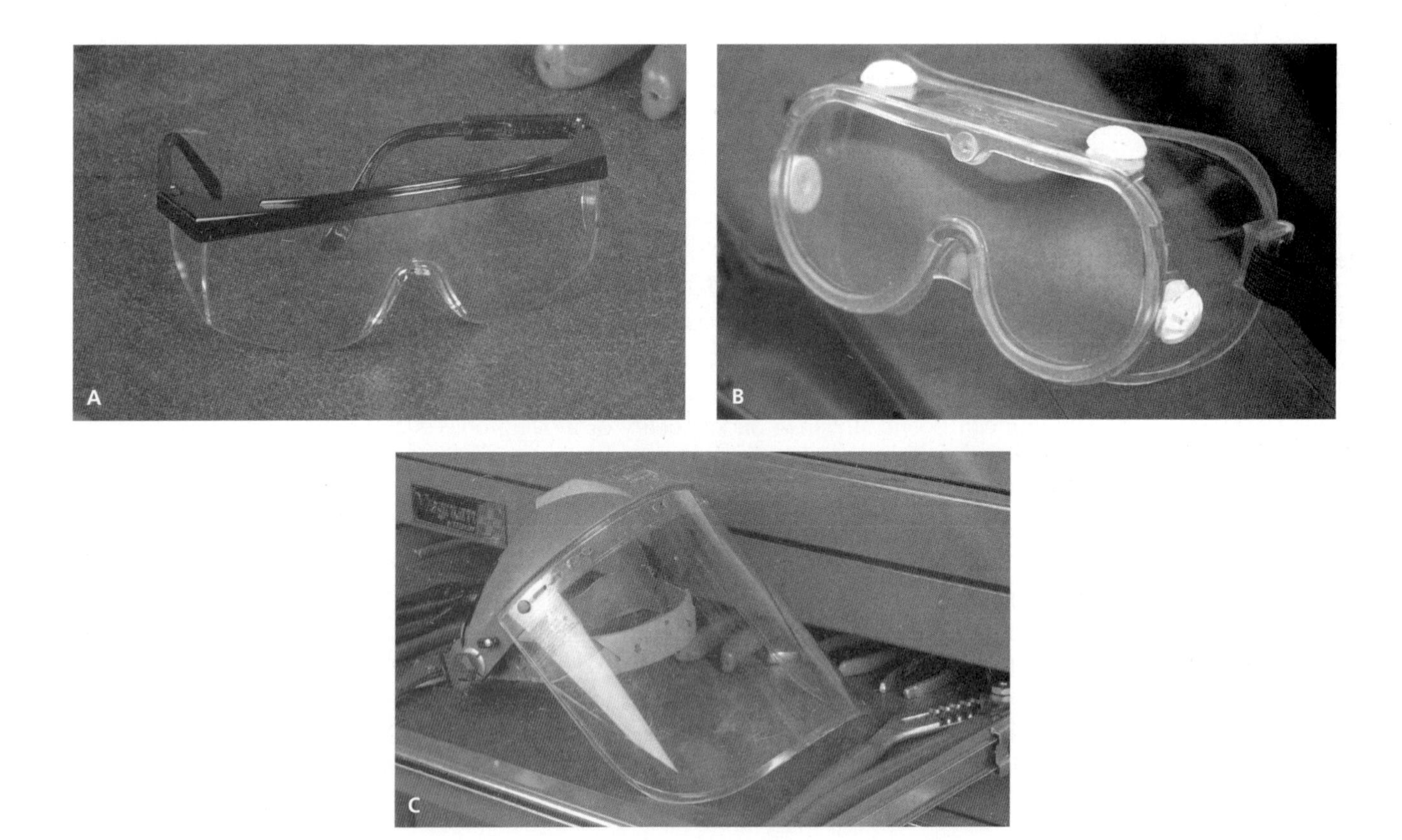

Figure 1-2 Different types of eye protection worn by truck technicians: (A) safety glasses with side shields; (B) safety goggles that can be worn over prescription glasses; and (C) a face shield that is worn over safety glasses or goggles and protects the face. (Courtesy of Goodson Shop Supplies, Winona, MN)

glasses or goggles should be worn when you are working under a vehicle and when you are using any machining equipment, grinding wheels, chemicals, compressed air, or fuels. Chemicals, such as battery electrolyte, can cause serious eye damage, which may lead to blindness. Always wear eye protection when working around or with batteries.

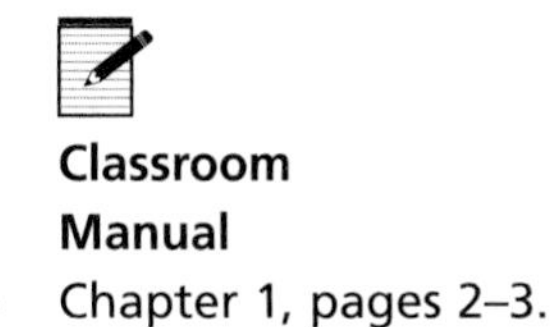

Classroom Manual
Chapter 1, pages 2–3.

The lenses of safety glasses should be made of tempered glass or safety plastic. Safety glasses should provide side protection as well. Common sense should tell you to wear safety glasses nearly all the time you are working in the shop. Some schools and shops make it a requirement to do so.

Work Area Safety

Work area safety needs to be as much a responsibility of the technician as is proper clothing. Most work area requirements are basic and deal with common sense. A good, safe work area is one that is clean, dry, and orderly. This prevents many slips and falls that can cause injury.

Dry floor absorbents are spread over a liquid spill such as oil or antifreeze for safety and environmental reasons.

Most shops require the use of dry floor absorbents to absorb oils and antifreeze from the floors, and they do not allow the use of water to clean floors. This prevents the contaminants from entering our water systems through sewers or flowing directly to groundwater sources.

Many shops require tools and machinery to be stored in certain areas of the shop; it is every technician's responsibility to follow through on their proper storage and placement. This prevents congestion of walkways. In many instances, specific tools are kept in certain dedicated areas in relationship to the type of work performed within those dedicated areas. This system contributes to a good shop layout, in which everyone knows where everything is located. Also, a good layout gives everyone access to safety areas such as exits, water, or eye flushing stations.

CUSTOMER CARE: Many shops do not allow customers in the shop area. However, a customer can often view the work environment from a distance. That first impression can make or break the customer's confidence in the caliber of service that is anticipated.

Inspecting Tools

Classroom Manual
Chapter 1, page 4.

Many shop accidents can be prevented by the proper use of hand tools. Not only is proper use important, but so also is proper care of hand tools.

Performing mechanical work on trucks may require prying, screwing, hammering, and chiseling. Using the incorrect tool for the job function can damage the tool, which in turn, can cause injury to yourself or another individual later on. For example, a screwdriver used as a chisel could shatter, causing you injury. The blade could also get worn, causing it to slip when turning a screw. No one cherishes a scraped knuckle.

The following are some common sense questions to ask yourself when checking your hand tools:

1. Are they clean? This will help to prevent slippage such as from a greasy hammer handle.
2. Are they bent, cracked, or weak? This may cause a breakage when force is applied.
3. Are they as sharp, pointed, or properly fitted as they were when originally purchased?

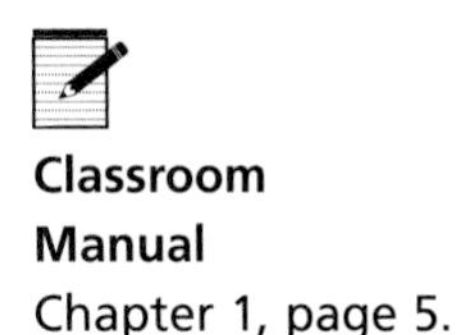

Classroom Manual
Chapter 1, page 5.

Power Tools

Many tools fall under the power tool category, and can be further classified as electric or pneumatic tools.

Power tools are powered by compressed air or electricity.

Electric power tools usually consist of drills and grinders. Both take electricity to perform work, usually in the form of rotation of a drill or grinding stone that is meant to cut. Use the following checklist before operating an electric power tool:

1. Is the tool clean?
2. Are all required safety guards in place?
3. Is the tool in good condition?
4. Is the electrical cord in good condition and properly grounded?

Do not use the tool if any of the above conditions are not met.

Pneumatic power tools include impact wrenches (Figure 1-3), drills, and grinders, to name a few. These tools, which are under great pressure, are used for many functions such as fastening or unfastening bolts, drilling, grinding, chiseling, and chipping. It is very important for your own safety that these tools are regularly inspected. They all have the potential to slip, break, or shatter, causing very serious damage. Use the following checklist before operating the pneumatic power tool:

1. Is the tool clean?
2. Are all required safety guards in place?
3. Does the tool appear to be in good condition?
4. Are the air hose and connections in good condition?
5. Are the attachments and/or tool bits in good condition?
6. Is the air pressure properly regulated?

Do not use the tool if any of the above conditions are not met.

Pneumatic tools are powered by compressed air.

SERVICE TIP: Use air tool lubricant to prolong the life of your tools. It also provides a more accurate speed or torque output from your tools.

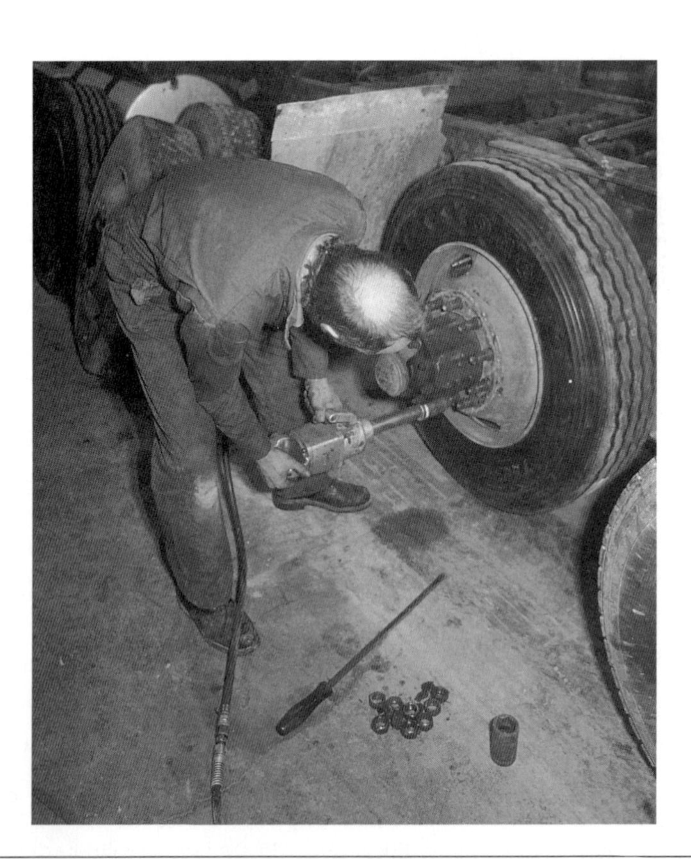

Figure 1-3 Impact wrench in use tightening tire lugs. (Courtesy of Heavy Duty Trucking/Ryder)

Compressed Air Safety

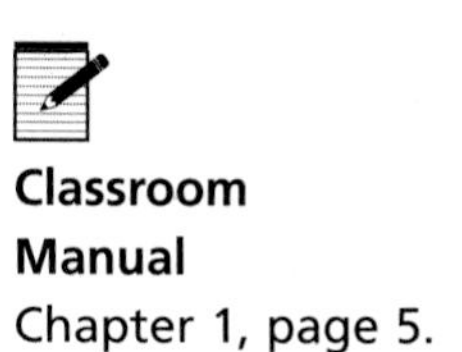

Classroom Manual Chapter 1, page 5.

Compressed air is used not only as a means of performing certain repair functions, but it is also found as a source of energy in heavy duty trucks for brakes and many other air-operated components.

The supply of air for shop use and in the heavy duty truck is in the form of high pressure. Serious injury can occur through careless use of this air. The following should be kept in mind when dealing with compressed air:

1. Proper protection, such as safety glasses, should be worn when performing tasks requiring compressed air equipment and tools.
2. Certain functions, such as riveting, air hammering, and chiseling, require the use of ear protection.
3. Integrity of air lines and connections need to be checked to prevent air hoses from "whipping off," which can cause serious injury.
4. Do not blow off clothing or hair for any reason.
5. Do not direct compressed air against the skin. This can be fatal if the air penetrates the skin and enters the bloodstream.
6. When working on trucks, beware of components such as air valves and air dryers that can exhaust air pressures at any given moment. Many technician's tasks are performed close to these components. The compressed air can come out at the same force as a blow gun.
7. When working on trucks, do not remove lines or fittings that are under high pressure, such as drain valves or plugs on air reservoirs.

Vehicle Operation

As a technician, you will be required to operate the vehicles on which you perform the tasks. This can include pulling the vehicles in and out of the shop or taking the vehicle on the road to duplicate a problem or verify a repair. Driving a commercial vehicle on the road requires a commercial driver's license (CDL). If it is necessary to operate a vehicle that you are unfamiliar with, make sure that you are knowledgeable of all its switches and controls. Medium and heavy duty vehicles react differently when they are empty versus when they are loaded. Tractor-trailers also react differently depending on whether they are hooked to a trailer or bobtailing.

Bobtailing is driving a tractor without the trailer hooked to it.

Most heavy duty trucks have manual transmissions with various ranges of shifting. Be sure that you understand and are comfortable with the shift patterns and/or sequences prior to a road test. Improper shifting can do serious damage to a transmission.

Many of the trucks will require the use of side mirrors to back up with no other means of visual access to what is directly behind the truck, tractor, or tractor-trailer. Prior to backing up, physically check behind the vehicle for any obstructions. It is not a bad idea to also sound the horn prior to backing up a vehicle. This is especially important when backing out of a shop bay when you are unable to see if a person or another vehicle is approaching.

When parking the vehicle, make sure that the proper park brake system is applied whether it be air or hydraulic. Using a transmission lever or leaving it in gear is not a park system. When performing a task on the vehicle, take the extra step and properly chock the wheels.

Lifting the Vehicle

Classroom Manual Chapter 1, page 5.

Many service tasks require lifting a medium or heavy duty truck. Some shops use lifts that are capable of lifting the whole unit with minimal amount of work. Still, the most basic means of raising a truck is using some form of a floor jack. Because of the weight involved when raising a truck or a portion of the truck, extra care must be taken to prevent injury and/or damage to the vehicle.

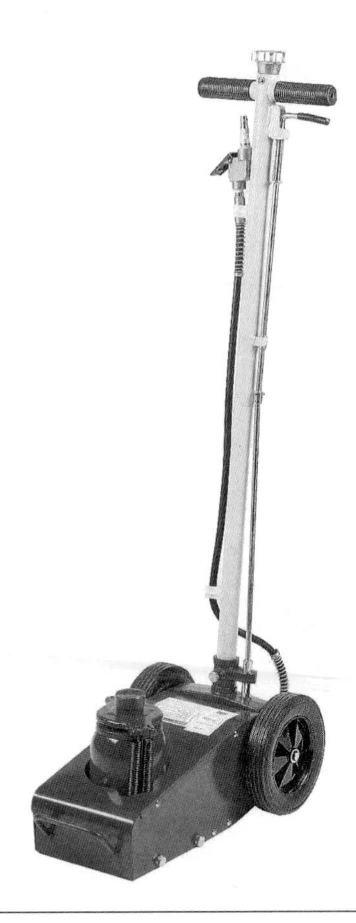

Figure 1-4 Hydraulic floor jack. (Courtesy of Mac Tools)

Special Tools

Floor jack

Safety stands

Wheel blocks

Service manual

A **floor jack** is a portable hydraulic tool used to raise and lower a vehicle.

Place the wheel chocks around the wheel(s) that remain on the ground.

Jack stands, or safety stands, are support devices used to hold the vehicle off of the floor after it has been raised by the floor jack.

Hydraulic Jack and Jack Stand Safety

The following procedures and precautions should be adhered to when using hydraulic jacks and jack stands:

1. Prior to using hydraulic jacks and jack stands, check the integrity of both pieces of equipment. There should be no leaks, or missing, bent, or broken parts.
2. Make sure the jack and stand weight capacity always exceed the weight of the vehicle portion being lifted.
3. The lifting and/or support pads have to be placed under part of the truck and/or axle where no slippage can occur.
4. When the truck is supported by jack stands, make sure that all the legs of the jack stand are touching the floor (Figure 1-5).
5. Whenever possible, maintain a balanced level of the truck from side to side in order to prevent a top heavy truck from slipping off the supports. This can occur when performing prying types of tasks on a unit.

WARNING: Never lift a portion of the vehicle or a component such as a wheel so high that it can slip off of the jack or stand; equipment damage and personal injury can result.

CAUTION: A worn or broken locking device being used on a lifting device or stand can be dangerous when trying to support the enormous weight of a medium/heavy duty vehicle. The vehicle can suddenly come down and cause personal injury.

Figure 1-5 Air spring suspension systems. (Courtesy of Freightliner Corporation)

Tilt Cab Safety

Many trucks today are of the tilt cab type (Figure 1-6). Certain precautions need to be taken when working on these vehicles. Some medium duty trucks do not use hydraulic jacks to tilt the cab and they can be quite heavy and cumbersome. Always ask for assistance whenever you deem it necessary to tilt those cabs. Do not take any unnecessary chances of hurting yourself. Prior to

Tilt cabs are the type of body style that allows the driver and passenger compartment to sit above the engine. This complete unit is tilted in order to gain access to the engine.

Figure 1-6 Do not raise or lower a tilt cab with the engine running. (Courtesy of General Motors Corporation, Service Operations)

tilting any cab, make sure there are no loose objects that can come through the windshield. When using hydraulics to raise a tilt cab, make sure you understand the operation of the control levers for raising, holding, and releasing.

CAUTION: All tilt cabs must be locked into place once they are tilted to the required level. Remember that you will be working under the cab and it is very heavy should it unexpectedly come down.

Air Quality

Carbon monoxide is an odorless, colorless, and toxic gas produced as a result of the combustion process.

Many tasks will require the running of the vehicle in the shop. This can be from the simple starting of the vehicle to build up of air to release the brakes (i.e., air-braked vehicles) so the vehicle can be moved within or out of the shop. The vehicle may also need to be run for an extended period for diagnostic purposes. This presents the possibility of health problems related to air quality. Carbon monoxide poisoning is at the top of the list. Not every shop has ventilation systems that will remove vehicle exhaust. In shops without ventilation systems, hoses should be routed to the outside from the exhaust when doing any prolonged running of the vehicle.

On occasion you will encounter a diesel truck that smokes excessively when it is started up, especially when it is cold. This might also continue for a prolonged period, especially if it is a truck that is slow in building up air pressure to release the brakes.

It is easy enough to make up an air coupler to connect to the vehicle's air system, utilizing shop air to fill the vehicle's air reservoirs. Once the air is built up, the vehicle can be started up, brakes can be released, and the vehicle can be moved or pulled out of the shop. This prevents excessive buildup of smoke and particulate matter in the shop building.

Fire Extinguishers

Classroom Manual
Chapter 1, page 8.

A **fire extinguisher** is a portable apparatus containing chemicals, water, foam, or a special gas that can be discharged to extinguish a small fire (Figure 1-7).

Tour the shop area and become familiar with the location of the fire extinguishers. Use a report sheet to record the locations. Also indicate the type of each extinguisher and what kinds of fires it will extinguish. Check the gauge and record the state of charge for each extinguisher.

The proper use of a fire extinguisher is very important. It is possible to deplete an extinguisher and still not put out even the smallest of fires if the extinguisher is used improperly. Procedures vary depending on the agents used. Technicians must become familiar with all the extinguishers equipped in the shop. Photo sequence 1 illustrates the proper use of a multipurpose dry chemical extinguisher. This type of extinguisher is the most widely used in the truck shop.

Figure 1-7 Typical fire extinguishers.

Photo Sequence 1
Using a Dry Chemical Fire Extinguisher

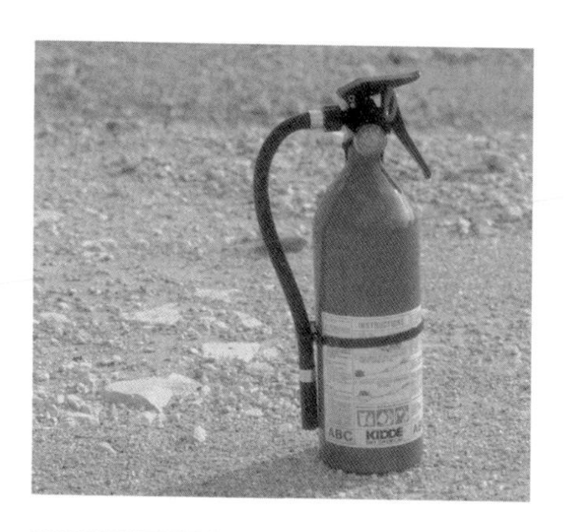

P1-1 Multipurpose dry chemical fire extinguisher.

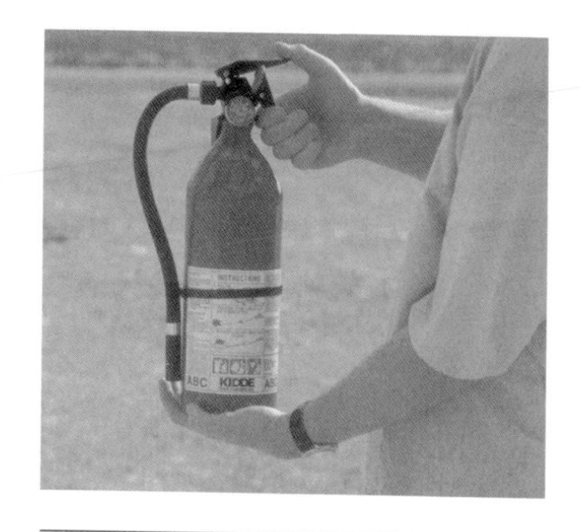

P1-2 Hold the fire extinguisher in an upright position.

P1-3 Pull the safety pin from the handle.

P1-4 Stand 8 feet from the fire. Do not go any closer to the fire.

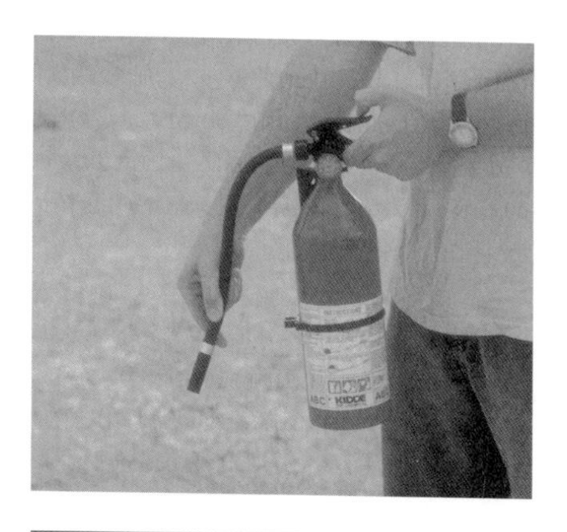

P1-5 Free the hose from its retainer and aim it at the base of the fire.

P1-6 Squeeze the lever while sweeping the hose from side to side. Keep the hose aimed at the base of the fire.

Battery Safety

Classroom Manual
Chapter 1, page 9.

Batteries are very dangerous components of the vehicle. It is important that you be able to demonstrate the ability to work around the battery in a safe manner. Throughout this manual there will be many instances where you will be required to perform a task involving the battery. Chapter 5 covers the subject of removing and testing the battery. Do not perform any tests or disconnect the battery until you have completed that chapter.

Antilock Brake System Safety

Eventually all commercial vehicles will be equipped with antilock brakes because of federal requirements. Certain precautions need to be taken when working on vehicles with antilock brakes. The following are a guideline:

Antilock braking systems (ABS) automatically pulsate the brake or brakes to prevent wheel lock up in order to maintain better steering and braking control of the vehicle.

1. Relieve any pressures prior to loosening or removing pressurized lines or components.
2. Be sure there is no power on when disconnecting any wiring connectors or components.
3. Disable the ABS system prior to welding on the truck. This can be accomplished by disconnecting the battery cables or wiring harness from the module.

Welding and Fabricating

Classroom Manual
Chapter 1, page 11.

On occasion, commercial vehicles require welding and/or fabricating to complete repairs or add accessories. Many components in a truck break more readily because of road and/or suspension condition. These repairs may also require welding and/or fabricating. The following are guidelines for those tasks:

1. Make personal safety a top priority. Wear proper clothing such as welding gloves. Shirt sleeves should be rolled down and buttoned. Wear safety glasses or welding glasses for fabricating and proper cutting (Figure 1-8).
2. Most vehicles today use some form of electronics. Disconnect the battery prior to welding.
3. Be aware of potential dangers when cutting or welding, such as pressurized lines or air bags that can be penetrated by sparks or hot components.
4. Metal retains heat for a long time. Be alert throughout your task and let others know of the potential of burns if they are approaching your work area.
5. Do not weld, drill, or heat on certain frame members. Always follow manufacturer's recommendations.

Welding gloves are special gloves that reduce burns and heat-related problems.

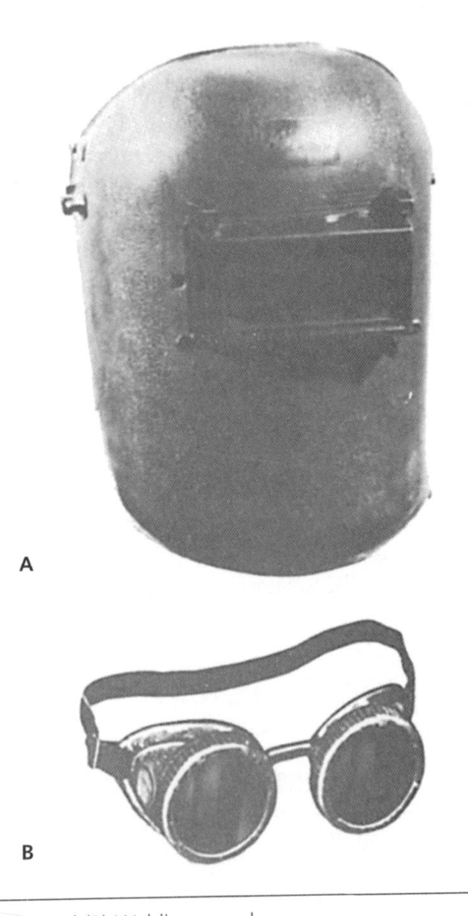

Figure 1-8 (A) Welding helmet and (B) Welding goggles.

CASE STUDY

A technician raised a fully loaded water truck to pull the rear wheels. The jack stands were placed under the rear axle on both sides for support. Unfortunately, the stands were not level. Upon bumping the vehicle on one side while performing a task, the truck slid off the stands and onto the ground! Fortunately, no one was underneath this truck. With no rear wheels on, the weight of the vehicle could have crushed anything underneath it. Needless to say, the technician learned an important lesson. The load was water. It could have been drained prior to raising the vehicle. Maintaining balance of the vehicle is a must. Also, always use an adequate number of stands and supports under the vehicle on which you are working.

Terms to Know

Antilock braking systems (ABS)	Fire extinguisher	Power tools
Bobtailing	Floor jack	Tilt cabs
Carbon monoxide	Jack stands	Tilt hood
Dry floor absorbents	Pneumatic tools	Welding gloves

ASE-Style Review Questions

1. *Technician A* says long hair is strictly a personal preference and has nothing to do with safety. *Technician B* says it is recommended that individuals wear eye protection before entering the shop area. Who is correct?

A. A only **C.** Both A and B
B. B only **D.** Neither A nor B

2. *Technician A* says the first check of a power tool is to plug it in and test its operation. *Technician B* says a complete inspection of the power tools is necessary before you use them. Who is correct?

A. A only **C.** Both A and B
B. B only **D.** Neither A nor B

3. Floor jack use is being discussed: *Technician A* says to place wheel blocks around the wheels that will be the first to be lifted from the ground. *Technician B* says the jack should be centered between the tires and positioned so that the lift will be straight up and down. Who is correct?

A. A only **C.** Both A and B
B. B only **D.** Neither A nor B

4. Jack stand (safety stand) use is being discussed: *Technician A* says to use a minimum of four jack stands if the entire vehicle is lifted. *Technician B* says to place the safety stands where they will not lean or slip. Who is correct?

A. A only **C.** Both A and B
B. B only **D.** Neither A nor B

5. A vehicle is about to be lifted on a frame contact hoist: *Technician A* says to locate the correct lift points in the service manual. *Technician B* says to test the pad contact by shaking the vehicle when it is just off the floor. Who is correct?

A. A only **C.** Both A and B
B. B only **D.** Neither A nor B

6. Dry chemical fire extinguisher use is being discussed: *Technician A* says to direct the agent to the base of the fire. *Technician B* says to stand 20 feet away from the fire when putting out a fire with an extinguisher. Who is correct?

A. A only **C.** Both A and B
B. B only **D.** Neither A nor B

7. Working with the battery is being discussed: *Technician A* says to disconnect the positive cable first. *Technician B* says to connect the negative cable last. Who is correct?

A. A only **C.** Both A and B
B. B only **D.** Neither A nor B

8. Work is being performed on a vehicle with a tilt cab. *Technician A* says the cab can be raised to any position to safely work on the vehicle. *Technician B* says the cab has to be raised to a specific lock position. Who is correct?

A. A only **C.** Both A and B
B. B only **D.** Neither A nor B

9. A vehicle was just pulled into the shop for repairs: *Technician A* says to put the manual transmission in gear and block the wheels to prevent the vehicle from moving. *Technician B* says to apply the parking brake and block the wheels to prevent the vehicle from moving. Who is correct?

A. A only **C.** Both A and B
B. B only **D.** Neither A nor B

10. Certain precautions need to be taken when welding on vehicles that use electronic components. *Technician A* says to disconnect the battery prior to welding to prevent damage to electronic components. *Technician B* says by lowering the current on the welder, damage can be prevented to electronic components. Who is correct?

A. A only **C.** Both A and B
B. B only **D.** Neither A nor B

Job Sheet 1

1

Name: ________________________ Date: ________________________

Shop Safety Survey

Upon completion of this job sheet, you should be able to identify safety issues in the shop.

Procedure

Shop and personal safety must always be a number one priority. It takes constant awareness of your surroundings to accomplish this goal. The following is a survey to heighten your awareness and instill good habits, which will help to prevent accidents and injuries.

1. Are you properly dressed for the task to be completed? Yes _____ No _____
2. Are you wearing the proper safety equipment, such as safety glasses, gloves, and ear protection? Yes ____ No _____
3. Have you removed jewelry that can pose a threat of personal injury? Yes ______ No ______
4. Is your work area free of clutter? Yes _____ No _____
5. Are the floors dry? Yes _____ No _____
6. Are all walkways clear? Yes _____ No _____
7. Are all tools stored in their proper places? Yes _____ No _____
8. Are all tools in a good and safe condition? Yes _____ No _____
9. Are all required guards in place? Yes _____ No _____
10. List the location of:
    - A. First-aid Kits: ________________________
    - B. Fire extinguishers: ________________________
    - C. Water sources: ________________________
11. List the proper procedures in the event of a fire. ________________________

    ____

    ____
12. List the things you can do to further enhance the safety of your shop environment.

    ____

    ____

☑ **Instructor Check** ________________________

Job Sheet 2

2

Name: ______________________ Date: ______________________

Batteries and Battery Safety

Upon completion of this job sheet, you should be able to identify the parts of a battery and work safely around them.

Tools and Equipment Needed

A medium/heavy duty truck battery

Procedure

1. Point out the following components of the battery to your instructor:
   - A. Negative terminal
   - B. Positive terminal
   - C. Vents
2. Answer the following questions concerning battery safety:
   - A. Why should you be concerned about battery acid? ______________________
   - B. What should be done if battery acid gets in your eyes? ______________________
   - C. What should you do if battery acid gets onto your skin? ______________________
   - D. What is meant by polarity? ______________________
   - E. Why is polarity a concern when connecting a battery? ______________________
   - F. Which terminal must be disconnected first? ______________________
   - G. When connecting battery cables, which cable is to be connected last? ______________________
   - H. Why should you never wear jewelry around the battery? ______________________
   - I. Why is smoking not allowed around the battery? ______________________
   - J. Why are tools not to be laid across the top of the battery? ______________________
   - K. What safety equipment should be worn while servicing or working around the battery? ______________________
   - L. What other safety precautions must be observed? ______________________

☑ **Instructor Check** ______________________

Electrical Diagnostic Tools

Upon completion and review of this chapter, you should be able to:

- ❑ Explain the proper use of analog volt/amp/ohm meters.
- ❑ Explain the proper use of digital volt/amp/ohm meters.
- ❑ Explain the proper use of test lights.
- ❑ Explain how and when to use a digital storage oscilloscope.
- ❑ Identify basic waveforms.

Introduction

Altough we cannot see electricity as it flows through a circuit, we can see the reaction of electricity. An example of this reaction is current flowing and encountering a resistance element, called a filament in a bulb. The heat generated by this high resistance causes the filament to glow and give off light. What happens if a problem occurs and the lightbulb doesn't come on when the control switch is turned on for that circuit? In most cases, it would be straightforward to determine if current is flowing within the circuit. Most likely it would be a bad bulb, broken wire, or bad ground. But what happens if the circuit involves flowing current through various conductors, connections, switches, and/or relay and the bulb is lit but it is dim?

The broken wire can usually be traced with a test light, although it is hard to accurately measure partial loss of pressure (voltage) with a test light. Today's trucks are composed of simple light circuits to complicated electronics circuits that require very accurate measurements to diagnose electrical problems. Today's technician needs to know how much current is flowing and/or the resistance to it. This requires circuit and component specifications and instruments with accurate measurement capabilities.

To diagnose the broad range of electrical problems a technician encounters, it is important to have the following common pieces of diagnostic test equipment:

- ❑ Jumper wire
- ❑ Test light
- ❑ Voltmeter
- ❑ Ammeter
- ❑ Ohmmeter

Basic Tools

Jumper wire

Nonpowered test light

Self-powered test light

Note that the voltmeter, ammeter, and ohmmeter are usually combined into a single meter called a multimeter. This allows the technician to measure electrical pressure in volts, current in amperage, and resistance in ohms with one hand-held meter.

An ampere is usually called an amp.

Nonpowered Test Light

The most common diagnostic tool that every technician should have is a nonpowered test light (Figure 2-1). The test light consists of a probe, a wire with a clamp on the end, and a lightbulb within a clear housing (handle). The wire end is clamped to a good ground and the probe is touched to various points of voltage in the circuit. When voltage is encountered, the light in the tester will come on (Figure 2-2). For this to occur, the circuit tested must be powered when using a nonpowered test light.

A **nonpowered test light** uses the vehicle's circuit for power.

WARNING: Do not use test lights on computer circuits or damage to components may occur. The test light can act like a jumper wire connected directly to ground, causing excess current.

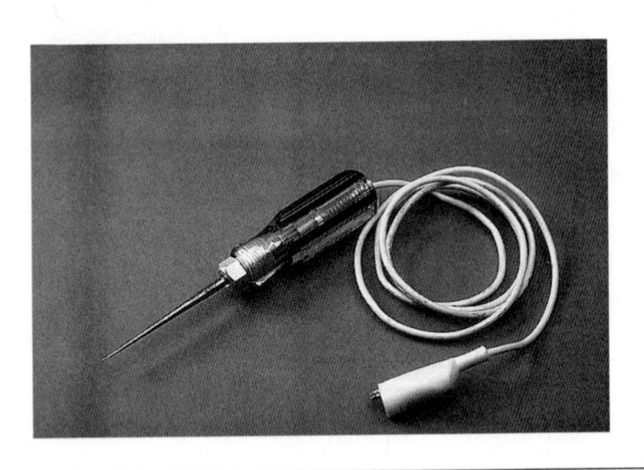

Figure 2-1 A typical test light, which is used to probe for voltage in a circuit.

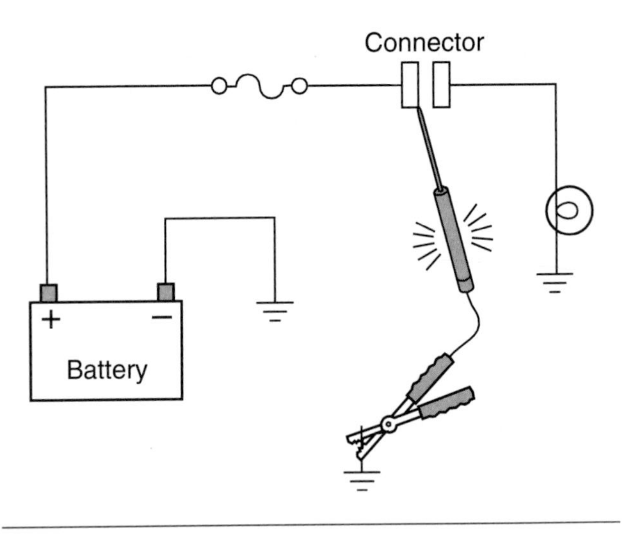

Figure 2-2 If voltage is present, the test light will light.

The test light is a very inexpensive tool to diagnose for breaks in wires or loss of contact within a circuit. However, the brightness of the bulb in the test light is not a measurement. Also, the test light itself needs approximately 1 amp to light its bulb, therefore making it useless on lower amperage electronic circuits. Sometimes a heavy duty truck technician will also select a 24-volt bulb for his or her test light. This allows for testing 24 volts or less. A 12-volt bulb will be ruined if used on a 24-volt system.

WARNING: A test light connected across a computer circuit can damage the components in the circuit. This occurs because the current draw of the test light overloads the circuit.

WARNING: Be careful not to puncture or penetrate wiring if you can avoid it. This allows moisture to enter the insulation, causing corrosion and eventually a failure.

Self-Powered Test Light

A **self-powered test light** has its own internal power to test the vehicle's nonpowered circuits.

This test light is different from the nonpowered test light in that it contains a small internal battery. It also has a bulb, probe, and a lead with an alligator clip. With its internal battery, this test light can be used to check circuits that are disconnected from the battery.

WARNING: A powered test light should only be used when the power to the component or circuit has been disconnected. Also, never use a powered test light on electronic circuitry. The voltage from the test light could cause damage.

Special Tools

DVOM
Oscilloscopes
Logic probes
Scan tools

The powered test light is a good tool for checking continuity in a circuit as well as a ground check. To use the test light for a continuity check, place the ground clamp on the negative side of the component (ground side) and touch the positive or power side with the probe. If the component or circuit has continuity, the bulb in the test light will light. If there is an open (break in the circuit), the light will not be illuminated.

Analog Meters

The test lights are very useful tools if you need to know if there is voltage present at a component or if the circuit and component are intact, allowing current to flow between the power (B+)

and ground. Many times though we need much more precise information. For example, how much current is flowing or being used by the circuit and/or component? Is the proper voltage or pressure being sent through the circuit or is there a pressure loss? Is there too much resistance in the circuit or is the resistance value wrong for a given component? These types of information are needed more often in today's medium/heavy duty trucks than ever before.

An **analog signal** is a direct current variable voltage signal.

Meters are used to retrieve certain kinds of electrical information. The meters selected by technicians should be accurate and dependable. Using an instrument that is not accurate could lead you to a wrong conclusion when diagnosing an electrical problem.

One of the first styles of meters technicians have used for years has been the analog meter. These are meters that use a sweeping needle over a sealed background (Figure 2-3). Analog signals change constantly. Most meters measure the amount of current flowing through them, but the analog meter can be calibrated to indicate any electrical quantity. Going back to Ohm's Law, we know that current flowing through a circuit is determined by the voltage applied and the resistance to overcome:

$$I = E/R$$

Applying this to a meter, we can say that the current flowing through a meter is determined by the applied voltage to the meter and the meter's resistance. Therefore, if a meter has a given resistance and different values of voltages are applied, it will cause specific values of current to flow. Although the meter measures current, its scale can be calibrated in voltage units. If different values of resistance are used in the meter for a given applied voltage, the result is a specific value of current flowing. Thus, the meter can be calibrated in units of resistance instead of units of current. This is Ohm's Law at work.

Analog meters need to be calibrated occasionally.

Most analog meters use a D'Arsonval (moving coil) meter movement. A D'Arsonval movement consists of a magnet, a frame and armature, coil of wire in the armature, a spiral spring, and a needle as an indicator. Basically it is a coil of wire in a magnetic field. Passing current through the coil of wire, a magnetic field is created that reacts with the permanent magnet's magnetic field, causing the coil to rotate. The pointer is attached to the coil and moves across a calibrated scale. The greater the current flow, the greater the magnetic field around the coil, which will cause the coil to rotate more and the pointer will swing further across the meter scale. However, meter construction limits the amount of current allowed through the meter movement. Fifty microamps is the maximum current allowed, and it would cause a full-scale deflection of the needle.

WARNING: Do not use analog meters on electronic circuits. They can load down a delicate electronic circuit and damage it or provide an inaccurate test result.

Analog meters need to be calibrated occasionally. Most analog meters have either an adjustment screw or zero adjustment knob. Adjustment is used to move the combined coil-needle to its base point.

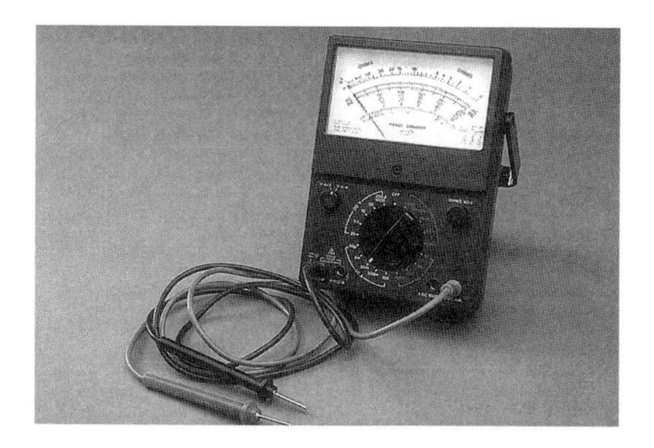

Figure 2-3 An analog meter.

Figure 2-4 A digital meter.

Digital Meters

Digital meters are more accurate than analog meters because of their internal electronic circuitry.

Digital meters are a much more accurate meter. They rely on electronic circuitry to measure values. They are much easier to read because the measurements are displayed numerically using light-emitting diodes (LEDs) or a liquid crystal display (LCD). See Figure 2-4.

In most cases the digital readout is accurate to four places. For example, the display could show 1250 ohms or 12.50 volts or 0.250 amps. In contrast, analog meters can be read inaccurately if viewed at an angle. This is known as a parallax problem.

Voltmeters

Voltmeters are valuable because they allow the measurement of **voltage drop** for each load in the circuit.

Of all the meters used, the voltmeter is the most useful. It is a very high resistance meter allowing it to be attached directly across the power source without being damaged. The voltmeter can be used to measure available voltage at the battery. It is an indicator of the voltage (pressure) sent through a circuit. The available voltage can be measured at the terminals of any component or connection within the circuit. If there is a difference of voltage potential at two points in a circuit, the meter needle will be deflected or a reading will be indicated depending on the type of meter.

There are two leads to the voltmeter: a red positive lead and a black negative lead. Always observe correct polarity. The most common hookup is called a shunt or parallel connection (Figure 2-5). In order for the voltmeter to take a reading, the circuit must be turned on and have current flowing.

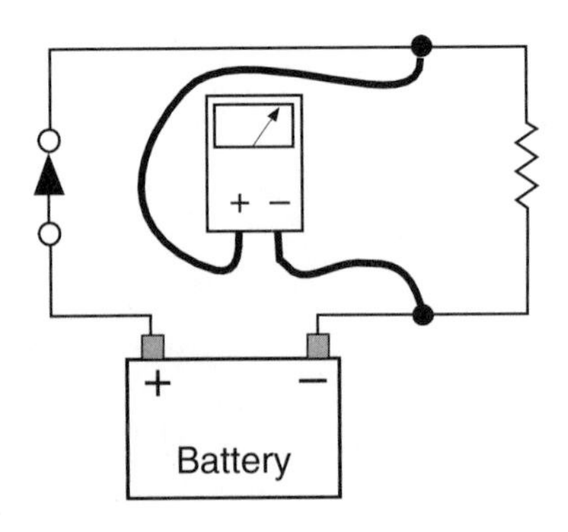

Figure 2-5 A voltmeter is connected in parallel or across the component or part of the circuit being tested.

The voltmeter becomes a very effective tool for locating a resistance in a live circuit. The voltmeter leads can be moved around a circuit from load to load. This allows you to measure the voltage drop for each load in the circuit.

By looking at voltage as pressure that pushes current through a circuit we can make sense of voltage drop. There will always be a voltage loss as current is pushed through a circuit (Figure 2-6). The resistance of a designed component will convert the dropped voltage to some other form of energy such as motion, heat, or light. Typically, the component in the circuit such as a lightbulb will require the applied voltage from the power source. If too much of the applied voltage is lost within the circuit due to poor connections, the bulb's brightness will be diminished. It should be noted that the resistance in the wire itself should be very low. Also remember that the total voltage dropped in a circuit must equal the source voltage.

As seen in Figure 2-6, all of the voltage applied to a circuit is converted to another form of energy. Most manufacturers allow a maximum of a 10% voltage loss of the system voltage in an electrical circuit across wires, connectors, and other conductors.

Figure 2-7 is a series circuit with three loads. The voltmeter is attached to the battery with the red positive lead to the positive post of the battery and the black negative lead of the meter to the negative post of the battery. This enables us to measure the voltage or pressure available from the battery. As you can see, point A is positive and point D is negative. To measure voltage drop across each of the loads requires a little bit of thought, because a point in the circuit can be either positive or negative depending on the load being measured. This is illustrated in Figure 2-8.

To measure voltage drop across load 1, the meter positive lead would be connected to point A and the negative lead to point B. The reading indicated by the meter is the difference in

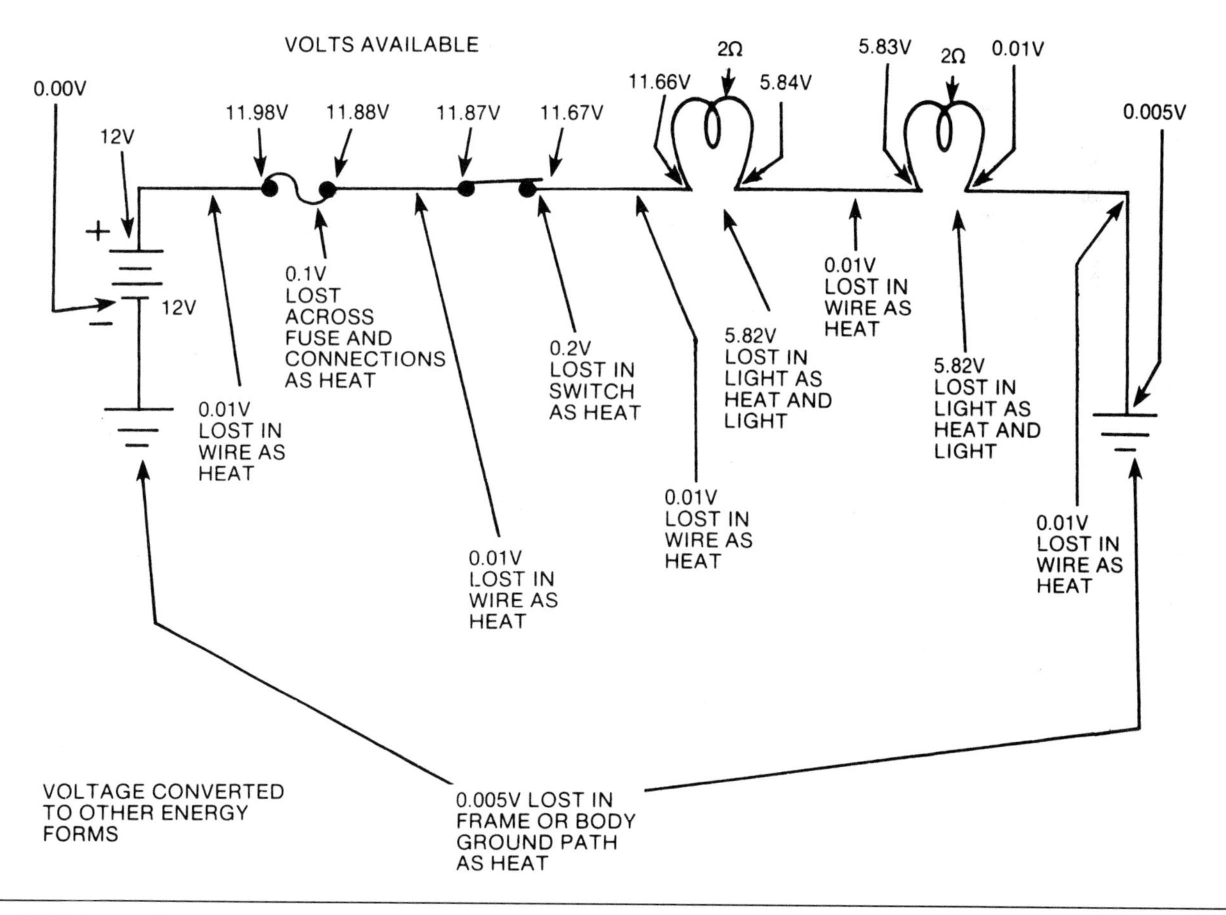

Figure 2-6 Voltage drop.

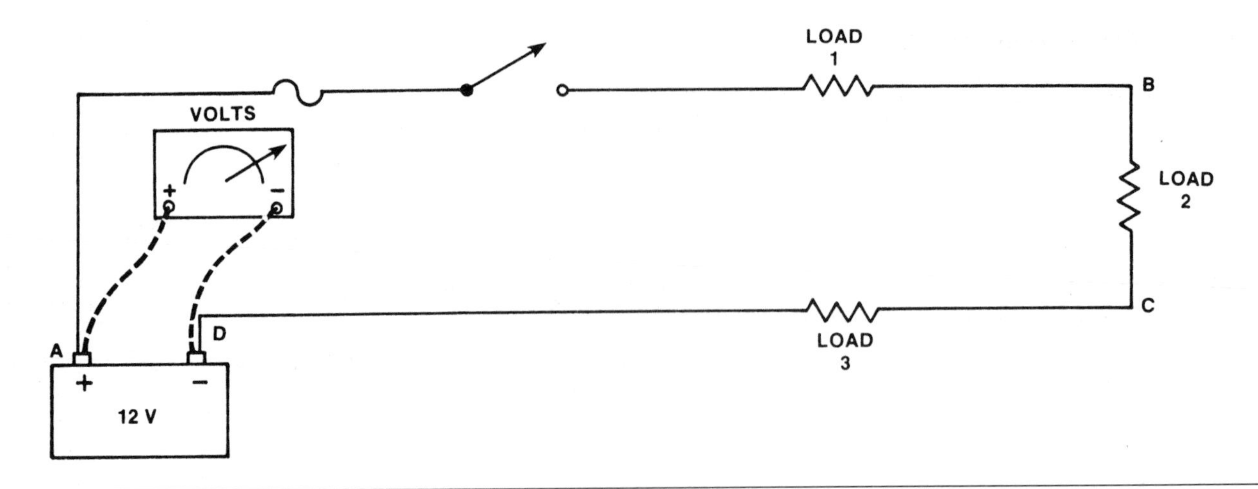

Figure 2-7 Voltmeter measuring battery voltage.

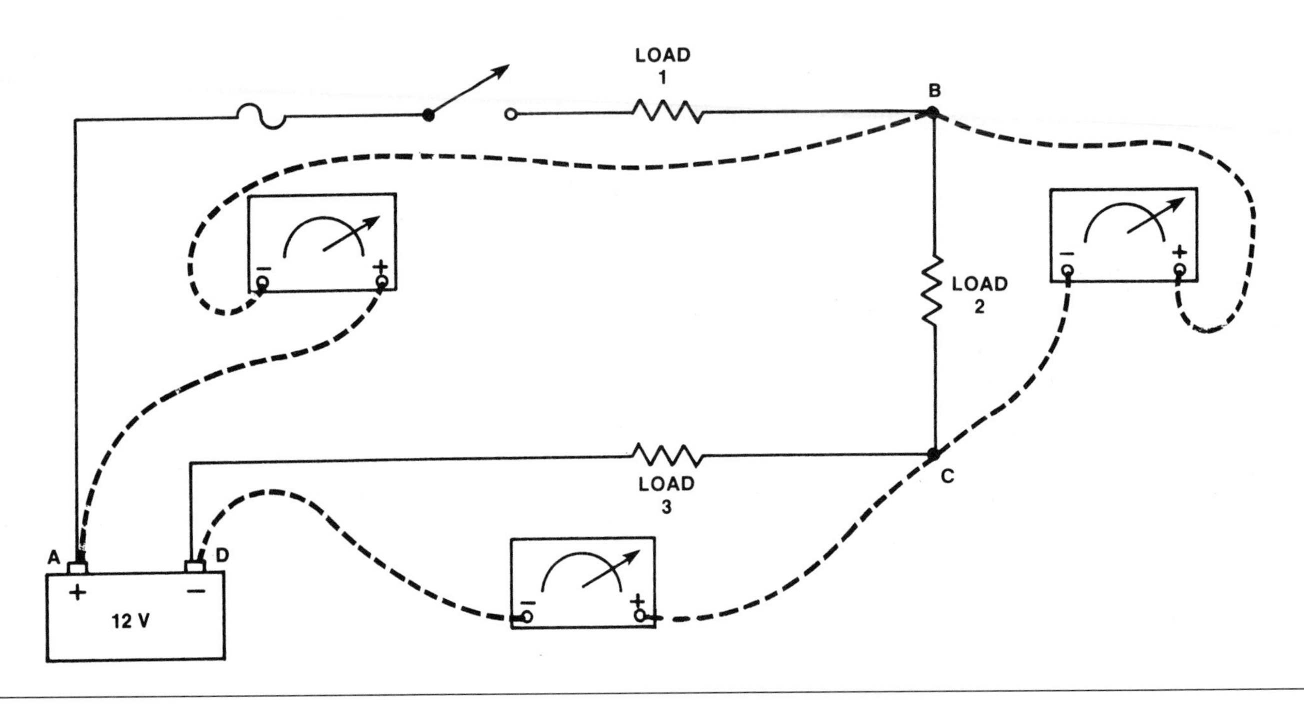

Figure 2-8 Voltmeter measuring voltage drops.

pressure between the two points. This is a voltage drop. To measure the voltage drop of load 2 in the same circuit, the polarity of point B will change from negative to positive in relationship to point C. When a meter is placed across the load, you need to stop and figure out which is the most negative point of the load and which is the most positive point of the load being measured.

By now you can see that measuring load 3, point C becomes the most positive point of the load and point D is the most negative point of the load. Don't forget that the load can be a working device, connector, switch, or anything that can provide resistance. This test is good for finding unwanted resistance due to dirty, corroded, or loose connections. When looking for unwanted resistance, don't forget to test both the feed and ground side of the load in the circuit. However in Figure 2-8 you will notice that the switch is open. There should be no reading. Remember that in order for the voltmeter to take a reading, the circuit must be turned on.

Photo sequence 2 shows a typical procedure for measuring voltage drops. The voltmeter will be used in other sections of the book, such as checking for battery voltage drop while cranking an engine. The voltmeter is also used for checking alternator output while the engine is running.

Photo Sequence 2
Performing a Voltage Drop Test

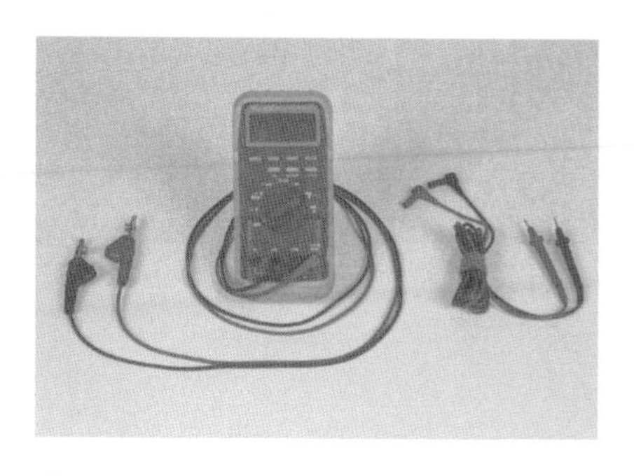

P2-1 The tools required to perform a voltage drop test are a voltmeter and leads.

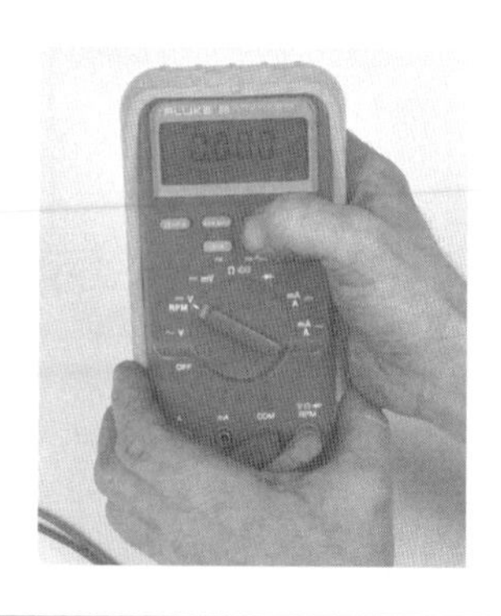

P2-2 Set the voltmeter to its DC voltage function. Depending upon the meter being used, either auto range the function or scale the function to aquire the most accurate reading.

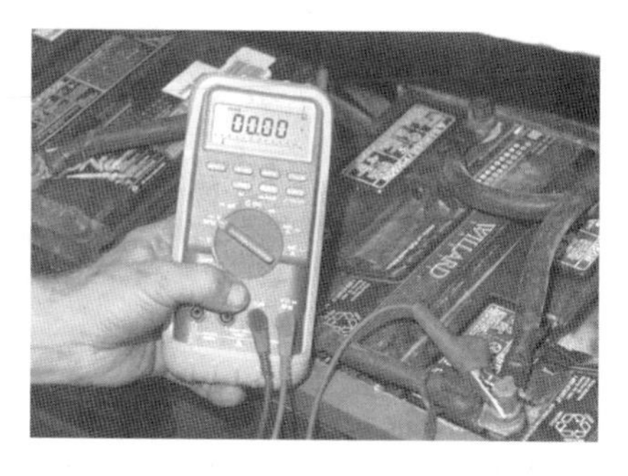

P2-3 To test the voltage drop of an entire power feed circuit, connect the positive (red) lead to the positive terminal of the battery.

P2-4 Connect the negative (black) lead to the power wire at the connector for the load. In this case a low beam headlight circuit is being used. Turn the circuit on. If there are no unwanted resistances in the circuit, the voltmeter will indicate a reading of less than 0.1 volts. A reading that is greater than 0.1 volts will indicate a wire or connector problem. To pinpoint the source of the problem, move the black test lead to another circuit connection closer to the battery.

Ammeter

Ammeters are used to measure the amperage or current that is being pushed through the load to do work. The more work the load will do, the more current (amps) the circuit will draw. By measuring this current flow, a technician can compare it to the specifications of a given load to determine if a problem exists. The ammeter is especially valuable for starting and charging system diagnosis.

Ammeters are also used to locate current drains. These can be minute drains of current from the battery. If sustained over a long period, such as overnight while the vehicle is off, the drain can be sufficient enough to cause a no start or slow cranking problem in the morning.

A **current drain** can be measured by using an ammeter.

Ammeters come in two types: the direct reading type and the inductive pickup type. In the direct reading type, all of the circuit current flows through the ammeter. The meter is placed in series with the circuit (Figure 2-9), and becomes part of it.

WARNING: Make sure that you don't exceed the amp limit of your meter. Exceeding the meter's amp limit will blow the meter's fuse or possibly damage it.

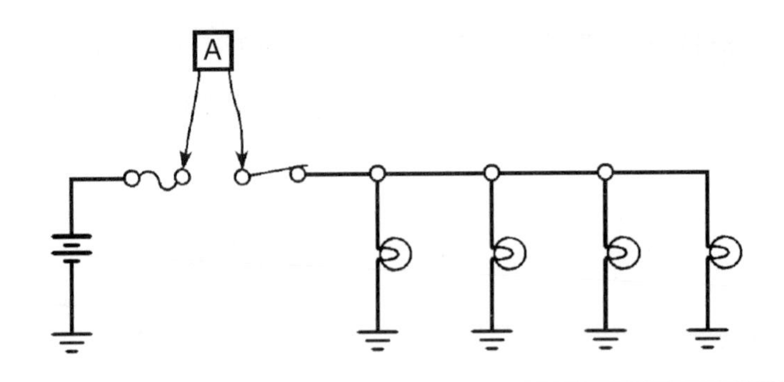

Figure 2-9 Measuring current with an ammeter. Notice this meter must be placed in series with the circuit.

A **current draw** is the term applied to the amount of amperes a load device uses to perform work.

To hook up the meter in series, it is a good idea to disconnect the circuit from the power source first. You must observe correct polarity. Place the red positive lead to the positive side of the circuit and the black negative lead to the ground or negative side of the circuit. When the circuit is turned on, a current draw will exist. The amount of draw is determined by the loads and the amount of work generated. For example, two headlights in series will require more current than one headlamp. You need to make sure that the ammeter is capable of handling the current flow; otherwise you will blow the ammeter's fuse.

Table 2-1 is meant to be a guideline to give you an idea of typical amp draws of some common devices. For very specific figures, you need to see manufacturer's specifications.

The other type of ammeter used is the inductive pickup type (Figure 2-10). To measure current flow, the pickup is clamped around the wire in the circuit. This eliminates the need to break

Table 2-1 Typical Amp Draw

Component	Amp Draw
A/C Compressor Clutch	4.0
Backup Lamp	2.1
Backup Alarm	0.65
Cab Marker Light	0.25
Defroster Fan	19.0
Drive Lights/Standard	4.5
Electric Horn	1.0
Front Turn Signal	2.0
Glow Plug Caterpillar	12.5
Headlamp Single (2)	9.0
Headlamp Dual (4)	14.5
Heated Mirror	5.0
Heater/A/C Motor	17.0
Lighted Mirror	1.15
Dome Light	1.45 to 2.75
Racor Fuel Heater	20.0
Stop Lights	2.25 per
Transistor Flasher	4.0
Warning Lamp	.25

It is important to note that the above are to be used only as a guideline.

Figure 2-10 An ammeter with an inductive pickup, which eliminates the need to connect the meter in series with the circuit.

into the circuit with the meter. A small winding in the pickup produces a voltage that is proportional to the magnetic field around the conductor when current flows. The voltage measured is calibrated to give an amperage reading.

▲ **WARNING:** When using a direct reading ammeter, never place the meter leads across a battery or a load. This can at a minimum, blow the meter's fuse or possibly destroy the meter.

Ohmmeters

Most load devices in an electrical circuit of a truck will have some sort of resistance. The specified resistance of the circuit or component can range from a fraction of an ohm to millions of ohms (mega ohms), or a value in between. Often you will have to measure for a specific value to verify component and/or circuit integrity.

Most load devices in trucks will have some sort of resistance.

An ohmmeter is used to measure this resistance to current flow. These measurements are often required to determine where an electrical problem exists. An example of this would be diagnosing the cause for a particular trouble code in an electronically controlled diesel engine.

An **ohm** is the term given to a quantity of resistance to current flow.

For example, a fault code retrieved on a Cummins Celect System is Fault Code 115. The reason given is: No engine speed signal detected at Pins No. 1 and 15 or 11 and 14 of S. H. (Sensor Harness). The effect would be no current to injectors. You will then be directed to a specific flow chart to find the majority of measurements you need to work with arc resistance readings.

An ohmmeter provides a means of effectively isolating parts of electrical circuits and components to pinpoint problems. The meter can accomplish this because it provides its own power supply to the load or circuit being tested and provides a return back to its internal power source. The meter becomes part of the circuit and provides the power. See Figure 2-11.

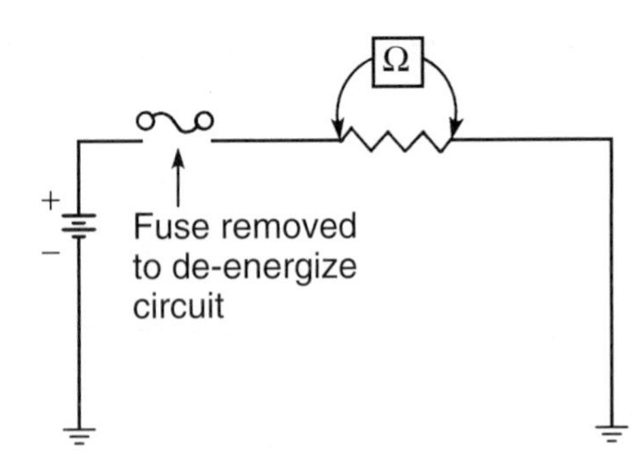

Figure 2-11 Measuring resistance with an ohmmeter. The meter is placed in parallel with the component after power is removed from the circuit.

WARNING: The circuit or load being tested must be disconnected from any power source when being tested or damage to meter or circuit may occur.

Figure 2-12 shows an analog ohmmeter. If you look closely, you will see the numbers getting smaller as the needle would be deflected. The meter is actually measuring the amount of current the internal battery was able to push through the circuit or device being tested.

If you remember Ohm's Law, you are looking at the relationship between current and resistance. More current relates to less resistance. When there is no resistance, you will have maximum current flow. That is why 0 ohms is at the far right on the scale. Maximum resistance is shown on the left side of the scale. Keep in mind that when resistance is at its maximum, there cannot be any current flow. This would be an open circuit or more resistance than can be measured on that scale or meter. Note the symbol for infinity at the very extreme left.

Looking at Figure 2-12 again, you will notice a scaling knob indicating 10×, 100×, and 1000×. Turning the knob from 10× to 100× allows you to measure a higher resistance. For example, if the reading on the scale were 50, you would multiply that by 100 if the pointer were on the 100× scale. The reading of 50 on the scale is actually 5000 ohms.

$$50 \times 100 = 5{,}000$$

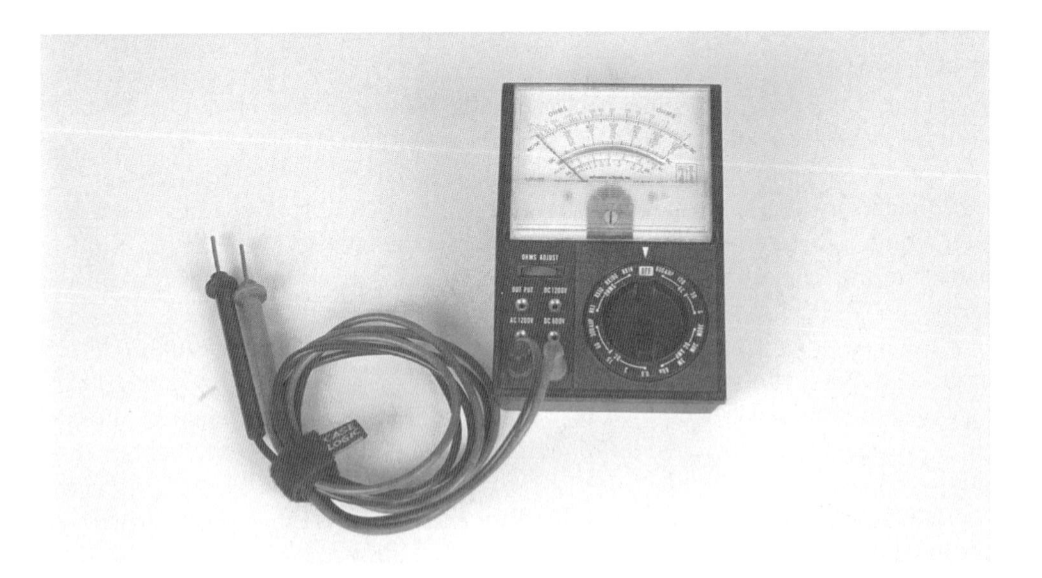

Figure 2-12 Analog ohmmeter.

If you don't know the resistance of the device or circuit you are testing, start on the lowest scale. If you read infinity, try the next scale. If the circuit or device is not open, you will be able to reach a resistance reading on a higher scale instead of infinity.

Looking at Figure 2-12 once again, you will notice an Ohm's Calibrate knob. This knob is used to calibrate the meter to measure resistance accurately. On this type of meter, you connect the test leads together and turn the calibration knob until the meter reads 0 ohms. Keep in mind that an inaccurate reading because of an improperly calibrated meter can mislead you when diagnosing electrical problems.

The ohmmeter is a very useful tool looking for a short (0 ohms) or an open circuit (infinity). It is also useful for comparing resistance values of components to manufacturer's specifications.

Digital Multimeters

Multimeters have been used for a long time by people who work with electricity. The advent of electronics in today's truck industry has brought the multimeter to the forefront. Today the digital multimeter can be found in most shops that work on medium and heavy duty trucks. Although analog meters are combined into a single meter to measure volts, amps, and resistance, the most popular meter in use today is the digital multimeter. The following are advantages of a digital, volt, ohmmeter (DVOM) (see Figure 2-13):

DVOM is an acronym for digital, volt, ohmmeter. It is a popular type of digital multimeter, or **DMM.**

1. The digital meter is very accurate.
2. It is easy to read the values of voltage, current, and resistance.
3. The meter has a very high internal resistance of 10 mega ohms (10 million ohms). This makes it ideal for electronics testing. With the high internal resistance, the internal amperage is low enough that it doesn't add a load to the circuit being tested.
4. The meter is not polarity sensitive. You cannot hurt the meter if the leads are hooked up backwards. Most DVOM will indicate a + or − on the screen to indicate polarity.
5. Most DVOM can be ranged manually and/or have an automatic ranger.
6. Besides reading volts, amps, and resistance, the meters can be purchased with the capability of reading frequency, duty cycle, and RPM, to name a few.

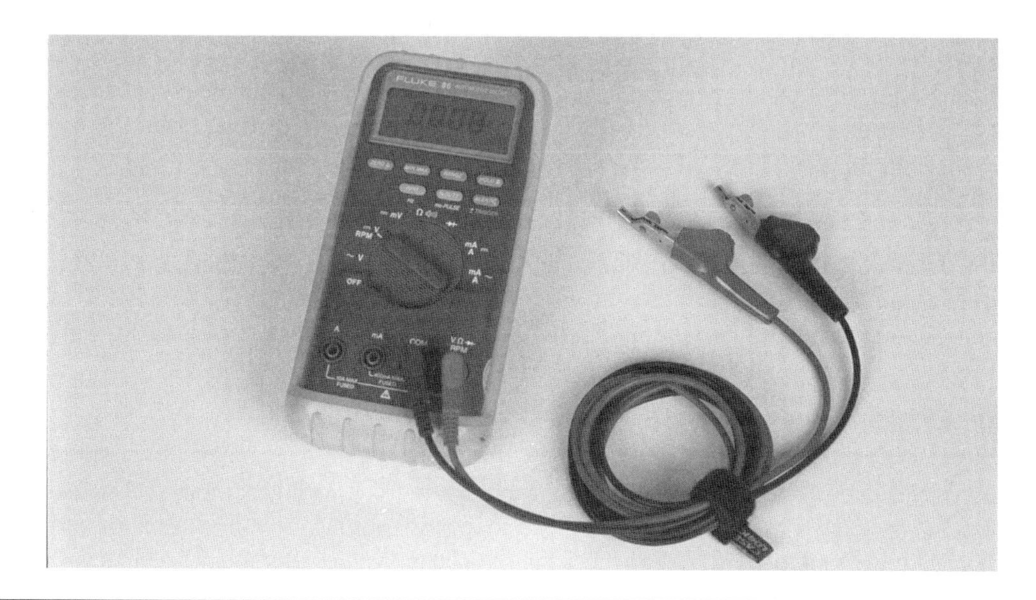

Figure 2-13 Digital multimeter.

PREFIX	SYMBOL	RELATION TO BASIC UNIT
Mega	M	1,000,000
Kilo	K	1,000
Milli	m	0.001 or $\frac{1}{1000}$
Micro	μ	.000001 or $\frac{1}{1000000}$
Nano	n	.000000001
Pico	p	.000000000001

Figure 2-14 Common prefixes used on meters.

0.345 KΩ = **345** Ω

1025 mAmps = **1.025** Amps

Figure 2-15 Placement of decimal and scale should be noticed when measuring with a meter in Auto Range.

Operating a DVOM is not that complicated. Hook up the test leads for the test you wish to perform. Through rotary switches or buttons, you can select a function and/or range. Many meters select the range automatically. After making a selection, most meters will display a prefix to indicate a specific range. Common prefixes are shown in Figure 2-14.

The ohmmeter function is a good example to use for ranging. Ohmmeters typically range in ohms, K-kilo or 1,000; M-mega or 1,000,000. For example, if the reading on the meter were 10.00 with a prefix of K ohms, the actual reading would be 10,000 ohms.

Another important aspect to take into consideration is that digital meters display four place numbers. You need to observe where the decimal point is placed (Figure 2-15). If you were to ignore the decimal point in Figure 2-15, you might very easily read the 345 ohms as 345,000 ohms.

Digital multimeters are fairly easy to use. You don't have to worry about polarity or over ranging. Like any tool you own, the more you use it, the more familiar you become with it.

Classroom Manual Chapter 2, page 27.

Oscilloscopes are commonly called scopes.

Lab Oscilloscopes

The lab oscilloscope is becoming a necessary tool for the truck technician today. Increased use of computers in today's trucks requires the need to see the outputs of various sensors and actuators. Digital meters are a good tool to check simple voltage values. But today's technician needs to see the rapidly changing values that occur in the electrical and electronic devices of today's trucks.

Scopes are said to be a visual voltmeter.

A lab scope can visually capture a momentary fault, which is hard to do with a digital meter. The following are some of the sensors and actuators used in trucks that are more easily diagnosed for problems by using a lab scope: vehicle speed sensors, magnetic camshaft and crankshaft position sensors, antilock wheel speed sensors, hall effect vehicle speed sensors, hall effect camshaft and crankshaft position sensors, EGR, turbo boost control solenoids, fuel injectors, and throttle position sensors.

At this point you may be saying to yourself that you are going to be a medium/heavy duty truck technician and no one in the truck industry really uses a lab scope, so why should you? This may be an appropriate answer—the following are some sensors and actuators from International's latest electronic control system for the DT466E and 530 E diesel engines:

- ❑ Thermistor type: engine coolant temperature sensor, engine oil temperature sensor, intake air temperature sensor
- ❑ Pressure sensors: manifold absolute pressure sensor, barometric pressure sensor
- ❑ Potentiometer: APS (Accelerator Position Sensor)
- ❑ Hall effect sensor: camshaft position
- ❑ Magnetic pickup sensor: vehicle speed sensor
- ❑ Fuel injectors

All of the sensors and actuators listed above are readable by a lab scope. Additionally, a lab scope is useful for any electrical signal dealing with voltage; most newer lab scopes can take amperage measurements through the use of inductive amp probes.

Oscilloscopes are valuable because they allow you to see the quantity and quality of a signal. It is a visual voltmeter that allows you to see voltage changes over time. A change in voltage will appear as a beam across the screen of the scope. This is illustrated by looking at a throttle position sensor (TPS) wave form (Figure 2-16).

The throttle position sensor is a potentiometer. The computer sends a 5-volt reference through the sensor. When the accelerator pedal is moved, the signal voltage going back to the computer changes. The computer uses the voltage it sees to determine various change requirements of the engine.

If you look at the horizontal (time) portion of the screen, you can see that accelerating (depressing the pedal) is a function of time. At the same time the pedal is depressed, voltage changes. That is why we refer to lab scopes as being able to measure voltage over time. If there had been an interruption of voltage during the sweep of the TPS, it would have shown up on

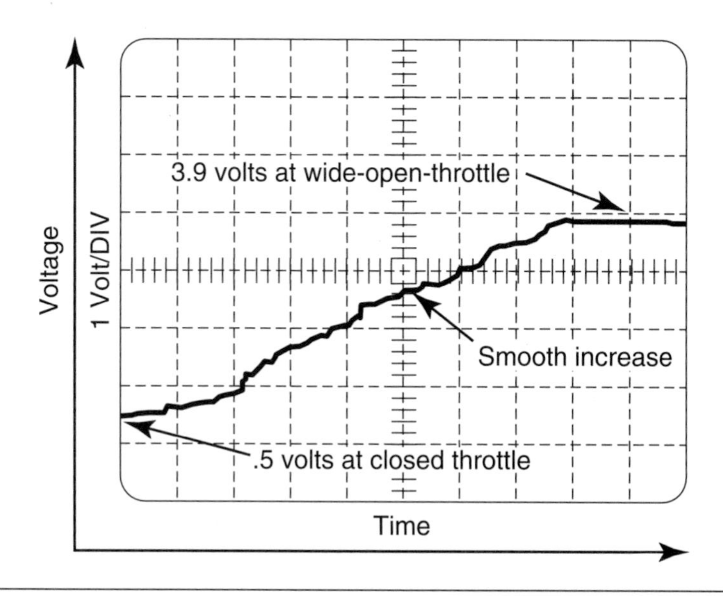

Figure 2-16 DC signal of a good TPS, measured from closed throttle to wide-open throttle.

the screen as a spike downward wherever the interruption of voltage would have occurred. We refer to this as a glitch.

Analog Versus Digital Scopes

The analog scopes have been around for a long time for truck use. They were mainly used to look at primary and secondary ignition systems. The signals we need to look at from various sensors are not as consistent as the ignition signals. To look at the variety of signals we need scopes that allow for more adjustments and parameters than an ignition scope would allow. Lab scopes fit all these needs.

Analog scopes show voltage signals in real time.

Digital storage oscilloscopes (DSO) sample voltage signals at many points of time.

The waveform on a scope is commonly called a trace.

The biggest difference between an analog and a digital oscilloscope is the way signals are acquired and displayed. Analog scopes show signals in real time. This means the signal you see on the screen is occurring at the same time. With a digital storage oscilloscope, the signal you see on the screen occurred a fraction of a second earlier.

The digital storage oscilloscope samples an incoming waveform signal at many points of time. They can freeze and store these waveforms to be more carefully analyzed. Most DSOs have a sampling rate of one million samples per second. The changes of amplitude over time can easily be observed. A loss or quick change during the sample rate period can also be observed. That makes a digital storage oscilloscope a valuable tool for the technician. The trace on the scope is repeated or refreshed at a fixed or variable rate.

Waveforms

The hardest part of using a DSO is setting it to display the various waveforms. To properly display a waveform, you must set the voltage scale and time base. Setting the trigger level or mode might also be required if the scope doesn't have an auto mode.

To help set the voltage scale and time base, the scope's screen has divisions (Figure 2-16). It might help to use the lines of the divisions as a ruler. Each division is adjustable. That means we can assign a value to each division to best reflect the waveform we are trying to see.

For example, we can assign a value of 1 volt for each vertical division. This would allow for a decent-looking waveform of a signal that stays within a 5-volt range. If the divisions were adjusted to 2 volts per division, the signal displayed would appear small even though it remained within a 5-volt range. In effect, we have changed the size of the window while looking at the same picture.

The time base (horizontal scale) can also be changed from milliseconds to seconds and minutes. This allows you to see changes in voltage that occur over a short or long time period. Also, the grids serve as a reference for measurement and time of the waveform.

The primary signals that are found and/or used in electronic vehicle systems are:

1. Direct current
2. Alternating current
3. Frequency modulated signals
4. Pulse width modulated signals
5. Serial data signals

Considering the number of signals that are used or generated in the vehicle's electronic systems, it is important to realize that these signals have certain dimensions.

Amplitude is a term that refers to the height of a waveform on a scope.

A **cycle** is one set of changes in a signal that repeats itself several times.

- **Amplitude** The voltage of the signal at a given time.
- **Frequency** Cycles or time between events. Usually given in cycles per second, or hertz (Figure 2-17). The higher the frequency, the more cycles occur in a second. One hertz is one cycle per second.
- **Duty cycle** The on time of the electronic signal during a cycle (Figure 2-18). It is a measurement of the amount of time something is on compared to the complete time of one cycle. It is measured in percentage. A 70% duty cycle means that the device was on 70% and off 30% of one cycle.

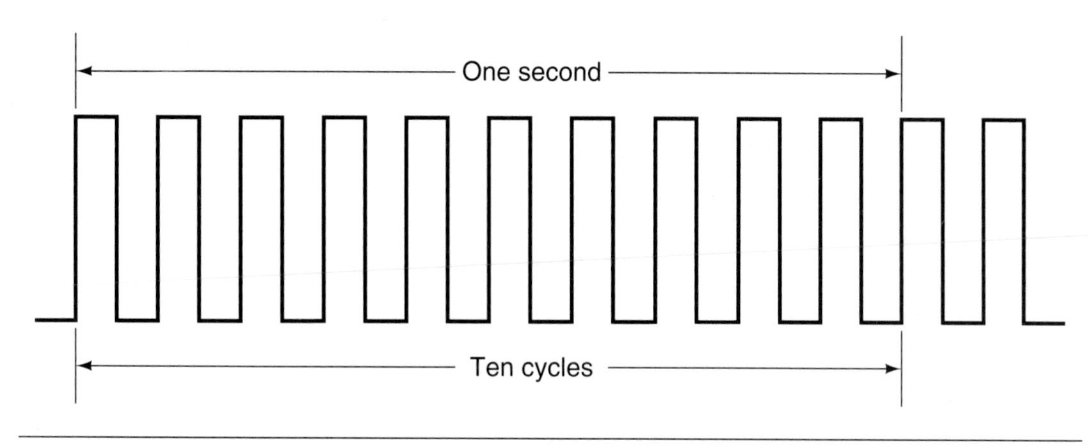

Figure 2-17 Signal frequency.

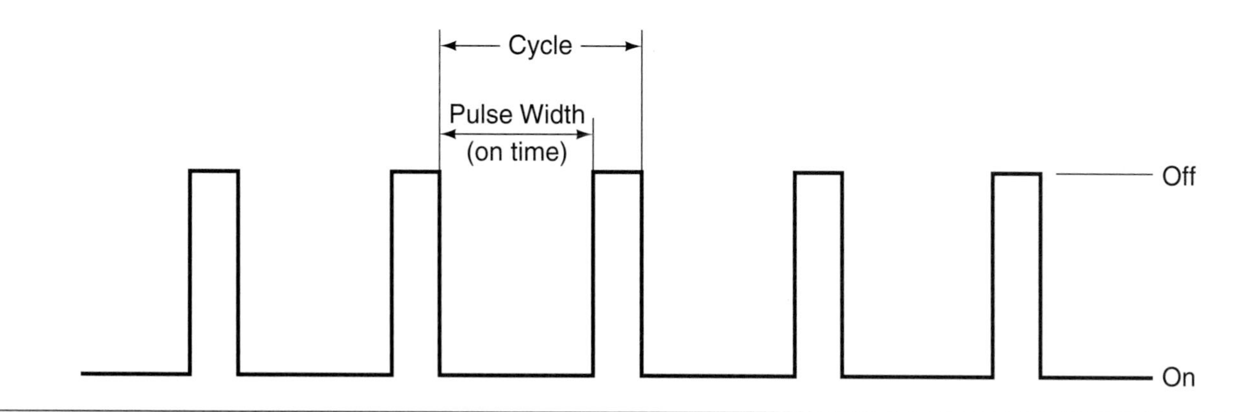

Figure 2-18 Duty cycle and pulse width.

- ❑ **Pulse width** Pulse width is a time function (Figure 2-18). It is similar to duty cycle except when you are measuring duty cycle, you are measuring the amount of time something is on during one cycle. Pulse width is normally measured in milliseconds. When measuring pulse width, you are looking at the amount of time something is on.
- ❑ **Shape** Shape is the signature of a signal such as certain curves, corners, etc.
- ❑ **Pattern** These are specific repeated patterns within a signal making up specific messages such as number 1 piston is at top dead center.

Most waveforms are composed of one or a combination of these dimensions.

A waveform represents voltage over time. When a trace is a straight horizontal line, the voltage is constant (Figure 2-19). This could be battery voltage and the divisions could be scaled to reflect 12.6 volts. The trace could be moving up and down in a straight line such as by an alternator output, showing an increase in voltage from 12.6 volts to 14.5 volts. The line could also drop very suddenly, showing a loss or drop of voltage.

A **waveform** represents voltage over time on an oscilloscope.

An alternating current (AC) waveform is shown in Figure 2-20. It reflects a consistent change of polarity. To look at one complete sine wave or cycle, start at the zero point, follow it to its positive peak, then move down through zero to its negative peak and return upward to the zero point. If the voltage of both the positive peak to zero is the same as the negative peak to zero, the wave is said to be sinusoidal. If the voltages are not the same, the wave is nonsinusoidal. The number of complete cycles per second is the frequency of the signal. Frequency of this type of a signal would be a good check for vehicle speed sensors or wheel speed sensors on ABS equipped vehicles.

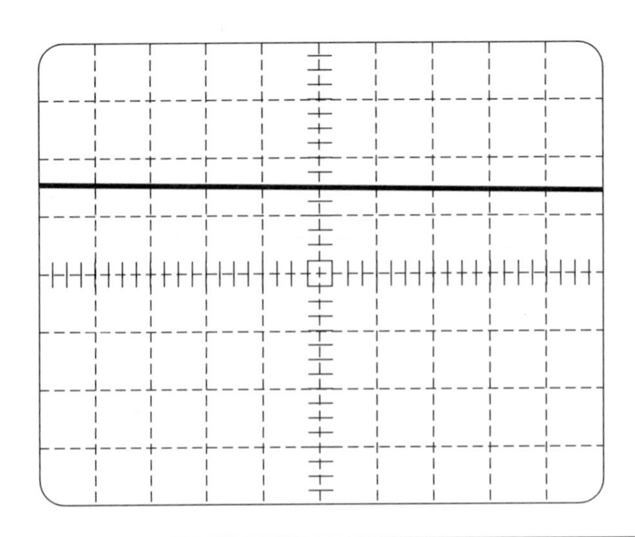

Figure 2-19 A constant voltage waveform.

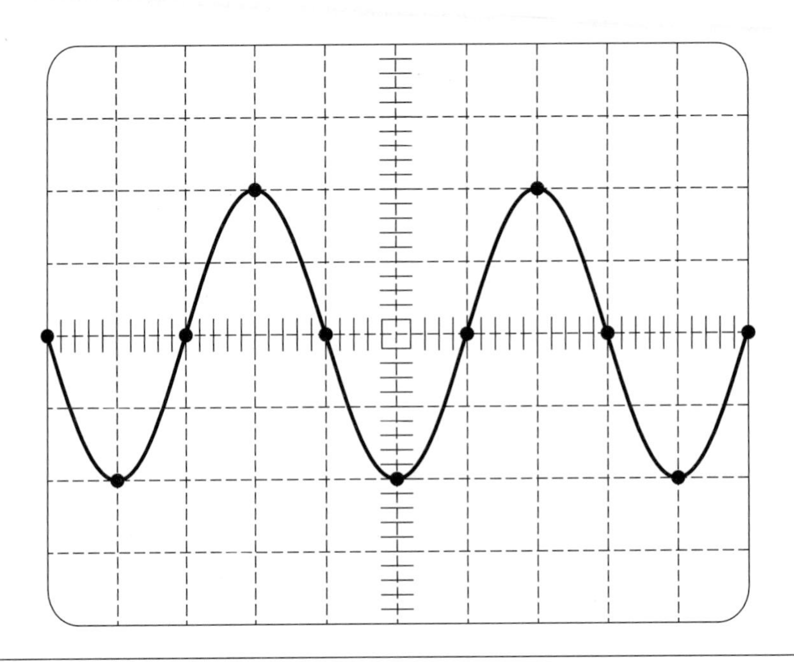

Figure 2-20 An AC voltage sine wave.

Direct current (DC) can also occur as a line showing a change in voltage over time. This can appear as a square wave, showing fast voltage changes (Figure 2-21). This figure shows the voltage being switched on and off with a fixed on/off time. The frequency of the signal can change with speed. This type of waveform is produced most commonly by a hall effect switch, and is called a fixed digital pulse signal.

It takes a lot of practice to become familiar with any waveform and to distinguish bad waveforms from good ones, but a picture is worth a thousand words. Knowing that every waveform has its own signature makes the digital storage oscilloscope a very valuable tool.

If a vehicle comes in with an electrical problem, you can take waveform readings prior to a repair. Once the repair is made, the lab scope can show you visually if a change was made and whether it is for the good or bad.

Hooking up the probes (leads) is straightforward. If you are comfortable hooking up the leads of a DVOM, you will see that you use the same hookups to test a circuit or device.

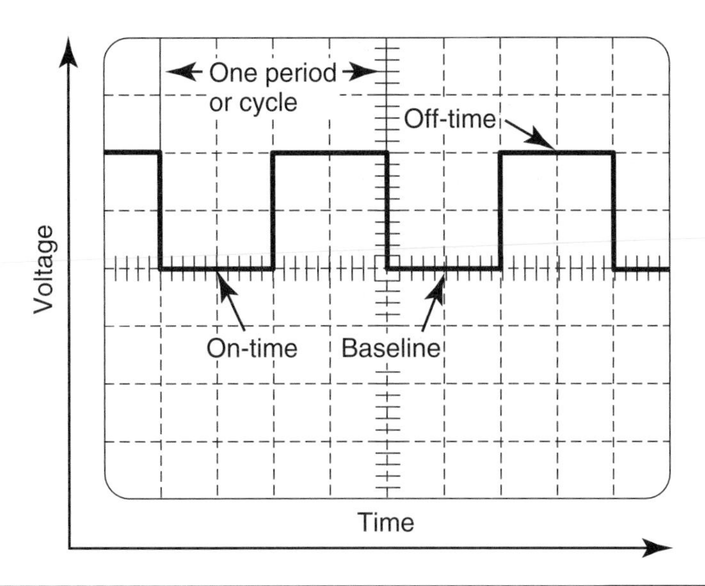

Figure 2-21 Fixed digital pulse-width signal waveform.

Other Test Equipment

A jumper wire is often overlooked as a piece of test equipment. Many times it is used as an extension of a circuit. It's value, though, comes when it is used to isolate a component, bypassing the circuit. The jumper wire has an alligator clip on each end. It should be fabricated with a fuse or circuit breaker to prevent damage to the circuit or component.

Connect one end of the jumper wire to the battery and the other end to the feed side of the component to check the operation of that component. If the component now works, you should look for a problem in the circuit.

CAUTION: Never use a jumper wire to bypass a fuse. Also, never connect a jumper wire across a battery. This will cause the wire to burn and could cause the battery to explode.

Another item of test equipment used by medium/heavy duty technicians is the scan tool. A scan tool is used to retrieve fault codes from the vehicle's computers. It can also receive serial data if the manufacturer offers the provision to do so. The serial data mode allows you to see the information that the vehicle's computer sees.

A **scan tool** is a test tool used to retrieve fault codes from the vehicle's computer(s) and/ or receive serial data.

Trouble codes are typically set by the vehicle's computer when a voltage signal is out of its normal range. By identifying the particular fault code, a technician can follow specific procedures to pinpoint the fault.

Becoming proficient with any test equipment takes a thorough understanding of the system and its components to be tested. It also requires access to various specifications.

CUSTOMER CARE: Servicing today's trucks without proper tools is a disservice to your customer. Also, accurate measurements taken today should be kept or recorded so they can be used as a reference in the future. Patterns or a change can be more accurately observed this way. Thus, this can be a form of preventive maintenance.

CASE STUDY

A vehicle was brought to the shop. The problem with the vehicle was that it didn't have any power and the engine would die out as if there were a lack of fuel; however, the problem was intermittent. This particular vehicle used an electronic engine control system.

The technician started with a good basic diagnosis to make sure there was fuel flow and all the filters were clean, including the air filter. After the initial diagnosis, a scan tool was hooked up to determine if any codes would be detected. The scan tool directed the technician to a throttle position sensor (TPS) problem.

The TPS was disconnected from the circuit so an ohmmeter could be used to determine its integrity. As the TPS was put through its motion, the resistance was varying. Since no problem was found with the component itself, the technician reconnected the TPS.

This presented her with the opportunity to try out her newest tool, the lab scope. She located the signal wire at the computer and connected the lab scope lead to it. This enabled her to look for any problems in the wiring and connection between the computer and TPS. She turned the ignition on and started to move the wiring and connections. Upon moving the connection at the TPS by the pedal, she noticed a glitch or a drop off of the signal.

The TPS on many trucks is at the throttle pedal. If the wiring is not properly fastened, the wires can be bumped and moved, and eventually the connection can be stressed enough to create a loose connection problem. In addition, environmental conditions that can occur in the winter, such as excess snow and salt around the pedal area, can lead to corrosive problems.

Because the connection looked corroded and worn, the technician replaced the complete plug assembly to ensure no comebacks. She rehooked her scope and moved the TPS through its full range of motion and moved all the wires and connections several times to ensure she made the proper repair. Road testing the truck verified that the repair cured the intermittent loss of power problem.

It should be noted that the same diagnosis of the electrical problem could be solved with an ohmmeter, voltmeter, lab scope, or any combination of these tools.

Terms to Know

Amplitude
Analog meter
Analog scope
Analog signal
Current drain
Current draw
Cycle
Digital meter
Digital storage oscilloscope (DSO)
DMM
Duty cycle
DVOM
Frequency
Nonpowered test light
Ohm
Pattern
Pulse width
Scan tool
Self-powered test light
Shape
Voltage drop
Voltmeters
Waveform

ASE-Style Review Questions

1. While discussing the use of a test light: *Technician A* says a test light can be used to check a switch while it is mounted. *Technician B* says a self-powered test light can be used to test a switch after it is removed. Who is correct?
   A. Technician A **C.** Both A and B
   B. Technician B **D.** Neither A nor B

2. Ohm's law is being discussed: *Technician A* says one known value will allow you to determine the remaining two. *Technician B* says Ohm's law is only used to determine resistance in a circuit. Who is correct?
   A. Technician A **C.** Both A and B
   B. Technician B **D.** Neither A nor B

3. While discussing measuring resistance: *Technician A* says an ohmmeter can be used to measure resistance of a component before disconnecting it from the circuit. *Technician B* says a voltmeter can be used to measure voltage drop. Something that has very little resistance will drop to zero or very little voltage. Who is correct?
   A. Technician A **C.** Both A and B
   B. Technician B **D.** Neither A nor B

4. When using a voltmeter: *Technician A* connects it across the circuit being tested. *Technician B* connects the red lead of the voltmeter to the most positive side of the circuit. Who is correct?
   A. Technician A **C.** Both A and B
   B. Technician B **D.** Neither A nor B

5. *Technician A* uses a DMM to test voltage. *Technician B* uses the same tool to test resistance. Who is correct?
   A. Technician A **C.** Both A and B
   B. Technician B **D.** Neither A nor B

6. *Technician A* says that in a series circuit the total resistance is the sum of all resistances. *Technician B* says that in a series circuit amperage is the same at any point. Who is correct?
   A. Technician A **C.** Both A and B
   B. Technician B **D.** Neither A nor B

7. *Technician A* says in a parallel circuit the total resistance is less than the lowest resistor. *Technician B* says all the voltage from the source must be dropped before it returns to the source. Who is correct?
   A. Technician A **C.** Both A and B
   B. Technician B **D.** Neither A nor B

8. While discussing electricity: *Technician A* says an open causes unwanted voltage drops. *Technician B* says high resistance problems cause increased current flow. Who is correct?
   A. Technician A **C.** Both A and B
   B. Technician B **D.** Neither A nor B

9. *Technician A* says if resistance decreases and voltage remains constant, amperage will increase. *Technician B* says Ohm's law can be stated as $I = E \div R$. Who is correct?
   A. Technician A **C.** Both A and B
   B. Technician B **D.** Neither A nor B

10. *Technician A* says voltage drop testing with a voltmeter is used to determine if excessive resistance is in the circuit. *Technician B* says voltage drop testing will determine the wattage rating of the headlights. Who is correct?
    A. Technician A **C.** Both A and B
    B. Technician B **D.** Neither A nor B

Job Sheet 3

Name: ______________________ Date: ______________________

Using Ohm's Law to Calculate Electrical Properties

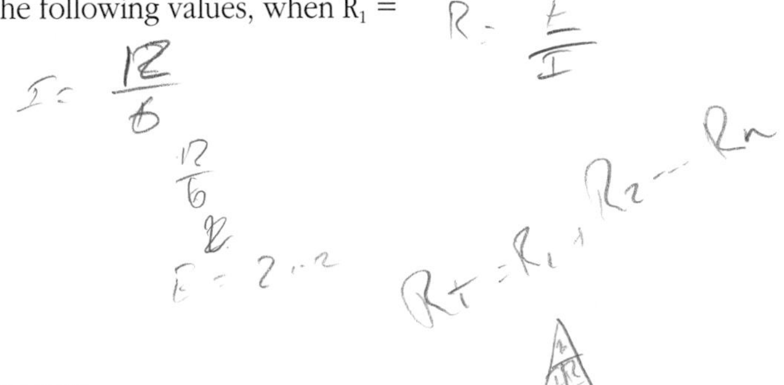

Upon completion of this job sheet, you should be able to use Ohm's law to calculate electrical properties.

Exercise 1—Series Circuit

Referring to the circuit in Figure 2-22, use Ohm's law to calculate the following values, when R_1 = 2 ohms and R_2 = 4 ohms:

Total circuit resistance = ______________ ohms
Circuit current = ______________ amps
Current through R_1 = ______________ amps
Current through R_2 = ______________ amps
Voltage drop across R_1 = ______________ volts
Voltage drop across R_2 = ______________ volts

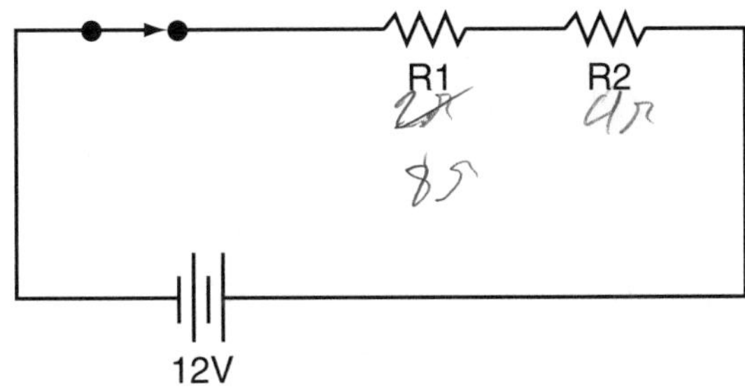

Figure 2-22 A series circuit.

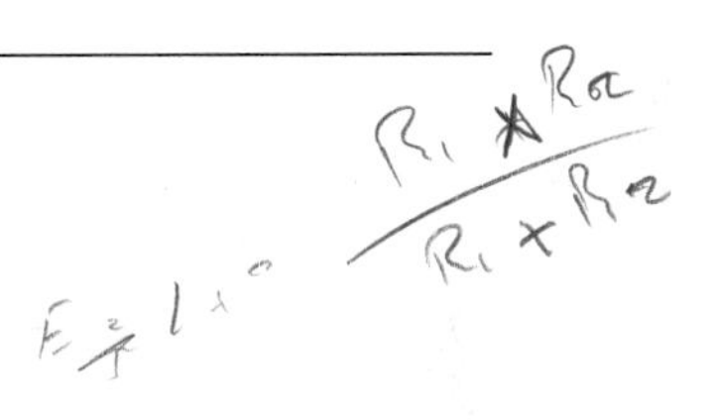

If the resistance of R_1 increases to 8 ohms, what are the new values?

Total circuit resistance = ______________ ohms
Circuit current = ______________ amps
Current through R_1 = ______________ amps
Current through R_2 = ______________ amps
Voltage drop across R_1 = ______________ volts
Voltage drop across R_2 = ______________ volts

Exercise 2—Parallel Circuit

Referring to the circuit in Figure 2-23, use Ohm's law to calculate the following values, when R_1 = 3 ohms and R_2 = 6 ohms:

Total circuit resistance = ______________ ohms
Circuit current = ______________ amps
Current through R_1 = ______________ amps
Current through R_2 = ______________ amps
Voltage drop across R_1 = ______________ volts
Voltage drop across R_2 = ______________ volts

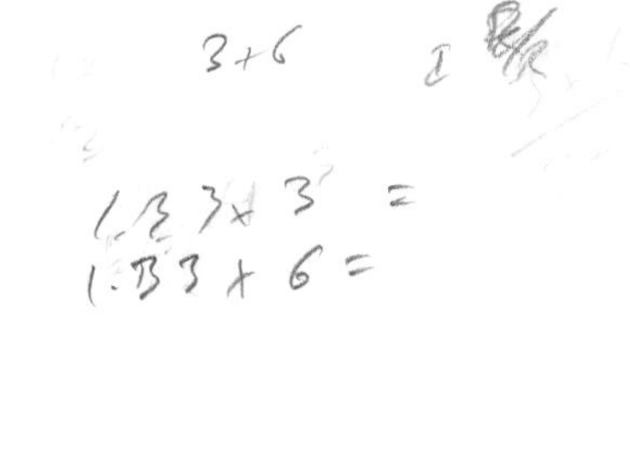

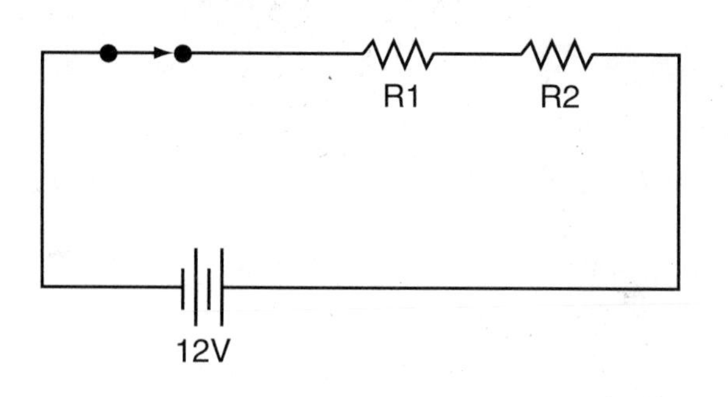

Figure 2-23 A parallel circuit.

Exercise 3—Parallel Circuit

Referring to the circuit in Figure 2-24, use Ohm's law to calculate the following values, when R_1 = 12 ohms and R_2 = 12 ohms:

Total circuit resistance = ___6___ ohms
Circuit current = ___2___ amps
Current through R_1 = ___1___ amps
Current through R_2 = ___1___ amps
Voltage drop across R_1 = ___12___ volts
Voltage drop across R_2 = ___12___ volts

$\frac{12 \times 12}{12 + 12} = \frac{144}{24}$

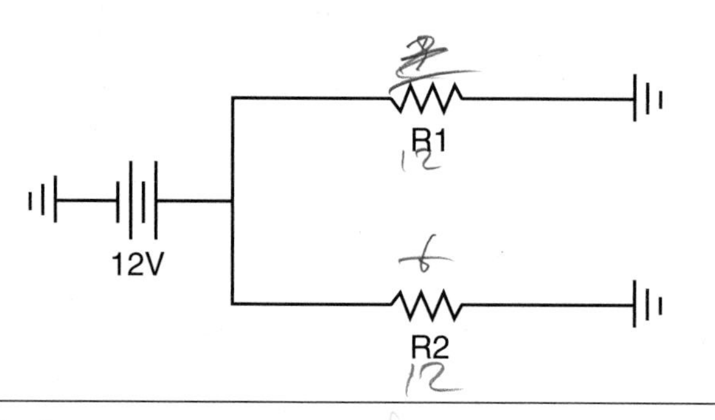

Figure 2-24 A parallel circuit.

Exercise 4—Parallel Circuit

Referring to the circuit in Figure 2-25, use Ohm's law to calculate the following values, when R_1 = 1 ohm, R_2 = 3 ohms, R_3 = 2 ohms, and R_4 = 2 ohms:

Total circuit resistance = ___2___ ohms
Circuit current = ___6___ amps
Current through R_1 = ___3___ amps
Current through R_2 = ___3___ amps
Current through R_3 = ___3___ amps
Current through R_4 = ___3___ amps
Voltage drop across R_1 = ___3___ volts
Voltage drop across R_2 = ___9___ volts
Voltage drop across R_3 = ___6___ volts
Voltage drop across R_4 = ___6___ volts

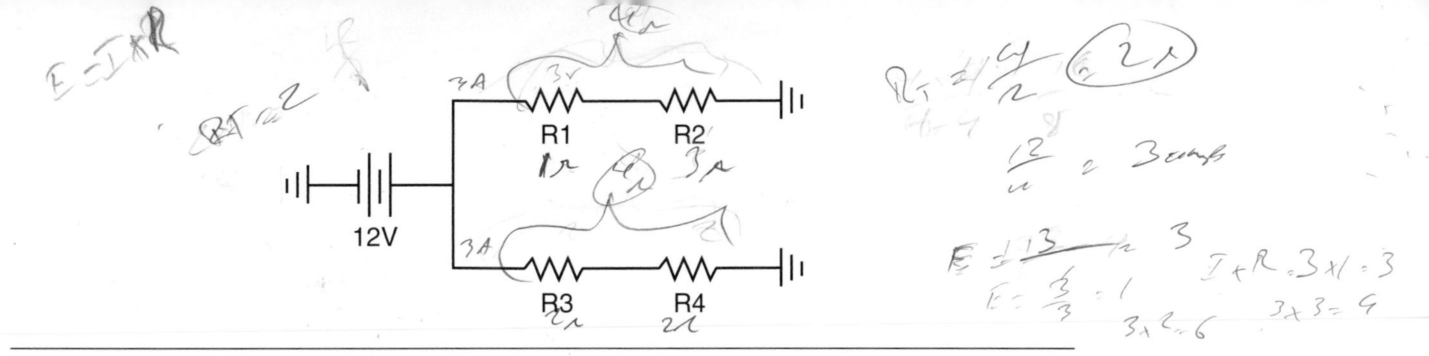

Figure 2-25 A parallel circuit.

Exercise 5—Series-Parallel Circuit

Referring to the circuit in Figure 2-26, use Ohm's law to calculate the following values, when R_1 = 1 ohm, R_2 = 2 ohms, R_3 = 3 ohms, and R_4 = 6 ohms:

Total circuit resistance = ____________ ohms

Circuit current = ____________ amps

Current through R_1 = ____________ amps

Current through R_2 = ____________ amps

Current through R_3 = ____________ amps

Current through R_4 = ____________ amps

Voltage drop across R_1 = ____________ volts

Voltage drop across R_2 = ____________ volts

Voltage drop across R_3 = ____________ volts

Voltage drop across R_4 = ____________ volts

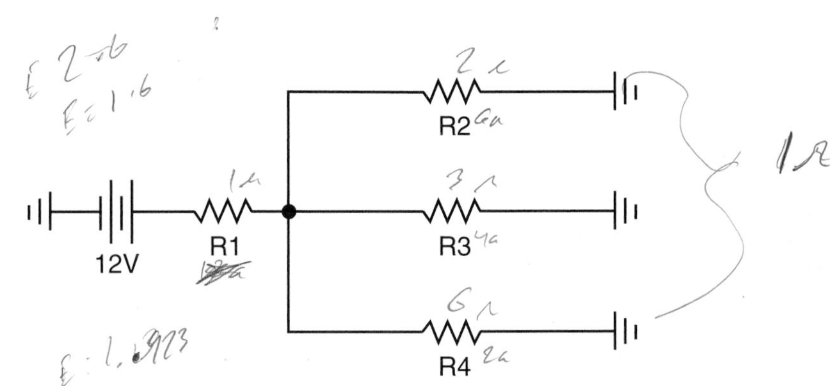

Figure 2-26 A parallel circuit.

Instructor Check ____

CHAPTER 3

Basic Electrical Troubleshooting and Service

Upon completion and review of this chapter, you should be able to:

- ❑ Describe proper troubleshooting techniques.
- ❑ Describe the various uses of test instruments.
- ❑ Test switches with various test instruments.
- ❑ Test relays and relay circuits for proper operation.
- ❑ Describe a potentiometer circuit.
- ❑ Test a potentiometer using a voltmeter and an ohmmeter.
- ❑ Diagnose diodes for opens and shorts.
- ❑ Locate opens in circuits.
- ❑ Locate shorts in circuits.
- ❑ Locate the cause of unwanted high resistance in a circuit.

Introduction

Troubleshooting electrical problems in trucks can involve simple problems such as a failed lightbulb to a more complicated problem within the lightbulb's circuit, such as a short to ground or a loose switch connection, which might involve more time in tracing the wiring.

The vehicle that comes into your shop often has a very tight schedule in order to get a time-committed load delivered. This means your diagnostic technique has to be sound and precise to accommodate a time schedule. One way of accomplishing proper diagnostics is to realize that electrical circuits, no matter where they are in the truck, have many common properties. One such property is that all electrical circuits must have voltage, current, and resistance. Another property is the relationship between voltage, current, and resistance.

Voltage is the electrical pressure needed for electron flows which is defined by current. Resistance is defined as opposition to current flow. The resistance is normally considered the load that can change electrical energy to light, heat, or movement. Of course the proper change of electrical energy cannot occur unless a "proper" path is provided for current flow, and all the electrical values of the circuit and its components stay within a predetermined range. The types of problems that can occur to change the electrical values are an open, a short, or an excessive voltage drop. With a solid understanding of basic electricity and the proper tools required to check the electrical values, the diagnostics of electrical problems can be accomplished regardless of where the circuit is in the truck or its use.

Basic Tools

Basic mechanics tool set

Service manual

Safety glasses

Classroom Manual
Chapter 3, page 44.

An **open circuit** is a circuit with a break that prevents current flow.

A **shorted circuit** alters the original path of current flow.

A high resistance in a circuit reduces current flow.

Basic Electrical Troubleshooting

Troubleshooting electrical problems requires logical steps to quickly lead to and pinpoint the problem. The following steps are guidelines to keep in mind when troubleshooting electrical problems:

1. Communicate with the driver if possible for the following information:
   - ❑ The nature of the problem.
   - ❑ When the problem occurs.
   - ❑ Is the problem intermittent or is it a complete system failure?
   - ❑ If the problem is intermittent, when does it tend to occur? For example, the circuit has to be on for awhile when the problem occurs. Another example is a failure of the circuit when another electrical circuit is turned on.
2. Verify the problem.
   - ❑ Operate the complete system to duplicate the problem, even if it is intermittent. This way you have a reference to verify the repair.
3. Narrow the problem. In order to narrow the problem so it can be isolated, you need to:
   - ❑ Know the system and how it operates.

- ❑ Know the circuit's limits and the effects if something goes wrong.
- ❑ Know the possibilities of certain conditions happening. This can save a lot of time. More conditions occur because of simple things rather than complex ones. Examples of simple things are loose connections, ground problems, and broken wires rather than components.
- ❑ Refer to wiring diagrams, schematics, and any other aid that will help in tracing and locating circuits and components.

4. Isolate the problem.

- ❑ Use proper tools and techniques to pinpoint the cause of the problem. This can start with simple checks of circuit protection devices, connectors, and switches and looking for broken or frayed wires. Isolating can also involve separating individual circuits from the main system by removing fuses, then observing symptoms.
- ❑ If it is a complex problem, you need to take electrical measurements so you have a baseline to refer to when any changes are made whether it is for the good or bad.

5. Make the repairs.

- ❑ Repair or replace the faulty portion of the circuit.
- ❑ Go one step further and make sure the cause of the fault is also taken care of. Examples of this include securing previously unsecured wiring, sealing connectors properly, and using properly specified wiring and components for that circuit.

Intermittent electrical problems take more patience to troubleshoot.

6. Verify the repair.

- ❑ Operate the system and other systems that might affect the problem circuit.
- ❑ If you used electrical measurements, now is the time to remeasure to make sure the repairs eliminated the failure.

SERVICE TIP: When repairing an electrical failure, a failure in another circuit can often be caused. Check the truck's other electrical circuits after the repair is finished.

A voltmeter is used to read electrical pressure, either available pressure or loss of pressure.

Because the failures of electrical circuits fall into few categories (opens, shorts, high resistance), the number of tools needed for diagnosing does not need to be extensive. For example, a voltmeter can be used to check for continuity, the presence of voltage, and the loss of voltage or voltage drop. Figure 3-1 shows the voltmeter being used to check for the presence of voltage. The meter is set to the volts direct current (VDC) setting and the negative lead to a good ground. With the circuit powered up, the positive lead is touched to various connections of the circuit to the load. Source voltage should be indicated anywhere in the circuit up to the load. A slight variation might exist because of resistances along the path. A loss of 1 volt or more will definitely indicate a high resistance. Zero volts indicates an open circuit.

A test light is used to indicate the presence of voltage.

A test light is used in the same fashion to indicate the presence of voltage. If the test light is on, voltage is present. If the lamp does not illuminate at a given point, it indicates an open or no voltage present. Keep in mind that for a circuit to operate properly:

- ❑ Voltage must be provided, which typically originates at the positive battery past.
- ❑ There should be no interruption of current through the conductors, controls, and loads.

WARNING: A test light should not be used on electronic components or damage to delicate electronic circuits and components may occur.

An **ammeter** is used to measure current flow.

An ammeter is used to measure current flow (Figure 3-2). The meter is placed in series so it becomes part of the circuit, unless an inductive ammeter probe is used. Either is a good indicator of current draw. A good example of the meter's use is to determine excess current draw by a failing load device such as a heater motor.

An **ohmmeter** is used to measure resistance and continuity.

An ohmmeter is used to measure resistance and continuity (Figure 3-3). To check the resistance of a component, disconnect it from the circuit. For accuracy, the resistance measurement needs to be compared with manufacturer's specifications whenever possible. Continuity is a common test performed by an ohmmeter. The following is a guide for continuity measurements:

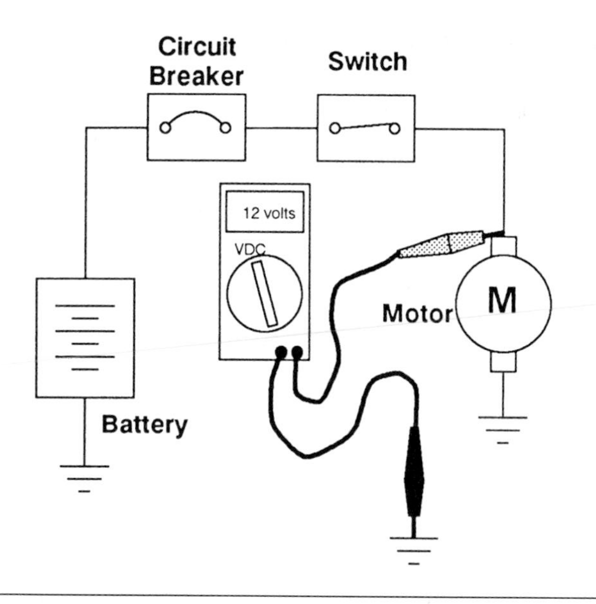

Figure 3-1 A voltmeter being used to check for the presence of voltage. (Courtesy of Mack Trucks, Inc.)

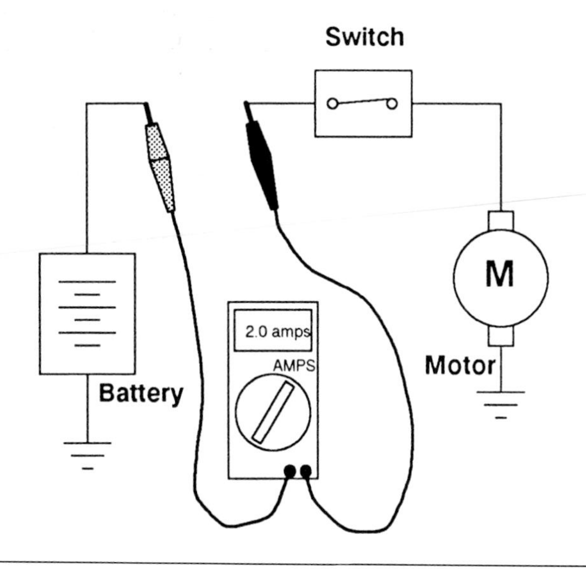

Figure 3-2 An ammeter being used to measure current flow. (Courtesy of Mack Trucks, Inc.)

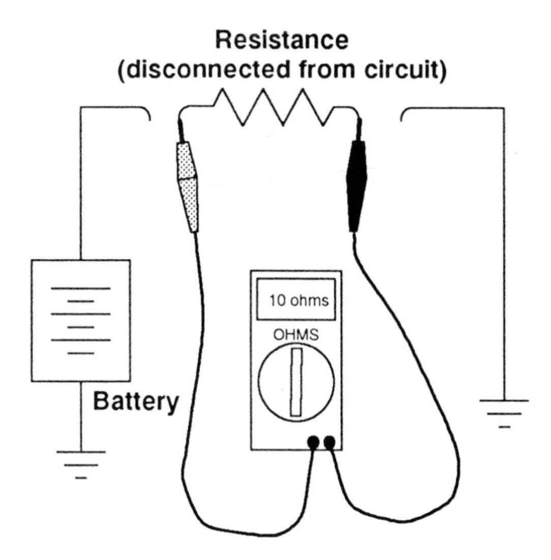

Figure 3-3 An ohmmeter being used to measure resistance and continuity. (Courtesy of Mack Trucks, Inc.)

- Low to zero resistance reading indicates continuity within the circuit or component.
- High resistance unless specified by a manufacturer indicates unwanted resistance such as poor connections or bad components.
- Infinity (an OL reading when using a digital meter) indicates an open circuit or the resistance is so high that current cannot flow.

Some meters might emit an audible "beep" when a circuit continuity problem is detected.

Infinity is used to define a resistance that is so high that current cannot flow.

WARNING: When using an ohmmeter, the power in the circuit must be turned off or damage to meter and/or circuit may occur.

CAUTION: The human body is a conductor of electricity; to prevent personal injury, observe all safety precautions when working with electricity.

Special Tools

Voltmeter

Ammeter

Ohmmeter

Test light

Testing Circuit Protection Devices

Classroom Manual
Chapter 3, page 45.

A fuse is a replaceable element that will melt if the current passing through exceeds the fuse rating.

Classroom Manual
Chapter 3, page 46.

Blade fuses use a code of either ATC or ATO.

Glass fuses use a code of SFE, AGA, AGW, or AGC. These codes are used to indicate various lengths and diameter of the fuses.

One advantage of a blade fuse is they can be pulled out of the fuse holder more easily and more safely than a glass type fuse.

Classroom Manual
Chapter 3, page 47.

A fuse link is commonly known as a fusible link.

Excess current flow in a circuit is called an **overload.**

As stated earlier, for a circuit to operate properly, voltage must be present to the load device. To protect the circuit or component, a circuit protection device is used to interrupt current flow whenever excessive current or overload condition exists. A voltmeter or ohmmeter would show a loss of continuity or an open circuit. When an open circuit occurs, the most obvious step to take is to see if the source of the open is the protective device, which indicates a problem in the circuit.

Fuses are one type of circuit protection device used in trucks. There are three basic types of fuses used: cartridge, blade, and ceramic. Common to these fuses are a metal strip that melts when an overload condition occurs, enabling a visual check of the fuse. Simply pull the fuse to see if the metal strip is intact or melted open. All types of fuses can also be checked with a test light, ohmmeter, or voltmeter. The test light and voltmeter show available voltage and continuity. The ohmmeter shows continuity. Pull the fuse out when using an ohmmeter.

All fuses are identified and rated at the current at which they are designed to blow. A three-letter code is used to identify the type and size of fuses. Blade fuses have codes ATC or ATO. All glass SFE fuses have the same diameter, but the length varies with current rating. Codes such as AGA, AGW, and AGC indicate the length and diameter of the fuse. Fuse lengths in each of these series are the same, but the current rating can vary. The code and current rating are usually stamped on the end cap of glass fuses. The current rating for blade fuses is usually marked on top and also indicated by the color of the plastic case (Table 3-1).

Table 3-1 Typical color coding of protective devices.

Blade Fuse Color Coding

Ampere Rating	Housing Color
4	pink
5	tan
10	red
15	light blue
20	yellow
25	natural
30	light green

Fuse Link Color Coding

Wire Link Size	Insulation color
20 GA	blue
18 GA	brown or red
16 GA	black or orange
14 GA	green
12 GA	gray

Maxi-fuse Color Coding

Ampere Rating	Housing color
20	yellow
30	light green
40	amber
50	red
60	blue

SERVICE TIP: To determine the correct fuse rating, use Watt's Law: Watts ÷ Volts = Amperes. For example, you are adding a 55-watt pair of fog lights to the truck: divide the 55 watts by the battery voltage, which is 12 volts. Since 55 ÷ 12 = 4.58, the current the circuit needs to carry is approximately 5 amperes. To allow for current surges, the correct in-line fuse should be rated slightly higher. In this case, an 8- or 10-ampere fuse would protect the circuit. *It is important to note that resistance values for lightbulbs are calculated hot. Solving for amps using resistance of a bulb can be difficult.*

Fuse links are used in circuits where a replaceable fuse might not be required. A typical application of the fusible link is in the starter circuit to protect the actual wiring harness. A fuse link is a short length of small-gauge wire installed in series with the conductor. Since the fuse link is a smaller gauge wire than the conductor, an overload condition will melt this wire, opening the circuit before further damage to it occurs. A special insulation is used that can bubble when it overheats, indicating a melted link. But if the insulation appears good, pull on the wire to see if it stretches, which also can indicate a melted link. The positive test, of course, would be using an ohmmeter or test light to check for continuity. To replace a fuse link, cut the protected wire where it is connected to the fuse link. Next, solder or tightly crimp a new fusible link with the same rating in its place. Be sure to mount fuse links as far as possible from combustible materials and components.

A visual inspection does not always validate a bad fuse. The melted strip might not be visible (Figure 3-4). For accuracy, use an ohmmeter, voltmeter, or test light. To check for continuity with a test light, remove the fuse from the fuse panel. Connect the ohmmeter's test leads across the protection device's terminals (Figure 3-5). On the meter's lowest scale, it should read between 0-1 ohms. If it reads infinity, the protection device is open. A fusible link can be tested the same way. Again, make sure there is no current flowing through the circuit. To be safe, disconnect the negative cable of the battery. Another method for testing circuit protection devices is using a voltmeter or test light to check for the presence of voltage on both sides (Figure 3-6). When using the test light or voltmeter, it is a good idea to also touch the fuse holder. There can often be a connection problem between the fuse and the holder.

SERVICE TIP: Prior to using a test light, verify that the tester's lamp functions. Connect the test light across a battery and the light should come on.

A **fusible link** is a special conductor with a heat-resistant insulation. When an overload occurs in the circuit, the conductor melts and opens it.

Measuring voltage drop is a useful check of the condition of the fuse and connections. A voltage drop of zero indicates a good fuse and/or connection. A reading of 12 volts indicates an open fuse. A reading above 0 volts indicates resistance. In this case, the circuit protection device is in poor condition and should be replaced.

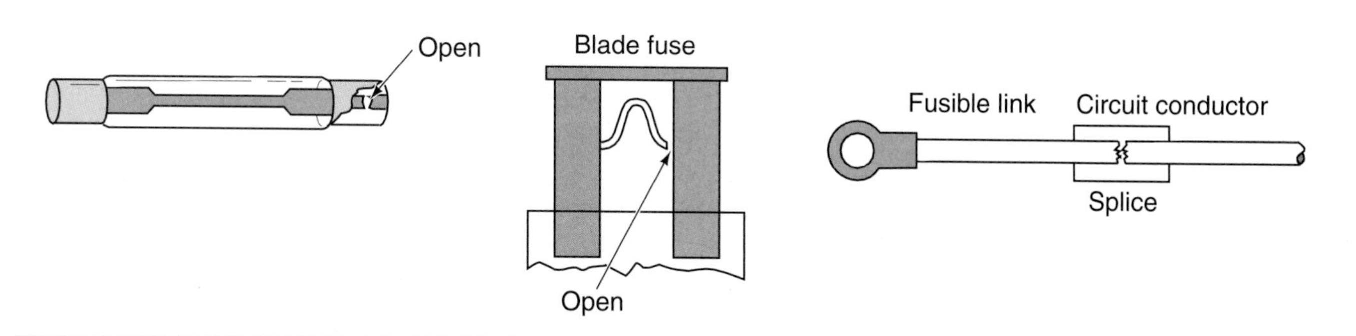

Figure 3-4 A fuse can have a hidden fault that cannot be seen by a technician. To make sure a fuse is good, test it.

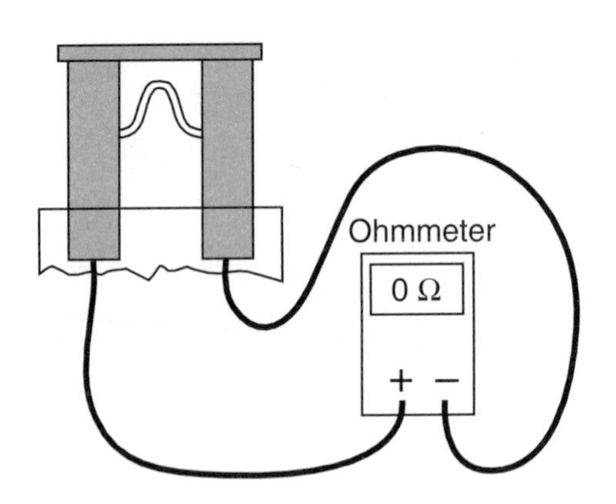

Figure 3-5 A good fuse will have zero resistance when tested with an ohmmeter.

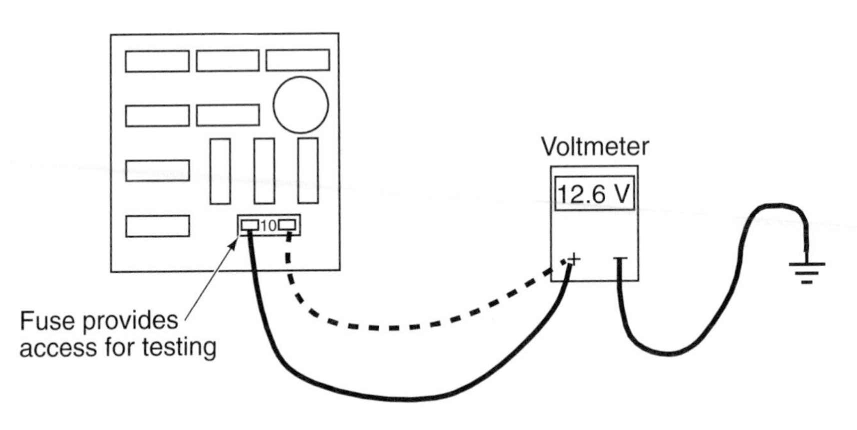

Figure 3-6 Circuit protection devices can be tested with a voltmeter. Make sure there is voltage present on both sides of the device.

Classroom Manual
Chapter 3, page 48.

Circuit breakers come in two basic types: manually resetting and automatic resetting.

SAE Type I circuit breakers are an automatic resetting "cycling" type of circuit protection devices.

SAE Type II circuit breakers are an automatic resetting "noncycling" type of circuit protection device. Power to the affected circuit breaker must be shut off before the circuit breaker will reset itself.

WARNING: Do not bypass a protection device with an unfused jumper. Circuit damage may result.

WARNING: Fuses are rated by amperage. Never install a higher amperage fuse into a circuit than the one designed by the manufacturer. Doing so may damage or destroy the circuit.

A very popular type of circuit protection device used in trucks is the circuit breaker. It is also rated by amperage, with the amperage designated on the top or side of the case. Circuit breaker styles can vary from a blade style to the style where the housing is used to hold the breaker in place and two threaded posts with nuts are used as connectors for the conductors.

Circuit breakers come in two basic types: manually resetting and automatic resetting. The SAE type 1 (automatic reset cycling) circuit breakers are most often used in trucks (Figure 3-7). When an overload condition occurs, the bimetallic strip heats and the arm flexes away, breaking the contact. The circuit remains open until the bimetallic strip cools. At this point the strip contracts, causing the contacts to connect, which reestablishes the circuit.

CAUTION: Continuous cycling of the breaker will make it very hot to the touch.

The breaker has two terminals labeled BAT and AUX. Battery voltage enters the breaker through the BAT terminal and is applied to the circuit through the AUX terminal. Circuit breakers are usually mounted in a central panel (Figure 3-8). This figure shows a typical installation of

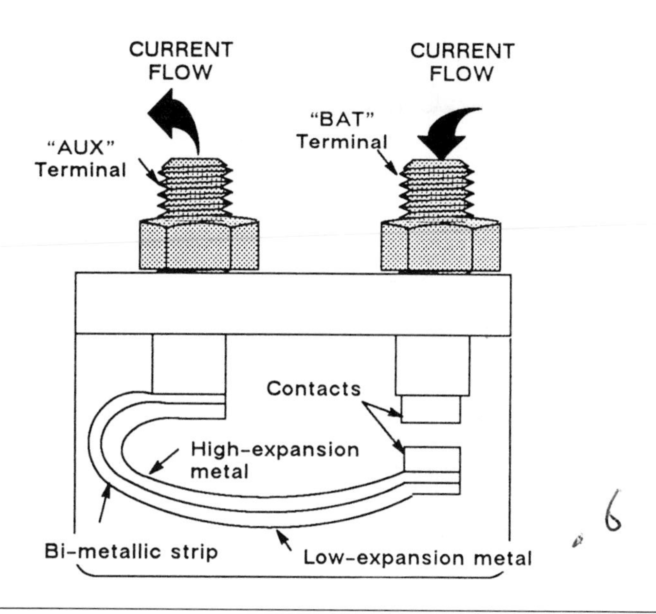

Figure 3-7 Type 1 (automatic reset cycling) circuit breaker. (Courtesy of Mack Trucks, Inc.)

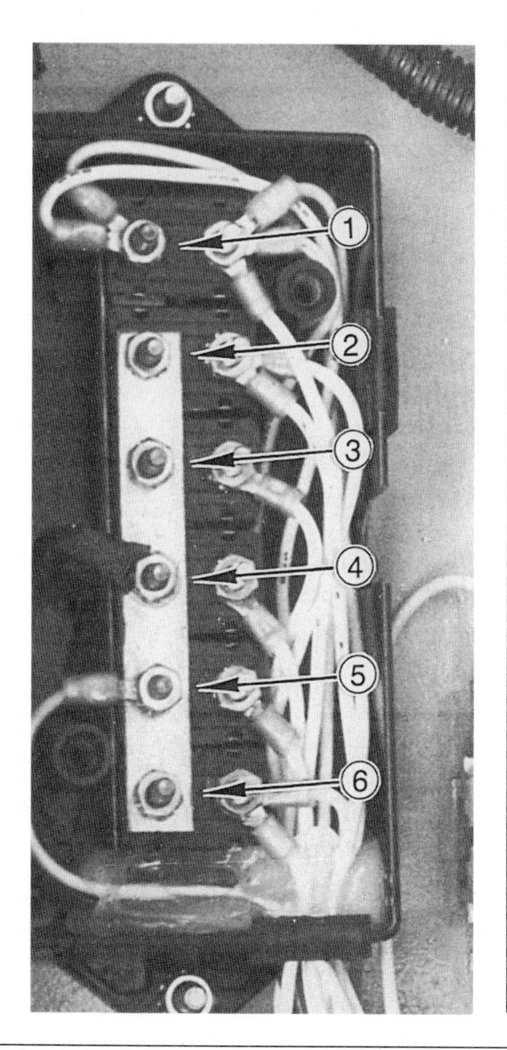

Circuit Breaker	Power	Amps	Function
#1	Keyed	15A	Backup Lights, Backup Alarm Option Equipment
#2	Battery	30A	Turn Signals, Horn
#3	Battery	30A	Taillights, Side Markers License Plate Light, Parking Lights, Clearance Lights, Panel Lights
#4	Battery	30A	Headlights- High and Low Beams
#5	Battery	30A	Key Switch, Starter Control, Windshield Washer, Windshield Wipers, Gauges, Drum Counter, Hydraulic Solenoids, Air Solenoids, Heater
#6	Battery	30A	Stoplights, Dome Light

Figure 3-8 A circuit breaker mounting panel. (Courtesy of Mack Trucks, Inc.)

Photo Sequence 3
Removing and Testing Circuit Breakers

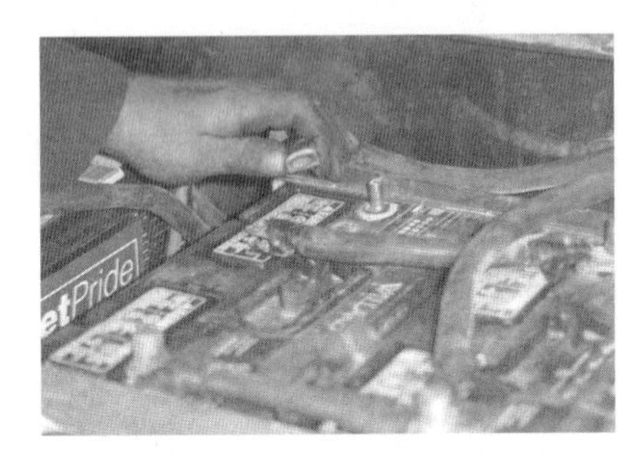

P3-1 Disconnect the negative battery cable.

P3-2 Remove the circuit breaker panel cover.

P3-3 Disconnect the suspected circuit breaker wires.

P3-4 This type of circuit breaker is removed by prying it out of the holder.

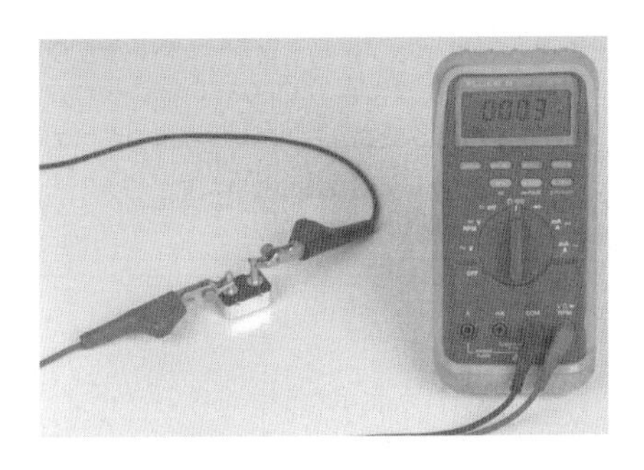

P3-5 This Type 1 circuit breaker can be tested by using the resistance (ohms) function of the meter and connecting the meter leads to the lugs of the breaker. A good circuit breaker will be indicated by a very low or no resistance reading. A defective circuit breaker will be indicated by a infinite resistance (OL on the digital meter) reading.

P3-6 To reinstall a circuit breaker, reverse the removal procedure. However, note that the terminal lugs are identified with the labels "BAT" or "AUQ." The proper connections have to be observed when reinstalling. The hot side of the circuit is connected to the "BAT" terminal lug.

the circuit breakers and a distribution legend. Notice how circuit breakers #2, #3, #4, #5, and #6 have a "bus-bar" used as a common connection for straight battery-powered circuits. Breaker #1 is not tied to the bus-bar because voltage is not available to that circuit until the key is turned on. Photo sequence 3 shows the proper methods used to remove and test circuit breakers.

A **bus-bar** is a common electrical connection to which many circuit breakers are attached. The bus-bar is connected to battery voltage.

Testing and Replacing Electrical Components

Failures of electrical circuits can often be attributed to electrical components. The first check to perform is to make sure there is available voltage to the component. If voltage is available, the next step is to test the component. The type of test is determined by what the component is meant to do and how it does it. A schematic can be a valuable tool for this if the component is complicated. The best way to test many components is to remove them from the truck and "bench-test" them.

Switches

A switch is a simple component that is normally open until acted upon to close the circuit. The easiest test of a switch is to jumper or bypass it (Figure 3-9). If the circuit operates with the switch bypassed, the switch is defective. Another test that can be performed is to check for voltage on both sides of the switch (Figure 3-10). Connect the voltmeter with the negative lead to a good ground and the positive lead to the power side of the switch. The voltmeter should indicate voltage to the input side of the switch. Move the positive lead of the meter to the output side of the switch. A normally open (NO) switch should have no voltage on the output side of the switch. Now close the switch and observe the meter. With the switch closed, there should be voltage present at the output side of the switch. A voltage drop is another good test to be performed when the switch is closed. Ideally, there should be no voltage drop. If a voltage drop is indicated, the switch should be replaced.

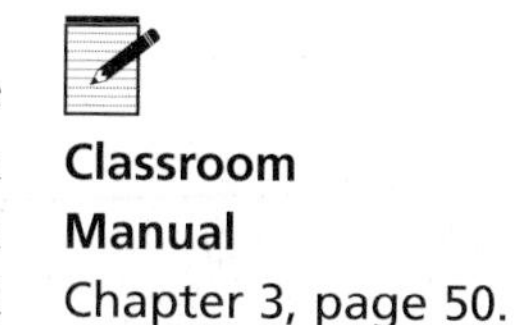

Classroom Manual
Chapter 3, page 50.

A **normally open (NO)** switch will not allow current flow in its rest position. The contacts are open until an outside force closes them to complete the circuit.

A **normally closed (NC)** switch will allow current flow when it is in its rest position. The contacts are closed until they are acted on by an outside force that opens them to stop current flow.

If a circuit being tested is powered through the ignition switch, it must be in the run position.

Any test performed with a voltmeter requires the battery or batteries to be fully charged in order to obtain accurate test results.

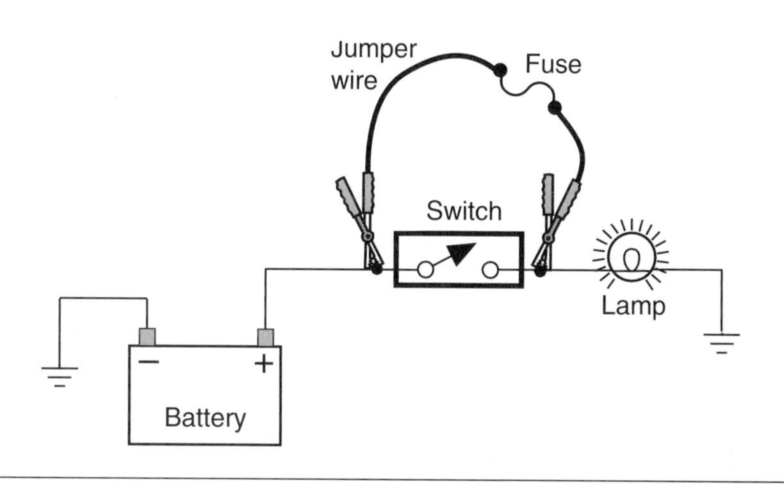

Figure 3-9 A switch can be checked by jumping around it with a fused jumper wire.

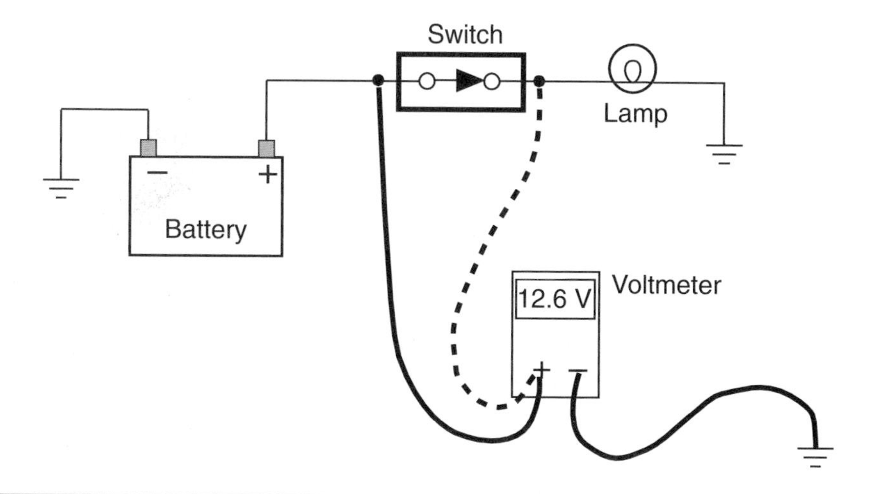

Figure 3-10 A switch can be checked with a voltmeter.

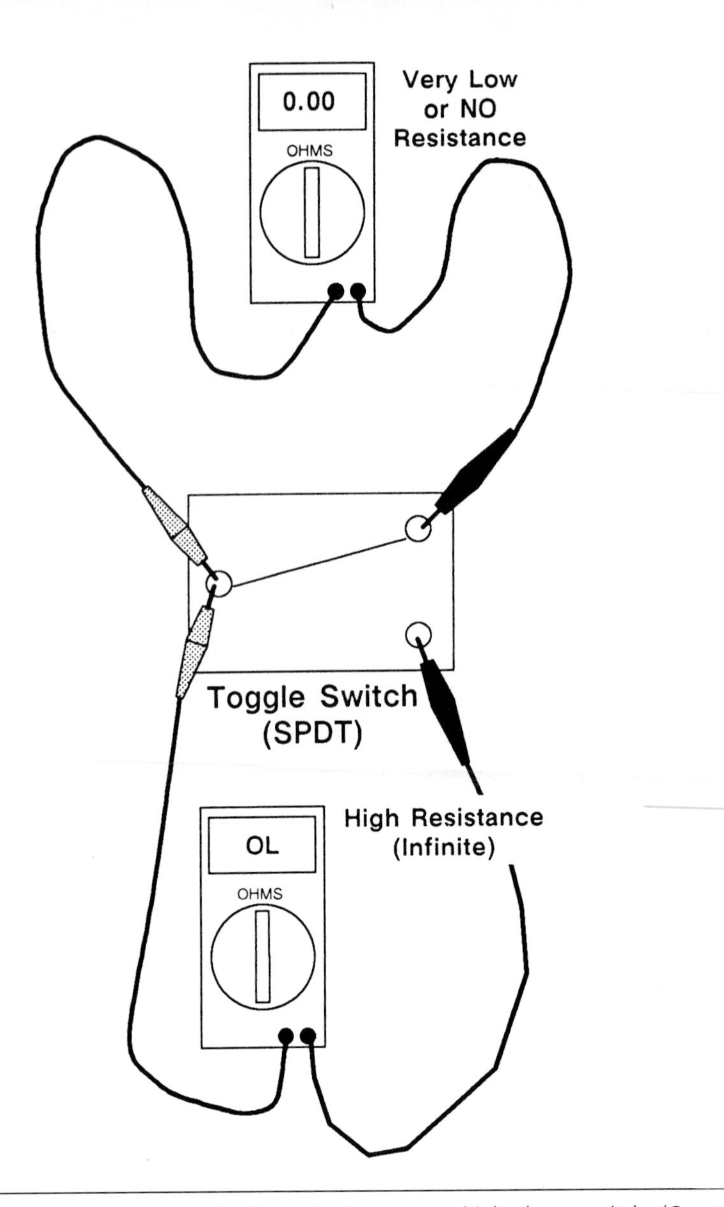

Figure 3-11 An ohmmeter being used to check continuity on a multiple throw switch. (Courtesy of Mack Trucks, Inc.)

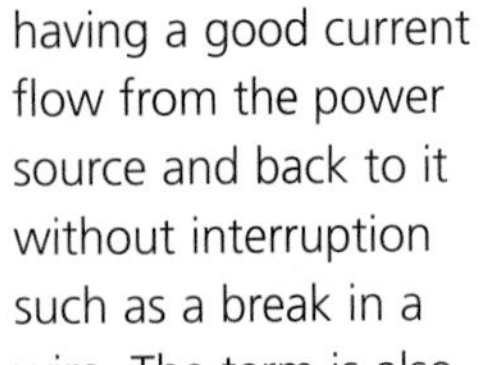

Classroom Manual
Chapter 3, page 51.

Continuity means having a good current flow from the power source and back to it without interruption such as a break in a wire. The term is also used when checking a component; for example, the wire has continuity or the switch has continuity, etc.

Using an ohmmeter to check continuity works well on multiple-throw switches (Figure 3-11). A single-pole double-throw (SPDT) toggle switch has three wire terminals: one common input and two output terminals. There should be continuity on one output and an open on the other output, dependent upon the direction of the pole toggled.

Classroom Manual
Chapter 3, page 51.

A relay is a device that uses low current to control a high-current circuit. The relay can be either a normally open or normally closed design.

Relays

A relay can also be checked using a jumper wire, test light, voltmeter, or ohmmeter. When it comes to relays, the type of test is often determined by the accessibility to the terminals. For example, a horn relay has its terminals accessible for various tests. The first step is to determine if the relay is controlled through a ground or insulated switch. To test a ground switch-controlled relay for a horn circuit, use Figure 3-12 as a guide.

In this illustration, the horn is powered through the fuse, which is at battery voltage so the horn can be used whether the key is on or off. The rest of the circuit is composed of the relay, horn switch, and a set of horns. When the horn button is depressed, a circuit to ground is completed. The current flow through the coil causes the relay contacts to close, which completes a circuit to the horn, causing the horns to operate. To diagnose this system, follow these steps:

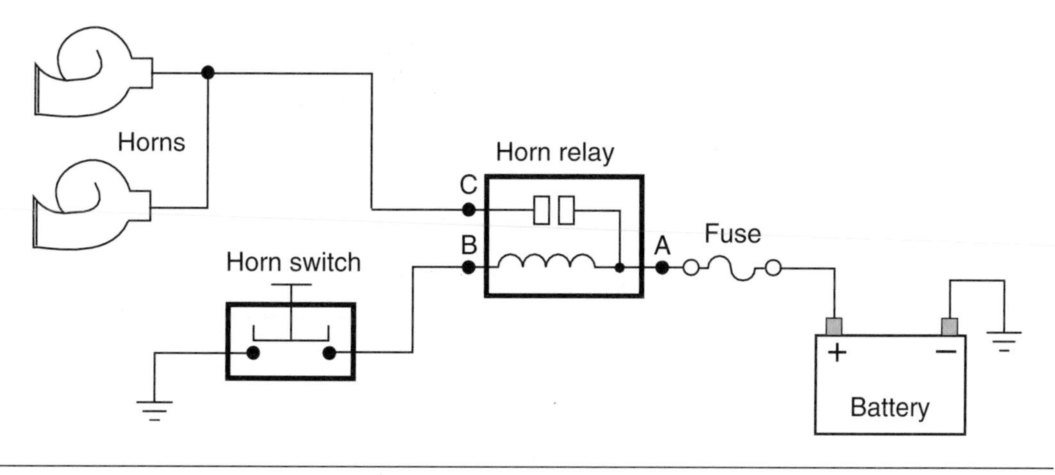

Figure 3-12 A relay circuit with a ground control switch.

1. Use a voltmeter to check for available voltage at terminal A. If voltage is not present, the problem is between terminal A and the battery. If voltage is present, continue testing.
2. Probe for voltage at control terminal B of the relay. If no voltage is present, the fault is in the relay coil. If voltage is present, continue testing.
3. This time a jumper wire can be used to connect terminal B to a good known ground. If the horn sounds, the fault is in the control side of the circuit, which can be the switch or conductors. If the horn does not sound, continue with the next step.
4. Connect the jumper wire from the battery positive to terminal C. If the horn does not sound, there is a fault in the circuit from the relay to the horn ground. If the horn sounded, the fault is in the relay.

Many times relays are computer controlled. A test light is not the recommended tool to use because the light might draw more current than the circuit is designed to carry and damage the computer. In these cases a digital voltmeter can be used to test a relay circuit (Figure 3-13). The

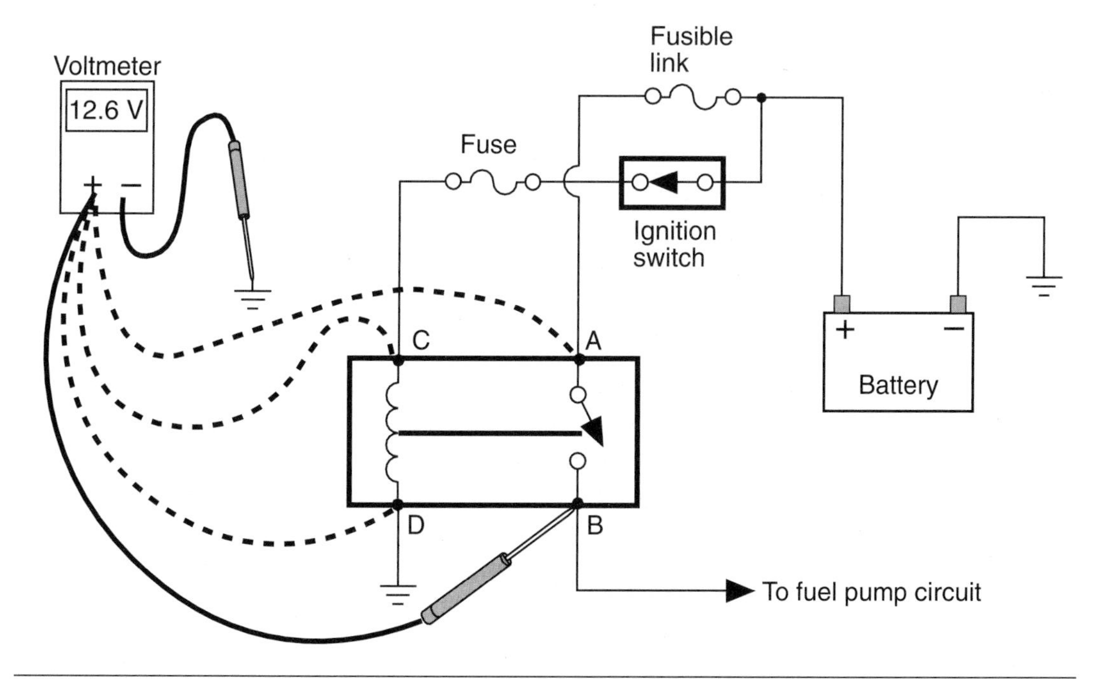

Figure 3-13 Testing relay operation with a voltmeter.

following are steps you can use with the voltmeter set on the 20-VDC, or volts DC, scale if it is auto-ranging:

1. Connect the negative voltmeter to a good ground.
2. Connect the positive voltmeter lead to terminal B. Turn on the ignition switch. If no voltage is present at this terminal, go to Step 3. If the voltmeter reads 12.0 volts or higher, turn off the control circuit. The voltmeter should then read 0 volts. If it does, the relay is good. If the voltmeter still reads any voltage, the relay is not opening and needs to be replaced.
3. Connect the positive meter lead to the power terminal A. The voltmeter should indicate at least 12 volts. If it is below this value, the circuit from the battery to the relay is faulty. If the voltage value is correct, continue testing.
4. Next, connect the positive meter lead to the control terminal C. The voltage should read 10.5 volts or higher. If it does not, check the circuit from the battery to the relay. If the voltage is 10.5 volts or higher, continue testing.
5. Connect the positive meter lead to the relay ground terminal D. If more than 1 volt is indicated on the meter, there is a poor ground connection. If the relay is less than 1 volt, replace the relay.

Sometimes a relay cannot be tested because the accessibility is not there. In those cases the relay can be tested with an ohmmeter after it is pulled out of the relay socket. A typical hookup of the ohmmeter to the relay for testing is shown in Figure 3-14. A schematic of a suppressed Hella relay is shown in Figure 3-15. If you look between terminals 86 and 85 of the relay, you will notice a resistor circuit parallel to the relay coil. It prevents voltage spikes from damaging electronic components in the vehicle. Take time to study the schematic closely to see the relationship of the relay terminals to the socket insertion end. Many relays will also have a circuit di-

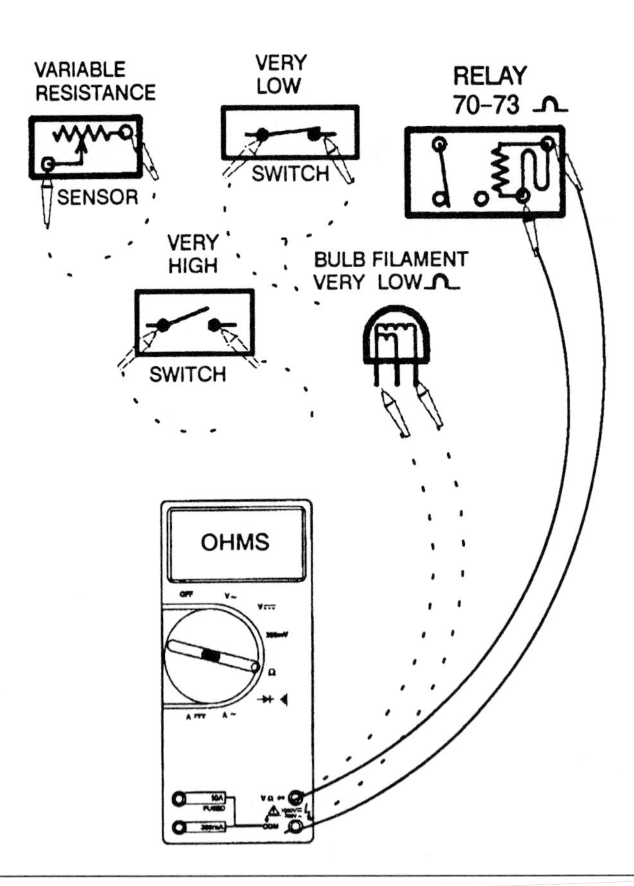

Figure 3-14 Place the ohmmeter leads across the component or circuit as shown for a resistance reading displayed in ohms. (Courtesy of Navistar International Engine Group)

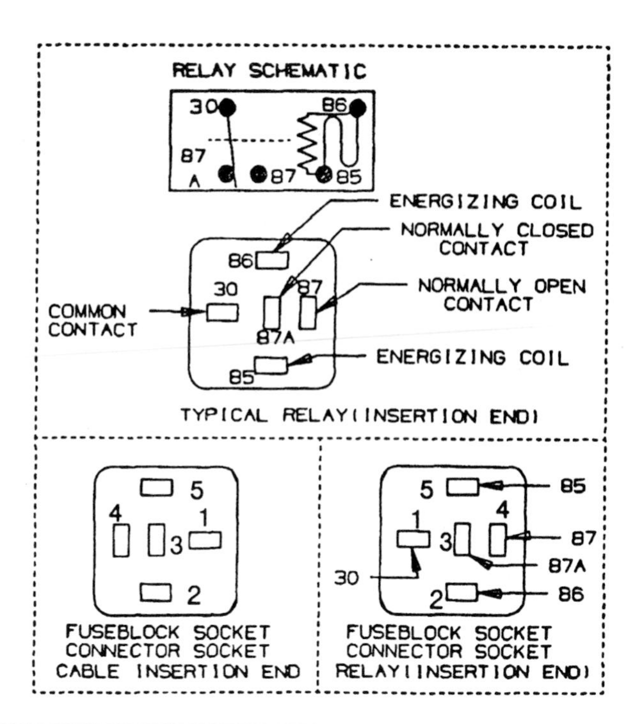

Figure 3-15 Relay function and wiring guide for a suppressed Hella relay. (Courtesy of Navistar International Engine Group)

agram located on the relay body. After you have become familiar with this relay schematic, the following steps can be used to troubleshoot the relay with an ohmmeter:

1. With the relay removed from the socket, place the ohmmeter leads between relay terminals 85 and 86 to read coil resistance. The resistance in this particular relay should be approximately 72 ohms. If the resistance is not correct, replace the relay. If the resistance is correct, go to Step 2.
2. Measure resistance between terminals 30 and 87A. The resistance should be less than 1.0 ohm. If resistance is less than 1 ohm, go to Step 3.
3. Use a 12V power source such as a battery and with test leads apply 12 volts across terminals 85 and 86. At this point, the relay should make an audible click. Use an ohmmeter and measure resistance between terminals 30 and 87 with the relay energized. Resistance should be 1.0 ohms or less. If continuity is not present, replace the relay. If there is continuity between terminals 30 and 87, go to Step 4.
4. With the relay still energized from Step 3, measure resistance across terminals 30 and 87A. The circuit resistance should be 100K ohms or more (no continuity). *It is important to note that some ohmmeters need to be ranged at this point if they do not have auto ranging.* If continuity is not present, the relay tests good. If there are less than 100K ohms, replace the relay.

The above relay test was taken from a manufacturer's flow chart with specific measurements, but the concept and test procedures are good for most types of relays with the same schematic. This emphasizes the need for a solid understanding of electrical circuitry and components.

Sometimes a relay needs to be powered up even though an ohmmeter is used for testing. Be careful not to touch the coil terminals with the ohmmeter test leads.

Classroom Manual
Chapter 3, page 54.

Testing Stepped Resistors

A stepped resistor is a control device with fixed resistances. The best method to test it is to use an ohmmeter. Connect the ohmmeter leads to the two ends of the resistor (Figure 3-16). Compare the results to manufacturer's specifications. A voltmeter can also be used to check the stepped resistor. It can check for applied voltage and output voltage after each resistor. By comparing the readings to specifications, you can tell if the resistor is good or not.

A **stepped resistor** has two or more fixed resistor values.

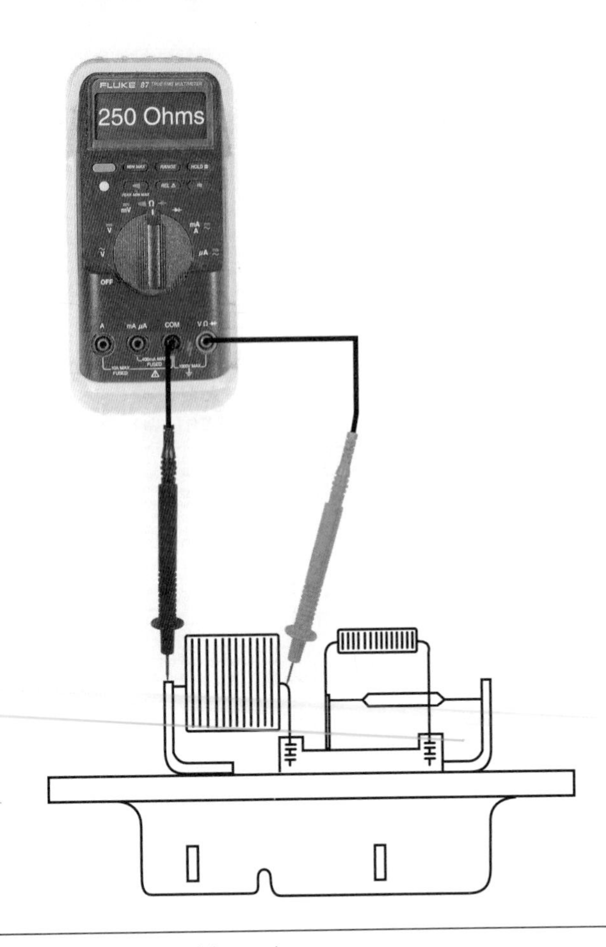

Figure 3-16 Testing a stepped resistor with an ohmmeter.

Classroom Manual
Chapter 3, page 54.

A **variable resistor** provides for an infinite number of resistance values within a range.

The **accelerator position sensor (APS)** is a potentiometer.

The **idle validation switch (IVS)** is incorporated as part of the APS.

A **potentiometer** is a three-wire variable resistor that acts as a voltage divider to produce a continuously variable output signal in proportion to a mechanical position.

Testing Variable Resistors

Most electronically controlled engines in today's trucks use a variable resistor device to check accelerator position. Many trucks use a combination device such as the one used for the Cummins Celect engine control system (Figure 3-17). The accelerator pedal accelerator position sensor/idle validation switch (APS/IVS) is a combination of a potentiometer and switch. The APS is the potentiometer attached to the accelerator pedal. Movement of the pedal results in a variable voltage signal being sent to the Electronic Control Module (ECM). A 5-volt signal is supplied to the APS at terminal C from the ECM. The variable signal, depending on the pedal position, is sent through terminal B to the ECM and the ground is through APS terminal A to the ECM.

SERVICE TIP: On this particular system, anytime the APS is disconnected from the ECM with the key switch on, they both must be recalibrated.

The IVS detects the position of the accelerator at low idle position. A 5-volt signal from the ECM is applied to the IVS terminal F. The IVS then sends a 5-volt signal to the ECM terminal #6, the accelerator is in low idle position, and a signal is sent to ECM terminal #9 when the throttle is not in the idle position.

For our purposes, we can check the APS portion using the illustration as a guide and an ohmmeter as the test tool in the following sequence:

1. With the key off, disconnect the APS from the cab harness at C306. Measure resistance at C306 (the APS assembly side) between pins A and C with the pedal in both the idle and full throttle position. If the resistance is between 2,000 and 3,000 ohms

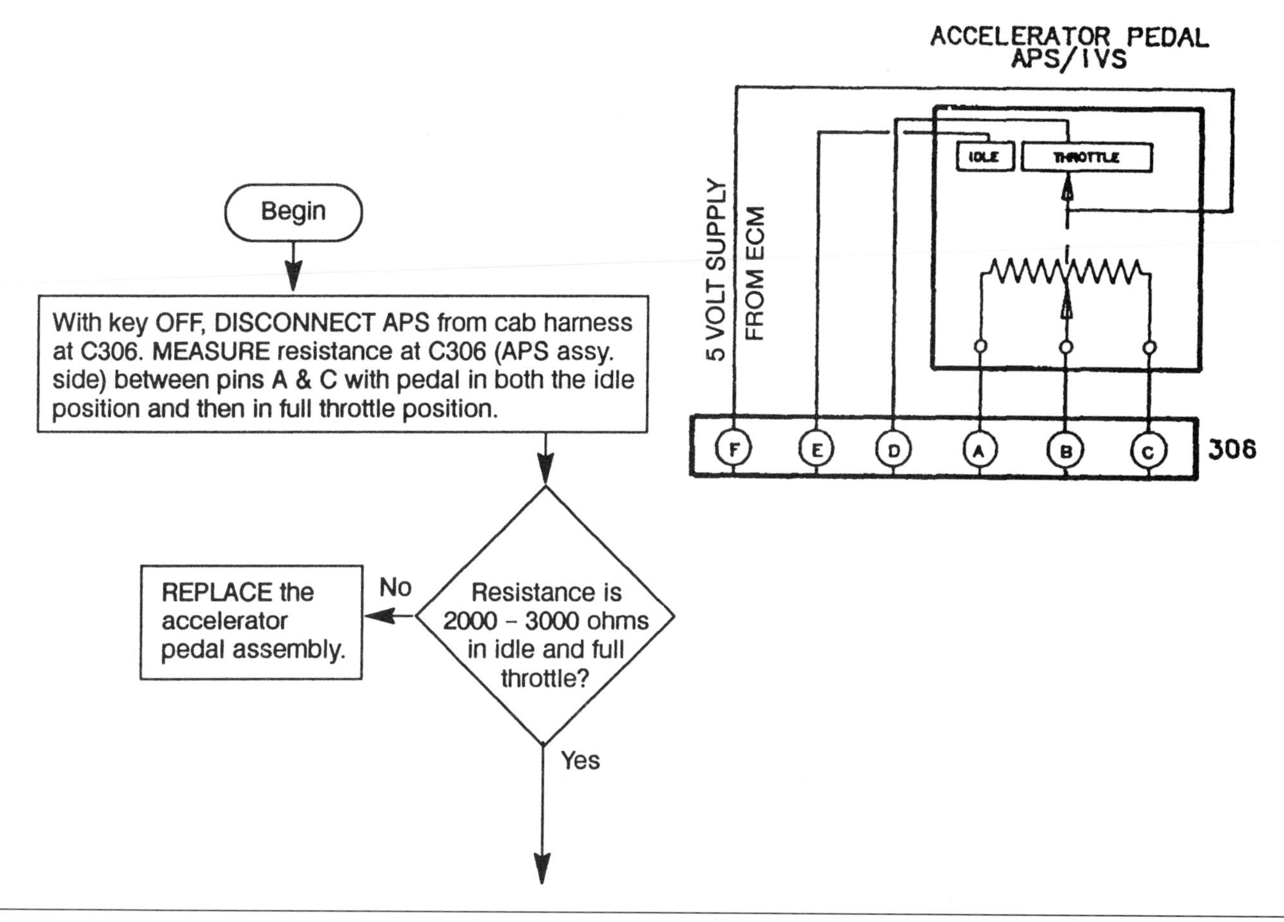

Figure 3-17 The APS/IVS is a combination device consisting of a potentiometer and a switch. (Courtesy of Navistar International Engine Group)

in both positions, go to the next step. If the resistance is not in that range, replace the accelerator pedal assembly.

2. In the idle position, measure resistance between pins C and B-. The resistance should be between 1,500 and 3,000 ohms. If the resistance is within that range, go to the next step. If the resistance is not within that range, replace the accelerator pedal assembly.
3. In the full throttle position, measure the resistance between pins C and B. The resistance must be at least 100 ohms less than the reading at idle and must be between 250 and 1,500 ohms. If the resistance is correct, go to the next step. If it is not correct, replace the accelerator assembly.
4. Measure resistance between the ground (metal part of the accelerator assembly) and connector terminals A, B, and C. If the resistance is 100 K ohms or more at A, B, and C, the accelerator position sensor portion of the assembly is all right. If the resistances were not correct, replace the APS/IVS assembly.

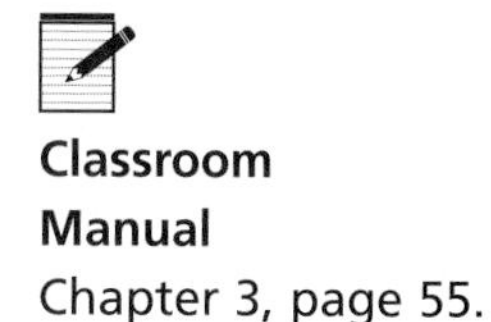

Classroom Manual
Chapter 3, page 55.

A voltmeter or labscope can also be used to check the potentiometer. Because the circuit needs to be on when using these meters, a set of jumper wires will have to be used or proper back-probe pins.

WARNING: Do not pierce the insulation to test the potentiometer. Piercing might break some of the strands of wire, resulting in voltage drops, not to mention that with the insulation broken moisture can enter and cause corrosion.

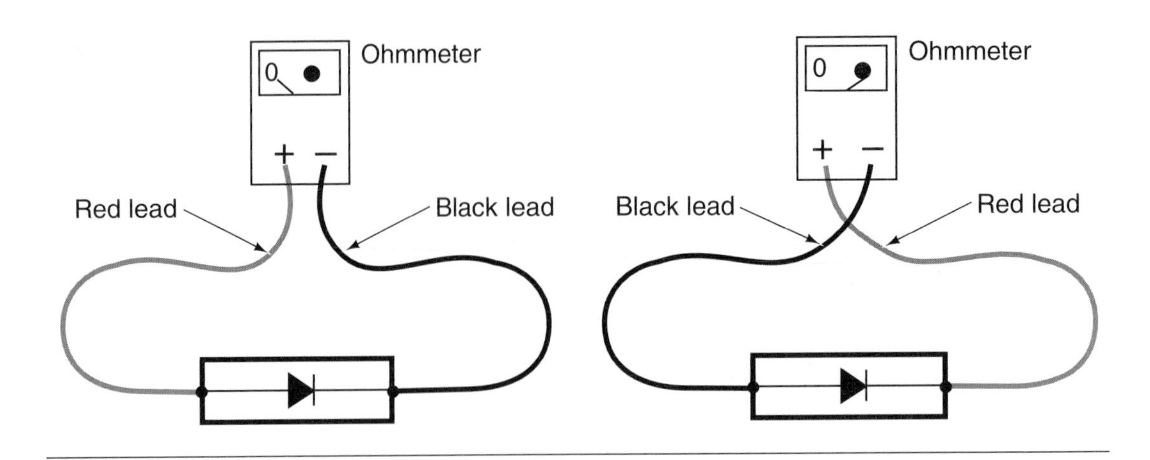

Figure 3-18 Use an ohmmeter to test a diode for an open or short.

Classroom Manual
Chapter 3, page 58.

A **diode** is a one-way electrical check valve that allows current to flow in one direction only.

Forward bias means a positive voltage is applied to the P-type material and negative voltage to the N-type material.

Reverse bias means that positive voltage is applied to the N-type material and negative voltage is applied to the P-type material.

Testing Diodes

Testing diodes is not complicated since a diode should allow current flow in one direction only. An analog ohmmeter or Digital Volt Ohm Meter (DVOM) with a diode testing function position are the best tools for testing diodes. Connect the meters leads across the diode (Figure 3-18). Observe the reading on the meter. Then reverse the leads and observe the reading again. In one direction, the reading should be very high or infinite and in the other direction, the resistance should be close to zero. If you observe any other readings, the diode is bad. For example, a diode that has low resistance in both directions is shorted and a diode that has a high resistance or an infinite reading in both directions is open.

The reason for using a DVOM with a diode function position is that some high-impedance digital meters might not have enough volts to forward bias the diode. Many diodes require at least 0.6 volts to allow current flow. If the meter cannot apply that voltage on the ohms function, the reading on the meter might indicate the diode is open, when in fact it may not be. Many DVOMs will show a reading between 500 and 600 ohms when forward biasing a diode on the diode check scale and "OL" when reverse biasing.

A diode-testing feature on a multimeter allows for increased voltage at the test leads. Some meters display the voltage required to forward bias the diode. If the diode is open, the meter will display "OL" or a reading indicating infinity or out of range. Some meters might beep when there is continuity during a diode check.

Classroom Manual
Chapter 3, page 65.

Testing for Circuit Defects

Common failures of electrical circuits that can occur in trucks are: an open, a short, a ground, or a high resistance problem.

Classroom Manual
Chapter 3, page 65.

An **open circuit** will not blow a fuse.

Testing for Opens

When testing for an open, you are trying to find a point within the circuit where current flow is not allowed to continue. Remember that in a complete circuit current flow originates and ends at the power source. To test for an open, it is possible to use a test light, voltmeter, ohmmeter, or a jumper wire. The test tool used will depend on the circuit being tested and the accessibility of components.

One method of isolating an open is to use a voltmeter and compare voltage readings at specific points. Figure 3-19 shows the voltmeter readings that should be obtained in a properly operating parallel circuit. When using a voltmeter, the circuit power must be ON. The idea is to

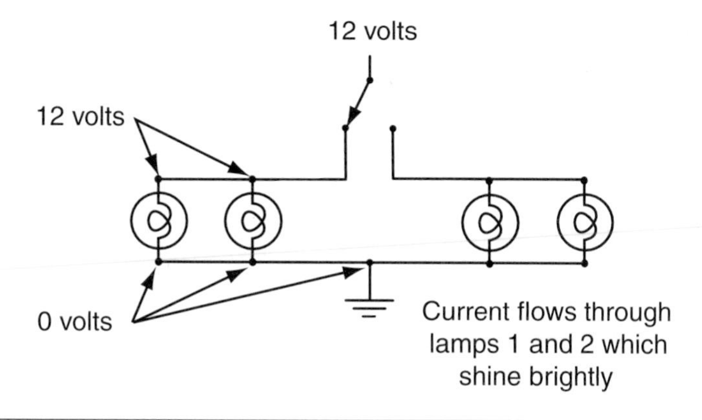

Figure 3-19 Voltmeter readings that would be expected in a properly operating parallel circuit.

move the positive (+) lead progressively toward the battery positive if no voltage is measured, or progressively toward battery negative if voltage shows.

Using Figure 3-20 as a guide and the following steps, you can easily locate an open if the load components are easily accessible:

1. Check for voltage at point A. If the voltage is 10.5 volts or higher, check the ground side (Point B). If the voltage is less than 1 volt, the load component is faulty. If more than 1 volt is present, there is excessive resistance or an open in the ground circuit. If the voltage at point A was less than 10.5 volts, continue testing.
2. Working towards the battery, test for voltage at all connections. If voltage is present at a connection, the open is between that connection and the previously tested connection (Figure 3-21). A jumper wire can then be used to verify the location of the open.
3. If battery voltage is present at point B, the open is in the ground circuit. Use a jumper wire to connect the ground circuit and then retest the component.

To test for an open circuit, it is possible to use a test light, voltmeter, ohmmeter, and/or a jumper wire.

A **series circuit** provides a single path for current flow from the electrical source through all the circuit's components and back to the source.

A **parallel circuit** provides two or more paths for electricity to flow.

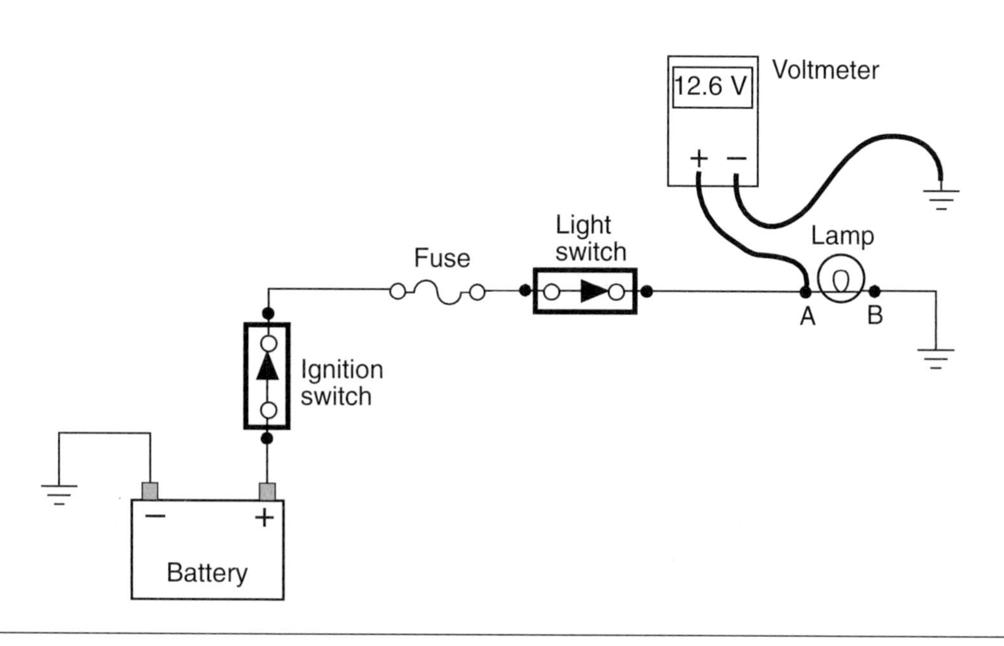

Figure 3-20 Locating an open by testing for voltage.

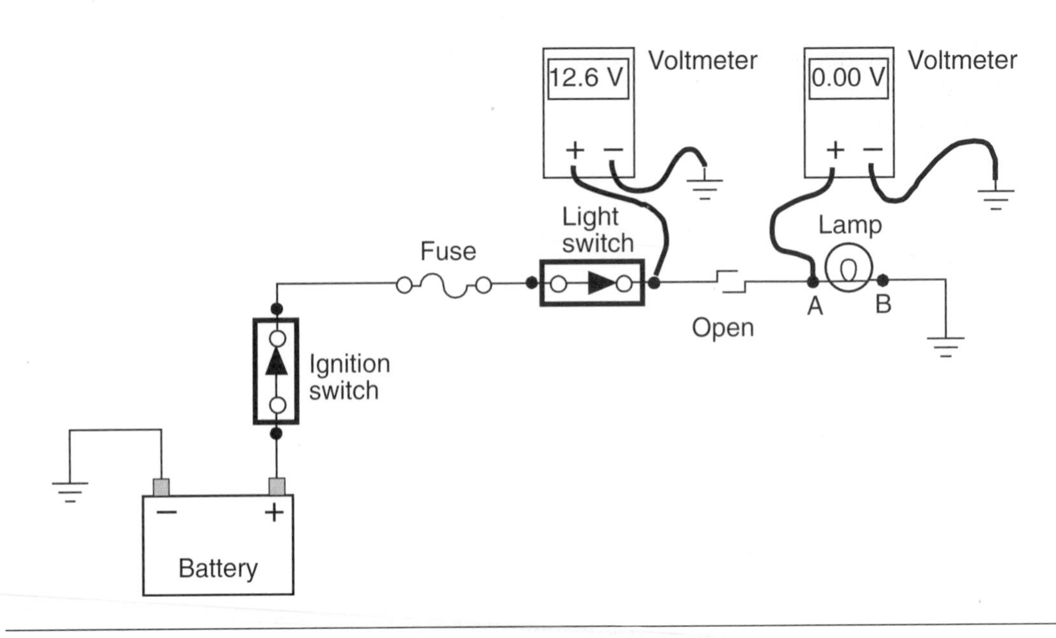

Figure 3-21 An open is present between the point where voltage was measured and where it was not.

Classroom Manual
Chapter 3, page 66.

A **shorted circuit** allows current to bypass part of the normal path.

A short may not blow a fuse, depending on the amount of current flowing.

Checking for Shorts, Grounds, and Opens

Another way to check for an open is to use an ohmmeter to check circuit continuity. This method, illustrated in Figure 3-22, is also useful for locating a short or ground. The following steps can be followed for this test:

1. Disconnect the load, such as a motor (shown in the illustration) and solenoids or remove the lightbulb(s), etc.
2. If there are any normally opened switches, they must be closed or jumped.
3. Connect one lead of the meter to the AUX terminal of the circuit breaker.
4. With the other lead, start probing various points or connections in the circuit while watching the meter. A reading of 0 ohms, or fractions of ohms, indicates a completed circuit. Infinite or OL on a digital meter indicates an opened circuit.

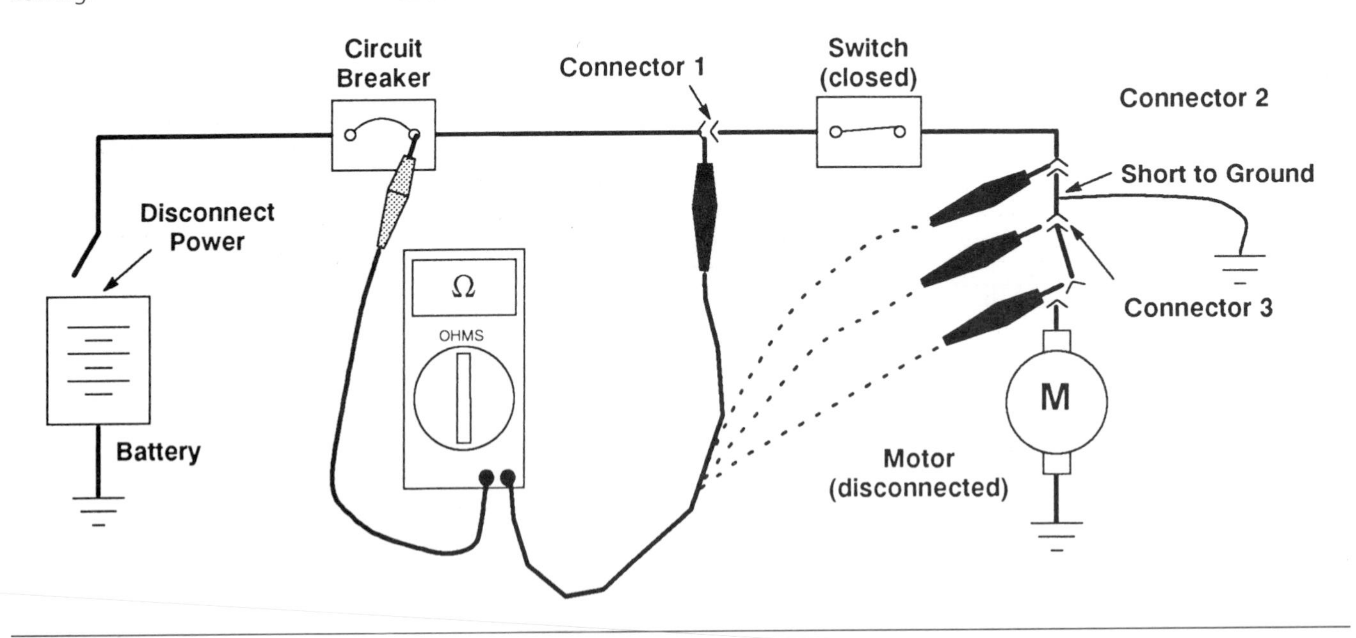

Figure 3-22 An ohmmeter being used to check for an open in a circuit. (Courtesy of Mack Trucks, Inc.)

WARNING: The power in the circuit must be turned off and the ground must be disconnected before performing any continuity test using the ohmmeter or damage to the meter and/or circuit may occur.

Sometimes the shorted, grounded, or opened circuit can be an intermittent problem. If the approximate area of the problem is known, you can perform the following steps:

1. Make sure the circuit is off and the ground is isolated.
2. Insert one lead of the ohmmeter into the connector of the suspected harness and connect the other lead to a good ground.
3. While watching the meter, start wiggling the wires every few inches.
4. When the resistance readings change—such as to 0 ohms from an infinite (OL) reading or to infinite (OL) from a 0 ohms reading—the problem is near that point.

If you don't know where the area of the problem is, the following steps can be performed using the same ohmmeter.

1. Connect the ohmmeter between a good ground and the AUX terminal of the circuit breaker. *It is important to note that the circuit is still isolated, with no power and ground.*
2. Starting at the circuit breaker, begin wiggling the harness at various points while watching the meter.
3. When the readings change, the approximate area of the problem has been found.

The following is another technique used to locate a shorted or grounded circuit. A circuit breaker that does not reset or continuously trip is an indicator of this type of circuit failure. This time a voltmeter is used to perform the following steps with Figure 3-23 as an illustrated guide:

1. Turn off all components that are powered through the affected circuit breaker.
2. Disconnect all loads powered through the circuit breaker by disconnecting connectors or leads from solenoids, motors, etc., or removing lightbulbs.
3. Set the multimeter to the VDC function. Connect the negative (black lead) to a good ground and the red lead (positive) to the BAT (battery) terminal of the affected circuit breaker. The meter should indicate battery voltage. This indicates that the circuit breaker is being powered. *It is important to note that some circuit breakers need the key on in order to be powered.*
4. Disconnect the black lead from ground and connect to the load side of the circuit breaker as shown in the illustration. *It is important to note that any normally opened switch or switches in this circuit should be either closed or jumped at this point.*

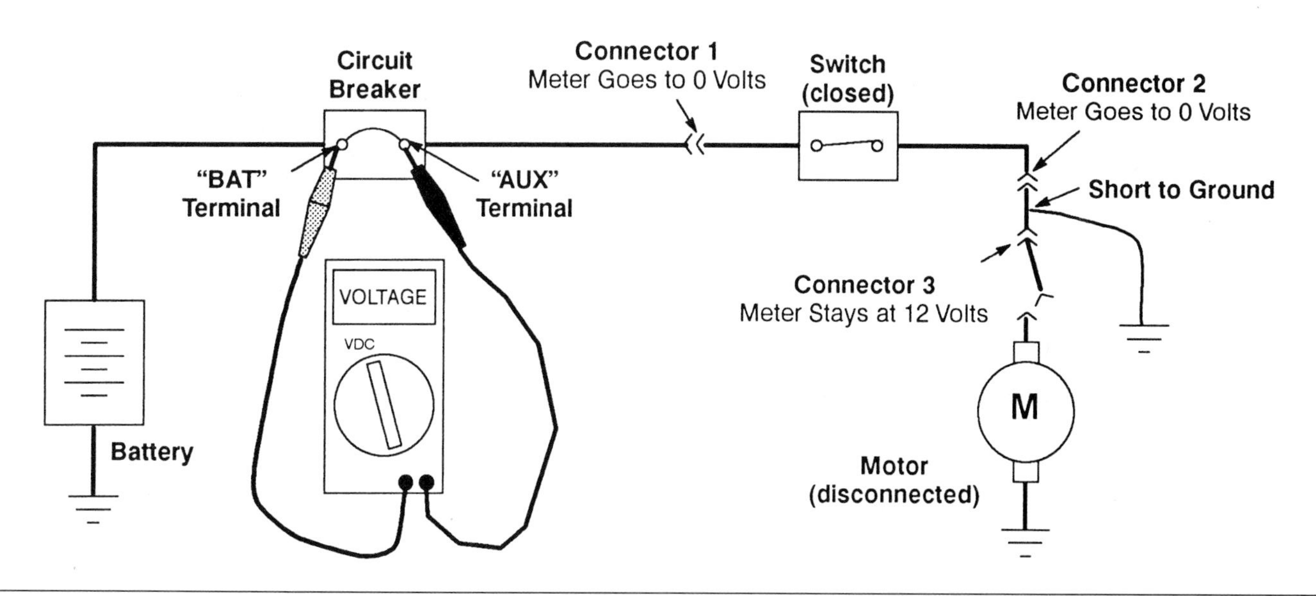

Figure 3-23 A voltmeter can also be used to check for a short. (Courtesy of Mack Trucks, Inc.)

5. If the meter reading shows no voltage, the short is located in one of the disconnected components.
6. If the meter indicates battery voltage, the short is located in the wiring. To further isolate the short, start disconnecting and connecting each connector found in the circuit while observing the meter. Start with the connector closest to the circuit breaker.
   - ❑ If the meter drops to 0 volts when a connector is disconnected, the wiring between the connector and the break is good.
   - ❑ If the meter remains at battery voltage when a connector is disconnected, the short is somewhere between that connector and the last connector disconnected.

CUSTOMER CARE: Troubleshooting and repairing the problem is not enough. Always look to see what can be done to prevent recurrence of the problem. Do not be afraid to make recommendations.

Voltage Drop Testing

Classroom Manual
Chapter 3, page 67.

For circuit problems, a voltage drop is used to define the loss of voltage due to unwanted resistance within a point or various points in a circuit. This unwanted resistance in the circuit reduces the amount of electrical pressure available beyond the unwanted resistance.

Voltage drop testing is used to locate high resistance in a circuit. Unwanted high resistance most often is the cause for dim lights, flickering lights, or slower than normal electrical motor speeds, etc. The unwanted resistance uses up some of the voltage that was required by the load components in the circuit. This excessive voltage drop can occur on either the insulated and/or ground return side of the circuit. When performing a voltage drop test, the circuit must be on. If testing the light circuit, the lights need to be turned on. If checking for a voltage drop in the starting system, the starter motor must be cranking the engine over. *As with any electrical test, make sure the source voltage is where it belongs before performing a test. When testing for a voltage drop, both sides of the circuit must be checked. It is important to note that in the case of the cranking system, the fuel should be shut off so the engine doesn't crank if it is a diesel. If it is a gas engine, you need to disable the ignition.*

Figure 3-24 illustrates a voltage drop test on a switch. Using a voltmeter, the positive lead must always be connected to the most positive portion of the circuit. In this figure, the positive lead is on the positive side of the switch and the negative lead is on the ground side of the switch. The voltmeter is "parallel" to the circuit. With the circuit on, the meter should indicate a minimal voltage drop. Consult the service manual for the maximum amount of voltage drop allowed. For example, one manufacturer's service manual shows the following for maximum voltage drop allowed:

0.1 or less—Wire, switch, or connector is all right.

0.2 or less—For solenoid is all right.

0.3 or less—For an insulated or ground "circuit" is all right.

Voltage drops exceeding these indicate poor connections or defective parts.

Figure 3-25 shows the location of the voltmeter leads to test the ground side of a circuit. Notice that the ground connection of the load is now the most positive location and the battery

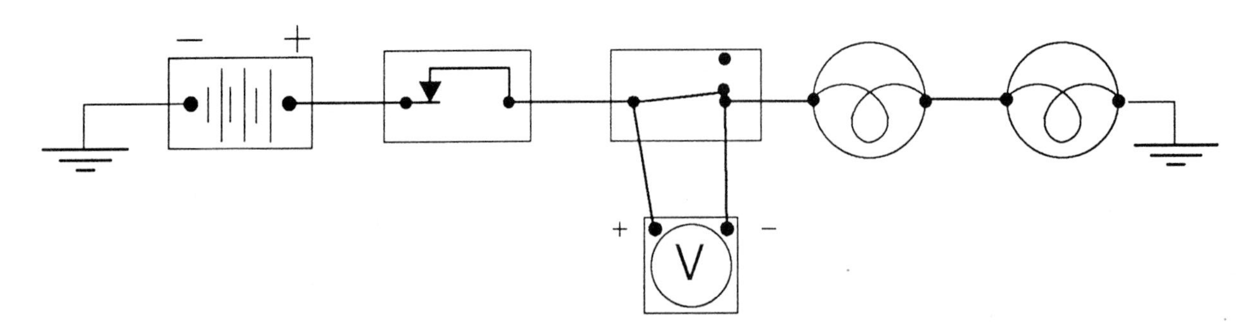

Figure 3-24 Proper placement of voltmeter leads to perform a voltage drop test. (Courtesy of Mack Trucks, Inc.)

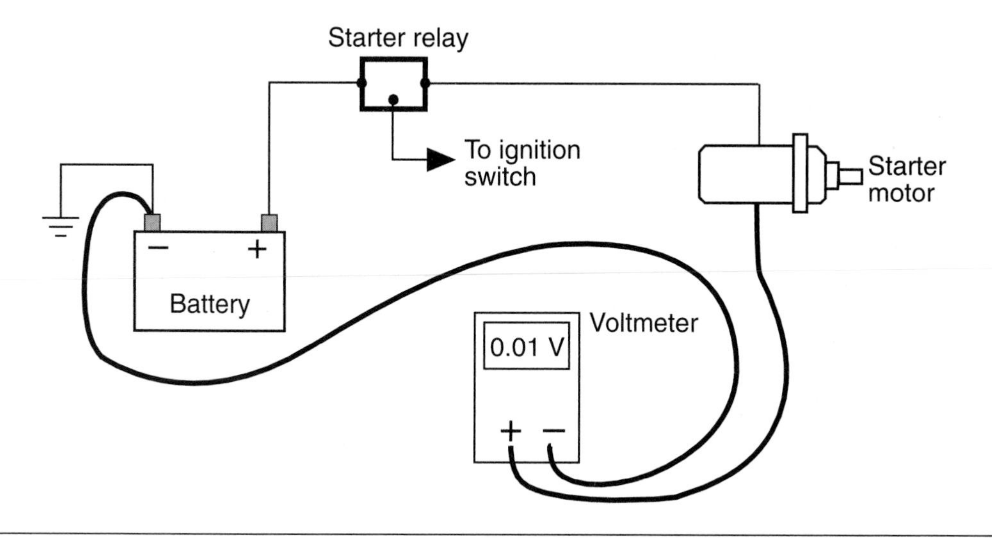

Figure 3-25 Testing the ground side of a starter for high resistance by measuring the voltage drop.

negative post is the most negative location. A reading of more than 0.1 volts indicates excessive resistance in the ground circuit.

CASE STUDY

A tractor-trailer is brought to the shop with no left turn signal on the trailer. The technician verifies that the turn signals work on the tractor, thus isolating the problem to the trailer lights circuit. The lens was taken off the affected light and the bulb was removed to visually check the filament and contact of the bulb socket. Everything seemed intact but the technician substituted a good bulb in place of the existing bulb. There was still no light. The technician pulled the bulb again and checked the contact in the socket for applied voltage with a test light and discovered no voltage present. The next logical step was to see if voltage was present at the trailer light cord using a test light. Voltage was present. Because the wiring is more exposed to the climate and corrosion in trailers, the technician proceeded to perform a visual inspection of the wiring. He noticed a lot of corrosion at one of the rear connector plugs. A test light verified that voltage was present at the plug but not going out towards the light. The turn signal contact was corroded to the point that no current could flow through. The plugs were cut out of the circuit and new plugs were spliced in and the repair was verified.

Terms to Know

Accelerator position sensor (APV) or throttle position sensor (TPS)
Ammeter
Bus-bar
Closed circuit
Continuity
Diode
Forward bias
Fusible link
Idle validation switch (IVS)
Infinity
Normally closed (NC)
Normally open (NO)
Ohmmeter
Open circuit
Overload
Parallel circuit
Potentiometer
Reverse bias
Series circuit
Shorted circuit
Stepped resistor
Variable resistor

ASE-Style Review Questions

1. Circuit protection devices are being discussed: *Technician A* says to figure the rated amperage required when installing a circuit breaker you have to know the resistance value of the load component. *Technician B* says the wire gauge size is needed to determine the proper amperage rating of a circuit breaker. Who is correct?
   A. A only **C.** Both A and B
   B. B only **D.** Neither A nor B

2. Testing of a switch is being discussed: *Technician A* says a switch can be tested with an ohmmeter. *Technician B* says a voltmeter is used to test a switch. Who is correct?
   A. A only **C.** Both A and B
   B. B only **D.** Neither A nor B

3. Voltage drop testing is being discussed: *Technician A* says an ohmmeter is used to the ground side of a circuit to test for a voltage drop. *Technician B* says a voltmeter is used on the insulated side of the circuit to test for a voltage drop. Who is correct?
   A. A only **C.** Both A and B
   B. B only **D.** Neither A nor B

4. Using an ohmmeter for continuity testing is being discussed: *Technician* A says infinity indicates an open circuit. *Technician B* says a low to zero resistance reading indicates an open circuit. Who is correct?
   A. A only **C.** Both A and B
   B. B only **D.** Neither A nor B

5. Troubleshooting techniques are being discussed: *Technician A* says one of the first steps should be to check for the presence of voltage. *Technician B* says a test light can be used to check for the presence of voltage. Who is correct?
   A. A only **C.** Both A and B
   B. B only **D.** Neither A nor B

6. Circuit continuity check using an ohmmeter is being discussed: *Technician A* says the ground must be isolated before performing a continuity test. *Technician B* says with the ground isolated there is no need to turn off the power in the circuit because the circuit is not complete with an isolated ground. Who is correct?
   A. A only **C.** Both A and B
   B. B only **D.** Neither A nor B

7. Checking for the presence of voltage at a load component is being discussed: *Technician A* says if 11 or more volts are present at the insulated input side of the load the circuit is okay. *Technician B* says less than 11 volts indicates an open circuit. Who is correct?
   A. A only **C.** Both A and B
   B. B only **D.** Neither A nor B

8. *Technician A* says a relay has to be connected in the circuit to test it. *Technician B* says to test a relay out of the circuit requires the use of an ohmmeter and possibly a jumper lead to power. Who is correct?
   A. A only **C.** Both A and B
   B. B only **D.** Neither A nor B

9. To test a potentiometer: *Technician A* says an ohmmeter can be used. *Technician B* says a voltmeter can be used. Who is correct?
   A. A only **C.** Both A and B
   B. B only **D.** Neither A nor B

10. Diode testing is being discussed: *Technician A* says some digital ohmmeters cannot properly test a diode. *Technician B* says a high or infinite reading in both directions of a diode indicates an open diode. Who is correct?
    A. A only **C.** Both A and B
    B. B only **D.** Neither A nor B

ASE Challenge

1. A high/low beam relay for a truck is being tested with an ohmmeter. *Technician A* says that only two of the switch (contactor) terminals should show continuity in a given position. *Technician B* says all terminals should show continuity when the coil is jumped with 12 volts. Who is correct?
   A. A only **C.** Both A and B
   B. B only **D.** Neither A nor B

2. *Technician A* says that a 12-volt test light should never be used on input wiring to a microprocessor, ECU, ECM, etc. *Technician B* says that poking or piercing microprocessor input wires with a sharp test probe for any instrument is a poor practice. Who is correct?
   A. A only **C.** Both A and B
   B. B only **D.** Neither A nor B

3. Using a DVOM, *Technician A* finds voltage to range from 0.5 to 4.5 volts from a TP sensor. She says this is way out of range. *Technician B* agrees but says that the source wire for the TP sensor should be jumped with a 12-volt positive voltage source. Who is correct?
   A. A only **C.** Both A and B
   B. B only **D.** Neither A nor B

4. The wire to a fuel tank sender is believed to be short-circuited because the instrument fuse no longer blows when this wire is disconnected from the gauge. *Technician A* says to check the wire using an ohmmeter with one lead to the wire end, the other to ground; a high reading will indicate a short. *Technician B* says to use a 12-volt test light with the clip to a 12-volt source, the probe to the wire; a dim light will indicate a short. Who is correct?
   A. A only **C.** Both A and B
   B. B only **D.** Neither A nor B

5. A battery drain problem is being diagnosed. *Technician A* says you should turn off all electrical components and systems then disconnect the positive battery cable and connect a good ammeter between the battery and cable end. A reading in excess of 250 ma indicates problems. *Technician B* says that a voltmeter can also be used in a similar manner, and any reading over 2 volts indicates a problem. Who is correct?
   A. A only **C.** Both A and B
   B. B only **D.** Neither A nor B

Table 3-1 NATEF and ASE Task

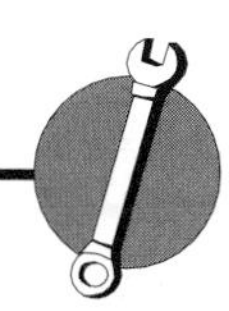

Read, interpret, and diagnose electrical/electronic circuits using wiring diagrams

Problem Area	Symptoms	Possible Causes	Classroom Manual	Shop Manual
ELECTRICAL/ ELECTRONIC FAULTS	Electrical/electronic component or systems fails to operate	1. Open or ground in circuit 2. Defective switch 3. Defective relay 4. Open in fuses or links 5. Incorrectly connected	65	58

Table 3-2 NATEF and ASE Task

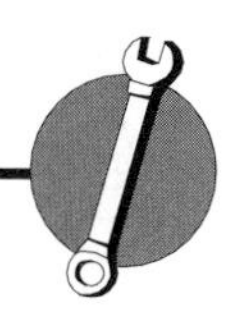

Check continuity in electrical/electronic circuits using appropriate test equipment

Problem Area	Symptoms	Possible Causes	Classroom Manual	Shop Manual
ELECTRICAL OPENS	Electrical/electronic component fails to operate	1. Broken conductor 2. Defective switch 3. Defective relay 4. Blown fuses or links 5. Defective circuit breakers	44	44

Table 3-3 NATEF and ASE Task

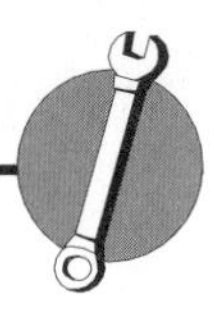

Check applied voltages, circuit voltages, and voltage drops in electrical/ electronic circuits using a digital multimeter (DMM)

Problem Area	Symptoms	Possible Causes	Classroom Manual	Shop Manual
EXCESSIVE RESISTANCE	Electrical/electronic component fails to operate or operates at reduced efficiency	1. Excessive resistance	48	60

Table 3-4 NATEF and ASE Task

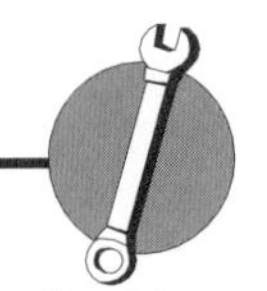

Check current flow in electrical/electronic circuits and components using an ammeter, digital multimeter (DMM), or clamp-on ammeter

Problem Area	Symptoms	Possible Causes	Classroom Manual	Shop Manual
A SHORT, OPEN, OR SHORT CIRCUIT IN THE CIRCUIT	Electrical/electronic component fails to operate or operates at reduced efficiency. Blown fuses	1. Broken conductor 2. Blown circuit protection device 3. Open or excessive current in the switch or relay 4. Short circuit	51	50

Table 3-5 NATEF and ASE Task

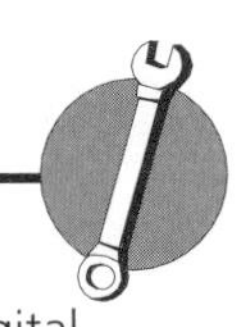

Check resistance in electrical/electronic circuits and components using an ohmmeter or digital multimeter (DMM)

Problem Area	Symptoms	Possible Causes	Classroom Manual	Shop Manual
EXCESSIVE RESISTANCE	Electrical/electronic component fails to operate or operates at reduced efficiency.	1. Open or excessive resistance in the switch or relay 2. Improper stepped resistor values 3. Open or shorted diodes	54	53

Table 3-6 NATEF and ASE Task

Find shorts, grounds, and opens in electrical/electronic circuits

Problem Area	Symptoms	Possible Causes	Classroom Manual	Sh Ma
COPPER-TO-COPPER SHORT OR SHORT TO GROUND	No electrical component operation or operation when another control is activated	1. Broken or burned insulation and/or connectors causing copper-to-copper short. 2. Broken or burned insulation and/or connectors causing a short to ground 3. Excessive ground or voltage drop across the ground.	44	60

Table 3-7 NATEF and ASE Task

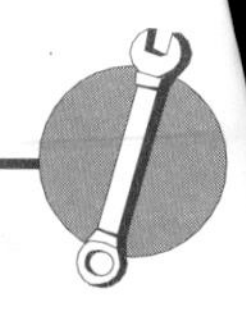

Inspect, test, and replace fusible links, circuit breakers, and fuses

Problem Area	Symptoms	Possible Causes	Classroom Manual	Shop Manual
BURNED FUSIBLE LINK OPEN CIRCUIT BREAKER BLOWN FUSE	Several electrical or electronic components fail to operate	1. Burned fusible link 2. Blown fuse 3. Open circuit breaker	45	45

Table 3-8 NATEF and ASE Task

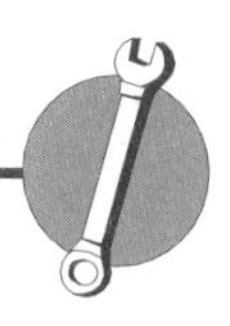

Inspect, test, and replace spike suppression diodes/resistors and capacitors

Problem Area	Symptoms	Possible Causes	Classroom Manual	Shop Manual
ECM/PCM OPERATION	Erratic operation of electronic components ECM, PCM, ECU failure	1. Failure suppression diode, resistor or capacitor	58	56

...TEF and ASE Task

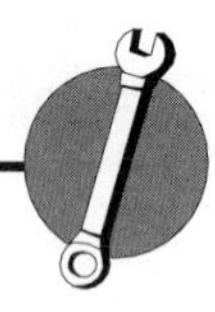

...ect, test, and replace relays and solenoids

Symptoms	Possible Causes	Classroom Manual	Shop Manual
System does not operate as intended	1. Failed relay or solenoid	51	50

Job Sheet 4

4

Name: ______________________________ Date: ______________________

Use of Various Testing Tools

Upon completion of this job sheet, you should be able to take amperage, voltage, and resistance measurements.

NATEF and ASE Correlation

This job sheet is related to NATEF and ASE Medium/Heavy Duty Truck Electrical/Electronic Systems List content area:

A. General Electrical System Diagnosis, ASE Task 1 and NATEF Task 2: Check continuity in electrical/electronic circuits using appropriate test equipment; ASE Task 2 and NATEF Task 3: Check applied voltages, circuit voltages, and voltage drops in electrical/electronic circuits using a digital multimeter (DMM); ASE Task 3 and NATEF Task 4: Check current flow in electrical/electronic circuits and components using an ammeter, digital multimeter (DMM), or clamp-on ammeter; ASE Task 4 and NATEF Task 5: Check resistance in electrical/electronic circuits and components using an ohmmeter or digital multimeter (DMM).

Tools and Materials

Test light/continuity checker
Voltmeter
Ammeter
Ohmmeter
12-volt battery
(2) 1157 park/stop lamps
(2) 6052-2B1 Head lamps
(6) 18+ gauge test leads with alligator clips

Procedure

This job sheet is quite extensive, so allow plenty of time to complete it. You will be required to make up various circuits and measure for certain values within the circuits. On this job sheet you will find instructions for making circuits and figures to be used as a guide when completing the exercises.

WARNING: Make sure you know your ammeter's maximum amperage limit. You can use ohm's law if needed to determine the approximate amp draw you will be working with.

1. Refer to Figure 3-26 to become familiar with the main components of the circuits. Holding the 1157 bulb so that the high locking tab is facing you, the right connector is the stop lamp, and the left connector is the park. Holding the 2B1 headlamp so it faces away from you, the leftmost connector is the ground, the top connector is the low beam, and the rightmost connector is the high beam. All measurements indicated are actual, however since measurements will vary with filament temperatures, use them as a guide only. Using the battery, lamps, and test leads, make up the following circuits and record your readings.

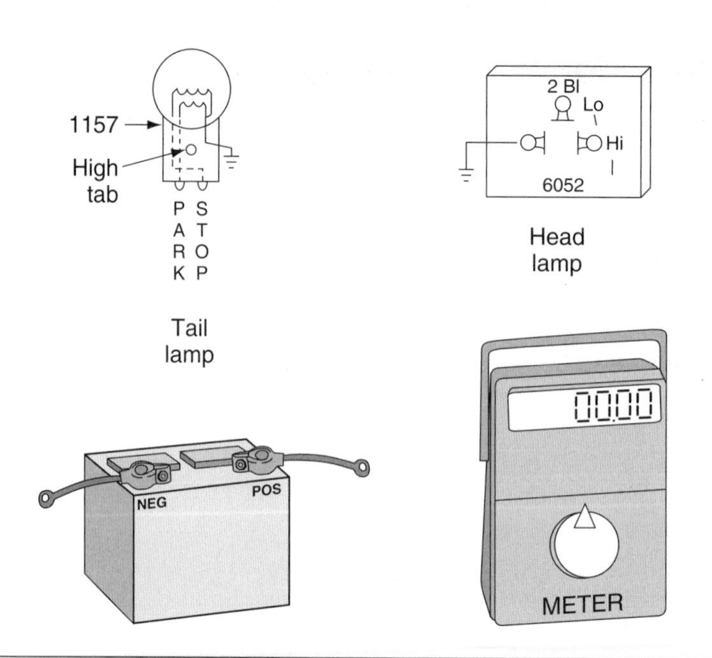

Figure 3-26

Exercise 1 (use Figure 3-27 as a guide)

A. Connect the 1157 ground connection to the positive side of the battery.

B. Connect the negative side of the ammeter to battery ground.

C. Connect the positive side of the ammeter to the park terminal of the bulb (____ amps).

D. Connect the positive side of the ammeter to the stop terminal of the bulb (____ amps).

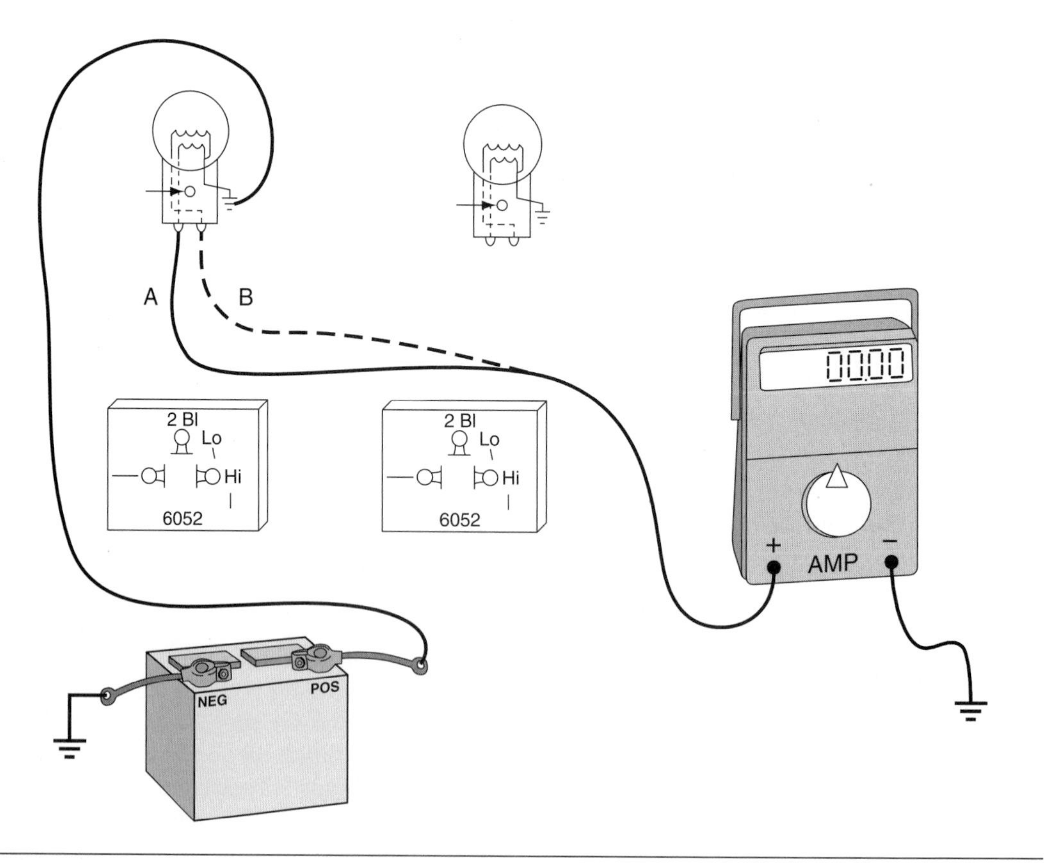

Figure 3-27

Exercise 2 (use Figure 3-28 as a guide)

A. Connect the 2B1 ground connection to the positive side of the battery.

B. Connect the negative side of the ammeter to the negative ground.

C. Connect the positive side of the ammeter to the high beam terminal of the bulb (______ amps).

D. Connect the positive side of the ammeter to the low beam terminal of the bulb (______ amps).

Exercise 3 (use Figure 3-29 as a guide)

A. Connect the negative terminal of two 2B1 headlights to the battery positive.

B. Connect the negative side of the ammeter to battery negative.

C. Connect the positive side of the ammeter to both high beam terminals (______ amps).

Exercise 4 (use Figure 3-30 as a guide)

A. Connect the negative side of an ohmmeter to the ground connection on the 1157 bulb.

B. Connect the positive side of the ohmmeter to the park terminal (______ ohms).

C. Connect the positive side of the ohmmeter to the stop terminal (______ ohms).

Exercise 5 (use Figure 3-31 as a guide):

A. Connect the negative side of an ohmmeter to the ground connection on a 2B1 bulb.

B. Connect the positive side of the ohmmeter to the high beam terminal (______ ohm).

C. Connect the positive side of the ohmmeter to the low beam terminal (______ ohm).

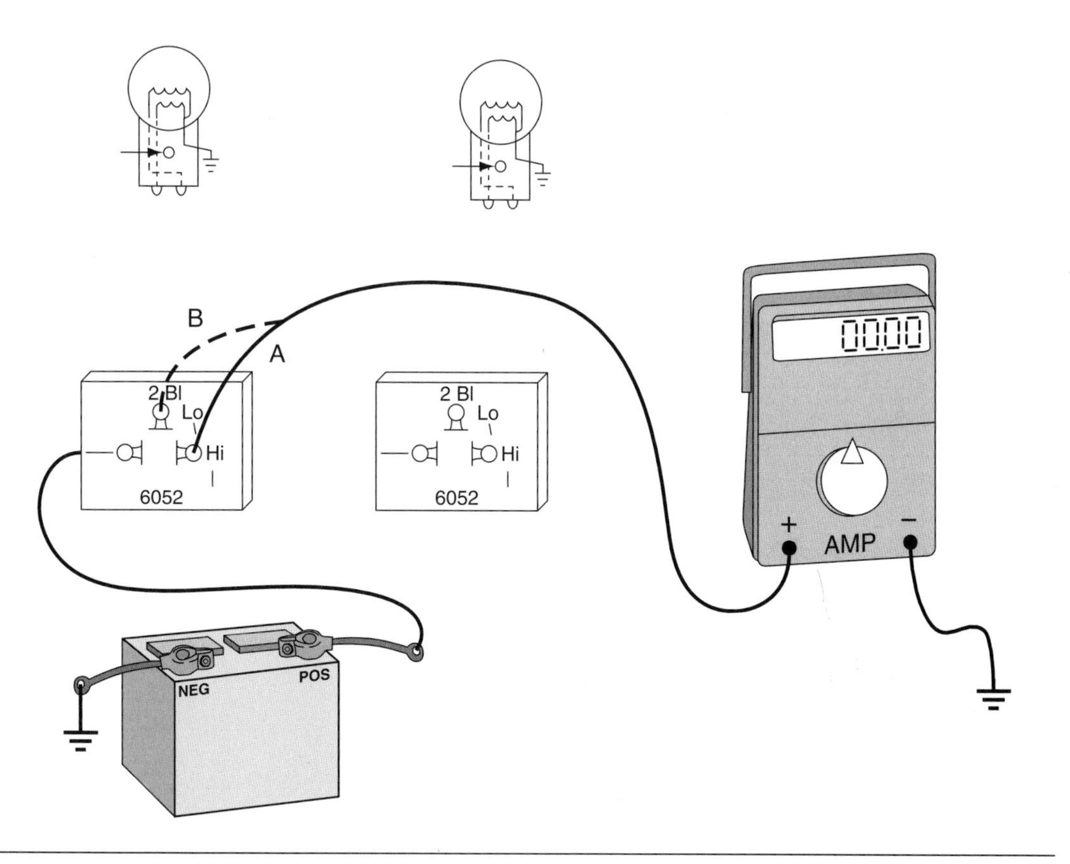

Figure 3-28

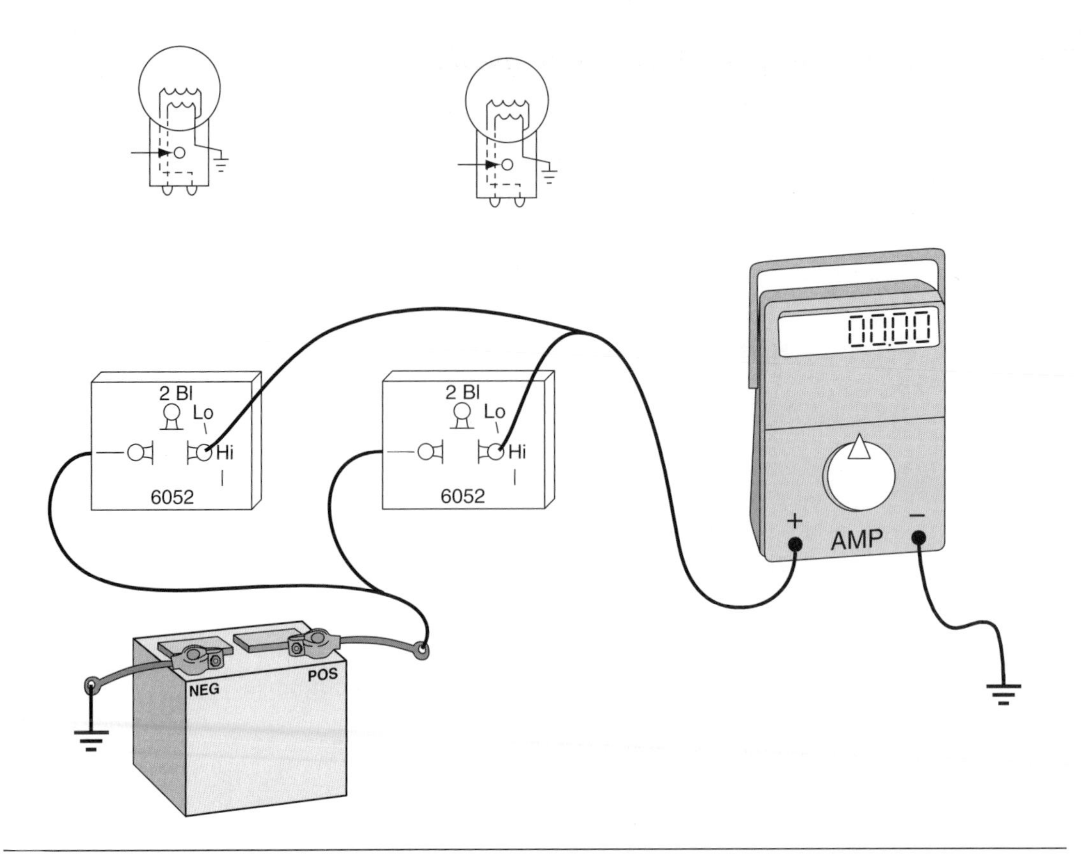

Figure 3-29

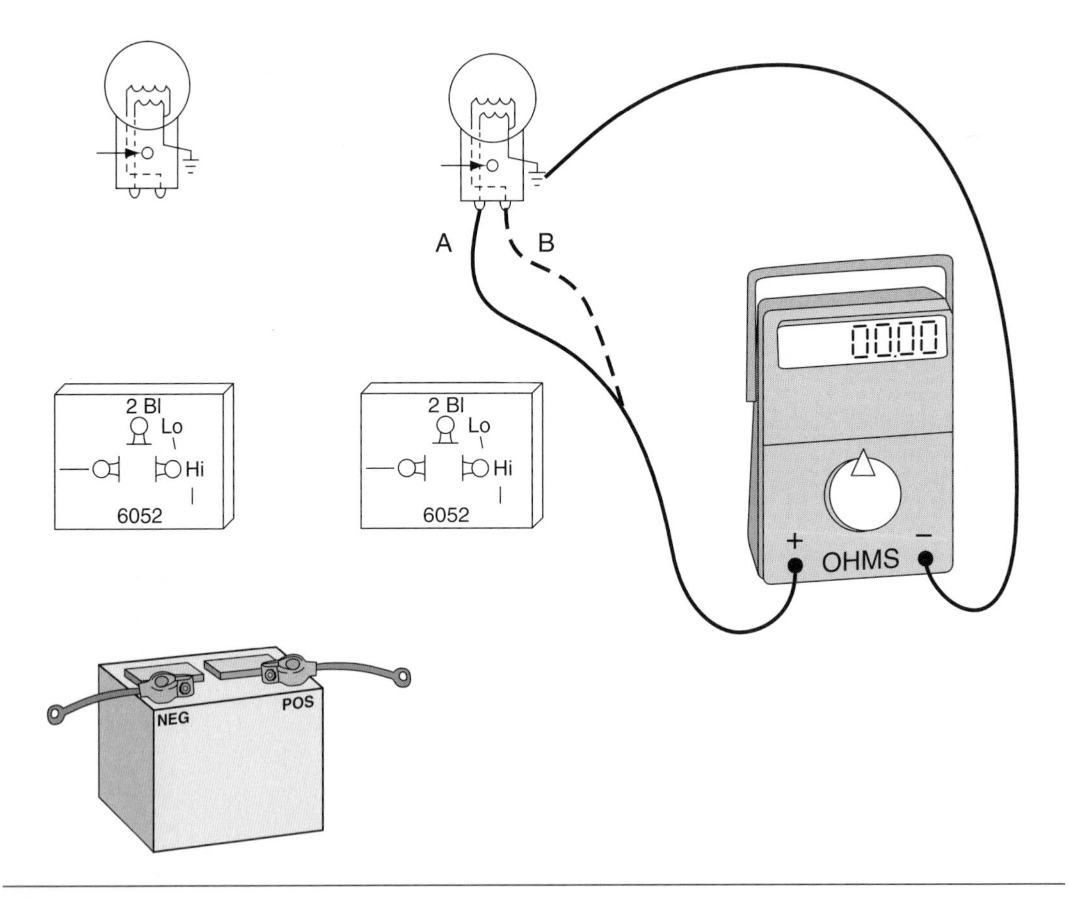

Figure 3-30

Exercise 6 (use Figure 3-32 as a guide):

A. Connect the positive side of an ohmmeter to the high beam terminals on two 2B1 lamps.

B. Connect the negative terminal of the ohmmeter to both negative terminals on the 2B1 lamps (______ ohms).

Exercise 7 (use Figure 3-33 as a guide)

A. With two 1157 light bulbs grounded, connect the battery positive lead to both park terminals.

B. Ground the voltmeter.

C. Connect the voltmeter to the ground terminal of one bulb (______ volts).

D. Connect the voltmeter to the ground terminal of the other bulb (______ volts).

Exercise 8 (use Figure 3-34 as a guide)

A. With one 1157 bulb grounded, connect the stop terminal to the ground connection of a second 1157 bulb.

B. Connect the stop terminal to the battery positive. Both bulbs should light dimly.

C. With the voltmeter grounded, connect the positive terminal to the stop terminal of the first bulb (______ volts).

D. Connect the negative terminal of the voltmeter to the connection joining the two bulbs. Then connect the positive terminal to the stop terminal of the second bulb (______ volts).

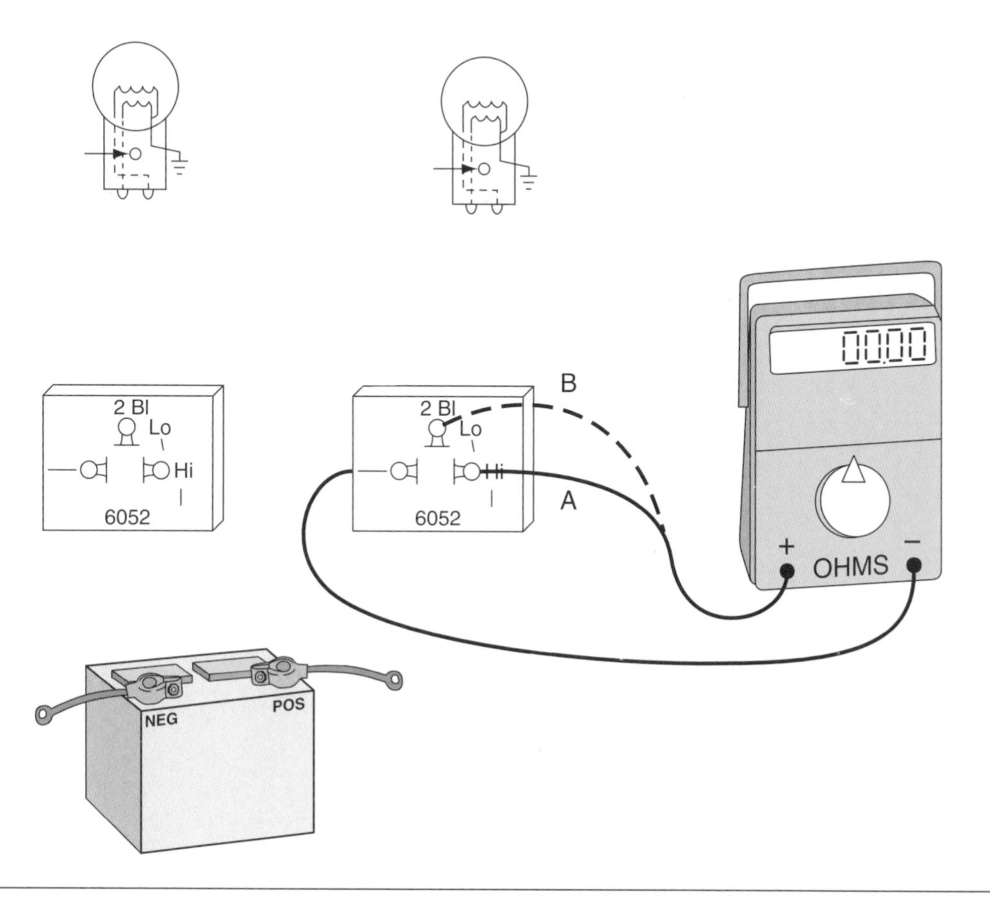

Figure 3-31

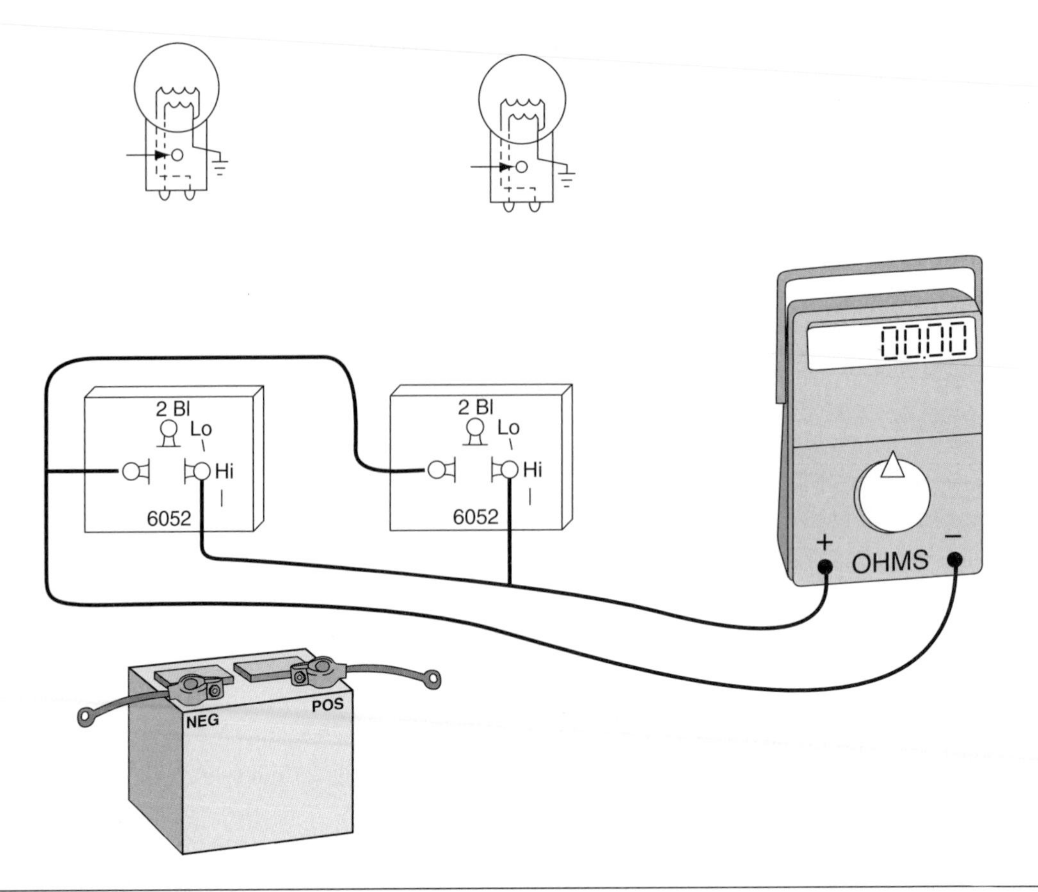

Figure 3-32

Figure 3-33

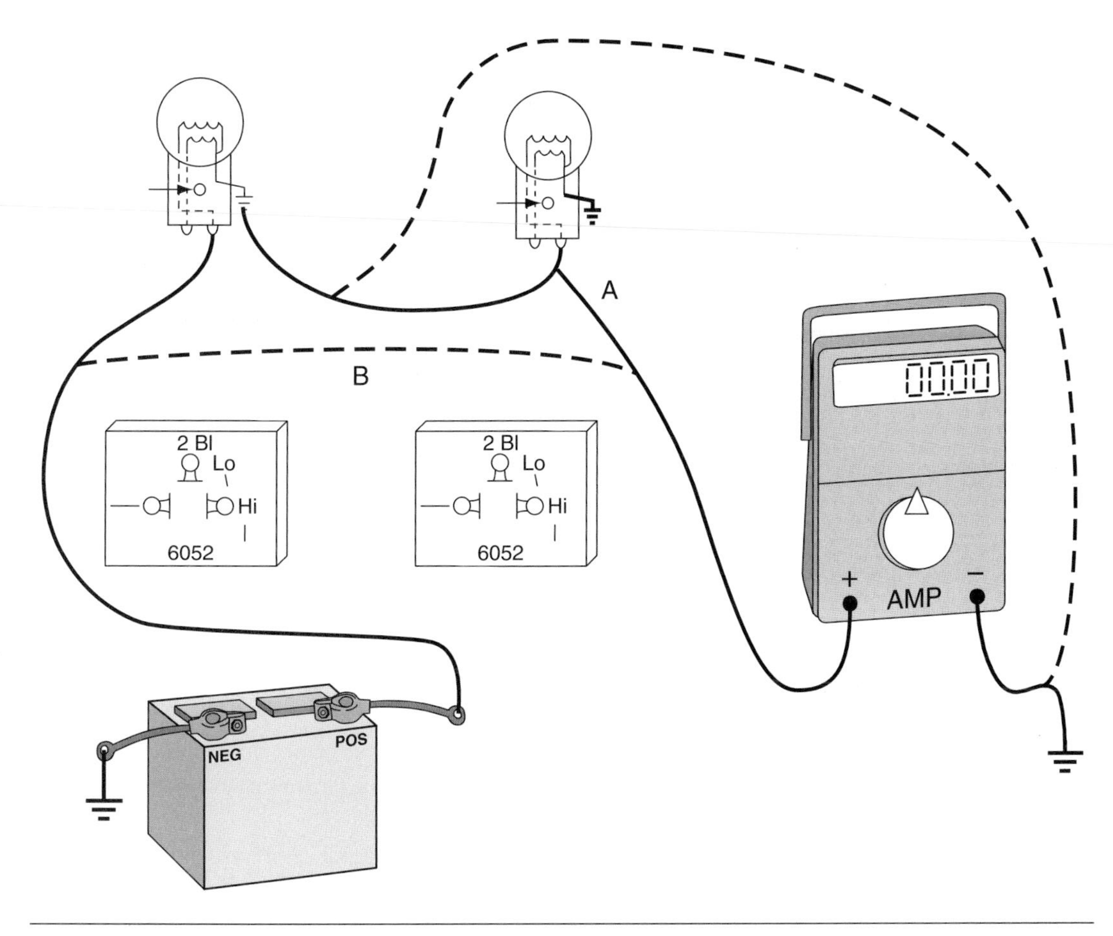

Figure 3-34

Exercise 9

A. Connect the battery positive to the ground terminal of one 1157 bulb.

B. Connect the stop terminal to the low beam terminal of a 2B1 lamp.

C. Connect the ground terminal of the 2B1 to the battery ground.

D. Connect the negative voltmeter lead to the park terminal of the 1157 and the positive lead to the ground terminal of the 1157 (______ volts).

E. Connect the negative terminal of the voltmeter to ground and the positive terminal to the low beam of the 2B1 (______ volts).

 **Instructor Check** ____

CHAPTER 4

Wiring Repair Practices and Diagram Reading

Upon completion and review of this chapter, you should be able to:

- ❑ Perform repairs to copper wires using solderless connections.
- ❑ Solder splices to copper wire.
- ❑ Repair twisted-shielded wires.
- ❑ Replace fusible links.
- ❑ Repair and/or replace terminals of a hard shell connector.
- ❑ Repair and/or replace terminals of various Packard connectors.
- ❑ Use a wiring diagram to diagnose a system failure.

Basic Tools

Basic mechanics tool set

Service manual

Safety glasses

Introduction

Electrical failure repairs in today's trucks involve many steps, beginning with proper diagnostics and troubleshooting techniques. Those, in turn, require a thorough understanding of electricity and its use in trucks. Because of the number of electrical and electronic circuits in trucks, you will always need to have access to reference materials such as wiring diagrams and schematics. Many manufacturers provide a lot of information on a single page to assist you (Figure 4-1). This figure, for example, shows a wiring diagram for an engine shutdown system, location of the harness connectors, and a view of some of the components and their location. With a thorough understanding of electricity, coupled with proper schematics and diagrams, you should be able to figure out how any circuit works. This is an important step in order to figure out why a circuit does not work.

But that is only part of a complete electrical failure repair. Once the problem is found, you have to make proper repair. Many times that involves taking connectors apart and using proper tools and procedures so more damage is not done in the process. Any repairs performed, whether they involve cutting, splicing, and/or rerouting, have to be performed in a manner that will prevent future failures of the circuit. After verifying the repairs you can consider the complete electrical failure repair process completed.

A **wiring diagram** is a schematic of electrical or electronic components using symbols and the wiring of the vehicle's electrical systems.

A **component locator** assists the technician in finding specific components in the vehicle. Both drawings and text are used to lead the technician to the desired component.

Wire Repair

One of the weakest links in a truck's electrical circuit tends to be where wires are connected, whether at a single connection, splice, or multiple plug connections. These connections need to stay tight and clean since corroded, loose, or damaged connections will cause unwanted high resistance that can prevent electrical load components to operate at their intended capacity.

The electrical failure often leads to a loose or damaged connector or you might have to make a splice in the wire, which involves adding a connector. Whatever the repair involves, it must be permanent and not increase the resistance of the circuit. The type of repair(s) performed will depend on many factors such as:

1. Type of conductor
2. Size of wire
3. Ease of access to the failed portion of the circuit
4. Circuit requirements
5. Manufacturer's recommendations

Most repairs are made to single wires, which can involve methods such as crimping, soldering, wrapping, and protecting damaged areas.

Troubleshooting is a diagnostic procedure to isolate a failure and its cause. It is a logical step-by-step process of elimination used for the proper determination of the failure.

Classroom Manual
Chapter 4, page 71.

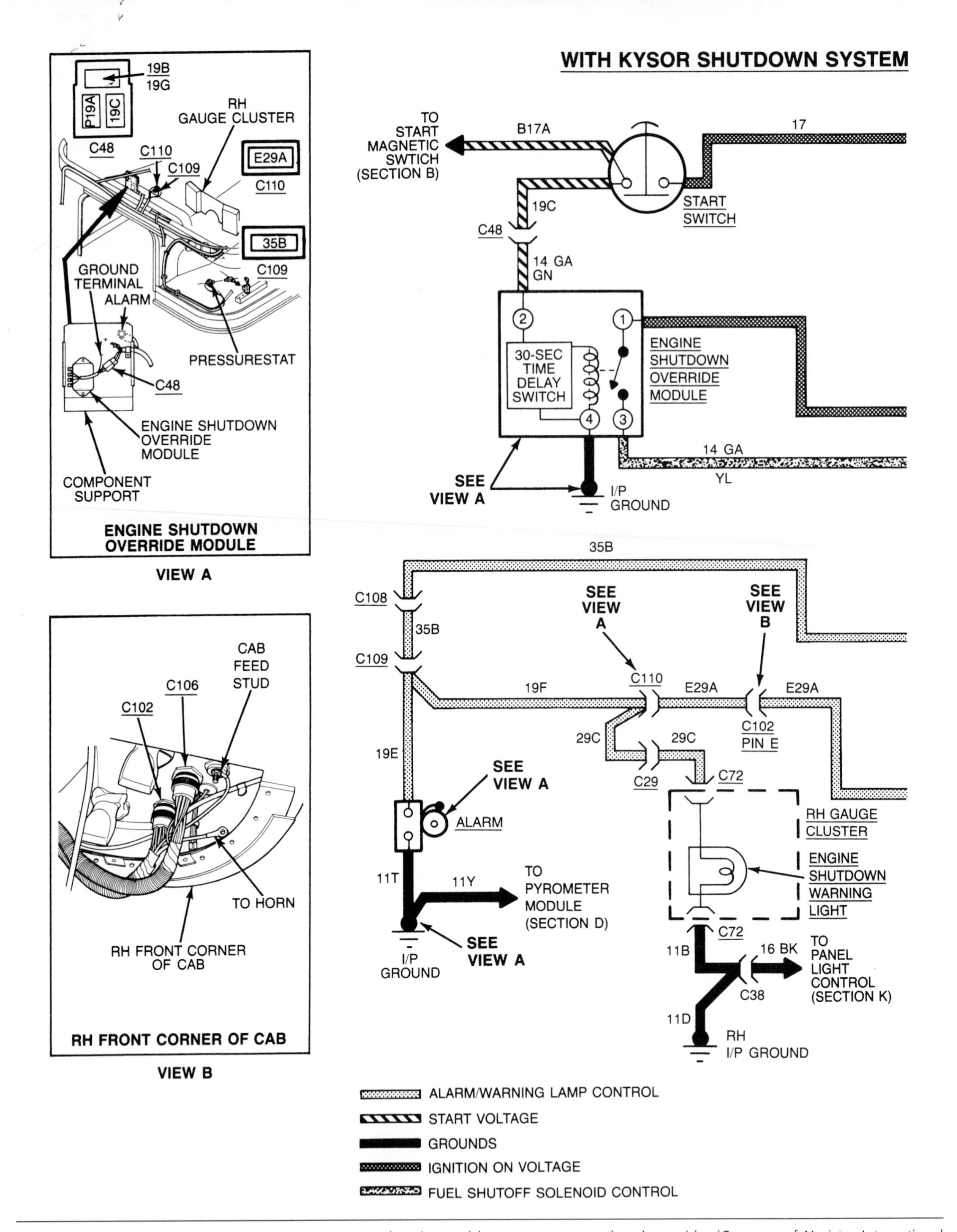

Figure 4-1 Combination wiring diagram, component location and harness connectors location guide. (Courtesy of Navistar International Engine Group)

Figure 4-2 Types of solderless connectors and terminals commonly used for wire repair.

Primary wires are conductors that carry low voltage and low current.

To **splice** means to join wire ends. Sometimes it can be more than two leads coming together at one point.

Copper Wire Repairs

Copper wire is the most common primary wire used in trucks. Because of stress and environmental conditions, the primary wire will occasionally fail due to a break, corrosion, damage to the insulation, or damage due to excessive current flow through it.

The location of the failure will dictate the type of repair that needs to be performed. This might involve replacing a whole section of a wire or simply replacing the end terminal. If you have to replace a long section of wire, make sure it gets properly protected, rerouted, and fastened. The two most common methods of splicing copper wire or replacing terminal ends are crimping and soldering. There are a variety of terminals that can be used for crimping type of repairs (Figure 4-2).

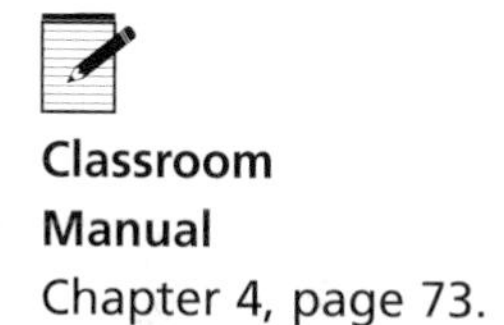

Classroom Manual
Chapter 4, page 73.

Crimping means to deform a connector around a wire so the connection is securely held in place.

CUSTOMER CARE: It is important to perform repairs properly the first time, so customers will not have to come back to your shop with the same problem. "Come backs" can cost you time, money, and your reputation as a quality repair shop.

Crimping

Crimping of solderless connections is the most common practice used to perform wiring repairs in trucks. To prevent future problems, the procedure of crimping has to be precise and correct, but we sometimes tend to take the basics for granted. A good crimping tool for this procedure is essential (Figure 4-3). Use the following steps to make a splice using solderless connections:

1. Determine the proper size of wire being used.
2. Use the correct size of the stripping opening of the tool matching the wire gauge.

Special Tools

Crimping tool
Solderless connector
Electrical tape or heat shrink tube
Safety glasses

The **crimping tool** is a special tool with different areas designed to perform different functions. This tool will cut the wire, strip the insulation, and crimp the connection.

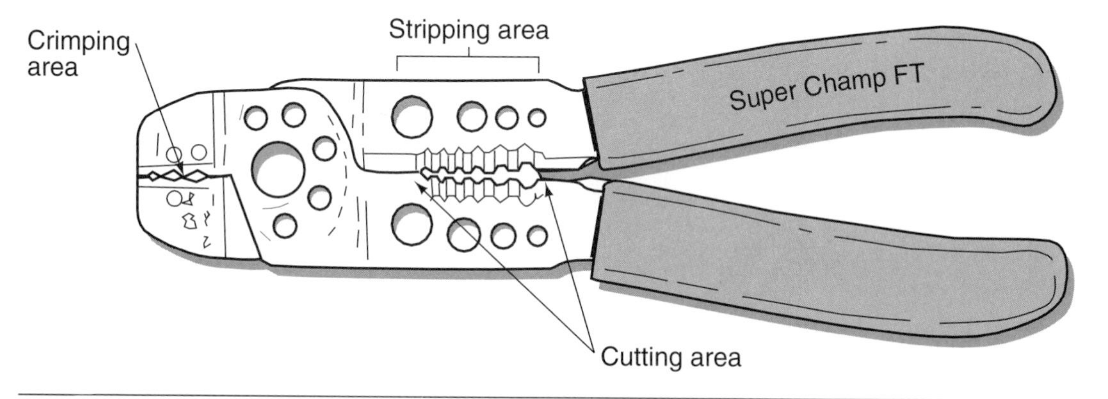

Figure 4-3 A crimping tool.

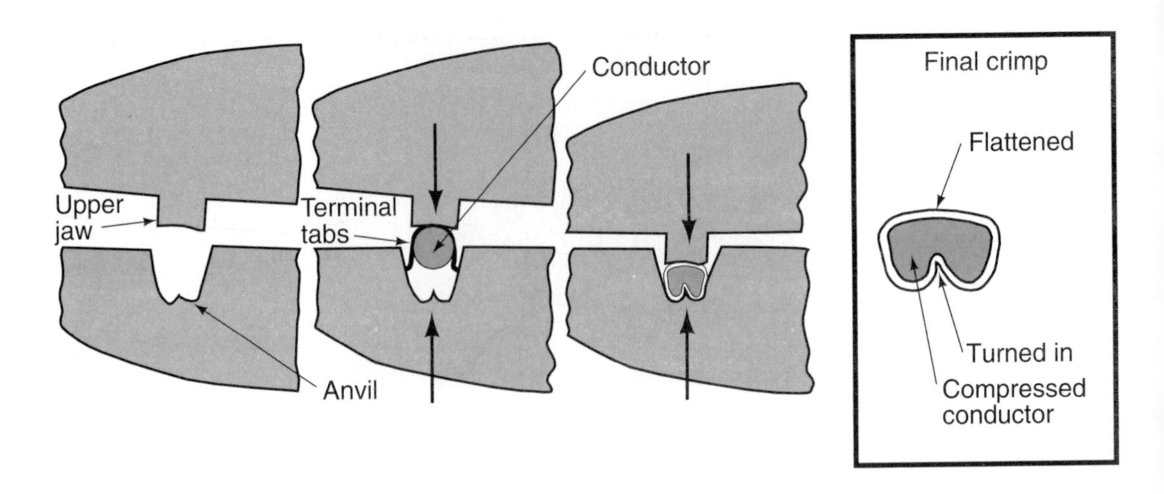

Figure 4-4 Properly crimping a connector.

Solderless connectors are hollow metal tubes covered with a plastic insulation. They can be butt connectors or terminal ends.

Heat shrink tube is a special plastic tube that shrinks in diameter when exposed to heat.

3. Remove enough insulation to allow the wire to completely penetrate the connector, but not so much as to leave exposed wire.
4. Place the wire into the connector and crimp it (Figure 4-4). For a proper crimp, place the open or slotted area of the connector facing toward the anvil. Verify that the wire is compressed under the crimp.
5. If the connector is the type that is used to splice a wire, insert the stripped end of the other wire into the butt connector and crimp in the same manner.
6. Use electrical tape or heat shrink tubing to provide additional protection. Make sure the covering material is watertight.

WARNING: When soldering or installing shrink tubing, the careless use of heat can do more damage than good.

SERVICE TIP: A good stripping tool has the gauge sizes marked on it.

WARNING: Using the wrong size connector for the wire can cause unseen problems or a failure of the circuit.

SERVICE TIP: Some connectors are manufactured with the shrink type tubing as part of the connector.

Sometimes a circuit might require the tapping of an extra wire into the circuit, such as when an extra marker light is added to the trailer. It is a convenient method of providing a connection without stripping wires (Figure 4-5). These connectors should not be used where precise power is critical for a component. A tap connector adds a circuit in parallel with another circuit, causing circuit resistance to decrease and circuit amperage to increase.

Soldering

Unfortunately, soldering is not a common practice for wiring repairs, but it is the best way to splice copper wires. Some repairs require the connection to be very secure with no added resistance because the circuit is sensitive to current variations, such as electronic circuitry. Soldered connections tend to stay secure longer. Solder itself consists of an alloy of tin and lead. It is melted over splice to hold the wire ends or a terminal to the end of the wire securely together. It takes a lot of practice to solder well.

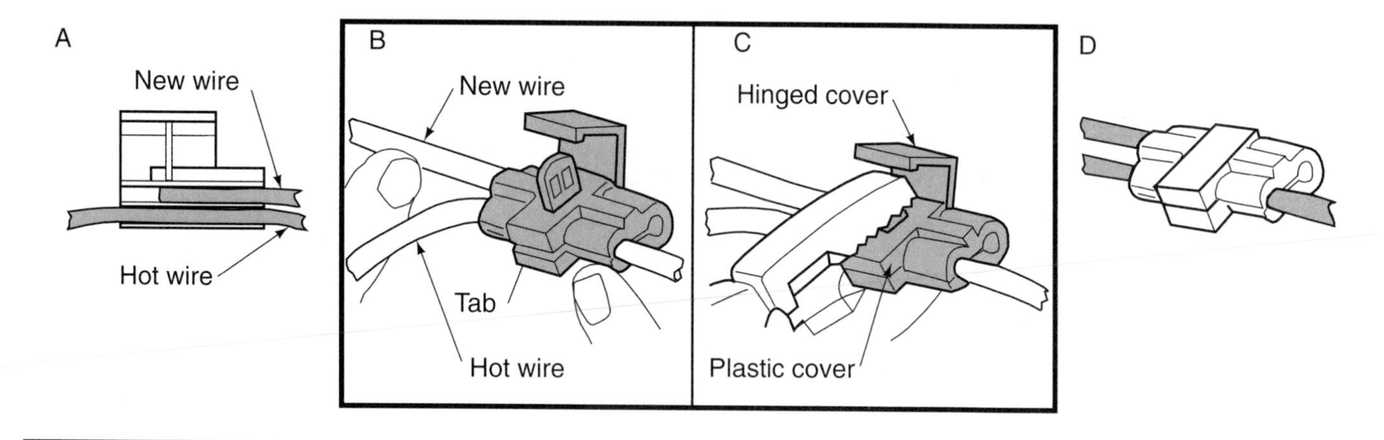

Figure 4-5 Using a tap connector to splice a new wire into a circuit.

CAUTION: The soldering iron and soldered connections get very hot, so prevent personal injury by avoiding contact with them.

Photo Sequence 4 shows the steps in a soldering process when using a splice clip. If a splice clip is not used, the wire ends should be braided tightly together. The key to soldering is heating the splice so the solder melts from contact with the splice, not from contact with the soldering gun's heating tip. For electrical repairs, you must use rosin core solder. Acid core solder can cause the wire to corrode, which can lead to high resistance.

SERVICE TIP: Before making a splice in a wire, check the rest of the wire to make sure there are not too many other splices in the line. Never use a wire that is smaller in size than the existing wire.

Wires can also be soldered together without using a splice clip. This involves removing about one inch of insulation from two or more wires. First join the wires using one of the methods shown in Figure 4-6. Next use the soldering gun to heat the splice, applying the solder to the wires at the splice, not the gun. Then allow the solder to melt and flow evenly among all the strands of wire (Figure 4-7). Let the soldered splice cool momentarily before moving it. Last, insulate the splice with electrical tape or heat shrinking tube.

SERVICE TIP: Before the new splice is insulated, perform a continuity and resistance test with an ohmmeter if possible or a voltage drop test across the connector if the circuit is powered up.

SERVICE TIP: Any new splice must be a minimum of 1.5 inches from a connector sleeve or another splice.

Splicing Twisted-Shielded Wires

With the advent of computers in vehicles, you will run across another style of wiring that consists of two wires twisted together within an outer shell and with or without a drain wire, as shown in Figure 4-8. This configuration of the wires protects the electronic circuits from electrical noise that could interfere with the operation of computer controls. Any splices made in these wires have to be perfect with no resistance. These wires may carry as low as 0.1 ampere of current. Any added resistance can end up causing a false signal to the computer or actuator. Note that the proper operation of the computer is dependent upon the integrity of its circuits.

Special Tools

100-watt soldering iron

60/40 rosin core solder

Crimping tool

Splice clip

Heat shrinking tube

Heating gun

Safety glasses

Sewing seam ripper

Electrical tape

Soldering is the process of using heat and solder (a mixture of lead and tin) to make a splice or connection.

A **splice clip** is a special connector used with soldering to ensure a good connection. The splice clip is not insulated and it has a hole for applying solder.

Photo Sequence 4
Soldering Copper Wire

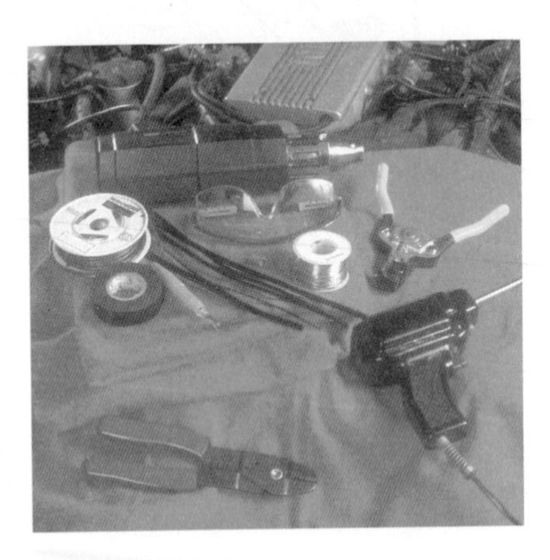

P4-1 Tools required to solder copper wire: 100-watt soldering iron, 60/40 rosin core solder, crimping tool, splice clip, heat shrink tube, heating gun, and safety glasses.

P4-2 Disconnect the fuse that powers the circuit being repaired. Note: If the circuit is not protected by a fuse, disconnect the ground lead of the battery.

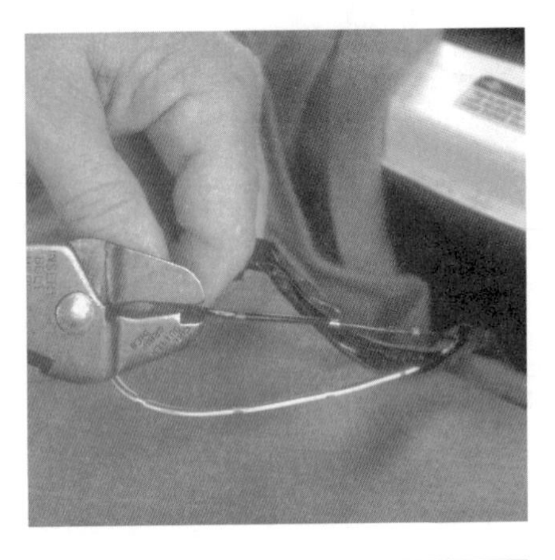

P4-3 Cut out the damaged wire.

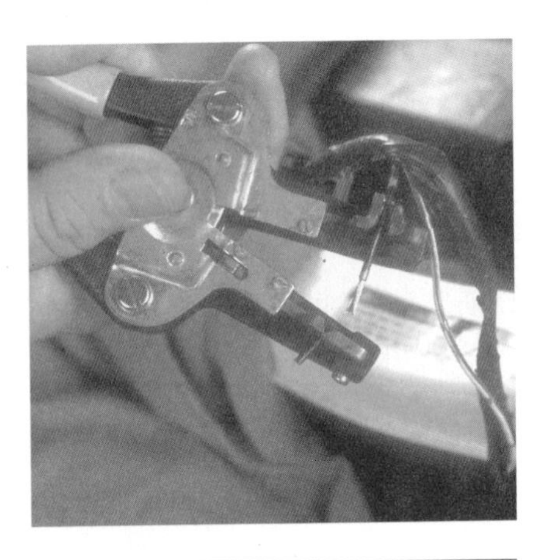

P4-4 Using the correct size stripper, remove about 1/2 inch of the insulation from both wires.

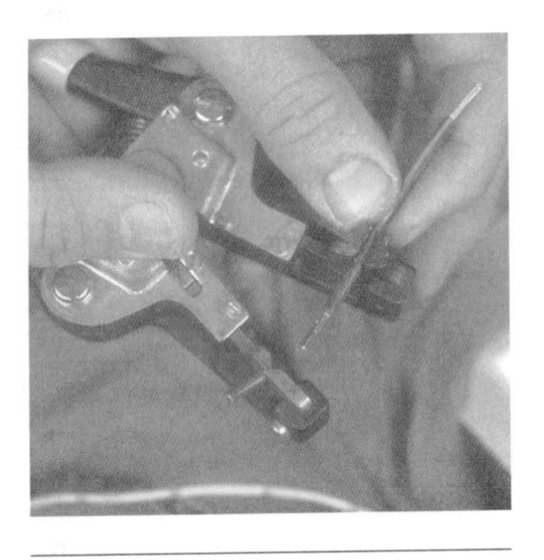

P4-5 Now remove about 1/2 inch of the insulation from both ends of the replacement wire. The length of the replacement wire should be slightly longer than the length of the wire removed.

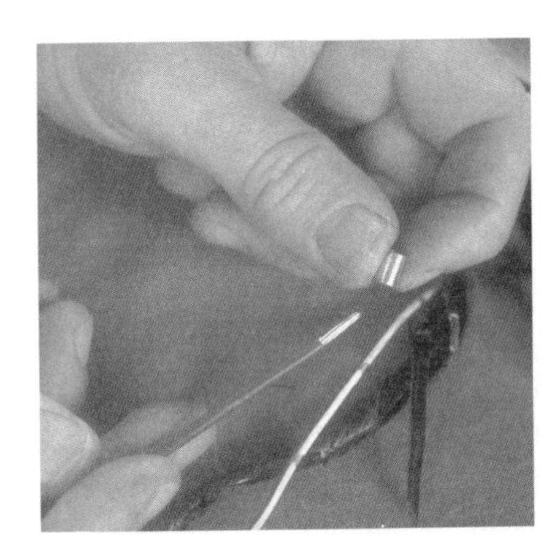

P4-6 Select the proper size splice clip to hold the splice.

Photo Sequence 4
Soldering Copper Wire (Continued)

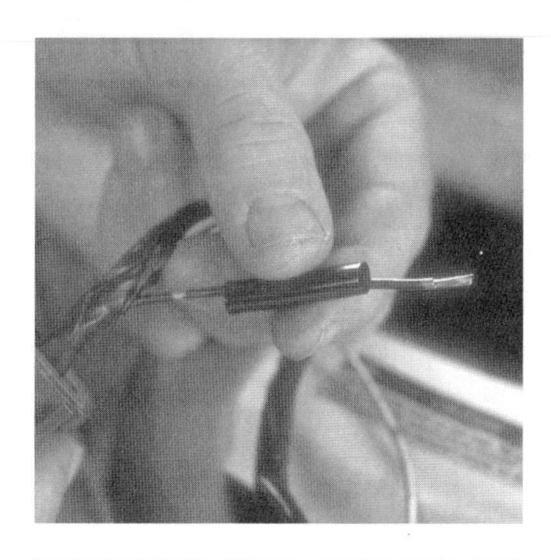

P4-7 Place the correct size and length of heat shrink tube over the two ends of the wire.

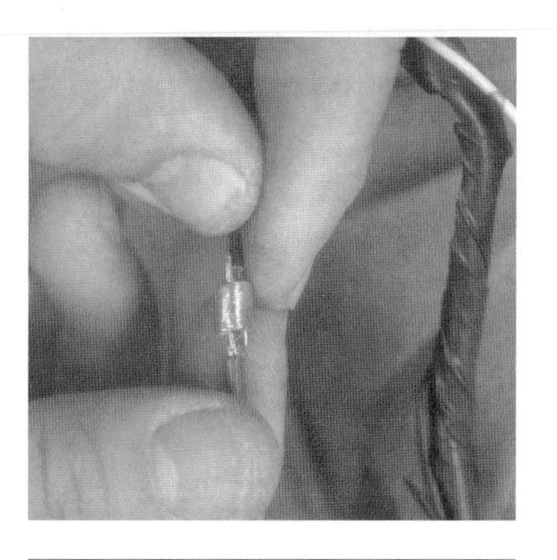

P4-8 Overlap the two splice ends and center the splice clip around the wires, making sure the wires extend beyond the splice clip in both directions.

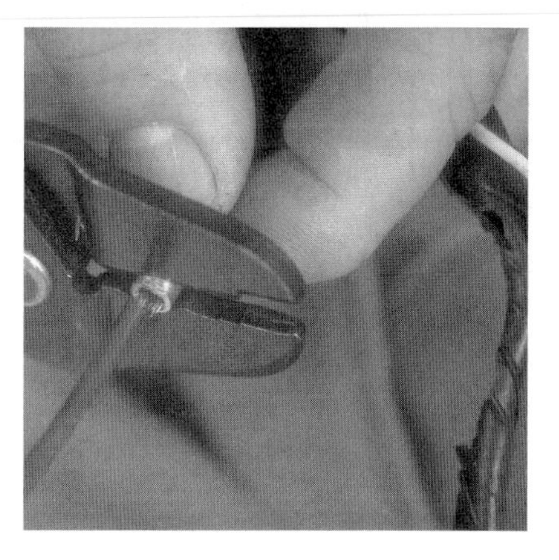

P4-9 Crimp the splice clip firmly in place.

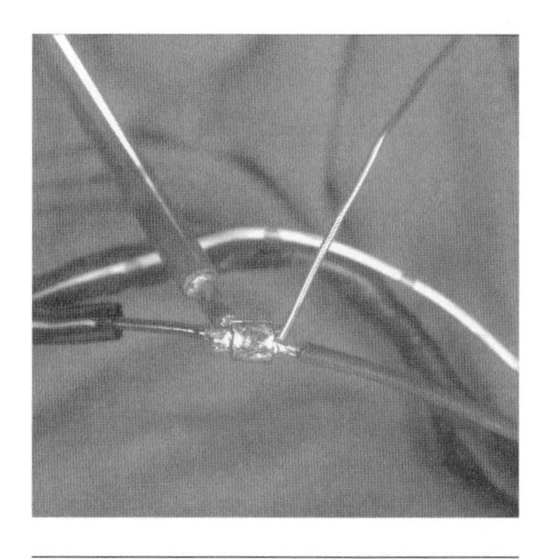

P4-10 Heat the splice clip with the soldering iron while applying solder to the opening of the clip. Do not apply solder to the iron. The iron should be 180 degrees away from the opening of the clip.

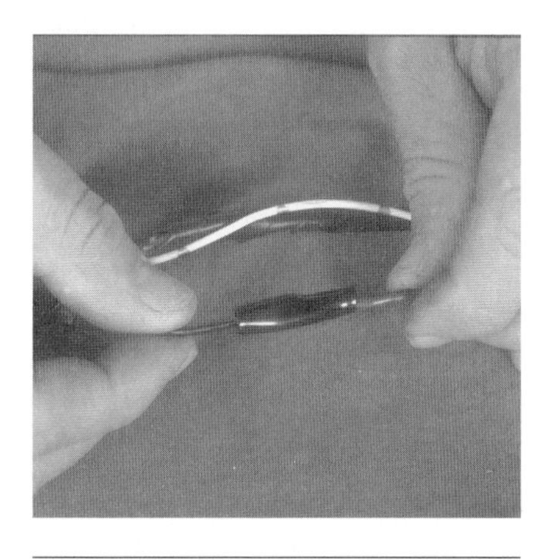

P4-11 After the solder cools, slide the heat shrink tube over the splice.

P4-12 Heat the tube with the hot air gun until it shrinks around the splice. Do not overheat the heat shrink tube.

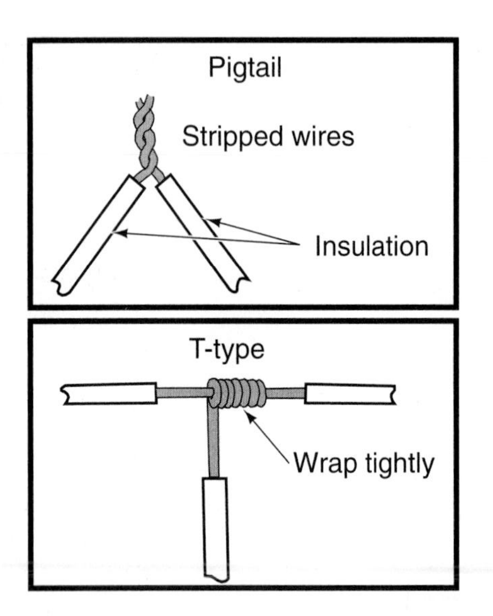

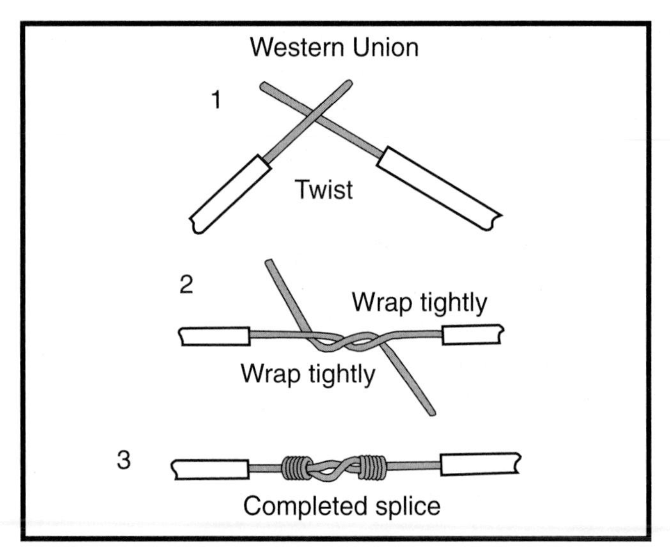

Figure 4-6 Methods for joining two wires together.

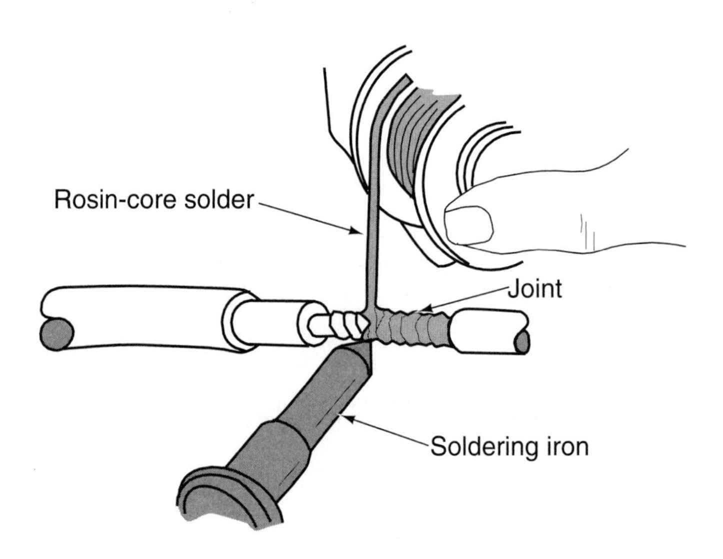

Figure 4-7 When soldering, apply the solder to the joint not to the tip of the soldering iron.

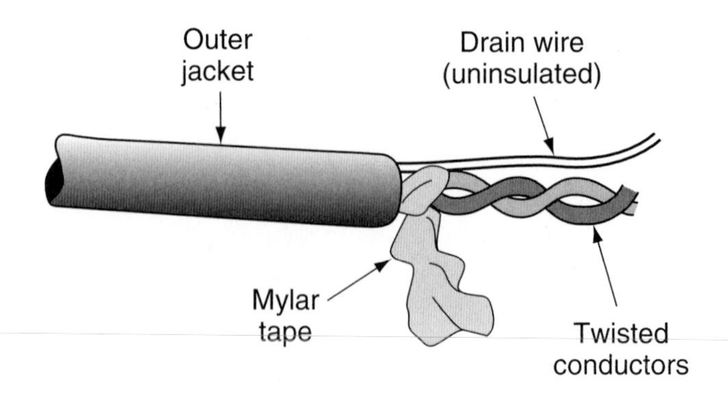

Figure 4-8 Twisted-shielded wire in computer circuits.

The following steps should be used when splicing twisted wires (Figure 4-9):

1. Locate and eliminate the bad section of the wire.
2. Cut back the jacket (1) leaving enough working room to retwist the wires (2).
3. Strip the proper length of insulation from the wires to connect them with a splice clip (3).
4. Crimp and solder the splice clip (4).
5. Wrap the individual splices with tape.
6. Wrap the outer jacket with electrical tape or use heat shrink.

WARNING: Stagger the splices so they cannot bundle up together and rub through. This could eventually cause a short.

SERVICE TIP: Never get the heat shrink hot enough to unsolder the splice. The proper "size" heat shrink tube will shrink very tight with minimal heat.

The following steps should be taken when splicing a twisted wire with a shielded cable (Figure 4-10):

1. Remove the outer jacket (1), being careful not to cut or damage the aluminum/Mylar tape (2) or drain wire (3) unless it is already damaged.
2. Once a good working area is exposed, you can make clean cuts and unwrap the Mylar tape (2).
3. Strip back enough of the wire insulation to insert the splice clips (4) on the twisted wires first.

Special Tools

Crimping tool

Splice clip

100-watt soldering gun

Safety glasses

60/40 rosin core solder

Electrical tape or shrink tube

Twisted-shielded wires are used to protect computerized circuits from electrical noise that can interfere with their operation.

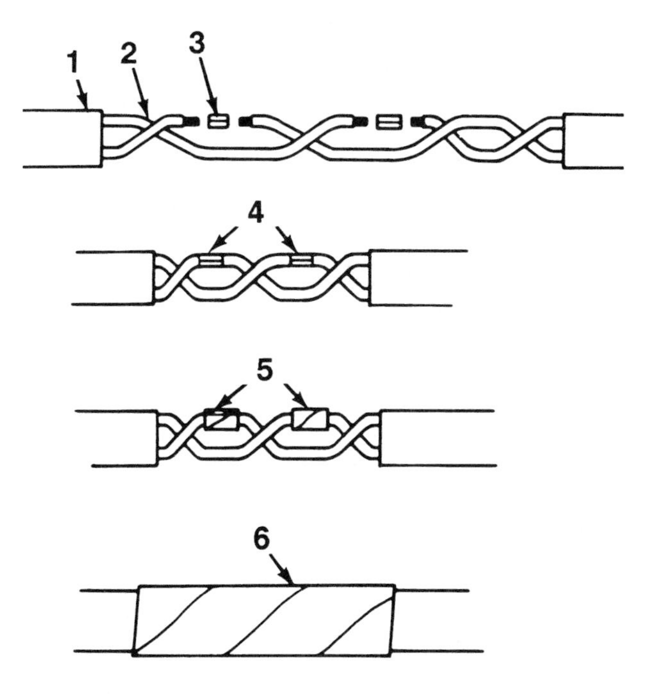

Figure 4-9 Proper twisted wire repair sequence. (Courtesy of General Motors Corporation, Service Operations)

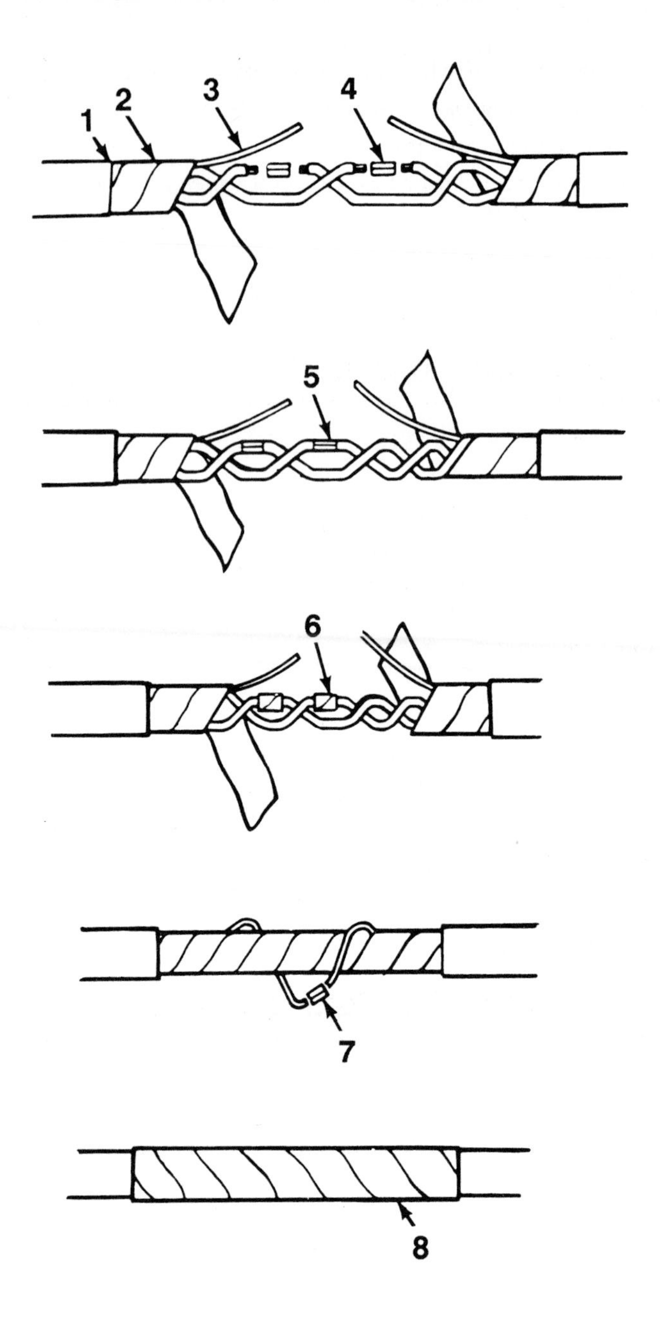

1. Jacket
2. Aluminum/Mylar Tape
3. Drain Wire
4. Splice Clip
5. Crimp and Solder
6. Electrical Tape Wrap
7. Drain Wire Splice Clip, Crimped And Soldered
8. Outer Electrical Tape Wrap

Figure 4-10 Proper twisted-shielded wire repair sequence. (Courtesy of General Motors Corporation, Service Operations)

4. After a proper twist, insert the wire ends into the splice clips (4).
5. Crimp and solder the splice clips to the wires.
6. Wrap electrical tape on the splices.
7. Wrap both splices with the unwrapped Mylar tape (2) leaving the drain wire leads (3) outside of the wrap.
8. Splice the drain wire (3) with a splice clip (7). Crimp and solder the splice clip.
9. Use electrical tape or heat shrink to complete the job.

Replacing Fusible Links

Verify a problem with the fusible link first. A visual inspection does not always indicate a bad fusible link. Test for battery voltage on both sides of the fusible link to confirm a failure. If the fusible link is bad, it is cut out of the circuit and replaced with a new one. A new fusible link is either crimped or soldered in place. The wire ends have to be cut back enough to allow a proper fit of the connector (Figure 4-11). The length of the new fusible link should never be more than 9 inches. A longer length might not provide sufficient overload protection. Sometimes a fusible link will feed two harness wires.

Special Tools

Crimping tool

100-watt soldering iron

60/40 rosin core solder

Safety glasses

WARNING: To prevent circuit damage, disconnect the battery negative cable before performing any repairs to the fusible link. Also, be sure to use the correct fusible link gauge required by the manufacturer.

A **fusible link** is a special conductor with a special insulation that is placed in a circuit to protect it from overload. It is usually located near the power source such as the battery to protect an entire circuit. During an overload, the conductor melts within the fusible link.

SERVICE TIP: Fuses and links do not blow unless there is a problem, and some problems can be borderline problems. Using a larger fuse is not the solution. The solution is to find the cause of the problem.

Repairing Connector Terminals

Connectors can be prone to a number of problems, including bent, broken, or corroded contacts. Heat generated by a loose contact can cause the connector shell to melt. Any of these problems warrant performing a connector repair. The method of repair depends on the type of problem and type of connector.

CUSTOMER CARE: Today's trucks rely on electrical and electronic systems to operate properly and safely. Be sure to include a visual inspection of major electrical connectors as part of a preventive maintenance program.

Molded Connectors

Some connectors such as a one-piece molded type connector cannot be taken apart for repairs (Figure 4-12). If the connector is damaged, it must be cut off and a new connector must be spliced in.

Molded connectors usually have one or more wires molded into a one-piece component.

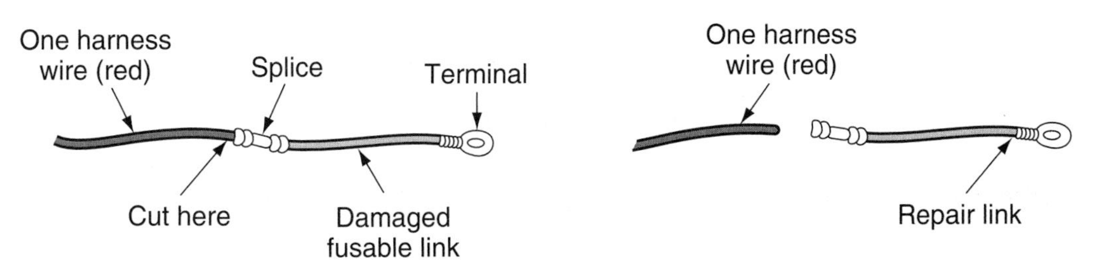

Figure 4-11 Replacing a fusible link.

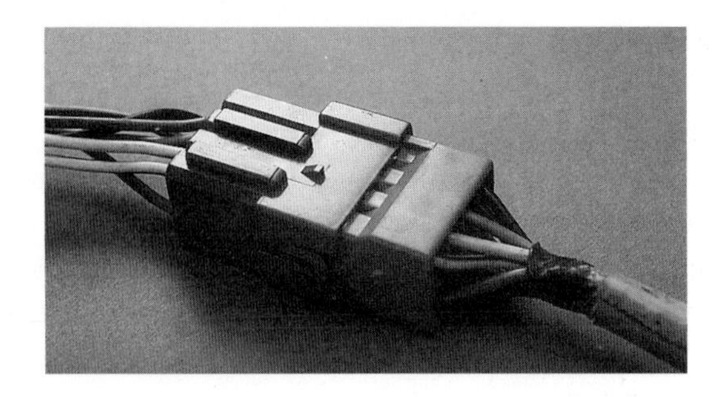

Figure 4-12 Molded connectors are a one-piece design and cannot be separated for repairs.

Classroom Manual
Chapter 4, page 75.

Hard shell connectors have a hard plastic shell that holds the connecting terminals of separate wires.

Hard Shell Connectors

Hard shell connectors usually have removable terminals. To remove the terminals from the connector usually requires a special tool or pick to depress the locking tang of the terminal (Figure 4-13). Pull the lead back far enough to release the locking tang from the connector. Remove the pick, and then pull the lead completely out of the connector. Make the repair to the terminal or replace it with a new one. Re-form the terminal locking tang before reusing to ensure a good lock into the connector (Figure 4-14). The pick is used to bend the tang back to its original shape. Insert the lead into the back of the connector. A catch should be felt as the lead is pushed halfway through the connector. Push it all the way through then gently tug back and forth on the lead to confirm that the terminal is locked in place.

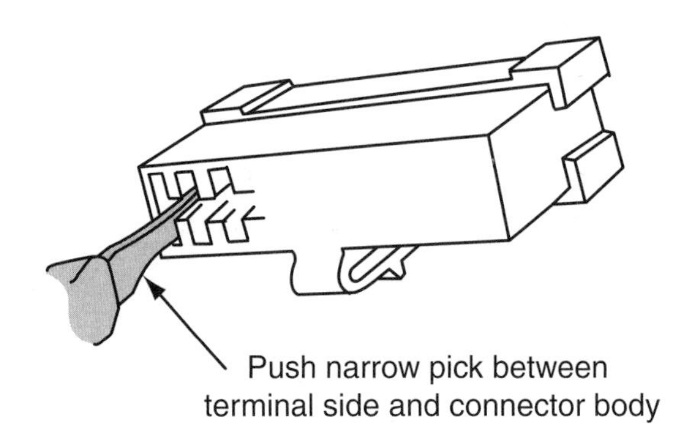

Figure 4-13 Depress the locking tang to remove the terminal from the connector.

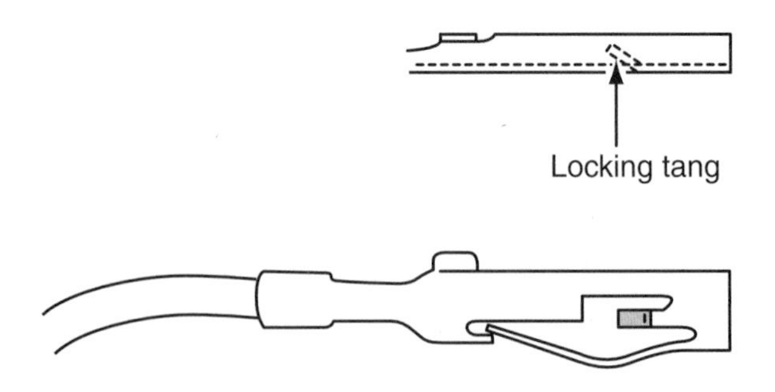

Figure 4-14 Re-form the locking tang to its original position before inserting the terminal back into the connector.

Packard Connectors

Packard connectors are commonly used in trucks. They come in various types and configurations (Figure 4-15). We will look at some of the more popular ones.

Repairing Weather-pack Series Connectors Weather-pack connectors use a primary lock to keep the two halves of the connector together (Figure 4-16) and a secondary lock to help retain the terminals in the connectors (Figure 4-17). The connectors are separated by pulling up on the primary lock while pulling the two halves apart. For terminal removal, the secondary lock is unlatched and allowed to swing open.

Weather-pack connector terminals use the flex-pin and lap-lock terminal design. Depressing the terminal tangs requires a special tool. First push the terminal tool into the terminal cavity from

Special Tools

Pick

Crimping tool

Safety glasses

Classroom Manual
Chapter 4, page 77.

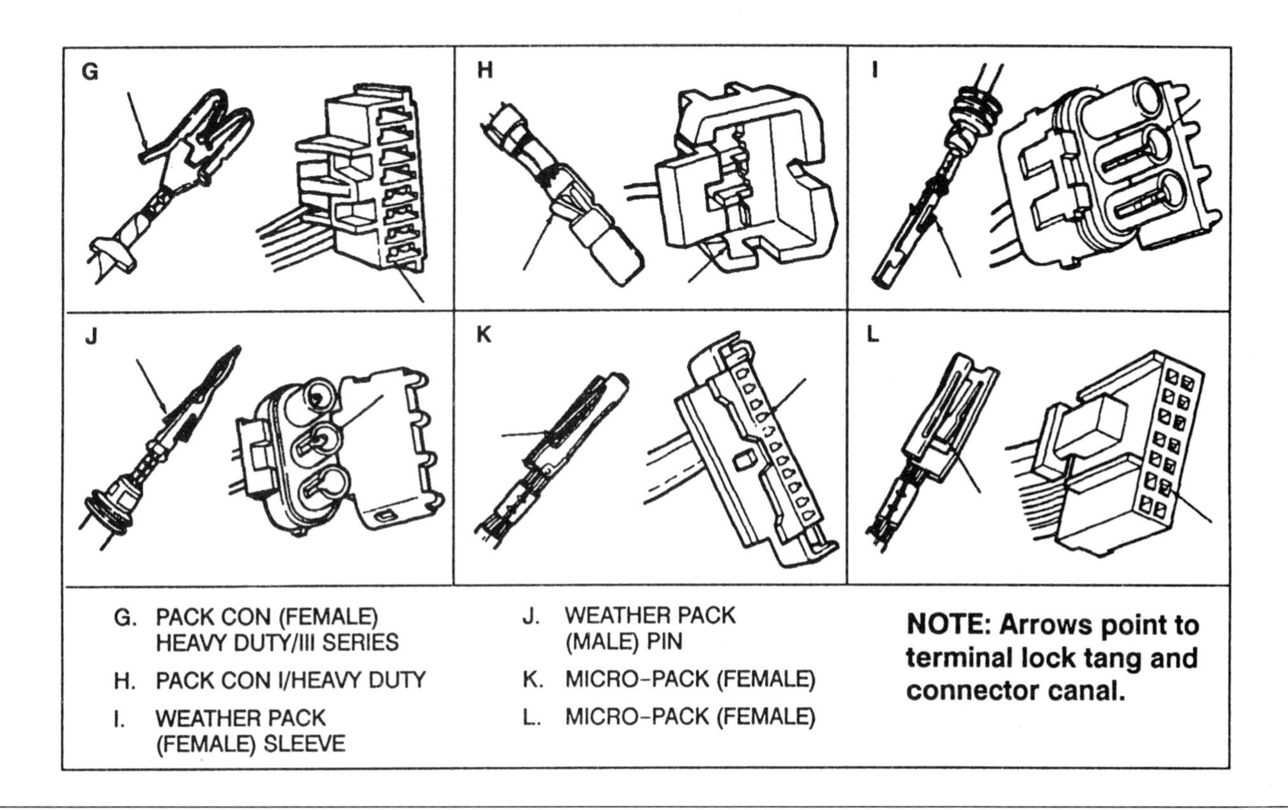

Figure 4-15 Various types of Packard connectors. (Courtesy of Navistar International Engine Group)

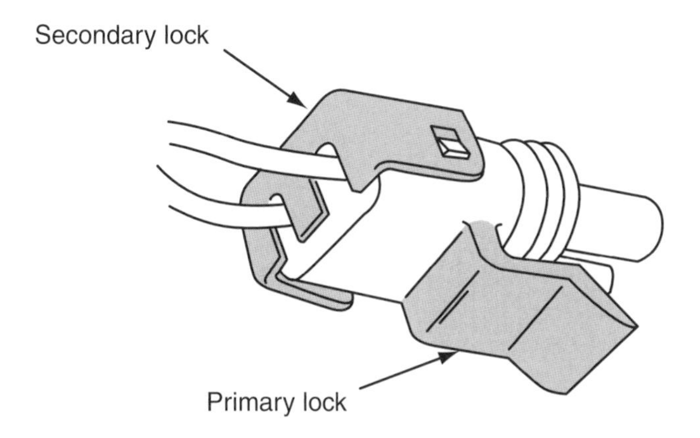

Figure 4-16 A weather-pack connector has two locks. Use the primary lock to separate the halves.

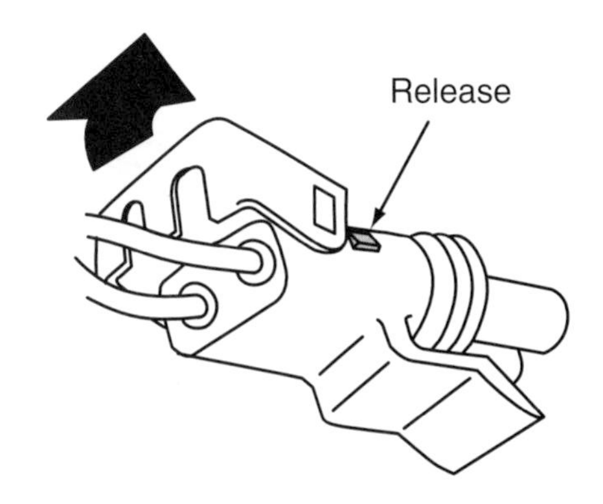

Figure 4-17 Unlock the secondary lock to remove the terminals from the connector.

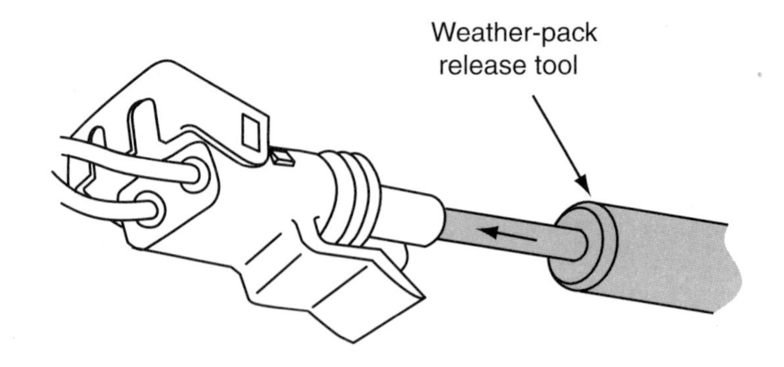

Figure 4-18 Use the recommended special tool to unlock the tang on the terminal.

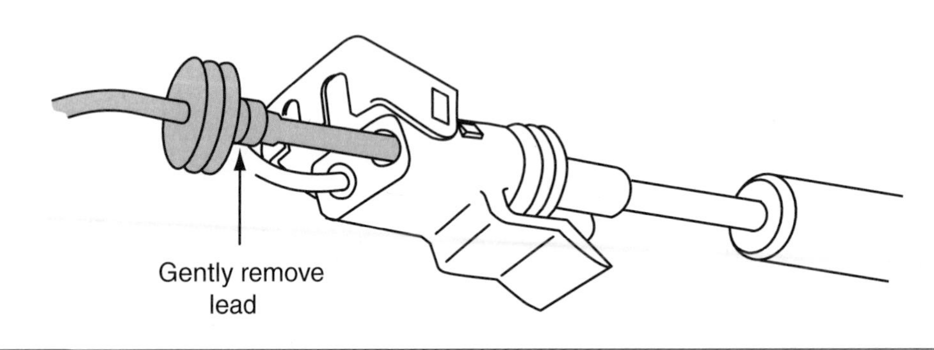

Figure 4-19 After the lock tang has been depressed, remove the lead from the back of the connector.

Weather pack connectors have rubber seals on the terminal ends and connector halves to protect the connections from corrosion.

Special Tools

Weather-pack terminal tool

Crimping tool

Safety glasses

Use a weather-pack repair kit with new seals for a complete repair.

the front until it stops (Figure 4-18). Then remove the tool and gently pull the lead out of the back of the connector (Figure 4-19). The terminal is either a male or female connector (Figure 4-20). Cut the wire behind the cable seal. Slip a new seal on the cable and strip the insulation back for a new terminal. Then crimp and solder the new terminal to the wire (Figure 4-21). Make sure the terminal lock tang is formed properly. If it is not, the tangs can be bent back (Figure 4-22).

Packard Series 56 Connectors Packard Series 56 connectors provide positive locking of both connectors and terminals (Figure 4-23). The female terminal has a spring-loaded tang that maintains a constant pressure against the male blade to provide a positive contact (Figure 4-24). To

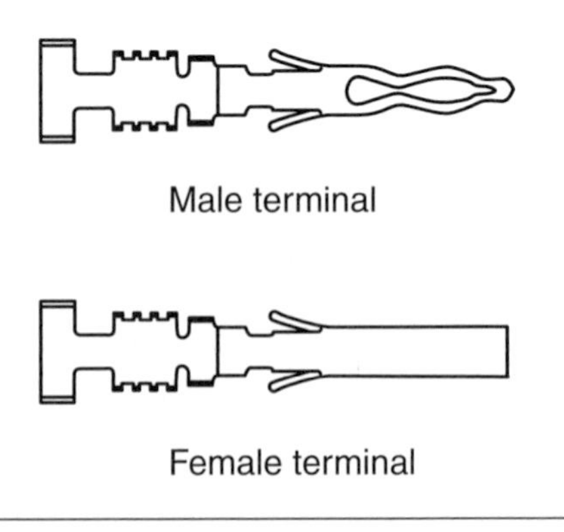

Figure 4-20 Male and female weather-pack terminals.

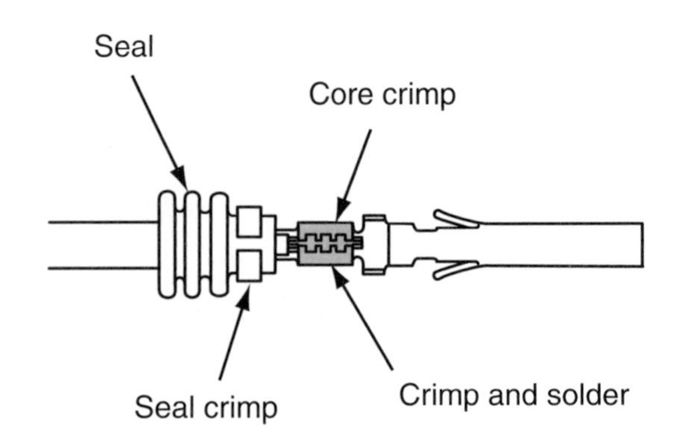

Figure 4-21 Crimp and solder the terminal to the lead.

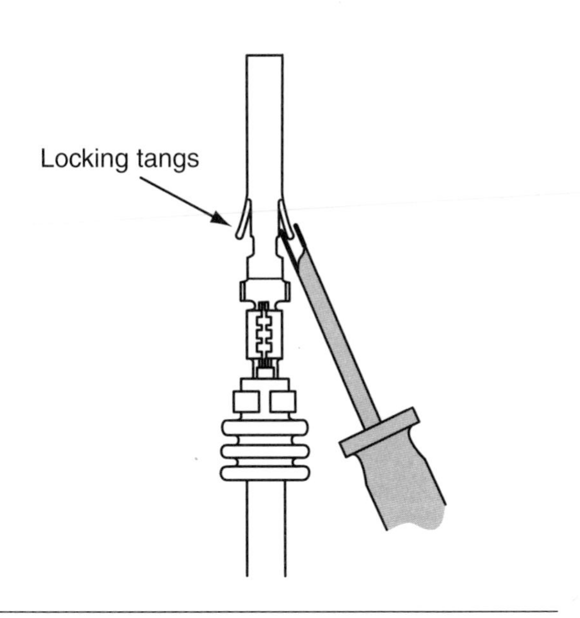

Figure 4-22 Reforming the locking tangs of the terminal.

Note: Female connectors may be used for panel-mount application.

Figure 4-23 Packard connectors. (Courtesy of Navistar International Engine Group)

remove the female terminal, insert a fine blade between the terminal's locking tang (Figure 4-25) and then pull on the lead to remove the terminal. To remove male terminals (Figure 4-26), insert a small blade between the locking tang and the insulator to compress the tang. Then pull on the lead to remove the terminal.

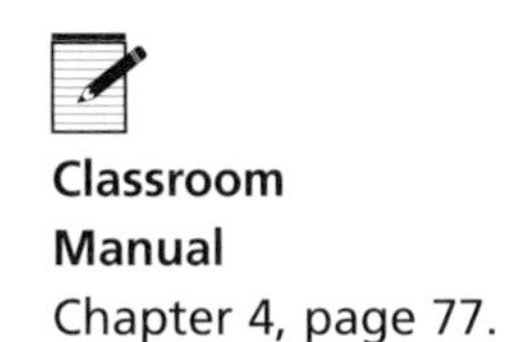

Classroom Manual
Chapter 4, page 77.

SERVICE TIP: Sometimes the wire is broken or breaks in the process of removing the terminal. When that happens, use needle nose pliers to remove the terminal.

To install a new terminal on the wire, refer to Figure 4-27 depicting a male terminal:

1. Strip approximately 1/4 inch of insulation from the end of the wire.
2. Insert the wire into the blade and core of the terminal.
3. Using a crimping tool, crimp the core wings so that the wire core is visible, both above and below the core wings.
4. Crimp the insulation wings over so that the wings cover the insulation of the wire. No wire core should be visible under the insulation wings.

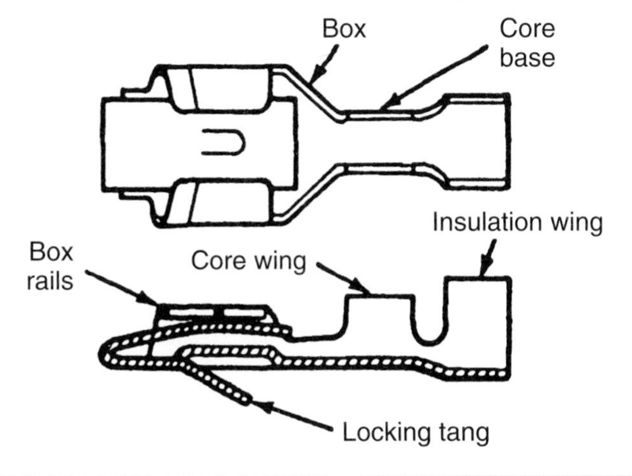

Figure 4-24 Female terminal. (Courtesy of Navistar International Engine Group)

Figure 4-25 Removing female terminal from connector. (Courtesy of Navistar International Engine Group)

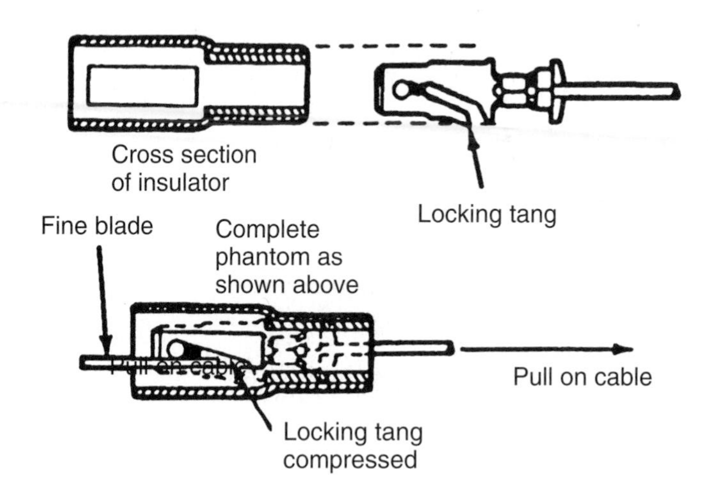

Figure 4-26 Removing male terminal from connector. (Courtesy of Navistar International Engine Group)

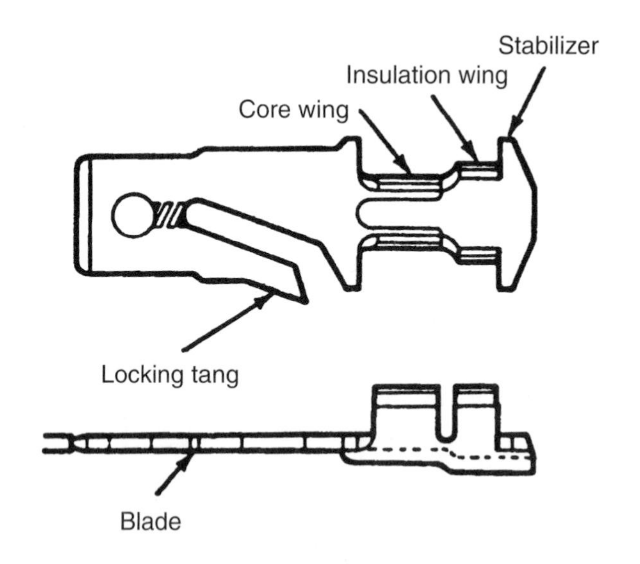

Figure 4-27 Male terminal. (Courtesy of Navistar International Engine Group)

Both male and female terminals are installed into the connector body from the back. When the wire/terminal assembly is inserted deep enough into the connector assembly, the locking tang will lock the terminal into place. Gently tug on the terminal to make sure it is seated. This is a push to seat type of connector.

Packard Metri-Pack Terminal Removal (Pull to Seat Type) To remove a Packard metri-pack terminal, a pick is inserted from the front of the connector and under the locking tang (Figure 4-28). Depress the locking tang to unseat the terminal. Gently push on the cable to remove the terminal through the front of the connector.

To replace a terminal or new wire, insert the wire from the back of the connector through the seal and proper connector cavity. Extend the wire far enough to work with. Strip the proper length of insulation and crimp the new terminal in place (Figure 4-29). Gently tug between the terminal and wire to make sure the terminal is on securely. To install the terminal, pull gently back into the connector until it locks in place (Figure 4-30).

Pull to seat and **push to seat** are methods used to install terminals into the connector.

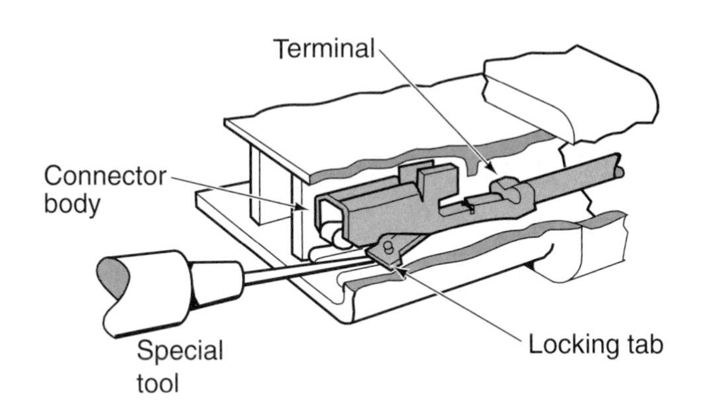

Figure 4-28 Pull down the lock tang to release the terminal from the connector.

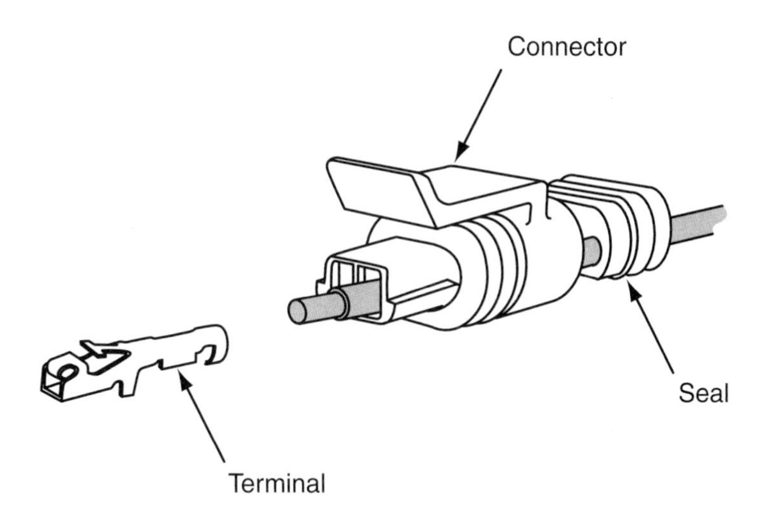

Figure 4-29 The wire lead must be installed into the seal and connector before attaching the terminal.

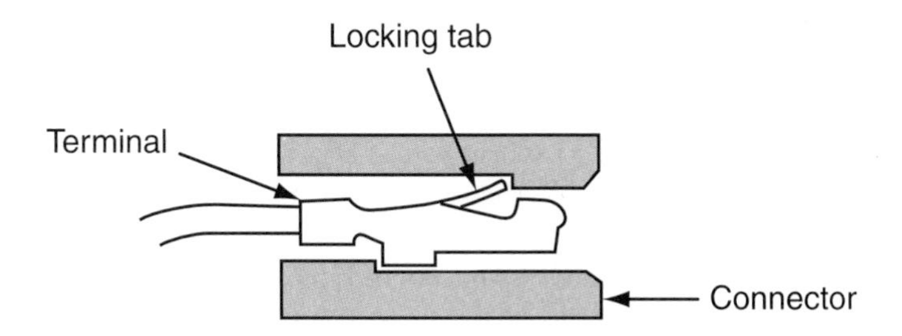

Figure 4-30 Make sure the terminal locks into the connector body.

Classroom Manual
Chapter 4, page 82.

Reading Wiring Diagrams

Many system malfunctions require the use of wiring diagrams to isolate the cause. When it comes to working on trucks, reading diagrams can sometimes seem complicated. For example, the same medium duty truck might specify different systems such as air brakes or hydraulic brakes. In this case, the warning systems will be wired differently. Any number of different manufacturer's engines could be specified for the same model truck. This would require additional or optional diagrams for the same model, but relating to different engines. The same is true with heavy duty trucks. In addition to different engines, we can add or delete other options such as engine shut down systems, automatic transmissions versus manual transmissions, or any number of environmental (comfort) options.

Most manufacturers take all this into consideration by offering electrical system troubleshooting manuals to cover a particular model of truck with any option system that is available for that model. For a truck technician that can be a blessing because manuals break up diagrams to cover portions of the system(s) rather than having to read and isolate a problem from a complete master diagram. However, at times it helps to look at the complete electrical system in order to understand the subsystems.

Before you use an electrical troubleshooting diagram system, you have to become familiar with the arrangement of the manual or diagram. Some manufacturers provide circuit diagram instructions (Figure 4-31). An important instruction found here is that switch and relay and solenoid positions as shown on circuit diagrams, indicate normal positions with the key switch in the OFF position, unless otherwise noted.

Classroom Manual
Chapter 4, page 83.

Circuit diagram instructions provide the reader with any special notes or requirements needed to read and understand the diagram.

Connector composites show pin configurations of the connector and which circuits are attached to the pins.

This diagram in Figure 4-32 shows a schematic picture of how a circuit is powered, the current path to circuit components, and how the circuit is grounded. Components that work together are usually shown together. The circuit components are also named, although sometimes abbreviations are used. Sectional diagrams also have page number references to tell you to where the rest of or next component or location of the wire leads.

Many diagrams also show connector composites (Figure 4-33). The composites show the pin configuration of the connector and which circuits are attached to the pins. Most manufacturers show the connectors as viewed from the mating end.

To get a better idea of how to read circuit diagrams, we will follow a circuit for electric windows (Figure 4-34). Before a problem is solved, the technician needs to understand how the circuit and its components work in the first place. Looking at the diagram, we see that power to the right and left electric window motors comes from separate circuit breakers. Circuit breaker M13 supplies the left hand (LH) window motor and M15 supplies the right hand (RH) window motor. *It is important to note that in this case, we would reference a power distribution diagram in the manual to determine when the circuit breaker would be powered up* (Figure 4-35).

This shows us that J2 is the ignition feed stud. Power is available at the window switch whenever the vehicle ignition key is in the IGN position. The windows do not operate with the key in accessory position. Going back to Figure 4-34, you should notice a difference between the LH window side circuit and the RH window side circuit. Each window has a switch on the control module but the RH window has an optional switch located on the passenger door.

In this particular case we will take the diagram and break it down into a more basic form to explain the circuit of the LH window operation (Figure 4-36). The LH window switch controls the driver's side window motor in the following manner:

1. With the switch in the up position, both pointers rotate to the right. Pointer "A" completes the circuit between switch terminals 2, 3, and 4. This action applies voltage to the BLUE motor wire. At the same time, pointer "B" connects terminals 1, 6, and 5, completing the ground path for the RED motor wire.
2. Going into the down position, both pointers ("A" and "B") rotate to the left. Pointer "A" completes the circuit between terminals 1 and 2, applying voltage to the RED motor wire. At the same time, pointer "B" completes the circuit between terminals 4 and 5, completing the ground path for the BLUE motor wire.

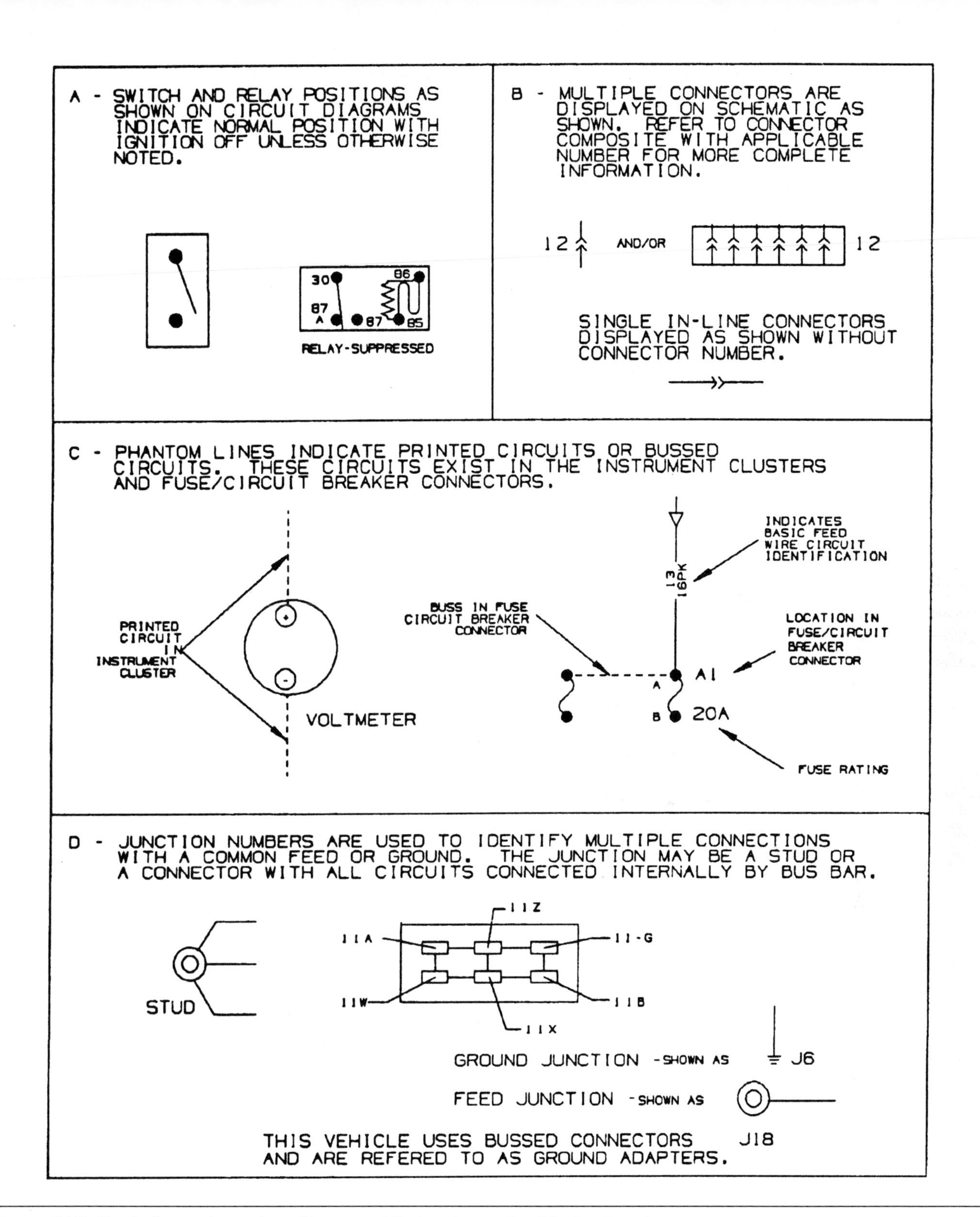

Figure 4-31 Circuit diagram instructions. (Courtesy of Navistar International Engine Group)

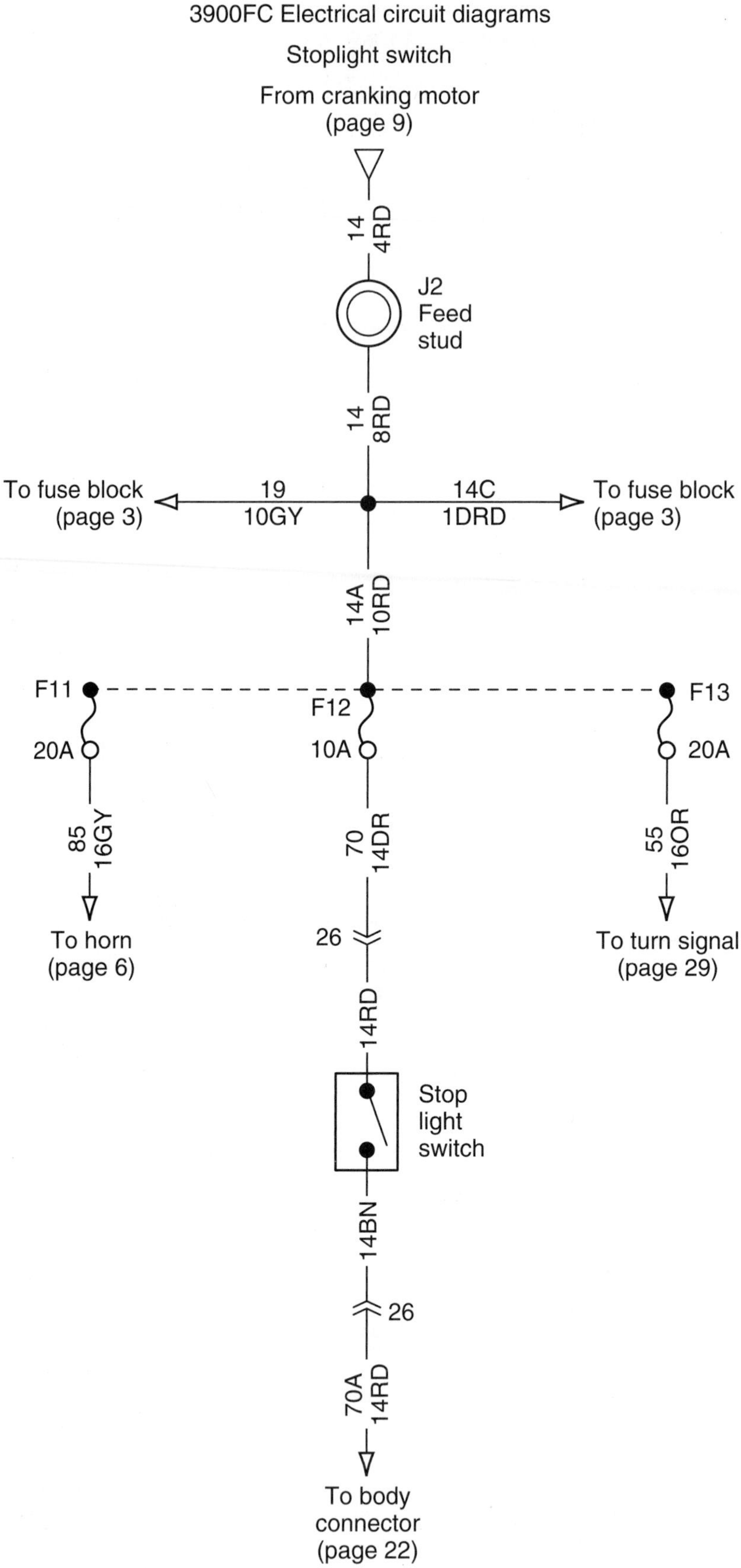

3900FC Electrical circuit diagrams

Stoplight switch

From cranking motor
(page 9)

14
4RD

J2
Feed
stud

14
8RD

To fuse block
(page 3)

19
10GY

14C
1DRD

To fuse block
(page 3)

14A
10RD

F11
F12
F13

20A
10A
20A

85
16GY

70
14DR

55
16OR

To horn
(page 6)

26

To turn signal
(page 29)

14RD

Stop
light
switch

14BN

26

70A
14RD

To body
connector
(page 22)

Figure 4-32 Typical circuit diagram.

TACHOMETER SENSOR
(36)

048A
N48
BK
BK

Figure 4-33 Typical connector end view. (Courtesy of Navistar International Engine Group)

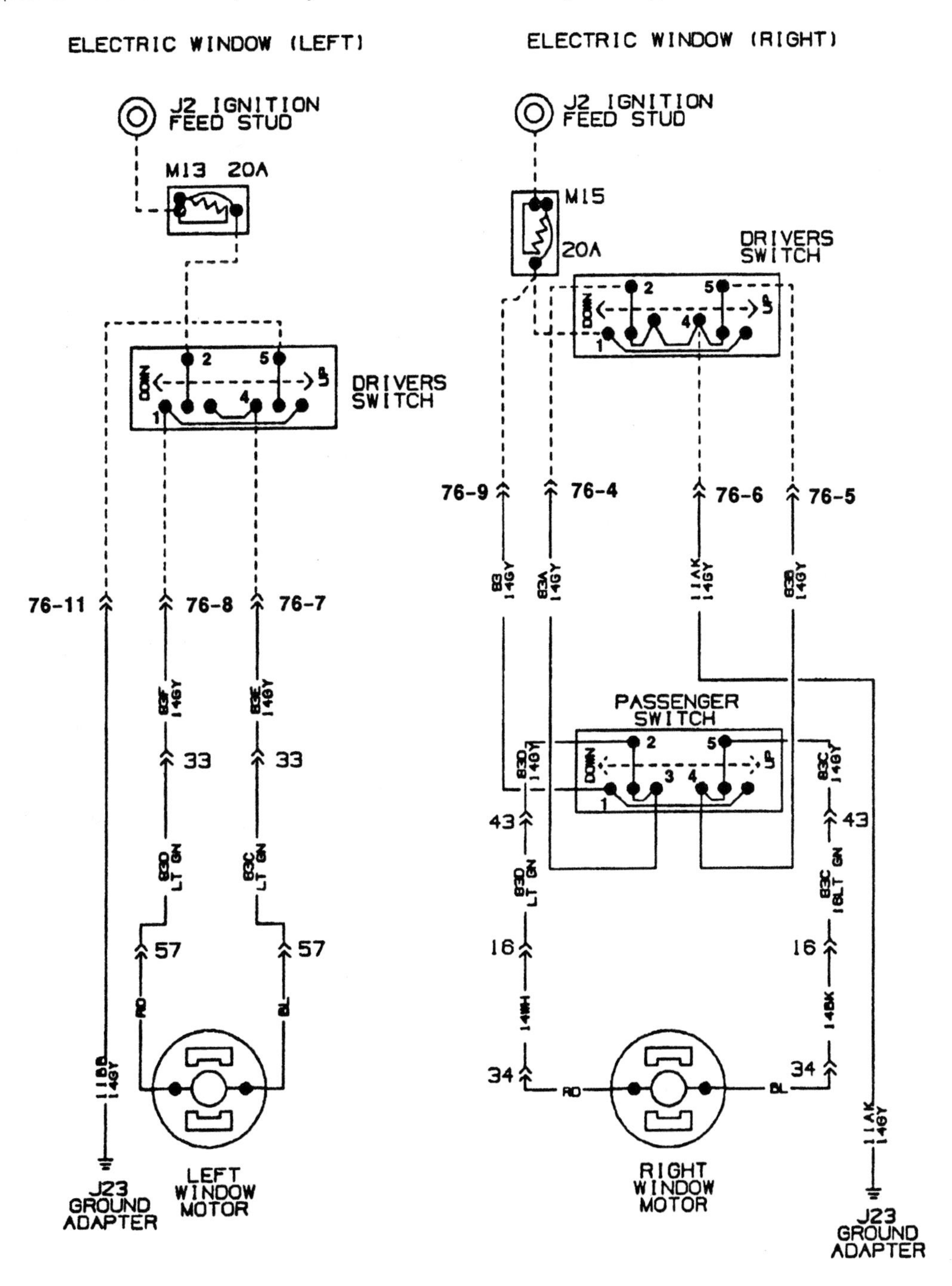

Figure 4-34 Circuit diagram of an electric window system. (Courtesy of Navistar International Engine Group)

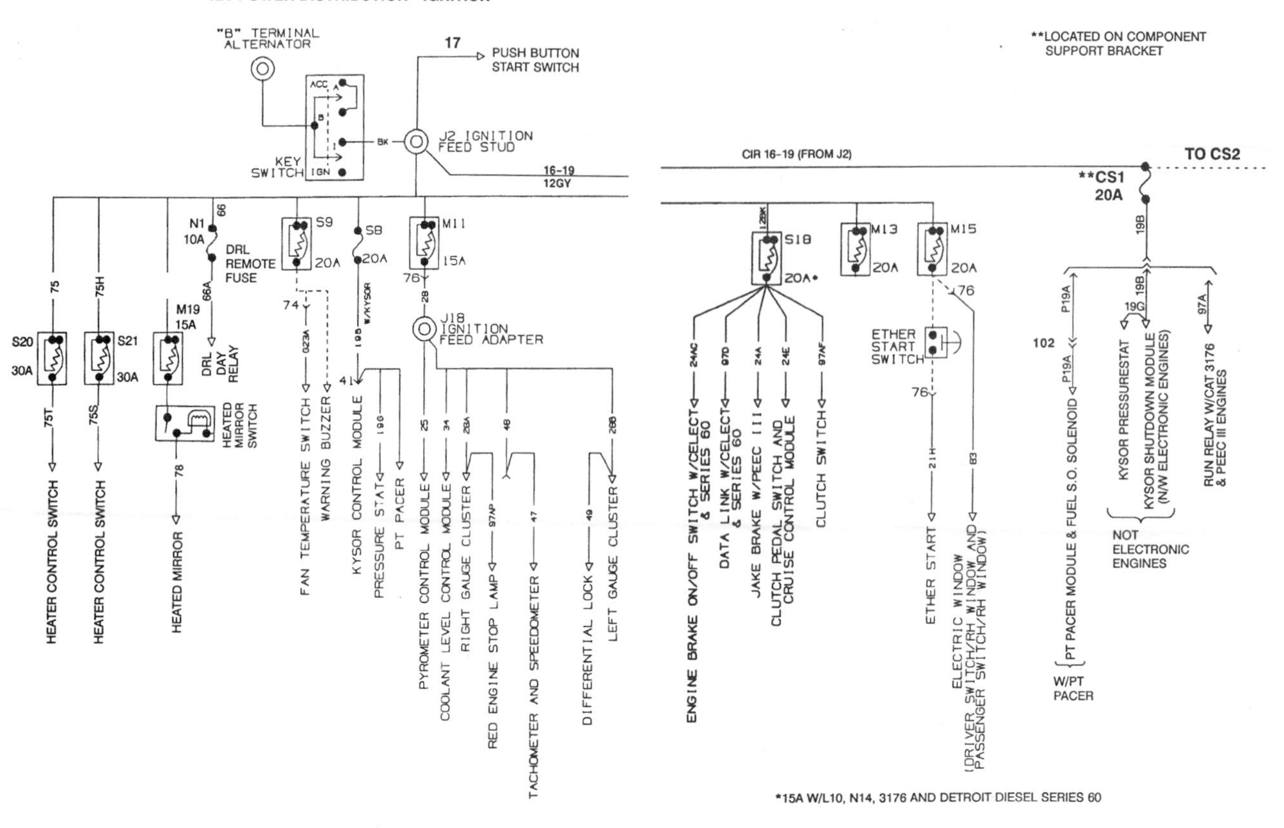

Figure 4-35 Power distribution diagram. (Courtesy of Navistar International Engine Group)

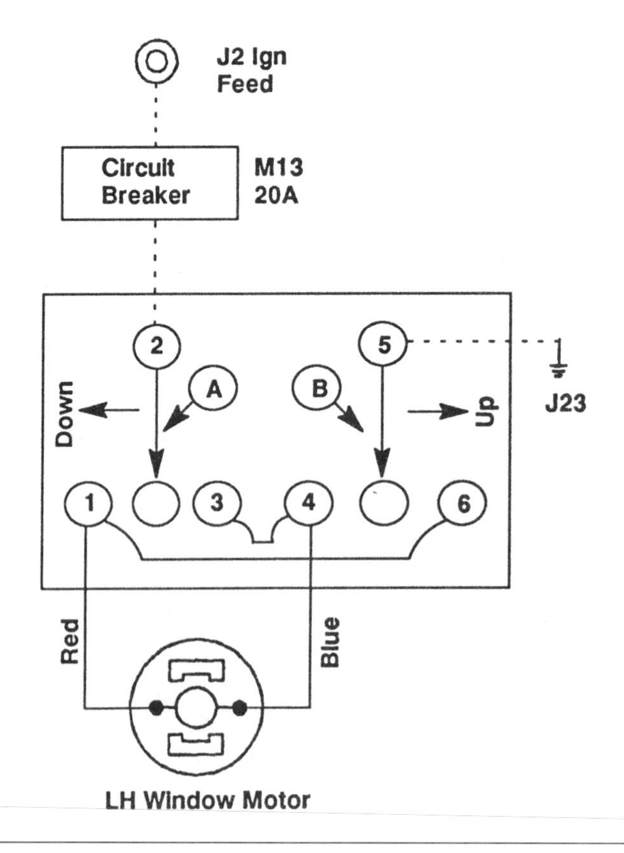

Figure 4-36 LH window switch (shown in neutral position). (Courtesy of Navistar International Engine Group)

Why would changing polarity to the motor make a window go up or down? The window motors are reversible DC motors. When battery voltage is applied to the red motor lead, the window goes down and when the voltage is applied to the blue motor lead wire, the window goes up.

Let's do the same thing with the RH window, using Figure 4-37 as a guide. In this case, the passenger window can be controlled by either of two switches. One switch is located on the driver's side and an optional switch is located in the right door panel for the passenger. These two switches operate in the following manner:

By changing the polarity to a **reversible DC motor**, the rotation of the motor is changed, making it useful to perform several functions with one motor.

1. The driver's RH window switch applies voltage from circuit breaker M15 through the passenger side window switch to the right window motor.
   A. With the driver's RH window switch in the down position, both pointers ("A" and "B") rotate to the left. Pointer "A" connects terminals 1 and 2, applying voltage through the passenger side switch to the RED motor wire. At the same

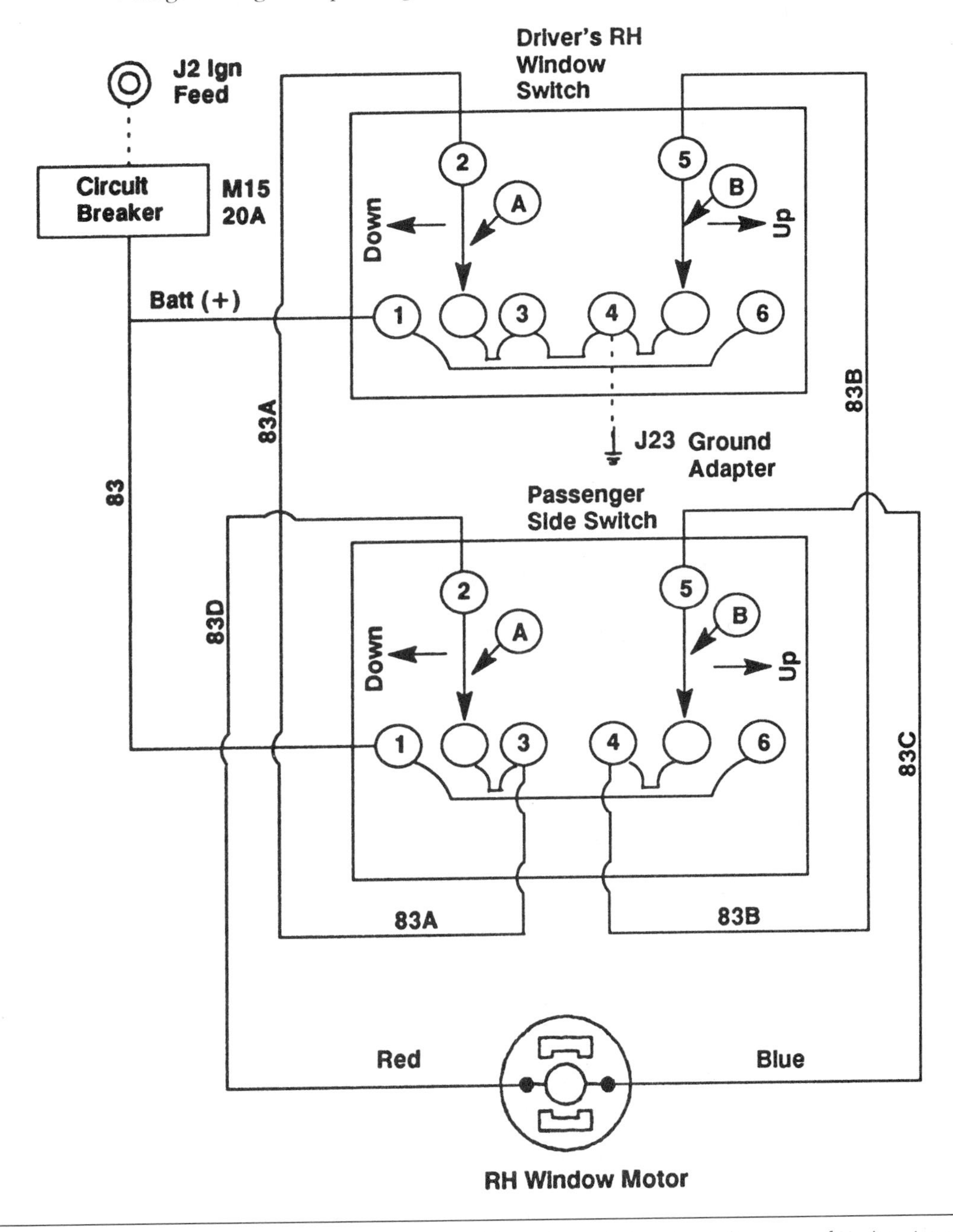

Figure 4-37 Passenger side electric window controls (shown in neutral position). (Courtesy of Navistar International Engine Group)

time, pointer "B" connected terminals 4 and 5, completing the ground path through the passenger side switch for the BLUE motor wire.

B. Putting the driver's side RH window switch in the UP position will have both pointers "A" and "B" rotate to the right. Pointer "B" connects terminals 5, 6, and 1, applying voltage through the passenger side switch to the BLUE window motor wire. At the same time, pointer "A" connects terminals 2, 3, and 4 to complete the ground path through the passenger side switch for the RED motor wire.

2. The passenger side window switch applies voltage from circuit breaker M15 directly to the window motor.

A. With the passenger side switch in the UP positions, both pointers "A" and "B" rotate to the right. Pointer "B" connects terminals 1, 6, and 5, applying voltage to the BLUE window motor wire. The "A" pointer connects terminals 2 and 3, completing the ground path for the RED motor wire. Note: The ground path is through the driver's RH window switch.

B. With the passenger side switch in the DOWN position, both pointers rotate to the left. Pointer "A" connects terminals 1 and 2, applying voltage to the RED motor wire. Pointer "B" connects terminals 4 and 5, completing the ground path through the driver's side RH window switch to the BLUE motor wire.

This window circuit is a perfect example in which a thorough understanding of the circuit operation is needed before an attempt is made to troubleshoot a circuit failure. For example, imagine that the complaint is that the driver's side window goes up, but it doesn't go down. First of all, if that occurs, we know now that the motor has the capability to cause the window to go up and down. We also know that both wires from the switch to the motor have continuity, because if the same wires can conduct current in one direction, they can conduct current in the other direction. We also know that the switch has received power and it also has received ground. This leads us to conclude that most likely the problem is in the switch. If we had not understood how the circuit works, we might have pulled the door panel apart and started probing for voltage at various terminals, etc.

CASE STUDY

A vehicle comes in with no low-beam lights but the high-beam lights work. The technician verifies the problem. At this point the technician concludes that it is very unlikely that both light assemblies can be bad. To save time, he gets a wiring diagram to better understand the headlight system and its circuit. The diagram indicates a dimmer switch controlling both high and low beams. A quick check of the switch showed no continuity in the switch on the low-beam side. The dimmer switch was replaced and the truck was once more safe for night driving.

Terms to Know

Component locator
Connector composite
Crimping
Crimping tool
Fusible link
Hard shell connectors
Heat shrink tube
Molded connectors
Primary wires
Pull to seat
Push to seat
Reversible DC motor
Soldering
Solderless connectors
Splice
Splice clip
Troubleshooting
Twisted-shielded wires
Wiring diagram
Weather-pack connectors

ASE-Style Review Questions

1. Splicing copper wire is being discussed: *Technician A* says using solderless connections is acceptable for wiring repairs. *Technician B* says soldering a splice is the preferred method for wiring repairs. Who is correct?
   - **A.** A only
   - **B.** B only
   - **C.** Both A and B
   - **D.** Neither A nor B

2. Soldering is being discussed: *Technician A* says that a rosin core solder should be used for electrical repairs. *Technician B* says acid core solder should be used for electrical repairs. Who is correct?
   - **A.** A only
   - **B.** B only
   - **C.** Both A and B
   - **D.** Neither A nor B

3. Soldering is being discussed: *Technician A* says to let the solder melt off the iron to the connection. *Technician B* says the solder should be placed where the tip of the iron and the conductor come together. Who is correct?
   - **A.** A only
   - **B.** B only
   - **C.** Both A and B
   - **D.** Neither A nor B

4. Twisted-shielded wires are being discussed: *Technician A* says twisted-shielded wires carry low current. *Technician B* says repairs to twisted-shielded wires need to keep resistance high so current can stay low. Who is correct?
   - **A.** A only
   - **B.** B only
   - **C.** Both A and B
   - **D.** Neither A nor B

5. *Technician A* says the proper length for a fusible link should not exceed 9 inches. *Technician B* says the longer the fusible link is, the better is the protection for the circuit. Who is correct?
   - **A.** A only
   - **B.** B only
   - **C.** Both A and B
   - **D.** Neither A nor B

6. Use of wiring diagrams is being discussed: *Technician A* says a wiring diagram cannot show how a circuit operates. *Technician B* says a wiring diagram will show the exact location of a component in a truck. Who is correct?
   - **A.** A only
   - **B.** B only
   - **C.** Both A and B
   - **D.** Neither A nor B

7. Repairing connectors is being discussed: *Technician A* says a pick is used to remove the terminal from the front of the connector. *Technician B* says a special tool is used to push the terminal towards the back of the connector to get it out. Who is correct?
   - **A.** A only
   - **B.** B only
   - **C.** Both A and B
   - **D.** Neither A nor B

8. Fuses are being discussed: *Technician A* says that a fuse for the parking light is blowing slowly so a higher rated fuse can cure the problem. *Technician B* says any blown fuse indicates a problem so the cause needs to be figured out first. Who is correct?
   - **A.** A only
   - **B.** B only
   - **C.** Both A and B
   - **D.** Neither A nor B

9. *Technician A* says troubleshooting is a diagnostic procedure of locating and identifying the cause of the fault. *Technician B* says troubleshooting is a step-by-step process of elimination of cause and effect. Who is correct?
   - **A.** A only
   - **B.** B only
   - **C.** Both A and B
   - **D.** Neither A nor B

10. Repairing multiple wires is being discussed: *Technician A* says keep the splice repairs in the same proximity for all the wires so future problems can be more readily found. *Technician B* says multiple wire splices need to be staggered to prevent future problems. Who is correct?
    - **A.** A only
    - **B.** B only
    - **C.** Both A and B
    - **D.** Neither A nor B

ASE Challenge

1. *Technician A* says that shielded wire can easily be repaired using solderless connectors. *Technician B* says that for a twisted pair special care should be used and that they usually require soldering. Who is correct?
   - **A.** A only
   - **B.** B only
   - **C.** Technician A if you use special connectors.
   - **D.** Neither A nor B

2. *Technician A* says that when making a fusible link, a wire that is one gauge size smaller than the main wire should be used. *Technician B* says that a wire for a fusible link must always have a rubber coating, never plastic. Who is correct?
   - **A.** A only
   - **B.** B only
   - **C.** Both A and B
   - **D.** Neither A nor B

3. *Technician A* says that all conventional and ladder type wiring diagrams show the wire number. *Technician B* says that only trunk or harness type diagrams show wire size and color. Who is correct?
   A. A only **C.** Both A and B
   B. B only **D.** Neither A nor B

4. A wiring diagram for a heavy truck uses several pages. *Technician A* says that to follow a given circuit from one page to the next you should always check wire numbers. *Technician B* says that alignment from one page to the next is another good way to trace a circuit. Who is correct?
   A. A only **C.** Both A and B
   B. B only **D.** Neither A nor B

5. A heavy truck has recently had the signal lights repaired at an out-of-town shop; the lights are flashing right turn when the switch is turned to the left or vice versa. *Technician A* says the problem is very likely the wrong signal light switch and that it should be replaced. *Technician B* says that the wiring diagram should be consulted for possible crossed signal light switch connections. Who is correct?
   A. A only **C.** Both A and B
   B. B only **D.** Neither A nor B

Table 4-1 NATEF Task

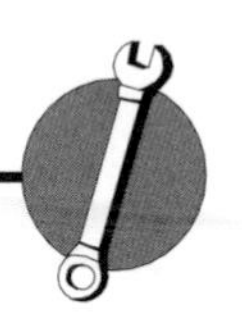

Read, interpret, and diagnose electrical/electronic circuits using wiring diagrams

Problem Area	Symptoms	Possible Causes	Classroom Manual	Shop Manual
ELECTRICAL/ ELECTRONIC FAULTS	Electrical/electronic component or systems will not operate	1. Open or ground in circuit 2. Defective switch 3. Defective relay 4. Open in fuses or links 5. Incorrectly connected	75	75

Table 4-2 NATEF and ASE Task

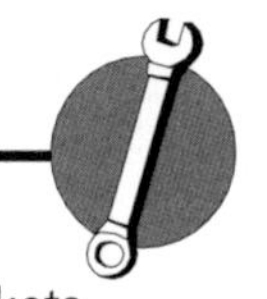

Test, repair, and replace headlight and dimmer switches, wires, connectors, terminals, sockets, relays, and control components

Problem Area	Symptoms	Possible Causes	Classroom Manual	Shop Manual
OPEN OR DEFECTIVE COMPONENTS	Improper or no headlight operation	1. Defective headlight switch 2. Defective relay 3. Defective dimmer switch 4. Open circuit 5. Poor connection	82	77

Job Sheet 5

5

Name: ____________________ Date: ____________________

Using a Wiring Diagram

Upon completion of this job sheet, you should be able to find the power source, ground connection, and controls for electrical circuits, and explain how they operate by using a wiring diagram.

NATEF and ASE Correlation

This job sheet is related to NATEF Medium/Heavy Duty Truck Electrical/Electronic Systems List content area:

A. General Electrical System Diagnosis, NATEF Task 1: Read, interpret and diagnose electrical/electronic circuits using wiring diagrams.

Exercise 1

Using the diagram in Figure 4-38, answer the following questions. (Note: Figure 4-35 in this chapter can also be referred to for power distribution.)

1. How is the circuit powered? ____________________

2. How is the circuit grounded? ____________________

3. How is the circuit controlled? ____________________

A. If it is a switch, is the switch normally open or normally closed? __________

B. Is the circuit power switched or ground switched? __________

C. Does the circuit have any other type of control devices? If so, name them. _____

____

4. How is the circuit protected? ____________________

5. Where does the power feed originate from? ____________________

6. What controls the power feed? ____________________

7. In this circuit, the circuit is highlighted to indicate the switch and circuit operation as ON and the headlight dimmer switch is high or low beam position.

A. Explain the headlight operation when the dimmer switch is on the low beam side. ____________________

____

B. Explain the headlight operation when the dimmer switch is on the high beam side. ____________________

____

Exercise 2

Using the wiring diagram in Figure 4-39, answer the following questions. (Note: Figure 4-35 in this chapter can be used if further reference is needed.)

1. How is the circuit powered? ____________________

2. How is the circuit grounded? ____________________

3. Where does J2 and J1 originate from? ____________________

4. How is the circuit controlled? ____________________

A. If it is a switch, is it normally open or closed? __________

B. Is the circuit power switched or ground switched? __________

C. Does the circuit have any other control devices? If so, name them. __________

____

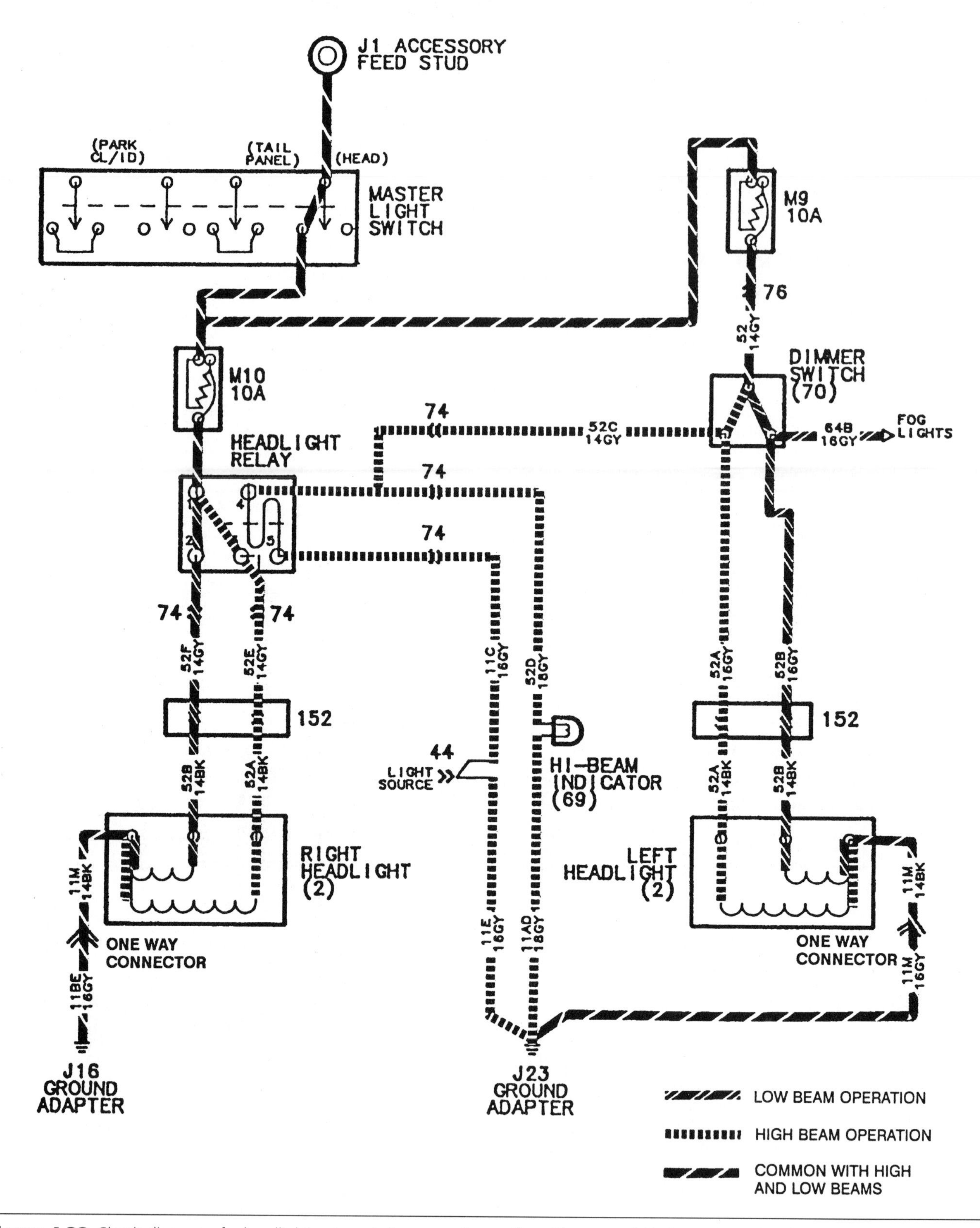

Figure 4-38 Circuit diagram of a headlight system during normal operation. (Courtesy of Navistar International Engine Group)

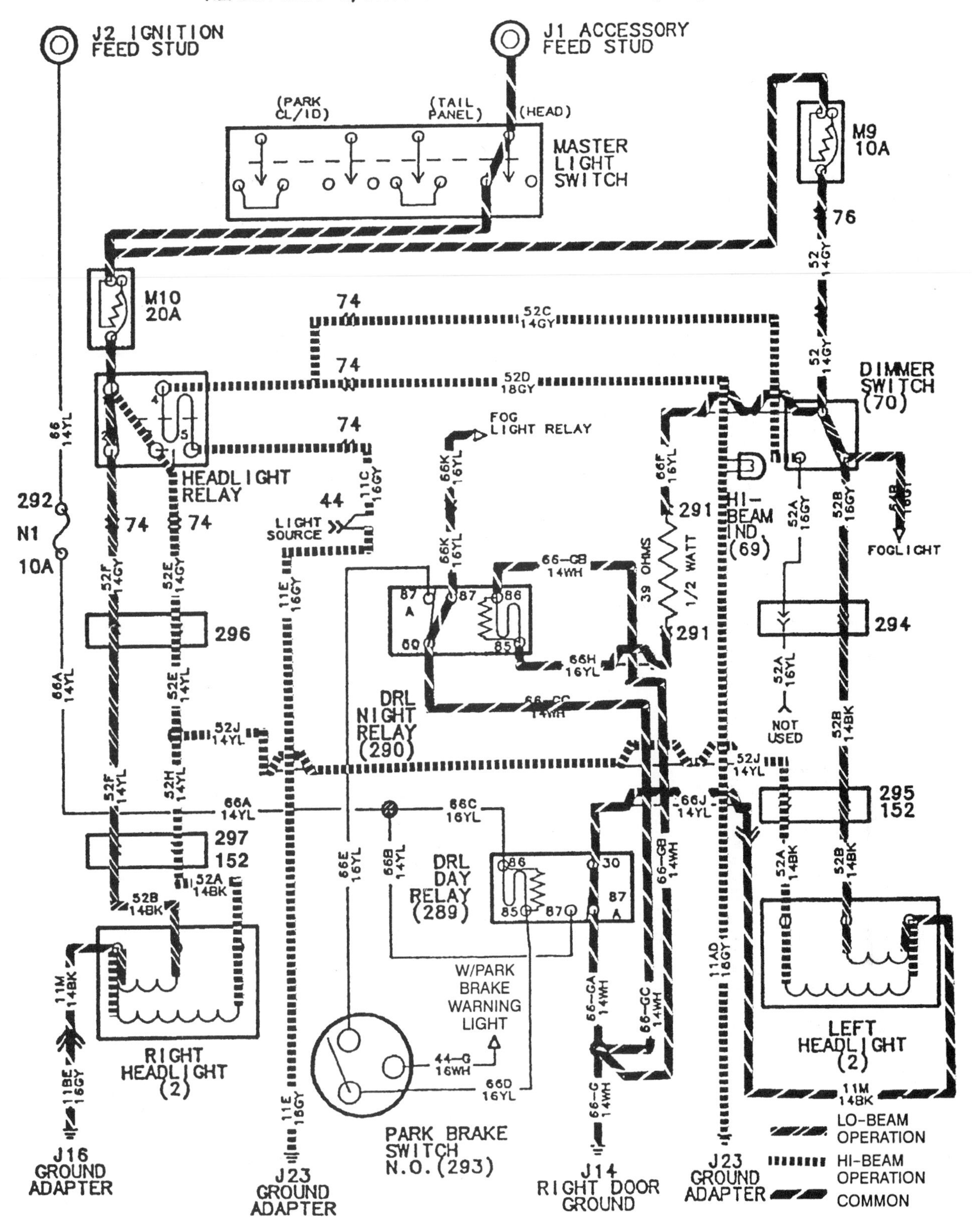

Figure 4-39 Circuit diagram of a headlight system with daytime running lights (DRL). (Courtesy of Navistar International Engine Group)

5. How is the circuit protected? ______________________________

6 How many relays are there in this circuit? ________ Name the relays. ____________

____

7. In this diagram, the three different circuit identification codes (high beam operation, low beam operation, circuits active in either mode) are used, indicating circuit operation when in the normal operating mode (not DRL). Those circuits that are not highlighted are not active in the normal operating mode (not DRL).

   A. In the normal mode, when is voltage available to the headlight switch?

   ____

   B. Where is the power from circuit breaker M9 applied to? There should be three different areas the power goes to. ______________________________

   C. In the low beam operation mode, is there any current going to any relays? If so, name the relays. ______________________________

   D. In the low beam mode, are there any relays that are energized? If so, name the relays. ______________________________

8. When does the daylight running light (DRL) not operate except when the key is off?

____

____

9. How many main ground leads are shown on the diagram? ______________

10. Name the load components that the grounds are used for or effect their operation. (Note: actual load components, not control devices) ______________

11. Explain what happens if the DRL night relay is de-energized? ________________

____

12. Is the fog light installation effected by the installation of daytime running lights? Yes ______ No _______

Instructor Check ______________________________

Battery Diagnosis and Service

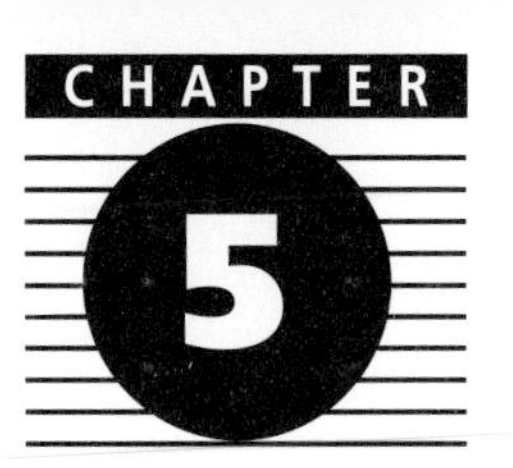

Upon completion and review of this chapter, you should be able to:

- ❑ Demonstrate and follow all safety precautions and rules associated with servicing the battery.
- ❑ Perform a visual inspection of the battery, cables, hold downs, and boxes.
- ❑ Perform a proper maintenance of the complete battery system.
- ❑ Test a conventional battery's specific gravity and interpret the result.
- ❑ Perform an open circuit test and interpret the results.
- ❑ Perform a load or capacity test of any battery.
- ❑ Perform a 3-minute charge test to determine battery sulfation or a possible charging system problem.
- ❑ Properly charge a battery.
- ❑ Perform several battery drain tests and determine the cause of the problem.
- ❑ Perform a voltage drop test to determine the condition of battery connections.
- ❑ Properly jump-start a vehicle whether it is a 12-volt or 24-volt system.

Basic Tools

Basic mechanic's tool set

Appropriate service manual

Introduction

The battery is the heart of the electrical system of the truck. Most of the truck's electrical circuits obtain their power supply from the battery in one way or another. Therefore, a battery with a problem can affect any of the truck's electrical circuits. If a problem is suspected in any of the truck's electrical systems, the battery condition must be checked and corrected if needed, prior to any other electrical system diagnosis.

The battery's function becomes much more difficult as the environmental conditions change from normal to extreme cold or even extreme heat. It is very important that the batteries in trucks are maintained properly. That is why every preventive maintenance (PM) schedule includes some form of battery checking. It makes more sense to find a possible problem before a failure occurs. Of course, the worst-case scenario would be for a truck not to start when it is scheduled to get to or be at a specific destination. Keep in mind that a commercial vehicle plays a role in commerce and it has to be profitable.

Preventive maintenance is a scheduled function performed on vehicles to maintain performance and prevent on-road breakdowns. A good fleet places top priority on this scheduled maintenance.

Most shops have tools for testing that range from simple hydrometers, hand-held draw testers, and multimeters to more sophisticated test equipment such as the VAT-40, manufactured by the Sun Electrical Corporation (Figure 5-1). Some tests in this chapter will require the procedures that are used with a VAT-40. However, there are many other testers that incorporate the same procedures. The trend is towards computer-based testers that conduct tests automatically after a test is selected (Figure 5-2). Test equipment manufacturer's procedures should always be followed very closely.

CUSTOMER CARE: Many computer-based testers determine the health of a battery. A test might indicate that a battery should be replaced, yet the engine starts with no problem. The customer might doubt your diagnosis. However, explain that environmental factors such as temperature can affect the battery's starting ability. Although the battery starts fine during the heat of the day, when the temperature drops overnight, the vehicle might not start in the morning. You will have to let the customer know that this is part of preventive maintenance.

General Precautions

The Consumer Product Safety Commission's National Eye Injury Surveillance System (NEISS) discovered some startling statistics. In a one-year period, they found that there were more than

Figure 5-1 A Sun VAT-40 battery, starting, and charging system tester.

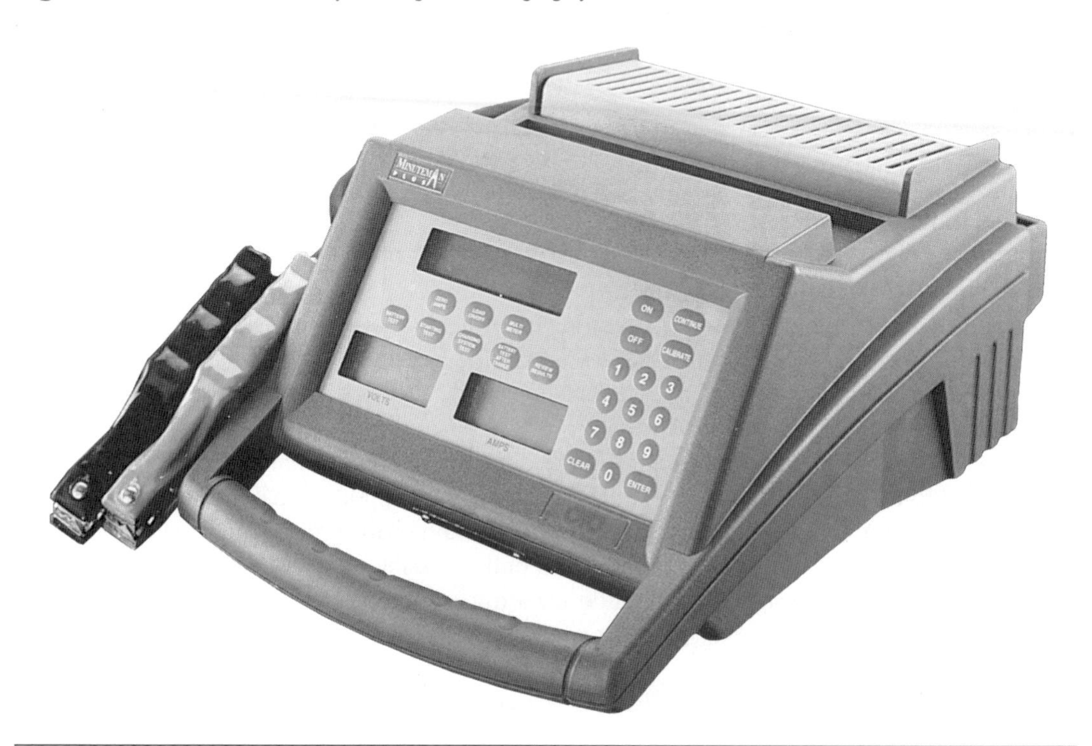

Figure 5-2 A microprocessor-controlled tester. (Courtesy of OTC Tool and Equipment, Division of SPX Corporation)

7,000 incidents of injuries from automotive batteries. These injuries included: battery explosions, chemical burns, muscle strains, and/or crush type injuries resulting from lifting or dropping a battery on a person's foot. Breaking it down to percentages, the commission found that 32% of the various injuries came from battery explosions, with approximately one-third exploding while the batteries were being charged and two-thirds while the batteries were being tested or serviced. According to NEISS, most people suffered eye injuries from exploding batteries (72%), chemical burns (62%), and injuries or lacerations due to flying shrapnel (21%).

Before moving on, it is important to review the following dangers posed by batteries and the precautions you need to take when working with them:

1. Batteries contain acid that is very corrosive and can cause blindness if it gets into your eyes. To prevent this, eye protection should be used. If battery acid does get into your eyes, rinse them thoroughly with clean water and receive medical attention immediately—you only have one set of eyes. If battery acid comes in contact with your skin, rinse it thoroughly with clean water. Baking soda added to water will help to neutralize the acid. Any acid that is swallowed requires the drinking of large quantities of water or milk, followed by milk of magnesia and a beaten egg or vegetable oil.

2. Batteries produce a lot of current. Care needs to be taken when dealing with connections and metal objects that can easily conduct. Conductive objects like jewelry can heat up very quickly when accidental contact is made between the battery positive and ground. Short circuits can cause damage to electronic components.
   - ❑ When making connections to a battery, be careful to observe polarity: positive to positive and negative to negative.
   - ❑ When disconnecting battery cables, always disconnect the ground cable first.
   - ❑ When connecting battery cables, always connect the ground cable last.
3. Batteries produce hydrogen and oxygen gases. If these gases get ignited, an explosion of the battery can occur, spreading acid and shrapnel over a large area.
   - ❑ Avoid any open flames or arcing near a battery.
   - ❑ Charge the battery in a well-ventilated area.
   - ❑ Do not connect or disconnect the charger's leads while the charger is turned on.
   - ❑ Never lay any tools or metal parts across the battery. The shorting of metal objects can create a spark or short out the battery and cause it to explode.

Battery Inspection and Cleaning

The following are some of the leading causes of battery failure in heavy duty trucks:

1. Vibration
2. Recharge-discharge cycling
3. Overcharging
4. Broken posts
5. Undercharging

To keep today's trucks rolling properly involves a good preventive maintenance (PM) program that includes a battery service. A battery inspection and cleaning should be performed whether it be a fleet maintenance function or as a part of a scheduled service function for private customers. The following parts should be included in a good visual inspection as shown in Figure 5-3:

Special Tools

Fender covers

Safety glasses

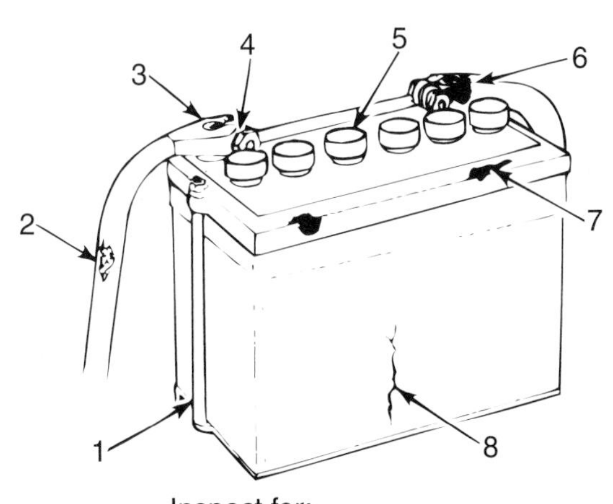

Inspect for:
1 Loose holddowns
2 Defective cables
3 Damaged terminal posts
4 Loose connections
5 Clogges vents
6 Corrosion
7 Dirt or moisture
8 Cracked case

Figure 5-3 Battery maintenance inspection points.

1. *Battery box condition:* A loose or deteriorated box can lead to vibration of the battery. This vibration can shake the active materials off the grid plates, thus shortening the battery's life. Plate connections and cable connection can also be loosened. The battery cover should be in place and properly fastened.
2. *Battery hold down condition:* Any loose or broken hold downs will allow the battery to vibrate and bounce, which also can cause the active material to shake off the grid plates and break the plate connectors and terminals. A noncoated broken metal hold down can come in contact with the positive terminal and if allowed to come in contact with the ground, a short will occur.
3. *Battery case condition:* Check the case for any cracks or bulging caused by overtightening of a battery hold down. A broken hold down can cause a cracked case due to excessive bouncing. The bulging of the battery can also result from excessive heat, buckled plates from an extended undercharged condition, freezing, or excessive charge rate.
4. *Battery corrosion:* This is typically caused by spilled electrolyte or electrolyte condensation from gassing. Corrosion of this type attacks the battery terminals, causing a chemical resistance to build up that reduces the applied voltage to the vehicle's electrical system. The corrosive material coupled with the dirt that is attracted to the moisture on top of the battery can create a path for current flow, allowing the battery to slowly discharge. Any corrosion that is allowed to spread beyond the battery connectors can physically damage the hold downs and the battery itself.
5. *Battery cables and terminals:* Check for corrosion, frayed or rubbed through cables, loose terminals, and broken clamps, all of which can reduce current flow or cause a short.
6. *Electrolyte level and condition:* A low level of electrolyte can dry out the material on the plates, which can more readily shake off. If the level is low, distilled water must be added. The level of the solution should be 1/2″ above the top of the plates. In addition, the electrolyte color should be checked. Discoloration of the electrolyte indicates an excessive charging rate, impurities in the electrolyte solution, or an old battery.
7. *Battery date code:* This tells the age of the battery. In the case of a borderline battery, the date on it can help in deciding whether to replace it. Keep in mind that in ideal conditions, a borderline battery can provide enough cranking power to start the truck. However, when the first cold spell occurs, the temperature might cause the battery to fail.
8. If the battery has a built-in hydrometer, check its color indicator.

▲ **WARNING:** Federal Motor Carrier Safety Regulations, or FMCSR, include battery installation as part of the parts and accessories necessary for safe operation of a commercial vehicle. For example, part of the regulation states: "Every storage battery on every vehicle, unless in the engine compartment, shall be covered by a fixed part of the motor vehicle or protected by a removable cover or enclosure. Removable covers or enclosures shall be substantial and shall be securely latched or fastened. The storage battery compartment and adjacent metal parts that might corrode by reason of battery leakage shall be painted or coated with an acid resistant paint or coating and shall have openings to provide ample battery ventilation and drainage." As can be seen from this quote, the way a medium and heavy duty truck technician performs his or her duties is often dictated by federal and/or local regulations.

▲ **WARNING:** If a major positive cable is allowed to make conduct with ground, major damage can occur. It causes the same result as connecting a piece of metal such as a wrench from the positive post to the negative post. There is very little resistance and the battery will totally drain, damaging the wrench. In the case of the truck, the chassis is the ground and it has very little resistance. If the positive cable is allowed to ground directly to the chassis, much damage can occur, especially with all the electronic components that are in today's trucks. *It is important to note that the positive or negative cable could be HOT, depending upon the ground system being used.*

SERVICE TIP: Before removing caps or cap assemblies, make sure the top of the battery is clean—you don't want to contaminate the inside of the battery from the battery top. This can easily occur during the battery filling process when water spilled on top flows into the cells, carrying with it dirt and residue.

Battery Leakage Test

A dirty battery can cause a battery drain, equivalent to leaving a light on. As mentioned previously, the dirt can actually allow current flow over the battery case. The **battery leakage test** is simple and quick, and should be performed prior to and after cleaning the battery to verify that if there was a leakage problem it has been eliminated. This test requires a voltmeter set at 12 volts DC scale. Connect the negative test lead to the negative terminal of the battery and move the red test lead across the top and sides of the battery case (Figure 5-4). There should be no voltage measurement anywhere on the case of the battery. If the meter does read voltage, a current path from the negative terminal to its positive terminal is completed through the dirt. This requires a good battery cleaning.

Special Tools

Safety glasses

Voltmeter

Battery Cleaning

A proper battery cleaning should include the hold downs and the battery box or platform. This requires the use of certain tools, depending on the type of battery terminals used. For example, Figure 5-5 shows a terminal puller being used on a top post type of battery. This prevents damage from occurring to the post that cannot be seen within the battery. If the battery is located where damage can occur to the vehicle's finish, it should be pulled and the entire battery, hold downs, and battery box should be washed with a baking soda and water solution. A cleaning brush is useful for removing heavy corrosion, but do not allow the baking soda and water mixture to enter the cells. After everything is clean, rinse off the battery, hold downs, and box with clean water. With the cables off, clean the cable ends and battery terminals (Figure 5-6).

Most battery tests require the battery to have a full charge. However, it is quite difficult to get a proper charge if the cables and terminals offer a resistance to the charging system's output current flow.

Classroom Manual
Chapter 5, page 99.

Special Tools

Safety glasses

Voltmeter

Terminal pliers

Terminal puller

Terminal and clamp cleaner

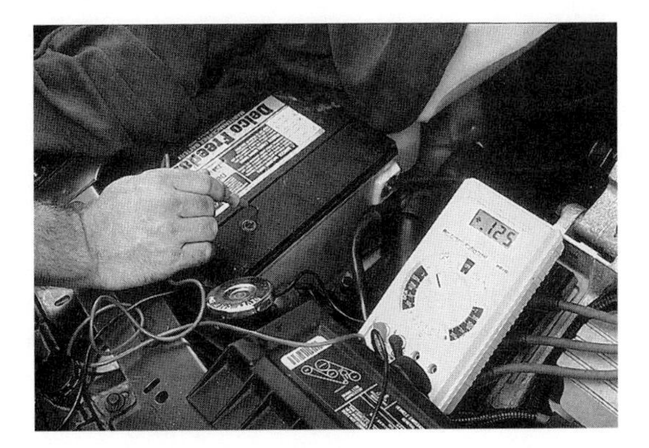

Figure 5-4 Checking a battery for leakage.

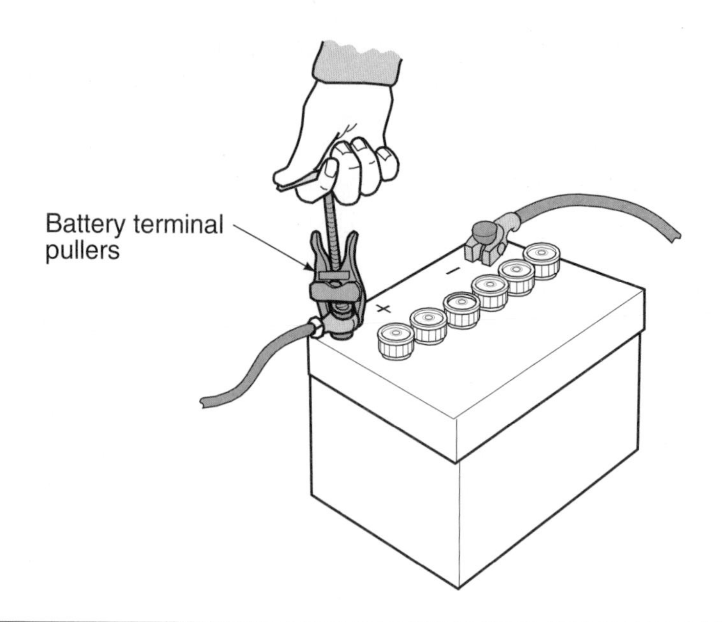

Figure 5-5 Use battery terminal pullers to remove a cable from a battery's terminal. Do not pry off the clamp.

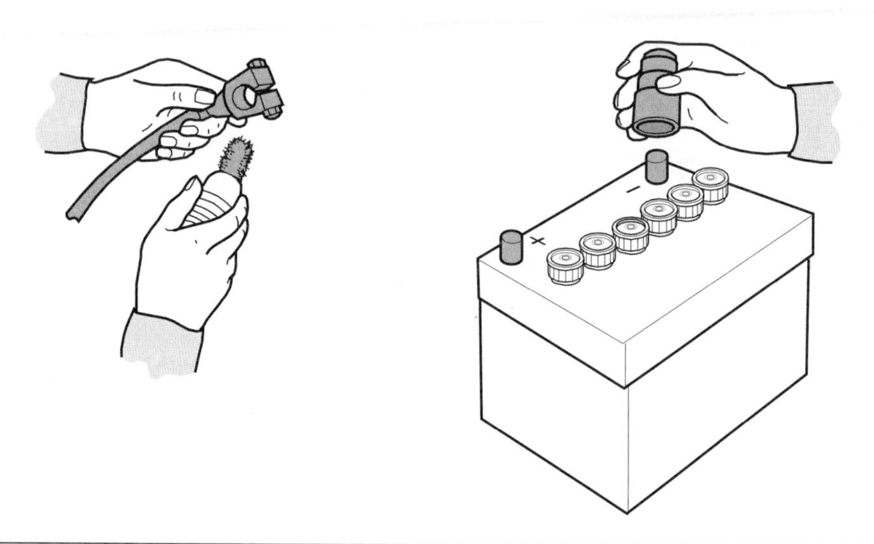

Figure 5-6 A terminal cleaning tool is used to clean the battery's terminal and the cables' clamps.

Classroom Manual
Chapter 5, page 102.

Charging the Battery

Prior to charging the battery, the electrolyte level should have been checked (as outlined in the visual inspection). The safest method of charging a battery is to remove it from the vehicle; however, that is not always feasible. If the battery is to be charged in the vehicle, it is always a good practice to disconnect the ground cable from the battery to protect the vehicle's electronic components.

Charging a battery means passing an electrical current through it in an opposite direction than during discharge.

CAUTION: To avoid battery explosion, all precautions should be taken at this point to prevent any sparks or flames near the charging area or the battery being charged.

CAUTION: A battery plate can short out a cell. This problem can occur when the grid plates start to grow to the point where they touch each other. A visual check can possibly catch this problem. If there is a normal electrolyte level in all cells but one, that cell is probably shorted out. The danger here is that the electrolyte has been converted to hydrogen gas. Be very cautious working with a battery with this condition because an explosion can occur.

CAUTION: Extreme cold weather can freeze a weak electrolyte solution, so check for ice crystals in the cells before charging. Charging a frozen battery may cause it to explode. Allow a cold battery to warm up at room temperature for a few hours before charging.

CAUTION: To avoid personal injury, always disconnect the ground cable first and use a lifting or carrying tool if the battery is removed from the vehicle.

There are two basic methods of recharging a battery: the slow charge method and the fast charge method. Either method has its advantages and disadvantages. Before we discuss them, let us define our objectives. First of all, the charge a battery receives is equal to the charge rate in amperes multiplied by the time in hours. For example, a typical truck battery will need at least 80-ampere hours of charge if it was completely discharged. Most chargers can be set for specific amperage rates. If you set the charger at a 20-amp rating, it will take 4 hours of charging time to produce 80 amp hours. Likewise, if you set the charger at 40 amps, it will take 2 hours of charging time to produce 80 amp hours.

Using that as a base, you can see that if the battery is 50% charged, you need to reduce the charging current or time by one half. Lowering the charger amp rating to extend the time period of charging is the preferred method because it reduces the chance of the battery overheating.

CAUTION: Never allow the temperature of a battery being charged to rise above 125° Fahrenheit or an explosion can occur. Also, look for signs of excessive gassing or liquid electrolyte escaping from the battery.

CAUTION: Do not overcharge batteries. Overcharging causes excessive loss of water, which is especially hard to see in maintenance-free batteries. It also causes the battery to produce explosive levels of hydrogen and oxygen.

Special Tools

Safety glasses

Battery charger

Voltmeter

Terminal adapters

Slow Charging

The slow charging method uses a low charge rate over a long time period. This method is often used in fleets where many batteries are kept on hand. The advantage of a slow charge is that it restores the battery to a fully charged state, and minimizes the chances of overcharging it. This enables the lead sulfate throughout the thickness of the plates to convert to lead peroxide. The recommended rate for slow charging is 1 ampere per positive plate per cell. Expect to see charging times of 24 hours or more to bring a battery to a full charge. Figure 5-7 shows various rates of charge according to the reserve capacity of the battery.

Slow charging uses a low current flow in the battery for a long time period. For example, a battery can be slow charged at the rate of 5 amperes for 20 hours.

Fast Charging

Many times you will have a time constraint preventing the battery being charged slowly. In those cases, fast charging is used to bring a battery to a **state of charge** that allows the vehicle to be started. Once the vehicle is started, hopefully the charging system will eventually bring the battery to a fully charged state. A slow charge converts the lead sulfate throughout the plate; however, in a fast charge, only the lead sulfate on the outsides of the plates gets converted. The fast charge method consists of a high charging rate for a short time period.

Fast charging uses a high charge current rate for a short time period. For example, a battery is charged at the rate of 40 amperes for 2 hours.

WARNING: Fast charging the battery requires that it be monitored at all times and the charging rate be controlled. To avoid battery damage, do not fast charge a battery for longer than 2 hours. Do not exceed a charging rate of 30 amperes. Do not let the voltage of a 12-volt battery exceed 15.5 volts and do not allow the temperature to rise above 125° Fahrenheit.

Battery Capacity (Reserve Minutes)	Slow Charge
80 minutes or less	10 hrs. @ 5 amperes 5 hrs. @ 10 amperes
Above 80 to 125 minutes	15 hrs. @ 5 amperes 7.5 hrs. @ 10 amperes
Above 125 to 170 minutes	20 hrs. @ 5 amperes 10 hrs. @ 10 amperes
Above 170 to 250 minutes	30 hrs. @ 5 amperes 15 hrs. @ 10 amperes

Courtesy of Battery Council International

Figure 5-7 Table showing the rate and time of slow charging a battery according to its reserve capacity.

A fast charging rate guide is shown in Figure 5-8. One method of monitoring the charging rate is by observing the voltage at the battery while it is being charged. With a voltmeter across the battery posts, adjust the charging rate to just below 15 volts. If the voltmeter reads over 15 volts, reduce the charging rate until it reads below 15 volts.

WARNING: To prevent damage to the AC generator and computer, disconnect the ground battery cable before fast charging the battery in the vehicle.

Three methods to determine if the battery is fully charged are:

1. An open circuit voltage test showing that the battery achieved 12.68 or higher volts after the battery has been stabilized.
2. A specific gravity test to see if it is 1.264 or higher after the battery is stabilized.
3. Using the ammeter on the battery charger to see if there is a drop of amperage to 3 amperes or less and it remains at that level for 1 hour.

BATTERY HIGH-RATE CHARGE TIME SCHEDULE

SPECIFIC GRAVITY READING	CHARGE RATE AMPERES	BATTERY CAPACITY—AMPERE HOURS			
		45	55	70	85
Above 1.225	5	★	★	★	★
1.200–1.225	35	30 min.	35 min.	45 min.	55 min.
1.175–1.200	35	40 min.	50 min.	60 min.	75 min.
1.150–1.175	35	50 min.	65 min.	80 min.	105 min.
1.125–1.150	35	65 min.	80 min.	100 min.	125 min.

★ Charge at 5-ampere rate until specific gravity reaches 1.250 @ 80°F.

Figure 5-8 Charging rates based on a battery's state of charge and capacity rating. Electrolyte temperatures should never be allowed to exceed 125° F during charging.

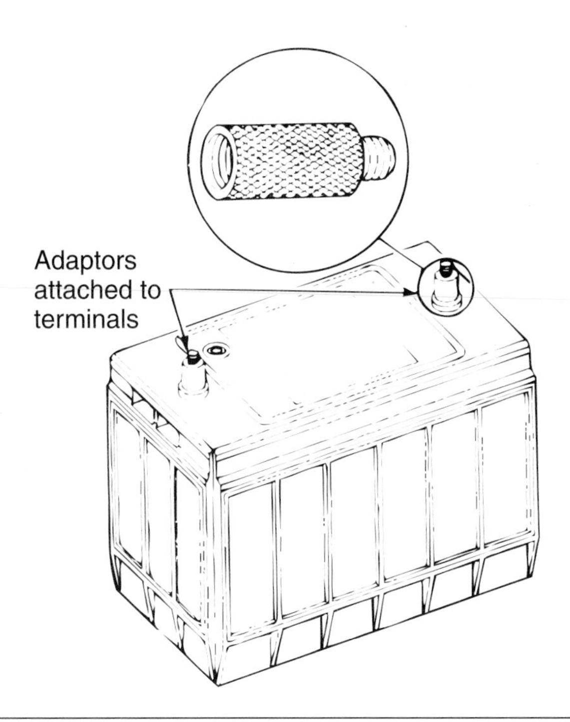

Figure 5-9 Batteries with threaded terminals might require the use of adapters for testing or charging.

Charging Instructions

The following is a guideline for charging batteries. Always follow manufacturer's instructions.

1. Before placing a battery on charge, ensure that the terminals are clean.
2. Verify that the electrolyte level is proper in all the cells. If not, add enough distilled water to cover the plates.
3. Connect the charger to the battery, observing proper polarity—the positive charger lead to the positive battery post and the negative charger lead to the negative post. Make sure the connections are tight.
4. Turn the charger on and slowly increase the charging rate until the recommended ampere value is reached.
5. When the battery is charged, turn the charger off and disconnect it.

If there is any evidence of smoke or dense vapor or liquid coming out of the battery, shut off the charger. The battery should be rejected or the charging rate reduced or temporarily halted.

SERVICE TIP: Most heavy duty batteries use the threaded terminals on their batteries. Adapters are available and should be used to ensure a good contact when charging or testing a battery (Figure 5-9).

Battery Test Series

As mentioned earlier, in most cases the battery should be fully charged in order to test it any further. Prior to any testing, though, you need to keep in mind the possible causes of battery failure. Many batteries might test good after a full charge, yet when reused in the vehicle, they might not perform satisfactorily. Therefore, the following factors need to be considered when testing and diagnosing a battery failure:

1. A vehicle accessory left on over an extended time period
2. A charging system failure such as a faulty generator or regulator, excessive wiring resistance, and a slipping belt

3. A vehicle's electrical load that exceeds the generator's output; this can happen if electrical accessories are added
4. A long period of idle or low speed driving with many accessories on at the same time, such as heaters, defrosters, lights, etc., in cold weather
5. A defect in the cranking system
6. Loose battery post and cable connection
7. A previously misdiagnosed or improperly charged battery

Classroom Manual
Chapter 5, page 105.

Batteries have to be fully charged to perform a battery terminal test.

The **battery terminal test** is a voltage drop test to check for poor electrical connections between the battery cables and terminals.

Special Tools

Safety glasses
Voltmeter
Jumper wire

The vehicle's engine needs to be cranked over without starting to perform a proper battery terminal test.

Special Tools

Safety glasses
Crimping tool
Soldering gun or torch
Terminal pliers
Terminal puller

Battery Terminal Test

This first test can be performed as part of the maintenance program or as part of the testing series. A terminal test is nothing more than performing a voltage drop check to test the integrity of the battery connection. It is a good check prior to disconnecting the battery cables and after cables are connected to verify a good connection. The check after reconnecting the cables will alleviate unnecessary come backs due to loose or faulty connections.

The test requires the vehicle to be cranked over without starting. With most diesel engines, you will have to disable the fuel system. On gasoline engine vehicles, disabling the ignition system is the easiest method to prevent the engine from starting. This can be done by removing the ignition coil secondary wire from the distributor cap and putting it to ground.

SERVICE TIP: Because of the variety of ignition systems used, always refer to the manufacturer's service manual for the correct procedure for disabling the ignition system.

Next, connect the negative voltmeter test lead to the cable clamp and connect the positive meter lead to the positive battery terminal. Looking at Figure 5-10, you can see that in this configuration, the polarities of the meter are correct because the battery post is the most positive part of the electrical circuit.

Have someone crank the engine while observing the voltmeter reading. If the voltage drop exceeds over 0.3 volts, there is a high resistance at the cable connection. Repeat the procedure on the negative side of the battery. Clean the connections and redo the voltage drop test to verify a proper connection.

This same test can also be performed to check for voltage loss between the cable and the cable end itself if the conductor of the cable is accessible. Many times a resistance due to corrosion develops between the cable and the cable end, requiring a repair of the cable or replacement of the cable assembly.

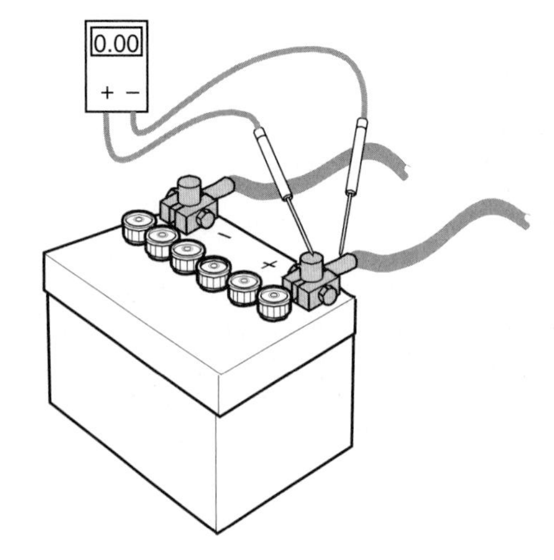

Figure 5-10 Voltmeter hookup for testing the condition of a battery terminal.

Many trucks require cables that are hooked to multiple batteries. In those cases, it makes sense to purchase pre-made cable assemblies (Figure 5-11). This type of assembly requires you to connect the two cable assemblies to the single starter cable. A butt splice is used for the connection and then covered with a heat shrink insulation. Remember that both cables of the assembly get hooked to the two positive terminals of the two batteries or the two negative terminals of the two batteries as seen in Figure 5-11. Four cable assemblies are also available to hook four positive or negative terminals of four batteries to a single cable.

Classroom Manual
Chapter 5, page 107.

At times only the cable end itself is all that needs to be replaced (Figure 5-12). If that is the case, cut off the old end and slide a correct length heat shrinkable tube over the cable. Cut off the proper length of insulation and install a new end terminal. Then crimp and solder the terminal to the cable. Next, slide the shrink tube over the new connection and apply heat to shrink the tubing. Make sure the routing of the cable is correct and it is properly clamped and protected. Always use proper crimping tools. Figure 5-13 illustrates some acceptable and not acceptable components and locations.

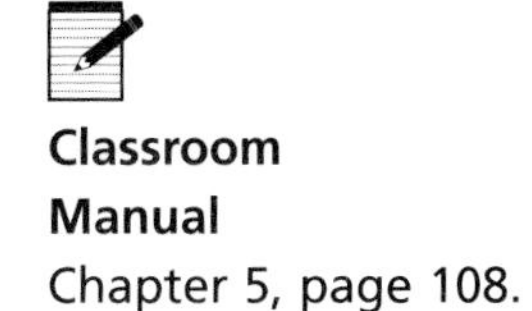

Classroom Manual
Chapter 5, page 108.

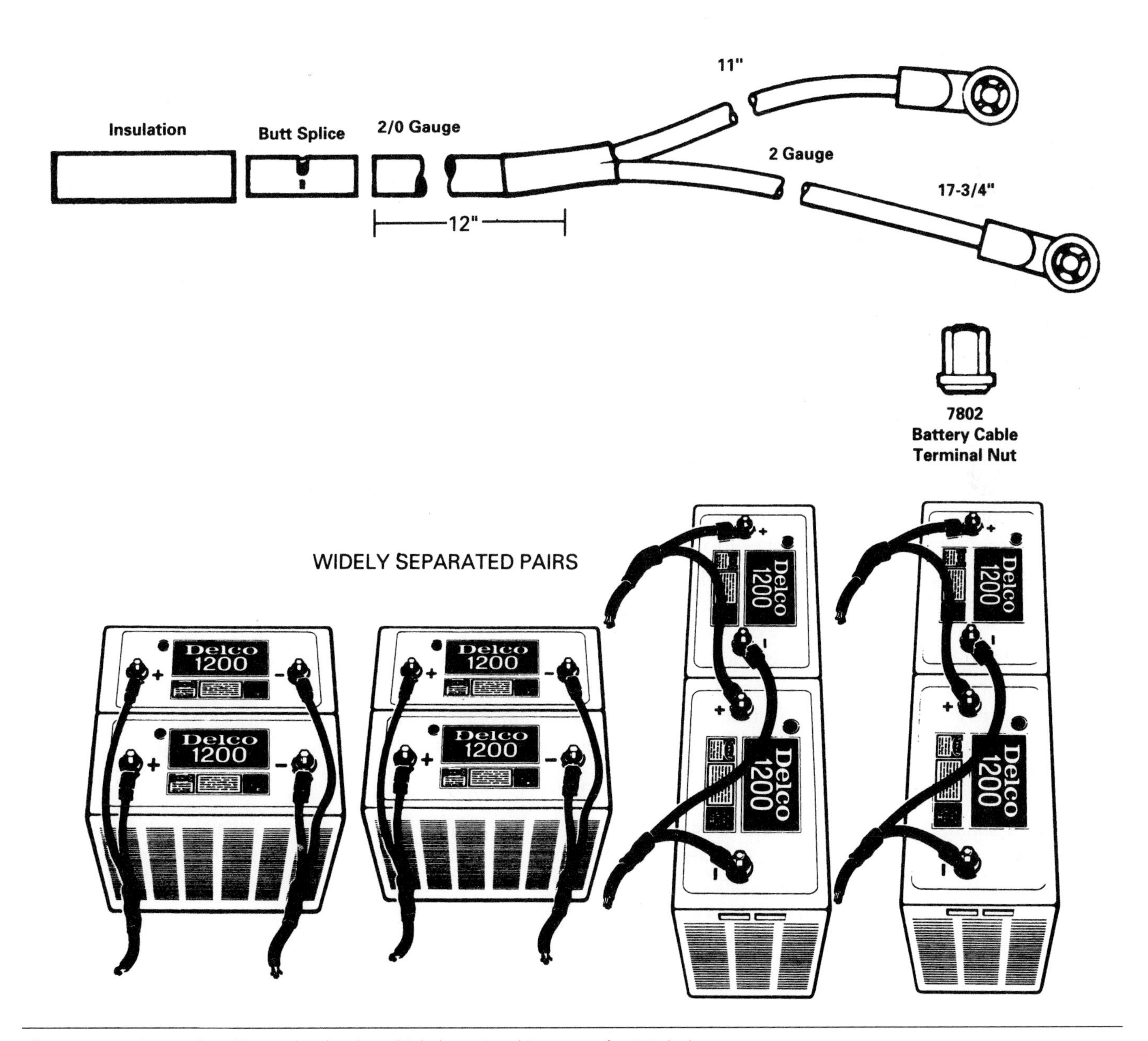

Figure 5-11 Pre-made cable used to hook multiple batteries. (Courtesy of AC Delco)

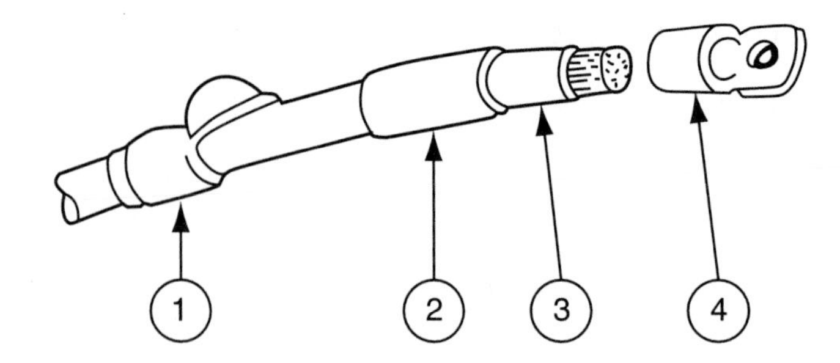

Figure 5-12 A battery cable with all the proper components. (Reprinted with permission. Courtesy of Volvo Trucks North America, Inc.)

A **hydrometer** measures the specific gravity of a liquid.

Special Tools

Safety glasses

Hydrometer

Classroom Manual
Chapter 5, page 104.

Specific gravity is a unit measurement for determining the sulfuric acid content of the electrolyte.

Electrolyte level has to be over the plates in order to perform a specific gravity test. If it is too low, add distilled water and charge the battery to mix the water and electrolyte solution.

State of Charge Test

Measuring the battery's state of charge can be accomplished by two methods. One method is to perform an open circuit voltage test and the other method, which we are going to look at first, is to measure the specific gravity of the electrolyte solution. This requires the use of a hydrometer (Figure 5-14). Using a hydrometer in truck batteries is very feasible because of the access available to the cells of the truck's battery, even if at times it might look like the battery is totally sealed (Figure 5-15).

The following steps show how to use a hydrometer:

1. Remove all battery caps.
2. Check the electrolyte level. There must be enough solution in the cell to be withdrawn by the hydrometer to get a proper reading.
3. Place the hydrometer in the cell vertically to prevent the float from sticking to the sides.
4. Squeeze the bulb to withdraw the solution (Figure 5-16). It might require a few squeezes to make sure that the float rises in a manner so that it isn't touching the bottom, the sides, or the top.
5. Hold the hydrometer in a vertical position with your eyes level to the surface of the liquid.

The specific gravity is read where the float scale intersects the top of the solution (Figure 5-17). Remember to make a temperature correction since hydrometers are calibrated to give a true reading at a temperature of 80° Fahrenheit (Figure 5-18). As you can see, a correction factor of 0.004 specific gravity is used for each 10° Fahrenheit in temperature change.

The state of charge versus specific gravity is shown in Figure 5-19. As you can see, any corrected reading below 1.265 would require recharging or replacing, depending on further diagnosis. One way of determining a defective battery is to compare the specific gravity readings of each cell of the battery. If there is a variation of more than approximately 0.050 between the highest and lowest specific gravity reading, chances are the battery will not hold a charge and it should be replaced.

SERVICE TIP: Do not take a chance with a battery. A major reason for checking a battery is to prevent a problem such as a no start when extreme conditions such as a cold snap require every ounce of power from the battery. This is called *preventive maintenance*.

If all cells are equal in specific gravity even though all are low, the battery is a good candidate for a recharge. Before concluding that a battery is good or bad, the voltage should be considered. For example, a battery with a good specific gravity reading but low voltage reading is bad and needs to be replaced.

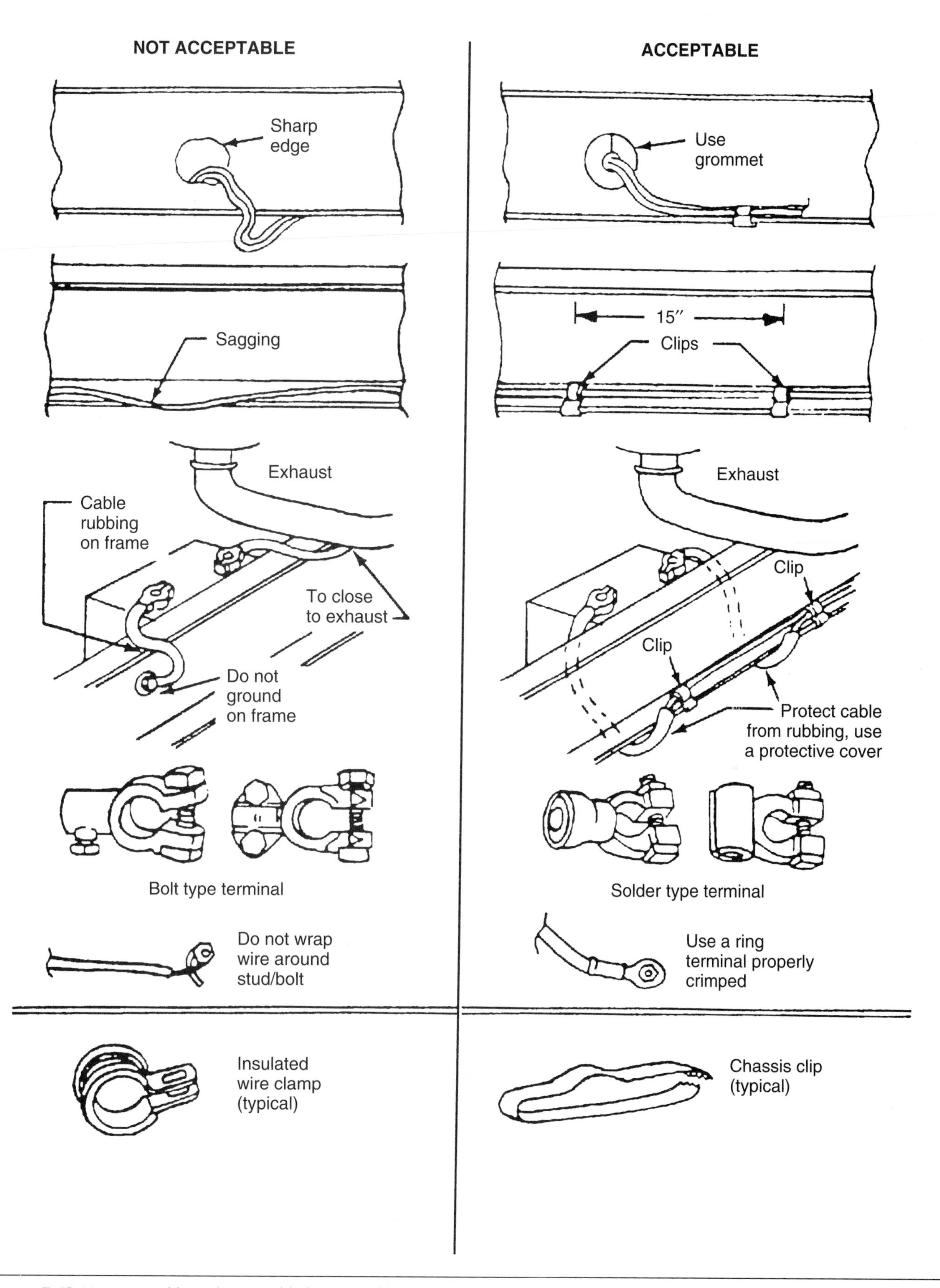

Figure 5-13 Not acceptable and acceptable battery cable service practices. (Reprinted with permission. Courtesy of Volvo Trucks North America, Inc.)

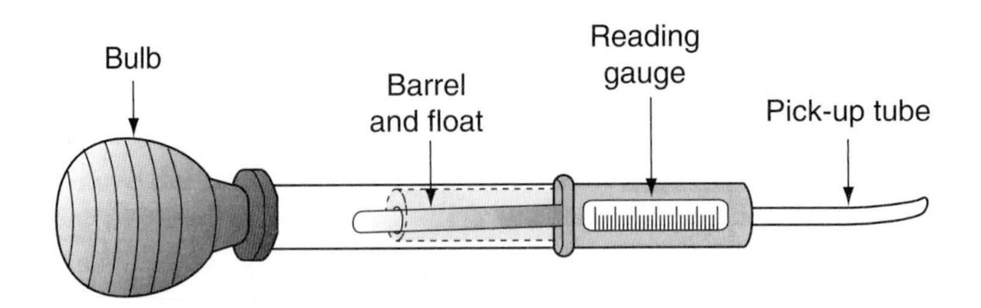

Figure 5-14 A hydrometer is used to measure the state of charge of a battery.

Figure 5-15 The electrolyte level can be checked in a maintenance-free battery by removing the decal to expose vent caps. (Courtesy of Navistar International Engine Group)

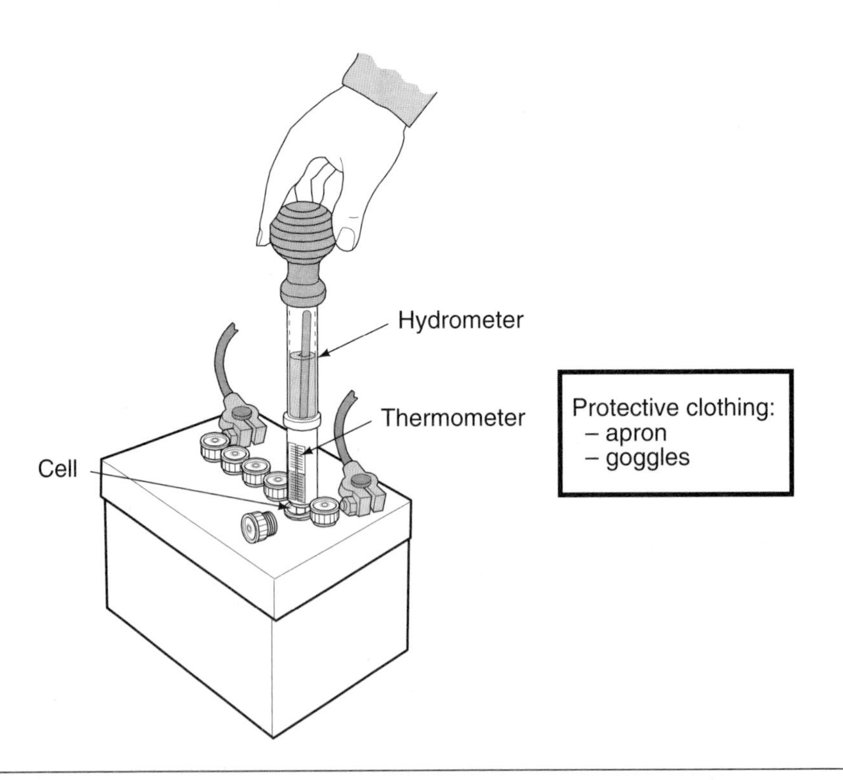

Figure 5-16 Drawing electrolyte out of the battery's cell into the hydrometer.

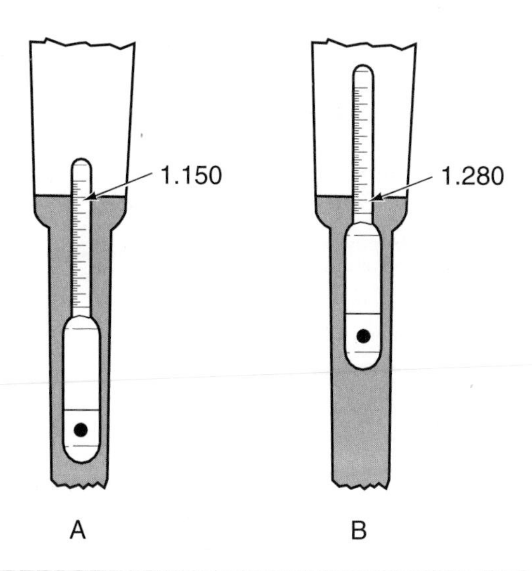

Figure 5-17 The specific gravity reading of electrolyte is read at the point where the electrolyte intersects the float. (A) A low reading and (B) a high reading.

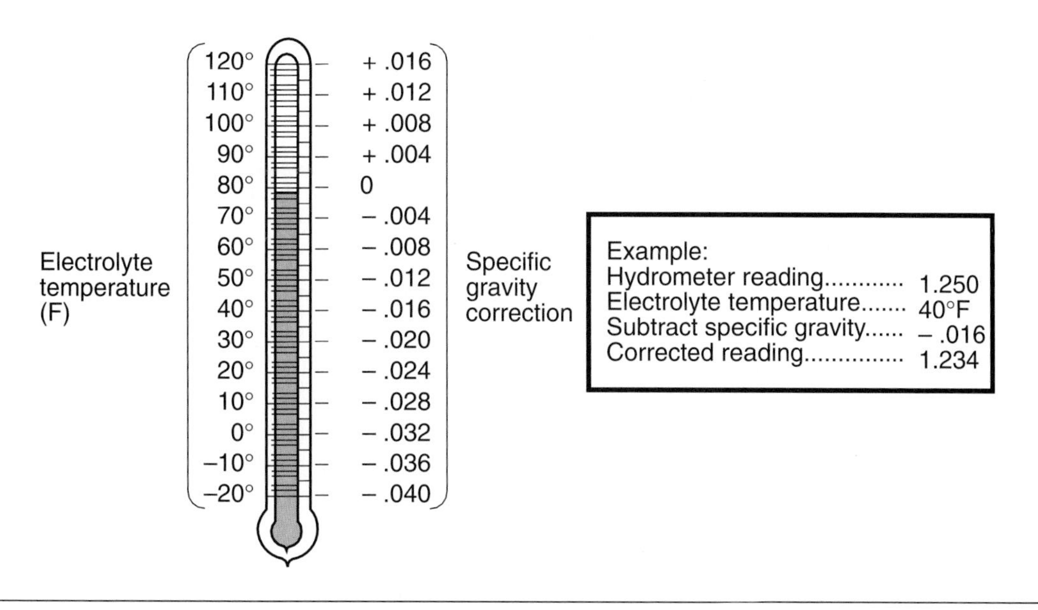

Figure 5-18 Correct the specific gravity reading according to the temperature of the electrolyte.

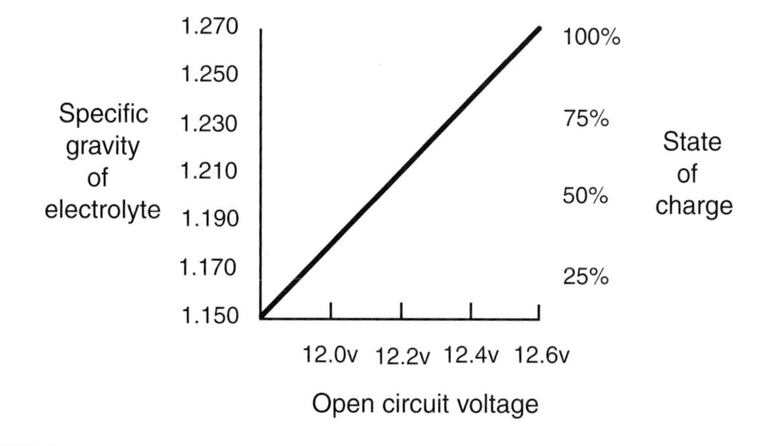

Figure 5-19 Open circuit voltage test results relate to the specific gravity of the battery cells.

Figure 5-20 Open circuit voltage testing.

The **open circuit voltage test** is used to determine the battery's state of charge. It is used when a hydrometer is not available or cannot be used.

Open Circuit Voltage Test

The location of batteries in some trucks makes it physically impossible to use a hydrometer to check the state of charge unless the batteries are pulled. Also, some batteries are totally sealed. An open circuit voltage test is especially useful in these circumstances.

Special Tools

Safety glasses

Digital voltmeter

To make this test accurate, the battery must be **stabilized**. This is especially true if the battery has just been charged either with an external charger or the truck's charging system. The surface charge has to be removed by cranking the engine for 15 seconds without starting the it. The same methods used in the voltage drop test for cables can be used to prevent the engine from starting.

Removing the surface charge stabilizes the battery for an open circuit voltage test.

The capacity test, or load test, is also useful for removing surface charge. Whatever the method used, the next step is to wait 10 to 15 minutes to allow the battery voltage to stabilize. Once the battery is stabilized, connect a voltmeter across the battery terminals, observing the proper polarity (Figure 5-20). Measure the open circuit voltage and compare it to the numbers in Figure 5-19. Notice that it only takes a few tenths of a volt to reflect a change in the state of charge of the battery. The relationship between specific gravity, open circuit voltage test results, and the state of charge is seen in Figure 5-19.

Special Tools

VAT 40

Carbon pile type load tester

Safety glasses

Capacity Test

This test is referred to as a load test. A load is placed on the battery to see how it functions under a demand condition. Prior to performing this test, the previous hydrometer and/or open circuit voltage test needs to be performed. If the state of charge is found to be low, the battery has to be charged in order for the load test to be accurate.

The **capacity test** checks the battery's ability to perform work if a heavy load is placed on it.

A battery tester with an adjustable carbon pile is used to perform this test (Figure 5-21). The tester usually has an ammeter to indicate the load placed on the battery and a voltmeter to indicate the result in volts. The result and specific numbers required for the test can be summed up in the following: A good 12-volt battery should maintain a voltage level of 9.6 volts or more when a load of 1/2 of its cold cranking amps (CCA) is applied to it for 15 seconds.

An adjustable carbon pile is used to place a load on the battery to determine the maintained voltage level.

To do this test with a load tester:

1. Check the state of charge of the battery and if necessary charge it to almost the full state of charge.
2. Determine the CCA of the battery so you can calculate 50% of it for the applied load.
3. Connect the cables across the battery terminals, observing the polarity. Depending upon the tester, there will usually be a set of large cables for the current draw and a set of smaller cables for the voltage reading.

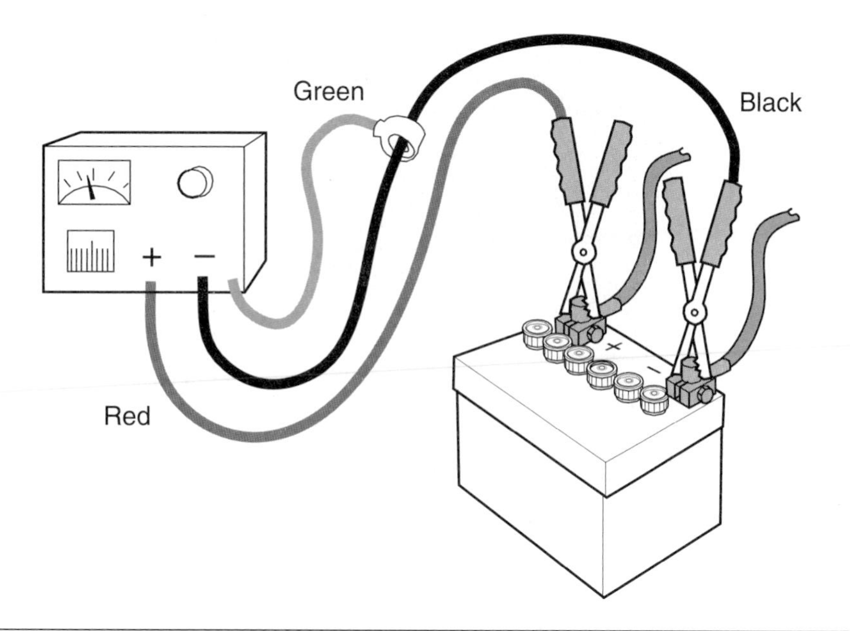

Figure 5-21 Capacity test connections using a battery tester with a carbon pile.

4. Zero the ammeter if required.
5. Some testers require an inductive pickup around one of the tester leads. Observe manufacturer's instructions.
6. Turn the load control knob slowly to apply the predetermined load.
7. While applying the load, read the voltmeter. Do not exceed the 15-second limit.
8. Memorize the reading and turn the off carbon pile.
9. Use the chart in Figure 5-22 to correct the reading.

If the recorded voltage is below the level as listed in Figure 5-22, the battery should be replaced. Also, keep in mind that if the battery was right on the minimum specification, the battery might still give a problem under adverse conditions. If the voltage reading exceeds the minimum specifications by a volt or more, it is an indication that the battery has sufficient reserve power to crank an engine over with a good margin of safety.

Three-Minute Charge Test

This test is performed if the voltage falls below 9.6 volts on the capacity test. As a battery ages, a condition called sulfation occurs in which sulfate crystals penetrate the plates. Sulfation inhibits the cells within the battery from delivering current and accepting a charge (Figure 5-23). This

The **three-minute charge test** is a method for diagnosing a sulfated battery.

Sulfation is a chemical action within the battery that inhibits the cells from delivering current or accepting a charge.

Electrolyte Temperature								
F°	70+	60	50	40	30	20	10	0
C°	21+	16	10	4	−1	−7	−12	−18
Minimum Voltage (12 volt Battery)	9.6	9.5	9.4	9.3	9.1	8.9	8.7	8.5

Figure 5-22 Correcting the results of a capacity test according to temperature readings.

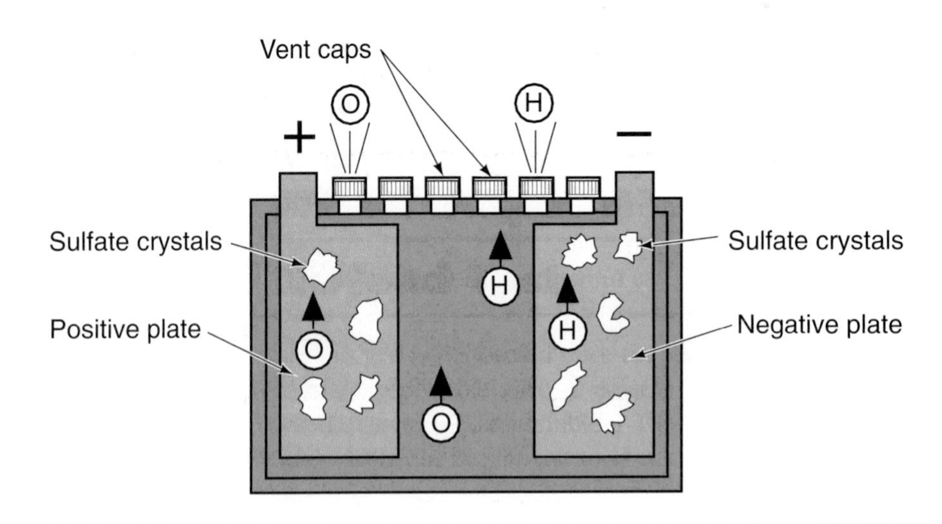

Figure 5-23 A sulfated battery is the result of sulfate crystals that penetrate the plates. The crystals become insoluble, not allowing the battery cell to deliver current or accept a charge.

Special Tools

Battery charger

Voltmeter

Safety glasses

condition often occurs in batteries that have been sitting around for extended periods. Another possibility is that the battery that failed the load test was not receiving an adequate charge from the vehicle's charging system over an extended period. The three-minute charge test is a method for diagnosing a sulfated battery and/or determining the battery's ability to accept a charge.

The following are the steps to conduct a three-minute test:

1. Remove the ground cable or disconnect the vehicle's computer if this test is performed with the battery still in the vehicle.
2. Connect a battery charger and voltmeter or volt/amp tester across the battery terminals observing polarity.
3. Turn on the charger to a setting of 40 amperes.
4. Maintain this rate for 3 minutes.
5. At 3 minutes, check the voltmeter reading: If the voltmeter reads less than 15.5 volts, the battery needs to be recharged and load tested again. If it passes this time, the battery can be placed back in service.

WARNING: Onboard computers are very sensitive to higher than normal voltages, which can damage them.

If the specified battery load cannot be obtained and the battery voltage is below 9.6 volts, perform the three-minute charge test.

A voltmeter reading over 15.5 volts indicates a high internal resistance of the battery due to sulfation or poor internal connections. It takes 1 volt more to push a charge into the battery than it is using to push out. Keep in mind that a sulfated battery can result from an undercharge condition. A charging system check should be performed to prevent future battery problems. Using the methods discussed so far, Figure 5-24 shows a logical troubleshooting flow chart for batteries.

Conductance Testing

As a battery ages, the battery material on the plates deteriorate and fall off. This means less material to pass current. Many times shorts or opens occur within the battery and are hard to detect. All of these reduce the ability of the battery to pass current. A battery with a dead cell can also be the cause of an explosion while the battery is being charged. Unless a hydrometer is used, a bad cell is very hard to detect. Most maintenance free batteries do not even have access to the cells and a battery with the "eye" only measures one out of six cells.

A conductance tester (Figure 5-25) determines a battery's state of health. **Conductance** is a measurement of the battery's ability to produce current using the power of the battery. The

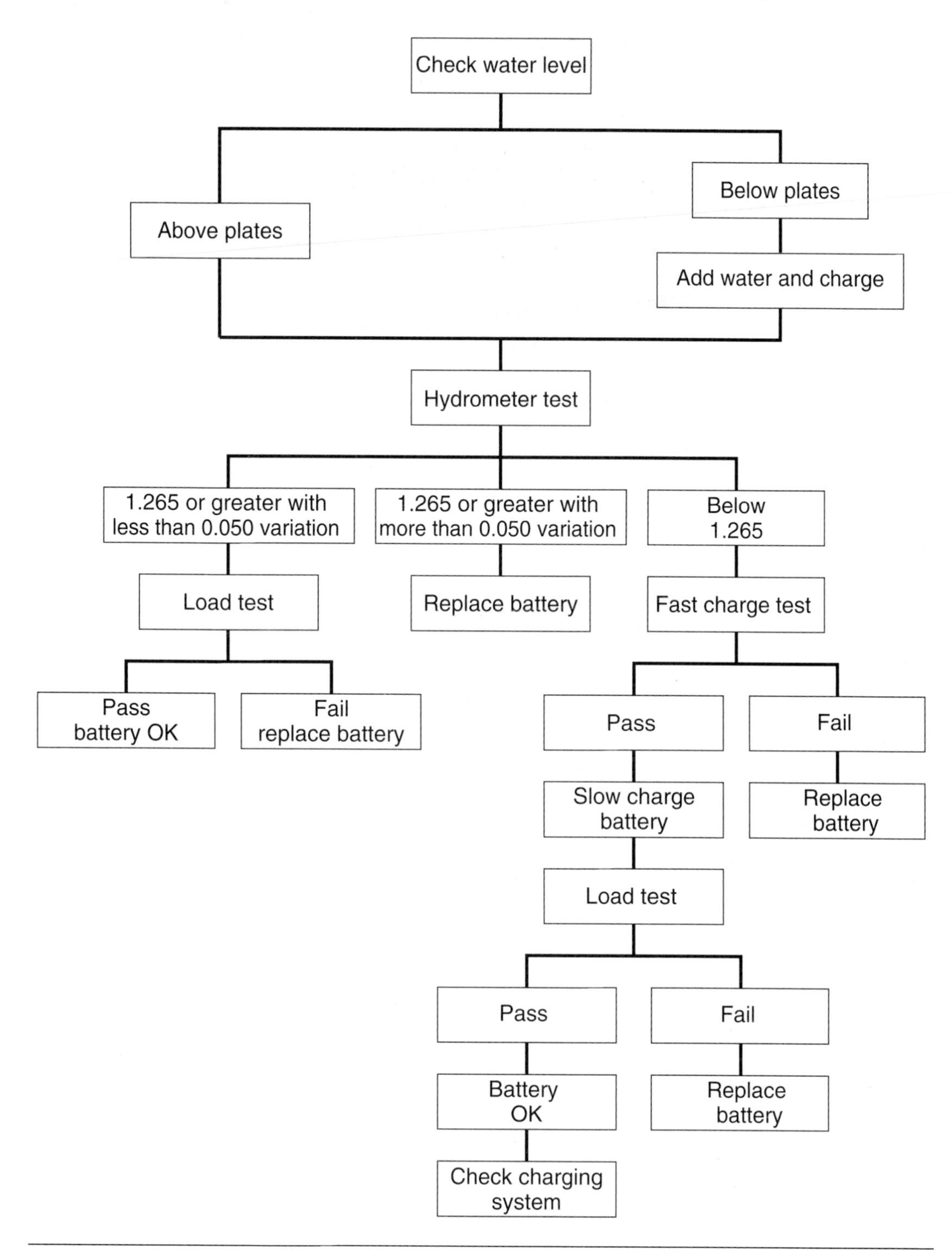

Figure 5-24 A battery troubleshooting flowchart.

tester creates a very small low-frequency AC voltage that is sent through the battery. Electronics inside the tester measure a portion of the AC current that passes through the battery. This AC current response is then compared to the battery's rating, which is entered manually into the tester by means of a keypad. Although multiple rating systems can be used, the most popular one is the battery's cold cranking ampere, or CCA, rating. The reading at the end of the test indicates voltage and the battery's relative power in CCAs. This value is the actual power available in the battery in relation to its potential power.

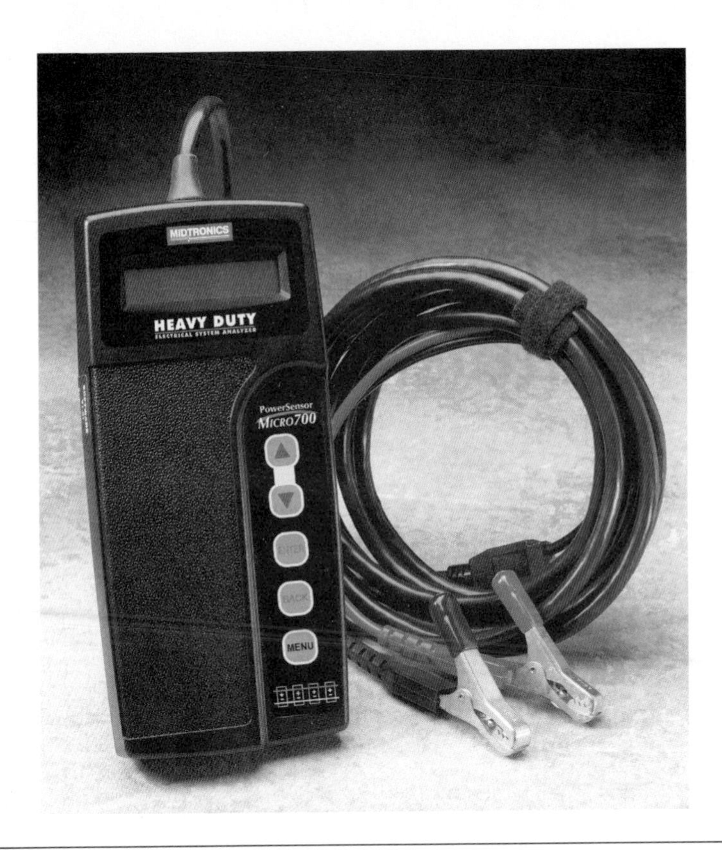

Figure 5-25 A hand-held battery and charging system diagnostic tool. (Courtesy of Midtronics)

The advantage of using this type of tester is that there is no need to place a load on the battery; it will test batteries that are down to 1 volt without having to charge the battery. Within a few seconds of the time it takes to test a battery, the conductance tester will tell if the battery is good, needs charging, is worn out, or has a bad cell. Tests are repeatable because there is no heat involved as there is with charging and load testing.

Battery Drain Test

A constant small current drain on the battery is called a **parasitic draw**.

Special Tools

- Test light
- Ammeter
- VAT-40 or equivalent
- Multiplying coil
- Voltmeter
- Terminal pliers
- Terminal pullers
- Safety glasses

If a battery is found to be dead, there has to be a reason for it. Good communication between you and the driver or customer can usually help you diagnose the problem. By asking the right questions you might discover that the battery is dead every time the driver attempts to start the vehicle after it has not been used for a while, such as over the weekend. The problem may be a current drain from one of the electrical systems, such as a light not turning off in the storage compartment, etc. Another good follow-up question is to find out if any add-on electrical devices have been installed (possibly improperly wired). Don't forget the possibility of a power cable, wire, or component shorting to ground, which can drain the battery.

There are several methods of checking for a battery drain. One method is to take an ammeter and connect it in series with the negative battery cable or place the inductive type ammeter pickup lead around the negative cable. If the meter reads 0.25 or more amps, there is an excessive drain. All the lights should be off. If any are on, remove the bulb to see if the battery drain is reduced or eliminated. If lights aren't the problem, the next step is to go to the fuse panel or distribution center and remove one fuse at a time while watching the ammeter. If the current drain decreases when a fuse is removed, chances are you will find the source of the problem within that particular circuit.

CAUTION: Do not allow the amperage draw to exceed your meter's amperage limit or damage to the meter may result.

The open circuit voltage reading must be 11.5 volts or higher to perform the battery drain test.

An amperage multiplier enables you to get a smaller current flow reading than with an analog type of ammeter.

Using an Amperage Multiplier

Parasitic draw refers to very low current drain on a battery. Some inductive ammeters such as a VAT-40 are not capable of accurately measuring very small current flow. By fabricating a 10X multiplier for use with an induction meter, the multiplier will allow the induction meter to read a much smaller current flow than would be visible without it. To fabricate a multiplier, use a 2 1/2-inch diameter cylindrical object and wrap a 16-gauge wire around it exactly 10 times. Allow one foot of extra lead of wire at each end of the coil. Then slide the coil off the round object and wrap the coiled wire in one place to keep the coiled wires together and retain the round shape. Then attach alligator clips to the two end leads.

Follow these steps to perform the test:

1. Turn off all the loads in the vehicle.
2. Disconnect the negative battery cable at the battery.
3. Connect the multiplier leads in series with the disconnected negative cable and the negative battery post (Figure 5-26).
4. Set your ammeter on the highest scale and clamp the inductive probe over a section of the multiplier.
5. Scale down the meter until the most accurate reading is achieved.

The way this method works is that the coil multiplies the reading by 10. So if you have an actual current flow of 1 amp flowing through the coil, the meter will read 10 amps. This means the reading that you get on the meter must be divided by 10 to get the actual amperage reading. As you can see, this enables you to get much smaller current flow readings than is possible with the analog type of meters such as used in the VAT-40 or other analog type of ammeters.

Using a Test Light for a Battery Drain Test

A test light can also be used to check for an unwanted battery drain, using the following steps:

1. Turn off all electrical loads.
2. Disconnect the negative battery cable.
3. Attach the test light between the negative battery cable and the battery terminal post (Figure 5-27).

If the test light is lit, there is a constant current drain. Check for any circuits or components that are on that should have been off. If the cause of the drain is not found, remove the fuses

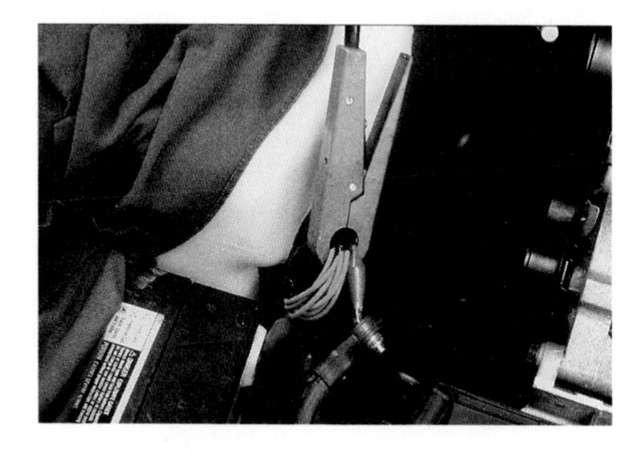

Figure 5-26 Using a multiplying coil to obtain accurate readings when measuring parasitic drains with a VAT-40 or similar tester.

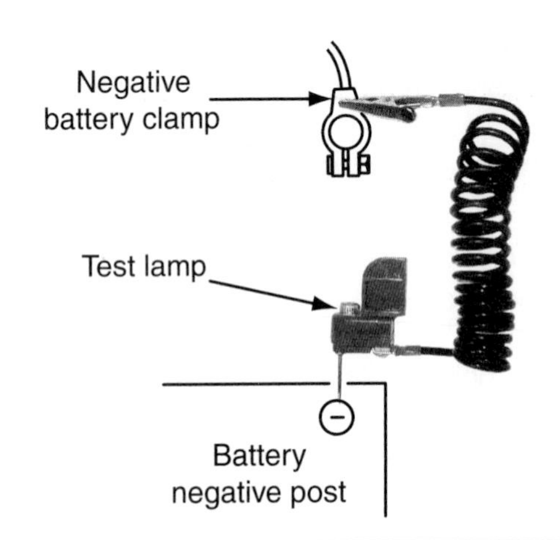

Figure 5-27 Using a test light to determine if there is a parasitic drain.

one at a time until the test lamp goes out. If the light goes out when a fuse is pulled, you need to trace this circuit to find the cause and repair it.

At times, pulling fuses is not enough. You might have to disconnect power leads at a power distribution center because someone might have wired a component into the vehicle without going through the fuse or circuit breaker system. You might also have to disconnect certain components that do not have a means of indicating whether they are on or off, such as thermostatic heaters on the purge valves of air dryers.

Short to Ground Battery Drain Test

A short to ground drain can occur also through any of the electrical circuits within the vehicle to cause the battery to go dead. To confirm this possibility, follow these steps:

A voltmeter is used to check for a short to ground type of battery drain.

1. Turn off all accessories and make sure the ignition is off.
2. Disconnect the battery negative cable.
3. Connect a voltmeter in series with the negative lead of the meter to the negative battery post and the positive lead of the meter to the battery negative cable (Figure 5-28).

If the meter reads less than 12 volts, there is no short to ground draining the battery. If the meter reads 12 volts, there is a short to ground draining the battery. When that occurs, start removing circuit breakers until the voltage drops, indicating the circuit with the short. Further diagnostics involve doing a short to ground circuit test to locate the short.

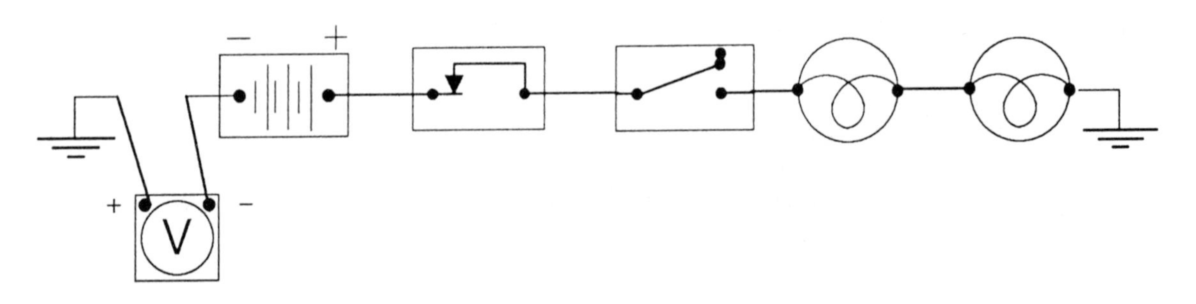

Figure 5-28 A voltmeter hooked up in series to check for a short to ground battery drain. (Courtesy of Mack Trucks, Inc.)

Jump-starting

There will be times, especially in cold weather, when you might have to start a vehicle by **jump-starting** it from another vehicle. This practice can cause electrical system problems or personal injuries if not performed properly.

Safety Precautions

Before attempting a jump-start, the following precautions should be learned and practiced:

1. Wear safety glasses or shield your eyes.
2. Make sure the tops of both batteries are clean of dirt and sulfation.
3. Assure both batteries have a proper level of solution and they are not frozen.
4. To lessen the risk of a short circuit or sparking, remove rings, jewelry, or any metal tools that might come in contact with the positive battery terminal.
5. Ensure that both the discharged battery and the booster battery are of the same voltage.
6. When it comes to heavy duty trucks, you have to determine the vehicle's starting system voltage, whether it is a 12-volt or a 24-volt starting system, and a negative ground electrical system.

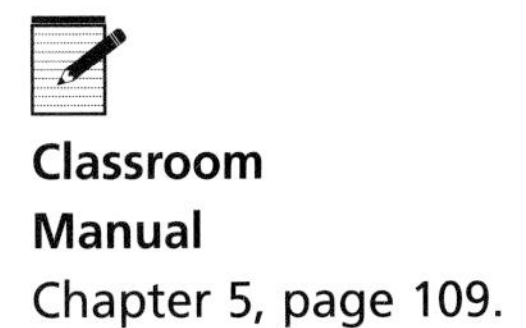

Classroom Manual
Chapter 5, page 109.

CAUTION: If unsure of the other vehicle's voltage or if the voltages and ground are different from your vehicle, do not try to jump-start. Serious personal injury and or damage to the vehicle's electrical systems can occur.

7. Do not allow the vehicles to come in contact with each other. This prevents excessive current from flowing through the vehicles' bodies, causing damage to low current circuitry.
8. Make sure both vehicles are properly set in park mode using their brake systems and the transmissions are in neutral or park.
9. Make sure all vehicle loads or accessories are off.
10. Do not allow jumper cable clamps to touch each other.
11. Do not smoke near the batteries or expose them to an open flame or electric sparks.

These precautions might seem lengthy but it only takes one mistake to cause an injury or catastrophe.

Procedures

The following are procedures for jump-starting a vehicle after following all of the above safety precautions. They are for a 12- or 24-volt starting system.

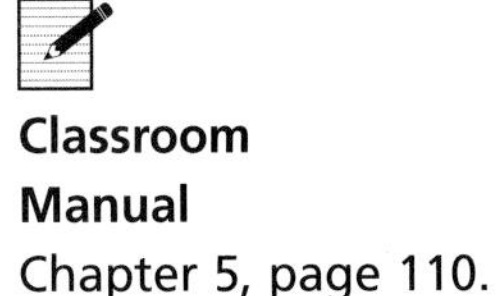

Classroom Manual
Chapter 5, page 110.

12-volt Starting System Many heavy duty vehicles use multiple batteries to start diesel engines. The following procedure can be used to start a single-battery vehicle from any of the diesel vehicle batteries. However, it may not be possible to start a diesel engine vehicle using multiple batteries from a single battery in another vehicle, especially at low temperatures.

1. Attach one end of the positive cable to the disabled battery's positive terminal (Figure 5-29).
2. Connect the other end of the positive jumper cable to the booster battery's positive terminal.
3. Attach one end of the negative jumper cable to the booster battery's negative terminal.
4. Attach the other end of the negative jumper cable to a ground at least 12 inches away from the battery.
5. Start the vehicle with the good battery and run the engine at a moderate speed. Next, try to start the discharged battery's vehicle. If it does not start within 15 to

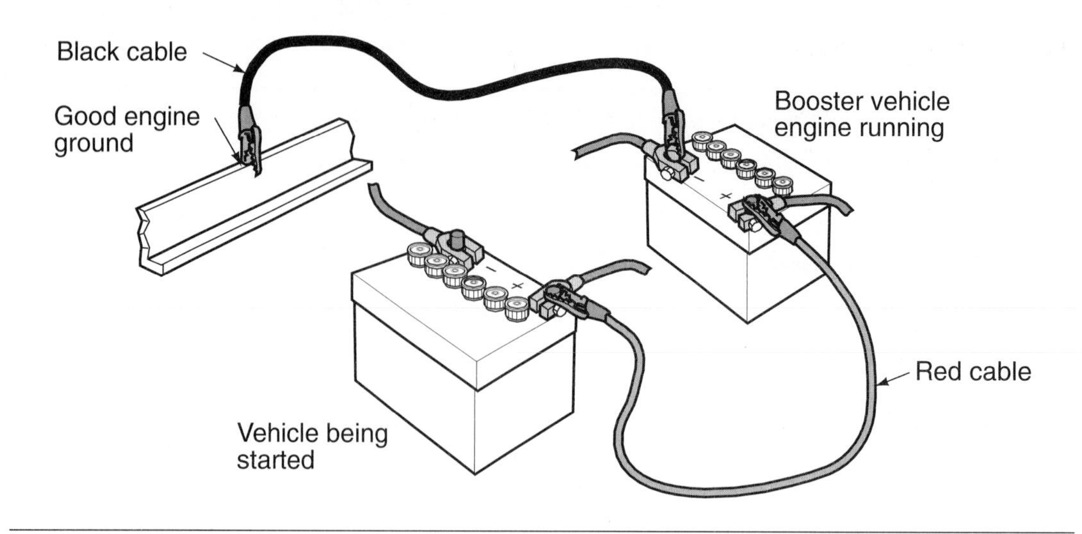

Figure 5-29 Proper jumper cable connections for jump-starting a vehicle.

20 seconds, stop and allow the starter time to cool down and the discharged battery to build up from the running vehicle's starting system. Patience is the key for cold weather jump-starting.

6. Once the vehicle has started, reverse the sequence in disconnecting the jumper cables.

CAUTION: Do not connect the negative booster cable clamp to the discharged battery. A spark might occur, causing the battery to explode.

Figure 5-30 shows the same sequence in hooking up jumper cables to vehicles with multiple batteries. As you can see, this is still a 12-volt system with the batteries wired in parallel for more power.

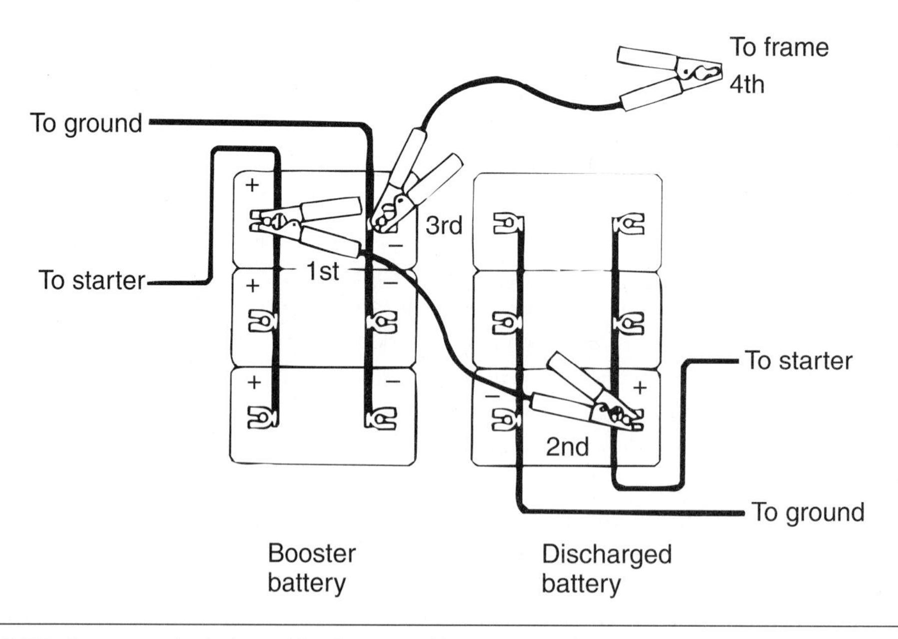

Figure 5-30 Correct method of attaching jumper cables to a 12-volt system.

24 Volt Starting System

WARNING: To prevent damage to electrical systems, both vehicles need to have the same starting systems.

Classroom Manual
Chapter 5, page 109.

Jump-starting 24-volt systems requires the use of four jumper cables.

The procedure to jump-start a 24-volt series-parallel system is similar to jump-starting a 12-volt system. However, it requires two sets of jumper cables. The following sequence is shown in Figure 5-31:

1. Attach one end of one jumper cable to the positive terminal of the booster battery that is connected directly to the starter (1st). The other end of the jumper cable is connected to the positive terminal of the discharged battery that is connected directly to the starter (2nd).
2. Attach one end of a second jumper cable to the negative terminal of the booster battery that is connected directly to the series-parallel switch (3rd). The other end of this same cable is attached to the negative terminal of the discharged battery, which is connected directly to the series-parallel switch (4th). *It is important to note that the booster cable ends are color coded—red for positive and black for negative.* This makes keeping the right polarity in context much easier when you add a second set of cables now.
3. Attach one end of a third jumper cable to the positive terminal of the booster battery that is connected directly to the series-parallel switch (5th). The other end of the cable is attached to the positive terminal of the discharged battery that goes directly to the series-parallel switch (6th).
4. Attach one end of the fourth jumper cable to the negative terminal of the booster battery that is connected directly to the starter (7th). The other end of this same jumper cable is connected to a ground at least 12 inches away from the battery of the vehicle being started.

Make sure your connections are good. Start the good vehicle. Next, start the vehicle with the discharged battery. Once the engine is idling, disconnect the ground connection from the vehicle with the discharged battery. Then disconnect the opposite end of the same cable. Disconnect the other cables in the same manner, the discharged battery ends first, then the opposite ends.

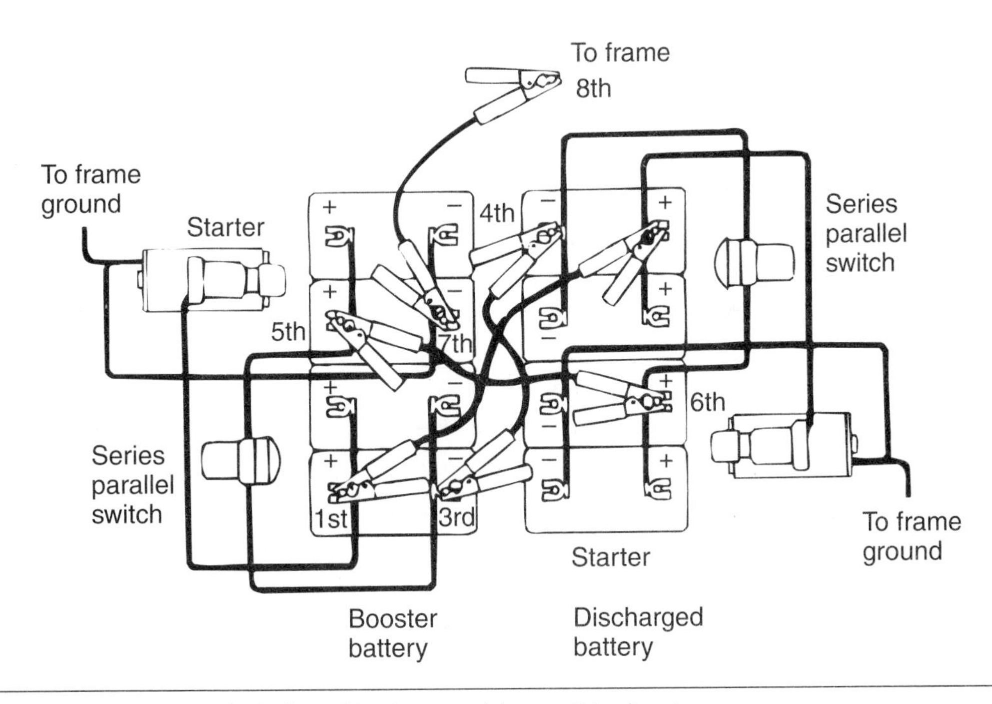

Figure 5-31 Correct method of attaching jumper cables to a 24-volt system.

SERVICE TIP: Many heavy duty vehicles use voltmeters on the dash to indicate battery voltage. Turning the key on in the discharged vehicle and observing the meter can tell you if you have made good connections. It can also indicate when the booster vehicle's charging system has brought the discharged battery's voltage to a level where it is safe to try to start the vehicle.

CASE STUDY

A customer's truck has to be jump-started every Monday morning after sitting for the weekend. During the rest of the week, the vehicle has no problem starting. After jump-starting the vehicle, it is brought to the shop. The half-hour drive to the shop is sufficient enough to charge the batteries so the vehicle starts by itself.

A visual inspection was performed first. The tops of the batteries were quite dirty, so the technician performed a battery leakage test. The technician found some leakage. After the batteries were cleaned, a leakage test was performed to verify that it was eliminated. A load test was then performed and the batteries passed.

The technician decided to go that one extra step and hook up an ammeter in series to check for any parasitic draw. Fortunately, the meter indicated a draw. The technician found a recently rewired air drier heater element that was on constantly because it was wired into a circuit that was constantly powered with or without the key on. The component was then properly wired and the battery system was rechecked for any other drains. The next Monday morning, the customer called to inform the shop that the vehicle had started with no problems.

Terms to Know

Ammeter
Battery leakage test
Battery terminal test
Capacity test
Conductance
Fast charging
Hydrometer
Jump-starting
Multiplying coil
Open circuit voltage test
Parasitic draw
Preventive maintenance
Slow charging
Specific gravity
Stabilized
State of charge
Sulfation
Three-minute charge test
Voltmeter

ASE-Style Review Questions

1. Battery testing is being discussed: *Technician A* says that an ammeter hooked in series by the battery will measure the state of charge. *Technician B* says using a hydrometer to measure conductance of the electrolyte will measure the state of charge. Who is correct?
A. A only **C.** Both A and B
B. B only **D.** Neither A nor B

2. Battery cleaning is being discussed: *Technician A* says that dirt on top of a battery can provide a path for current flow. *Technician B* says a voltmeter is used to determine if the top of the battery is the cause of a current drain. Who is correct?
A. A only **C.** Both A and B
B. B only **D.** Neither A nor B

3. Battery terminal connections are being discussed: *Technician A* says when disconnecting battery cables, always disconnect the negative cable first. *Technician B* says when connecting battery cables, always connect the negative cable first. Who is correct?
   A. A only **C.** Both A and B
   B. B only **D.** Neither A nor B

4. Specific gravity of a tested battery is being discussed. The test was performed during a cold spell with the ambient temperature of 40° Fahrenheit. *Technician A* says that the specific gravity was 1.250. The corrected reading would be lower. *Technician B* says that the corrected specific gravity reading would be higher. Who is correct?
   A. A only **C.** Both A and B
   B. B only **D.** Neither A nor B

5. *Technician A* says that disconnecting the battery cable from the battery and connecting an ammeter in series between the two will indicate if the connection was good or bad. *Technician B* says that placing the voltmeter leads in parallel to the battery terminal and the cable end is a method used to determine the battery connection's integrity. Who is correct?
   A. A only **C.** Both A and B
   B. B only **D.** Neither A nor B

6. The open circuit test is being discussed: *Technician A* says that the surface charge must be removed before performing this test. *Technician B* says that a reading of 12 volts indicates a fully charged battery. Who is correct?
   A. A only **C.** Both A and B
   B. B only **D.** Neither A nor B

7. Jump-starting is being discussed: *Technician A* says that a 24-volt system boosting a 12-volt system makes sense in cold weather because the extra power from the 24-volt system helps to overcome the cold weather cranking problem. *Technician* B says that a multiple 12-volt battery system can be used to jump-start a single 12-volt battery system. Who is correct?
   A. A only **C.** Both A and B
   B. B only **D.** Neither A nor B

8. The results of a three-minute charge test are being discussed: *Technician A* says if the voltmeter indicates fewer than 15.5 volts, the battery must be replaced. *Technician B* says if the voltmeter reading is above 15.5 volts, the battery is good. Who is correct?
   A. A only **C.** Both A and B
   B. B only **D.** Neither A nor B

9. *Technician A* says that a load applied to a battery during a load test should be 1/2 the A/H rating. *Technician B* says that during a load test, the battery voltage must not fall below 9.6 volts. Who is correct?
   A. A only **C.** Both A and B
   B. B only **D.** Neither A nor B

10. Sulfated batteries are being discussed: *Technician A* says that a sulfated battery is the result of a water and baking soda mix getting into the cells during the cleaning process. *Technician B* says that sulfation can occur in batteries that sit around for extended periods. Who is correct?
    A. A only **C.** Both A and B
    B. B only **D.** Neither A nor B

ASE Challenge

1. *Technician A* says that overcharging a battery can make it more susceptible to explosion. *Technician B* says that any arcing or sparks near a battery could cause an explosion. Who is correct?
   A. A only **C.** Both A and B
   B. B only **D.** Neither A nor B

2. The owner of a large dump truck complains about the vehicle's batteries always being low in charge. Upon inspection, shop personnel notice the lid to the battery box is missing and that the batteries are covered with dirt. *Technician A* says that the top of batteries must be cleaned along with the battery post and connectors and that a new cover should be installed. *Technician B* says that modern battery construction eliminates the worries regarding dirt in the lower half of a battery box. Who is correct?
   A. A only **C.** Both A and B
   B. B only **D.** Neither A nor B

3. Battery containment within the battery box for a heavy truck is being discussed. *Technician A* says that good hold down straps and bolts are essential. *Technician B* says that good batteries have separator plates and internal supports that allow for severe vibration and jarring of battery components. Who is correct?

A. A only　　**C.** Both A and B
B. B only　　**D.** Neither A nor B

4. Correct battery charging procedure is being discussed. Technician A and B agree that a slow, low charge is best but that the time often limits such a procedure. *Technician A* says that a high-rate, fast charge is OK if excessive gassing can be avoided. *Technician B* says that a high-rate charge should never be used for more than five minutes. Who is correct?

A. A only　　**C.** Both A and B
B. B only　　**D.** Neither A nor B

5. A couple of technicians are attempting to locate the reason for a slow cranking starting system on a heavy truck. The batteries have been tested and found acceptable. *Technician A* says that a voltage drop of 2 volts at battery cable connections is acceptable. *Technician B* says that a large voltage drop of 6 to 8 volts between the solenoid battery and motor post is acceptable. Who is correct?

A. A only　　**C.** Both A and B
B. B only　　**D.** Neither A nor B

Table 5-1 NATEF Task

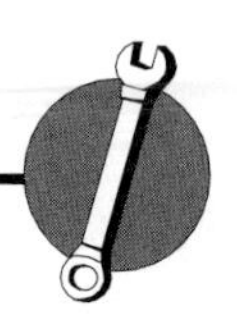

Perform battery hydrometer test; determine specific gravity of each cell

Problem Area	Symptoms	Possible Causes	Classroom Manual	Shop Manual
LOW BATTERY OUTPUT	Inadequate current to start engine under heavy load conditions	1. Low Specific gravity 2. Defective battery	101	107

Table 5-2 NATEF and ASE Task

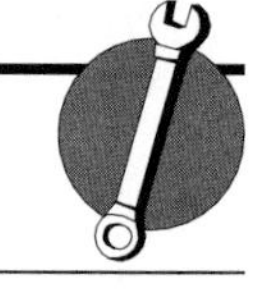

Perform battery load test; determine needed service

Problem Area	Symptoms	Possible Causes	Classroom Manual	Shop Manual
LOW BATTERY VOLTAGE	Battery requires jump-starting after the engine shuts down	1. Undercharged battery 2. Defective battery	103	120

Table 5-3 NATEF and ASE Task

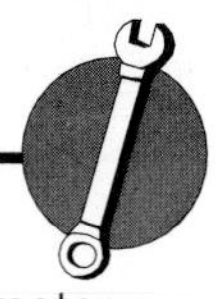

Determine battery state of charge by measuring terminal post voltage using a digital multimeter (DMM)

Problem Area	Symptoms	Possible Causes	Classroom Manual	Shop Manual
LOW BATTERY OUTPUT	Inadequate current to start engine under heavy load conditions	1. Undercharged battery 2. Defective battery	110	120

Table 5-4 NATEF and ASE Task

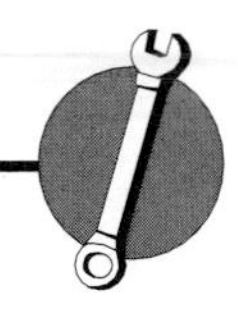

Inspect, clean, service, or replace battery and terminal connections

Problem Area	Symptoms	Possible Causes	Classroom Manual	Shop Manual
REDUCED STARTING AND LOAD CAPABILITIES OF THE BATTERY	Starter fails to crank the engine or cranks slowly	1. Contaminated terminal clamps 2. Defective battery cables	105	107

Table 5-5 NATEF Task

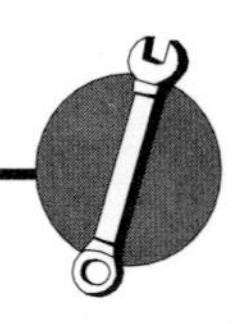

Inspect, clean, and service battery; replace as needed

Problem Area	Symptoms	Possible Causes	Classroom Manual	Shop Manual
DEFECTIVE OR DIRTY BATTERY		1. Dirty battery case resulting in constant current draw 2. Contaminated battery terminals 3. Defective battery	107	107

Table 5-6 NATEF and ASE Task

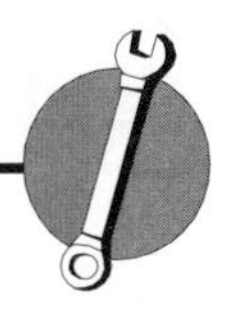

Inspect, clean, repair, and replace battery boxes, mounts, and hold downs

Problem Area	Symptoms	Possible Causes	Classroom Manual	Shop Manual
LOW STATE OF CHARGE	Starter cranks engine slowly or fails to crank engine	1. Discharged battery	103	107

Table 5-7 NATEF and ASE Task

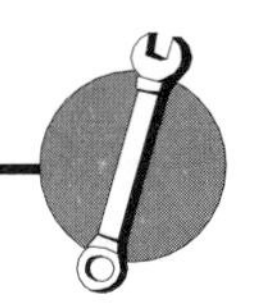

Charge battery using slow or fast charge method as appropriate

Problem Area	Symptoms	Possible Causes	Classroom Manual	Shop Manual
LOW STATE OF CHARGE	Starter is unable to crank engine	1. Low state of charge	102	110

Table 5-8 NATEF and ASE Task

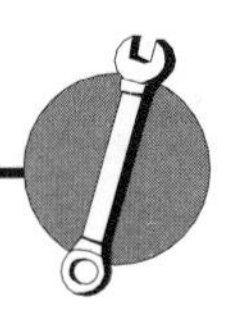

Jump-start a vehicle using jumper cables and a booster battery or auxiliary power supply.

Problem Area	Symptoms	Possible Causes	Classroom Manual	Shop Manual
LOW STATE OF CHARGE	Starter is unable to crank engine	1. Low state of charge	109	127

Job Sheet 6

6

Name: ______________________ Date: ______________________

Inspecting a Battery

Upon completion of this job sheet, you should be able to visually inspect a battery.

NATEF and ASE Correlation

This job sheet is related to ASE and NATEF Medium/Heavy Duty Truck Electrical/Electronic Systems List content area:

B. Battery Diagnosis and Repair, ASE Task 3 and NATEF Task 4: Inspect, clean, service, or replace battery and terminal connections.

Tools and Materials

A medium/heavy duty truck with a 12-volt battery or batteries

A digital multimeter (DMM)

Procedure

1. Describe the general appearance of the battery. ______________________
2. Describe the general appearance of the cables, routing and terminals. ______________________
3. Describe the general appearance of the battery hold downs and battery box. ______________________
4. Check the tightness of the cable at both ends. Describe the condition. ______________________
5. Connect the positive lead of the meter (set on DC volts) to the positive terminal of the battery. Move the negative lead of the meter around the top and sides of the battery case. What readings do you get on the voltmeter? ______________________
6. What is indicated by the readings? ______________________
7. Measure the voltage of the battery with the voltmeter. What is the reading and what is indicated by the reading? ______________________
8. Sum up the condition of the battery based upon this visual inspection and tests. ______________________

☑ **Instructor Check** ______________________

Job Sheet 7

7

Name: ______________________ Date: ______________________

Testing the Battery's State of Charge and Capacity

Upon completion of this job sheet, you should be able to test a battery's capacity and state of charge.

NATEF and ASE Correlation

This job sheet is related to ASE and NATEF Medium/Heavy Duty Truck Electrical/Electronic Systems List content area:

B. Battery Diagnosis and Repair, ASE Task 1 and NATEF Task 2: Perform battery load test; determine needed service; ASE Task 2: Determine battery state of charge by measuring terminal post voltage using a digital multimeter (DMM); NATEF Task 3: Determine battery state of charge using an open circuit voltage test.

Tools and Materials

A medium/heavy duty truck with a 12-volt battery
Starting charging system tester with a carbon pile
Ammeter and voltmeter (or similar equipment)
Hydrometer
Battery charger

Procedure

1. Describe the truck being worked on: Year: ______________ Make: ______________
VIN: ______________________ Model: ______________________
Engine size: ______________________

2. Record the battery specifications:
Installation date: ______________ Cold cranking amps: ______________

3. Perform a battery state of charge using a hydrometer.

A. Record the readings of each cell:
Pos. (1) ______ (2) ______ (3) ______ (4) _____ (5) _____ (6) _____ Neg.

B. Did the reading need to get temperature corrected? Yes _______ No ________

C. If yes, by how much? ________________

D. Are all cells within 25 points? Yes ____________ No __________

E. Are the cell conditions good enough to perform a load test? Yes ________ No _________

4. Open circuit voltage test:
This test is optional to relate to the specific gravity test or if the battery is a maintenance-free battery.

A. Use a voltmeter and record the reading. __________________

B. Locate the chart to determine the battery's state of charge.

5. Review the battery's state of charge and perform any necessary service to the battery in order to perform a load test. Describe service performed. __________________
____

6. Connect the starting system tester to the battery.
   A. Describe the function of the amperage portion of the test. ____________________
   ____
   B. Describe what is expected of the voltmeter portion of the test. ________________
   ____
7. Determine the correct load and record one or the other. ____________________
   A. Cold cranking amps ÷ 2 = Load
   A. Battery Amp/Hour × 3 = Load
8. Conduct the load test and record the readings.
   Battery voltage decreased to __________ volts after ________ seconds at an applied load of ________ amps.
9. 3-minute charge test:
   This test is optional for a battery that has failed the load test and is suspected of being sulfated.
   A. Disconnect the ground battery cable.
   B. Connect a voltmeter across the battery terminals (observe polarity).
   C. Connect battery charger (observe polarity).
   D. Turn the charger on to a setting of 40 amps.
   E. Maintain a 40 amp rate for 3 minutes while observing voltmeter.

   Did the voltmeter read less than 15.5 volts? Yes _______ No _______
   If yes, recharge slowly and reload test.
   If no, battery is sulfated and should be replaced.

☑ **Instructor Check** ____

CHAPTER 6

Starting System Diagnosis and Service

Upon completion and review of this chapter, you should be able to:

- ❑ Describe safety precautions and perform safe procedures when working with the starting system.
- ❑ Perform in-vehicle starting system checks.
- ❑ Determine the cause of a no-crank or slow-crank condition.
- ❑ Perform quick-tests to determine problem areas.
- ❑ Perform a starter system current draw test and interpret the results.
- ❑ Perform voltage drop test of both the control circuit and the starter motor circuit.
- ❑ Perform a no-load test and interpret the results.
- ❑ Perform starter motor component checks and tests.
- ❑ Remove and reinstall the starter.
- ❑ Disassemble, clean, inspect, repair, and reassemble a starter motor.

Introduction

One system in a truck that can be counted on to pose a potential problem, is the starting system. There is nothing worse than being committed to a timed delivery or pickup and the vehicle won't start. Think about the trucks and tractors with tankers that haul gasoline and the problems that can occur if a starting system fails. These vehicles are not allowed to run during the loading process at the racks. What would happen if the vehicle was loaded and it would not start? Think of the effects that sparks could create during jump-starting at the rack area where gasoline vapors are very heavy. Jump-starting is not an option. As you can see, a truck that doesn't start can become quite an expensive ordeal to the trucker.

Basic Tools

Basic mechanic's tool set

Service manual

Starting problems are not necessarily related to the starting system. Other problems such as ignition problems in gasoline engines or fuel problems in diesel engines can prevent the vehicle from starting. If everything else is operating properly in an engine, the starter motor should be able to rotate the engine fast enough to start the engine so it runs under its own power. A typical starter system has five basic components:

A starter system uses two connected circuits to reduce resistance: the starter circuit and the **control circuit.**

1. Battery
2. Key switch and/or starting switch
3. Battery cables
4. Magnetic switch
5. Starter motor

In heavy duty trucks, the starter motor might require 400 amps or more of current to perform work. This requires integrity of the whole system, especially the battery's cables and connections. To reduce resistance of the cables, the starting system is designed with two connected circuits: the starter circuit and the control circuit.

Starting System Service Precautions

Before any test or service is performed on the starter system, some precautions need to be observed. Keep in mind that part of the starting system is the battery. Any precautions given in

Chapter 5 that apply to the batteries also apply to the starting system. The following are further precautions that should be observed:

1. Refer to manufacturer's manuals for correct procedures. Today, vehicles have electronics that are directly or indirectly related to the battery portion of the starting system.
2. If the starter is required to be disconnected or removed for service, disconnect the battery ground cable first.
3. Always make sure the vehicle is properly parked.
4. Always make sure the transmission is in neutral (manual), or neutral or park (automatic) before performing any cranking tests.
5. Follow manufacturer's directions for disabling the ignition or fuel system whenever required for a test.
6. When testing the system with the engine running make sure:
   A. Test leads and tools are clear of any moving parts.
   B. Loose clothing is clear of any moving parts.
   C. Hands and other body parts are clear of any moving parts.
7. Use a helper if the removal of a heavy starter is too much weight for one person or the position is too awkward for one person to handle.
8. Wear safety glasses, especially when working underneath the vehicle.

A **no-crank** condition means the ignition switch is placed in the start position and the starter motor does not turn the engine. However, at times a clicking noise can be heard from the solenoid or relay, or there is a buzzing noise indicating the drive has engaged the flywheel but the engine is not turning.

Starting System Troubleshooting

Most starter problems fall into four categories: the engine does not crank, the engine cranks slowly or unevenly, the starter motor spins but does not crank the engine, or there is noisy starter cranking. These are simply symptoms. The technician's job is to diagnose the cause of these symptoms. One thing is for certain, starting an engine takes a team effort and a minimum cranking speed. Attaining that speed depends on many factors such as oil viscosity, temperature, battery capacity, and cable resistance. They all play a role as to whether or not an engine will crank fast enough to start.

Many starting system complaints wind up being battery-related problems. Any starting system test performed with a weak battery can result in a possible misdiagnosis that can be unnecessarily time consuming and costly. A thorough visual inspection of the entire starting system is a good place to start with a diagnosis. The chart in Figure 6-1 is a good troubleshooting reference to start with to determine the possible causes of the various symptoms. Go through this chart to familiarize yourself with the possible directions to take and/or tests to be performed when troubleshooting a problem.

Slow cranking means the starter drive engages the flywheel and the starter motor turns the engine but at a reduced speed. Sometimes not enough speed is reached to start the engine.

Keep in mind that all the terminals and contacts, switches, and connecting wiring are potential starter system problem areas. You must be comfortable with the basic starting system because some flow charts often refer to specific terminals in order to do pinpoint testing. Figure 6-2 shows a set of flow charts pertaining to an engine cranking slowly. Looking at these two charts, you can see that a majority of this test involves checking for a **voltage drop** at various points. That is why it is so important to check the battery and starter cables, wiring, and grounds for clean, tight connections before beginning the test procedure. These individual voltage drop tests will be covered in more depth later in this chapter.

Flow charts are specific procedures in sequence to help a technician pinpoint a problem. A good flow chart also supplies specific test values.

Sometimes a starter spins but the engine does not turn over. The most likely cause of this is the starter drive portion of the starter. Also, sometimes the starter drive attempts to engage but cannot do so because of bad teeth. Both cases would require the starter to be pulled to install a new starter drive. Before faulting the drive, though, make sure you check the flywheel ring gear teeth for wear or breakage. Manually turn the crankshaft so you can inspect the complete circumference of the ring gear.

Problem	Possible Cause	Tests and Checks	Remedy
Engine cranks slowly or unevenly	1. Weak battery	1. Perform battery open circuit voltage and load voltage tests. Perform battery load tests (capacity). Check capacity and voltage ratings against engine requirements.	1. Service, recharge, or replace defective battery.
	2. Undersized or damaged cables	2. Perform visual inspection.	2. Replace as needed.
	3. Poor starter circuit connections	3. Perform visual inspection for corrosion and damage.	3. Clean and tighten. Replace worn parts.
	4. Defective starter motor caused by high internal resistance	4. Perform cranking current test and no-load test.	4. If cranking current is under specs, proceed with no-load bench testing.
	5. Engine oil too heavy for application	5. Check oil grade.	5. Change oil to proper specs.
	6. Seized pistons or bearings	6. Check compression and cranking torque.	6. Repair as needed.
	7. Overheated solenoid or starter motor	7. Check for missing or damaged heat shields.	7. Replace shield. Service as needed.
	8. High resistance in starter circuit	8. Use cranking current test. Insulated circuit test, and ground circuit test to pinpoint area of high resistance.	8. Replace defective components.
	9. Poor starter drive/ flywheel engagement	9. Perform visual inspection of drive and flywheel components.	9. Replace damaged items.
	10. Loose starter mounting	10. Perform visual inspection.	10. Tighten as needed.
Engine does not crank	1. Discharged battery	1. As listed above	1. As listed above
	2. Poor or broken cable connections	2. As listed above	2. As listed above
	3. Seized engine components	3. As listed above	3. As listed above
	4. Loose starter mounting	4. As listed above	4. As listed above
	5. Open in control circuit	5. Perform control circuit test to determine "open" condition or areas of high resistance.	5. Repair or replace components as needed.
	6. Defective starter relay	6. Perform starter relay bypass test.	6. Replace starter relay if engine cranks when these components are bypassed.
	7. Defective starter motor caused by internal motor malfunction	7. Perform starter relay bypass test.	7. Replace starter motor if engine will not crank when starter relay is bypassed.
Starter motor spins but does not crank engine	1. Defective starter drive	1. Perform starter drive test.	1. Replace starter drive.
	2. Worn or damaged pinion gear	2. Perform visual inspection of components.	2. Replace starter drive.
	3. Worn or damaged flywheel gears	3. Perform visual inspection of flywheel.	3. Replace as needed.

Figure 6-1 Troubleshooting a starting system.

Problem	Possible Cause	Tests and Checks	Remedy
Starter does not operate or Movable pole shoe starter chatters or disengages before engine has started	1. Battery discharged	1. Perform battery load test.	1. Recharge or replace.
	2. High resistance in starting circuit	2. Perform cranking current test, insulated circuit test, and ground circuit tests to pinpoint area of high resistance.	2. Replace defective components.
	3. Open in solenoid or movable pole shoe hold-in winding	3. —	3. Replace solenoid or movable pole shoe starter.
	4. Worn solenoid unable to overcome return spring pressure	4. —	4. Replace solenoid. Install lighter return spring.
	5. Defective starter motor	5. —	5. Replace motor.
Noisy starter cranking	1. Loose mounting	1. Perform visual inspection.	1. Tighten mounts. Correct alignment.

Figure 6-1 Continued.

Testing the Starting System

One of the most valuable tools for performing a starter test is a combination battery, starter, charging system tester as discussed in Chapter 5. These testers are able to load down batteries and read current draw and voltages. Battery performance and starter performance go hand in hand—that is why a full battery test series is performed as part of a starter system test.

The **quick test** is performed to isolate a problem area to determine whether the control circuit, starter motor, or solenoid is at fault.

If no lights come on and the engine doesn't crank, look for a main power distribution problem.

Special Tools

Hand tools

Battery terminal pliers

Battery terminal cleaning tools

Jumper wire

Quick Testing

A quick test can be performed to help you locate the problem area if the starter does not turn the engine at all. To perform this test, make sure the vehicle is in the park position with the transmission in neutral. Turn on the headlights. Next, turn on the ignition switch to start the engine while observing the headlights. One of the following three things will happen to the headlights during this test:

1. They will go out.
2. They will dim.
3. They will remain at the same brightness level.

If the lights go completely out, the most likely cause is a connection problem at one of the battery terminals. Check all the cables for a good connection. To check for and ensure a proper connection, pull the cables off and clean the terminals and cable ends. Reinstall them and retest again.

If the headlights dim when the ignition is turned to the start position, the battery may be discharged. Check the battery or its condition. If it checks out good, there might be a mechanical problem in the engine preventing it from rotating. If that is suspected, try turning the engine manually with a socket and ratchet by the crank pulley. Some trucks have access to the flywheel teeth by the bell housing. If that is the case, a large screwdriver can be used as a lever to pry against the access hole, with the tip of the screwdriver in the teeth to rotate the crankshaft. If the engine turns, the starter motor may have internal damage.

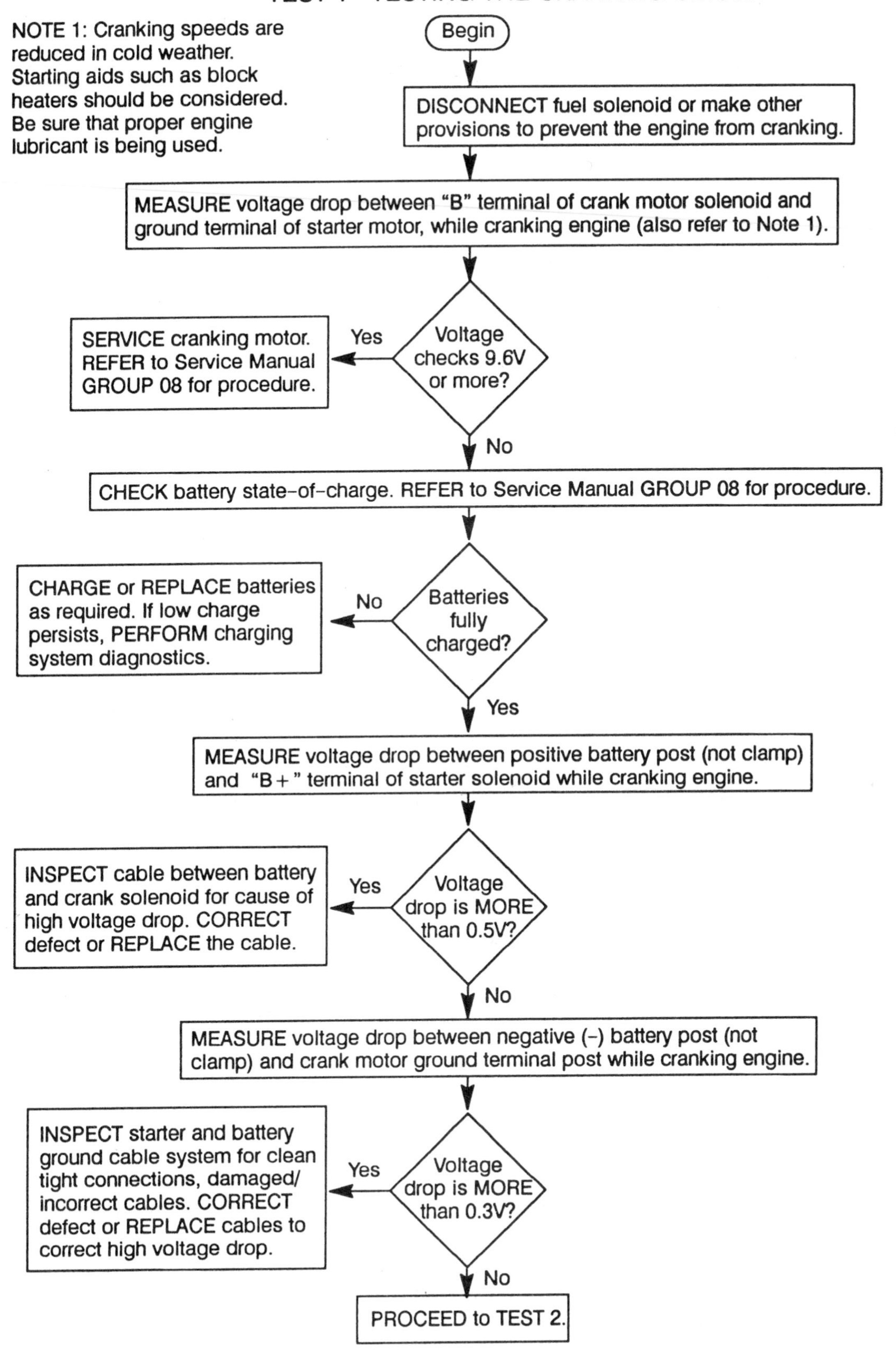

Figure 6-2 A flowchart pertaining for an engine cranking slowly. (Courtesy of Navistar International Engine Group)

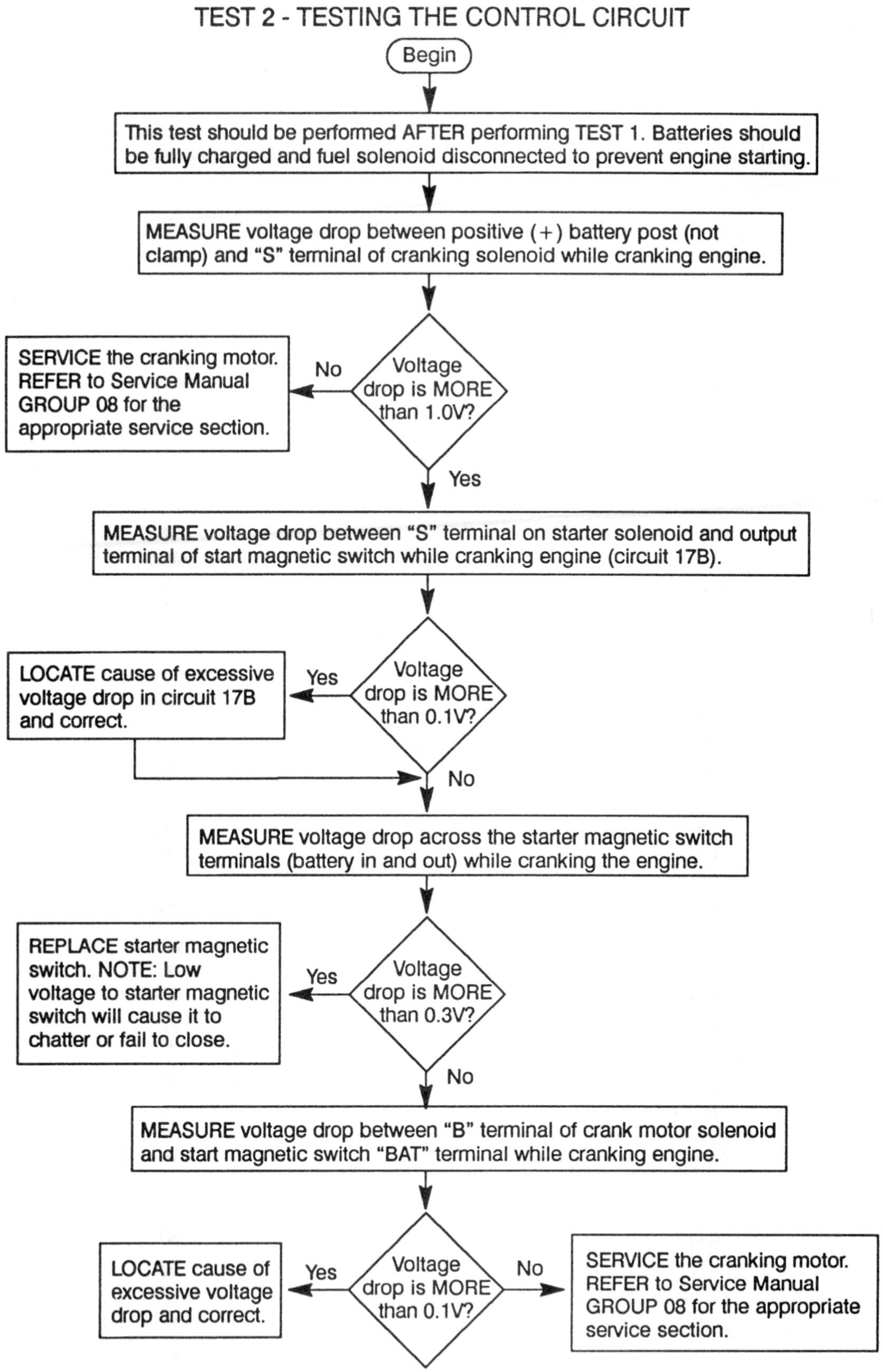

TEST 2 - TESTING THE CONTROL CIRCUIT

Begin

This test should be performed AFTER performing TEST 1. Batteries should be fully charged and fuel solenoid disconnected to prevent engine starting.

MEASURE voltage drop between positive (+) battery post (not clamp) and "S" terminal of cranking solenoid while cranking engine.

Voltage drop is MORE than 1.0V?

No → **SERVICE** the cranking motor. **REFER** to Service Manual GROUP 08 for the appropriate service section.

Yes ↓

MEASURE voltage drop between "S" terminal on starter solenoid and output terminal of start magnetic switch while cranking engine (circuit 17B).

Voltage drop is MORE than 0.1V?

Yes → **LOCATE** cause of excessive voltage drop in circuit 17B and correct. (then continue ↓)

No ↓

MEASURE voltage drop across the starter magnetic switch terminals (battery in and out) while cranking the engine.

Voltage drop is MORE than 0.3V?

Yes → **REPLACE** starter magnetic switch. NOTE: Low voltage to starter magnetic switch will cause it to chatter or fail to close.

No ↓

MEASURE voltage drop between "B" terminal of crank motor solenoid and start magnetic switch "BAT" terminal while cranking engine.

Voltage drop is MORE than 0.1V?

Yes → **LOCATE** cause of excessive voltage drop and correct.

No → **SERVICE** the cranking motor. **REFER** to Service Manual GROUP 08 for the appropriate service section.

Figure 6-2 Continued.

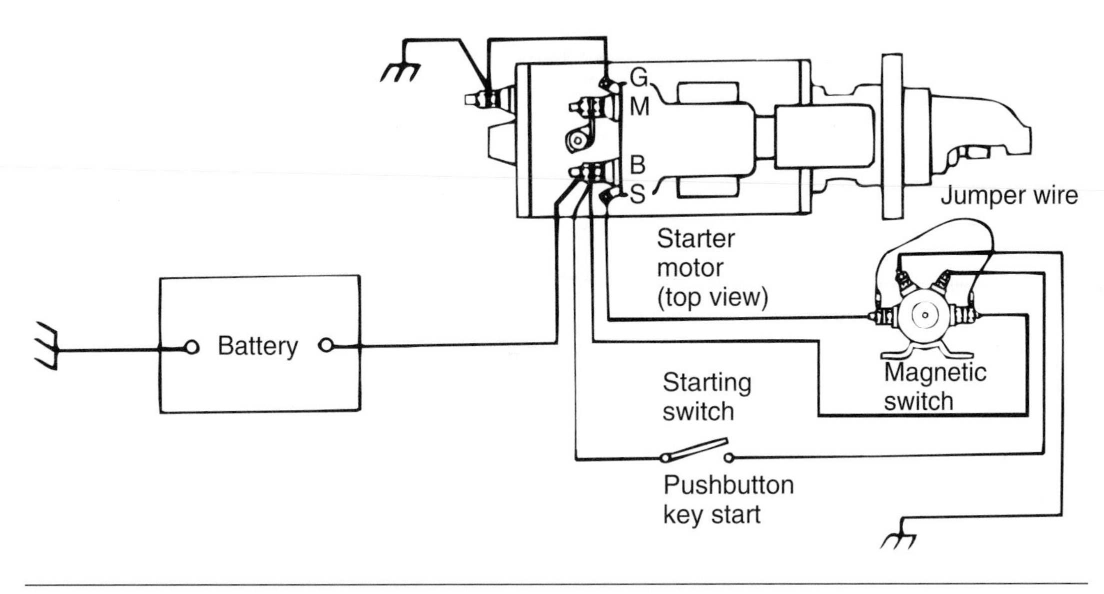

Figure 6-3 Check the starter relay operation by bypassing the switch with a jumper wire. (Courtesy of Freightliner Corporation)

If the light stays brightly lit and the starter makes no clicking noise, there is an open in the circuit and the fault is most likely in the solenoid or the control circuit. A quick check can be performed using a jumper wire. If the starter system has a magnetic switch, jumper around the heavy terminals as shown in Figure 6-3 to see if the motor cranks. If it cranks, the relay is defective assuming that control current from the starting switch is available at the small terminal of the relay. If the starter system doesn't have a magnetic switch, the same process can be used but this time you will jumper between the large "B" terminal and the "S" terminal of the solenoid. This bypasses the control circuit to see if the motor cranks. If the motor cranks, there is a problem in the control circuit such as wiring, connections, or switch. A test light can also be used to check for available voltage at all the various terminals.

CAUTION: You are working with high currents. Personal injury and damage to your tools can occur if you arc a power connection to ground.

Special Tools

Starting/charging system tester

Jumper wires

WARNING: When jumping circuits and components, make sure the wires or tools used are able to withstand the large current draws of the starter system or circuit and/or component damage may occur.

Current Draw Test

Not all starter problems are a no-crank type of a problem. The symptom of a possible problem is often a slow cranking speed. In those instances, a cranking test is a good first test to perform. The cranking current test is used to measure the amount of current in amperes that the starter circuit draws to crank the engine. This information is used to determine the next step needed to pinpoint a starter system problem.

Figure 6-4 shows a typical battery-starting-charging system tester. This tester has an ammeter to measure amps, a voltmeter to measure volts, and a carbon pile to load a battery or system down. The information from this tester is important because of the relationship between battery performance and starter performance. It is important to note that a battery can cause a starting problem just as it can be caused by the starting system. Remember that in Chapter 5 we learned

A VAT-40 is one example of a starting/charging system tester.

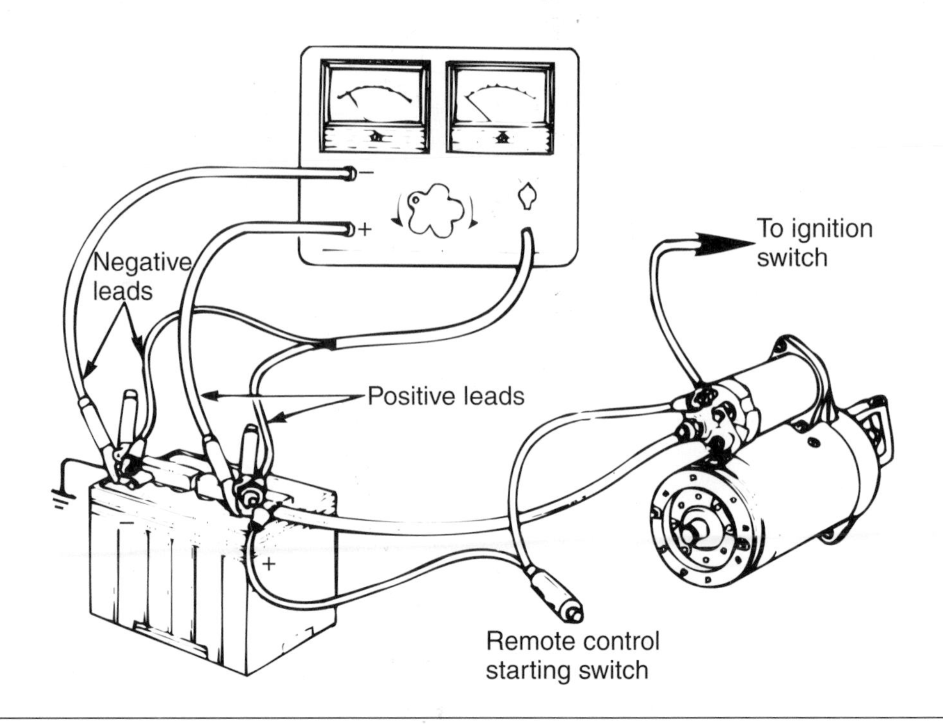

Figure 6-4 Test connections for a current draw test. (Courtesy of Navistar International Engine Group)

The **starter current draw test** measures the amount of current drawn by the starter to turn the engine over.

that a battery must have the ability to keep its voltage above 9.6 volts and still deliver the necessary current to crank the engine over.

Using wattage, we can see that if a battery cannot keep its voltage up, the starter will draw more current. The formula for wattage is: Volts × Amps = Wattage. Wattage is the actual work that is used to crank the engine. For example, if a starter cranks at 2,000 watts and 10 volts is applied to it, 200 amps will flow: 10V × 200A = 2,000W. What happens if the battery cannot keep its voltage up? If 8 volts is applied to the starter it will draw 250 amps: 8V × 250A = 2,000W. Keep in mind that more current drops voltage lower, which increases the current, which then drops more voltage, increasing the current, and so on.

For these examples you can see why it is important to deliver amperage at a reasonable voltage level. For this reason, a battery has to pass a load test prior to a starter current draw test. A weak battery will throw off this test. To perform this test the hook-up shown in Figure 6-4 is needed, or if using a tester such as a VAT-40 with an inductive pick-up, see Figure 6-5. The following procedure is used:

1. Connect the leads as shown in the figures.
2. Zero the ammeter if required.
3. Make sure the carbon pile is not on.
4. Make sure all vehicle loads are off.
5. Disable either the ignition system or fuel system to prevent the engine from starting.
6. If using the VAT-40, select INT 18V and place the test selector in the starting position.
7. Crank the engine and note the voltmeter reading.
8. Stop cranking and make sure the ignition key is off.

WARNING: Do not allow the motor to crank for more than 15 seconds. Allow the motor to cool down between cranking or motor damage may occur.

9. Now, turn the carbon pile or load control knob slowly until the voltmeter reading matches the reading taken in Step 7.
10. Read the ammeter scale to determine the amount of current draw.

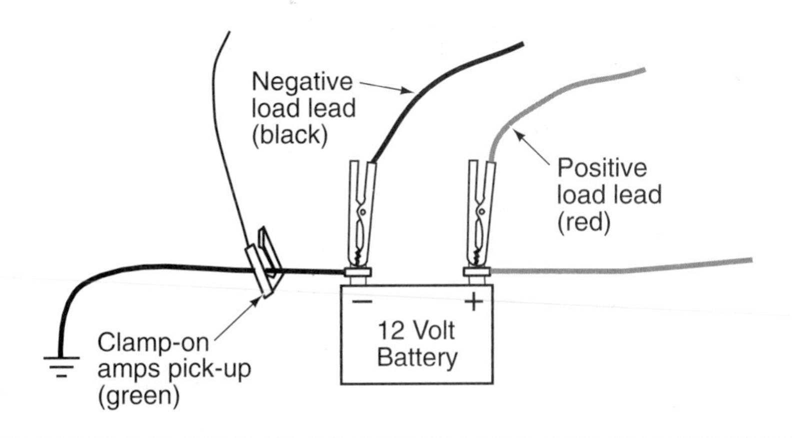

Figure 6-5 Connecting the tester's leads to conduct a starter current draw test. (Courtesy of Sun Electric Corporation)

Compare the reading to the manufacturer's specifications. For example, a V8 gasoline engine has a current draw of approximately 200 amperes, while diesel engines can require over 400 amperes of current. These readings are meant to determine if the starter draw is normal, less than normal, or excessive.

A higher than normal current draw is an indication of a resistance problem internal to the starter, either electrical or mechanical and also, the possibility of a mechanical problem with the engine which impedes rotation. A lower than normal current draw usually indicates a resistance problem in the starter system circuit, causing an excessive voltage drop.

Wattage is the actual work that is used to crank the engine.

The specification for current draw is the maximum allowable and the specification for cranking voltage is the minimum allowable.

A battery that is weak will affect the current draw of a starter motor with a specific wattage determined to start an engine. This results in a higher starting system current draw.

Starter Voltage Test

The next series of tests requires the use of a digital multimeter with a Volts DC function. A digital meter is preferred because it is accurate to four places. The advantage of using voltage readings is that we can easily locate excessive resistance by measuring the amount of voltage lost anywhere within the starter circuit. The starter circuit requires a great deal of current to operate, so that any unwanted resistance will hinder the proper operation of the starter motor.

SERVICE TIP: Starting system problems can be easily confused with charging system problems. A slow cranking starter may be the result of a charging system not keeping batteries fully charged. Before performing any starter tests, verify that the charging system is operating properly and the battery is fully charged and passes all the battery tests.

This following starter voltage test is an alternative to the starter current draw test to determine whether a problem is internal to the starter or an external problem within the starter system circuit. It is meant to isolate problems related to slow cranking speeds or no cranking at all. Cranking of the engine is required in this test without the engine starting; therefore, you must first disable the ignition system or fuel system following the manufacturer's procedures:

1. Set the multimeter to the volts DC function.
2. Connect the negative (−) lead to the negative battery terminal and the positive (+) lead to the positive battery terminal (Figure 6-6).
3. Make sure proper steps were taken to prevent the engine from starting. Turn the starter switch to the start position and energize the starter.

WARNING: Limit the cranking period to 15 seconds or less or cranking motor damage can result.

Figure 6-6 Checking starting voltage at batteries.

Special Tools

Basic hand tools

Voltmeter

A starter voltage test is performed to determine if there is excessive resistance.

A voltage test is performed on both the **insulated side** and the **ground side** of the starter system.

4. Observe the voltmeter scale, note the reading, then release the switch.
5. Disconnect the voltmeter from the battery and move the leads to the starter (Figure 6-7).
   - ❑ The meter's negative (−) lead goes on the starter ground terminal.
   - ❑ The meter's positive (+) lead goes on the starter motor power terminal—the connection on the starter that comes from the solenoid's "M" terminal.
6. Turn the switch to the start position and energize the starter.
7. Again observe the voltmeter scale, note the reading, then release the switch.

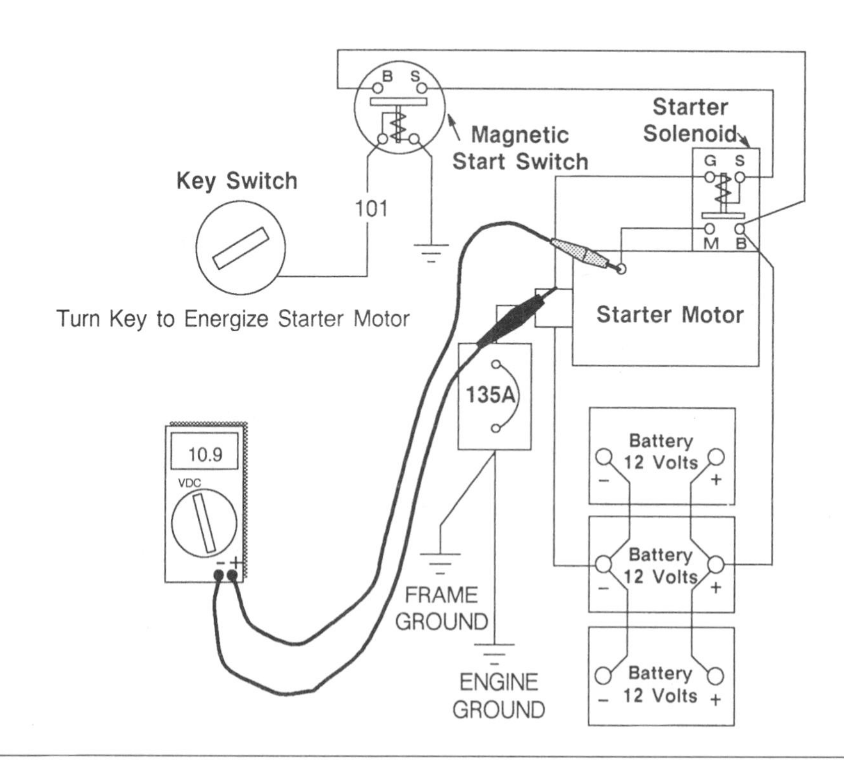

Figure 6-7 Starter voltage test. (Courtesy of Mack Trucks, Inc.)

Compare the two readings. A good circuit has the same amount of voltage (within 0.8 volts) at the starter as was measured at the battery. If that was the case and the starter motor does not crank at all or cranks too slowly, the most probable cause is a high internal resistance within the starter motor. If less voltage was measured at the starter motor than at the battery, suspect a voltage loss problem somewhere in the starter cranking circuit. The voltage loss indicates a resistance problem.

Voltage drop tests performed at cables, terminals, and the starter solenoid can pinpoint the voltage loss.

A **starter solenoid** is a relay device used to control current to the starter motor and cause the armature to turn. It is also used to engage the starter drive into the flywheel.

SERVICE TIP: Verify that a slow-crank condition is not due to the fault of the engine before pinpointing the starter.

SERVICE TIP: The voltage difference between the starter and batteries should be greater than 0.8 volts when there is a suspected voltage loss problem somewhere in the starter cranking circuit.

Battery Cable Tests

Battery cables are the most logical place to look for resistance problems. Chemical resistance, especially between lead terminals that are exposed to vapors from the battery and other environmental contaminants, make battery connections a likely candidate for resistance. Other problems can occur due to improper fastening of cable that allows too much movement or no flexibility, which is just as bad due to the stress that can be exerted at terminals.

The following steps are used to perform a **battery cable voltage drop test** on an engine that is still disabled to prevent it from starting:

1. Set the multimeter to the Volts DC function.
2. Connect the positive (+) meter lead to the positive battery post. Make this connection on the post, not on the cable clamp or terminal end (Figure 6-8).

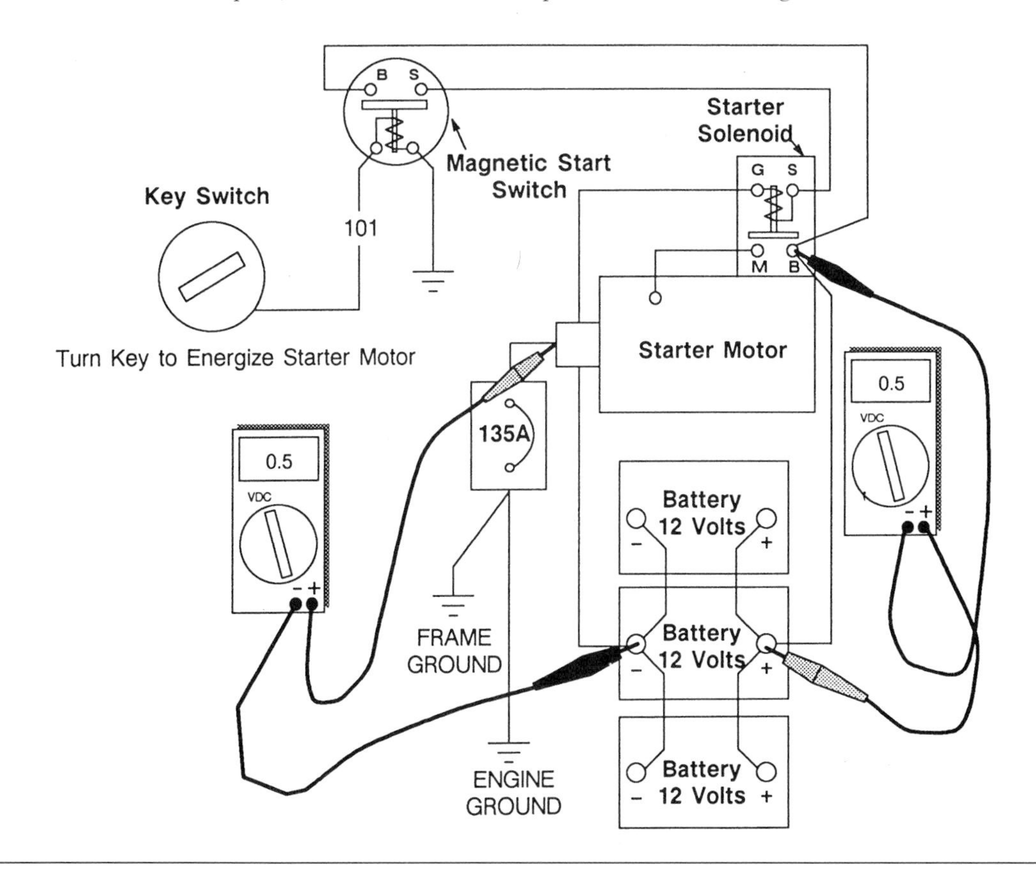

Figure 6-8 Battery cable tests. (Courtesy of Mack Trucks, Inc.)

3. Connect the negative (−) meter lead to the starter solenoid "B" terminal.
4. With the engine disabled not to start, energize the starter motor.
5. Observe and note the reading on the meter, then turn the switch off.
6. Next, move the meter negative (−) lead to the negative battery post and the meter positive (+) lead to the starter motor ground connection.
7. Energize the starter motor again and observe and note the voltage reading indicated on the meter, then turn the key off.

On trucks, the voltage loss should not exceed 0.5 volts through the positive battery cable and the negative battery cable. If an excessive voltage loss is indicated through either or both cables, locate and repair the cause. Again, this could be a loose or corroded connection anywhere between the points tested.

Classroom Manual
Chapter 6, page 123.

Classroom Manual
Chapter 6, page 124.

A **solenoid circuit test** and **magnetic start switch test** determine the condition of the control circuit.

Special Tools

Hand tools
Jumper wire
Voltmeter

Starter Solenoid and Magnetic Start Switch Voltage Drop Test

This test is used on the control circuit of the starter system. *(It is important to note that in this test the engine is disabled to prevent it from starting.)* The following steps are used to check for voltage losses in the control circuit:

1. Set the multimeter to the Volts DC function.
2. Connect the positive (+) lead of the meter to the starter solenoid "B" terminal and the negative (−) lead of the meter to the starter solenoid "M" terminal (Figure 6-9).
3. With the engine disabled not to start, turn the key to the start position and energize the starter.
4. Observe and note the reading on the meter, then turn the key off.
5. If a magnetic key switch is used, move the meter leads to the magnetic start switch "B" and "S" terminals as shown in the figure.
6. Again, with the engine disabled not to start, turn the key to the start position and energize the starter.
7. Observe and note the reading on the meter, then turn the key off.

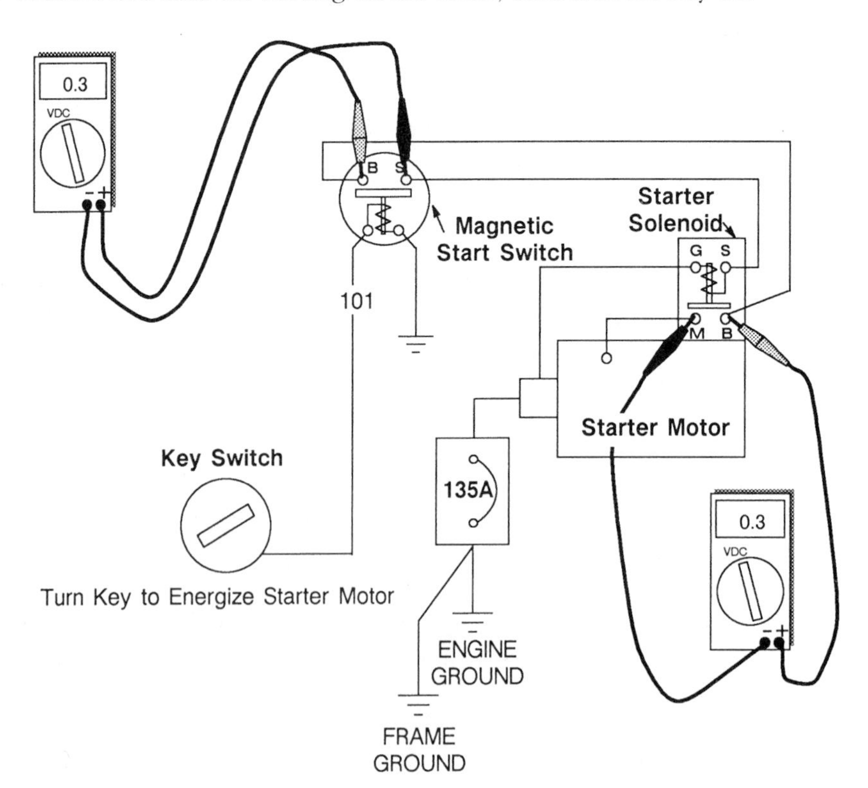

Figure 6-9 Starter solenoid and magnetic start switch voltage drop test. (Courtesy of Mack Trucks, Inc.)

Voltage drop through the solenoid or the magnetic switch should be 0.3 volts or less. A voltage drop that is greater than 0.3 volts indicates a high resistance inside the component. Study the figure and think about why the leads of the meter are placed where they are in regard to polarity. Refer also to Figure 6-8 where the batteries are also shown. Now try to determine which connection winds up being the most positive to the battery. Is not the "B" terminal of the solenoid the most positive point to the battery when the meter positive lead is hooked to it? What about when the meter is hooked to the magnetic switch? Now the "B" terminal of the magnetic switch is the most positive point to the battery. Where would you connect the meter's positive lead if you wanted to perform a voltage drop between terminal "S" of the magnetic switch and the terminal "S" of the starter solenoid? With the magnetic switch energized by the key first, the "S" terminal of the magnetic switch becomes the most positive, therefore, the meter's positive lead would be hooked to the magnetic switch "S" terminal and the negative lead would be hooked to the starter solenoid's "S" terminal. Crank the motor and observe and note the meter's reading for any excessive voltage drop. As you can see, the voltmeter can be a very powerful tool to pinpoint resistance within any portion of a circuit. This same concept also applies to any portion of the ground sides of the circuit.

Ground problems can be attributed to many starting system symptoms.

SERVICE TIP: Ground sides develop just as many problems as the insulated sides of the circuit. Corrosion easily develops where a ground connection is made to a frame. Remove the connection and sand or wire brush the frame or engine area where the terminal makes connection to ensure proper connection.

Magnetic Start Switch/Key Switch Test for a No-Crank

Classroom Manual
Chapter 6, page 125.

This test is also very useful if the starter does not energize when the key is turned to the start position. A voltmeter is used to test for certain voltage at the magnetic start switch. For this test, disconnect the wire from the starter solenoid "S" terminal before performing the following:

1. Set the multimeter to the Volts DC function.
2. Connect the meter leads across the magnetic start switch coil windings. These are the two small terminals on the switch (Figure 6-10).
   - ❑ Connect the positive (+) meter lead to the magnetic start switch key switch connection.
   - ❑ Connect the negative (−) meter lead to the magnetic start switch ground connection.
3. Turn the key to the start position.
4. Observe and note the voltage reading, then turn the key off.

A voltage reading of 0 volts indicates an open circuit between the key switch and the magnetic start switch. Check for anything that could have prevented total current flow such as broken or damaged wires, disconnected or broken connectors, or a faulty key switch. A voltage reading that is less than 11 volts indicates a high resistance in the starter control circuit. For resistances, check for loose or corroded connections and damaged wires. If after repairs the voltage is still less than 11 volts, replace the magnetic start switch.

Most starting system tests require the engine to be disabled so it won't start. Follow manufacturer's procedures. Either the ignition is disabled or the fuel solenoids get disabled.

A **fuel solenoid** is an electric solenoid used to shut off fuel to the diesel engine so the engine is prevented from running.

SERVICE TIP: A magnetic start switch should have an audible click coming from it if the key is turned on. If no click is heard, the switch is most likely defective if voltage was found to be available to the coil terminal. To quickly verify solenoid integrity, disconnect the two small terminal wires and place an ohmmeter across the two terminals. Then measure the resistance through the coil wires. There should be a very small resistance through the coil. If the meter indicates a very high resistance or infinite resistance the magnetic start switch is defective.

Special Tools

Ohmmeter
Hand tools

All these tests are used to pinpoint problems within a starting system. Typically one or two of these tests will quickly lead you to the cause of the problem without having to go through all of them. However, you will become more comfortable with the procedures and more familiar

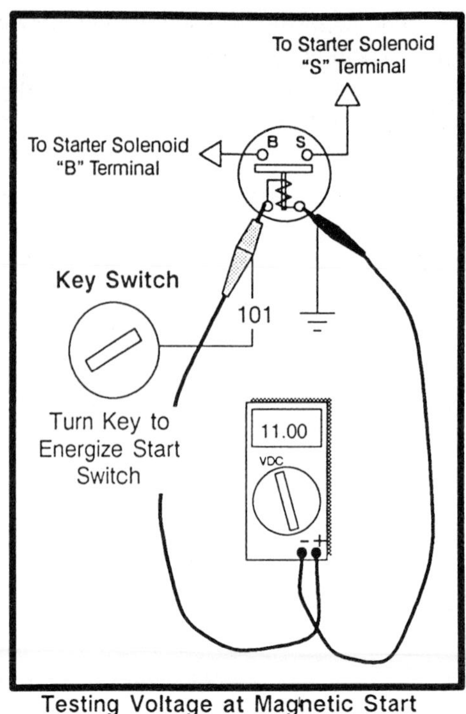

Figure 6-10 Testing voltage at magnetic start switch. (Courtesy of Mack Trucks, Inc.)

with relating certain readings to different problems by performing all these tests periodically, whether or not they are all needed.

Starter Motor Removal

Special Tools

Basic hand tool set

Jack stands

Battery terminal pliers

Battery terminal puller

Occasionally the on-vehicle tests indicate that the starter motor has to be removed. The following steps and procedures should be followed and observed when removing the starter motor:

1. Disconnect the battery from the system. If left connected, you would be working in close quarters with high battery current. It is very easy to make contact with the positive cable at the starter to the frame or engine ground via the wrenches or sockets and ratchet.

CAUTION: Remove the ground battery cable. Wrap the cable end with tape or any other insulating material to prevent accidental contact with the battery terminal. Such contact can result in personal injury.

2. If the vehicle needs to be lifted off the ground to gain proper height, use proper equipment. If a hydraulic jack is used for lifting, make sure the jack stands are also placed under the vehicle.
3. Disconnect the wires leading to the solenoid. Keep the wires together for easy identification later when the starter is reinstalled. For example, there might be two, three, or four wires connected to the "B" terminal of the starter solenoid. For easy identification, wire wrap, zip tie, or tape those wires together so there will be less confusion. The same can apply to the battery cables at the starter. Some heavy duty starter systems have more than one heavy battery cable going to the "B" terminal of the starter solenoid and also to the ground terminal of the starter because of the use of multiple batteries.

SERVICE TIP: Sometimes the nut will not come off the heavy "B" terminal of the starter solenoid and the stud assembly wants to turn instead. If you are able to squeeze a narrow wrench on the nut between the cable and the solenoid housing try tightening it, which might keep the stud from turning.

CAUTION: Do not use a torch to heat the outer solenoid nut. You can easily destroy the cable and terminal this way. Also, a fire can occur because of the close proximity of the flame to any leaking oil or other materials that can easily burn. In the long run, you will be better off cutting the cable(s) or disconnecting it from the other end and bringing it out with the starter.

4. Make sure nothing will hinder you from removing the starter. Truck starters are quite heavy so get assistance if you need it. Also, watch for any spacer shims or plates and support brackets that will need to be reinstalled in the proper position.

Installing Starter

The procedure is reversed when installing the starter. Make sure all connections are tight and clean. If a new or rebuilt starter from an outside source is used for replacement, make sure that the starter solenoid is positioned so that it is free of obstructions. The starter nose opening also has to be positioned correctly for proper engagement of the pinion gear of the starter to the ring gear of the flywheel. This might require repositioning the nose piece but make sure the scallops will provide a wrench clearance for the mounting bolt (Figure 6-11A). However, in cases where the scallops do not provide sufficient clearance for a wrench (Figure 6-11B), a 12-point head cap screw is used (Figure 6-12).

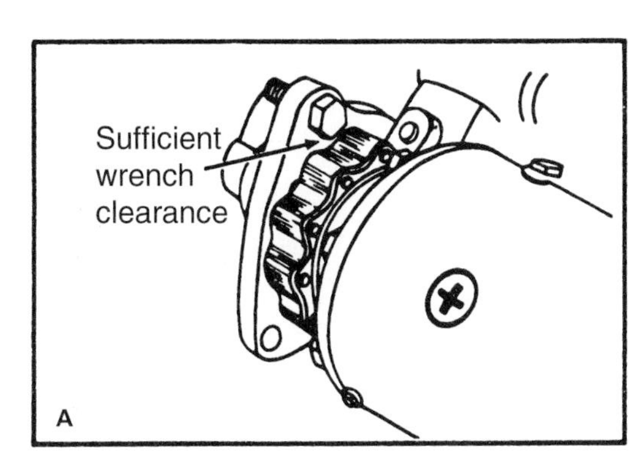

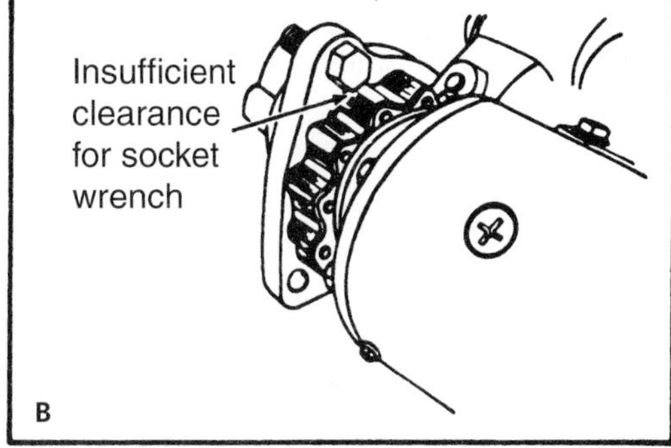

Figure 6-11 (A) Sufficient wrench clearance; (B) Insufficient wrench clearance. (Courtesy of AC Delco)

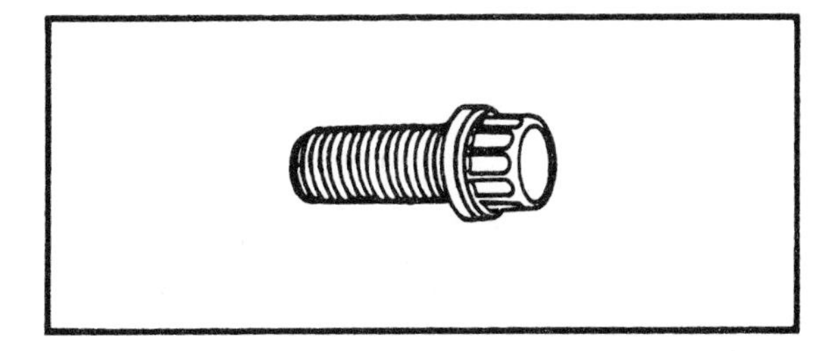

Figure 6-12 A twelve-point head cap screw is used as a starter mounting bolt. (Courtesy of AC Delco)

SERVICE TIP: Before installing a heavy duty truck starter, clean all the threads of the mounting bolts and apply never-seize or anti-seize to them. Also, run a thread cleaner through the mounting screw holes in the bell housing. This will make your job much easier, because you are trying to hold the starter up, align the holes, and finger turn the mounting bolt in as far as possible so it can help hold up the starter motor.

Starter Motor No-load Test

Special Tools

Special starter tachometer

Jumper cables

Starting/charging system tester

Remote starter switch

Soft jaw vise

A no-load, or free spinning test, is performed on a starter that is out of the vehicle. This determines the free rotational speed of the armature.

An **armature** is an arrangement of looped coils wound on an iron core that is free to rotate within a magnetic field.

This test is not an in vehicle test. This test is performed once the starter is removed from the vehicle. A **no-load test** is used to identify possible specific defects within the starter that can be verified with further testing once the starter is disassembled. The no-load test is also useful in identifying open or shorted fields, which can be difficult to check when the starter is disassembled. If a shop is equipped with the proper equipment to perform this test, it can be useful to verify the normal operation of a starter before it is installed in the vehicle.

To perform this test you will need a starter/charging system tester and a tachometer as shown in Figure 6-13. Perform the test as follows:

1. Place the starter motor in a secure vise.
2. Attach a revolutions per minute (RPM) indicator to the armature shaft, using a rubber bushing or cup.
3. Connect a remote starter switch between the "BAT" and "S" terminals of the starter solenoid.
4. Connect the jumper cables as shown in Figure 6-13 using a VAT-40 tester or as shown in Figure 6-14 using a tester setup without an inductive pickup.
5. With the VAT-40, select Int 18V.
6. Zero the ammeter.
7. With the VAT-40, place the selector to the starting position.
8. Load the battery to a desired voltage. (Note: it is not necessary to obtain the exact voltage specified, because an accurate interpretation can be made by recognizing that if the voltage is slightly higher the RPM will be proportionately higher with the current remaining essentially unchanged.)

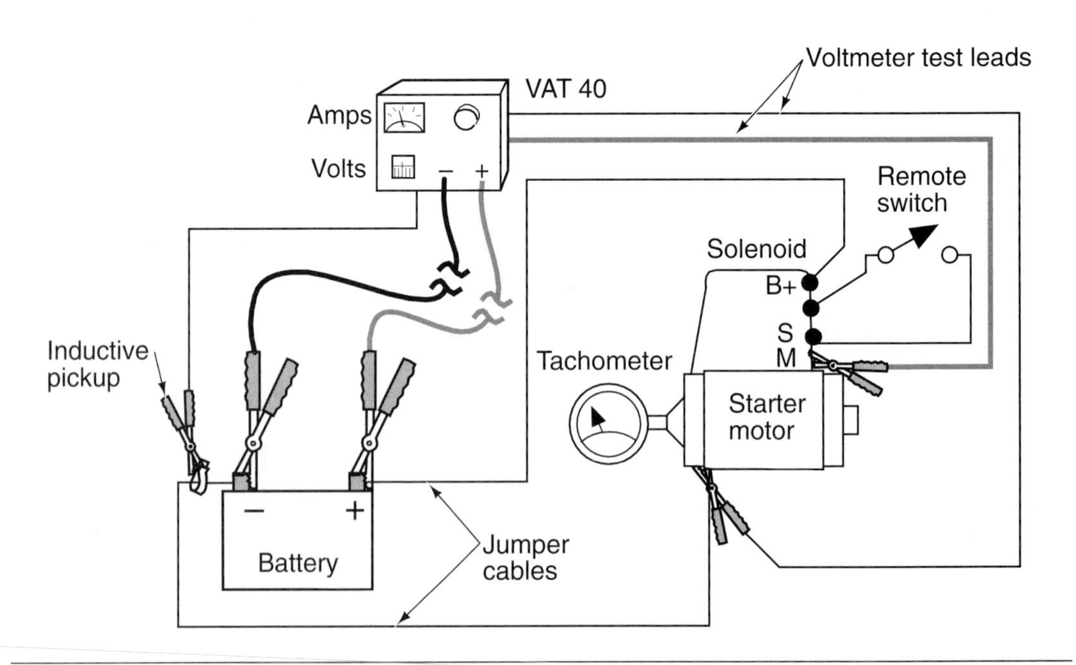

Figure 6-13 Starting/charging system tester (VAT-40) connections for a free speed test.

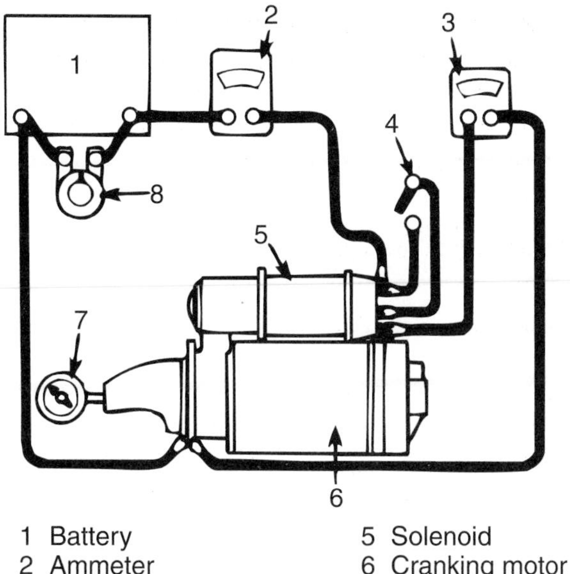

Figure 6-14 No-load test connections. (Courtesy of Delco-Remy Co.)

9. If using the VAT-40, switch to the EXT 18V position.
10. Close the remote starter switch while reading the ammeter, voltmeter, and tachometer.

Compare the test results to manufacturer's specifications. For example, the following are specifications for a typical 40MT DELCO-Remy starter.

No-load Test

VOLTS	Minimum Amps	Maximum Amps	Minimum RPM	Maximum RPM
9	140	190	4,000	7,000

Interpret the test results as follows:

1. If the rated draws and no-load speeds are as specified, the starter is functioning properly. Reminder: This is why the test is also useful to perform before installing a newly rebuilt starter.
2. Low free speed and high current draw indicate:
   - **A.** Too much friction—tight, dirty, or worn bearing; bent armature shaft; or loose poles allowing armature to drag
   - **B.** Shorted armature—This can be further checked with a **growler** after disassembly.
   - **C.** Grounded armature or fields—This can also be checked further after disassembly.

SERVICE TIP: Prior to performing a no-load test, try turning the armature on the pinion end by prying with a screwdriver. If the armature is tight, preventing it from turning freely, disassemble the starter instead of doing a no-load test. Look for tight bearings, bent armature shaft, or a loose pole shoe screw that is preventing the armature from turning freely.

3. Failure to operate with high current draw indicates:
   A. A direct ground in the terminal or fields.
   B. Frozen bearings (this should have been determined by turning the armature by hand).
4. Failure to operate with no current draw indicates:
   A. Open field circuit. This can be checked after disassembly by inspecting internal connections and tracing the circuit with a test lamp.
   B. Open armature coil. Inspect the commutator for badly burned bars.
   C. Broken brush springs, worn bushings, high insulation between the commutator bars, or anything that would prevent good contact between the brushes and the commutator.
5. Low no-load speed and low current draw indicate high internal resistance due to poor connections, defective leads, dirty commutator, and other causes listed under Number 4.
6. High free speed and high current draw indicate shorted fields. If shorted fields are suspected, replace the field coil assembly and check for improved performance.

As can be seen, most test results require the starter to be disassembled for further checking and testing.

Starter Motor Disassembly

In today's truck shops, if a starter is found to be defective, a replacement starter is purchased and is immediately installed. However, it is feasible for a shop to replace the pinion drive or perform a quick internal repair such as of broken connections if another starter is not immediately available. Also, you might possibly choose a career as a unit rebuilder specializing in starter and charging system troubleshooting and repairs. We seem to have nurtured an industry of parts changers. It is important to grasp all aspects of components and systems in order to troubleshoot and diagnose. Perhaps that is why consumers have so little faith in us.

Prior to disassembling a starter to check or rebuild it, a technician should study the manufacturer's service manual for proper procedures. In the next section, we will look at a smaller starter such as is used in medium vehicles and a larger starter such as is used in heavy duty trucks. Sometimes they will not be distinguished from each other because certain tests and procedures apply to all starters. Photo Sequence 5 shows a DELCO Remy starter that is found in light and medium duty trucks. A disassembled view of this starter is shown in Figure 6-15.

A cross section of a heavy duty 40MT type 400 starter is shown in Figure 6-16. Using this heavy duty starter as an example, we will use the following precautions and steps for disassembly:

CAUTION: To prevent personal injury, safety glasses must be worn for starter service work.

1. Note and scribe the relative position of the solenoid, lever housing, and nose housing so the motor can be reassembled in the same manner.
2. Disconnect the field coil connector from the solenoid motor terminal, and the lead from the solenoid ground terminal.
3. On motors that have brush inspection plates, remove them and then remove the brush lead screws. This will disconnect the field leads from the brush holders.
4. Remove the attaching bolts and separate the commutator end frame from the field frame.
5. Separate the nose housing and field frame from the lever housing by removing attaching bolts.
6. Remove the armature and clutch assembly from level housing.
7. Separate the solenoid from the lever housing by pulling them apart.

Photo Sequence 5
Typical Procedure for DELCO Remy Starter Disassembly

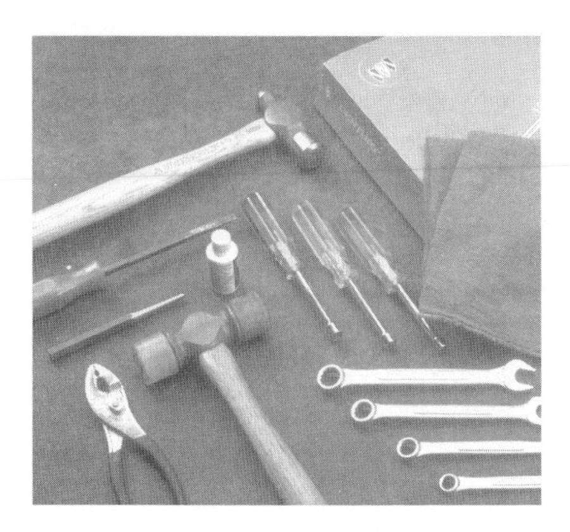

P5-1 Provide a clean and organized work area. Have the following tools on hand; rags, assorted wrenches, snap ring pliers, flat blade screwdriver, ball-peen hammer, plastic head hammer, punch, scribe, safety glasses, and arbor press.

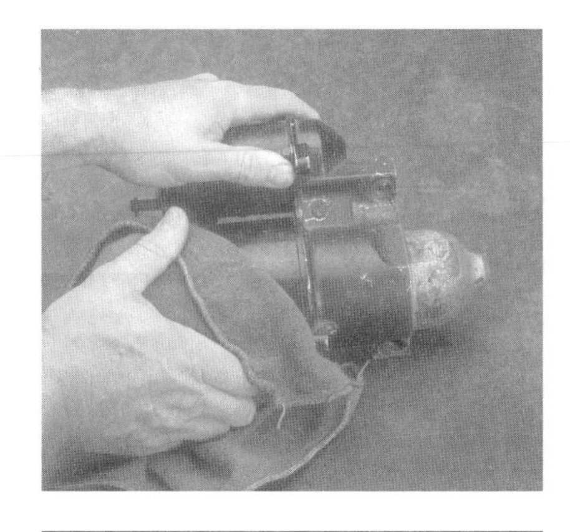

P5-2 Clean the case.

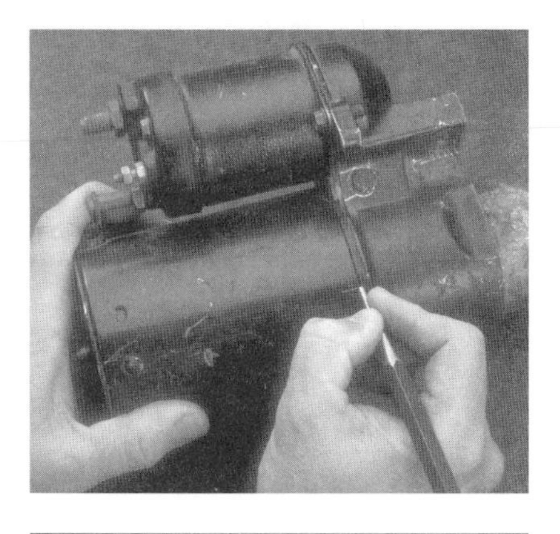

P5-3 Scribe reference marks at each end of the starter end housings and the frame.

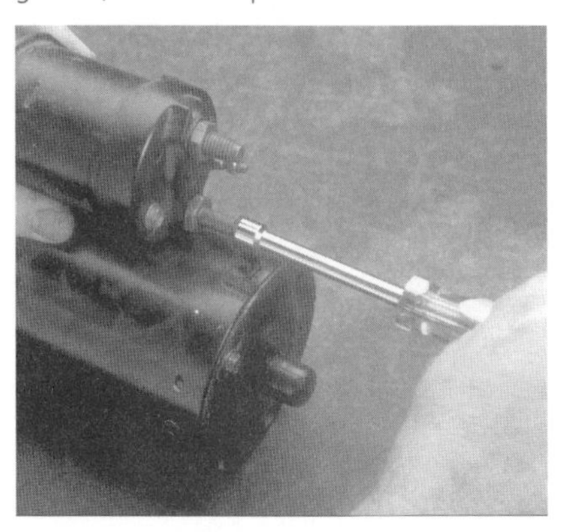

P5-4 Disconnect the field coil connection at the solenoid M terminal.

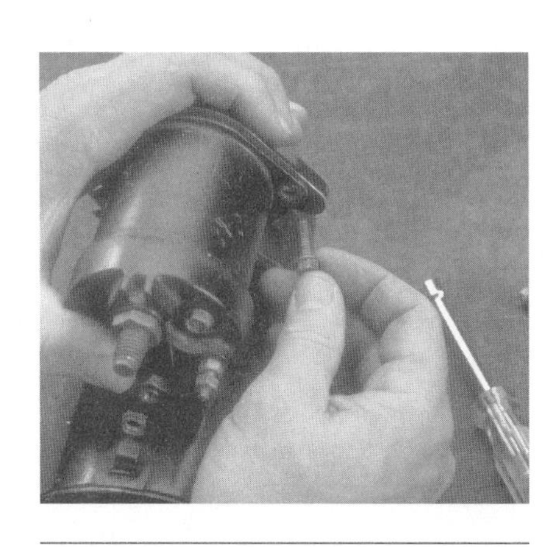

P5-5 Remove the two screws that attach the solenoid to the starter drive housing.

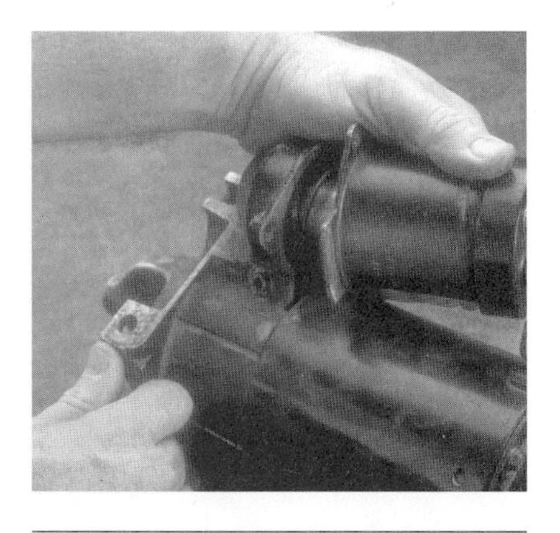

P5-6 Rotate the solenoid until the locking flange of the solenoid is free. Then remove the solenoid.

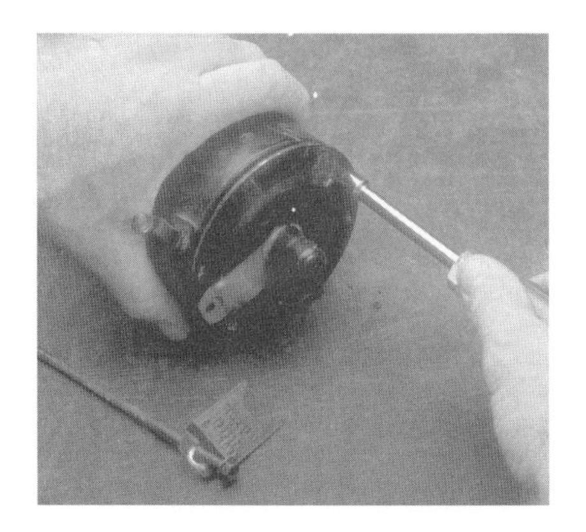

P5-7 Remove the through bolts from the end frame.

P5-8 Remove the end frame.

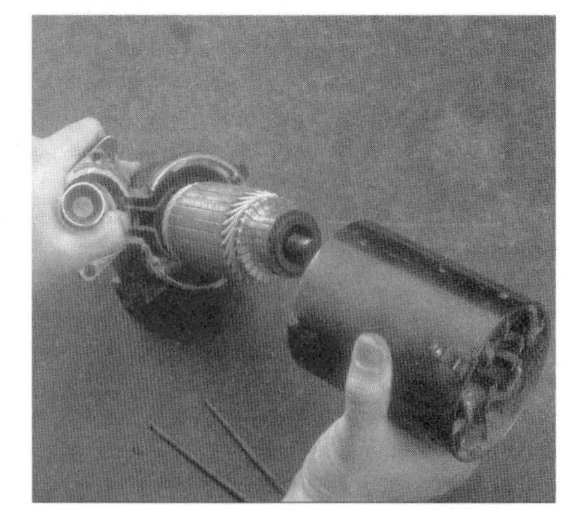

P5-9 Remove the frame.

Photo Sequence 5
Typical Procedure for DELCO Remy Starter Disassembly

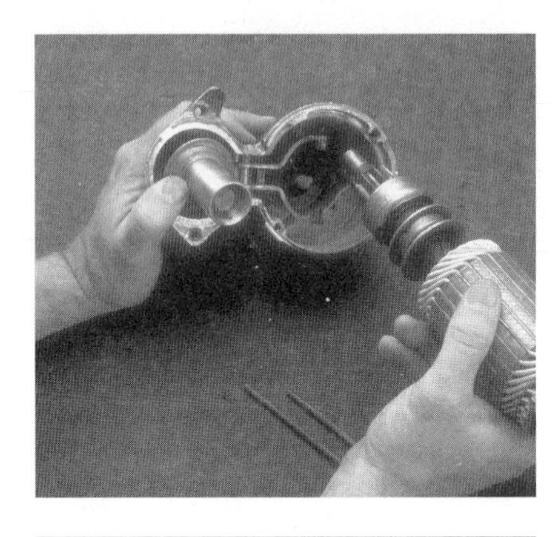

P5-10 Remove the armature from the drive housing. Note: On some units it may be necessary to remove the shift lever from the drive housing before removing the armature.

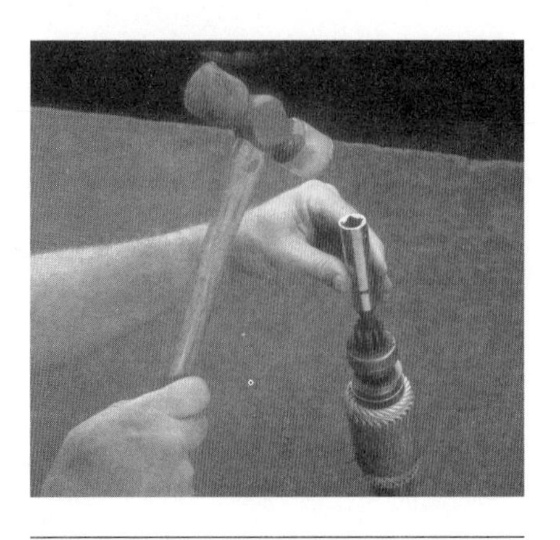

P5-11 Place a 5/8" deep socket over the armature shaft until it contacts the retaining ring of the starter drive.

P5-12 Tap end of socket with a plastic hammer to drive the retainer toward the armature. Move it only far enough to access the snap ring.

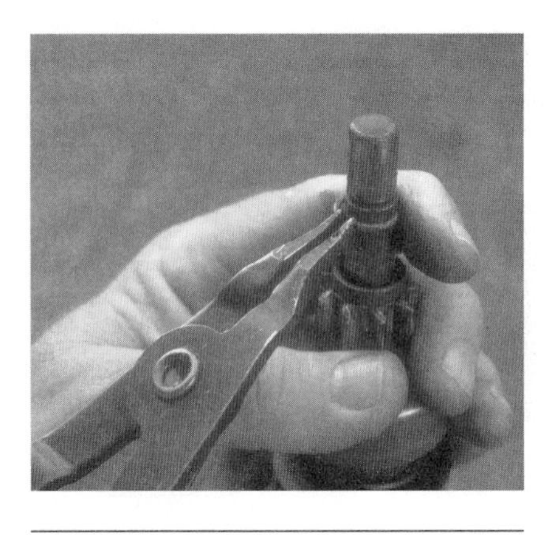

P5-13 Remove the snap ring.

P5-14 Remove the retainer from the shaft and remove the clutch and spring from the shaft. Press out the drive housing bushing and the end frame bushing.

The starter motor can now be cleaned and inspected.

WARNING: The drive, armature, and field coils should not be cleaned using solvents, gasoline, or a degreasing tank with grease-dissolving solvents. All of these can dissolve the lubricant in the drive and damage the insulation in the armature and field coils. All parts except the drive should be cleaned with mineral spirits and a brush and then wiped with clean rags.

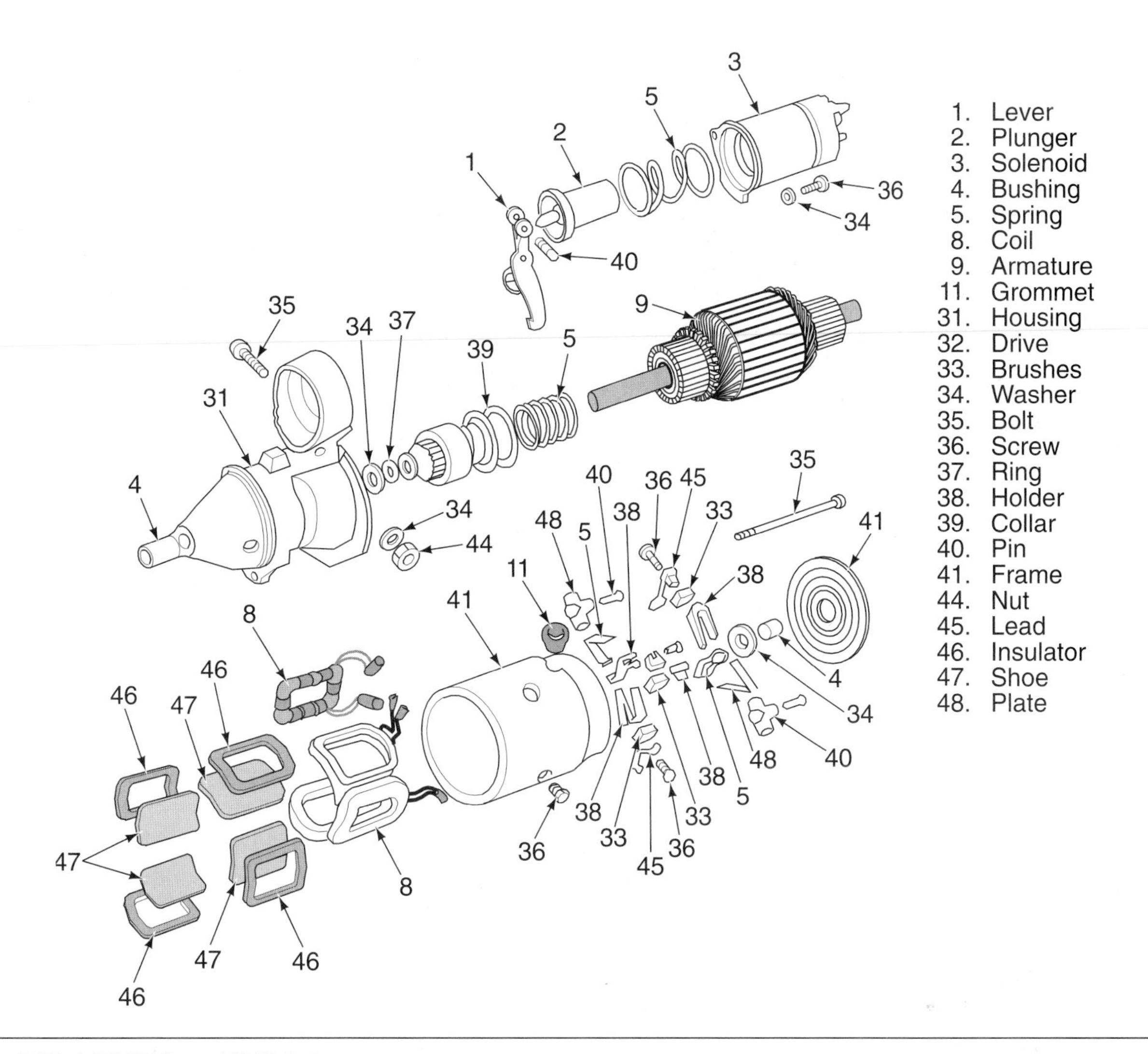

Figure 6-15 A DELCO-Remy 10MT starter.

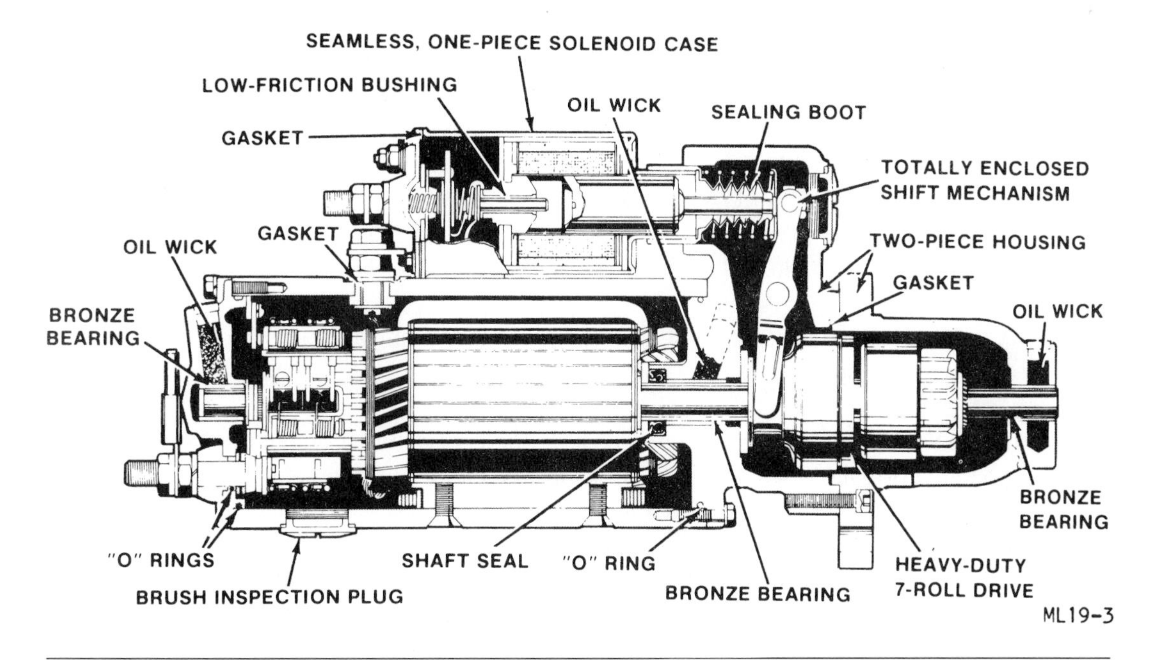

Figure 6-16 Cross-section of a 40MT type 400 starter. (Courtesy of Mack Trucks, Inc.)

The most common procedure with starters is to clean and undercut the commutator.

SERVICE TIP: Never use emery cloth to clean the commutator. If the commutator is dirty it may be cleaned with a very fine sandpaper.

Inspection

Inspect all housing components such as the end frame and drive housing for cracks. Check the frame assembly for loose pole shoes and broken or frayed wires. Inspect the drive gear for worn teeth and proper overrunning clutch operation. Inspect the brushes for wear, comparing them to new brushes and replacing them if they are excessively worn. Make sure the brush holders are clean and not damaged, and that the brushes are free to move in the holders. They need to make full contact with the commutator. Check the brush springs to make sure they can exert enough force to allow full surface contact. Replace any weak, distorted, or discolored springs. Check end housing and nose cone bearing for wear or other damage and replace them if there is any doubt about their condition.

Starter Motor Component Test

Next perform pinpoint tests of certain components to isolate the reason for the starter failure and also to verify if a component is functioning properly so it doesn't have to be needlessly replaced.

Field Coil Testing

Field coils are electromagnets used to produce the magnetic field of a generator or starter motor.

Classroom Manual
Chapter 6, page 118.

Special Tools

Safety glasses

110-volt test light

Ohmmeter

The field coils and frame assembly should be tested for opens and shorts to ground. In order to perform this test accurately you need to know how they are wired. By looking at Figure 6-17, you can tell where the coils get their power and where they ground. Always refer to a service manual or specific wiring diagram when testing field coils.

One method of checking for grounds and opens is to use a 110-volt test lamp as follows:

1. Grounds: If the motor has one or more coils normally connected to ground, the ground connections must be disconnected during this check. Connect one lead of the 110-volt test lamp to the field frame and the other lead to the field connector. If the lamp lights, at least one field coil is grounded and must be repaired or replaced. This check cannot be made if the ground connection cannot be disconnected.
2. Opens: Connect test lamp leads to the ends of the field coils. If the lamp does not light, the field coils are open.

A concept picture of a 110-volt test lamp is shown in Figure 6-18. As you can see, it is a test light using household AC current as its power. Remember that the starter uses alternating current to provide alternating magnetic fields to produce rotation. This same test light is often used to check stator windings in alternators and also for grounds and continuity between phases.

WARNING: Careless use of this type of test lamp can cause electrical shock and possible injury.

Another method of checking for grounds and opens is using an ohmmeter. With the ohmmeter on the lowest scale, place one lead on the starter motor input terminal and the other lead on the insulated brushes. The ohmmeter should indicate zero resistance. If there is resistance in the field coil, replace the coil and/or frame assembly.

The field coils on many heavier starters can be removed from the field frame by using a pole shoe screwdriver. A pole shoe spreader should also be used to prevent distortion of the field frame. Careful installation of the field coils is necessary to prevent shorting or grounding of the field coils as the pole shoes are tightened into place.

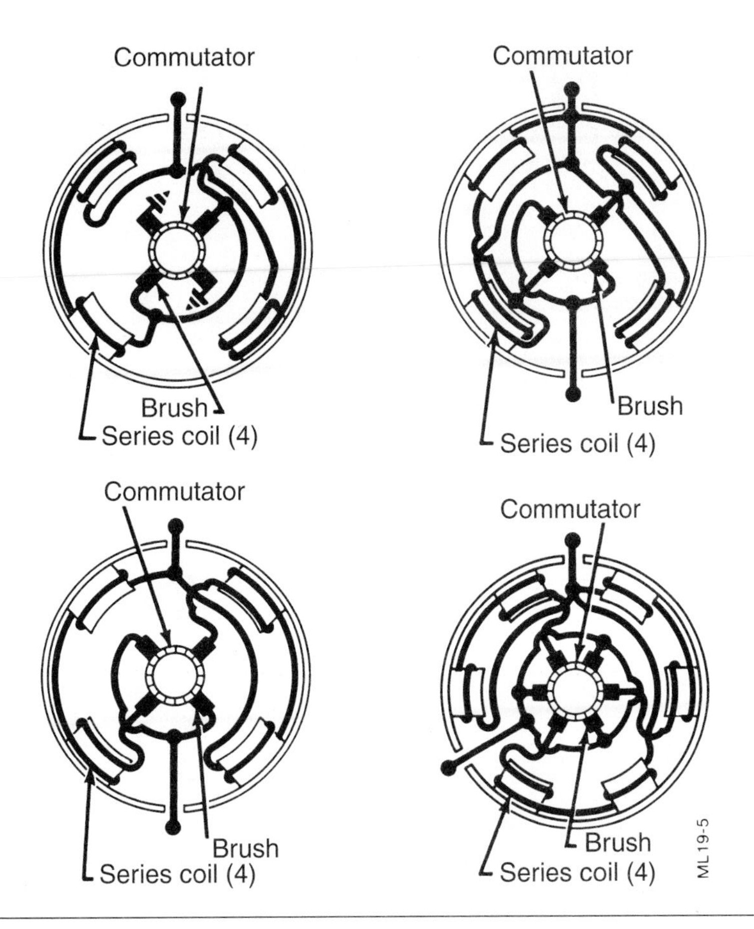

Figure 6-17 Internal motor circuits. (Courtesy of Mack Trucks, Inc.)

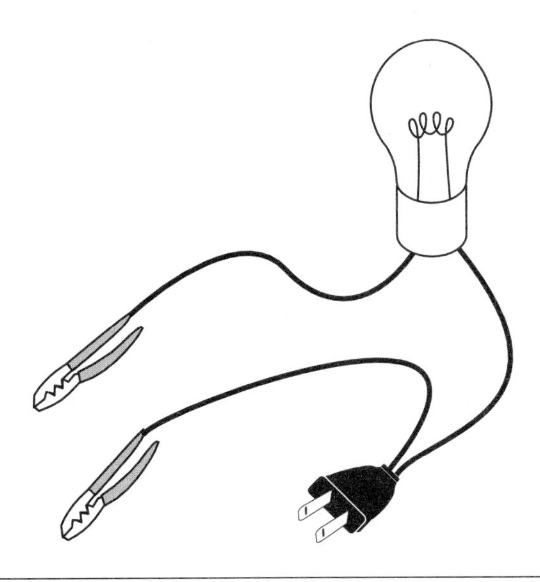

Figure 6-18 A 110-volt test lamp.

Armature Testing

Armatures are checked for opens, short circuits, and grounds.

1. Opens: Opens are usually caused by excessively long cranking periods. The most likely place for an open to occur is at the commutator riser bars. Inspect the points where the conductors are joined to the commutator bars for loose connections.

Classroom Manual
Chapter 6, page 118.

Figure 6-19 A growler is used to test an armature for shorts.

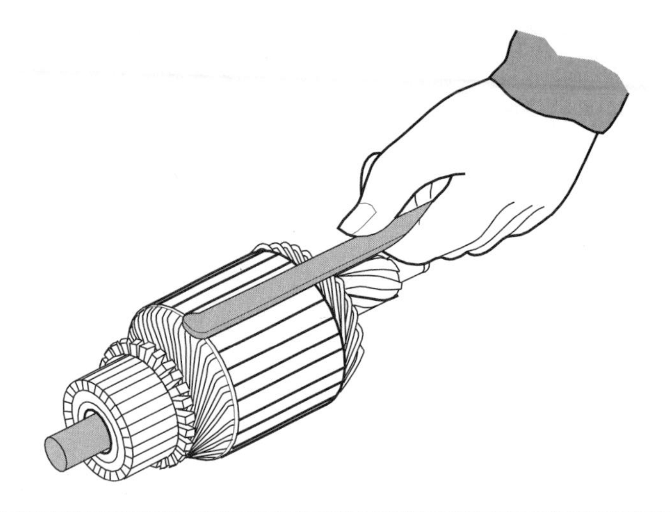

Figure 6-20 The growler generates a magnetic field. If there is a short, the hacksaw blade will vibrate over the area of the short.

Special Tools

Safety glasses

Growler

Hacksaw blade

110-volt test light

Ohmmeter

The **commutator** is the part of the armature assembly used to connect the brushes to the rotating armature windings. It also converts AC voltage to DC voltage in DC generators.

2. Short Circuits: Short circuits in the armature are located by using a growler (Figure 6-19). This growler produces a very strong magnetic field that is capable of inducing a current flow and magnetism in a conductor. When the armature is revolved in the growler with a steel strip such as a hacksaw blade held above it, the blade will vibrate above the area of the armature core in which the short circuit is located (Figure 6-20). Sometimes, these shorts are produced by copper or brush dust between the bars. Try cleaning out the slots and then retest the armature.
3. Grounds: A growler has a built-in continuity tester or a 110-volt test lamp can be used. With either tester, one point of it is placed on the commutator and the other lead is placed on the shaft. If the lamp lights, the armature is grounded. Most likely the cause will be breakdown of the insulation caused by overheating of the cranking motor, caused in turn by excessively long cranking periods.

Commutator Tests

If a growler is not available, the commutator can be tested for opens or grounds using an ohmmeter. To check for continuity, place the ohmmeter on the lowest scale. Connect the test leads to any two commutator sections (Figure 6-21). There should be zero ohms of resistance. If there is resistance, the armature has to be replaced. Place the ohmmeter on the 2K scale and connect one of the test leads to the armature shaft. Connect the other lead to the commutator

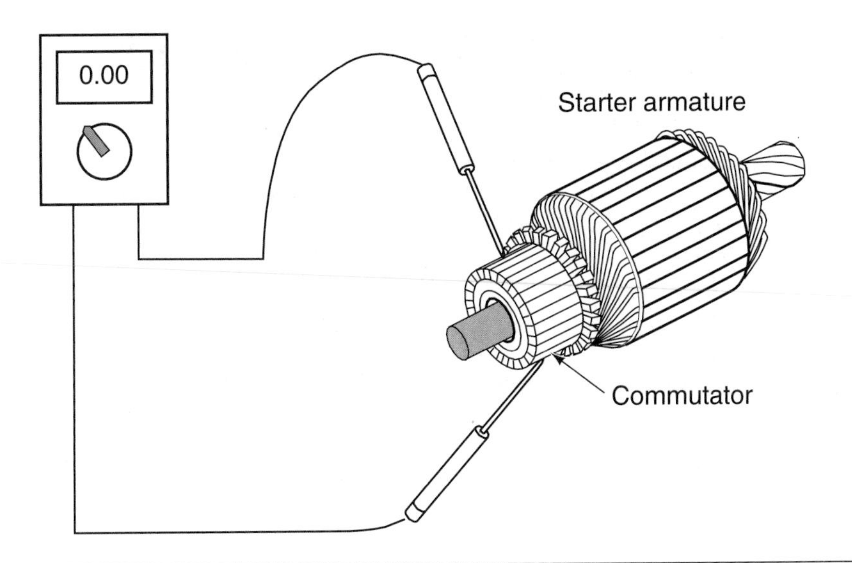

Figure 6-21 Testing the armature for opens. There should be zero resistance across the segments of the commutator.

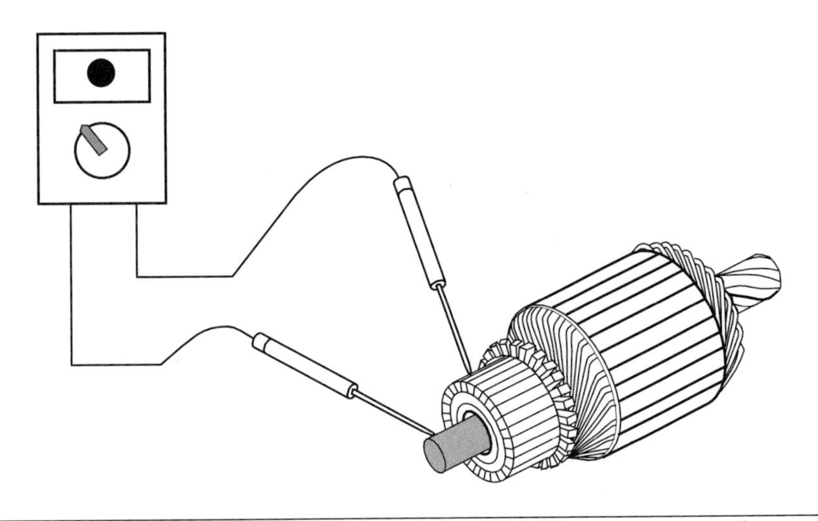

Figure 6-22 Testing the armature for shorts to ground. The meter should read infinite when placed on the shaft and the different segments of the commutator.

(Figure 6-22) and check each segment. There should be no continuity to ground. If there is, replace the armature.

Special Tools

Battery

Starter/charger tester or equivalent

Solenoid Checks

The solenoids of heavy duty trucks can also be checked for excessive resistance or a shorted hold-in winding. Although some solenoids differ in appearance, they can be checked electrically by connecting a battery of the specified voltage, a switch, an ammeter, and a voltmeter. A carbon pile is used to decrease voltage to a specified value. The hookup to check the hold-in winding is shown in Figure 6-23. A VAT-40 or similar starter charger tester with an inductive lead can also be used. With all leads disconnected from the solenoid, make the test connections as shown.

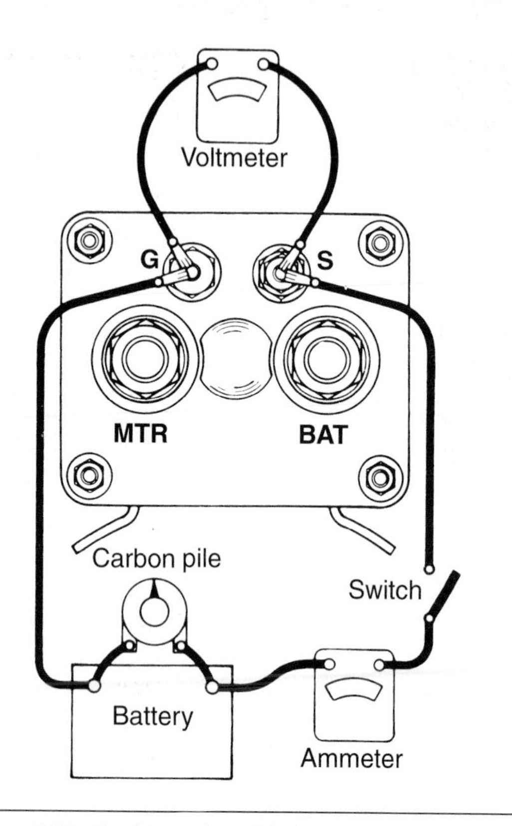

Figure 6-23 Checking solenoid hold-in winding. (Courtesy of Mack Trucks, Inc.)

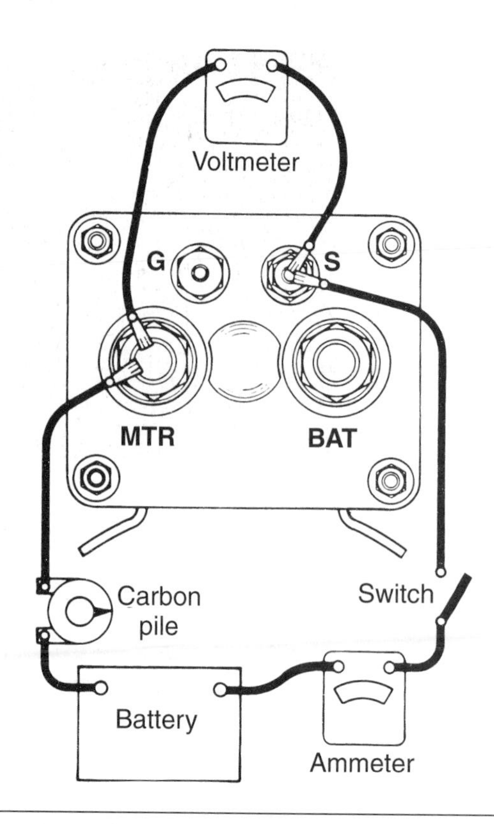

Figure 6-24 Checking solenoid pull-in winding. (Courtesy of Mack Trucks, Inc.)

Use the carbon pile to decrease the battery voltage to the specified value and compare the ammeter readings with the specifications:

- ❑ A high reading indicates a shorted hold-in winding.
- ❑ A low reading indicates excessive resistance.

Classroom Manual
Chapter 6, page 123.

To check the pull-in winding, connect from the solenoid switch terminal (S) to the motor (M) terminal (Figure 6-24). Close the switch and observe the readings. The results are interpreted the same as the hold-in windings. To check for grounds, move the battery lead from "G" as shown in Figure 6-23 or from "MTR" as shown in Figure 6-24 to the solenoid case (not shown). The ammeter should read zero. If it does not, the winding is grounded.

WARNING: Do not leave the pull-in winding energized for more than 15 seconds or damage can occur due to overheating. The current draw will decrease as the winding temperature increases.

In the above tests of the solenoid, if the carbon pile is not needed to achieve a certain voltage, a jumper wire can be used in place of it.

Reassembly

The most likely component to be replaced in a starter is the brushes. Manufacturers use two methods of connecting the brushes: they are either soldered to the coil leads or screwed to terminals. If they are soldered to the coil leads, cut the old leads (Figure 6-25). Place a piece of insulation or shrink tube over the brush connector. Crimp the new brush lead connector to the coil lead and solder them with rosin core solder. Slide the insulator over the connector and if using shrink tubing, use a heat gun to shrink the tube.

Figure 6-25 Removing brushes.

Special Tools

Feeler gauge set

100-watt soldering iron

Rosin core solder

Safety glasses

Heat shrink tubing

Two blocks of wood

Basic hand tool set

Jumper cables

Starting/charging system tester

Remote starter switch

Classroom Manual
Chapter 6, page 121.

To reassemble the starter motor, reverse the disassembly procedure. Prior to assembly, make sure all components are clean and proper lubrication is used wherever required. One of the first subcomponents that needs to be assembled before a complete reassembly of the starter motor is to install the pinion drive gear assembly to the armature shaft. The steps are as follows:

1. Stand the commutator end of the armature on a block of wood.
2. Install the drive gear assembly onto the shaft.
3. Install the snap ring onto the shaft as shown in Figure 6-26. Position the snap ring at the end of the shaft and use a block of wood and hammer to drive the snap ring onto the shaft.
4. Some starters use a snap ring retainer (Figure 6-27). This retainer is installed first with the cupped surface facing the end of the shaft. Force the ring on with the hammer and block of wood. Slide the ring down into the shaft groove. To force the retainer over the snap ring, place a suitable washer over the shaft and squeeze with pliers.

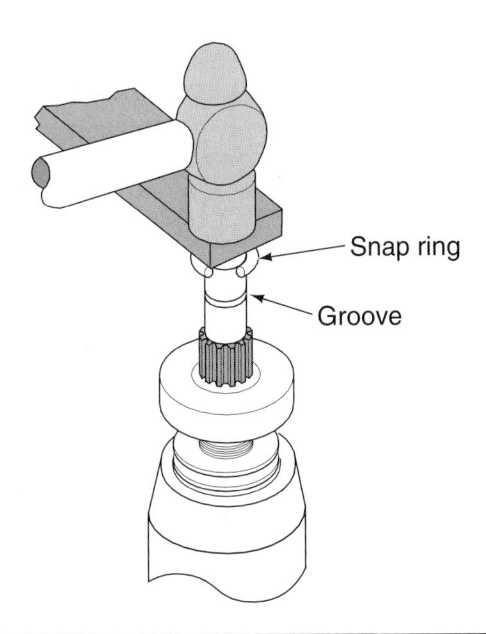

Figure 6-26 Once the snap ring is centered on the shaft, a hammer and block of wood can be used to install the ring onto the shaft.

Figure 6-27 Installing a snap ring retainer.

Pinion clearance is a specified clearance between the pinion gear and drive housing with the starter engaged to ensure proper engagement between the pinion drive gear and flywheel gear.

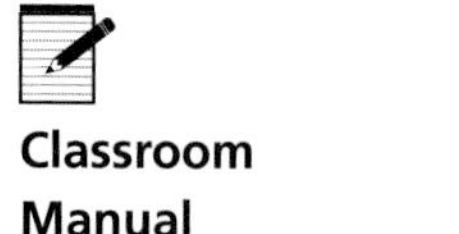

Classroom Manual
Chapter 6, page 122.

To reassemble the end frame with brushes onto the field frame, pull the armature out of the field frame just far enough to permit the brushes to be placed over the commutator. Then push the commutator end frame and the armature back against the field frame. Align all frame and housing components with the disassembly scribe marks.

Some smaller starters do not have provisions to adjust pinion clearance, but many heavy duty starters do. In either case, the pinion clearance should be checked for excessive clearance and adjusted if possible. Excessive clearance can be an indication of excessive wear of the solenoid linkage or shift lever.

To check for pinion clearance, perform the following:

1. Make connections as shown in Figure 6-28.
2. Using a jumper lead, momentarily flash it between terminal "G" to terminal "MTR." The drive will now shift into cranking position and remain so until the battery is disconnected.
3. Push the pinion or drive back towards the commutator end to eliminate slack movement.
4. Measure the distance between the drive housing or drive stop (Figure 6-29).
5. Adjust clearance by removing the plug and turning the adjusting nut.

Always refer to the manufacturer's specifications for specific starter models.

Smaller starters' pinion clearance is checked similarly; however; the "M" terminal to the starter motor's field coils is disconnected (Figure 6-30). A jumper cable from the positive terminal is connected to the "S" terminal of the solenoid. Connect the other jumper cable from the battery negative terminal to the starter motor frame. Connect a jumper wire from the "M" terminal and momentarily touch the other end of the jumper wire to the starter motor frame. This will shift the pinion gear into the cranking position and hold it there until the battery is disconnected. Push the pinion back towards the armature to remove any slack. With a feeler gauge, check the clearance (Figure 6-31). Compare the clearance with manufacturer's specifications. A typical clearance is 0.010 to 0.140 inches.

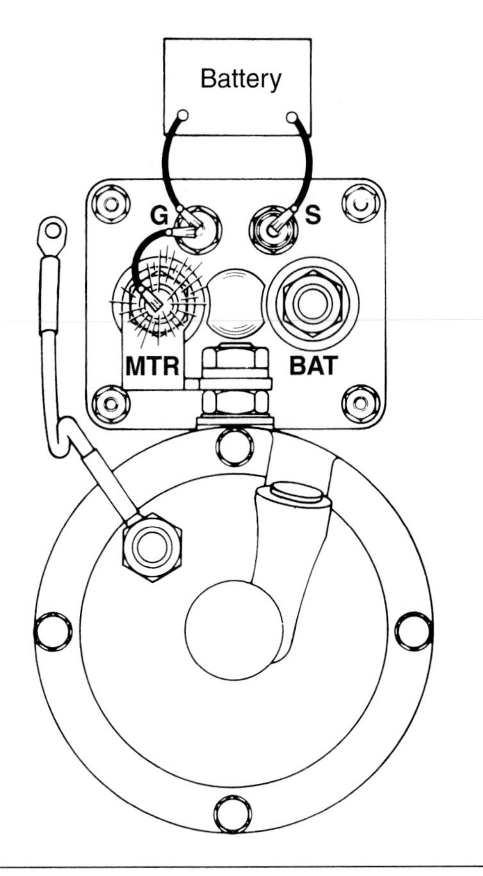

Figure 6-28 Checking pinion clearance circuit. (Courtesy of Mack Trucks, Inc.)

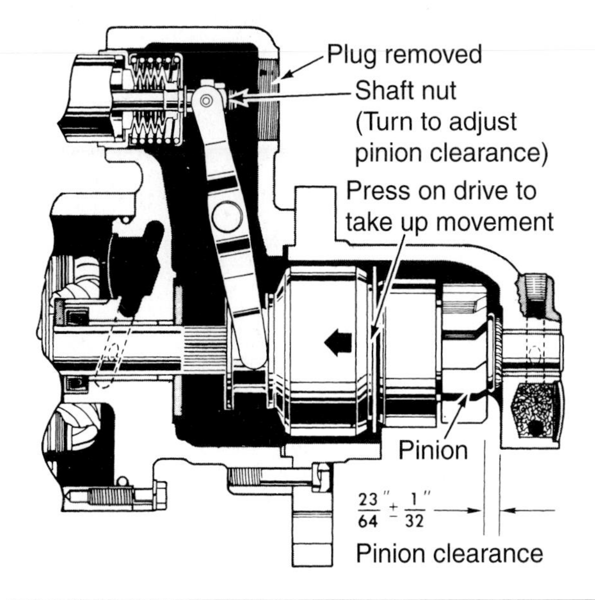

Figure 6-29 Measuring pinion clearance. (Courtesy of Mack Trucks, Inc.)

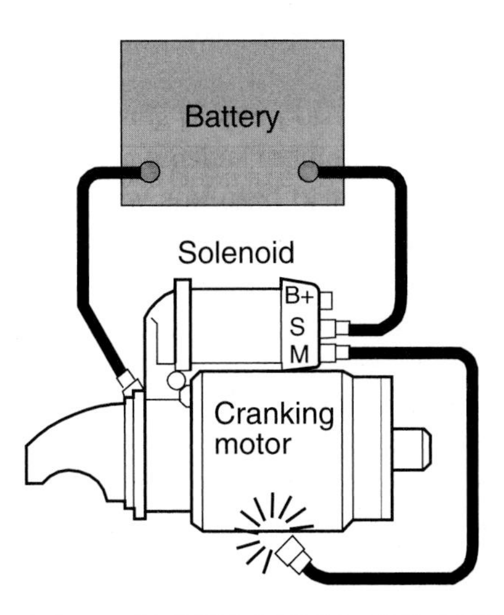

Figure 6-30 Jumper cable connections for checking pinion gear clearance.

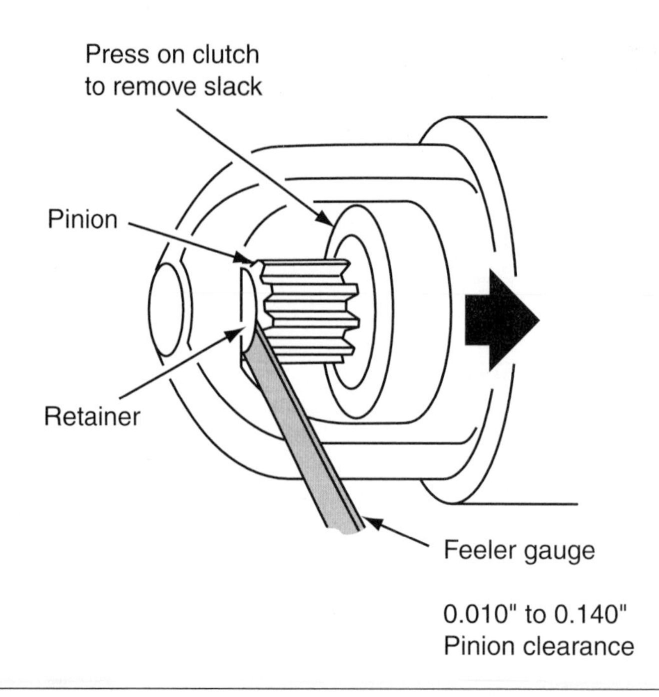

Figure 6-31 Checking the pinion gear to drive housing clearance.

CASE STUDY

A customer on the road had to shut off his vehicle. Upon restarting, it would not crank over. A road service tried jump-starting the vehicle. Wires and cables were moved but nothing would get the starter to crank. The customer decided to get it pull started and did not shut the truck off until he got it to our shop. He was told on the road that he needed a starter.

The technician had the driver hit the key switch, and she could hear the solenoid click. Using a test light, she quickly verified that voltage was available at the solenoid "B" terminal. She further verified that the "S" terminal was being activated with the key. The test light was put on the stud of the "M" terminal of the solenoid and it showed voltage there. Moving the test light to the starter power terminal she noticed there was no voltage there. A voltmeter was used next. The positive lead was hooked up to the stud of the solenoid and the negative lead to the connector strap. The meter indicated an 11 volt drop. The nuts holding the strap on the starter and solenoid were pulled off and the strap was removed. A visual inspection showed that the area where the strap was connected to the solenoid was corroded and burned. A new strap was installed with new nuts. Before installing the strap, the hold down nuts were tightened first. The battery cables were then reinstalled and the vehicle was cranked over. A starter current draw test was performed to assure the starter motor integrity.

The customer was happy that he did not have to invest in a starter. Many shops might have condemned the starter immediately and installed a new starter, which would have worked because the new starter comes with the connector strap already on it. However, you can bet this customer will keep coming back to this shop for future problems.

Terms to Know

Armature
Battery cable voltage drop test
Commutator
Control circuit
Current draw test
Field coil
Fuel solenoid
Ground side
Growler
Insulated side
Magnetic start switch
No-crank
No-load test
Pinion clearance
Quick test
Slow cranking
Solenoid circuit test
Starter current draw test
Starter solenoid
Wattage
Voltage drop

ASE-Style Review Questions

1. A no-load test is being discussed: *Technician A* says that a no-load starter test is used to confirm whether the starter is drawing the proper amperage. *Technician B* says that a no-load starter test is used to confirm whether the starter is rotating at the recommended speed. Who is correct?

A. A only **C.** Both A and B
B. B only **D.** Neither A nor B

2. *Technician A* says that if a starter motor spins but does not crank the engine, the most likely cause is a defective starter drive. *Technician B* says the most likely cause of that symptom is high resistance in the starter circuit. Who is correct?

A. A only **C.** Both A and B
B. B only **D.** Neither A nor B

3. Use of a growler is being discussed: *Technician A* says that a growler is used to check the continuity of the field coil. *Technician B* says that a growler is used to check for shorts and grounds in the armature. Who is correct?

A. A only **C.** Both A and B
B. B only **D.** Neither A nor B

4. Voltage drop tests are being discussed: *Technician A* says on a no-crank complaint, perform a voltage drop test to find the location of the resistance. *Technician B* says the circuit must be activated in order to obtain a proper voltage drop reading. Who is correct?

A. A only **C.** Both A and B
B. B only **D.** Neither A nor B

5. *Technician A* says that a weak battery can throw off the results of a starting system test. *Technician B* says that a weak battery can be the cause of an engine cranking slowly. Who is correct?

A. A only **C.** Both A and B
B. B only **D.** Neither A nor B

6. Starter solenoid testing is being discussed: *Technician A* says that a 110-volt test light can be used to check for opens, shorts, or grounds on a starter solenoid. *Technician B* says that a starter/charger tester and a battery can be used to check a starter solenoid. Who is correct?

A. A only **C.** Both A and B
B. B only **D.** Neither A nor B

7. Brush replacement is being discussed: *Technician A* says that if the new brushes are soldered on, insulation or heat shrinking tube should be used to insulate the connection. *Technician B* says that not all brushes are soldered—some use screw terminals. Who is correct?

A. A only **C.** Both A and B
B. B only **D.** Neither A nor B

8. Starter solenoid and magnetic start switch voltage drop is being discussed: *Technician A* says that the voltage drop through the solenoid or the magnetic start switch should be 0.3 volts or less. *Technician B* says that the voltage drop through the solenoid or the magnetic start switch should be 0.8 volts or less. Who is correct?

A. A only **C.** Both A and B
B. B only **D.** Neither A nor B

9. Pinion gear to starter drive housing is being discussed: *Technician A* says if there is excessive clearance, the problem could be starter drive lever wear. *Technician B* says that all starter drive pinion gear to starter drive clearance is nonadjustable. Who is correct?

A. A only **C.** Both A and B
B. B only **D.** Neither A nor B

10. Starter motor installation is being discussed: *Technician A* says that on heavy duty starters, the starter nose housing is adjustable. *Technician B* says that on heavy duty starters, the starter solenoid has to be in a proper position in relationship to the engine prior to installing the starter. Who is correct?

A. A only **C.** Both A and B
B. B only **D.** Neither A nor B

ASE Challenge

1. The correct current draw for a starter on a heavy truck is being discussed. *Technician A* says that the actual reading can vary due to engine battery and starter temperature. *Technician B* says that all identical makes and models of truck engines should exhibit identical readings if operating temperatures are the same. Who is correct?
   A. A only **C.** Both A and B
   B. B only **D.** Neither A nor B

2. A solenoid switch is suspected of causing a slow and erratic cranking problem. *Technician A* says that you should check for voltage drop across the main solenoid terminals (battery and motor) with a voltmeter while the engine is being cranked. *Technician B* says that the main solenoid terminals should be checked with an ohmmeter while the starter is not engaged. Who is correct?
   A. A only **C.** Both A and B
   B. B only **D.** Neither A nor B

3. A starter on a heavy duty truck makes considerable whirring or spinning noise, but fails to crank the engine when engaged. *Technician A* says that the problem could be a faulty magnetic relay. *Technician B* says the problem could be a poor connection at the solenoid ground terminal. Who is correct?
   A. A only **C.** Both A and B
   B. B only **D.** Neither A nor B

4. Starter installation is being discussed. *Technician A* says that on most heavy duty starters, support brackets are unnecessary and can be left off to save time. *Technician B* says that all medium duty starter support brackets should be reinstalled regardless of the difficulty. Who is correct?
   A. A only **C.** Both A and B
   B. B only **D.** Neither A nor B

5. A starter is being rebuilt and is found to have an oil-soaked interior. *Technician A* says the starter parts could be cleaned with a strong solvent then dried with compressed air. *Technician B* says that hot water and detergent make a good cleaning solution for starter components if dried with compressed air. Who is correct?
   A. A only **C.** Both A and B
   B. B only **D.** Neither A nor B

Table 6-1 NATEF and ASE Task

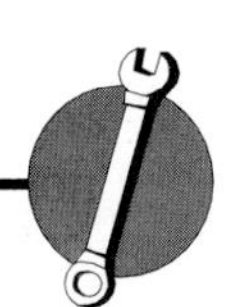

Perform starter circuit voltage drop tests; determine needed repairs

Problem Area	Symptoms	Possible Causes	Classroom Manual	Shop Manual
EXCESSIVE VOLTAGE DROP	Starter fails to crank engine or operates at reduced efficiency	1. Excessive starter-circuit voltage drop or loss	123	150

Table 6-2 NATEF Task

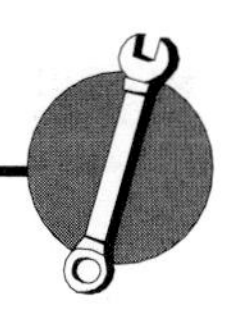

Perform starter current draw test; determine needed repairs

Problem Area	Symptoms	Possible Causes	Classroom Manual	Shop Manual
EXCESSIVE CURRENT DRAW	Starter fails to crank engine or operates at reduced efficiency	1. Shorted armature 2. Worn starter bushings 3. Bent armature 4. Thrown armature windings 5. Loose pole shoes 6. Grounded armature 7. Shorted field windings	120	145
CURRENT DRAW TOO LOW	Reduced or no starter operation	1. Worn brushes 2. Excessive circuit voltage drop		147

Table 6-3 NATEF and ASE Task

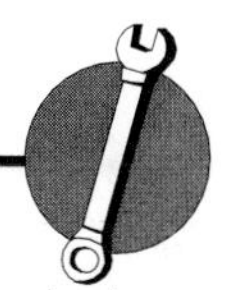

Inspect, test, and replace components (key switch, push button, and/or magnetic switch) and wires in the starter control circuit

Problem Area	Symptoms	Possible Causes	Classroom Manual	Shop Manual
STARTER WILL NOT OPERATE	Starter fails to operate	1. Failed key switch 2. Open in circuit between starting switch and magnetic switch	123	150
	Solenoid makes noise	1. Defective magnetic switch. 2. Defective push button		
REPLACE THE KEY SWITCH, MAGNETIC SWITCH, OR PUSH BUTTON	Failed component testing	1. Faulty key switch, magnetic switch, or push button	125	150

Table 6-4 NATEF and ASE Task

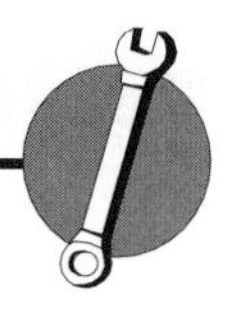

Inspect, test, and replace starter relays and solenoids/switches

Problem Area	Symptoms	Possible Causes	Classroom Manual	Shop Manual
STARTER WILL NOT OPERATE	Starter fails to operate Solenoid makes noise	1. Failed solenoid 2. Open in circuit between starting switch and magnetic switch 3. Defective magnetic switch. 4. Defective relay	125	150
REPLACE THE RELAY MAGNETIC SWITCH, OR SOLENOID	Failed component testing	1. Faulty relay, solenoid, or magnetic switch	125	150

Table 6-5 NATEF and ASE Task

Remove and replace starter; inspect flywheel ring gear or flex plate

Problem Area	Symptoms	Possible Causes	Classroom Manual	Shop Manual
STARTER FAILS TO OPERATE	Starter fails to crank or rotates at reduced efficiency	1. Defective starter requiring repair or replacement	125	140

Table 6-6 NATEF and ASE Task

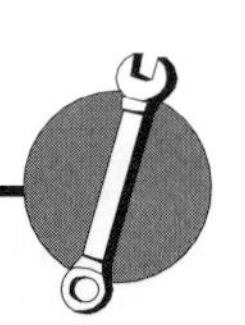

Inspect, clean, repair, and replace battery cables and connectors in the cranking circuit

Problem Area	Symptoms	Possible Causes	Classroom Manual	Shop Manual
REDUCED STARTING AND LOAD CAPABILITIES OF THE BATTERY	Starter fails to crank the engine or cranks slowly	1. Contaminated terminal clamps 2. Defective battery cables	123	149

Job Sheet 8

8

Name: [illegible] Date: [illegible]

Testing the Starting System Circuit

Upon completion of this job sheet, you should be able to visually inspect the starting circuit and test it for excessive resistance.

NATEF and ASE Correlation

This job sheet is related to ASE and NATEF Medium/Heavy Duty Truck Electrical/Electronic Systems List content area:

C. Starting System Diagnosis and Repair, ASE Task 1 and NATEF Task 2: Perform starter circuit voltage drop tests; determine needed repairs; ASE Task 2 and NATEF Task 3: Inspect, test, and replace components (key switch, push button, and/or magnetic switch) and wires in the starter control circuit; ASE Task 3: Inspect, test, and replace starter relays and solenoids/switches.

Tools and Materials

A medium/heavy duty truck with a 12-volt starting system
Wiring diagram for the above truck
DVOM

Procedure

1. Describe the truck being worked on:
   Year: 95 Make: KW VIN: ______
   Model: T600
2. Locate all components and wiring that are related to the starting system and list them. cables, batteries, ing switch, clutch, starter
3. Perform a visual inspection of all the cables, connections, and components. List all of the components and connections checked and describe their condition.
   cables - good, ~~battery~~ good, starter - good
   button - broken post
4. Perform a quick-test and record the result. good
5. Disable the engine so that it will not start. Describe how it was disabled. remove wire from fuel shutoff solenoid
6. Connect a multimeter set at volts DC across a battery. The positive meter lead to the positive post and the negative meter lead to the negative post.
7. Crank the motor, observe, and note the meter reading. ~~10.9~~ 12.4 - 10.9
8. Move the meter leads to the starter.
   A. The negative meter lead on the starter ground terminal.
   B. The positive meter lead on the starter motor power terminal.
9. Energize the starter, observe, and note the voltage reading. 12.2 - 6.2
10. Compare the voltages and describe your findings. voltage drop thru cables

11. Based upon the above, what are your recommendations? check and clean cables

12. Gain access to the starter solenoid and identify the various terminals.

13. With the multimeter set at the volts DC function, make the following hook-up:

    A. Connect the positive meter lead to the starter solenoid "B" terminal.

    B. Connect the negative meter lead to the starter solenoid "M" terminal.

14. With the engine still disabled so that it will not start, energize the starter.

15. Observe and note the voltmeter reading. ______________________

16. What does this reading indicate? ______________________

17. Based upon the above reading, what are your recommendations? ______________________

____

____

✔ Instructor Check ______________________

Job Sheet 9

9

Name: ______________________ Date: ______________________

Measuring the Current Draw of a Starter Motor

Upon completion of this job sheet, you should be able to measure the current draw of a starter motor and interpret the results of the test.

NATEF Correlation

This job sheet is related to NATEF Medium/Heavy Duty Truck Electrical/Electronic Systems List content area:

C. Starting System Diagnosis and Repair, NATEF Task 1: Perform starter current draw test; determine needed repairs.

Tools and Materials

A medium/heavy duty truck with a 12-volt starting system
Starting/charging system tester with carbon pile
Service manual for the above truck
Hydrometer

Procedure

1. Describe the truck being worked on:
   Year: ____________ Make: ____________ VIN: ______________________
   Model: ______________________
2. Perform a battery state of charge and condition check. Describe the condition and determine whether a starter current draw test can be performed using these batteries.
   ____
   ____
   Note: Before continuing with this job sheet, be sure the batteries are healthy enough to perform a proper starter current draw test.
3. Disable the ignition or fuel system to prevent the engine from starting.
4. What is the expected starter current draw? ____________ amps
5. Voltage should not drop below ____________ volts.
6. Connect the starting charging system tester cables to the truck.
7. Zero the ammeter on the tester.
8. Observe the amperage when the engine begins to crank and while it is cranking. Also, note the voltage when you stop cranking the engine.
   The initial current draw was __________ amps. After __________ seconds, the current draw was __________ amps and the voltage dropped to __________ volts.
9. Compare your measurements to the specifications. What is indicated by the test result? ____
   What is your recommendation? ____
   ____
10. Reconnect the ignition or fuel system and start the engine.

☑ **Instructor Check** ____

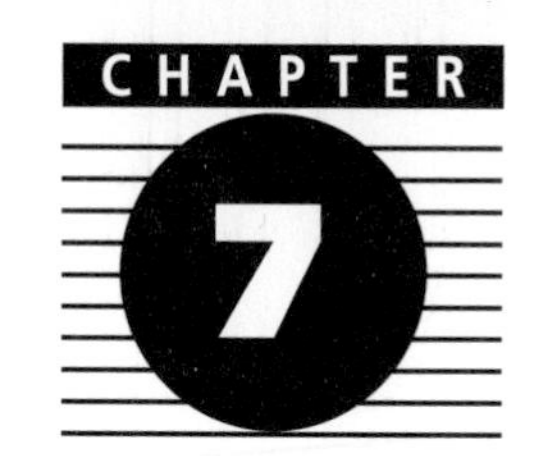

Charging System Testing and Service

Upon completion and review of this chapter, you should be able to:

- ❑ Perform a preliminary inspection of the charging system to detect any problems.
- ❑ Check for symptoms of undercharging and overcharging.
- ❑ Diagnose problems that cause an undercharge or no-charge condition.
- ❑ Diagnose problems that cause an overcharge condition.
- ❑ Perform charging system output tests and determine the proper repairs.
- ❑ Perform charging system circuit voltage tests and determine the needed repairs.
- ❑ Perform many in-vehicle tests of the AC generator itself.
- ❑ Perform a proper voltage adjustment.
- ❑ Disassemble a typical AC generator.
- ❑ Perform bench tests on various AC generator components.
- ❑ Remove and reinstall an AC generator pulley.
- ❑ Properly remove and reinstall an AC generator and adjust the belts.

Basic Tools

Basic mechanic's tool set

Service manual

Introduction

The charging system consists of the AC generator, voltage regulator, batteries, charging indicator, and all of the associated wires to complete the charging system circuit (Figure 7-1). Many different testers can be used to test the charging system and AC generators. A multimeter is a very useful tool. Many multimeters accept an inductive ammeter to check for current flow. Electronics and the use of microprocessors have brought on a new generation of hand-held testers that are specific to battery, starting, and charging system testing, as mentioned in Chapters 5 and 6. Those test units provide prompts and easy-to-read displays with preselected test sequences to take a lot of the complication out of testing battery starting and charging systems. Some units use waveform analysis when testing diodes and also give RPM readings. Some of the tests in this text use the VAT-40 with the carbon pile type of multitester. Regardless of the type of tester that is being used, you always have to follow the operating procedures for the specific tester.

Before any charging system tests are performed, you have to make sure the battery is checked first. Remember that the battery supplies the electrical power for the charging system and provides the reference voltage for the regulation of the AC generator. If the battery is dead, the charging system cannot work properly. AC generators are designed to maintain a specific battery voltage, not to charge a dead battery.

Before performing a charging system test, a thorough preliminary inspection can detect a problem that is the cause or fault of the charging system failure. Check the following items:

1. Condition of the belt, whether it is the AC generator belt or any other belt that indirectly causes the generator to turn. Figure 7-2 shows various conditions of a belt. If a belt is worn or glazed, it will prevent the AC generator from attaining the proper RPM due to slippage. Keep in mind that when a full electrical demand is placed on the generator, it takes more effort to turn the generator.
2. Drive belt tension is critical also to prevent slippage.
3. Electrical connections at the generator, regulator, batteries, starter, and grounds.
4. Excessive current drain caused by an electrical component staying on after the ignition is switched off.

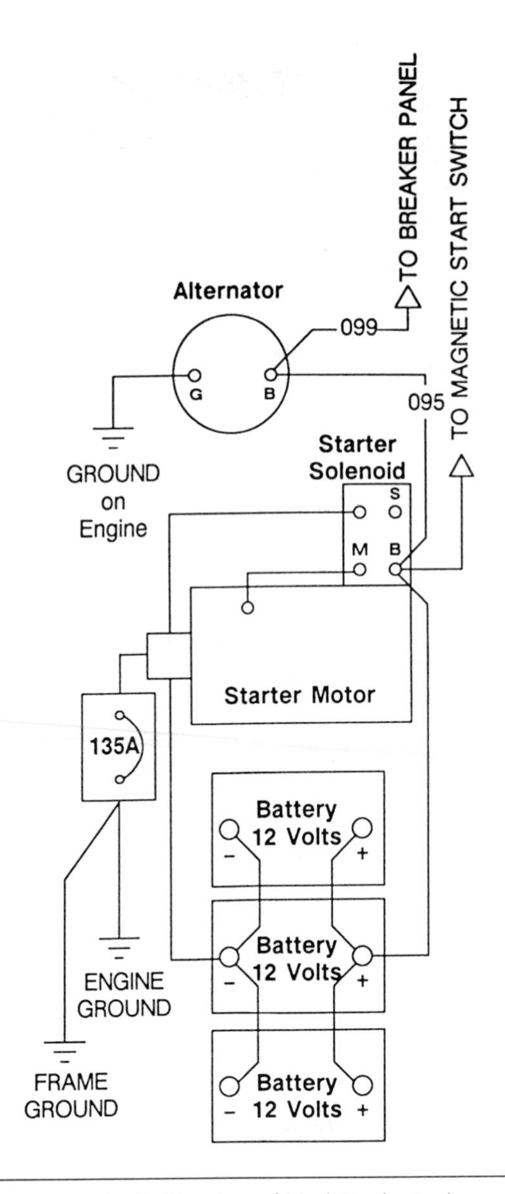

Figure 7-1 A typical charging system circuit. (Courtesy of Mack Trucks, Inc.)

5. Check for symptoms of an overcharging condition. This symptom can be indicated by batteries that experience rapid water loss or overheating and/or the ammeter indicating a higher than normal reading.

   It is important to note that the battery itself could be defective, making it seem like there is a problem in the charging system.

6. Check for symptoms of an undercharging condition. This is the symptom that normally prompts checking, testing, and repairing the charging system. The typical indicator is low ammeter and/or voltmeter readings, a slow cranking starter motor, and/or discharged batteries.

CAUTION: Many charging systems require the vehicle to be running. Be aware of moving pulleys and the fan, which can cause personal injury. If the vehicle is running inside the shop area, make sure it is well ventilated. Also, make sure the vehicle is properly "parked."

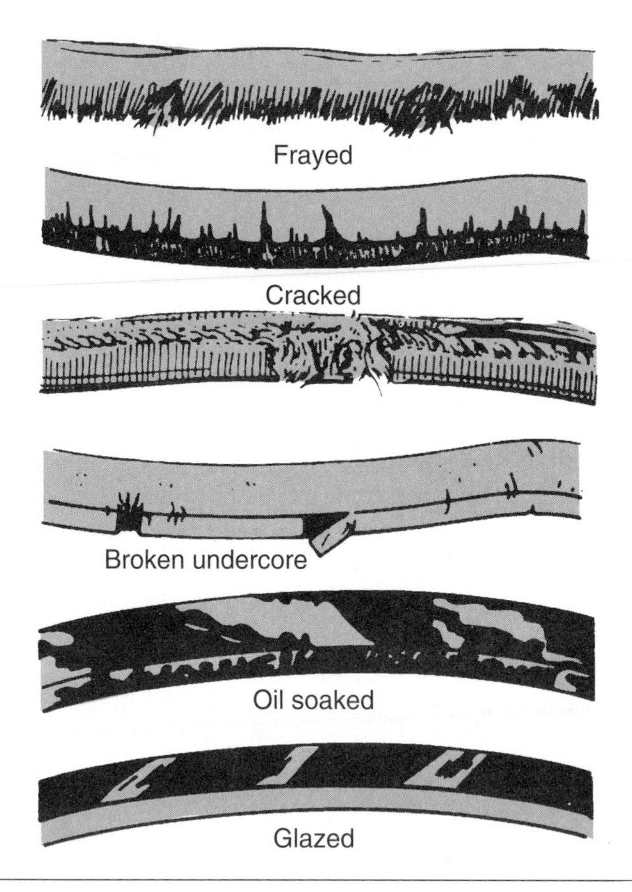

Figure 7-2 The generator's drive belt must be replaced if any of these conditions exist.

Charging System Service Precautions

Charging systems can easily be damaged in the process of testing or servicing. For example, in the process of charging, jumping, or rehooking cables if the polarity is reversed, damage can occur to the diodes. The following are some general guidelines that should be observed when working with charging systems:

1. Do not disconnect the AC generator output lead while it is running.
2. Do not disconnect the battery while the vehicle is running. The battery is a reference for voltage monitoring and it also stabilizes any voltage spikes that can damage a vehicle's delicate electronics.
3. Do not allow output voltage to increase over 16 volts when performing charging system tests.
4. Do not allow anything to accidentally touch the output terminal of the AC generator in a manner that would also allow contact to ground.
5. Do not attempt to polarize an AC generator.

    Note that this process is very rarely performed. It is only required if the rotors' residual magnetism is lost, which might occur during disassembly of the generator for repairs. If polarization is required, you must follow manufacturer's procedures.
6. Do not remove the AC generator from the vehicle until the battery's ground cable is removed first.

Polarizing a generator is performed to reintroduce residual magnetism to the field.

Additionally,

1. Observe all polarities whether with the batteries or AC generator.
2. Make sure that the test equipment is accurate.
3. Make sure all connections are tight and clear before proceeding with any testing or charging.
4. Check drive belt(s) integrity before performing any charging system tests.
5. Make sure the ignition switch is off before making any electrical connections or disconnections.
6. Make sure the battery(s) is fully charged and in good condition before performing, testing, and diagnosing.

AC Generator Noises

A charging system can be the cause of certain noises that develop from the engine compartment. The most common noise that can be attributed to the AC generator is a squealing sound. The first component to be checked for that noise is the belt(s) tension and condition.

SERVICE TIP: With the engine off, rub a bar of soap on the pulley surfaces of the drive belts. Do this and run the engine with one belt at a time until the noise stops. This is a quick way of determining which belt is the culprit.

Faulty bearings can also cause a squealing noise. The bearings are used to support the rotor in the housing ends. A long screwdriver, length of hose, or a technician's stethoscope can be used to test for these noises. Simply place the listening device near the bearings to pinpoint the noise.

SERVICE TIP: There will always be some noise coming from the AC generator when listening with a stethoscope. It doesn't hurt to compare the noise to a good alternator in another truck if possible. This can help prevent misdiagnosing. Bearing replacement will require disassembling the AC generator.

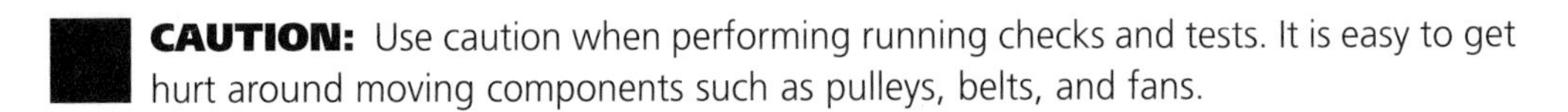

CAUTION: Use caution when performing running checks and tests. It is easy to get hurt around moving components such as pulleys, belts, and fans.

A whining noise can also emanate from inside the AC generator due to shorted diodes or stator. A quick way to test for this type of noise is to disconnect the wiring to the generator, then start the engine. If the noise is gone, the cause could be a magnetic whine due to shorted diodes or stator windings. If the noise remains, the cause is mechanical. A scope or some of the newer charging system testers can be used to verify the condition of the diodes and stator.

Charging System Troubleshooting

Troubleshooting charts are valuable tools for technicians. A chart such as the one shown in Figure 7-3 gives several possibilities for each type of complaint, which is useful for quickly finding a direction for the next step that needs to be taken towards pinpointing the fault. For example, if the basic type of trouble is that the ammeter registers constant discharge, the service procedure tells you to perform a voltage output test as one of the procedures. In that case, a more direct flowchart becomes useful (Figure 7-4).

BASIC TYPES OF TROUBLE	PROBABLE CAUSES	SERVICE PROCEDURES
Generator noise Note: Water pump A/C compressor, P/S pump or any other belt driven component noise is sometimes confused with alternator noise. A sound detecting device, such as a stethoscope, will eliminate indecision in this respect.	Generator drive belt (squealing noise)	Adjust or replace belt, as required. (An application of belt dressing may eliminate noise caused by minor surface irregularities.)
	Generator bearing (squealing noise)	Replace bearing if found to be out-of-round, worn, or causing shaft scoring.
	Generator diode (whining noise)	Test generator output. (A shorted diode causes a magnetic whine and a reduction in output.) Test diodes and replace, as required. Replacing rectifier assembly may be most feasible fix.
Indicator gauge fluctuates —or— Indicator light flickers.	Charging system wiring	Tighten loose connections. Repair or replace wiring, as required.
	Regulator contacts	Oxidized or dirty regulator contacts. Replace regulator, if necessary.
	Brushes	Check for tightness and wear. Replace, if necessary.
Indicator light stays on.	Broken, loose, or slipping drive belt	Adjust or replace, as required.
	Battery cables, charging system wiring	Clean battery cables and terminals. Tighten loose connections. Repair or replace as required.
Ammeter or voltmeter registers constant discharge.	Battery specific gravity	If unsatisfactory, replace battery.
Battery will not hold charge.	Generator output too low, or no generator output	Perform the voltage tests to determine if trouble is in the regulator, wiring harness or the alternator itself.
	Generator drive belt	Adjust or replace belt, as required.
	Battery cables, charging system, wiring	Clean battery cables and terminals. Tighten loose connection. Repair or replace, as required.
Battery low in charge. Headlights dim at idle. Note: A history of recurring discharge of the battery, which cannot be explained, suggests the need for checking and testing the complete charging system.	Electrolyte (specific gravity)	Test each cell and evaluate condition: All readings even at 1.225 or above—Battery OK All readings even, but less than 1.225—Recharge and retest. High-low variation between cells less than 50 gravity points—Recharge and retest. High-low variation between cells exceeds 50 gravity points—Replace battery.

Figure 7-3 Charging system troubleshooting chart.

BASIC TYPES OF TROUBLE	PROBABLE CAUSES	SERVICE PROCEDURES
	Battery (capacity)	Test capacity and evaluate condition: Minimum voltage 9.6 for 12 volt battery or 4.8 for 6 volt battery. (Both values under specified test load conditions.) If capacity is under minimum specifications, perform 3 minute charge test. If below maximum (15.5 volts for 12 volt battery or 7.75 volts for 6 volt battery at 40 and 75 amps, charge rate, respectively) battery is OK—recharge. If above maximum, battery is sulfated. Slow charge at 1 amp/positive plate. Replace battery if it doesn't respond to slow charge.
Lights and fuses fail prematurely. Short battery life. Battery uses excessive water. High charging rate.	Charging system wiring, including regulator ground wire.	Tighten loose connections. Repair or replace wiring, as required.
	Voltage limiter setting.	Perform the voltage output tests to verify the condition of the regulator.

Figure 7-3 Charging system troubleshooting chart. (Courtesy of Freightliner Corporation)

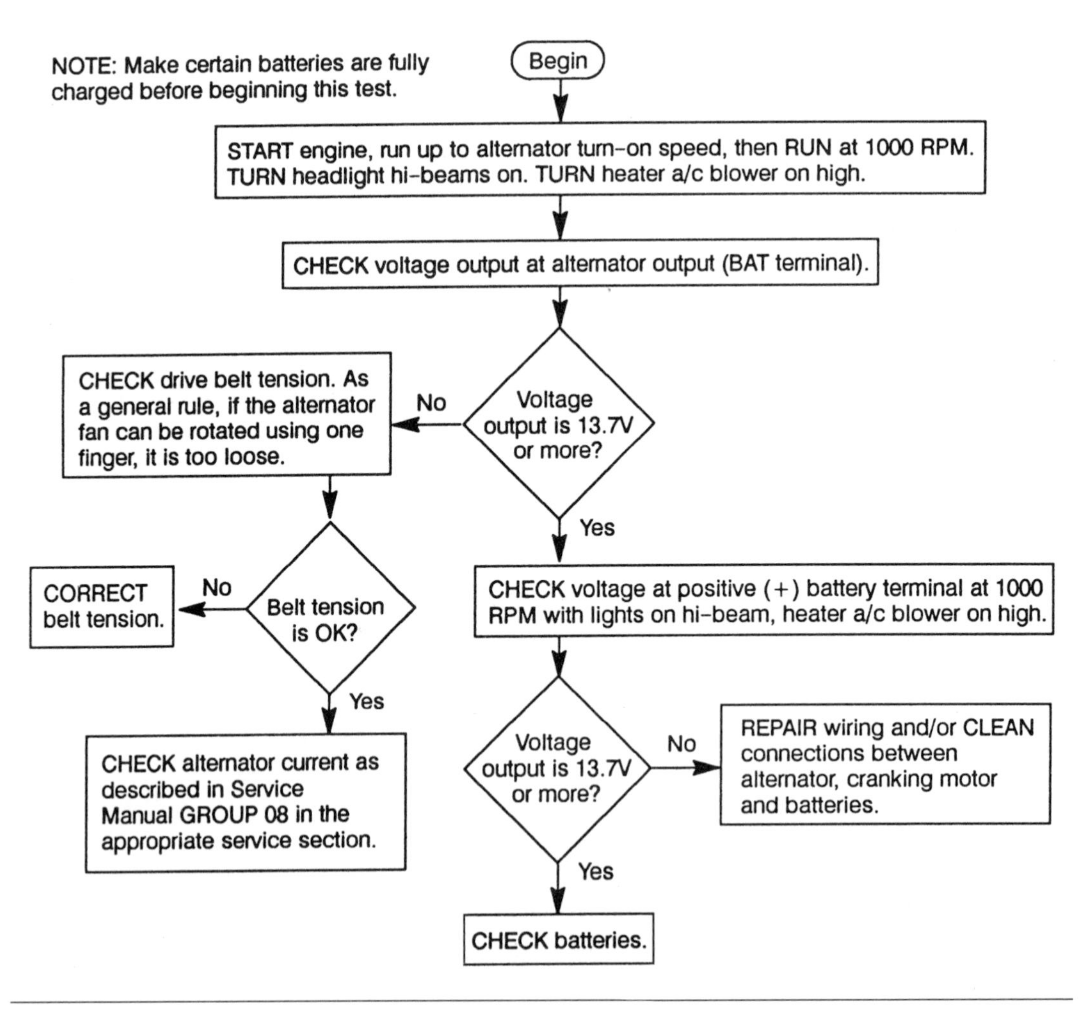

Figure 7-4 A flowchart for checking alternator output. (Courtesy of Navistar International Engine Group)

Voltage Output Testing

Special Tools

Voltmeter

A **voltage output test** is one of the first alternator tests performed whenever there is a charging system problem, assuming a visual inspection was performed first. It is a fairly quick and simple test to verify whether the AC generator is working properly. This test is also performed at the output of the AC generator and at the batteries to compare the two voltages to see if there is a problem.

SERVICE TIP: Before proceeding with a voltage output test, make sure the batteries are in good condition, fully charged, and all connections are clean and tight.

Unloaded Output Test

Unloaded tests are performed with no loads on at both idle and high RPMs.

The following procedure is for performing this test using a multimeter:

1. Set the multimeter to the VDC function.
2. Connect the meter's positive lead to the alternator BAT terminal and the meter's negative lead to a good ground (Figure 7-5).
3. Make sure all electrical loads are off and start the engine.
4. Bring the engine speed to approximately 1200 RPM or more to obtain a maximum voltage reading.

   It is important to note that this higher than idle RPM ensures that enough was reached to turn on the regulator. No output can be obtained until the regulator turns on.
5. Observe and note the voltage reading.
6. Move the voltmeter to the batteries and hook up the meter leads across the positive and negative battery terminals, observing polarity.
7. With the engine running at the same speed when the meter was on the alternator, observe and note the voltage reading.

Use the following figures as a guide:

System Voltage	Rated Voltage	Operating Range
12	14.0	13.0–15.0V
24	28.0	26.0–30.0V

WARNING: The voltage should never be allowed to rise above the operating range or circuit, or component damage can occur.

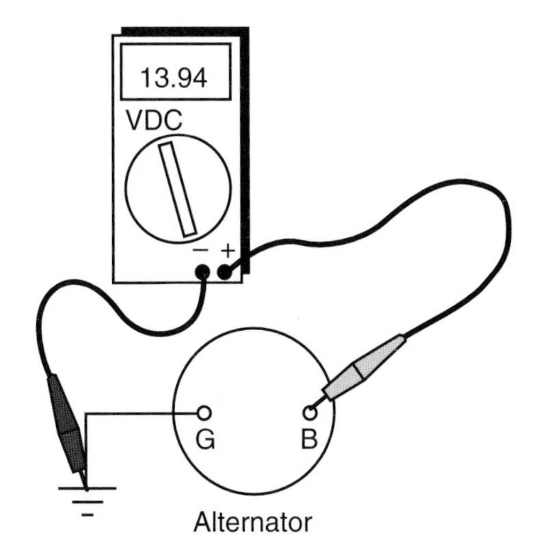

Figure 7-5 Checking alternator output at the alternator. (Courtesy of Mack Trucks, Inc.)

If both the alternator and battery's voltages were below the operating range, this would indicate an **undercharge** condition leading to the following areas for the fault:

1. Defective battery
2. Damaged or corroded wires and connections
3. Loose or glazed belts
4. Defective regulator or regulator adjustment, if applicable
5. Defective AC generator
6. Defective diode trio

It is important to note that many of the Leece-Neville AC generators provide an easy access to the diode trio without having to pull the alternator. This makes it convenient to check the diode trio while the generator is still in the vehicle.

Classroom Manual
Chapter 7, page 165.

If the diode trio is in good shape, the next test is a full field test.

If the test result showed both the AC generator and battery's output voltage to be over 14.2 volts (for most 12 volt systems), an overcharging condition is indicated. The following areas should be checked for an overcharging condition:

1. Defective battery
2. Defective or improperly adjusted regulator (if applicable)
3. Defective diode trio
4. Defective wiring between the voltage regulator and the AC generator

It is important to note that most AC generators have integral voltage regulators today, which eliminate a lot of external wiring problems.

Loaded tests require that all the electrical loads are on to determine if the AC generator can meet the load demands of the vehicle's electrical components.

If the voltages are correct, you can go one more step with the voltmeter still hooked to the batteries and perform a *loaded AC generator voltage output test*:

1. With the engine running, turn on as many electrical components as possible, such as headlights and heater motor fan(s).
2. Perform this test at idle and at 2,000 RPMs.
3. Observe and note the readings.

At idle, the voltage should be at least at the minimum operating range of 13.0 volts and increase at least 0.5 volts or more as the RPM is increased.

Why did we check the output voltage at the AC generator and then again at the batteries? What would be the problem if the AC generator output voltage was satisfactory, but at the battery the voltage indicated was less?

Special Tools

Voltmeter

Safety glasses

Voltage Drop Testing

Voltage drop test determines if all the components, wires, and connections are operating at the same potential.

The AC generator's output must get to the batteries and the chassis electrical components with a minimum loss of voltage or the batteries will not adequately recharge. The voltage regulator is designed to control maximum system voltage, which should be available at the AC generator's output terminal. However, that doesn't mean that the maximum voltage always reaches the battery and electrical components. A problem in the wiring can cause some of the maximum voltage to be lost. The greatest voltage loss tends to occur when the charging system output is at its maximum rated amperage. The following test is performed to find the lost voltage. (Note that with this test, the engine has to be running with as many electrical loads on as possible.)

CAUTION: This test requires the testing of various components and areas within the engine compartment. Be aware of all rotating components to prevent personal injury.

Using a multimeter set at Volts DC, check for voltage drops at the following locations and refer to Figure 7-6 as a guide for meter lead locations.

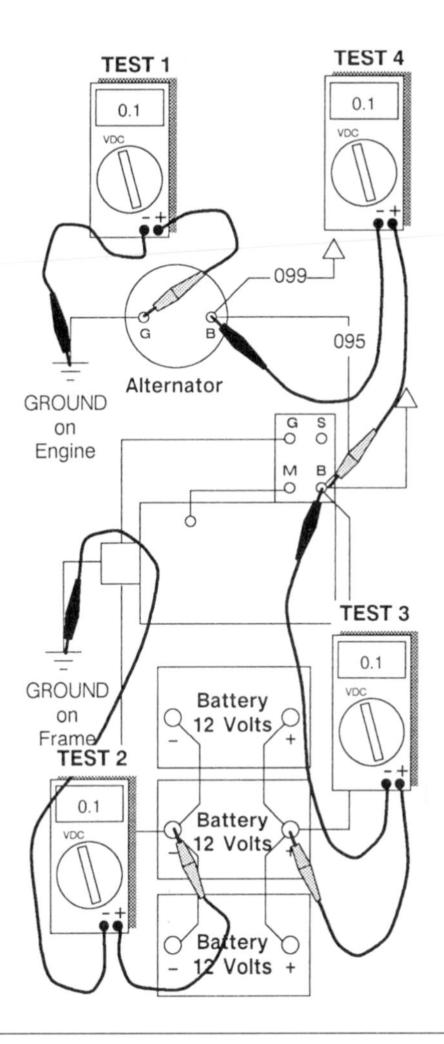

Figure 7-6 A voltmeter is being used to check for voltage drops at various points of the circuit. (Courtesy of Mack Trucks, Inc.)

Test 1. Connect meter negative lead to ground and the positive lead to the "G" (ground) terminal of the AC generator. Observe and note the voltage reading.

Test 2. Connect the negative lead of the meter to the ground on the frame and the positive lead of the meter to the battery negative terminal. Observe and note the voltage reading.

Test 3. Connect the negative lead of the meter to the starter solenoid "B" terminal and the positive lead of the meter to the positive terminal of the battery. Observe and note the voltage reading.

Test 4. Connect the negative lead of the meter to the "B" (+) terminal of the AC generator and the positive lead of the meter to the "B" terminal of the starter solenoid. Observe and note the voltage reading.

Voltage loss should not exceed 0.1 volts through any cable. If the voltage drop is excessive, look for damaged cables or loose or corroded connections and repair as necessary. Then verify the repair with another voltage drop test. Remember that this procedure can be used in any part of the circuit as long as the meter is hooked up properly. It should also be noted that current has to flow in order to see a voltage drop, which is why the charging system is loaded down.

A voltage drop of the complete positive (+) side can be also performed by placing the meter's positive lead to the B terminal of the AC generator and the negative meter lead to the battery positive terminal. Why this type of connection? If the alternator is charging, what is the most positive point?

This test can also be performed with a battery/starter system tester utilizing the carbon pile to regulate the load from 9 to 20 amps. As you can see, there is a constant relationship between the battery, starting, charging system, and the connecting wires and cables.

Determining Current Output Needs

AC generator RPM is referred to as the speed of the AC generator shaft.

A current output tests the AC generator's capability to put out its rated current output. Figure 7-7 shows manufacturer's specifications for a DELCO-Remy 25S1 Series AC generator. As you can see, the current output in amperes varies with different RPMs. Notice the high RPM is at 6,500 RPM, three times the approximate full RPM of a diesel engine. What you are interpreting is the AC generator's shaft rotation RPM, because the current output test is most often performed as a bench test, but can also be performed as an in-vehicle test.

The AC generator's RPM is determined by the generator's pulley diameter in relation to the engine's drive pulley diameter, known as the pulley ratio. A typical pulley ratio is selected to allow an AC generator to carry approximately 50% of the electrical load at idle. This output can be adjusted by changing the pulley ratio. A desired RPM to generator output can be determined by looking at a manufacturer's AC generator's performance curve (Figure 7-8). Note that the bottom

GENERATORS

Delcotron Generator Model	Series	Type	Grd	Service Bulletin	Rotation Viewing D. E.	Spec. No.	Field Ohms (80°F)	FIELD CURRENT (80°F) Amps	Volts	Specified Volts	COLD OUTPUT Amps	Approx. RPM	Amps	Approx. RPM	Rated Hot Output‡ (Amps)
1117485	30SI	400	N	1G-280	CW	4567	2.7–3.4	3.6–4.3	12	13	68	2500	86	6500	90
1117611	20SI	300	N	1G-276	EITHER	4577	1.7–2.1	5.7–7.1	12	14	28	2600	62	6500	60
1117612	20SI	300	N	1G-276	EITHER	4577	1.7–2.1	5.7–7.1	12	14	28	2600	62	6500	60
1117613	20SI	300	N	1G-276	EITHER	4577	1.7–2.1	5.7–7.1	12	14	28	2600	62	6500	60
1117614	20SI	300	N	1G-276	EITHER	4578	6.0–7.3	3.3–3.8	24	26	12	2500	45	6500	45

Figure 7-7 Sample of manufacturer's specifications. (Courtesy of Alternator Service Manual)

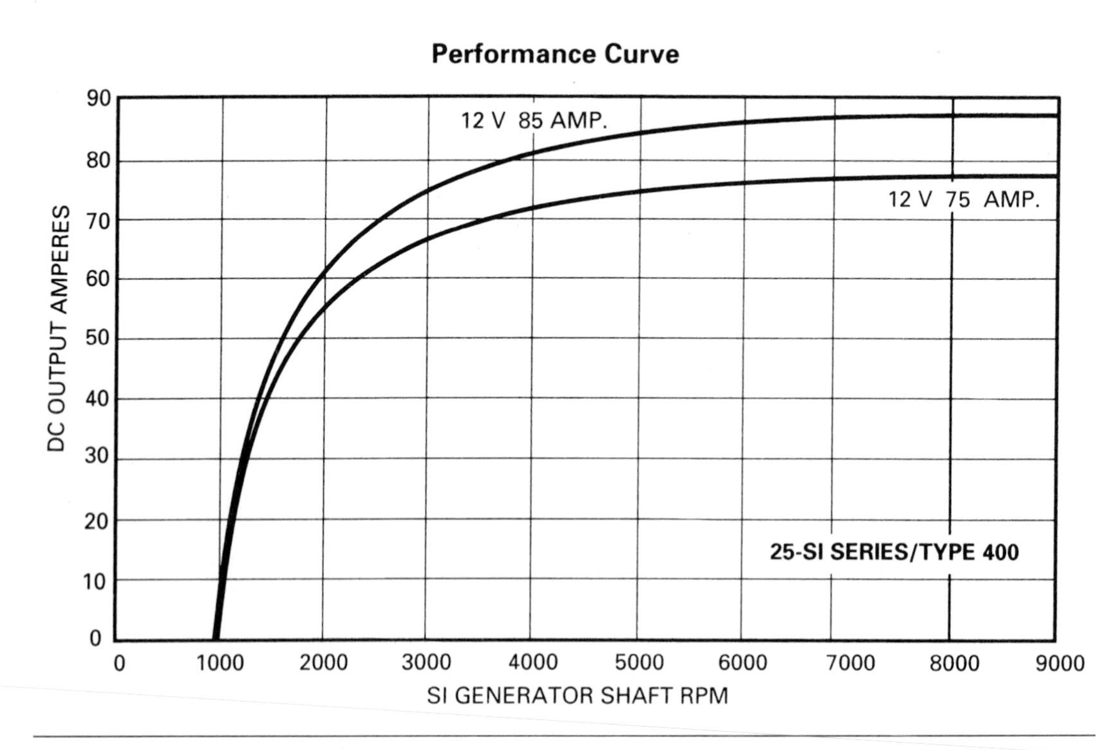

Figure 7-8 AC generators performance curve. (Courtesy of AC Delco)

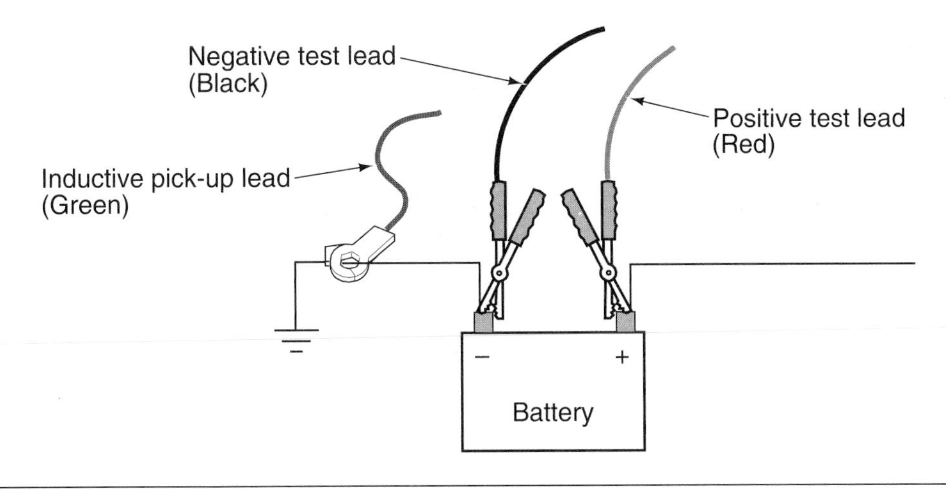

Figure 7-9 Common tester connections for performing most charging system tests.

of the graph shows the generator shaft RPM. For example, a higher output at idle may be required to obtain maximum battery life by reducing battery cycling. This can be beneficial if a truck is being run during long or low idle periods with various loads on.

To determine the proper pulley ratio, divide the required AC generator idle RPM by the engine idle RPM to get the pulley ratio. For example, 1600 ÷ 650 = 2.5. This means the AC generator must turn 2.5 times faster than the engine, therefore, the pulley ratio is 2.5:1, which is considered the minimum. The pulley diameter is determined by the engine drive pulley diameter. Divide the engine pulley diameter by the pulley ratio as determined above. If the engine drive pulley is 9″, the AC generator pulley diameter would be: 9″ ÷ 2.5 = 3.6″ (3–5/8″) pulley diameter. AC generator manufacturers provide the performance curves for their AC generators to enable the truck owner to better match the electrical needs to the driving condition.

AC generator's RPM is determined by the generator's pulley diameter in relation to the drive pulley's diameter, which is known as the **pulley ratio**.

Another important aspect to consider is the charging system requirement. This can be derived by using an ammeter placed in series with the battery or if using an inductive pickup such as in the VAT tester, the hookup illustrated in Figure 7-9. To perform this check, perform the following:

Performance curves are used to determine the AC generator's output at certain RPMs.

1. With the engine stopped (VAT tester selector will be in charging position), turn on all loads—all accessories, ignition switch, headlights (high beams), heater and A/C blower motor(s), etc.—at their highest settings.
2. Note the ammeter reading. This is the total accessory load, which is important for specifying an alternator and also when performing an AC generator current output test. Total alternator output current readings should exceed accessory load readings by 5 amps or more.

Current Output Test

Special Tools

Starting and charging system tester

Safety glasses

In this test, the charging system is loaded down using a carbon pile to obtain the AC generator output current. The carbon pile regulates the load to obtain the highest reading, within a volt reading of 12 volts or higher. One method of measuring the AC generator's output current is shown in Figure 7-10. The following steps are performed if using a VAT style of tester with an inductive pickup lead as shown in Figure 7-9.

Current output test determines the maximum output of the generator.

1. Connect the large red and black cables across the battery, observing polarity, and connect the green amps pickup around the vehicle's battery ground cable(s). In this manner, the ammeter will not sense the portion of the AC generator's output that is used to run the vehicle while the output test is performed.
2. Select charging.

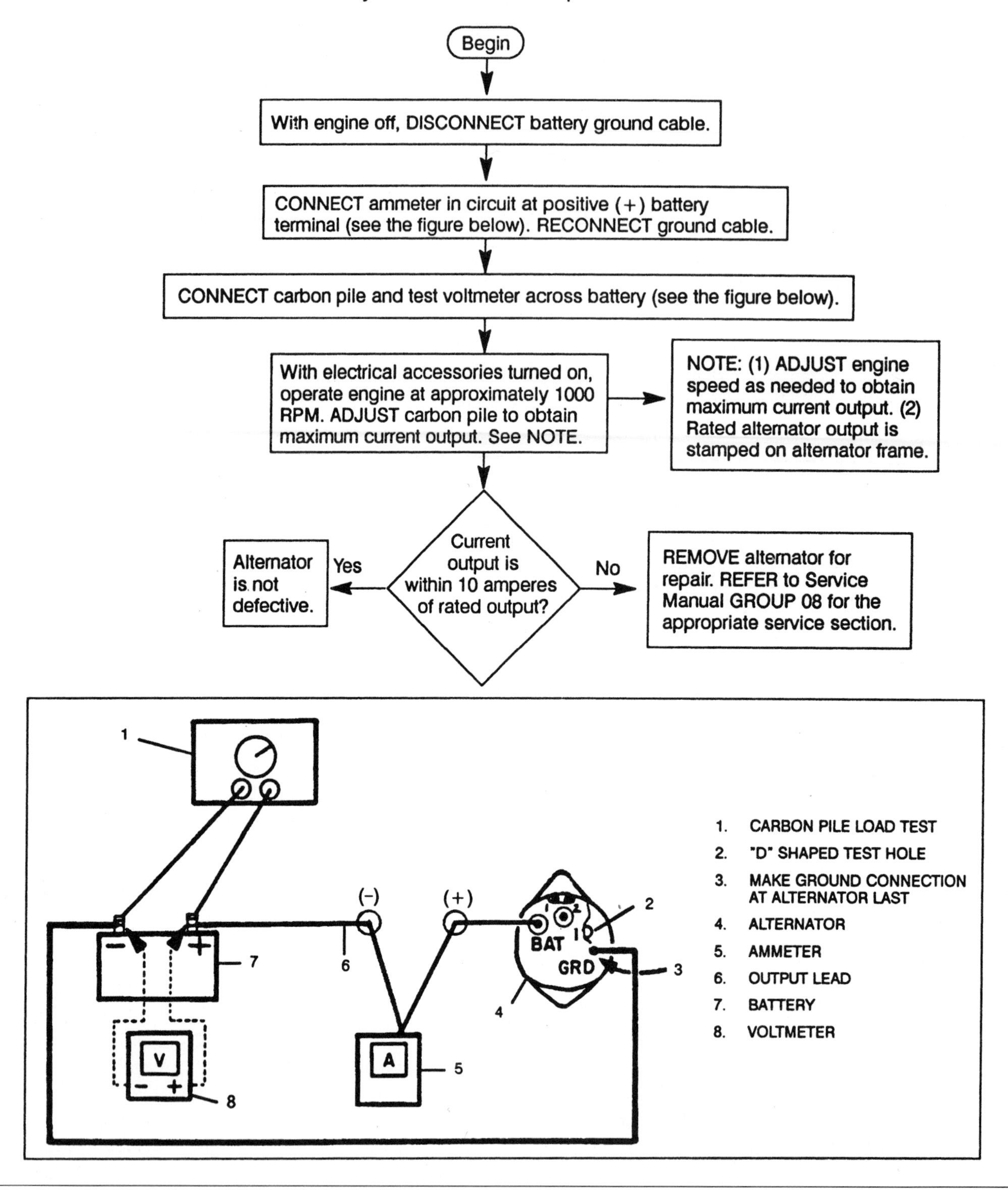

Figure 7-10 One method of measuring an AC generator's output current. (Courtesy of Navistar International Engine Group)

3. Zero the ammeter.
4. Turn the ignition key on (engine not running) and observe the ammeter reading. This measures the ignition and accessory draw that would be used to run the engine and is not seen by the tester.
5. Start the engine and keep it between 1,500 and 2,000 RPM.
6. Adjust the load knob to obtain the highest reading on the ammeter without causing the voltage to drop lower than 12 volts. Applying a load causes the voltage regulator to stop regulating and permits the alternator to produce its maximum output.
7. The highest reading observed indicates maximum current output.
8. Return the load knob to OFF.
9. Add the ammeter reading obtained in Steps 4 and 7.
10. This final combined reading should be within 10% of the manufacturer's specification.

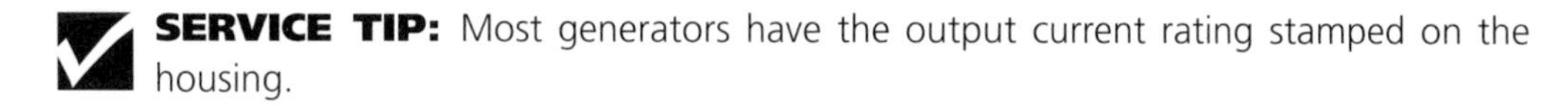

SERVICE TIP: Most generators have the output current rating stamped on the housing.

If the output is not within 10% of specifications, a field circuit test has to be performed next. This is assuming that some basic checks were performed already, such as checking for slipping belts.

WARNING: The current output test can burn up the AC generator if not performed properly. The ammeter reads the load placed upon the battery with the carbon pile. But what happens if there is a 15-amp draw on when the engine is running, such as from a fuel solenoid draw, fuel pump draw (if electrical), or any other accessory that could be on when the engine is running? For example, the AC generator is rated at 85 amps and you place a load of 80 amps on the charging system. All of a sudden the amps drop to almost zero (no output). What happened? The most likely explanation is that the 15 amps not seen on the ammeter combined with the loaded 80 amps, totaling 95 amps. This meant the generator worked beyond its maximum output rating, generating enough heat to cause it to fail.

WARNING: A generator that is constantly working to produce current because there is a charging system problem will get very hot. This excessive heat will fry an AC generator. After installing an AC generator, let it run awhile and monitor for overheating.

Full Fielding Test

Special Tools

Starting and charging system tester

Jumper wire

Small screwdriver

The full fielding test tries to pinpoint whether a charging problem is caused by either the AC generator or the voltage regulator. Some alternators with internal voltage regulators do not have the capability of being full field tested while on the vehicle. A full fielding test bypasses the voltage regulator, allowing full field current to energize the AC generator's field coil. If AC output is present, the voltage regulator is the suspected component. If the output is still lower than specified output, the AC generator is the cause of the problem.

Classroom Manual
Chapter 7, page 151.

WARNING: When full fielding the system, the battery should be loaded to protect vehicle electronics. With the voltage regulator bypassed, there is no control of voltage output and the AC generator is capable of producing high voltages. This increased voltage will damage circuits not designed to handle high voltages.

In the next section we will look at the procedures for full fielding some of the more common AC generators.

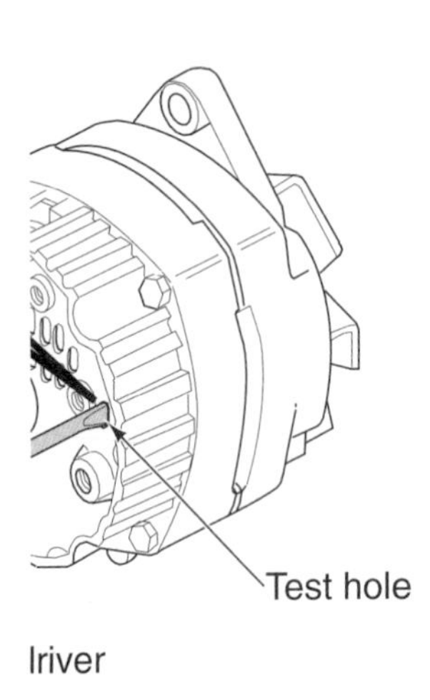

Figure 7-11 Full fielding the GM 10SI generator by grounding the tab in the "D" test hole.

Full fielding means the field windings are constantly energized with full battery voltage. If the AC generator is working properly it will produce maximum generator output.

Full fielding tests are used to determine if a problem exists in the regulator or generator.

Classroom Manual
Chapter 7, page 151.

General Motors Full Field Testing

Many of the DELCO-Remy S1 type of generators with internal regulators provide a D-shaped test hole, or D hole, for full fielding (Figure 7-11). This is considered an A circuit as described in Chapter 7 of the Classroom Manual. The D holes lines up with a small tab attached to the negative brush. A small metal probe such as small screwdriver is placed into the D hole approximately 1/2" to contact the tab. As the probe also contacts the case at the D hole, the AC generator is forced into a full field operation. Observe the ammeter or voltmeter for maximum output. If the output is within specifications, the regulator is at fault.

WARNING: Do not full field for longer than 10 seconds. Damage to electrical systems can occur.

When the rotor's field coil is energized, a magnetic field is produced. If the same test is performed by grounding the D hole, the rear bearing cap area should be magnetized. When performing a full field test, the rear bearing should attract a small screwdriver or paper clip indicating the brushes and rotor are good.

Ford External Regulator Full Field Testing

Special Tools

Voltmeter or starting/charging system tester

Small screwdriver

Jumper wire

The following steps are used to full field an external voltage regulator system:

1. Disconnect the wiring plug from the regulator. (Remember these tests are meant to bypass the voltage regulator.)
2. Use an ohmmeter to measure the resistance between the "F" terminal of the removed plug and ground (Figure 7-12).
3. The reading should be higher than 2.4 ohms. If not, the field circuit is grounded and needs to be repaired before continuing. The ground can be in the wiring harness or the AC generator field circuit.

WARNING: Removal of all electrical plugs and connections must be done with the key off to prevent damage to electrical system components.

Classroom Manual
Chapter 7, page 154.

4. If the resistance is higher than 2.4 ohms, connect a wire between the "F" and "A" terminals of the regulator wiring plug (Figure 7-13). This jumps the B+ voltage from the "A" terminal to the "F" terminal to energize the rotor's field coil.

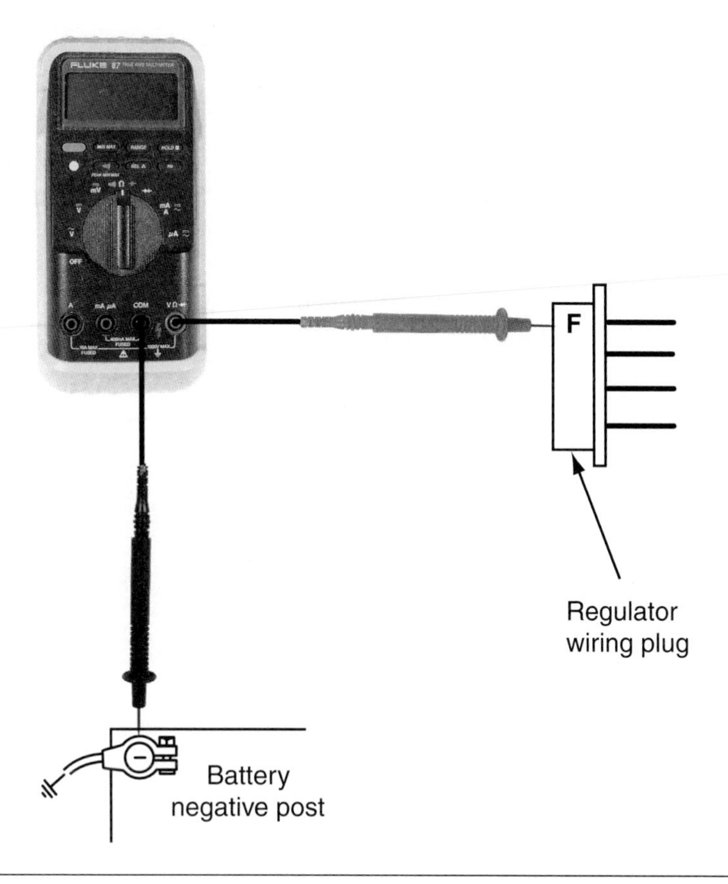

Figure 7-12 Ohmmeter connections to test field wire resistance.

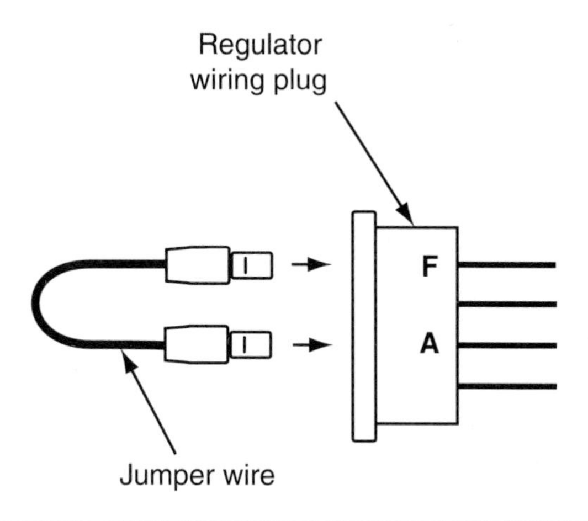

Figure 7-13 Jumper wire connections to bypass regular and full field the generator.

5. Run the engine between 1,500 and 2,000 RPM.

WARNING: To prevent circuit and/or component damage, do not allow output voltage to increase beyond 16 volts. Load the battery by turning on the headlights to prevent excessive voltage output.

6. Compare the test results with manufacturer's specifications.

SERVICE TIP: Since you need to get voltage to the field, it makes sense to make sure that voltage is available at the "A" terminal with the key on before performing the full fielding test.

Special Tools

Voltmeter

Jumper wire

Stiff wire or 1/32″ drill bit

Classroom Manual
Chapter 7, page 145.

Leece-Neville

To perform a full field test requires the use of a digital voltmeter for accuracy, a jumper wire and a piece of stiff wire such as a paper clip or 1/32″ drill bit. Follow these steps to perform this test:

1. Connect the voltmeter across the AC generator's output terminals, observing polarity. The meter function is set at Volts DC.
2. Make sure all electrical accessories are off and run the engine at approximately 1,000 RPM.
3. Observe and note the output voltage for later reference.
4. Insert a paper clip into the full field access hole (Figure 7-14).

   It is important to note that the paper clip or piece of stiff metal needs to be at least 2″ long.
5. Attach one end of the jumper wire to the paper clip and the other end to the negative (−) terminal of the AC generator.
6. With the voltmeter still across the output terminals of the generator, run the engine at approximately 1,000 RPM.

   Make sure the wire is held firmly against the brush terminals inside the housing.
7. Observe and compare this voltage reading with the voltage reading obtained in Step 3.

   In a Leece-Neville alternator, the individual stator windings can also be accessed easily and used as part of the full fielding test final determination. This requires the voltmeter to be set to the Volts AC function and placed on the terminals shown in Figure 7-15.
8. With the jumper still hooked up and the wire in the full field access hole, connect the AC voltmeter across Terminals 1 and 2, 1 and 3, and 2 and 3, and note the voltages. If all of the voltages are approximately the same, they are considered balanced.
9. Remove the jumper wire, clip, and meter leads.

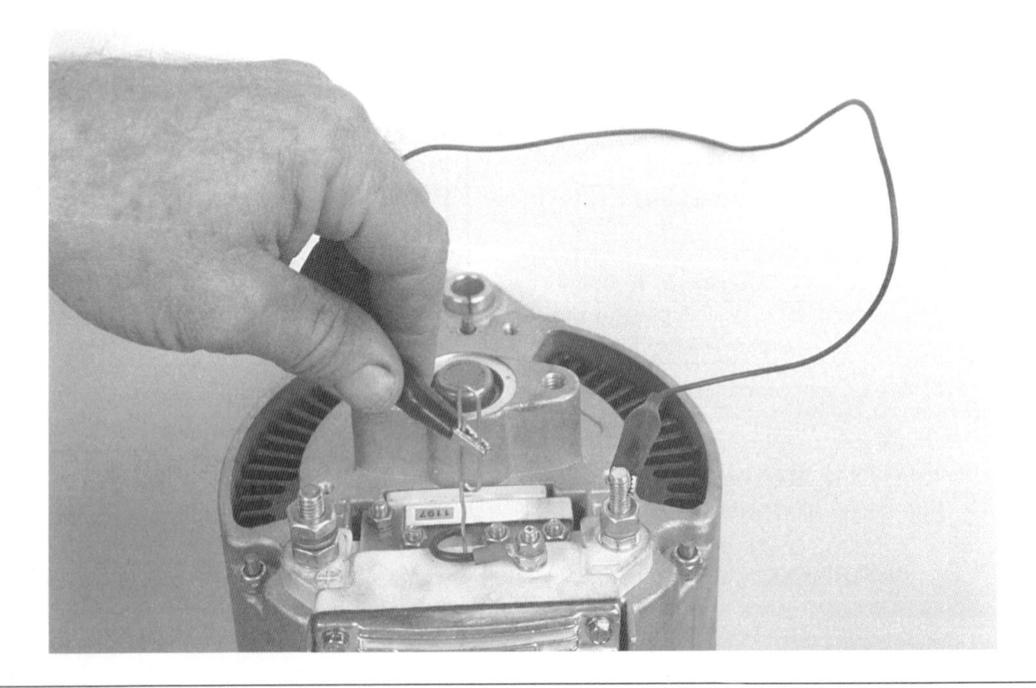

Figure 7-14 Inserting a paper clip into the full fielding access hole to perform a full field test.

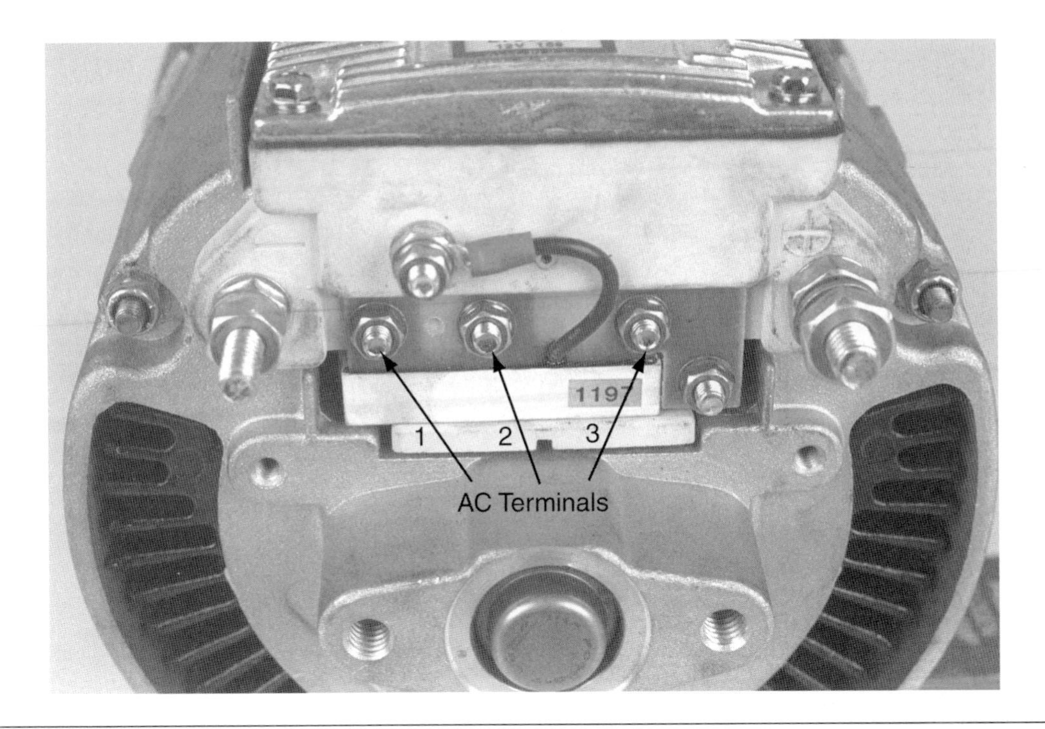

Figure 7-15 Stator winding terminals.

If the voltage reading in Step 7 is higher than that in Step 3, and the voltages measured in Step 8 are balanced, the stator and alternator are okay and the next step is the voltage regulator adjustment procedure.

If the voltage reading in Step 7 is higher than that in Step 3 and the voltages in Step 8 are not balanced, the alternator stator or rectifier(s) is defective. Repair or replace the alternator.

If the voltage reading in Step 7 is lower or equal to that in Step 3 and the voltages in Step 8 are balanced, the alternator is defective.

If the voltage reading in Step 7 is lower or equal to that in Step 3, and the voltages measured in Step 8 are not balanced, the alternator stator or rectifier(s) are defective.

Diode Trio Test

On the Leece-Neville alternators, the **diode trio** can be checked without removing the alternator from the vehicle. However, the diode trio has to be removed from the AC generator by first removing the black diode trio lead from the outer brush housing. Then take off the three nuts from the stator winding terminals that were previously used to test for stator winding balance. Next, lift the diode trio assembly off the AC terminal studs.

Use the following procedure to test the diode trio using Figure 7-16 as a guide:

1. Use a DVOM with a diode check function or an ohmmeter having an R times 1,000 or an R times 10,000 scale.
2. Connect the negative (−) meter lead to the diode trio lead terminal as indicated in the diagram. Connect the positive (+) meter lead to each of the three copper contact pads one at a time. Observe and mark down the resistance reading for each contact.
3. Reverse the leads so the positive (+) meter lead is connected to the diode trio lead terminal. Connect the negative (−) meter lead to each of the contact pads. Again, observe and mark down the resistance reading for each contact.

Special Tools

Basic hand tools

Ohmmeter or DVOM with diode function

Classroom Manual
Chapter 7, page 150.

Diode tests are used to determine that the diodes are capable of rectifying AC to DC.

Readings Interpretation The diode trio is okay when a "low" resistance reading is observed in the other direction. If using an analog ohmmeter, the low resistance is observed by the full

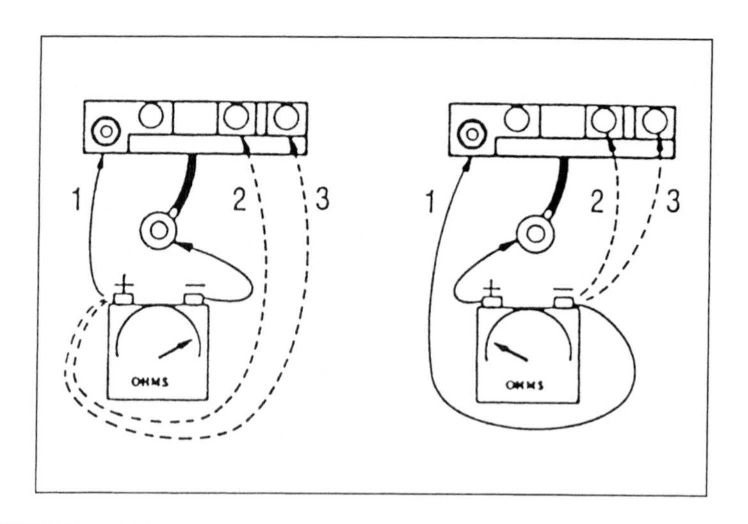

Figure 7-16 Using an ohmmeter or diode checker to test the diode trio. (Courtesy of Leece-Neville)

deflection of the needle. This means there is a voltage flow. The high resistance is indicated by infinity or very little or no needle deflection. Infinity resistance means there is no voltage flow.

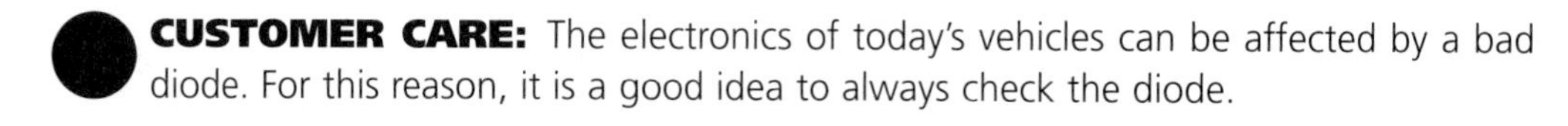

CUSTOMER CARE: The electronics of today's vehicles can be affected by a bad diode. For this reason, it is a good idea to always check the diode.

Voltage Regulator Adjustment

Classroom Manual
Chapter 7, page 150.

Some AC generators have **adjustable voltage regulators**.

Some integral charging systems have adjustable regulators to fine-tune the AC generator's output voltage.

WARNING: Adjustable regulators are not meant to compensate for bad batteries, connections, or AC generator defects. For example, a battery with a bad cell or low state of charge can cause any voltage regulator adjustment to be misleading. Remember that one function of the battery is to act as a voltage reference for the charging system.

Leece-Neville

Special Tools

Voltmeter

Small screwdriver

Leece Neville AC generator units are equipped with two types of regulators. One type is the fully adjustable regulator that is recognized by a flat cover plate. The other type is a three-step regulator distinguished by its finned, curved cover plate. In the Leese-Neville full fielding segment, we are directed to perform a voltage regulator adjustment procedure if the full field output reading is higher than the non full-field output reading, providing the stator outputs are balanced. The following procedure is used for the fully adjustable regulator (Figure 7-17):

SERVICE TIP: Before making any adjustments, the battery has to have at least a 95% state of charge. Also make sure the belt tension is okay and all wire connections are tight and in good condition.

1. Shut off all electrical accessories and run the engine at approximately 1,000 RPM.
2. Connect a voltmeter to the alternator outputs observing proper polarity and set the function at Volts DC.
3. Remove the protective plastic screw from the regulator and insert a small screwdriver in the hole. The blade of the screw needs to engage the adjustment screw slot inside the regulator.

Figure 7-17 Adjusting the voltage regulator.

WARNING: To prevent damage to the voltage regulator, do not force the adjustment beyond the low and high stops of the adjustment potentiometer screw. Exert only enough pressure to turn the screw.

4. Turn the screwdriver clockwise to raise the voltage and counterclockwise to lower it. Set the voltage between 14.0 and 14.2 volts.
5. Remove the screwdriver and voltmeter and install the plastic screw in the adjustment screw access hole.

If the voltage is excessively high and can't be lowered, the regulator is probably faulty and should be replaced. If the output voltage is low and can't be increased by adjustment, the regulator, alternator, or diode trio may be faulty.

WARNING: Before removing the regulator, be sure that the battery ground cable is disconnected to prevent circuit and/or component damage.

Three Step Adjustable Regulator

To adjust this type of regulator requires that the voltage regulator be lifted out of the housing (Figure 7-18). The following procedures are used for this type of regulator:

1. Disconnect the battery ground cable.
2. Remove the #10-32 nuts and lockwasher from the regulator terminal and disconnect the diode trio lead if the alternator is equipped with a diode trio.
3. Remove the four regulator hold down screws from the regulator cover. Lift the regulator out of the housing and move it out of the way as far as the leads will permit.
4. Inspect the two regulator brush contact pads. If dirty or corroded, clean the pads with #600 or finer paper.

SERVICE TIP: A charging system problem such as a low charge condition can be remedied by simply cleaning these pads. In this case, a voltage adjustment may not even be necessary.

Figure 7-18 Voltage regulator lifted off to perform a regulator adjustment.

5. Inspect and reinstall the brushes if they are good.
6. To adjust the voltage, remove and reinstall the adjustment strap in one of three positions:
   - ❑ For low—between terminals A and B
   - ❑ For medium—between terminals A and C
   - ❑ For high—between terminals B and C

   It is important to note that each change in the strap will result in an increase or decrease in the alternator output voltage of approximately 0.4 volts.
7. Reinstall the regulator.
8. Rehook the battery ground cable.
9. Start the truck and with a voltmeter hooked to the alternator's output terminal, verify proper output voltage.

Classroom Manual
Chapter 7, page 159.

DELCO-Remy

Some DELCO-Remy AC generators also have a provision to allow for minor voltage adjustment. Figure 7-19 shows a wiring diagram with the voltage adjustment. It is a network of connections and resistors that can easily be manipulated in segments to be a part of the battery voltage monitor section of the voltage regulator. To accomplish this, a voltage adjustment cap is used (Figure 7-20). When rotated, this cap can either add or reduce resistance to the monitoring circuit of the regulator. The cap is placed into an adjustment cube. Look at the wiring diagram in Figure 7-19. If the voltage adjustment cube becomes an open circuit, TR3 and TR1 will turn off, thus preventing high voltage.

To determine if voltage adjustment is required, you have to first perform a current output test. If the ampere output is within 10 amperes of rated output as stamped on the AC generator frame or manufacturer's specifications, the generator is a good candidate for voltage adjustment. To perform the adjustment, remove the adjuster cap and rotate it in increments of 90°, then reinsert it into the cube. In Figure 7-20, the cap is set for medium high voltage. If Position 2 were aligned with the arrow, it would be set for medium low. Position "LO" is low and position "HI" is the highest regulator setting.

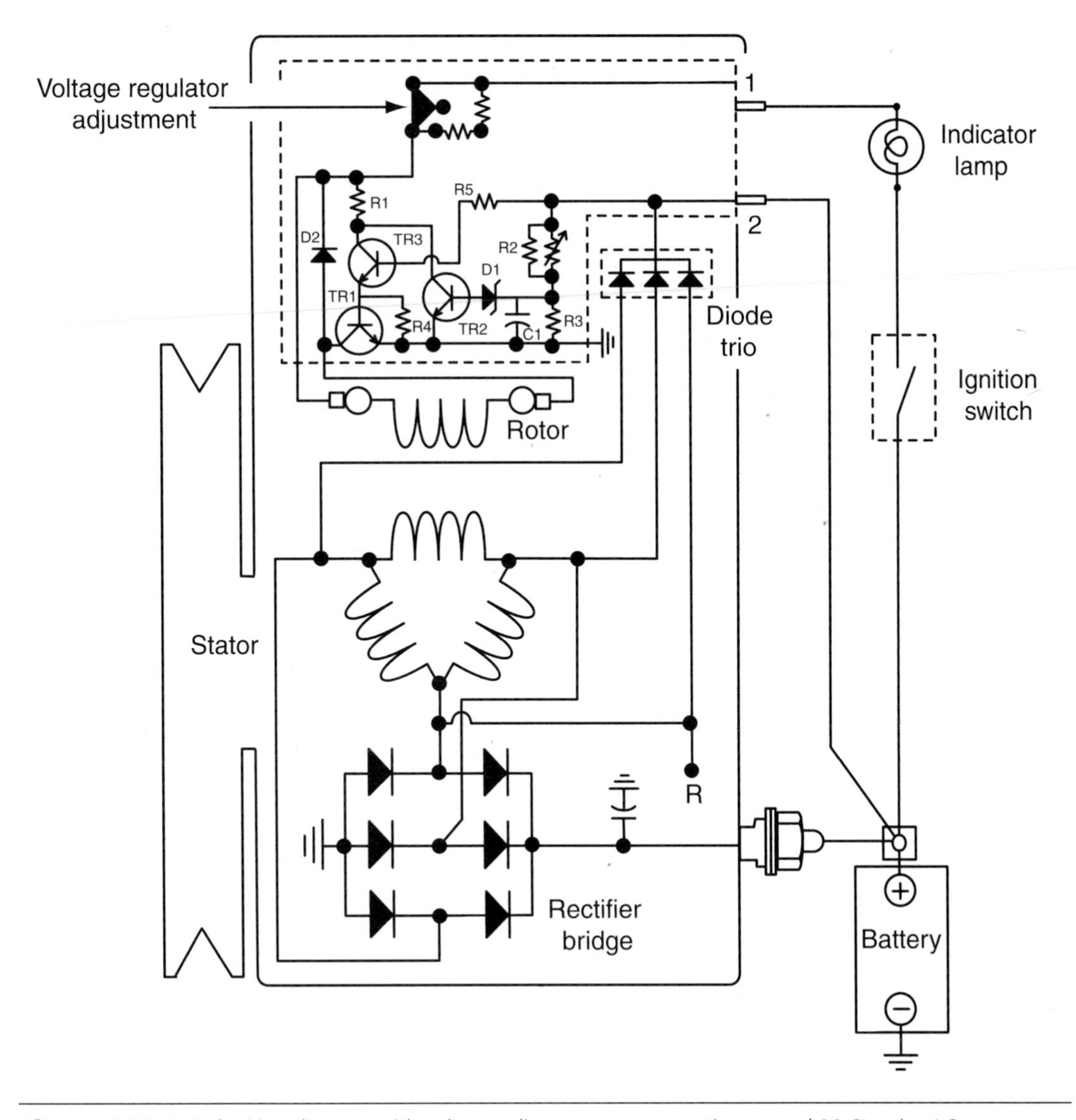

Figure 7-19 Typical wiring diagram with voltage adjustment on a negative ground 30-SI series AC generator.

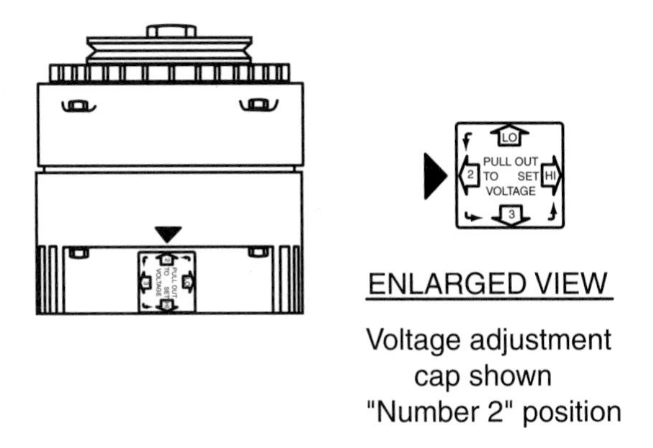

Figure 7-20 Voltage adjustment cap.

Special Tools

Basic mechanic's hand tool set

AC Generator Removal and Replacement

Removal of the AC generator varies somewhat from one vehicle to another. Some generators are easy to get at while others require the removal of other components to get at it or to make room so the generator can physically be maneuvered out of a tight area.

WARNING: Never attempt to disconnect any wires to the AC generator or remove it without first disconnecting the battery(s) negative cable, or circuit and/or component damage may occur.

Removal Procedure

The following is a typical removal procedure of an AC generator (Figure 7-21):

1. Disconnect the battery(s) ground cable.
2. Disconnect the wires from the alternator. Note their position for reassembly.
3. Loosen the mounting bolt assembly(s).
4. Loosen the adjustment bolt to pivot the alternator so the belt(s) can be removed.
5. Remove the upper adjustment bolt from the alternator.
6. Remove the lower pivot bolt(s) and remove the alternator from the vehicle.

Pulley Replacement Procedure

Whether the AC generator is rebuilt or a new one is installed, chances are the pulley has to come off.

WARNING: Do not attempt to prevent the shaft from turning by using a screwdriver or prybar in the cooling fan fins. Damage to the fan and front housing of the AC generator will occur.

WARNING: Sometimes a different manufacturer's AC generator is installed in place of the defective one. Make sure that the pulley distance in relation to the proper alignment of the belts remains the same. If the belt does not rotate in alignment from the drive pulley to the AC generator pulley, it can be ruined or thrown off.

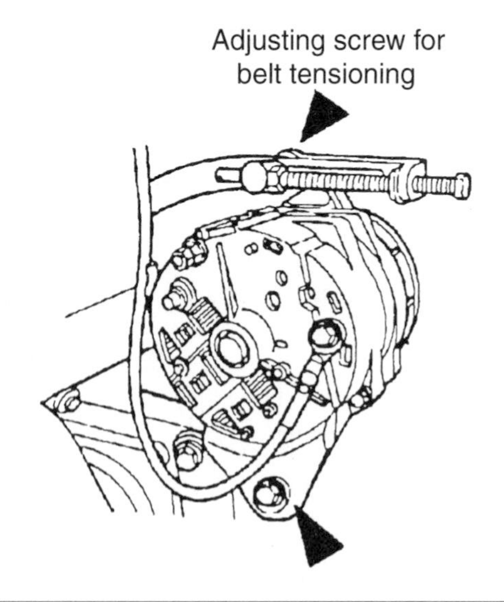

Figure 7-21 Mounting and belt adjustment bolts. (Reprinted with permission. Courtesy of Volvo Trucks North America, Inc.)

To remove a pulley, follow this typical procedure:

1. Place the AC generator into a bench vise. Use jaw protectors to prevent damage to the body of the generator.
2. Insert an Allen key into the end of the pulley shaft (typically, 7/16″).
3. Place a wrench over the pulley retaining nut and loosen it.
4. Remove the pulley and cooling fan from the shaft.

SERVICE TIP: Impact with a socket can often be used to spin the nut off. Even though the shaft spins with the impact gun, there is enough resistance in the alternator so that the impact gun speed is greater than the rotating shaft speed.

To install the pulley, follow this typical procedure:

1. Place the AC generator into the vise with jaw protectors.
2. Place the removed or new pulley and cooling fan onto the new alternator.

WARNING: Some AC generators use spacer washers to maintain proper cooling fan distance. Make sure the proper components and sizes are used when installing a cooling fan and pulley, or damage may occur.

3. Insert the proper size Allen key into the end of the pulley shaft.
4. Using a wrench, tighten the pulley nut to specifications. A typical torque is 60–70 foot pounds.

AC Generator Installation Procedure

1. Position the AC generator to the mount bracket and install the lower pivot bolt(s) and nut(s). Do not tighten this all the way.
2. Install the upper adjustment bolt. Do not tighten this.
3. Place the belt onto the pulley.

WARNING: Do not pry the belts to get them into place on the pulley. Doing this can damage the belt(s) and/or pulley.

4. Adjust the belt tensioner or apply pressure to the AC generator front housing to get the proper belt tension.
5. Tighten the alternator mounting and adjuster bolts and nuts to the proper torque specifications.
6. Reconnect the wires, observing proper polarity and position.

Figure 7-22 shows a table of proper tension required for different widths of belts. As you can see, the table has a figure for new and used belts. A belt is considered used if it is operated more than 10 minutes. A tension gauge is the proper tool to measure the belt's tension (Figure 7-23).

Special Tools

Basic hand tools

Belt tension gauge

Pry bar

SERVICE TIP: If a **belt tension gauge** is not available, belt tension can be determined by depressing the belt at the center of its span. As a rule, 3/8″ is the approximate distance that the belt should be allowed to move.

CUSTOMER CARE: A new belt tends to stretch when it is initially used. If possible, allow the truck to run a while and then readjust the belt(s). Also, it is a good practice to advise the customer to periodically check a new belt for tension until it settles down to a stable condition. This can prevent a needless "comeback."

BELT WIDTH (INCHES)	NEW BELT TENSION MIN–MAX	USED BELT INSTL. TENSION MIN–MAX
.380	140–150	60–100
.440	140–150	60–100
1/2	140–150	60–100
11/16	160–170	60–100
3/4	160–170	60–100
7/8	160–170	60–100
6 rib	160–170	70–120

Figure 7-22 Belt tension chart. (Reprinted with permission. Courtesy of Volvo Trucks North America, Inc.)

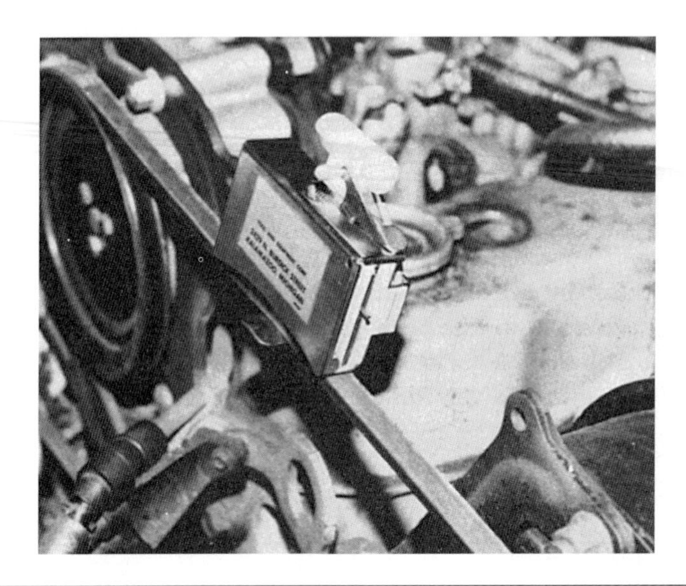

Figure 7-23 Using a belt tension gauge is the only way to be sure there is proper belt tension.

Bench Testing the AC Generator

Bench testing is a set of tests and procedures used to check the integrity of a generator that is out of the vehicle.

Bench testing is performed to check the integrity of the many components within the AC generator that are difficult or impossible to do while still in the vehicle. These tests are usually performed by people who rebuild these generators to determine which component(s) needs to be repaired or replaced to make the generator functional again. To become an all around proficient technician, it is important to be able to perform these tasks in case the need ever arises. We will use a Leece-Neville AC generator as a guide to perform various component tests. Many of the tests are similar to other manufacturers' bench testing procedures.

Special Tools

Basic hand tools

Press

Various pullers

Leece-Neville

The following are typical steps used to disassemble some of the more popular Leece-Neville AC generators such as the 2000JB, 2500JB, 2700JB, 2800JB, and 2805JB.

1. Remove the pulley nut, pulley, fan, key, and spacer.
2. Remove the four regulator hold down screws and carefully lift the regulator free of the housing. Remove the red and black leads from the regulator and note their position for reassembly (Figure 7-24).

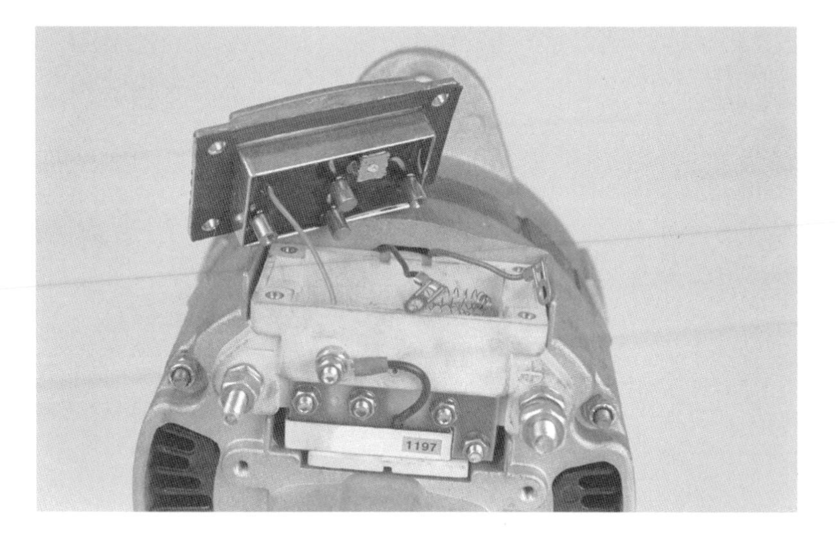

Figure 7-24 Accessing voltage regulator and brushes.

3. Remove the diode trio lead to terminal or regulator housing. Loosen the inner nut, which will allow the blue regulator lead to be withdrawn from under the head of the terminal screw. Remove the regulator (Figure 7-24).
4. Lift the brush and spring assemblies out of the housing (Figure 7-24).
5. Remove the three nuts and lift the diode trio off of the AC terminal studs (Figure 7-25).
6. Remove three self-locking nuts and through bolts.
7. Remove the rotor and drive housing assembly from the stator and slip ring end housing assembly. If drive end housing binds on the stator, loosen by tapping gently on the mounting ear with a fiber hammer. Be sure that the drive end housing separates from the stator and that the stator remains attached to the slip ring end housing to avoid damage to stator leads (Figure 7-26).
8. Remove the three nuts that secure stator leads to terminals and remove the stator.
9. Remove nuts from positive and negative output terminal bolts and then remove the bolts. Note the location of the red and black regulator leads on the heat sinks (Figure 7-27).

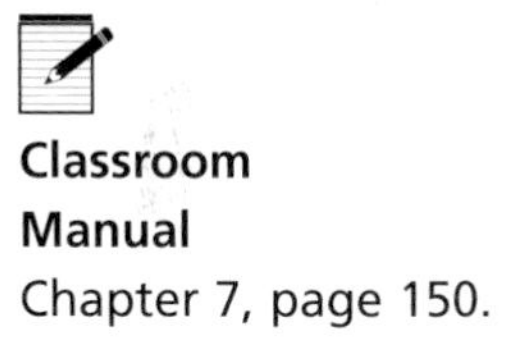

Classroom Manual
Chapter 7, page 150.

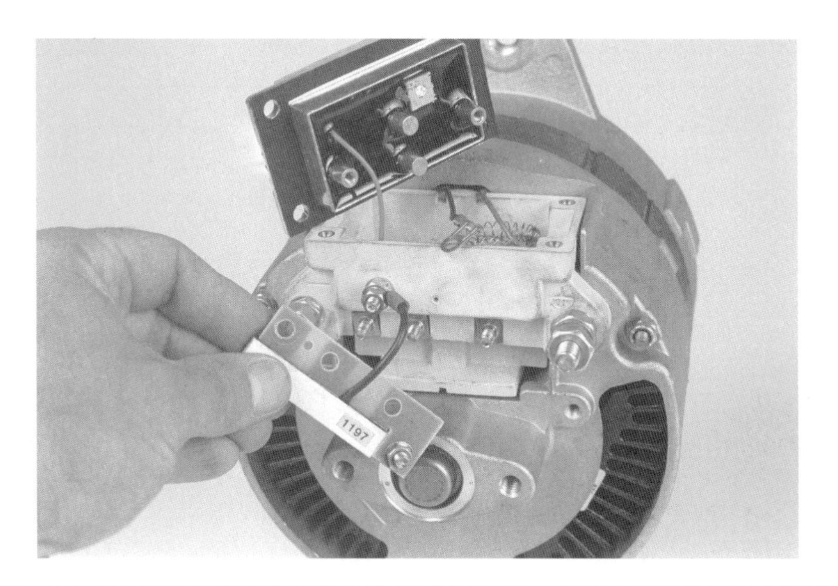

Figure 7-25 Remove the three nuts and lift the diode trio off the AC terminal studs.

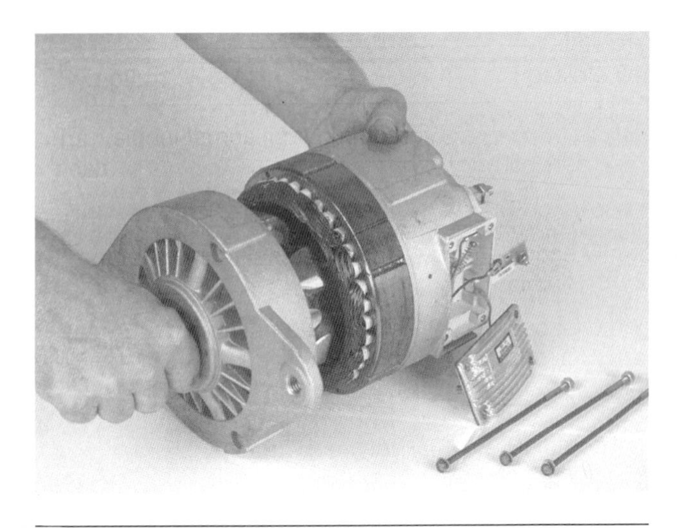

Figure 7-26 Separating the housings from the stator.

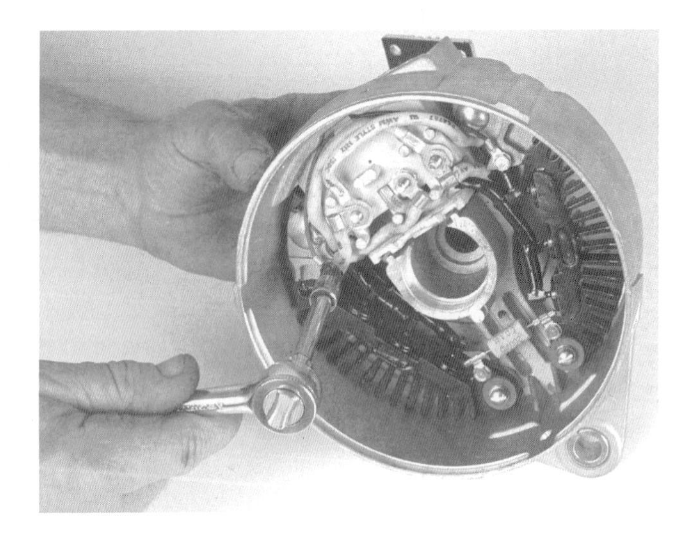

Figure 7-27 Remove nuts from positive and negative output terminals and remove the bolts.

Classroom Manual
Chapter 7, page 165.

10. Remove three hex head screws and remove the capacitor connected between the heat sinks (Figure 7-27).
11. Remove regulator housing. Note the location of the gasket that seals the brush compartment (Figure 7-28).
12. Remove the terminal stud insulating bushings from the housings. There are two bushings in each terminal hole (Figure 7-28).
13. Remove the two screws, lock washers, guard washers, and insulating washers, which retain the lower end of the heat sinks. Remove the heat sinks. Note the location of the insulating washers and bushings (Figure 7-29).
14. Pry the flanged dust cap out of the housing (Figure 7-30).
15. Slip ring bearing replacement should seldom be required. However, if it is, press the bearing from the inside of the housing outward (Figure 7-31).

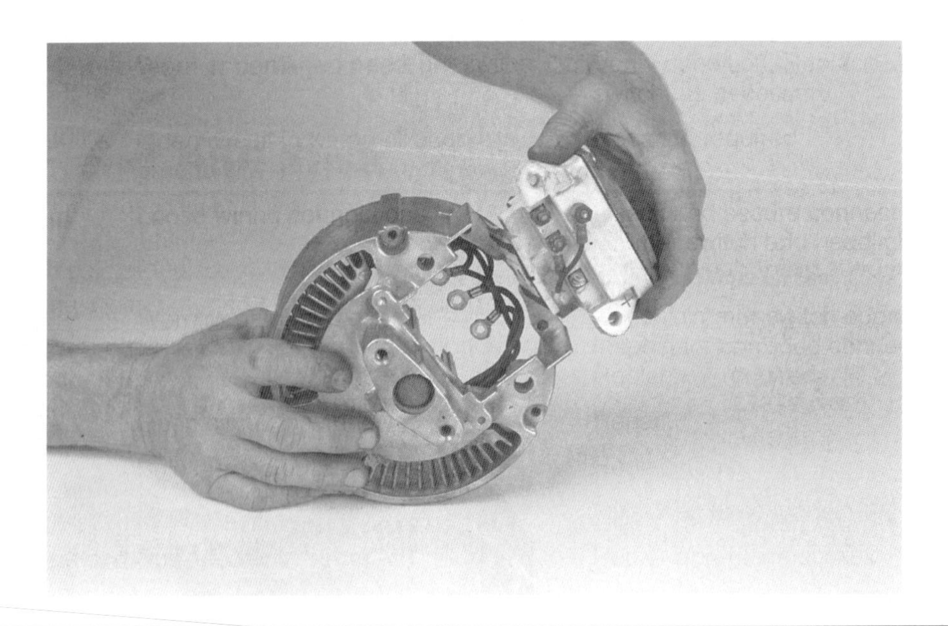

Figure 7-28 Remove the terminal stud insulating brushes from the housings.

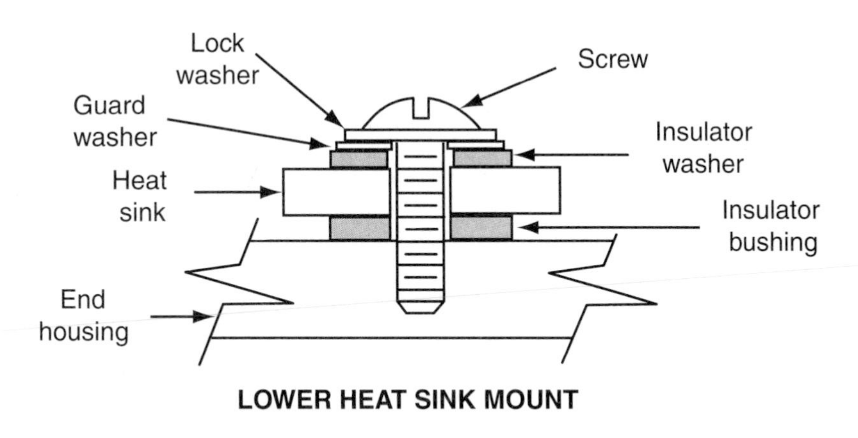

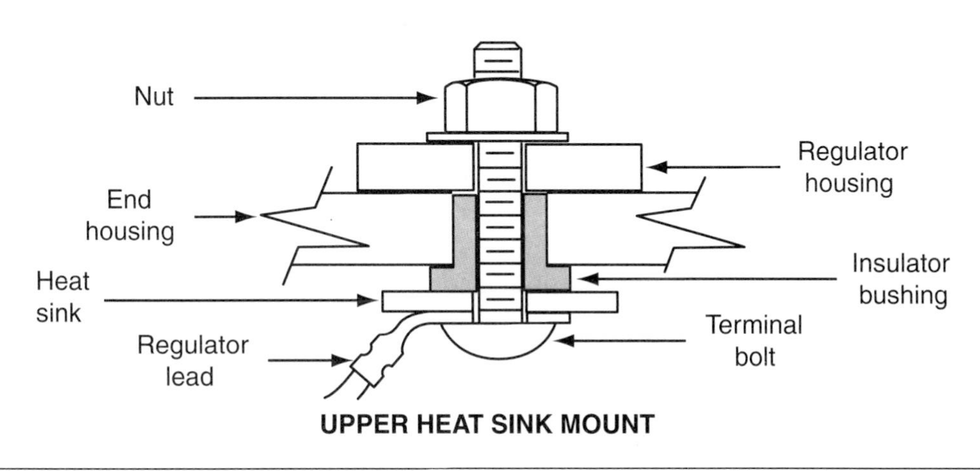

Figure 7-29 Note the location of the insulating washers and bushings when removing the heat sinks.

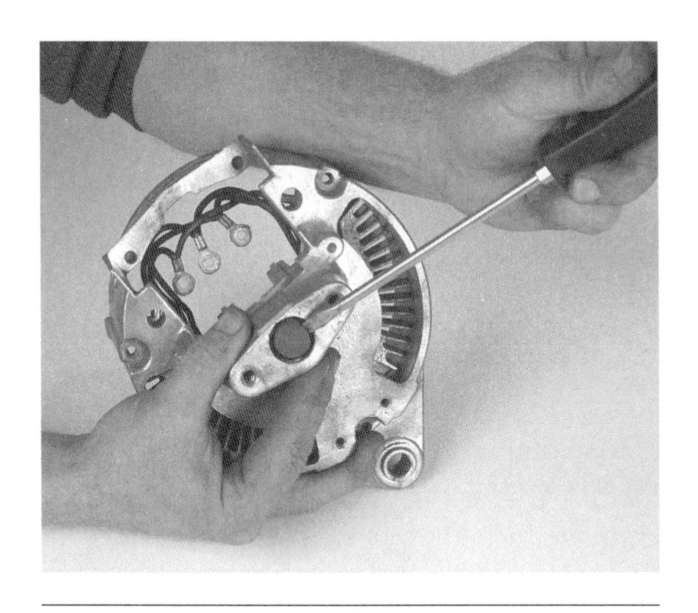

Figure 7-30 Pry the flanged dust cap out of the end housing.

Figure 7-31 Slip ring bearing replacement. Note: press the bearing from the inside of the housing outward.

Figure 7-32 Removing the drive end housing and bearing assembly using a puller.

Figure 7-33 Remove four screws and bearing retainer and press the bearing out of the drive end housing.

16. Using a puller or arbor press, remove the drive end housing and bearing assembly from the rotor shaft (Figure 7-32).
17. Remove four screws and the bearing retainer and press the bearing out of the drive end housing (Figure 7-33).

Once the AC generator is disassembled, the parts should be carefully inspected for any wear, cracks, breakage, or any other defects. Discard all damaged parts. Next we can proceed with component testing.

Special Tools

Ohmmeter or DVOM with diode function

115–220 test lamp

Diode Tests

Diode tests may be performed on heat sink assemblies without removing them from the end housing. If they are tested in this manner, remove the stator and be sure that the red and black leads are disconnected from the regulator and are not touching each other. Be sure that the diode trio has been removed from the AC studs, and disconnect the capacitor across the lower end of the heat sink.

Diodes are tested to insure that they pass current in one direction only. Diodes that do not allow current flow in either direction are said to be open, while those that pass current in both directions are said to be shorted. The best method to test diodes is with a meter that has a diode testing function. However, an ohmmeter or a battery-powered test light may be used.

Classroom Manual
Chapter 7, page 146.

Positive Heat Sink Tests

The positive heat sink is the one to which the positive output terminal is connected. Also, the square hole in the terminal end of the positive heat sink is larger than the terminal hole of the negative heat sink. To perform the test:

1. Connect the positive lead of the meter to the positive heat sink and touch the negative lead to each of the three diode terminals (Figure 7-34A). A high resistance should be indicated. If a test light is used, it should not light. If any of the three diodes show a low resistance or the test lamp lights, the diode is shorted.
2. Reverse the leads so the negative test lead is connected to the heat sink and touch the terminal of each diode with the positive lead (Figure 7-34B). Now a low resistance reading should be obtained. A high resistance or a test lamp that fails to light indicates an open diode.

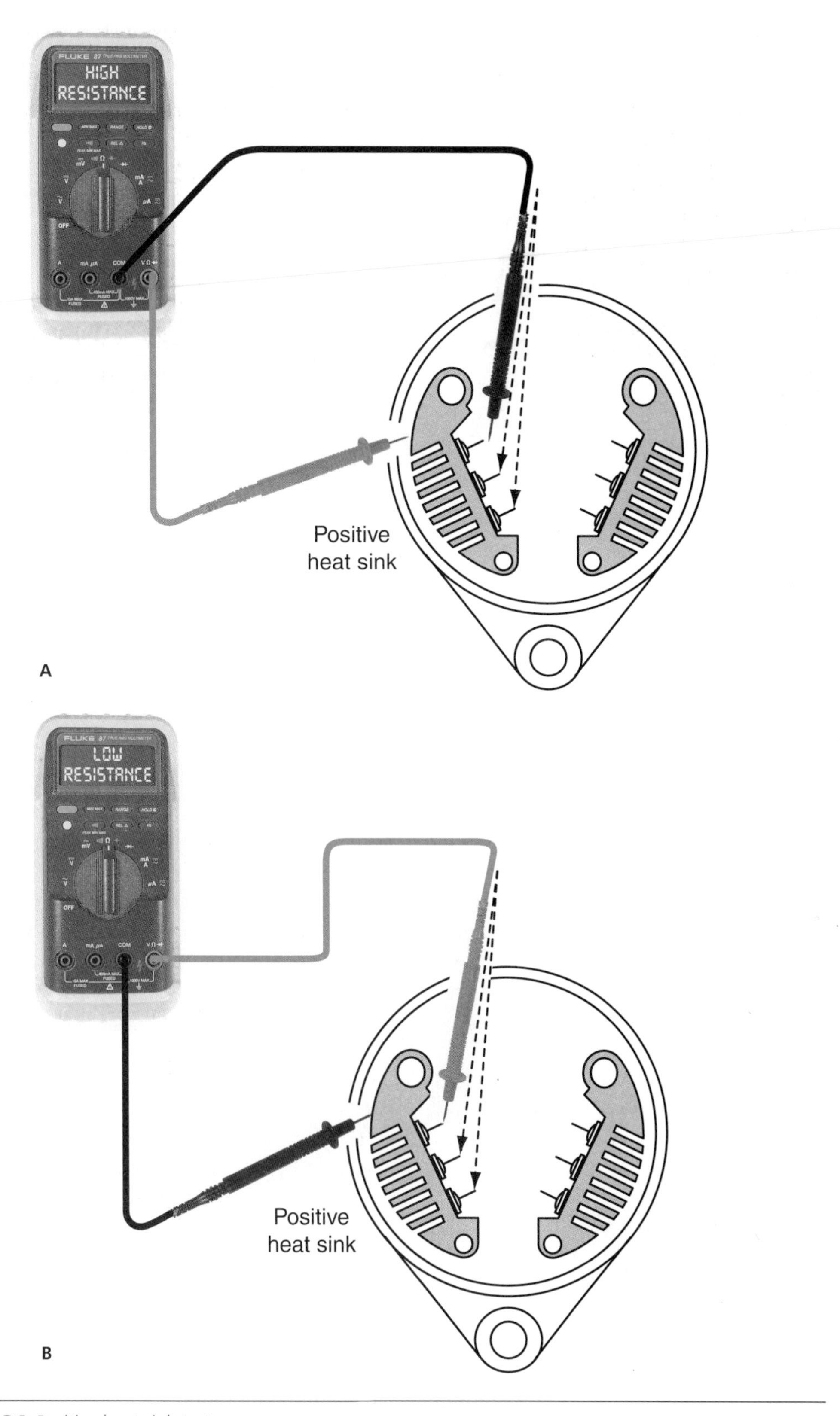

Figure 7-34 Positive heat sink test.

Classroom Manual
Chapter 7, page 146.

Negative Heat Sink Test

1. Connect the negative test lead to the negative heat sink and touch the positive test lead to each of the diode terminals (Figure 7-35A). If a low resistance reading is obtained or if the test lamp lights, the diode is shorted.
2. Reverse the test leads so that the positive lead is connected to the negative heat sink and the negative terminal is touched to each diode terminal (Figure 7-35B).

A low resistance reading should be obtained, or the test light lamp should light. If a high resistance is indicated or if the test light lamp is off, the diode is open. If a bad diode is detected, the entire heat sink assembly should be replaced.

Classroom Manual
Chapter 7, page 150.

Diode Trio Test

The same meter or test equipment is used to check the diode trio. The same test is performed as shown in the on-vehicle test. See the full fielding section Leece-Neville AC generator diode trio test.

Capacitor

The **capacitor** connected across the heat sinks may be tested in one of two ways. If a capacitor tester is available, the value derived should be 0.5 MFD and 200 working volts DC. However, if a capacitor tester is not available, an ohmmeter can be used in its place (Figure 7-36). Connect

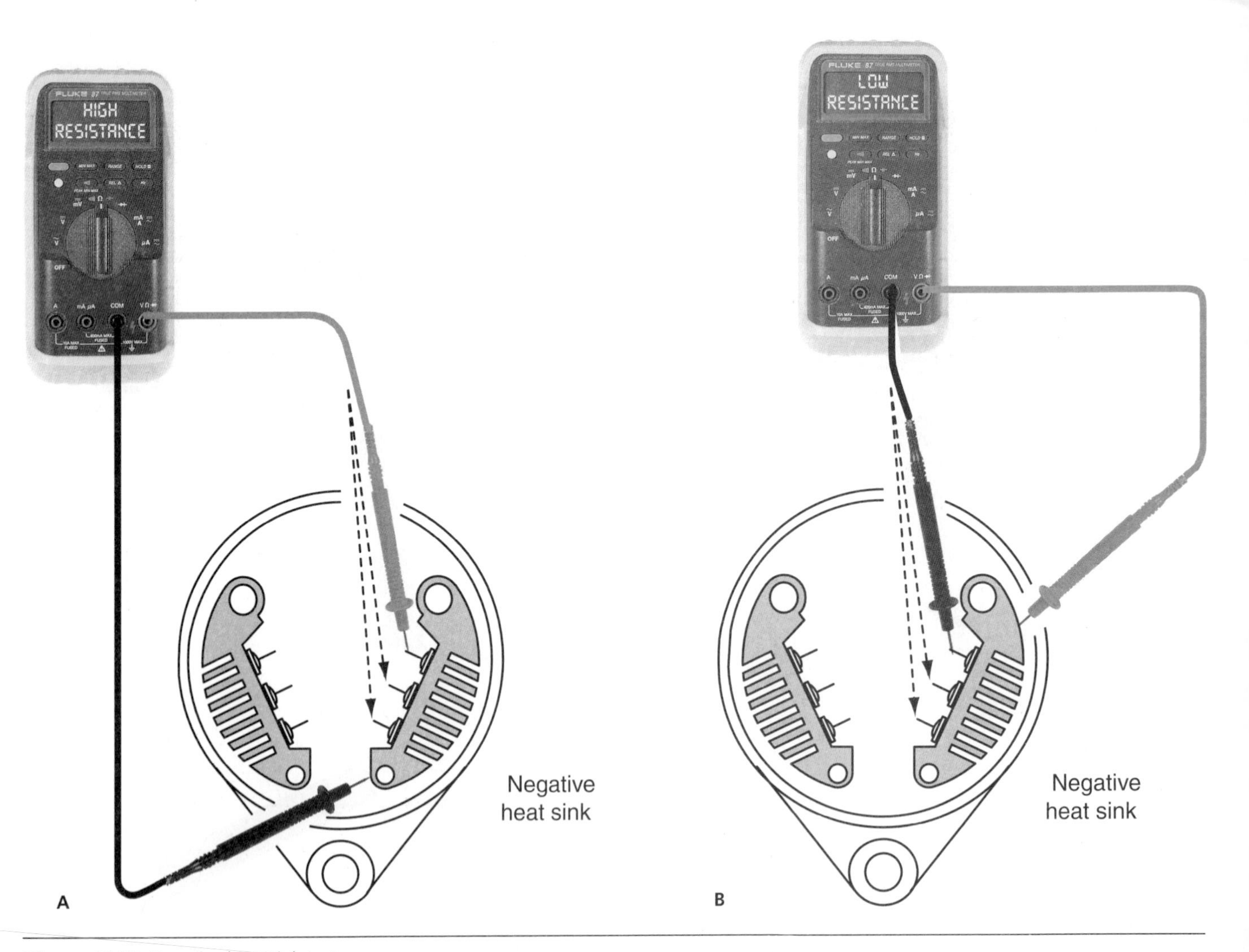

Figure 7-35 Negative heat sink test.

the meter leads across the capacitor that was removed in the disassembly procedure. The reading should indicate a very high resistance. Low resistance indicates a shorted or leaking capacitor, which should be replaced.

Rotor Test

In the **rotor test**, the rotor is checked for grounds and proper resistance using an ohmmeter. Prior to performing the test, a good visual inspection of the rotor should be done. The windings should show no signs of overheating or discoloration. Also, the slip rings should be flat, smooth, and free of any damage. If any of these conditions are indicated, the rotor should be discarded.

Classroom Manual
Chapter 7, page 140.

To check a rotor for a grounded coil, connect the ohmmeter leads between the rotor shaft and either slip ring (Figure 7-37). It should have a very high resistance reading (infinity). If any other reading is obtained, the rotor coil is grounded and must be replaced. To check for rotor coil resistance, connect the ohmmeter leads across the two slip rings (Figure 7-38). The readings obtained need to match manufacturer's specifications. The following are some examples:

Leece-Neville 2500JB: 2.2–5.5 ohms
2700JB: 1.9–2.3 ohms

DELCO-Remy uses a method whereby the specified resistance value can be determined by dividing the voltage by the field current, as shown previously in Figure 7-7. For example, in that figure if we picked generator model 1117485, we would get the following reading:

$$12V \div 2.7A = 4.4 \text{ ohms}$$
$$12V \div 3.4A = 3.52 \text{ ohms}$$

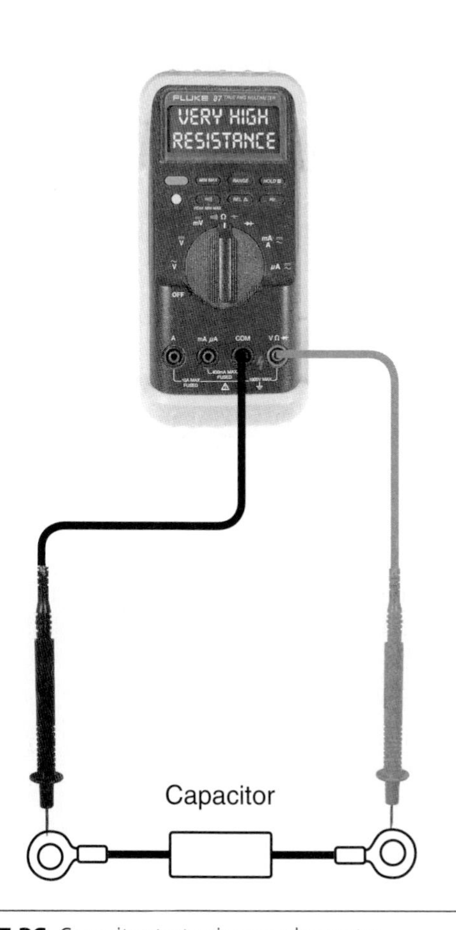

Figure 7-36 Capacitor test using an ohmmeter.

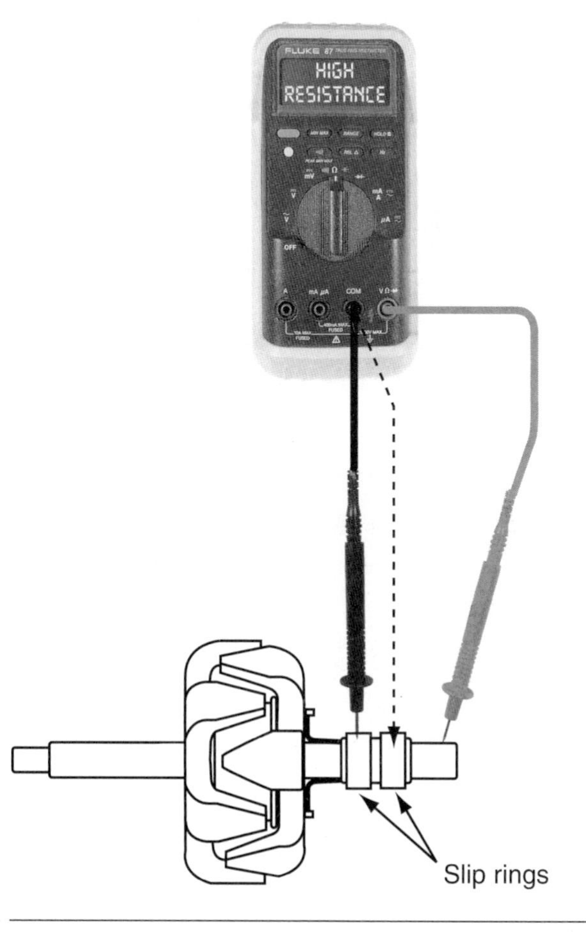

Figure 7-37 Rotor coil ground test.

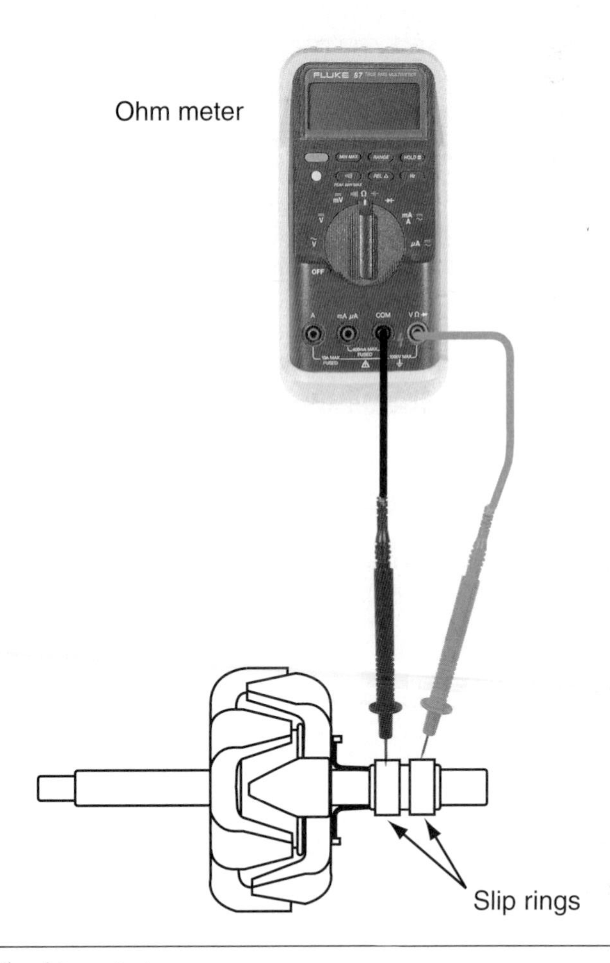

Figure 7-38 Rotor coil resistance test.

As you can see, the values fall in place with the field ohms reading of 2.7–3.4 ohms. To attain this small of a resistance reading in accuracy requires the use of a digital ohmmeter.

SERVICE TIP: Before condemning a rotor for an open coil, check the solder points at the slip ring leads. If a poor connection is suspected, resolder and retest before discarding the rotor. If proper resistance is not obtained after soldering, then discard the rotor.

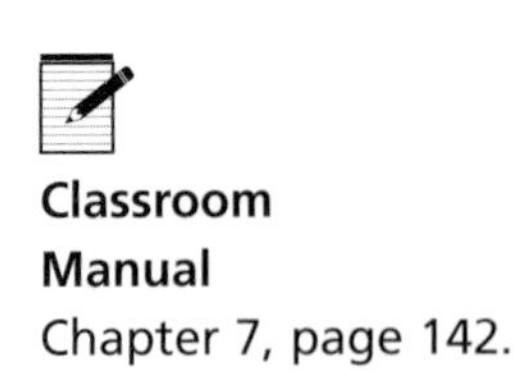

Stator Test

In the **stator test**, a good visual inspection of the stator should also be performed. Heat damage can show up by discoloration. Also, check for damaged windings. Note that the windings of the stator have very low resistance, making it hard to measure with a conventional ohmmeter. For that reason a 115–220 volt test light is used (Figure 7-39).

The leads are connected alternately (Figure 7-40) between the center lead and each outer lead and then between the two outer leads. For continuity of the stator, the lamp should light on all three phases. For a stator ground test, connect one lamp test lead to the stator body (Figure 7-41). Alternately touch the other lamp lead to each of the three stator leads. The lamp should not light. Discard any stator that fails the continuity or ground test. However, if an ohmmeter is used, the hookup in Figure 7-42 is used to test the stator for opens. Connect the leads in combination as previously performed using the test light. If the resistance is high (infinity) between any two leads the stator has an open and it must be replaced.

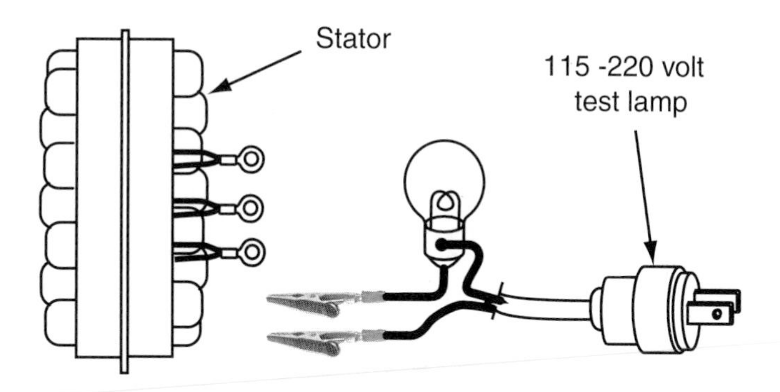

Figure 7-39 Using a 110-volt test light to test stator windings.

Amp should light on all 3 phases

Stator continuity test

Figure 7-40 Stator continuity test.

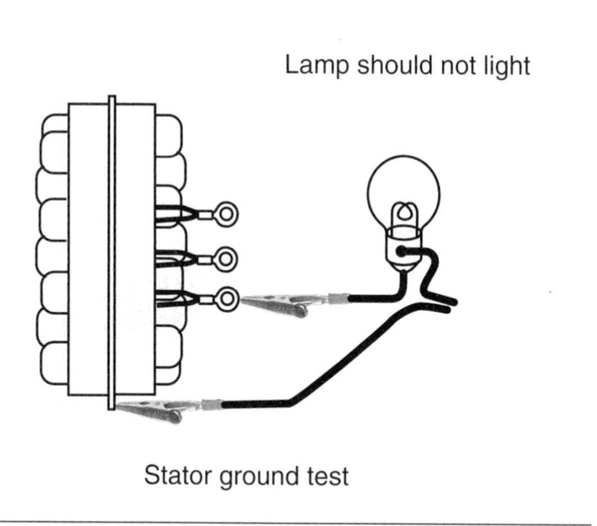

Figure 7-41 Stator ground test.

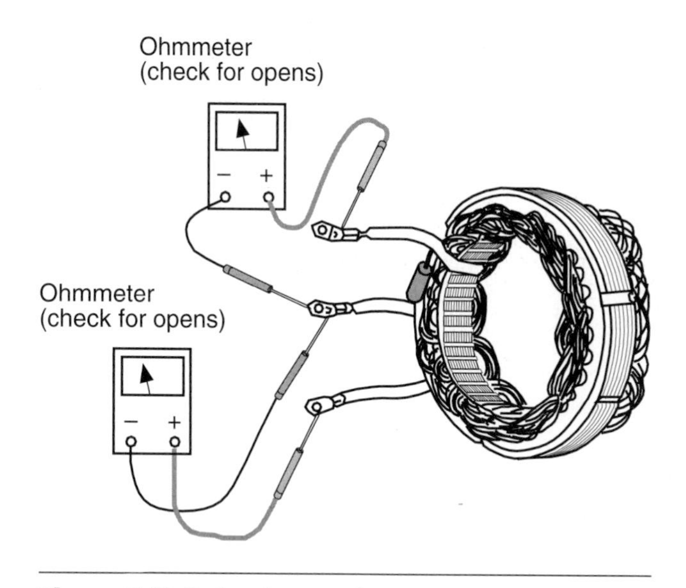

Figure 7-42 Testing the stator for opens.

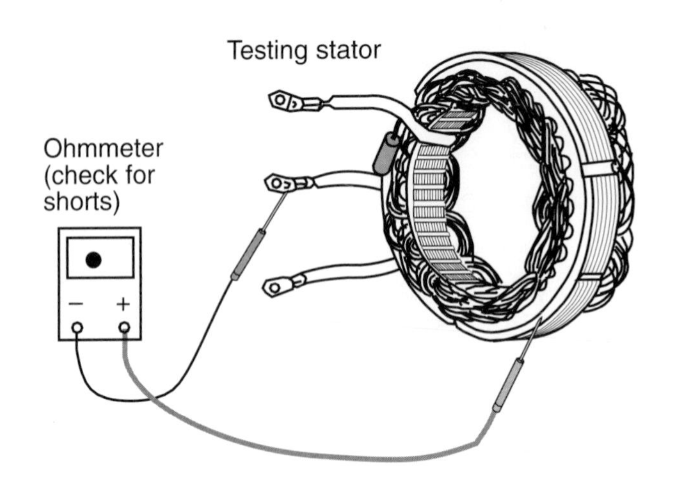

Figure 7-43 Testing the stator for shorts to ground.

An ohmmeter can be used to test the stator for a short to ground by connecting the meter leads as shown in Figure 7-43. The ohmmeter should read infinity on all three stator leads. If the reading is less than infinity, the stator is shorted to ground and it must be replaced.

Reassembly Once all components have been tested and appropriate action has been taken whenever required, the AC generator needs to be reassembled. This procedure is the reverse of the disassembly procedure. The following are highlights to ensure proper assembly.

1. If a slip ring end bearing has been removed, press a new bearing in place from the outside of the housing.

   It is important to note that the bearing should seat against the lip on the inside end of the bearing bore and that the manufacturer's part number stamped on it will face the outside of the housing.

2. Reinstall heat sinks, making sure that the upper and lower insulating washers are in their proper location.
3. Install terminal bolts and regulator lead wires.

   It is important to note that the red wire goes on the positive heat sink and the black wire on the negative heat sink.

4. Install the regulator housing. Make sure you install the two insulating bushings on each terminal bolt.

   Be sure that the regulator leads are properly routed through the cut-away section of the end housing.

5. Reinstall the capacitor.
6. Reinstall the stator. Align the stator and housing by temporarily installing the through bolts.
7. Press drive the end housing and bearing onto the rotor shaft.

WARNING: Use a sleeve around the shaft; press on the inner race to avoid brinnelling the bearing. (Brinnelling is a condition in which the needle rollers wear grooves in the surface or surfaces they ride on.)

8. Install the rotor and housing assembly into the stator and slip ring end housing assembly, being sure that the mounting ears are aligned.
9. Install the three through bolts and self-locking nuts. Torque to 50–60 inch pounds. Reinstall the metal cap after placing a small amount of proper grease in the housing.
10. Install the diode trio.

Figure 7-44 Using a small pin through a hole in the rear of housing helps keep the brush springs compressed.

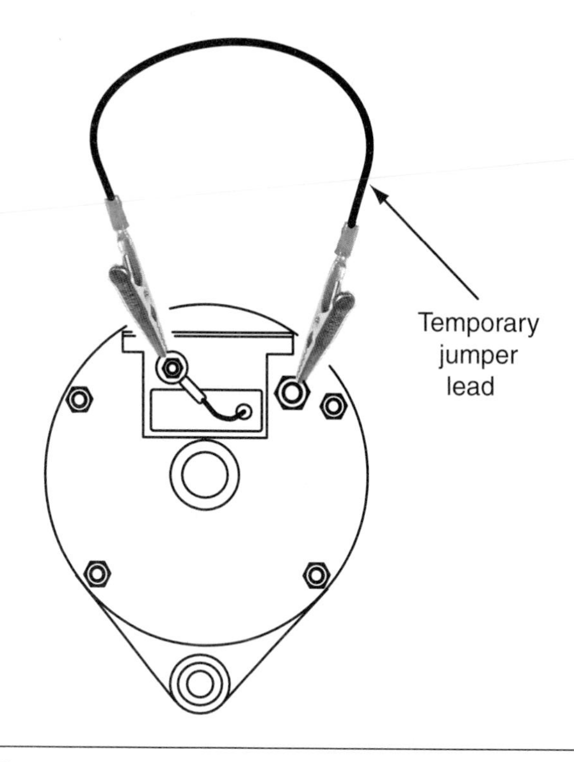

Figure 7-45 Restoring residual magnetism.

11. Insert the brush and spring assembly into the housing and compress the brush spring, using a small screwdriver. While holding the spring compressed, insert a small pin such as a 1/16″ drill bit through the hole in the rear of the housing and over the compressed brush spring to keep the spring compressed. Install the other brush and spring, pushing in the pin so both brush spring assemblies will stay compressed (Figure 7-44).
12. Attach the regulator leads to the regulator and install it. Withdraw the brush retaining pin.
13. Install the spacer, key, fan, pulley, and nut.

WARNING: Do not force or pound the pulley on the shaft or damage can result.

The next step is to test the AC generator for proper output using a proper test bench. Because the AC generator has been apart, it might have lost its residual magnetism. Before performing an output test, momentarily flash the field by connecting a jumper between the diode trio terminal and the alternator positive output terminal (Figure 7-45). This will restore the residual magnetism.

Diode Pattern Testing

The **diode pattern test** is often called a diode **ripple voltage test**. Ripple voltage is the AC component of the DC voltage after rectification. What we are looking for is any excess AC voltage that comes out with the DC. Sometimes this is called AC riding on the DC. Problems such as shorted or open diodes can result in high ripple. If the AC voltage riding on the DC line is high enough, it can affect the many electronic components in the vehicle. There are many sensors and components that work with AC-generated signals. If an unwanted AC signal reaches these com-

Special Tools

DVOM with volts AC function

Oscilloscope

Starter/charging system tester or equivalent with carbon pile

ponents, the computer can misinterpret a signal. In addition, many DC electronic components can be damaged if they cannot accept AC current.

Any AC voltage that is over 0.10VAC is excessive. One method of checking for this is to use a DVOM with a Volts AC function. The meter leads are placed on the output of the AC generator and ground. Many of the newer conductive type of battery, starter, and charging system tests perform a diode ripple test. **Oscilloscopes** are most often used to observe the AC generator's waveform for diode problems. Set an oscilloscope on a low scale or select a built-in diode pattern function. Connect the signal test lead to the AC generator output terminal and the ground lead to ground. Start the engine and place a moderate load on the charging system. Different loads placed on the charging system can affect the appearance of the pattern. Figure 7-46 illustrates a good diode pattern as does Figure 7-47, except that in the latter, the AC generator is under a full load. Patterns with high resistance, shorted, or open diodes are illustrated in Figures 7-48 through 7-51.

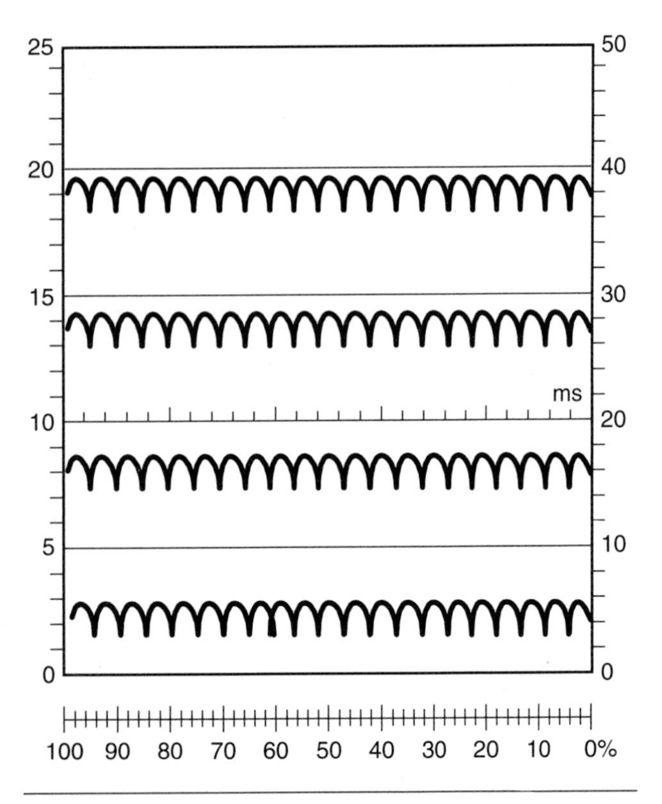

Figure 7-46 A good diode waveform. (Courtesy of Sun Electric Corporation)

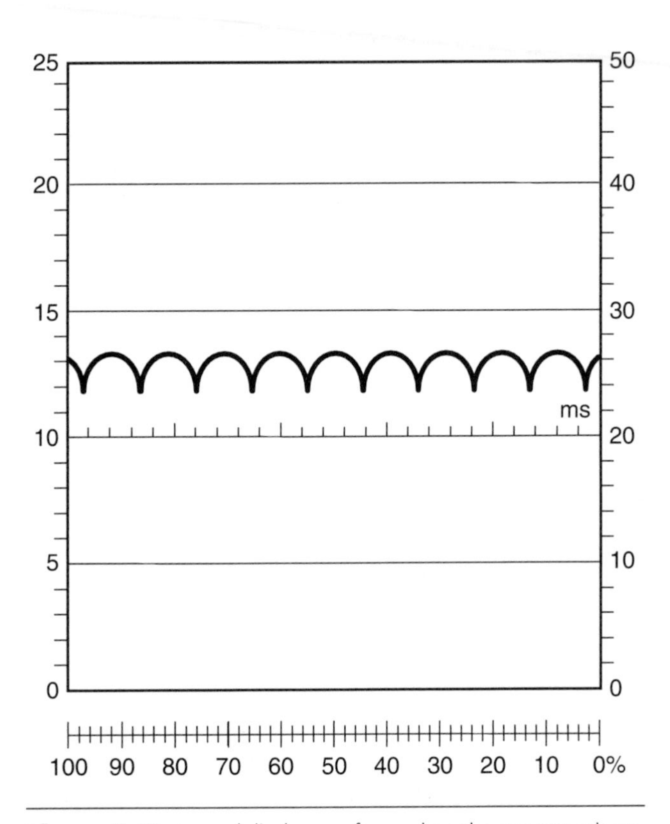

Figure 7-47 A good diode waveform when the generator is operating under full load. (Courtesy of Sun Electric Corporation)

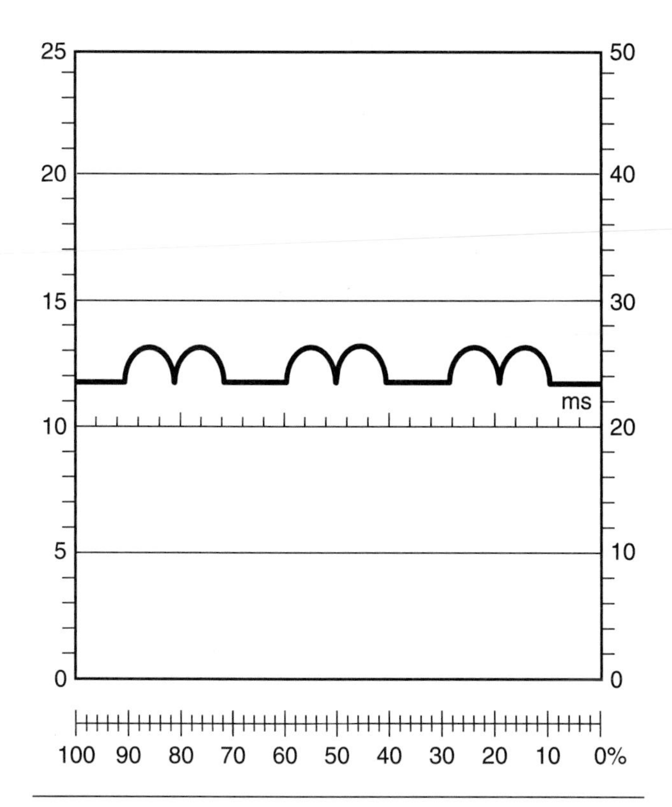

Figure 7-48 A waveform of the generator's output when under full load showing shorted diodes or a shorted stator. (Courtesy of Sun Electric Corporation)

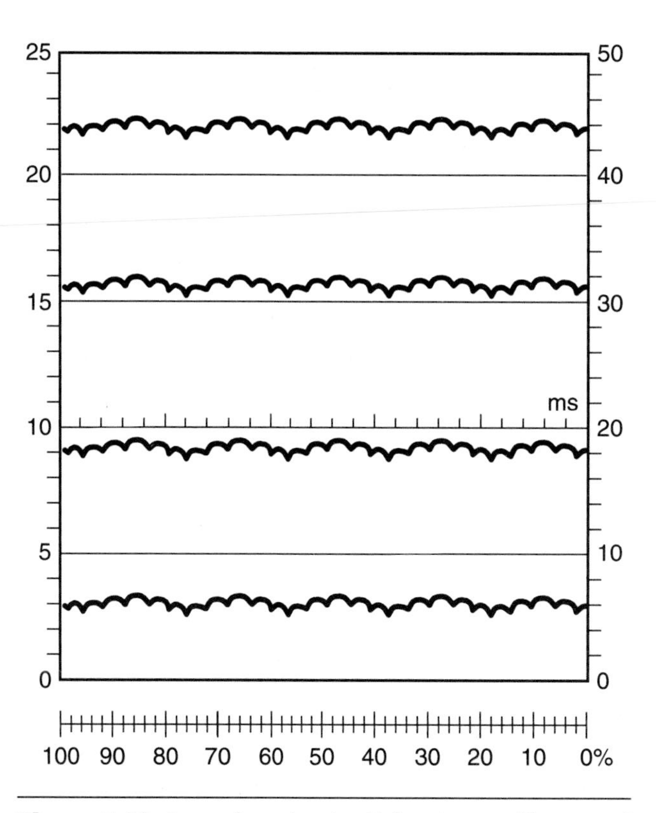

Figure 7-49 A waveform showing high resistance. (Courtesy of Sun Electric Corporation)

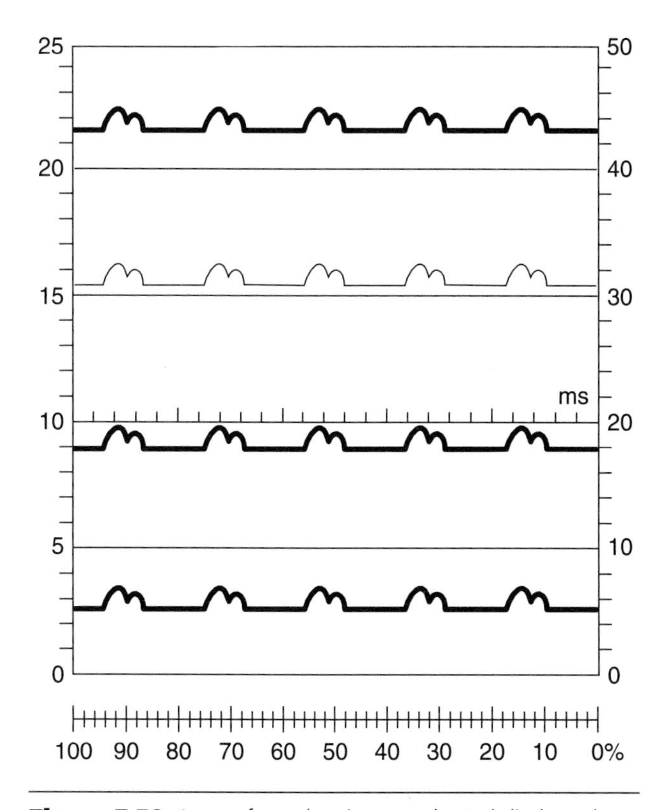

Figure 7-50 A waveform showing one shorted diode and one open diode. (Courtesy of Sun Electric Corporation)

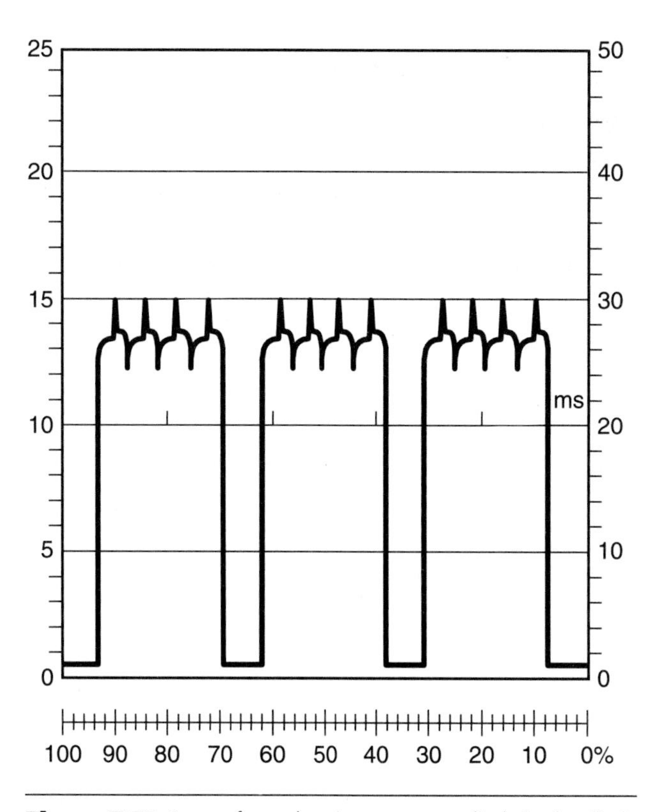

Figure 7-51 A waveform showing one open diode in the diode trio. (Courtesy of Sun Electric Corporation)

CASE STUDY

A customer brought her truck to a shop complaining that his vehicle's traction control system came on at approximately 40 miles per hour. This system is an option on some trucks that prevents slippage between the road surface and the tire during acceleration mode. The spinning of tires can do damage to the truck's drive components. Traction control systems take advantage of the existing antilock brake systems (ABS) in today's trucks. When an anti-spin-equipped vehicle accelerates on a slippery surface, the brakes are intermittently applied to any drive wheel that is about to exceed its specific acceleration speed as determined by the computer. The ABS/traction control computer can also communicate with the engine's computer to reduce power if too much is applied by the driver. This system relies on many signals and frequencies to ensure that the appropriate action is always taken.

The customer had taken her vehicle to another shop, where numerous tests were performed on the system and numerous components, including the computer, were replaced. However, the traction control would still come on intermittently at 40MPH, so the technician decided to check the alternator for the cause of the problem. He hooked up a labscope and found an open diode that at 40MPH set up a frequency and signal that the ABS computer somehow misinterpreted as a traction problem. The technician replaced the alternator and the problem disappeared.

Terms to Know

Adjustable voltage regulator
Belt tension gauge
Capacitor
Carbon pile
Current output test
Diode pattern test (ripple voltage test)
Diode trio
Full fielding test
Oscilloscope
Pulley ratio
Rotor test
Stator test
Undercharge
Voltage drop test
Voltage output test

ASE-Style Review Questions

1. *Technician A* says: Battery current on a brush type of AC generator is supplied to the field coil by a zener diode. *Technician B* says: Battery current on a brush type of AC generator is supplied to the field coil by the slip rings. Who is correct?

A. A only **C.** Both A and B
B. B only **D.** Neither A nor B

2. Full fielding is being discussed: *Technician A* says that on a Leece-Neville alternator a jumper wire is connected to a stiff wire that has made contact with the brushes and the negative terminal of the AC generator. *Technician B* says that on the same alternator a jumper wire is connected to the stiff wire and the positive terminal of the AC generator. Who is correct?

A. A only **C.** Both A and B
B. B only **D.** Neither A nor B

3. Diode testing is being discussed: *Technician A* says that a diode that has a high resistance in both directions is shorted. *Technician B* says that a diode that has a low resistance in both directions is open. Who is correct?

A. A only **C.** Both A and B
B. B only **D.** Neither A nor B

4. Replacing pulleys on AC generators is being discussed: *Technician A* says that an Allen wrench and a regular wrench should be used to loosen the pulley nut. *Technician B* says that placing the pulley itself tightly in a vise will prevent the shaft from turning, thus making it easier to get the pulley nut off. Who is correct?

A. A only **C.** Both A and B
B. B only **D.** Neither A nor B

5. *Technician A* says that if an AC generator that wasn't outputting puts out current when it is full fielded, there is a problem in the AC generator. *Technician B* says that if an AC generator that wasn't outputting puts out current when it is full fielded, there is a problem in the voltage regulator. Who is correct?
   A. A only **C.** Both A and B
   B. B only **D.** Neither A nor B

6. Charging systems are being discussed: *Technician A* says that a bad battery can be the cause of a charging system not working. *Technician B* says that a charging system that is not working properly will cause a battery to fail. Who is correct?
   A. A only **C.** Both A and B
   B. B only **D.** Neither A nor B

7. A squealing noise attributed to the charging system can be caused by all of the following except:
   A. A only **C.** Both A and B
   B. B only **D.** Neither A nor B

8. Voltage output test is being discussed: *Technician A* says that this test can be performed loaded and unloaded. *Technician B* says that engine speed should be approximately 1200 RPM or more to make sure the voltage regulator turns on. Who is correct?
   A. A only **C.** Both A and B
   B. B only **D.** Neither A nor B

9. Voltage drops are being discussed: *Technician A* says that if a voltmeter positive lead is hooked to the negative terminal or case of the alternator and the negative lead is hooked to ground on the engine, a voltage drop test on the ground side is being performed. *Technician B* says that the positive lead on the AC generator output and the negative lead on the "B" terminal of the starter solenoid indicate a voltage drop test is being performed on the power side of the charging system. Who is correct?
   A. A only **C.** Both A and B
   B. B only **D.** Neither A nor B

10. A rotor test is being performed: *Technician A* says to check for a grounded coil put the ohmmeter leads across the two slip rings. *Technician B* says to check for a grounded coil put the ohmmeter leads between the shaft and either slip ring. Who is correct?
    A. A only **C.** Both A and B
    B. B only **D.** Neither A nor B

ASE Challenge

1. A truck driver has pulled a truck into a shop complaining of a very high charge reading on the instrument panel voltmeter. *Technician A* says that if the batteries were extra warm and low on electrolyte, this would be a further indication of an overcharging condition. *Technician B* says you should check battery voltage with the engine at a fast idle; if the voltage is in excess of 14.6, there is definitely a problem. Who is correct?
   A. A only **C.** Both A and B
   B. B only **D.** Neither A nor B

2. A truck has been diagnosed as having an overcharging AC generator. *Technician A* says the condition could be due to a short in the output wiring. *Technician B* says the problem could be due to a short in the rectifiers. Who is correct?
   A. A only **C.** Both A and B
   B. B only **D.** Neither A nor B

3. A truck with a low charge rate problem is being diagnosed. No "D" hole, test hole, or exterior field terminal is evident. *Technician A* says a good way to test this system is to load the battery with the engine running at a fast idle and watch for a good generator output increase. *Technician B* says that when testing a generator, an increase in output amperage not voltage should be observed when either "loading" or "full fielding." Who is correct?
   A. A only **C.** Both A and B
   B. B only **D.** Neither A nor B

4. A truck has an intermittent charge/no charge condition. *Technician A* says this is likely due to a shorted stator winding. *Technician B* says it could be a loose wiring connection at the back of the alternator. Who is correct?
   A. A only **C.** Both A and B
   B. B only **D.** Neither A nor B

5. A truck driver complains that his rig runs with dim lights and that he has trouble starting the engine. Upon initial inspection it is determined that the battery is undercharged but otherwise okay and that the battery wiring and connections are okay. The charging system is then tested with a load on the electrical system. The amperage shows slightly low. With no load and a fast idle the voltage shows well below specs at 13.4 volts. *Technician A* says that the voltage regulator is likely the problem. *Technician B* says that if possible a full field test should be run. Who is correct?
   A. A only **C.** Both A and B
   B. B only **D.** Neither A nor B

Table 7-1 NATEF and ASE Task

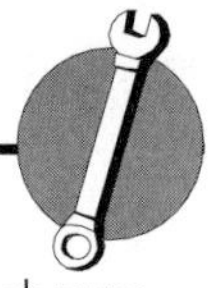

Diagnose dash mounted charge meters and/or indicator lights that show a no charge, low charge, or overcharge condition; determine needed repairs

Problem Area	Symptoms	Possible Causes	Classroom Manual	Shop Manual
CHARGING SYSTEM SHOWING NO CHARGE, LOW CHARGE, OR OVERCHARGE	Run-down battery Engine will not crank Battery electrolyte low Low battery open-circuit voltage	1. Voltage regulator too high or too low 2. Faulty regulator 3. Open stator 4. Shorted stator 5. Open field coil 6. Shorted field coil 7. Bad dash gauge 8. Worn or slipping belt 9. Worn brushes	170	180

Table 7-2 NATEF and ASE Task

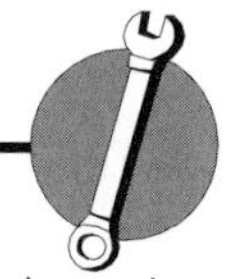

Diagnose the cause of a no charge, low charge, or overcharge condition; determine needed repairs

Problem Area	Symptoms	Possible Causes	Classroom Manual	Shop Manual
CHARGING SYSTEM SHOWING NO CHARGE, LOW CHARGE, OR OVERCHARGE	Run-down battery Engine will not crank Battery electrolyte low Low battery open-circuit voltage	1. Voltage regulator too high or too low 2. Faulty regulator 3. Open stator 4. Shorted stator 5. Open field coil 6. Shorted field coil 7. Worn or slipping belt 8. Worn brushes	149	177

Table 7-3 NATEF and ASE Task

Inspect, adjust, and replace alternator drive belts/gears, pulleys, fans, and mounting bracke

Problem Area	Symptoms	Possible Causes	Classroom Manual	Sh M
REPLACE AND ADJUST DRIVE BELT	Charging system output below specifications	1. Slipping or worn drive belt	137	177
REPLACE AC GENERATOR PULLEY	Belts wear prematurely	1. Pulley bent		
REPLACE AC GENERATOR FAN	Noises	1. Bent fan blades	148	198

Table 7-4 NATEF and ASE Task

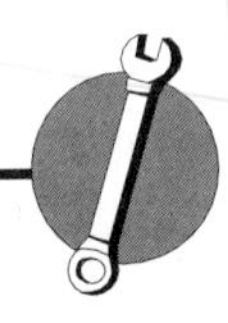

Perform charging system voltage and amperage output tests; determine needed repairs

Problem Area	Symptoms	Possible Causes	Classroom Manual	Shop Manual
OUTPUT TESTING	Charging system producing low or no output	1. Faulty regulator 2. Open stator 3. Shorted stator 4. Open field coil 5. Shorted field coil 6. Worn or slipping belt 7. Worn brushes	134	187

Table 7-5 NATEF and ASE Task

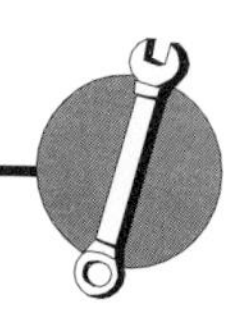

Perform charging circuit voltage drop tests; determine needed repairs

Problem Area	Symptoms	Possible Causes	Classroom Manual	Shop Manual
VOLTAGE DROP TESTING OF THE CHARGING SYSTEM	Low or no output	1. Excessive resistance in system circuit. 2. Open in the circuit 3. Improper ground connection 4. Loose or corroded connection		184

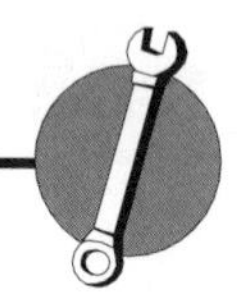

...EF and ASE Task

...ve and replace alternator

Problem Area	Symptoms	Possible Causes	Classroom Manual	Shop Manual
	AC generator fails output test	1. Faulty AC generator	159	198

...-7 NATEF and ASE Task

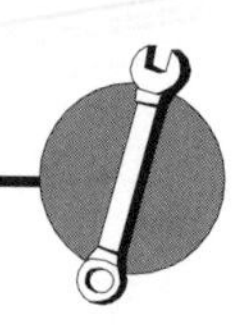

Inspect, repair, or replace connectors and wires in the charging circuit

...em Area	Symptoms	Possible Causes	Classroom Manual	Shop Manual
...RING IN ...NTROL ...RCUIT TO ...C GENERATOR	AC generator not charging	1. Open circuit to generator 2. Faulty connector	149	184

Table 7-8 NATEF and ASE Task

"Flash"/Full field alternator to restore residual magnetism

Problem Area	Symptoms	Possible Causes	Classroom Manual	Shop Manual
AC GENERATOR	AC generator not charging	1. Faulty voltage regulator	151	189

Job Sheet 10

10

Name: ______________________ Date: ______________________

Testing Charging System Output

Upon completion of this job sheet, you should be able to perform a charging system output test.

NATEF and ASE Correlation

This job sheet is related to ASE and NATEF Medium/Heavy Duty Truck Electrical/Electronic Systems List content area:

D. Charging System Diagnosis and Repair, ASE and NATEF Task 4: Perform charging system voltage and amperage output tests; determine needed repairs.

Tools and Materials

A medium/heavy duty truck
Service manual for the truck
Starting/charging system tester (VAT-40 or similar)
Inductive amp-probe equipment

Procedure

1. Describe the truck being worked on:
   Year: ______________ Make: ______________ VIN: ______________________
   Model: ______________________
2. Identify the type and model of AC generator. ______________________
   ____
3. What are the output ratings for this AC generator?
   ______________ amps and __________ volts at __________RPM.
4. Hook up an ammeter to the battery circuit and determine the draw requirement of the truck's electrical system.
   A. What is the draw? __________ amps.
   B. Is this within the limit of the AC generator's rated output? Yes ______ No ______
   C. If not, are there any electrical add-ons installed on the truck? Yes ______ No ______ If yes, identify ______________________
   ____
   D. What is your recommendation? ______________________
   ____
5. Connect the starting/charging system tester to the truck.
6. Start the engine and run it at the specified engine speed.
7. Rotate the load control until the proper load reading is obtained.
   Record: __________ amps at __________ volts.

WARNING: Do not leave the load on.

8. Compare the readings to specifications and give recommendations. ______________
   ____

9. If readings are not to specifications, perform a full field test following proper procedure for that specific AC generator. Describe the full fielding method to be performed.

____

____

10. Full field the AC generator.

    The readings are ________________.

11. What is your recommendation and why? ________________________________

____

☑ **Instructor Check** ________________________________

Job Sheet 11

11

Name: ______________________ Date: ______________________

Testing the Charging System Circuit

Upon completion of this job sheet, you should be able to test voltage output and voltage drop of both the insulated and ground side of the charging system circuit.

NATEF and ASE Correlation

This job sheet is related to ASE and NATEF Medium/Heavy Duty Truck Electrical/Electronic Systems List content area:

D. Charging System Diagnosis and Repair, ASE and NATEF Task 2: Diagnose the cause of a no charge, low charge, or overcharge condition; determine needed repairs; ASE and NATEF Task 5: Perform charging circuit voltage drop tests; determine needed repairs.

Tools and Materials

A medium/heavy duty truck

Wiring diagram for the truck

A DMM

Lead or jumper long enough to connect the meter from the battery to the AC generator

Procedure

1. Describe the truck being worked on:
Year: ____________ Make: ____________ VIN: ______________________
Model: ______________________

2. Measure the open circuit voltage of the battery. __________ volts

3. With the voltmeter still at the battery(s) and all electrical accessories turned off, start the engine.

4. Increase the engine speed as necessary to obtain a maximum voltage reading and record here. __________ volts.

5. Reconnect the voltmeter to the AC generator: the positive meter lead to the "BAT" output terminal and the negative lead to ground. (preferably the AC generator ground terminal.)

6. With all electrical accessories turned off, start the engine and increase the engine speed to obtain the maximum voltage reading and record here. __________ volts.
Was the reading the same as in step number 4? ____________ If not what are your recommendations? ____

____

7. For the following test, you will need the engine running and as many electrical accessories turned on as possible. Using the voltmeter, check for voltage drop at the following locations, observing proper polarity:

A. From AC generator "G" (ground terminal) to ground on engine. Record the reading. __________ volts.

B. From battery negative terminal to ground on frame. Record the reading. __________ volts.

C. From positive battery terminal to starter solenoid "B" terminal. Record the reading. __________ volts.

D. From starter solenoid "B" terminal to AC generator "B" terminal. Record the reading. __________ volts.

8. Compare the readings and record your recommendation. ______________________

____

Charging System Diode Check

9. Connect the DVOM to the AC generator output terminals observing polarity.
10. Set the function to volts AC.
11. With the engine at fast idle, measure the AC current at the output terminal. Record the reading. __________ volts AC.
12. What is your determination and/or recommendation? ______________________

____

Instructor Check ____

Lighting Troubleshooting and Repairs

Upon completion and review of this chapter, you should be able to:

- ❑ Explain regulations relating to lighting systems.
- ❑ Replace and aim sealed-beam and composite headlights.
- ❑ Diagnose the cause of dimmer than normal lights.
- ❑ Diagnose lighting systems that do not work.
- ❑ Test a master headlight switch.
- ❑ Diagnose problems with taillight systems with stop and turn signal lamps.
- ❑ Replace a dimmer switch.
- ❑ Remove and reinstall sealed-beam taillights.
- ❑ Diagnose problems with a turn signal circuit.
- ❑ Diagnose problems with a trailer light system.

Basic Tools

Hand tools

Service manual

Test light

Safety glasses

Introduction

Classroom Manual
Chapter 8, page 176.

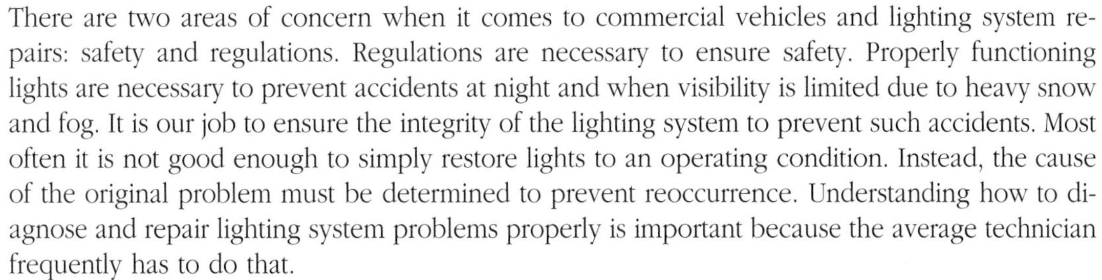

There are two areas of concern when it comes to commercial vehicles and lighting system repairs: safety and regulations. Regulations are necessary to ensure safety. Properly functioning lights are necessary to prevent accidents at night and when visibility is limited due to heavy snow and fog. It is our job to ensure the integrity of the lighting system to prevent such accidents. Most often it is not good enough to simply restore lights to an operating condition. Instead, the cause of the original problem must be determined to prevent reoccurrence. Understanding how to diagnose and repair lighting system problems properly is important because the average technician frequently has to do that.

The lighting systems in trucks tend to be the longest power carrying circuits, because power has to reach lights in the rear, on top, on the sides, in the front, and within the cab, and any other inside vehicle area requiring lighting. The circuits may easily consist of more than 50 bulbs and thousands of feet of wire. Also included within the circuits are switches, circuit protection devices, relays, and connectors. As you can see, any failures require a systematic approach in order to diagnose, locate, and perform a timely, effective repair. For instance, the simple act of replacing a bad bulb requires the technician to be able to properly identify bulbs in order to use the proper type and size bulb for the application.

CUSTOMER CARE: The industry spends a lot of money in needless replacement of bulbs. It is a common practice for technicians to automatically replace a bulb without checking its or the circuit's integrity. Once a new bulb or light is used, the old one is rarely returned to the shelf even though there was nothing wrong with it. This practice can contribute to an expensive maintenance cost for a fleet. That is not to say that when a single bulb is not operating the cause is usually the bulb itself. When troubleshooting lighting systems, start with the most obvious cause first and don't lose sight of the basics of electricity.

Regulations

Classroom Manual
Chapter 8, page 177.

Many repairs and inspections that you perform on commercial vehicles fall under a number of federal, state, and possibly local laws and regulations. One guideline that is used by many operators in the industry is the **Federal Motor Carrier Safety Regulations** (FMCSR) (Figure 8-1).

Part 382	Controlled Substances and Alcohol Use and Testing
Part 383	Commercial Drivers License Standards; Requirements and Penalties
Part 387	Minimum Levels of Financial Responsibility for Motor Carriers
Part 390	General
Part 391	Qualifications of Drivers
Part 392	Driving of Motor Vehicles
Part 393	Parts and Accessories Necessary for Safe Operation
Part 394	Removed and Reserved
Part 395	Hours of Service of Drivers
Part 396	Inspection, Repair, and Maintenance
Part 397	Transportation of Hazardous Materials
Part 399	Employee Safety and Health Standards
Part 400	Procedures for Transportation Workplace Drug and Alcohol Testing Programs
Appendix G	Minimum Periodic Inspection Standards

Figure 8-1 Federal Motor Carrier Safety Regulations (FMCSR).

Federal Motor Carrier Safety Regulations (FMCSR) need to be adhered to on an ongoing basis. These regulations exist to insure the safety of everyone on the road.

CAUTION: PART 382 and PART 40 of the FMCSR pertain to controlled substances and alcohol use. The use of controlled substances and alcohol can jeopardize the health of your fellow workers or the general public when it affects the quality of your work. Considering the lives and property that can be placed in jeopardy by negligence, the industry is committed to abiding by all the regulations.

We are not going to cover all the regulations; however, we will look at parts of them that pertain to lighting. After all, the vehicle that you are working on might someday receive a "random roadside inspection." The first part we will look at is PART 396—Inspection, Repair, and Maintenance.

- ❑ 396.1 Scope.
  General—Every motor carrier, its officers, drivers, and agents, representatives, and employees directly concerned with the inspection and maintenance of motor vehicles shall comply and be conversant with the rules of this part.
- ❑ 396.3 Inspection, repair, and maintenance.
  "(a)General—Every motor carrier shall systematically inspect, repair, and maintain, or cause to be systematically inspected, repaired, and maintained, all motor vehicles subject to its control.

  (I) Parts and accessories shall be in safe and proper operating conditions at all times. These include those specified in Part 393 of this subchapter and any additional parts and accessories which may affect safety of operation, including but not limited to, frame and frame assemblies, suspension systems, axles and attaching parts, wheels and rime, and steering systems."

  One area that commercial vehicle owners are concerned with is the possibility of having the vehicle and/or driver placed "out of service." The "Out of Service Criteria" is a document that is under constant revision. It is published (and updated annually) by the **Commercial Vehicle Safety Alliance (CVSA)** and it abides with the Federal Motor Carrier Safety Regulations.

  The inspections, referred to as road side inspections, are usually performed by law enforcement officials. When a vehicle goes through an inspection, not all of its defects will cause it to be put out of service. Also many law enforcement officials have the right to come to the shop or fleet's facility to perform different levels of vehicle inspections.

Many states require that annual safety inspections be performed on any vehicle registered within the applicable state.

Federal Annual Inspection requirements are met by annual inspections performed in certain states and some provinces of Canada.

Let's look at PART 393 and how it affects the lighting system.

- ❑ 393.1 Scope of rules in this part.
  Every employer and employee shall comply and be conversant with the requirements and specifications of this part. No employer shall operate a commercial motor vehicle or cause or permit it to be operated, unless it is equipped in accordance with the requirements and specifications of this part.

- ❑ 393.5 Definitions

  Note that we will refer to lighting only. As used in this part, the following words and terms are construed to mean:
  - ❑ Clearance lamp—A lamp used on the front and rear of a motor vehicle to indicate its overall width and height.
  - ❑ Grommet—A device that serves as a support and protection to that which passes through it.
  - ❑ Hazard warning signals—Lamps that flash simultaneously to the front and rear on both the right and left sides of a commercial motor vehicle to indicate to an approaching driver the presence of a vehicular hazard.
  - ❑ Head lamps—Lamps used to provide general illumination ahead of a motor vehicle.
  - ❑ Identification lamps—Lamps used to identify certain types of commercial motor vehicles.
  - ❑ Lamp—A device used to produce artificial light.
  - ❑ License plate light—A lamp used to illuminate the license plate on the rear of a motor vehicle.
  - ❑ Reflective material—A material conforming to Federal Specifications L-S 3000, "Sheeting and Tape, Reflective; Non-exposed lens, Adhesive backing" (Sept. 7, 1965) meeting the performance standard in either Table 1 or Table 1A of SAE Standard J594f, "Reflex Reflectors" (January 1977).
  - ❑ Reflex reflector—A device used on a vehicle to give an indication to an approaching driver by reflected light from the lamps on the approaching vehicle.
  - ❑ Side marker lamp (Intermediate)—A lamp shown to the side of a trailer that indicates the approximate middle of a trailer 30 feet or more in length.
  - ❑ Side marker lamps—Lamps used on each side of a trailer to indicate its overall length.
  - ❑ Stoplamps—Lamps shown to the rear of a motor vehicle to indicate that the service brake system is engaged.
  - ❑ Taillamps—Lamps used to designate the rear of a motor vehicle.
  - ❑ Turn signals—Lamps used to indicate a change in direction by emitting a flashing light on the side of a motor vehicle towards which a turn will be made.

Subpart B—Lighting devices, reflectors and electrical equipment

- ❑ 393.9 Lamps Operable.

  All lamps required by this subpart shall be capable of being operated at all times.
- ❑ 393.11 Lighting devices and reflectors.

Note: This part sets forth the required color, position, and required lighting devices by type of commercial motor vehicles and was shown in Chapter 8 of the Classroom Manual.

Figures 8-2, 8-3, 8-4, and 8-5 relate to regulation in PART 393 pertaining to lights.

392.30

All vehicles on the road must have their lamps on during the period of one-half hour after sunset to one-half hour before sunrise. At any time of the day, vehicles must use lamps when there is not enough light to clearly see persons and vehicles up to 500 feet away.

COMBINED LAMPS OR REFLECTORS: **393.22(a)**

Two or more lamps and reflectors may be combined if:
1. Each required lamp and reflector conforms to the regulations.
2. The means of mounting does not impede visibility of the lamp or reflector.
3. A nonrequired lighting device does not impede the visibility of the lamp or reflector.

Prohibited combinations: **393.22(b)**

1. A turn signal lamp may not be combined with a head lamp or any device that produces a greater intensity of light than the turn signal.
2. Clearance lamps cannot be combined with tail lamps nor identification lamps.
3. Turn signal lamps may not be combined with stop lamps unless the stop lamp function is deactivated when the turn signal function is activated.

Figure 8-2 Regulations pertaining to lamps or reflectors.

INSPECTING LIGHTING DEVICES - A CHECKLIST

This manual addresses the necessary lighting devices for straight trucks, tractors, and semi- and full trailers. ONLY THOSE LAMPS INCLUDED IN THE NORTH AMERICAN STANDARD AND WALK AROUND INSPECTIONS ARE LISTED.

Component	Possible Defects	
1. ALL LAMPS:	• Missing • Inoperative • Improper color • Unsecure mounting • Poor electrical connection (flickering, blinking, etc.)	393.25 393.9
2. ELECTRICAL WIRING or "PIGTAIL": (Wiring for lamps that extends from towing to the towed units)	• Crimping • Cracks • Abrasions in protective coating • Exposed, twisted-together wiring • Wires resting on other components • Loose connections	393.28
3. PROJECTING LOADS: (Objects extending four feet beyond the side or the rear of the vehicle.)	• Day or night, must have at least one operating red lamp on each side of the projecting load. • One red flag, 12 inches square; must be placed wherever lamps are required. (This flag is used IN ADDITION to the red lamp.)	393.18 393.87
4. STRAIGHT TRUCKS: (80 inches or more in width)	• STRAIGHT TRUCK FRONT: a. At least two head lamps, an equal number on each side. (White) b. Two turn signals, one on each side. (White or amber) c. Two emergency flashers, combined with turn signals. EMERGENCY FLASHERS ARE NOT INCLUDED IN THE NORTH AMERICAN STANDARD.	393.12

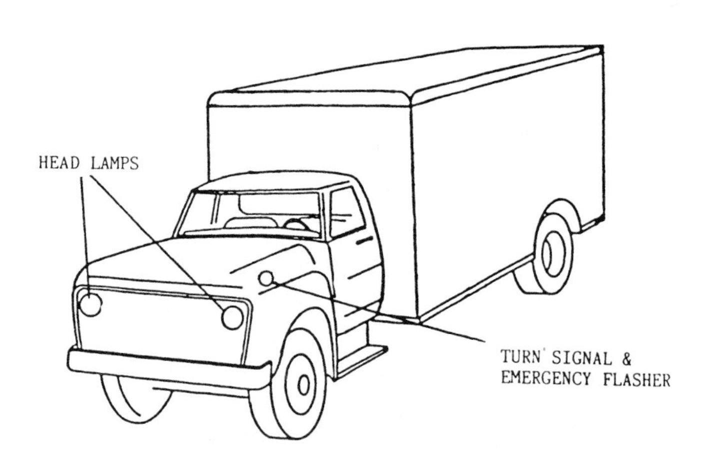

STRAIGHT TRUCK - FRONT AND SIDE VIEW

Figure 8-3 Lighting checklist.

- STRAIGHT TRUCK REAR:
  a. Two tail lamps, one on each side. (Red)
  b. Two stop lamps, one on each side. (Red)
  c. Two turn signals, one at each side. (Red, yellow, or amber)
  d. Two emergency flashers, combined with turn signals. EMERGENCY FLASHERS ARE NOT INCLUDED IN THE NORTH AMERICAN STANDARD.

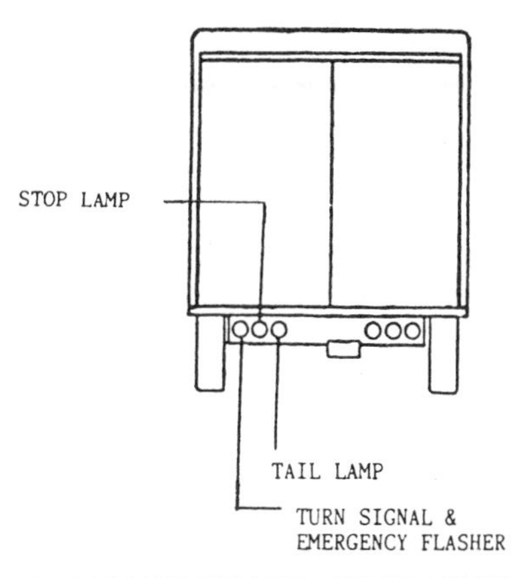

STRAIGHT TRUCK - REAR VIEW

5. TRUCK-TRACTOR: (80 inches or more in width)

- TRUCK-TRACTOR FRONT: **393.13(a)**
  a. At least two head lamps; equal number on each side. Must have upper and lower beams. (White)
  b. Two turn signals; one on each side. (White or amber) **393.25(e)(f)**
  c. Two emergency flashers, combined with turn signals. EMERGENCY FLASHERS ARE NOT INCLUDED IN THE NORTH AMERICAN STANDARD. **393.19**

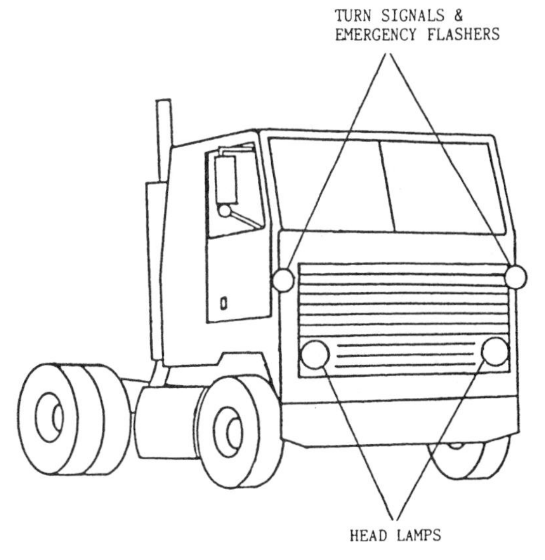

TRUCK-TRACTOR - FRONT AND SIDE VIEW

Figure 8-4 Regulations pertaining to truck-tractor lighting.

393.13(b)

- TRUCK-TRACTOR REAR:
  a. One tail lamp minimum on back of frame. (Red)
  b. One stop lamp minimum on back of frame. (Red)
  c. Unless the turn signals on the front are double-faced and visible to passing drivers, two turn signals on the rear of the cab, one at each side. (Red, yellow, or amber)
  d. Two emergency flashers, combined with turn signals. EMERGENCY FLASHERS ARE NOT INCLUDED IN THE NORTH AMERICAN STANDARD.

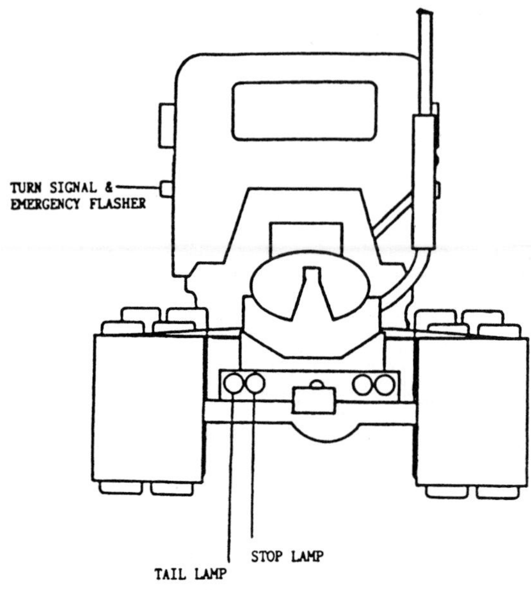

TRUCK-TRACTOR - REAR VIEW

393.14(b)

6. LARGE SEMITRAILERS & FULL TRAILERS: (80 inches or more in width)

- REAR OF TRAILERS
  a. Two tail lamps, one at each side. (Red)
  b. Two stop lamps, one at each side. (Red)
  c. Two turn signals, one at each side. (Red, yellow, or amber)
  d. Two emergency flashers, combined with turn signals. EMERGENCY FLASHERS NOT INCLUDED IN THE NORTH AMERICAN STANDARD.

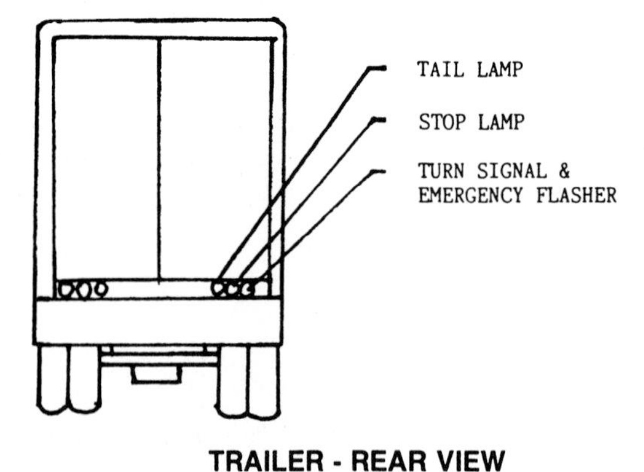

TRAILER - REAR VIEW

Figure 8-5 Lighting checklist.

CAUTION: Follow regulations very carefully. As a rule, they are very specific so do not read anything into them.

CUSTOMER CARE: There is nothing worse than sitting on the side of the road because a roadside inspection revealed that the vehicle's required lights weren't working. When checking lights be sure to check the wiring as well. The lights can illuminate at the moment, but the first time the vehicle goes over a bump, a bad connection or wire can lose its integrity, causing the lights to fail. Checking the wiring prevents conversations with angry customers.

Common Lighting Problems and Repair Practices

Lights in trucks can fail for various reasons. For example, a blown bulb can be caused by water hitting a hot bulb that had a cracked lens. However, water reaching the bulb socket area has also caused the socket and pigtail to corrode. Will replacing the bulb cause the light to work again? Possibly, if there is enough contact between the bulb base and the contacts. Will the problem reoccur if the lens is not changed? Probably. Will the next bump or shock impact cause the bulb to socket contact to lose its connection? Probably. A proper repair would require the replacement of the lens, bulb, socket, and pigtail or a complete light assembly to prevent the recurrence of the light failure.

The following are major reasons that lights fail in a truck, bus, tractor, or trailer:

1. Physical damage—Light assemblies, sealed lamps, and lenses are manufactured to meet many strict standards. However, a strong enough impact can indirectly or directly damage the bulb, or **filament**.
2. Shock and vibration—These can be caused by poorly mounted light assemblies or even a loosely fitted bulb in a socket. Through new designs today's light manufacturers have come a long way in eliminating these types of problems.
3. Corrosion—This problem occurs due to lack of proper sealing, not only in the lens area, which allows contaminants to enter the bulb or filament area, but also in the wiring and all its connections. To prevent these types of problems, many fleets and shops use a nonconductive corrosion-resistant compound to seal the lighting system's problem areas. The use of heat shrink tubing on wires and connections is highly recommended.
4. Wiring related problems—These problems have decreased with the use of sealed wiring systems. However, they can occur after improper repairs and wiring routing have been performed. It is not uncommon to see exterior light wiring with a small length of the wire having several butt connectors rubbing against another object and not properly sealed.

Special Tools

Test light

WARNING: Many technicians use the sharp tip of the test light probe to pierce the wire when looking for power or loss of power to the light. This practice can lead to failure of the circuit later. Piercing of the insulation exposes the wire strands to moisture and corrosion. The moisture can actually travel along the wire, making it hard to see the location of the actual circuit problem.

Several manufacturers offer test equipment that allows for quick testing of circuits without puncturing the wire (Figure 8-6).

Simply plug the tester into the end of the cable nearest the trailer. Bring the cable end with the tester into the cab of the truck and operate each circuit of the lighting system. The LEDs on the tester will light, indicating a properly functioning circuit. The tester can be moved to the lamp receptacle on the tractor for further testing if one or more circuits appear faulty. If all the circuits appear normal, the connector cord is faulty. Otherwise, the fault is in the tractor.

Special Tools

Basic mechanics tool set

Service manual

LED type circuit tester

Figure 8-6 Lamp, trailer cable, and circuit tester. (Courtesy of Trucklite)

SERVICE TIP: Most lighting problems occur with combination units. If it is necessary to probe a wire or harness, be sure to seal the puncture completely before the repair is considered done.

Testing

Testing should be performed with a meter or power supply to determine if power is reaching a lamp. This ensures that the circuit is live. On some circuits with multiple components and lights, you might have to walk through the circuit first using a wiring diagram. This helps to locate the various connectors and components involved in the circuits giving you a better idea of the relationship of all the components to each other. You might discover that the circuit has a number of loads on one fuse (Figure 8-7).

Two major problem areas with truck lights and wiring are bad grounds and shorts.

- ❑ Grounds—An improper ground is a major cause of lamp failure due to the environmental exposure of many of the lights and connections. This is more prevalent when a trailer is used as a ground. In this case, make sure there is a fifth wheel ground strap for added protection. Also, many lights are grounded through the lamp housing so make sure the connections are clean. Testing the system properly can determine if the failure is the result of a poor ground or a bad connection.
- ❑ Shorts to ground—Pinpointing the location of this type of failure can be quite frustrating at times. One concept of diagnosing this type of failure is to control the current with a false load. A short to ground will always open a fuse or circuit breaker due to the increase of current associated with the decrease of resistance (no load). What is needed is the "time" to diagnose or pinpoint this short while it is occurring. This is where a device such as a short detector or loud test light comes in handy. One such device is composed of a headlight and warning buzzer wired to a **flasher** as shown in Figure 8-8. When this device is connected to B+ and B−, it will buzz and light. The most important aspect of this device is that it can control the current if it is wired in place of the burned-out fuse or opened circuit breaker. The light and buzzer are forms of resistance that when wired in series will buzz and light because of the B+ available at the fuse box and the short to ground. A diagram of this setup and system is shown in Figure 8-9.

Special Tools

Short detector

Loud test light

Service manual

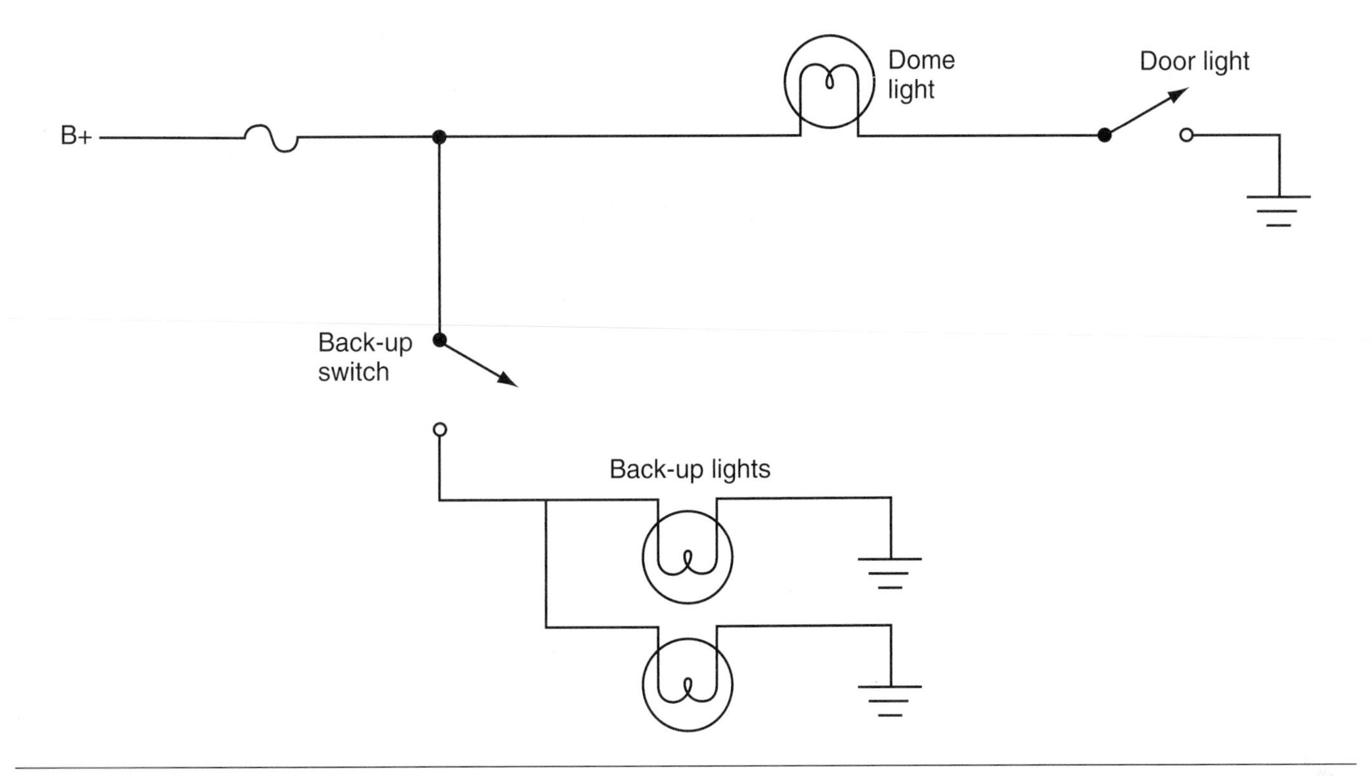

Figure 8-7 Multiple loads on one fuse.

Figure 8-8 Short detector wired in place of a blown fuse.

Figure 8-9 Schematic of short detector.

Once the tester is hooked up, there is an audible and visual indicator of the short. To pinpoint the problem area, start separating connectors in succession, moving away from the fuse. As you open the connector, if the loud test light stops, you have eliminated the short. The short is beyond the connector you have opened. If the test tool is still making a noise, the short is ahead of the open connector. This system allows the technician to pinpoint the short usually to a small section of the wiring harness.

The filament is a fine wire tungsten that is heated electrically to incandescence.

CAUTION: Do not substitute a higher amperage fuse or items such as tinfoil or bolts in place of the circuit protection device to keep the circuit live for a "smoke" test. This will damage wiring and components, not to mention that you might start a fire.

Clues

Sometimes it is possible to find out why a bulb failed by looking at it closely.

- ❑ If the bulb's filament is broken or stretched, it probably was subjected to vibration or shock.
- ❑ If the glass contains a yellow, white, or blue glaze, it usually indicates there was a rupture in the glass envelope.
- ❑ A black or sooty appearance on the glass indicates that a poor seal existed in the bulb. Note that the quality of the bulb varies between manufacturers and is most often reflected in its cost.
- ❑ A dark metallic-looking finish around the head of the bulb usually indicates a failure due to old age.

A troubleshooting chart for lighting systems is shown in Figure 8-10. This is a very general troubleshooting chart. Keep in mind that the type of components, switches, and controls varies between types of trucks, busses, trailers, and manufacturers.

LIGHTING SYSTEM DIAGNOSIS

Symptom	Possible Cause	Remedy
Hazard flasher lamps do not flash.	Fuse or circuit breaker burned out	Replace fuse or circuit breaker. If fuse or circuit breaker blows again, check for short circuit.
	Worn or damaged hazard flasher	Substitute a known good flasher. Replace flasher, if damaged.
	Worn or damaged turn signal operation	Repair turn-signal system.
	Open circuit in wiring	Repair, as required.
	Worn or damaged hazard flasher switch	Repair or replace turn-signal switch and wiring assembly, which includes hazard flasher switch.
Backup lamps—one lamp does not function.	Bulb burned out	Replace bulb.
	Loose wiring connections	Secure connections, where accessible.
	Open circuit in wiring	Repair, as required.
Backup lamps—both lamps do not function.	Fuse or circuit breaker burned out	Replace fuse or circuit breaker. If fuse or circuit breaker blows again, check for short circuit.
	Backup lamp switch out of adjustment	Adjust switch.
	Worn or damaged backup lamp switch	Replace switch.
	Loose wiring connections	Secure connections, where accessible.
	Open wiring or poor ground	Repair, as required.
	Bulbs burned out	Replace bulb.

Figure 8-10 Lighting system troubleshooting chart.

LIGHTING SYSTEM DIAGNOSIS

Symptom	Possible Cause	Remedy
Instrument panel lamp does not light.	Bulb burned out	Replace bulb.
	Fuse burned out	Replace fuse.
	Open circuit in wiring rheostat, or printed-circuit board	Check for short circuit. Repair, as required.
Dome lamp does not come on when door is opened.	Connector loose	Secure or replace.
	Blown fuse	Replace fuse. If fuse blows again, check for short circuit.
	Bulb burned out	Replace bulb.
	Open circuit in wiring	Replace, as required.
	Worn or damaged door jamb switch	Replace switch.
Dome lamp stays on.	Worn or damaged door jamb switch	Replace switch.
	Worn or damaged main lighting switch	Replace main lighting.
Map lamp does not come on when switch is actuated.	Bulb burned out	Replace bulb.
	Blown fuse	Replace fuse.
	Open circuit in wiring	Repair, as required.
	Worn or damaged switch in lamp assembly	Replace lamp assembly.
Map lamp stays on.	Worn or damaged switch in lamp assembly	Replace lamp assembly.
Side or roof marker lamp does not light.	Bulb burned out	Replace bulb.
	Open circuit or poor ground	Check socket for corrosion and good ground. Repair, as required.
Turn signal lamps do not light.	Fuse or circuit breaker burned out	Replace fuse or circuit breaker. If fuse or circuit breaker blows again, check for short circuit.
	Worn or damaged turn signal flasher	Substitute a known good flasher. Replace, if required.
	Loose wiring connections	Secure connections, where accessible.
	Open circuit in wiring or poor ground	Repair, as required.
	Damaged turn-signal switch	Check continuity of switch assembly. Replace turn signal switch and wiring assembly, as necessary.

Figure 8-10 Continued.

LIGHTING SYSTEM DIAGNOSIS

Symptom	Possible Cause	Remedy
Turn signal lamps light but do not flash.	Worn or damaged turn signal flasher	Substitute a known good flasher. Replace, if required.
	Poor ground	Repair ground.
Front turn signal lamps do not light.	Loose wiring connector or open circuit	Repair wiring, as required.
Rear turn signal lamps do not light.	Loose wiring connector or open circuit	Repair wiring, as required.
One turn signal lamp does not light.	Bulb burned out.	Replace bulb.
	Open circuit in wiring or poor ground	Repair, as required.
Headlights do not light.	Loose wiring connections	Check and secure connections at headlight switch and dash panel connector.
	Open circuit in wiring	Check power to and from headlight switch. Repair as necessary.
	Worn or damaged headlight switch	Verify condition. Replace headlight switch, if necessary.
One headlight does not work.	Loose wiring connections	Secure connections to headlight and ground.
	Sealed-beam bulb burned out	Replace bulb.
	Corroded socket	Repair or replace, as required.
All headlights out; park and taillights are okay.	Loose wiring connections	Check and secure connections at dimmer switch and headlight switch.
	Worn or damaged dimmer switch	Check dimmer switch operation. Inspect for corroded connector. Replace, if required.
	Worn or damaged headlight switch	Verify condition. Replace headlight switch, as necessary.
	Open circuit in wiring or poor ground	Repair, if required.
Both low-beam or both high-beam headlights do not work.	Loose wiring connections	Check and secure connection at dimmer switch and headlight switch.
	Worn or damaged dimmer switch	Check dimmer switch operation. Inspect for corroded connector. Replace, if required.
	Open circuit in wiring	Repair.
One taillight out.	Bulb burned out	Replace bulb.
	Open wiring or poor ground	Repair as necessary.
	Corroded bulb socket	Repair or replace socket.

Figure 8-10 Continued.

LIGHTING SYSTEM DIAGNOSIS

Symptom	Possible Cause	Remedy
All taillights and maker lamps out; headlights okay.	Loose wiring connections	Secure wiring connections, where accessible.
	Open wiring or poor ground	Check operation of front park and marker lamps. Repair as necessary.
	Blown fuse	Replace fuse.
	Damaged headlight switch	Verify condition. Replace headlight switch, if necessary.
Stop lights do not work.	Fuse or circuit breaker burned out	Replace fuse or circuit breaker. If fuse or circuit breaker blows again, check for short circuit.
	Worn or damaged turn signal circuit	Check turn signal operation. Repair as necessary.
	Loose wiring connections	Secure connection at stop-light switch.
	Worn or damaged stop-light switch	Replace switch.
	Open circuit in wiring	Repair as required.
Stop lights stay on continuously.	Damaged stop-light switch	Disconnect wiring connector from switch. If lamp goes out, replace switch.
	Switch out of adjustment	Adjust switch.
	Internal short circuit in wiring	If lamp stays on, check for internal short circuit. Repair as necessary.
One parking lamp out.	Bulb burned out	Replace bulb.
	Open wiring or poor ground	Repair as necessary.
	Corroded bulb socket	Repair or replace socket.
All parking lamps out.	Loose wiring connections	Secure wiring connections.
	Open wiring or poor ground	Repair as necessary.
	Bad switch	Replace switch.

Figure 8-10 Continued.

Headlights

Classroom Manual
Chapter 8, page 183.

Special Tools

Basic mechanics tool set

Safety glasses

Torx drivers

Trucks use a number of different types of headlights: standard sealed-beam, halogen sealed-beam, and composite type of lights. We will look at common replacement procedures first because they are the most frequent services performed. Also, the diagnostics of the headlight system often require the removal of the sealed units to get at the connectors and wiring.

Headlight Replacement

To determine the need for headlight replacement the technician must remove the light. This allows direct access to the headlight plug and a voltmeter or test light can be used to detect the presence of voltage at the light. If voltage is present, it usually indicates a bad lamp. The

The conventional **sealed-beam headlight** is a self-contained glass unit. **Halogen** is the term used to identify a group of chemically related nonmetallic elements. **Composite** is the term that is used to describe the headlight system that uses separate components to make up the unit. For example, in some systems the lamp is separate from the lens assembly and is replaceable.

Classroom Manual
Chapter 8, page 183.

A

B

Figure 8-11 Bulb removed from (A) front and (B) back. (Courtesy of Freightliner Corporation)

A **bezel** is the retaining trim around the headlight unit.

procedure for headlight replacement depends on the type of bulb used. With standard sealed-beam lights it usually requires the removal of some type of bezel first (Figure 8-11).

SERVICE TIP: Because of the construction and placement of the prisms in the lens, it is important that the headlight be installed in its proper position. The lens is usually marked "TOP."

The following is a typical replacement procedure for sealed-beam headlights:

1. Remove the bezel.
2. Remove the retaining screws that hold the sealed-beam retaining trim (Figure 8-12A and Figure 8-12B). *Do not turn the headlight adjusting screws, which are shown more clearly in Figure 8-13.*
3. Remove the headlight from the housing.
4. Disconnect the wire connector from the back of the lamp.
5. Check the wire connector for corrosion and integrity. Clean as needed.
6. Coat the connector terminals and prongs of the new headlight with dielectric grease to prevent corrosion.
7. Install the connector to the new headlight and place the headlight in the housing. Ensure proper positioning of the headlight.

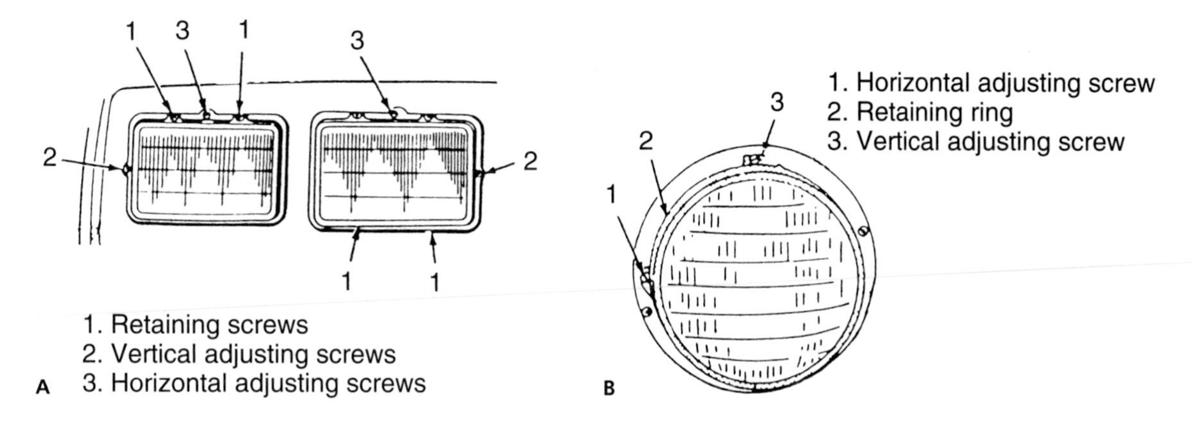

Figure 8-12 (A) Double-sealed beam unit. (Courtesy of Volvo); (B) Single-sealed beam unit. (Reprinted with permission. Courtesy of Volvo Trucks North America, Inc.)

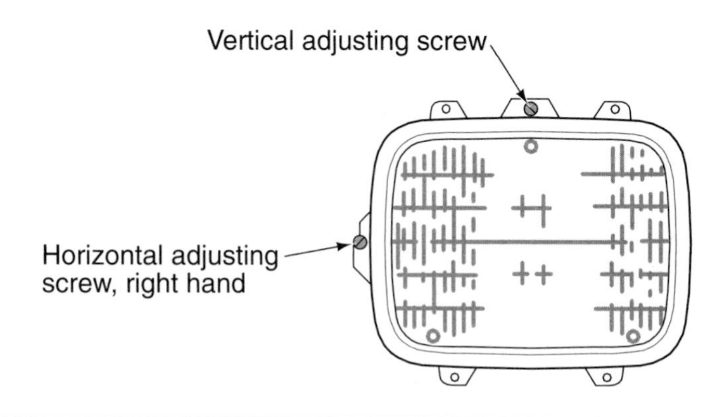

Figure 8-13 Typical locations for the adjustment screws.

8. Install the retainer trim and fasteners.
9. Check the headlight operation before proceeding.
10. If the headlights require adjusting, perform this next as described in the next section.
11. Install the headlight bezel.

WARNING: It is not recommended mixing standard and halogen headlights on a vehicle. If the original equipment manufacturer (OEM) had halogen lights on, it is not wise to replace them with standard beam headlights. Doing so will result in poor light quality.

To replace a composite headlight, use the following procedure:

Because the housings of some composite headlight systems are vented, condensation may develop inside the lens assembly. This condensation is not harmful to the bulb and does not affect headlight operation.

Removal

1. Be sure that the **headlight switch** is off.
2. Open the hood (conventional trucks) and locate the bulb installed from the rear of the headlight body.
3. Remove the electrical connector from the bulb by releasing the locking tab and pulling the connector in a rearward direction.

Classroom Manual
Chapter 8, page 186.

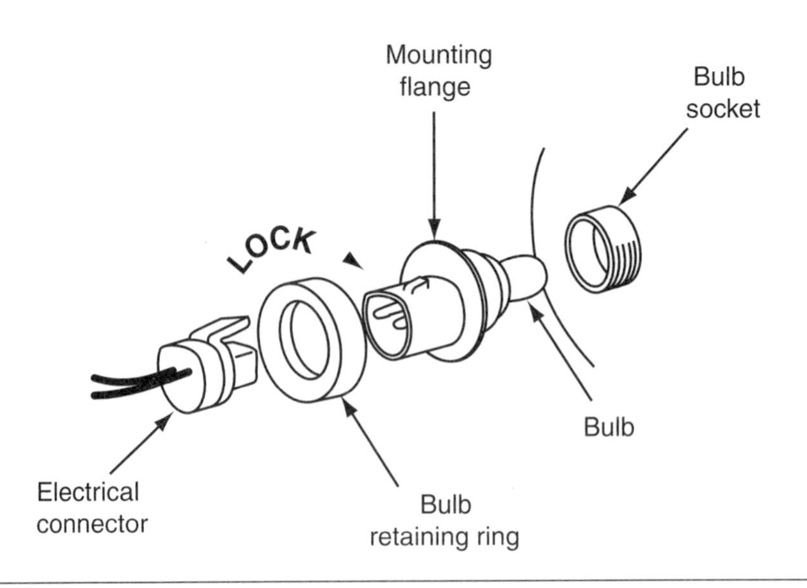

Figure 8-14 Composite headlight bulb replacement.

Classroom Manual
Chapter 8, page 185.

4. Remove the bulb retaining ring by rotating it counterclockwise when viewed from the rear (Figure 8-14). It takes approximately 1/8 turn to unlock it.
5. Slide the ring off the base and save it to retain the new bulb.
6. Carefully remove the halogen bulb from its socket in the reflector by pulling gently backwards. Do not rotate the bulb.
7. Check the wire connector for corrosion and clean as needed.
8. Coat the connector terminals and the prongs of the new headlight with dielectric grease to prevent corrosion.

Classroom Manual
Chapter 8, page 186.

Installing Headlamp Bulb

1. With the flat side of the plastic base of the bulb facing upward, insert the glass envelope of the bulb into the socket. The base may have to be turned slightly to the left or right to align the grooves in the forward part of the plastic base with the corresponding locating tabs inside the socket. When the grooves are aligned, push the bulb firmly into the socket until the mounting flange on the base contacts the rear face of the socket. Note: An angled version is shown in Figure 8-15.
2. Slip the retaining ring over the plastic base and against the mounting flange. Lock the ring into the socket by rotating the ring clockwise. It will be fully locked when you feel a detent (or stop).
3. Push the electric connector into the rear of the plastic base until it snaps and locks into place.
4. Turn on the headlights and check for proper operation.

CAUTION: The halogen bulb contains gas under pressure. The bulb may shatter when it is mishandled or dropped. Handle the bulb carefully—grasp it by its plastic base only.

WARNING: Avoid touching the envelope of the halogen composite lamp. Staining the bulb with normal oil skin can shorten the life of the bulb. Handle the lamp by its base only.

WARNING: Leaving a bulb out of the headlamp reflector assembly for a prolonged period will allow moisture and contaminants to enter the assembly, which can affect the performance of the headlamp.

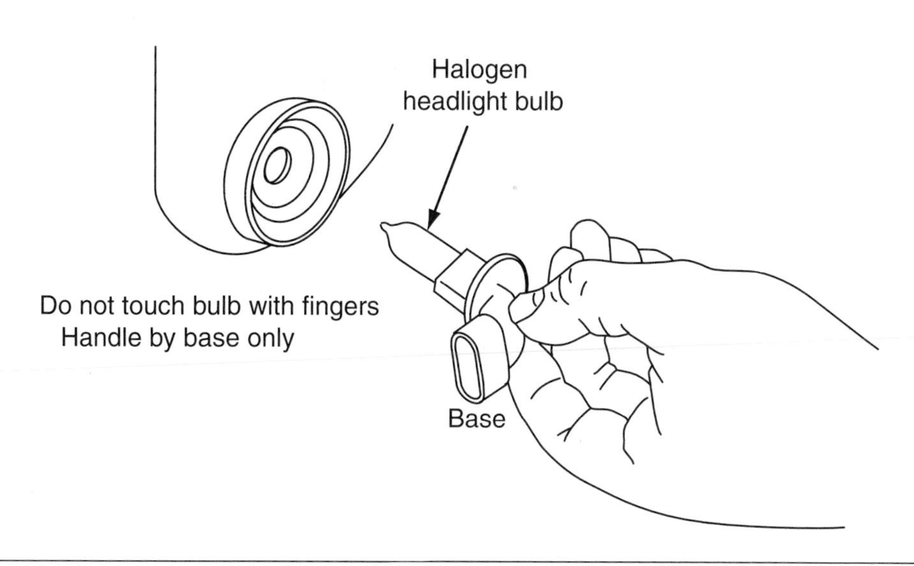

Figure 8-15 The correct method for handling a composite bulb during replacement.

WARNING: To prevent early failure, do not energize the bulb unless it is installed into the socket.

Headlight Adjustment (Aiming)

Special Tools

Headlight aiming unit

Basic mechanics hand tool set

Safety glasses

To obtain maximum illumination during low-beam and high-beam operation of the headlights requires the lights to be kept in adjustment. Not only is proper aiming important for good light projection onto the road, but also discomfort and dangerous conditions can be created for oncoming drivers if the headlights are not properly aimed.

Two methods that can be used to check headlight aiming are: the screen method and the portable mechanical aiming unit method. Regardless of the method, the vehicle's tire inflation, springs, and proper ride height should be checked first.

Screen Method

Headlight aiming by the screen method requires a level area in a darkened condition sufficient for the vehicle and an additional 25 feet from the lamps to the screen. The vehicle must be located accurately in front of the screen. The screen should be marked in the following manner (Figure 8-16): first, measure the distance between the centers of the matching headlights and draw two vertical lines on the screen, with each line corresponding to the center of the headlight. Next, draw a vertical centerline halfway between the two previous vertical lines. Then measure the distance from the floor to the centers of the headlights. Subtract 2 inches from this height and then draw a horizontal line on the screen to correspond to this new height.

With the headlights on high beam, the hot (brightest) spot of each headlight should be centered on the point where the corresponding vertical and horizontal lines intersect on the screen for each headlight. The headlight adjusting screws are turned in or out to adjust the headlights vertically and/or laterally to obtain a proper aim.

Portable Mechanical Aiming Units

Most shops use portable aiming units to make it easier to check headlight aim on the variety of headlights used today (Figure 8-17). These units are secured to the headlight lens by

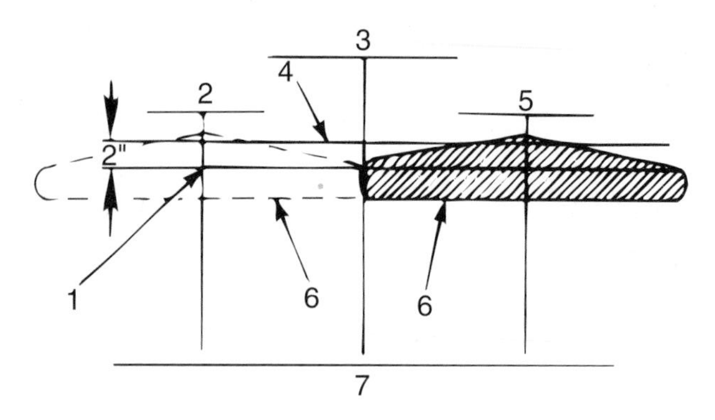

Figure 8-16 Headlight aiming pattern. (Courtesy of Navistar International Engine Group)

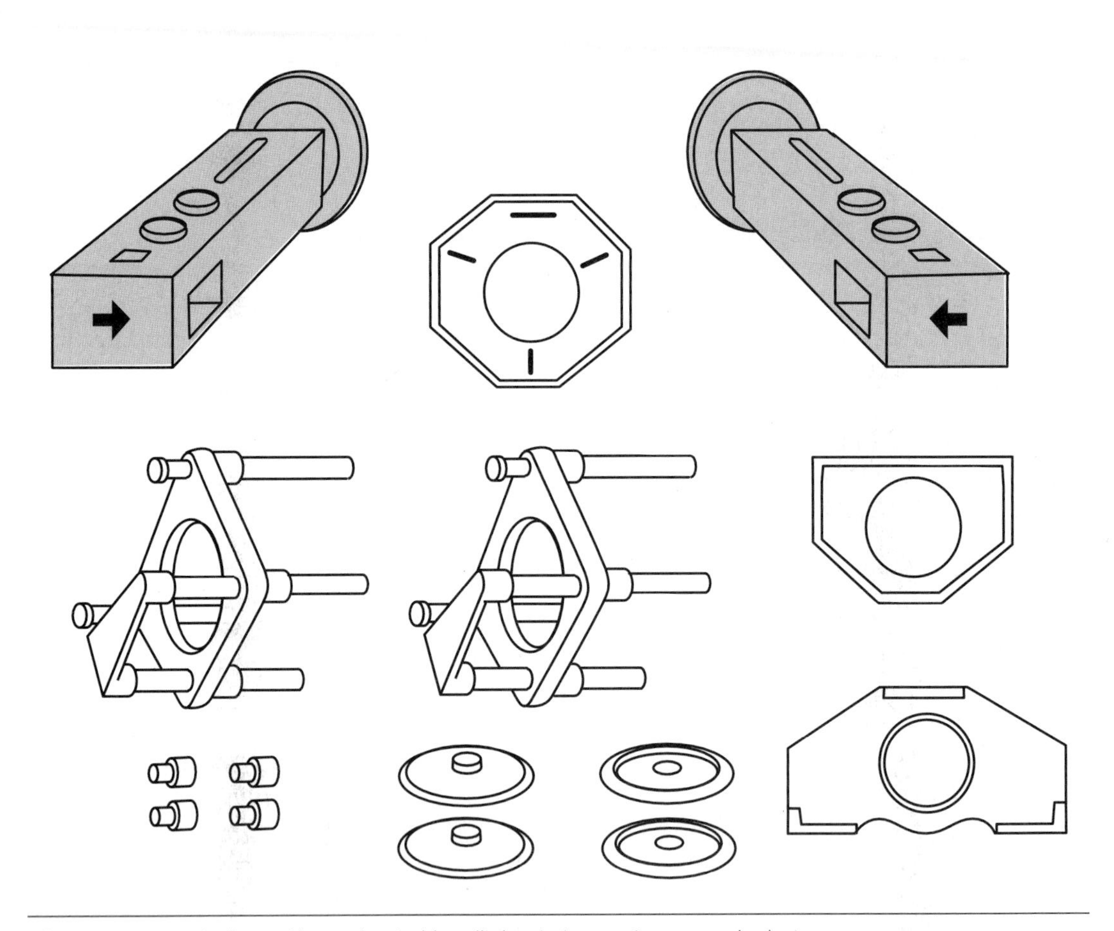

Figure 8-17 Typical portable mechanical headlight aiming equipment and adapters.

suction cups (Figure 8-18). The aiming units usually come with a variety of adapters to enable proper attachment to the various styles of headlights. Always follow manufacturers' procedures. The next sections discuss procedures used for sealed-beam and composite headlight aiming.

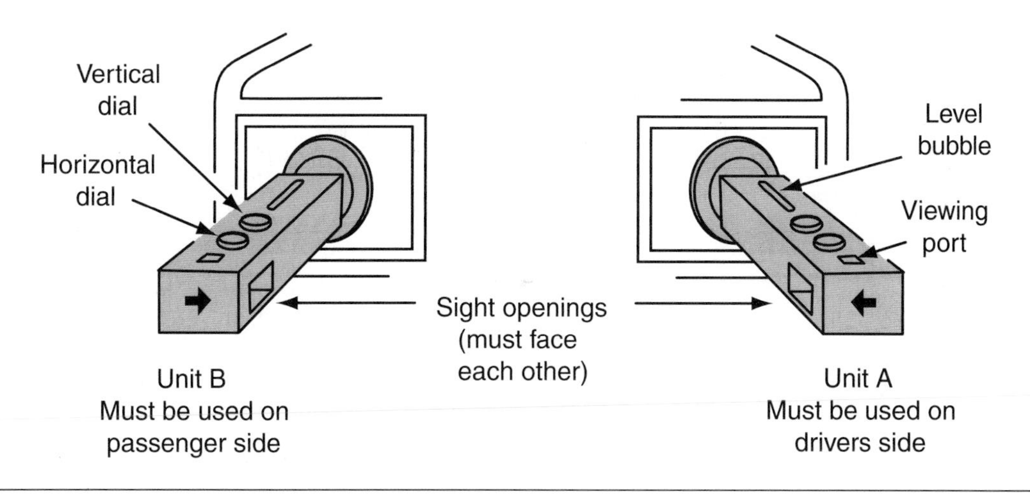

Figure 8-18 The aiming units attach to the headlight lens with suction cups.

Sealed Beam Park the vehicle on level ground. Most units will have leveling devices and adjustments for the units to compensate for floors that are not level. See manufacturers' procedures.

Using the correct adapter, connect the calibrated aimer units to the headlights. It might be necessary to remove the bezels on some headlights first. Be sure the adapters fit the headlight aiming pads on the lens (Figure 8-19). Zero the horizontal adjustment dial and confirm that the split image target lines are visible in the view port (Figure 8-20). If the target lines are not seen, rotate the aimer unit. Turn the headlight horizontal adjusting screw until the split image target lines are aligned. Repeat for the headlight on the other side.

To set the vertical aim of the headlight, turn the vertical adjustment dial on the aiming unit to zero. Turn the headlight vertical adjustment screw until the level bubble is centered (Figure 8-21). Recheck the horizontal aim throughout this procedure because the vertical adjustment may have altered the original adjustment.

If the vehicle is equipped with a four headlight system, repeat the procedure for the other pair of lamps.

WARNING: Make sure the headlight aimers are properly fastened. They are very easily damaged if dropped.

Classroom Manual
Chapter 8, page 186.

Composite Headlight Aiming Composite headlight designs are often referred to as aerodynamic design with unique contours that require a special adapter to check the aim and adjust (Figure 8-22). These types of headlights and lenses have aiming pads in order to use a

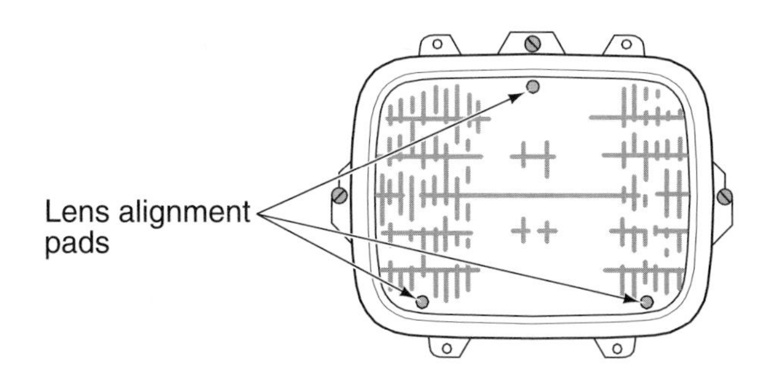

Figure 8-19 Headlight aiming pads.

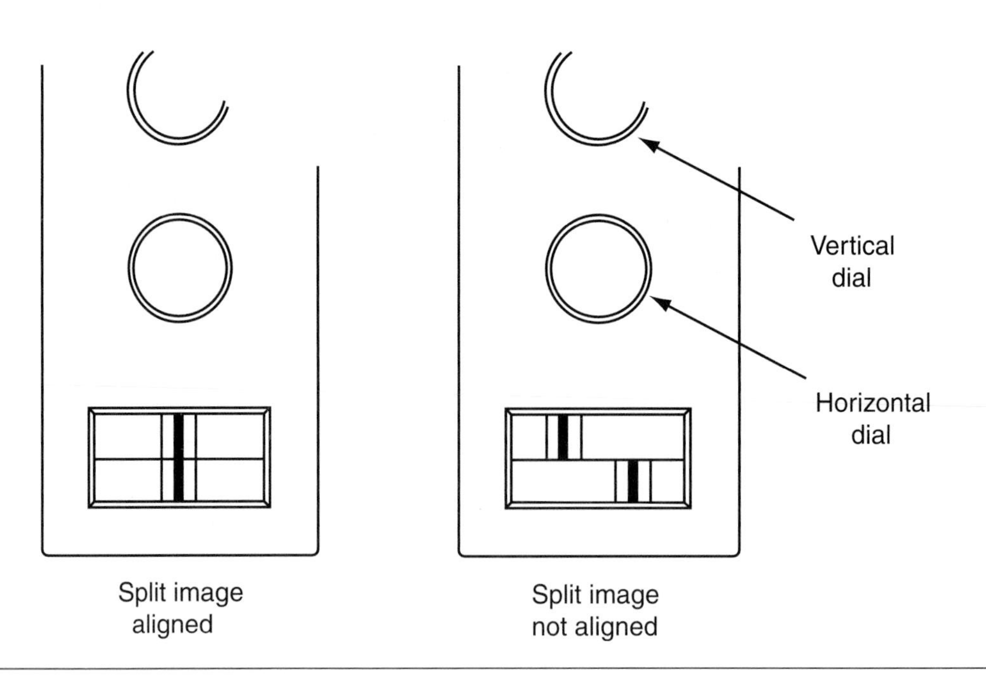

Figure 8-20 Split image target.

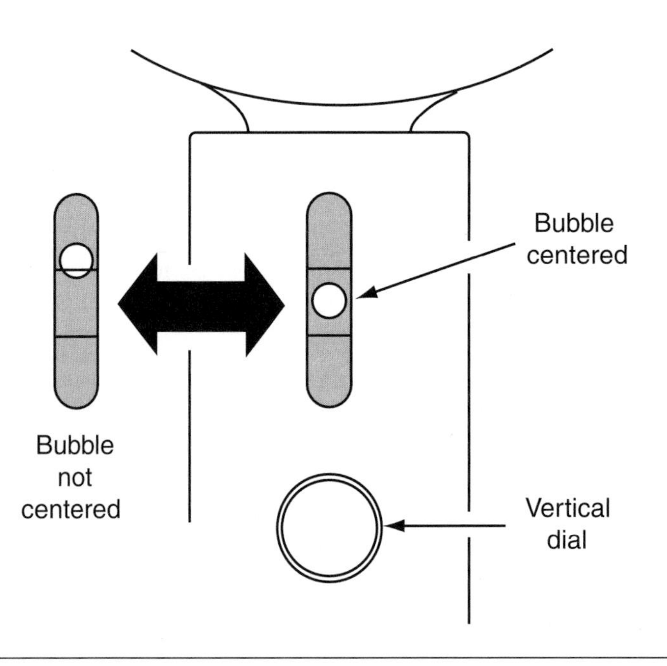

Figure 8-21 Center the spirit level by turning the vertical aiming screw.

mechanical aiming unit. The lens also has a number molded on it. The unit's adjustment rod must be set to correspond with that number and locked in place. The aiming unit is attached to the headlight lens in the same manner as the sealed-beam headlights (Figure 8-23). Articulating vacuum cups are used for composite light aiming. If the articulating cup does not touch the lens of the headlamp, then add both of the extension studs to the vacuum cup assembly (Figure 8-24).

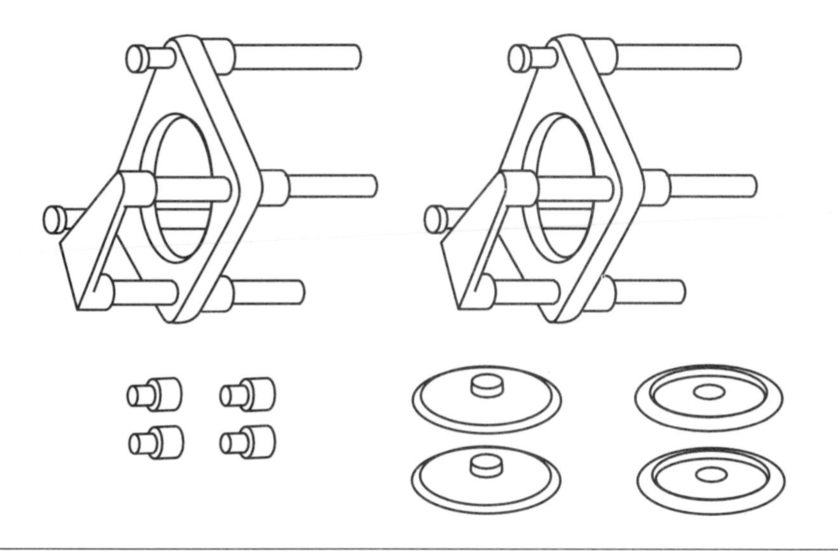

Figure 8-22 Special adapter for aiming composite headlights.

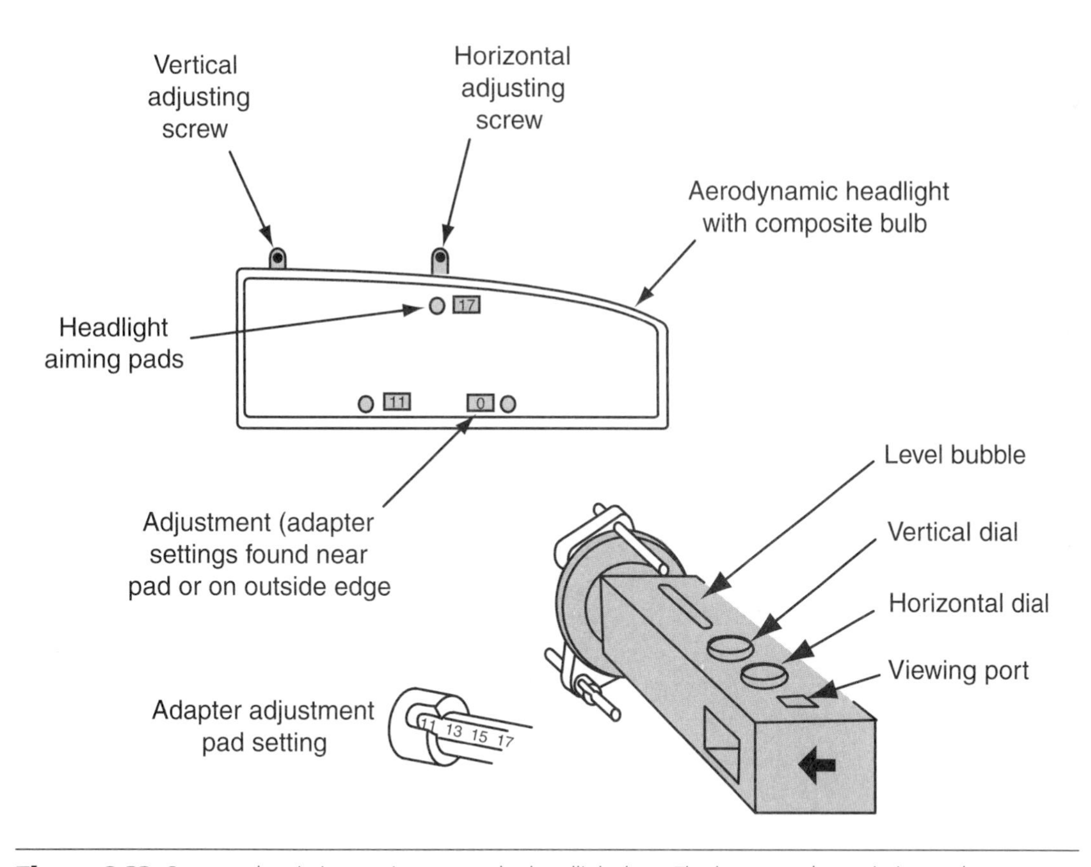

Figure 8-23 Connect the aiming equipment to the headlight lens. The lens must have aiming pads.

The adjustment procedure is identical to that of the sealed-beam headlights. The illustration (Figure 8-25) shows the location of one manufacturer's headlight adjusting screws. As you can see, it is at the rear of the headlight assembly.

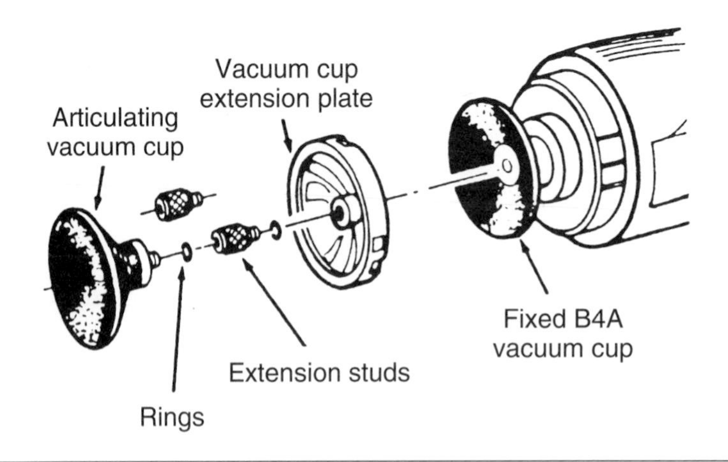

Figure 8-24 Fitting the vacuum cup. (Reprinted with permission. Courtesy of Volvo Trucks North America, Inc.)

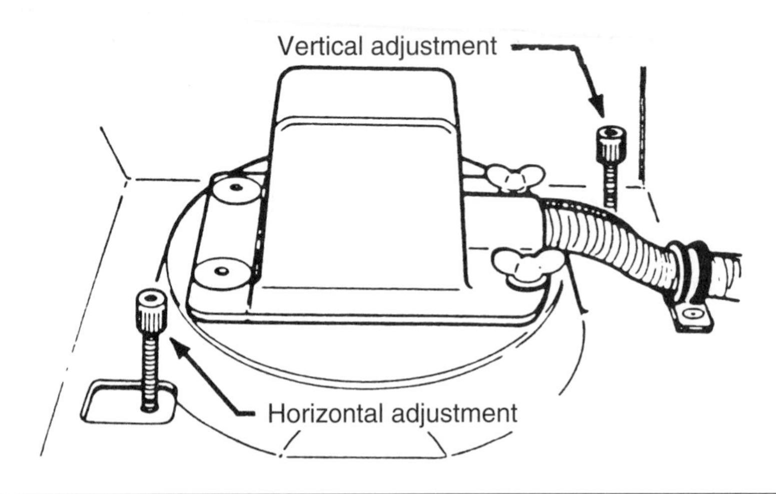

Figure 8-25 Adjustment screw location. (Reprinted with permission. Courtesy of Volvo Trucks North America, Inc.)

Special Tools

Basic mechanic's tool set

Test light

Voltmeter

Safety glasses

Jumper wire

Voltage drop testing is used to find unwanted resistances in an electrical circuit.

Headlight Troubleshooting

Headlight problems usually fall into two categories: complete failure of one or both headlights and dimmer or brighter than normal lights. Both symptoms require a logical approach. We will use Figure 8-26 as a reference for a typical headlight system.

In the low beam operation, power for low beams comes from the headlight relay via Circuit 286. A left and right junction box is used to connect the circuit to the headlights.

High beams are operated by lifting the turn signal stick in this truck. Lifting the indicator stick on the turn signal switch activates the high beams by completing a path for current flow to ground through the headlight relay coil. The headlight relay is connected to the turn signal/dimmer switch by Circuit 284. Before going any further, let us create a failure in which the high beams do not work but the low beams do work. By verifying that the low beams work, you should be able to figure out by looking at the diagram that the headlight switch must be okay because the relay gets its power from that switch. If both high beam lights are out, we can narrow the problem to the circuit between the dimmer switch and a common point in Circuit 285 before the circuit splits to the two junction blocks.

A test light is a useful tool in this case to check for available voltage at the relay where circuit 285 comes out. If voltage is available, the problem is towards the junction block and head-

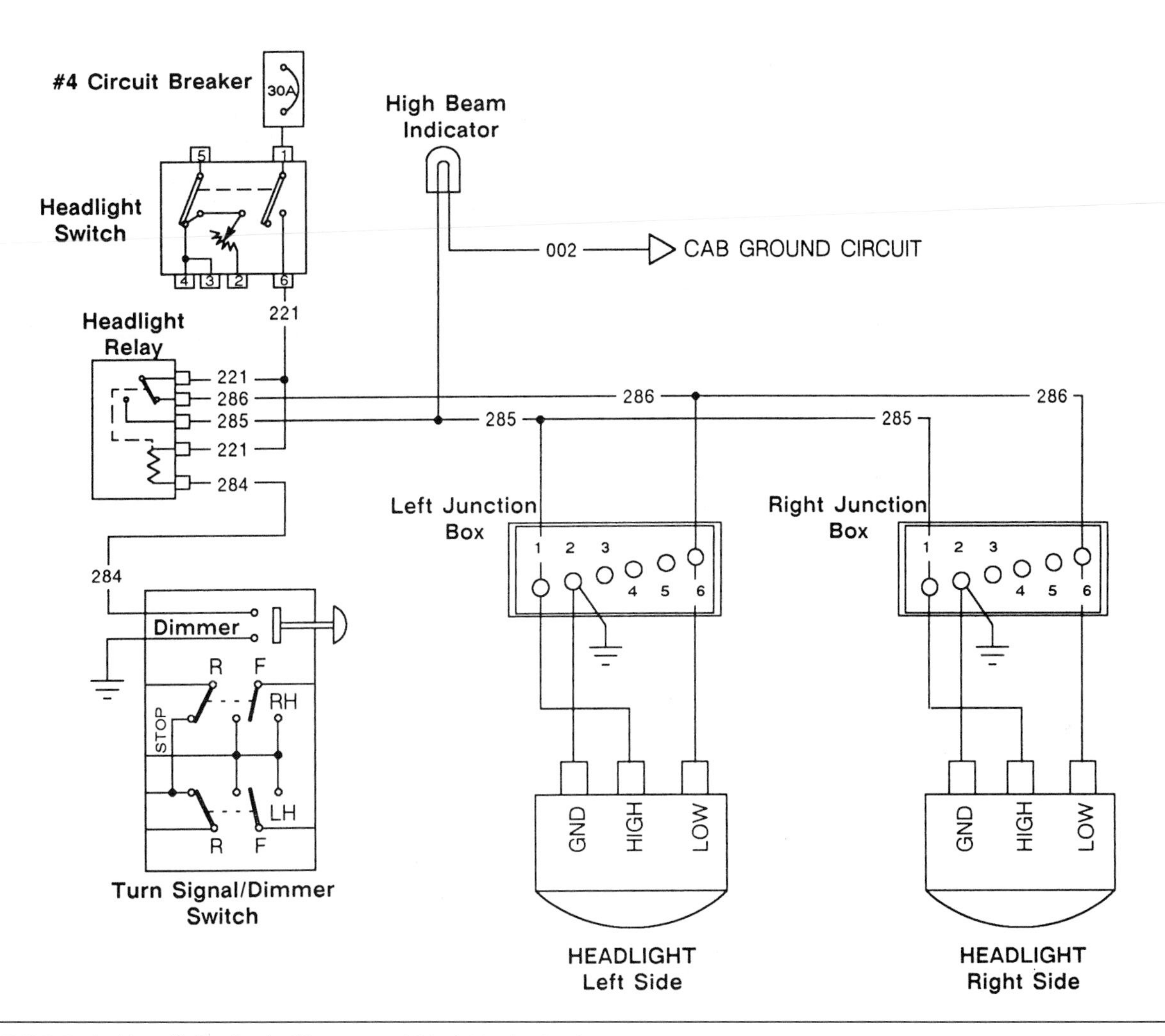

Figure 8-26 Typical headlight circuit. (Courtesy of Mack Trucks, Inc.)

lights. If the test light doesn't come on, the problem would be the relay itself, Circuit 284, or the dimmer switch. Take a good look at the figure and visualize how the relay operates. We know that it takes current flowing through the relay coil to energize it and move the relay contacts into position to feed voltage to the high beam Circuit 285. However, what is needed to complete this circuit? Hopefully, you answered power to Circuit 221 at the relay and a ground through Circuit 284 controlled by the dimmer switch. To further pinpoint the problem, you will need to ensure there is power and there is ground. If you have any doubts about the ground side, simply connect a jumper to ground the relay to verify the problem.

To diagnose a dim light or lights, the technician should start with the premise that excessive resistance in a part of the circuit could be the culprit. The extra resistance can be on the insulated side or the ground side of the circuit. To locate the excessive resistance, perform a voltage drop test (Figure 8-27). This requires the use of the vehicle's electrical diagram to identify the various components and connects of that particular system. Start at the light and work towards the battery. Don't forget to check both the power and ground side. Also remember that the circuit has to be on in order to perform a voltage drop test. Keep in mind that headlights are wired in parallel. If both headlights are dim, the excessive resistance is probably at a point common to both lights, such as an alternating current generator that isn't putting out right.

A higher than specified generator output can cause lights to be brighter than normal as well as improper lamp application.

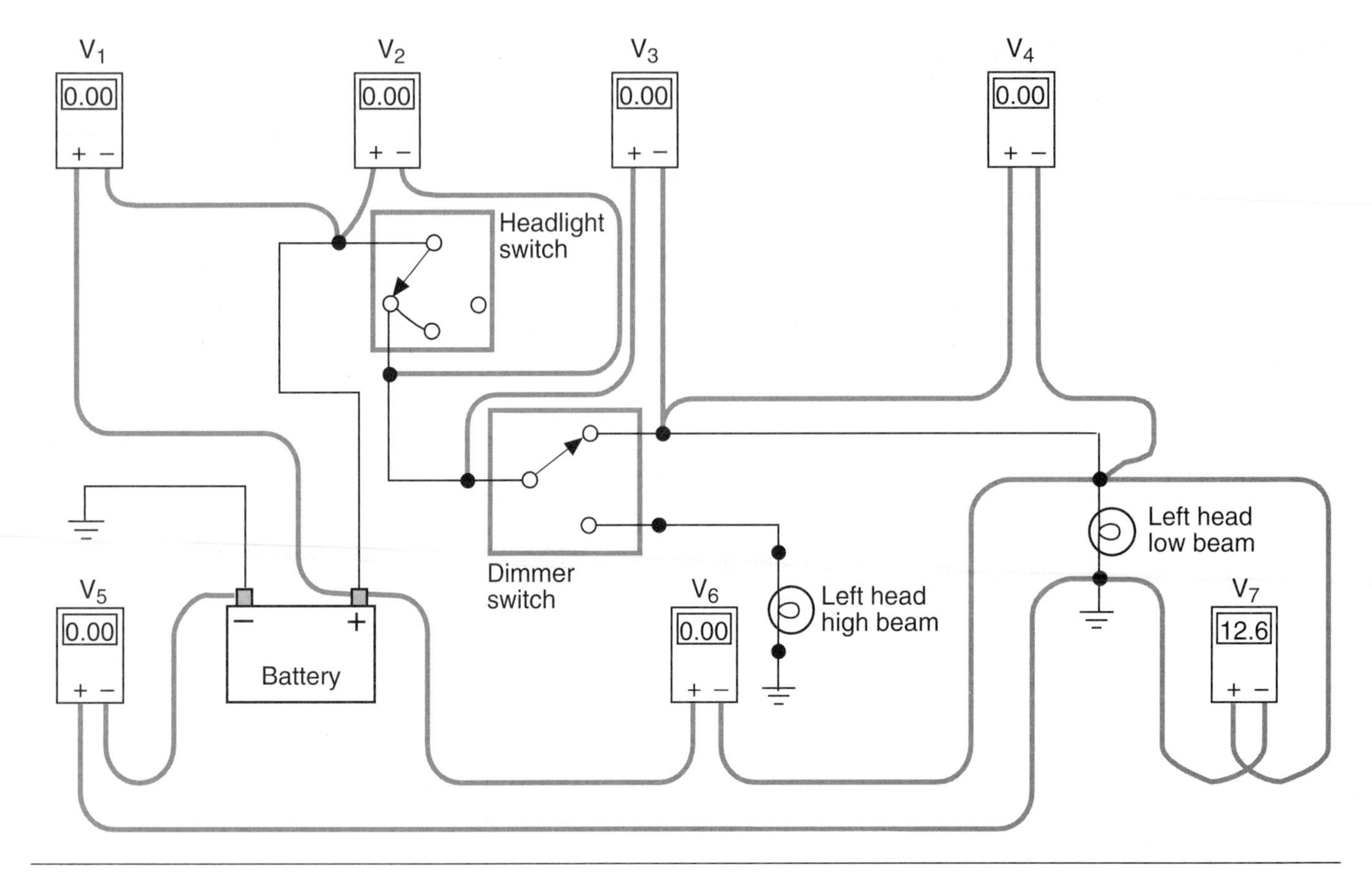

Figure 8-27 Voltage drop testing the headlight system.

SERVICE TIP: Headlights do not wear out with age and become dimmer. Therefore, if one headlight is dimmer than the other, there is an excessive voltage drop in that circuit and the headlight plug is the problem. Sometimes disconnecting and connecting the plug to the lamp is enough to clean the contacts. This typically occurs when a new headlight is installed. A technician might think the problem is the headlight when in reality, the fault lies with the connection.

We can summarize the troubleshooting of headlights in the following logical steps, depending on the severity of the problem:

1. Have a clear understanding of the symptom.
2. Analyze the appropriate circuit diagram to get a clear understanding of the system operation and identify possible causes. Then use the appropriate tool such as a test light and/or digital multimeter to check for expected voltages and grounds.

Headlight Switch Testing and Replacement

Special Tools

12-volt test light
Ohmmeter or self-powered test light
Jumper wire
Battery terminal pliers
Terminal pullers
Safety glasses

Headlight switches used in trucks can differ from one vehicle to another. Some vehicles use toggle switches with separate rheostats for panel lights while others use a master headlight switch that controls most of the vehicle's lighting system. Many headlight problems can be attributed to the headlight switch. If a headlight switch is suspected, it might have to be removed from the dashboard to access the wiring in order to perform a visual check or perform pinpoint testing of the circuits.

Photo Sequence 6 shows the removal of a master headlight switch. As you can see, this type of headlight switch has a knob assembly that has to be removed in order to remove the

Photo Sequence 6
Headlight Switch Removal

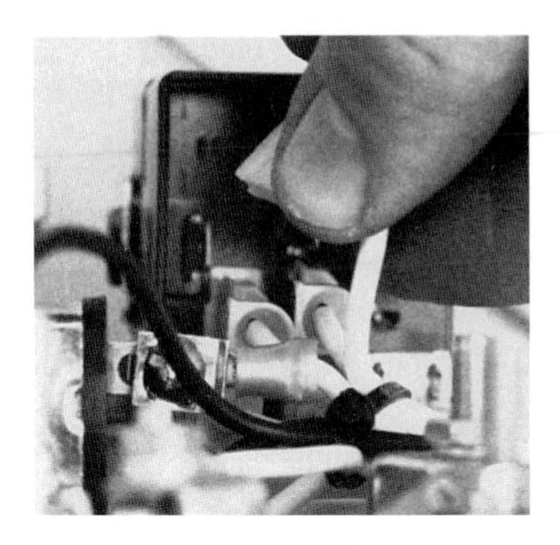

P6-1 Disconnect the wires from the headlight switch terminals.

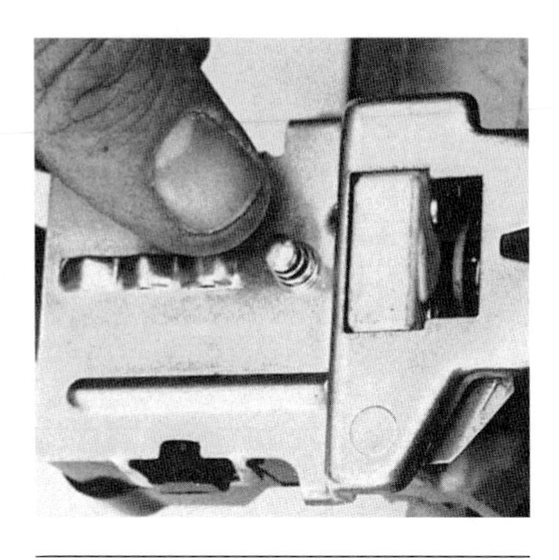

P6-2 Locate the spring-loaded release button on the bottom of the switch body.

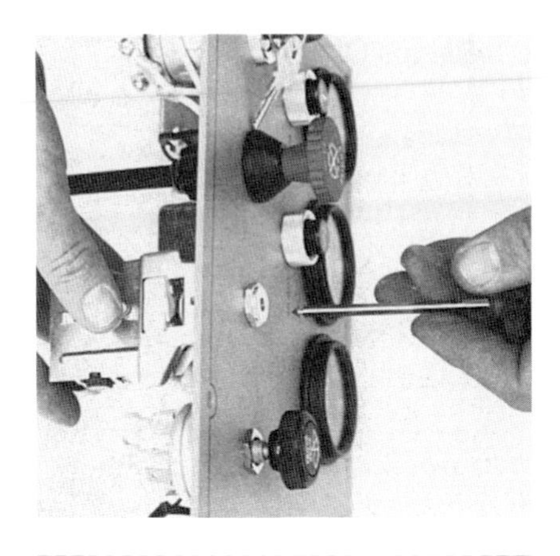

P6-3 Push the button in to release the shaft and knob assembly. Pull the shaft and knob assembly out.

P6-4 Remove the switch assembly retaining nut securing the switch to the instrument panel. Then remove the switch. Note: To reinstall the headlight switch, reverse the removal procedure.

switch retaining nut so that the switch itself can be removed from the dash panel. Once the switch is removed, it can be tested for voltages and continuity.

WARNING: Disconnect the battery ground cable first before working in areas where you might ground out or arc and otherwise damage circuits and components.

A diagnosis chart for the headlight switch is shown in Figure 8-28.

If the diagnosis leads towards the headlight switch, remove the switch from the dash so the connector plug can serve as a test point for the lighting circuits. First, test at the connector. Consult the vehicle's service manual for terminal identification. In this procedure, the terminals are identified as follows:

- ❏ Terminal B—battery
- ❏ Terminal A—fuse
- ❏ Terminal H—headlights

Classroom Manual
Chapter 8, page 186.

HEADLAMP SWITCH

NOTE: Replace burned out bulbs or fuses before proceeding.

CONDITION	POSSIBLE SOURCE	ACTION
Headlamps do not work. Park and taillamps OK.	• Open or shorted wiring	• Check wiring and connections between power source and headlamp switch, and between headlamp switch and lamps. Service as necessary.
	• Poor ground connections	• Check and service as necessary.
	• Damaged dimmer switch	• Check dimmer switch; replace if necessary.
	• Damaged headlamp switch	• Check headlamp switch; replace if necessary.
All exterior lamps do not work.	• Open or shorted wiring	• Check wiring and connections between power source and headlamp switch.
	• Damaged headlamp switch	• Check headlamp switch; replace if necessary.
Headlamps flash on and off.	• Shorted circuit	• Check wiring and connections between headlamp switch and headlamps.
	• Damaged headlamp switch	• Replace headlamp switch.
Park and taillamps do not work. Headlamps OK.	• Open wiring or poor ground	• Check wiring and connections between power source and headlamp switch, and between headlamp switch and lamps. Service as necessary.
	• Blown fuse	• Service as necessary.
	• Damaged headlamp switch	• Check headlamp switch; replace if necessary.
Instrument panel lamps do not work, or will not dim.	• Open or shorted wiring	• Check wiring between headlamp switch and lamps.
	• Damaged headlamp switch	• Check headlamp switch; replace if necessary.
Dome lamps will not work.	• Open or shorted wiring	• Check wiring and connections between headlamp switch and dome lamp, and between head-lamp switch and fuse panel.
	• Damaged headlamp switch	• Check headlamp switch; replace if necessary.

Figure 8-28 Headlight switch diagnostic chart. (Courtesy of Freightliner Corporation)

- ❑ Terminal R—rear park and side markers
- ❑ Terminal I—instrument panel lights
- ❑ Terminal D1—dome light feed
- ❑ Terminal D2—dome light

1. Connect a 12-volt test light across Terminal B and ground (Figure 8-29). The test light should light. If not, there is an open in the circuit back to the battery.
2. Connect the test light across Terminal A and ground (Figure 8-30). The test light should come on. If not, repair the circuit back to the fuse panel.
3. Connect a jumper wire between Terminals B and H (Figure 8-31). The headlights should come on. If they don't, trace the H circuit to the headlights. Don't forget to check the ground circuit side from the headlights.
4. Connect a jumper wire between Terminals A and R (Figure 8-32). The rear lamps should come on. If not, trace the circuit to the rear lights. Also check the ground return path.
5. Connect a jumper wire between Terminals A and I (Figure 8-33). The instrument panel lights should come on. If not, trace the circuit to the panel lights.

The chart in Figure 8-34 indicates the test result that should be obtained when testing the headlight switch itself for continuity. For this test, use an ohmmeter or a self-powered test light.

The previous test's concept can be also used for toggle type of switches. Remember that you are looking for available voltage and ground. Jumpers are used to provide voltage to determine the direction of diagnosis.

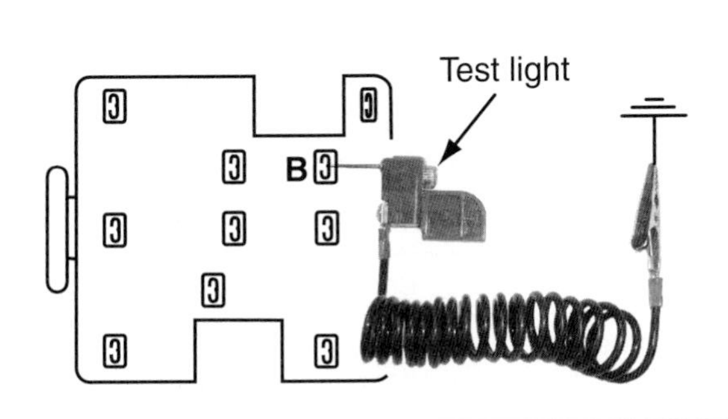

Figure 8-29 Testing for battery voltage at the connector.

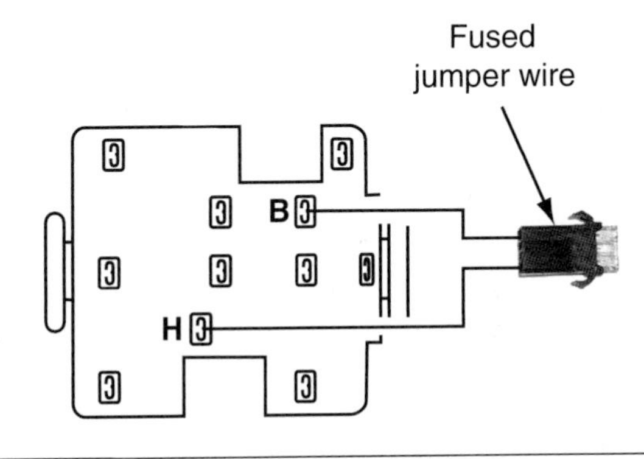

Figure 8-31 Use a jumper wire to test the circuits from the battery to the headlights.

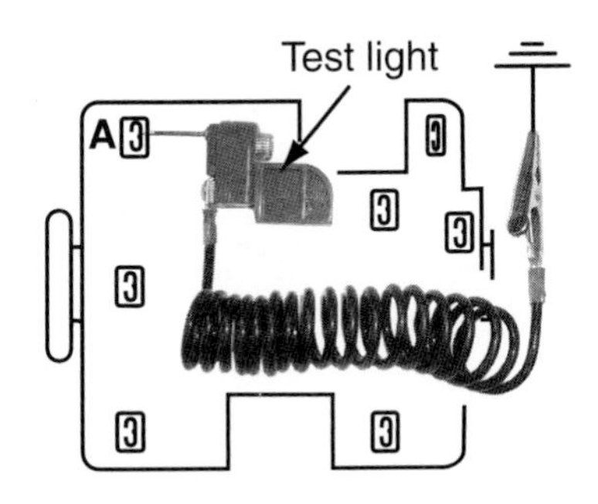

Figure 8-30 Testing for battery voltage through the fuse to the connector.

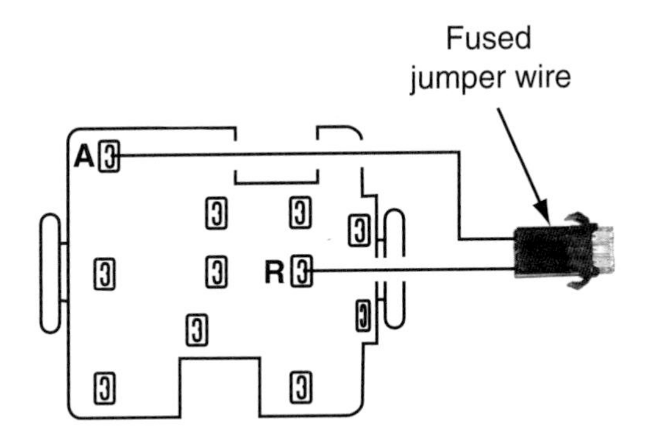

Figure 8-32 Use a jumper wire to test the circuit from the fuse to the rear lights.

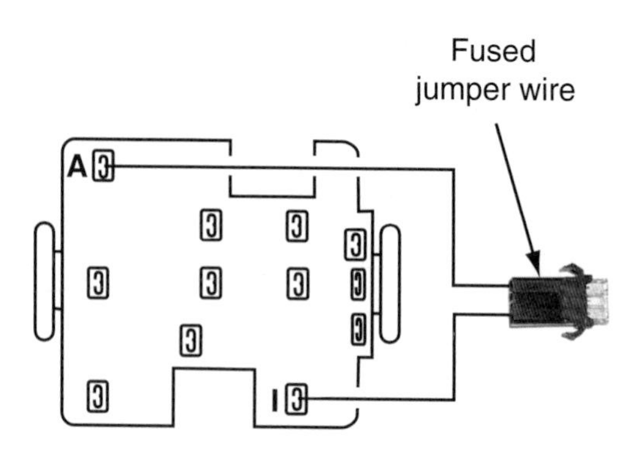

Figure 8-33 Testing the circuits between the fuse and the instrument panel lamps.

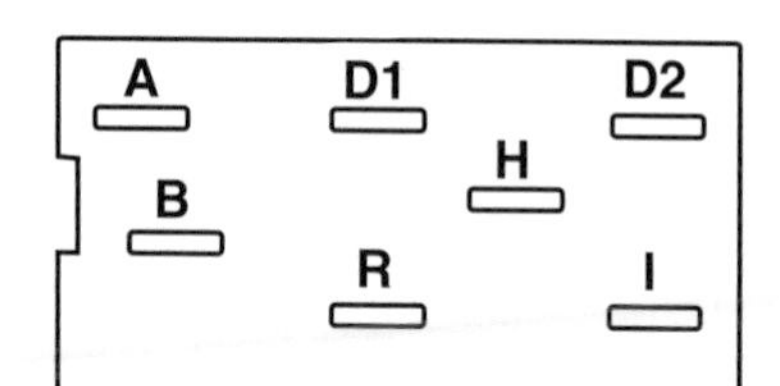

A self-powered test light or ohmeter will be required.

Terminal identification used on test proceedure corresponds with actual identification on headlight switch.

Switch terminals	Headlight switch positions		
	Off	Park	Headlight
B to H	No continuity	No continuity	Continuity
B to R	No continuity	No continuity	No continuity
B to A	No continuity	No continuity	No continuity
R to H	No continuity	No continuity	No continuity
R to A	No continuity	Continuity	Continuity
H to A	No continuity	No continuity	No continuity
D1 to D2	Continuity with rheostat in full counterclockwise position		
I to R	Continuity with rheostat in full counterclockwise position Slowly rotate rehostat clockwise and test lamp should dim		

Figure 8-34 Continuity test chart for a headlight switch.

CAUTION: Use proper size jumpers and avoid excessive arcing, or personal injury can result.

SERVICE TIP: Some light switch problems occur in trucks because too many lights are added and wired to a single switch. This can overload the switch. Don't repair the failure without taking care of the cause of the failure. Poor contact problems can be visually detected by looking for contacts that are discolored or loose.

Classroom Manual
Chapter 8, page 188.

Dimmer Switch Testing and Replacement

A dimmer switch is connected in series with the headlights to control the current path for high and low beams. In trucks, it is either located on the floorboard next to the left kick panel or on

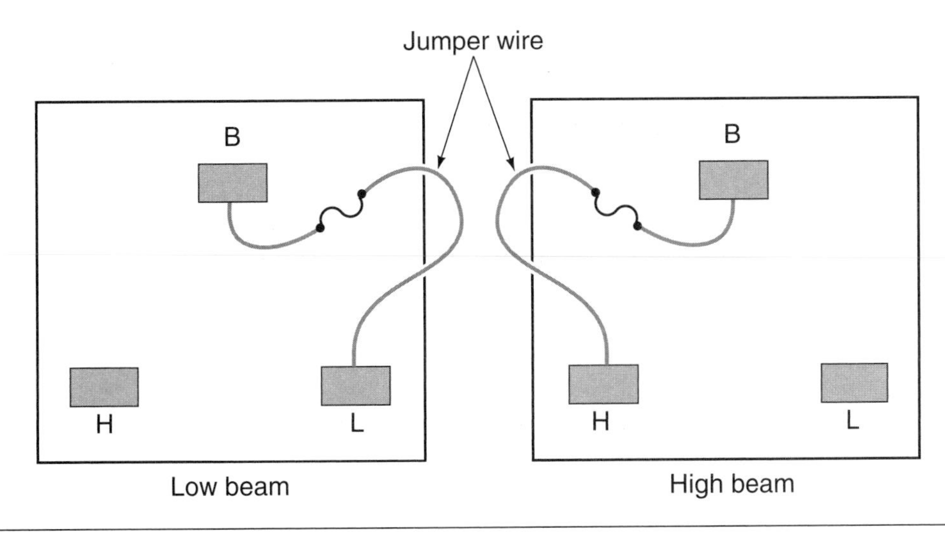

Figure 8-35 Jumper wire connections to bypass the dimmer switch.

the steering column, usually as part of the turn signal switch. Testing of the switch is fairly straightforward. One method is to look for voltage at the battery or power terminal and then look for voltage at the high or low beam terminal when the switch is switched. Another method is to use a jumper wire to bypass the switch (Figure 8-35).

If the headlights operate with the switch bypassed, the switch is faulty. Older vehicles tend to have problems with the floor mount switches due to their location, which subjects the switch and connector to rust, corrosion, etc.

The **dimmer switch** is also known as a courtesy switch. It provides the means for the driver to select either high- or low-beam operation.

Replacing Headlight Dimmer Switches

To replace a floor-mounted dimmer switch, use the following procedures:

1. Remove the two screws that hold the dimmer switch to the floorboard or mounting bracket (Figure 8-36).
2. Disconnect the dimmer switch wiring connector. Note that sometimes it is easier to remove the connector from the switch when the switch is off the floorboard. Sometimes the connector needs to be removed first to allow more working room with your tools.
3. Position the new switch and refasten.
4. Connect the wiring harness if it hasn't been done yet.
5. Check the dimmer switch operation.

Special Tools

Jumper wires

Safety glasses

Basic hand tool set

Battery terminal pliers

Terminal pullers

Classroom Manual
Chapter 8, page 189.

WARNING: The mounting screws tend to freeze to the switch and/or floorboard or mounting bracket. Patience is the key if that is the case. It is easier to use a good quality penetrating oil and allow it to penetrate than it is to break the mounting screw. A broken screw or bolt can involve more time in the long run.

SERVICE TIP: The connectors plugs are prone to corrosion. After-market plugs are very common and inexpensive. If a problem is suspected, replace the plug assembly using the wiring repair procedures you have learned.

Trucks also use various styles of dimmer switches that are steering column mounted. The following procedure is used for one of the common types of steering column-mounted dimmer switch:

1. Make sure the key switch is off and disconnect the battery negative cable.
2. Disconnect the switch wiring connector from the cab wiring harness connector (Figure 8-37).

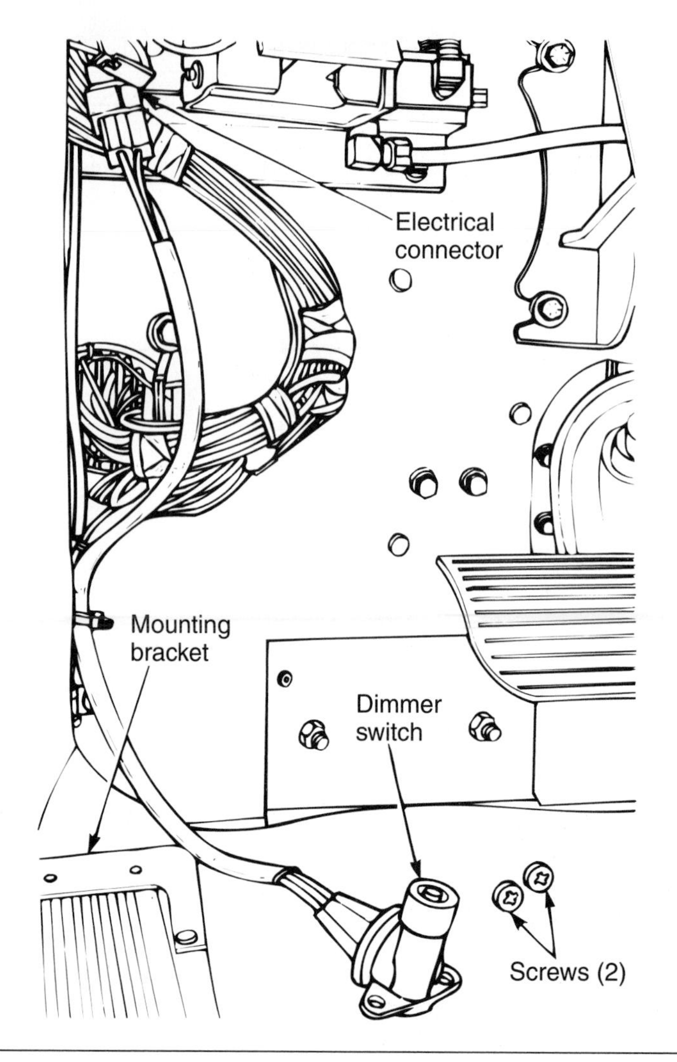

Figure 8-36 Typical components of the dimmer switch. (Courtesy of Navistar International Engine Group)

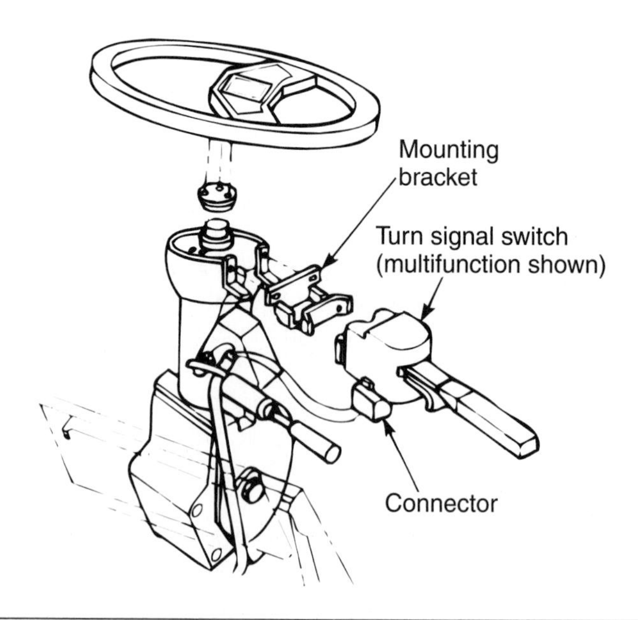

Figure 8-37 Typical parts of a turn signal/hazard switch. (Courtesy of Navistar International Engine Group)

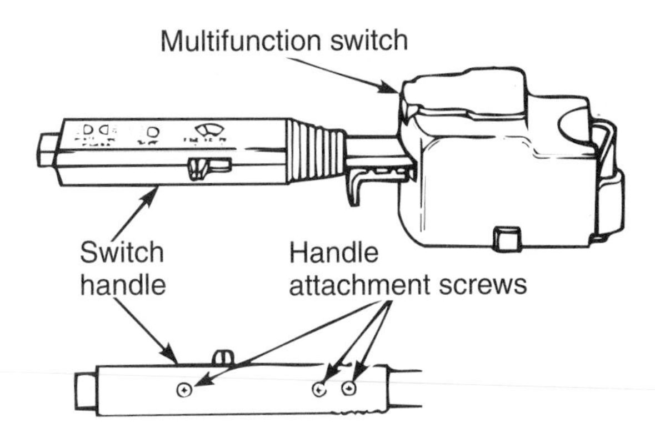

Figure 8-38 Three screws holding a turn signal grip together. (Courtesy of Navistar International Engine Group)

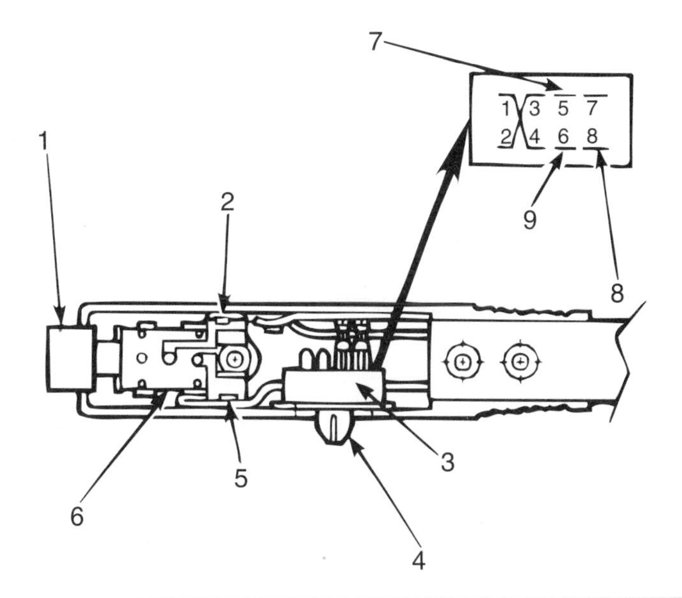

1 Courtesy switch button cover
2 Yellow wire terminal
3 Windshield wiper switch
4 Wiper switch button cover
5 Red wire terminal
6 Switch with circuit board facing up
7 Yellow wire from courtesy switch to no. 5 position
8 Green wire (high speed) to no. 8 position
9 Brown wire (low speed) to no. 6 position

Figure 8-39 Courtesy switch detail. (Courtesy of Navistar International Engine Group)

3. Remove the switch mounting screws and then the switch assembly from the steering column.
4. Remove the three screws holding the plastic grips together (Figure 8-38). Lift off the grip with the function symbols embossed on it.
5. Lift out the courtesy switch, disconnecting the edge terminals of the yellow and red wires (Figure 8-39).
6. Carefully reassemble the edge terminals to the new switch.
7. Position the new courtesy switch inside the handle.
8. Attach the grip with the embossed symbols to the mating grip and refasten it with the three screws.
9. Reconnect the battery cable and check the switch operation.

Special Tools

Basic hand tool set

Wiring diagram

12-volt test light

Jumper wires

Safety glasses

Voltmeter

Taillight Assemblies

Taillights can be configured in a number of ways. In a three-bulb taillight system, the brake lights are controlled directly by the stoplight switch (Figure 8-40). Most light systems use dual filament bulbs that perform multifunctions. In this type of circuit, the stoplights are wired through the turn signal switch (Figure 8-41).

Classroom Manual
Chapter 8, page 191.

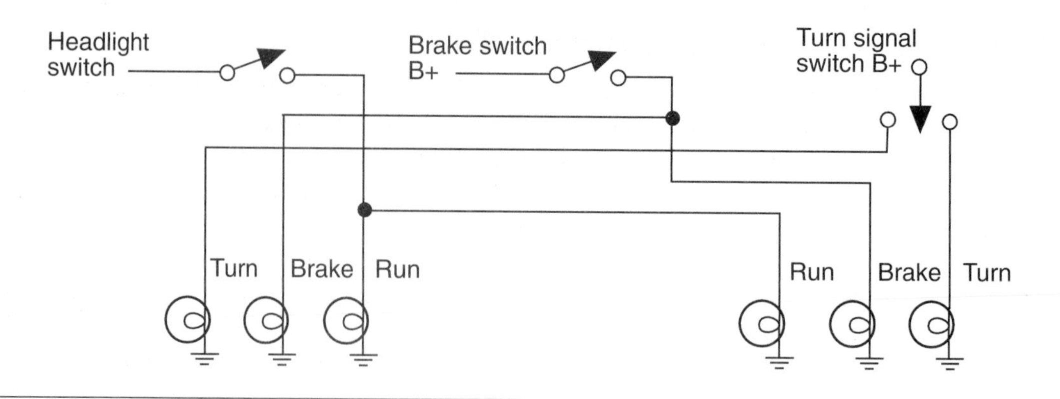

Figure 8-40 A three-bulb taillight circuit with an individual control for each bulb.

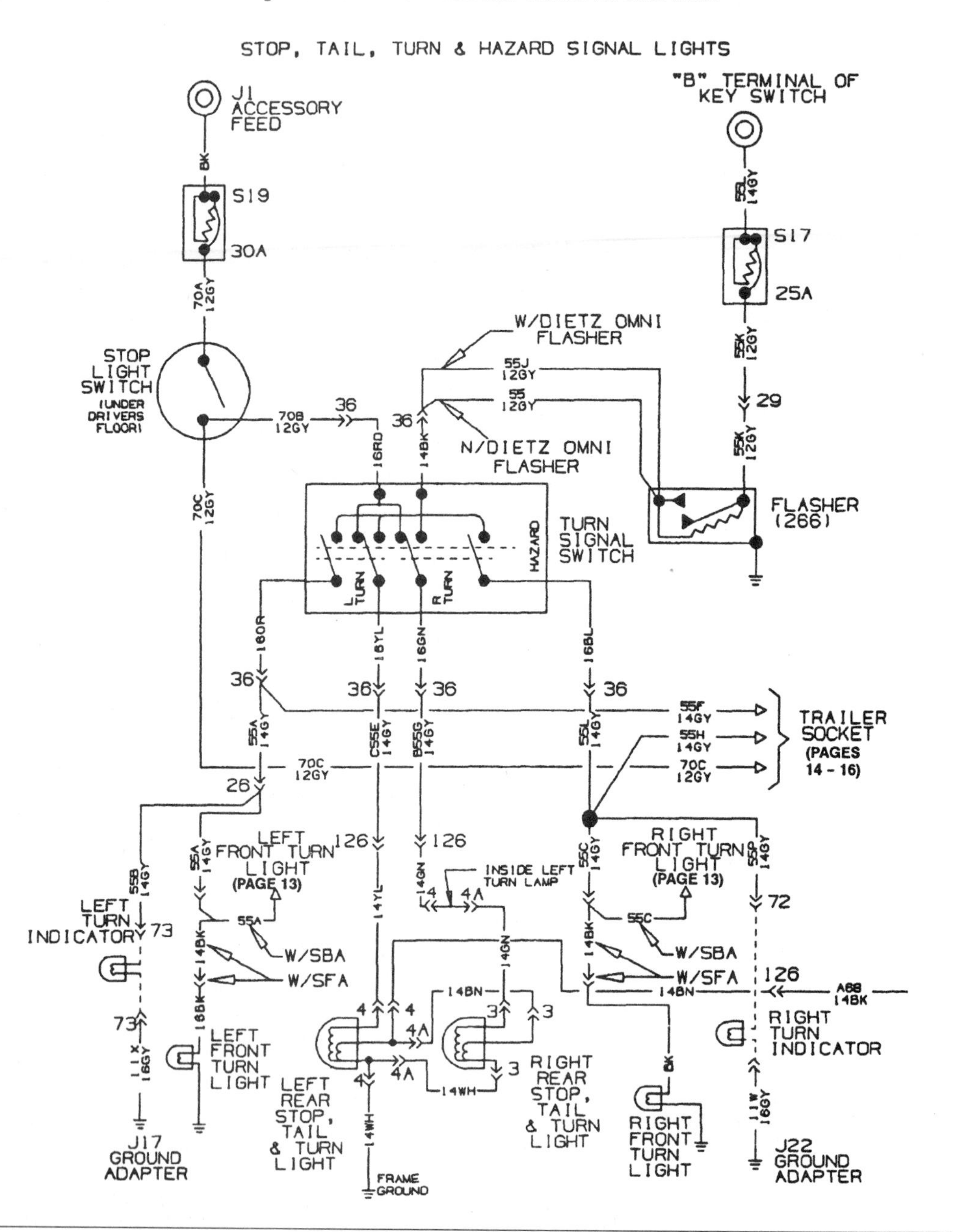

Figure 8-41 Circuit diagram of a stop, tail, turn, and hazard signal lights system. (Courtesy of Navistar International Engine Group)

Turn Signals and Brake Lights

Classroom Manual
Chapter 8, page 196.

When the service brakes are applied, the stoplight switch supplies power to the brake circuit (70B) in the turn signal switch. The power flows through the turn signal switch to illuminate the left and right stoplights. Looking at the circuit diagram, you notice that the stop and turn lights share the same filament in the taillight assembly. In Figure 8-41, the schematic of the turn signal switch, you notice that the four switch arms are shown with turn signals not applied.

To help in diagnosing turn signal or stoplight problems, study Figure 8-42. This figure illustrates the switch arm positions in the right or left turn mode or with hazard switch on. Compare this figure with the circuit diagram.

Not all flashers are located in the fuse box. Use the component locator to find where the flasher is installed.

For example, closing either the RH or LH turn signal switch closes the circuit to the turn signal lamp on the applicable side of the vehicle. The rear turn signals on the tractor use the same lamp filaments as the stoplights. When the turn signals are turned on, the rear bulb filaments on the applicable side of the vehicle are fed through the turn signal switch.

Classroom Manual
Chapter 8, page 197.

Flashers

The flashing signal for the turn and hazard lights is provided by the turn signal flasher unit. A common complaint of this unit is that it flashes too slow or too fast. This can happen if the flasher is

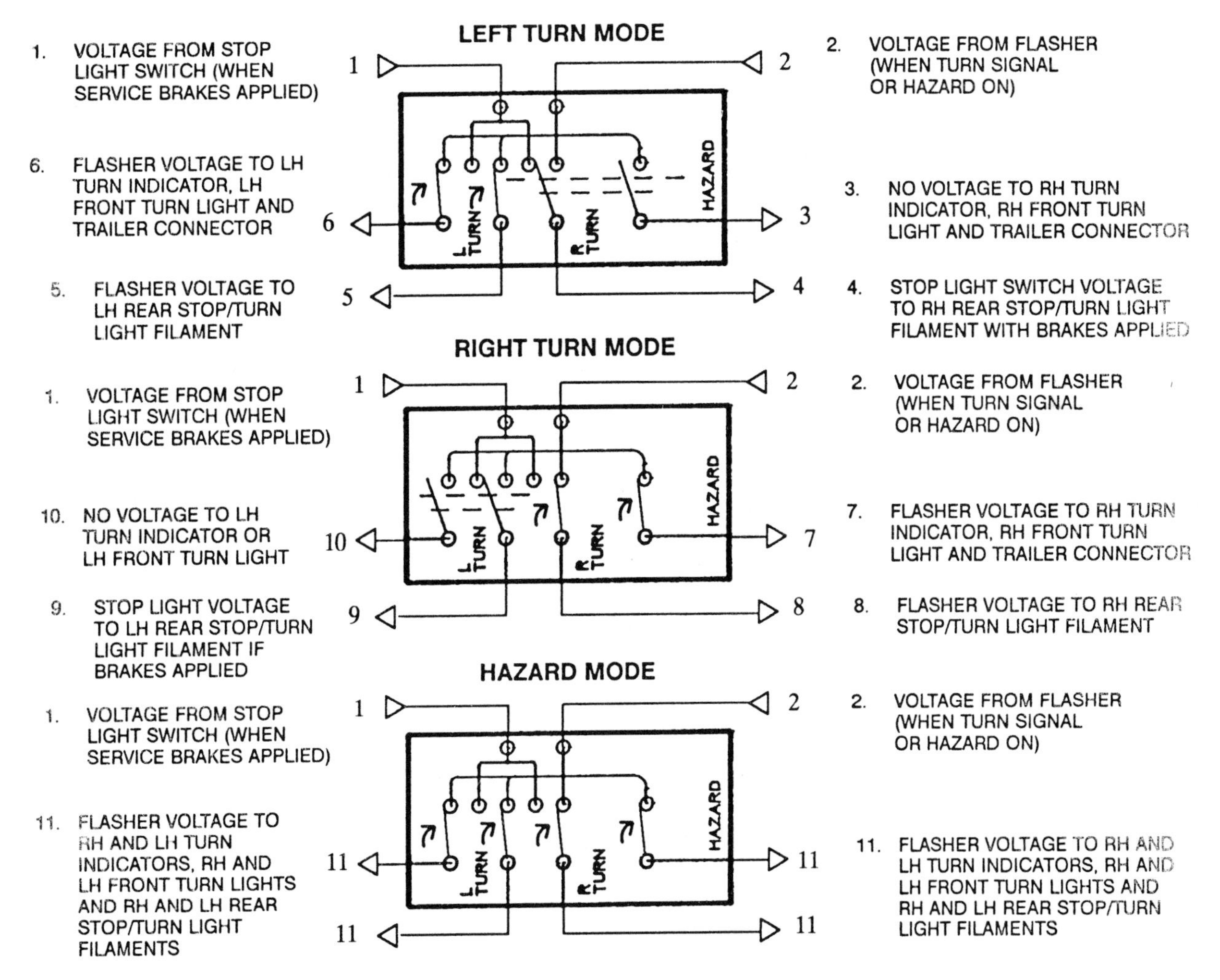

Figure 8-42 Turn signal switch operation shown in different modes. (Courtesy of Navistar International Engine Group)

the wrong type or has the wrong rating. Also, newer electronic circuit type of flashers flash at an increased speed if one or more of the turn signal bulbs is burned out or the circuit is defective.

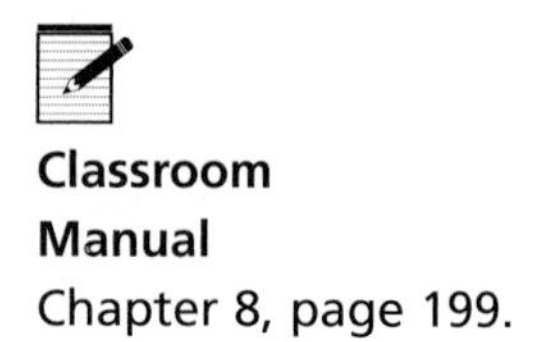

Classroom Manual
Chapter 8, page 199.

Troubleshooting Stop and Turn Lights

Figures 8-43, 8-44, and 8-45 show charts for troubleshooting stop and turn lights as shown in the wiring diagram in Figure 8-40.

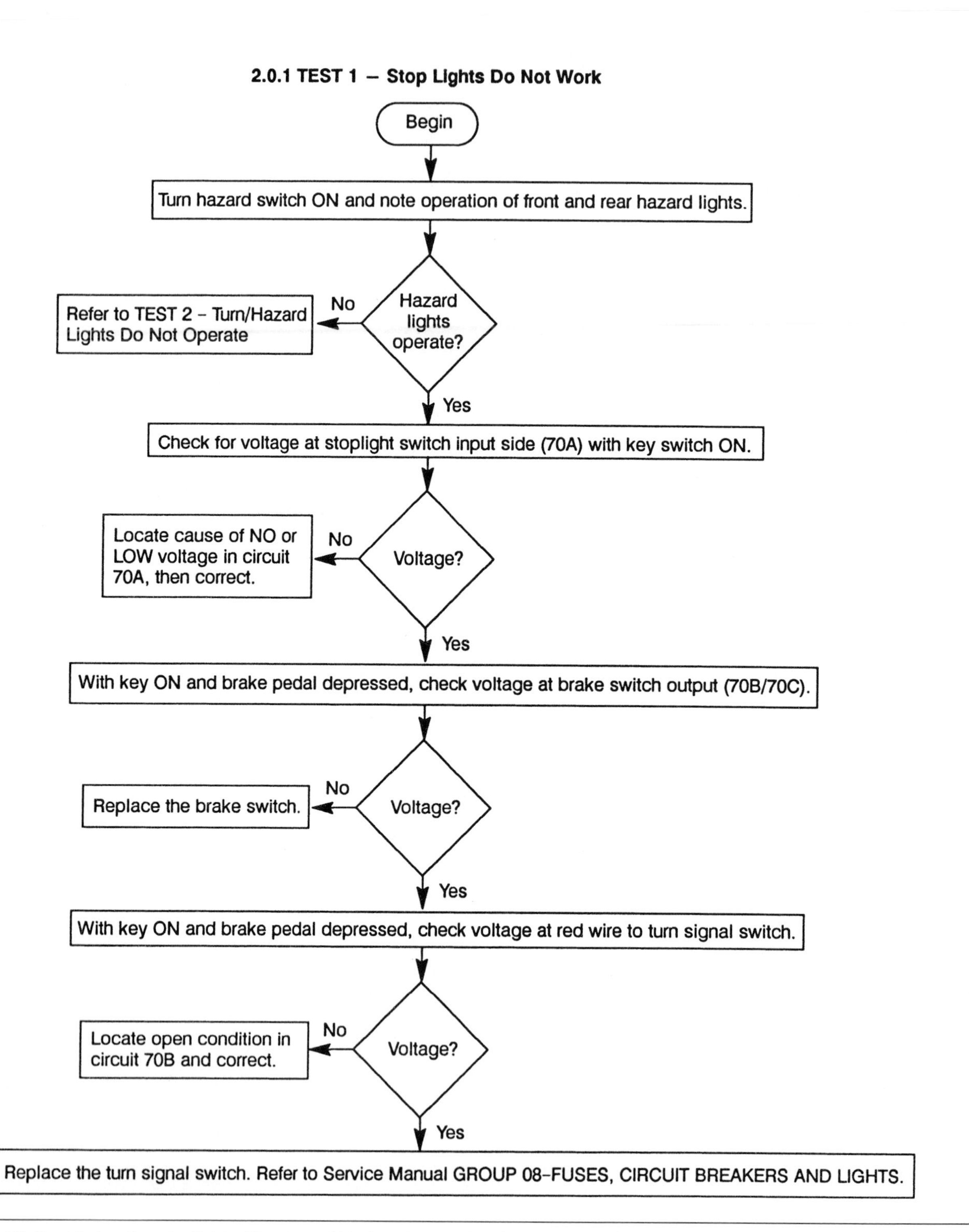

Figure 8-43 Troubleshooting stoplights. (Courtesy of Navistar International Engine Group)

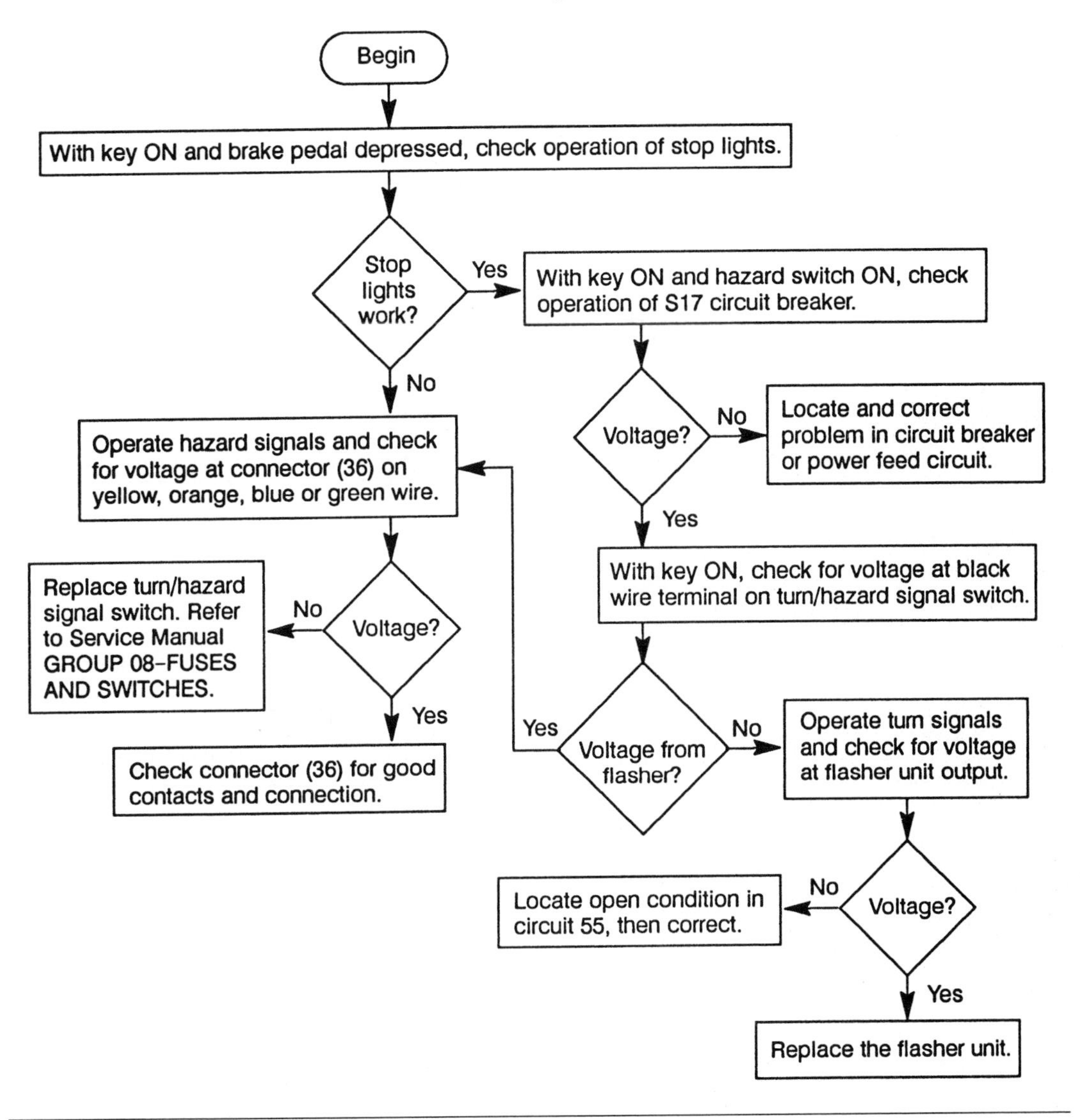

Figure 8-44 Troubleshooting turn/hazard signals. (Courtesy of Navistar International Engine Group)

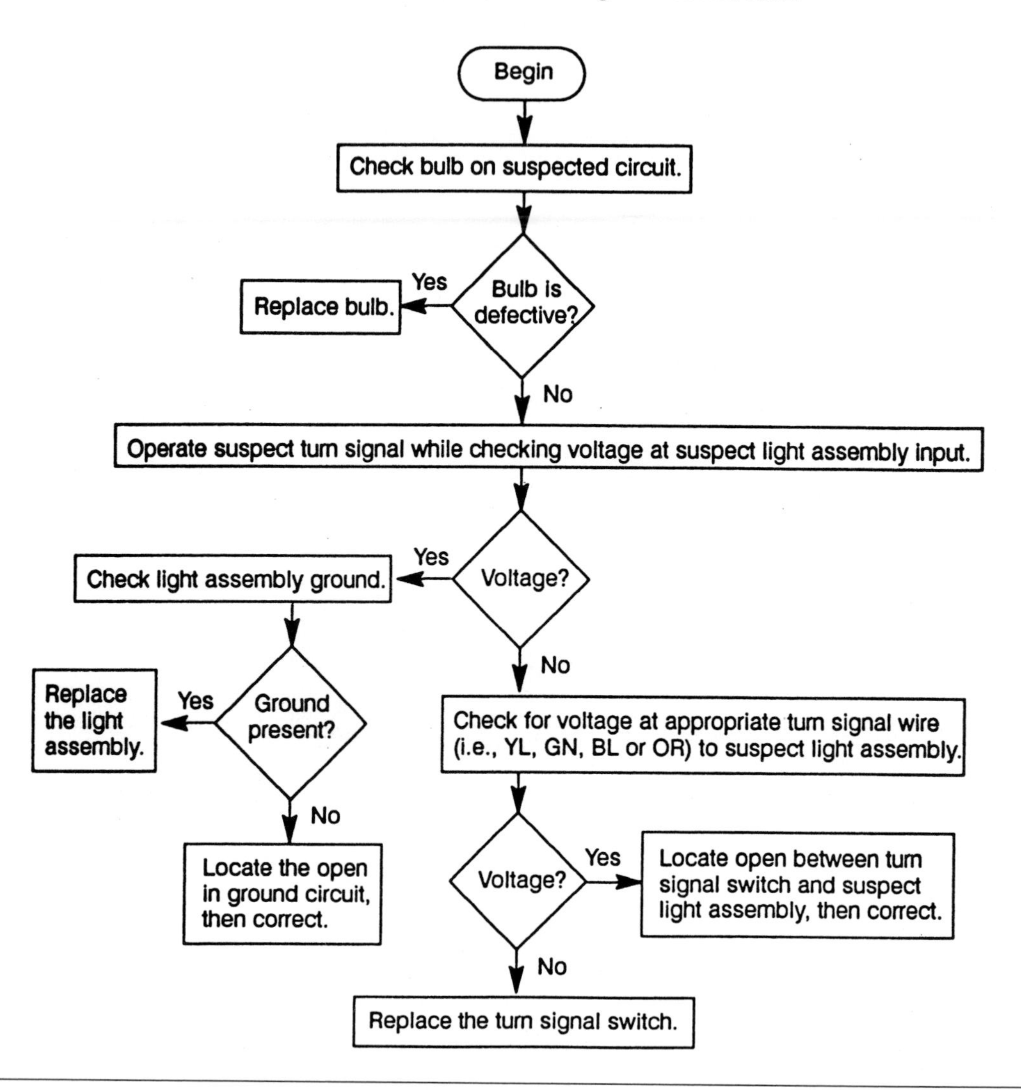

Figure 8-45 Troubleshooting for one turn/hazard light not working. (Courtesy of Navistar International Engine Group)

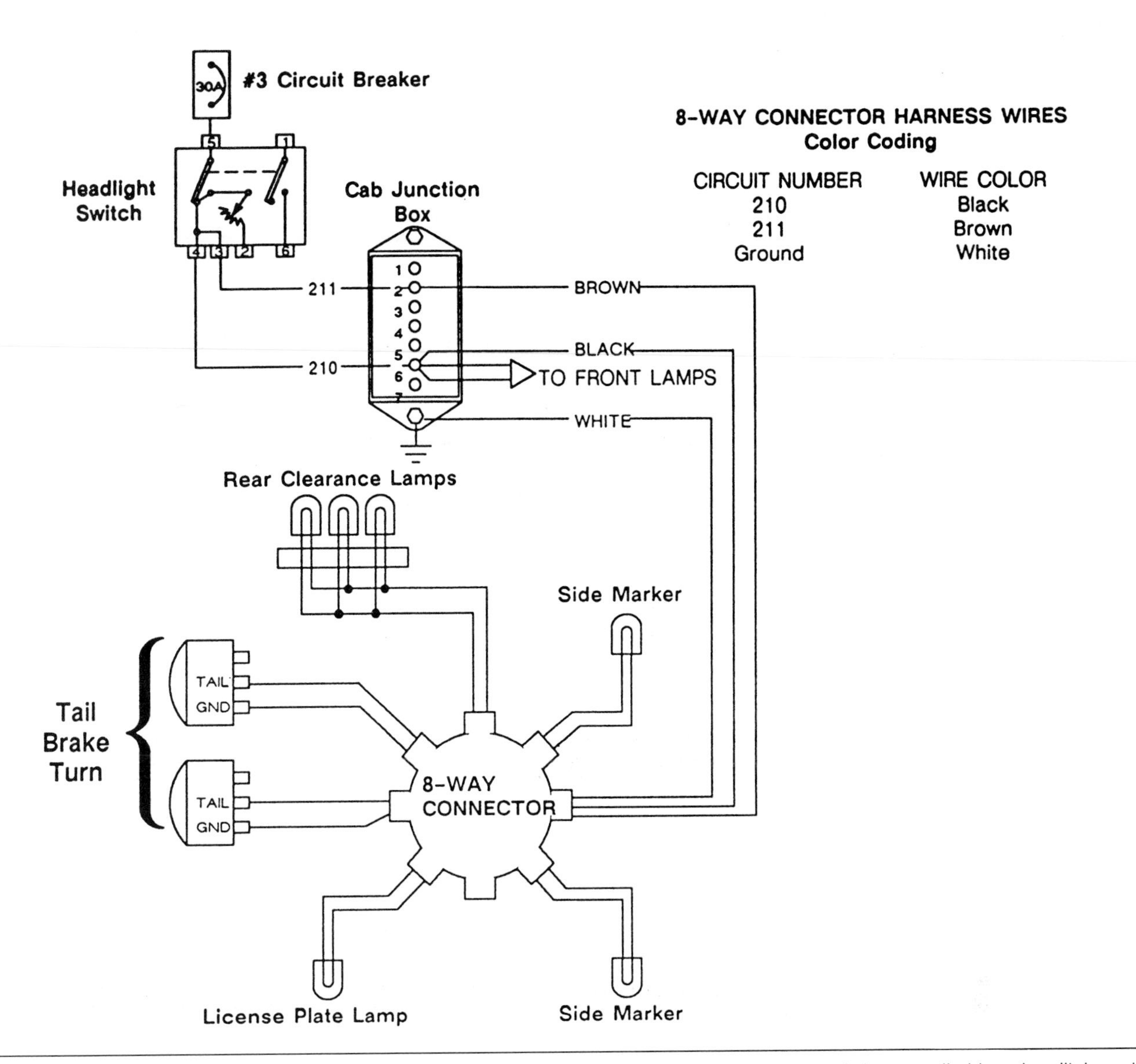

Figure 8-46 Circuit diagram showing rear taillights, clearance lights, side markers, and license plate light controlled by a headlight switch. (Courtesy of Mack Trucks, Inc.)

Taillights

The taillight or rear running lights are on a circuit with the headlights (Figure 8-46). You will notice that also the markers and clearance lamps are on the same circuit system. As with any light system, you need to check for voltages at the loads and grounds.

Classroom Manual
Chapter 8, page 193.

Replacing Sealed Tail, Brake, and Clearance Lights

Classroom Manual
Chapter 8, page 181.

Special Tools

Basic hand tool set

Safety glasses

Many vehicles are now incorporating the sealed type of light systems. Photo Sequences 7 and 8 show the procedures for removing and installing these types of lights.

These style of lights are used in many trucks, tractors, and trailers. The white wire is the ground wire. Some circuits use a common ground, while others require the ground to be separate

Photo Sequence 7
Brake/Tail/Backup/Marker Lamp Removal

P7-1 Place both thumbs on each side of the lamp assembly.

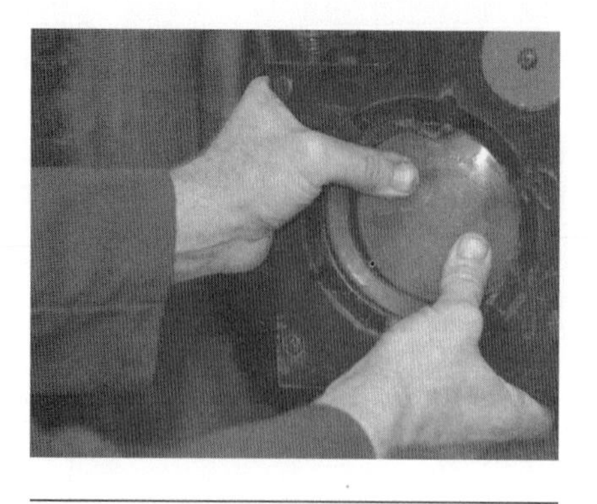

P7-2 With a firm steady pressure, push inward until the lamp assembly pushes through the rubber grommet and mounting hole. Note: The rubber grommet will often push through also with the lamp.

P7-3 Separate the lamp from the rubber grommet, then remove the lamp.

P7-4 Disconnect the harness connector.

Feedback is common on ground problems. One example of this is the marker lights blinking when the turn signals are turned on. This occurs because electricity always wants to seek a path of least resistance. This alternate path operates another component than the one intended. Feedback can be classified as a short.

and connected to the chassis near the light that is being served. Always be sure there is a proper ground.

Trailer Connector and Lights

The trailer lights require an electrical connection to the tractor. The tractor supplies the applicable voltages to the trailer from a 7-way receptacle that is usually mounted to or at the rear of the tractor cab.

SERVICE TIP: Trailers using ABS require the dedicated wire for it to be larger. Use the appropriate trailer cord with ABS.

The first step is to see if the lighting system problem is in the trailer or the tractor. It is easy to pull the **trailer cord plug** from the tractor receptacle (Figure 8-47). Simply probe the failed light corresponding connector on the tractor receptacle to see if the test light illuminates. Figure

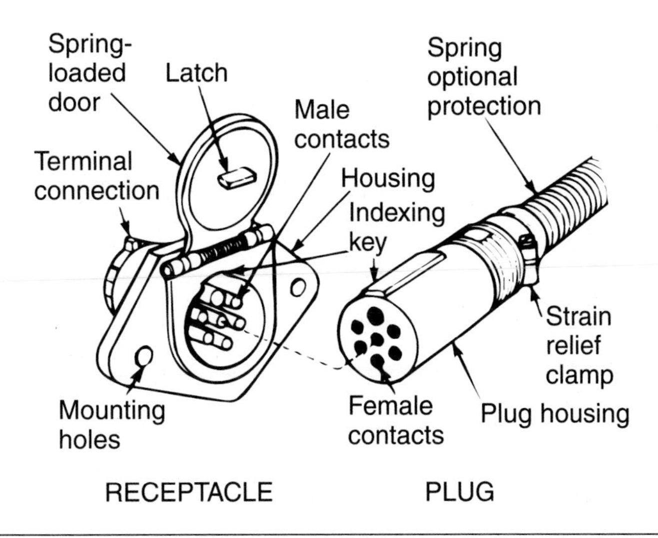

Figure 8-47 Light cord plug connector. (Courtesy of Navistar International Engine Group)

Wire Color	Light and Signal Circuits
White	Ground return to towing vehicle
Black	Clearance, side marker, and identification lights
Yellow	Left-hand turn signal and hazard signal
Brown	Tail and license plate lights
Blue	Auxiliary circuit

Figure 8-48 Wire colors are used to identify specific circuits.

8-48 shows the corresponding wire color to light function. This can be compared to the pin configuration shown in Figure 8-46. If the test light comes on, the problem could be in the trailer cord and/or in the trailer light circuits. If the light failed to illuminate, the problem is most likely in the tractor light circuit. This requires a diagram of the vehicle's lighting system as it is related to the semi-trailer (Figure 8-49). Take time to study this figure.

Notice that the vehicle stoplight switch not only feeds power to the turn signal switch, but it also has a separate circuit (70C). When the tractor brakes are applied, the stoplight switch supplies power through that circuit to the trailer socket (90) at the RD terminal, designating the color red for stoplights. Take time to compare the rest of the colors to the turn signal functions, taillights, and markers. These are standard colors used by vehicle manufacturers for trailer wiring.

This system also uses a normally closed, momentarily open interrupter switch. Depressing the interrupter switch momentarily de-energizes the trailer relay (5) turning off the power to the trailer socket BN and BK terminals. For example, the interrupter switch is used to signal other vehicles that it is okay to re-enter a lane after passing a vehicle.

Special Tools

Basic hand tool set

12-volt test light

Digital multimeter

Safety glasses

Classroom Manual
Chapter 8, page 207.

Troubleshooting Summary

Diagnosing lighting systems does not have to be complicated. Always have as much information as possible available to you. Have a good grasp on how the system functions. Start with the most obvious part first such as burned out lamps. If the lamps are functional, check the connectors and ground connections of the failed circuit. Many lighting systems share common power sources and grounds. If more than one circuit is not functional, start looking for these common sources.

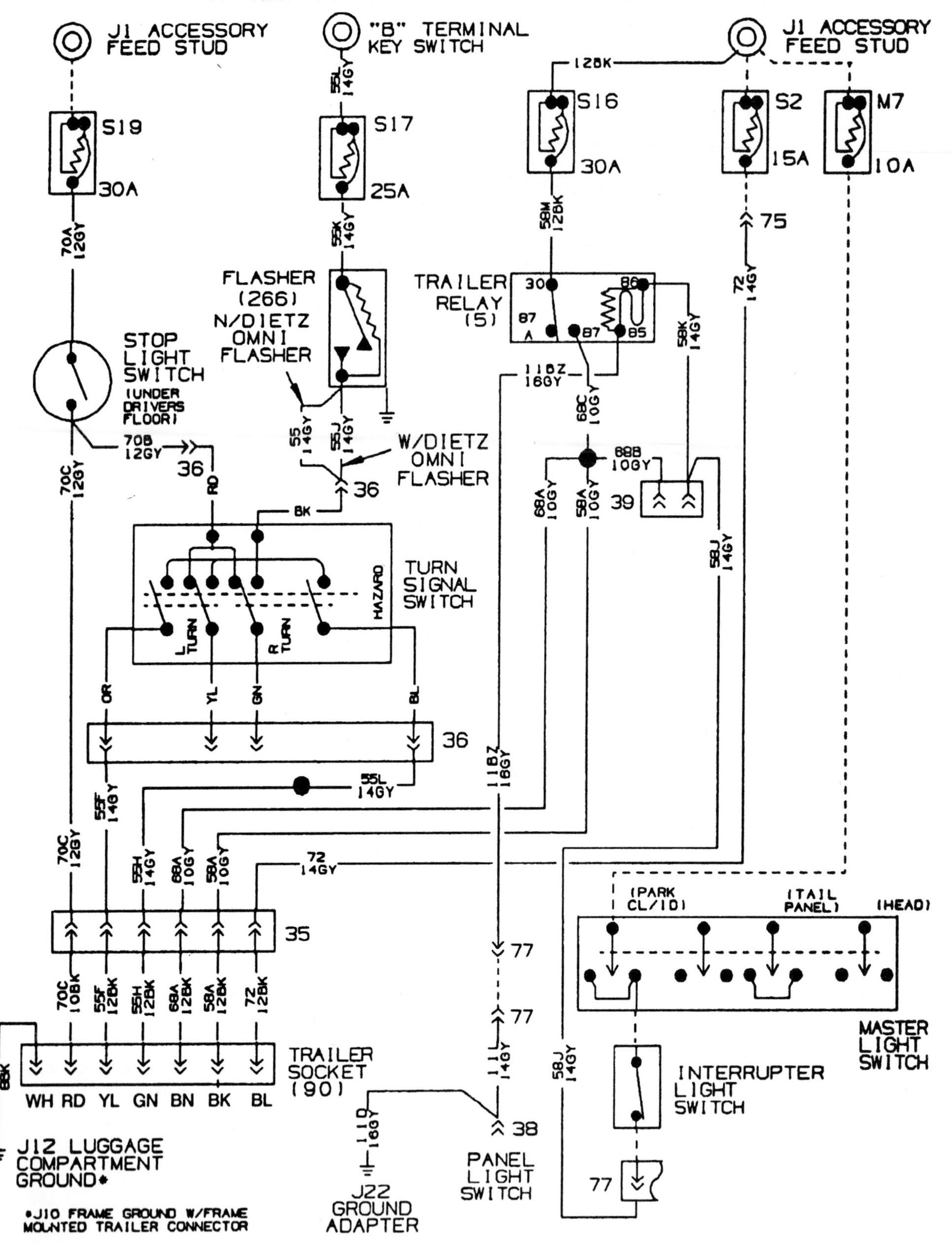

Figure 8-49 Tractor's lighting circuit pertaining to the trailer lights. (Courtesy of Navistar International Engine Group)

Check harnesses for breaks and deteriorated or frayed wires and then repair and/or replace them as necessary. If these basic checks do not solve the problem, use a test light or multimeter to determine current flow through the circuits. When the faulty component or problem area has been located, repair or replace it as necessary. Don't forget to take any action if deemed necessary to prevent future problems.

CASE STUDY

One cold winter day a customer came in with no taillights on the trailer. One of the technicians in the shop verified that the fuse was blowing every time the running lights were turned on. However, every time he tried to find the problem, the new fuse would blow. After poking around some wires, he got discouraged. He substituted a small bolt in place of the fuse. This gave him enough time to see the smoke from the wire that was rubbed through and touching the crossmember. He moved the wire away from the crossmember, taped up the bare spot, and verified that the lights would stay on with the new fuse.

The next day, the customer was back with the same trailer and a ticket from the DOT inspector for not having any taillights. Needless to say, the truck driver's company was not pleased. The original technician was out of the shop, so another technician in the shop was given the job order. He noticed that the circuit protection was intact. A thorough check showed there were several harness connections. A few quick checks with the test light brought him to the section of the harness where no current was flowing. He noticed the tape job his co-worker had done the day before. He took the tape off and noticed that the area seemed to have some strands still left. He decided to pull out that section of harness. In a section that couldn't be seen before, he noticed a section that was burned or heated to a point that any vibration could open up that wire, preventing the flow of current.

He replaced the section of harness following proper wiring procedures. He then reconnected the lights and performed a voltage drop test to ensure there were no other problems in the circuit indicating a high resistance. The shop did not charge for the corrective work, hoping that the driver and his company would be understanding and return to the shop for future work.

Terms to Know

Bezel
Commercial Vehicle Safety Alliance (CVSA)
Composite
Dimmer switch
Federal Motor Carrier Safety Regulations (FMCSR)
Filament
Flasher
Halogen
Headlight switch
Sealed-beam headlight
Trailer cord plug

ASE-Style Review Questions

1. *Technician A* says commercial vehicle lighting systems are regulated by the federal government. *Technician B* says commercial vehicle lighting system requirements are often enforced by certain state and local entities. Who is correct?
   A. A only **C.** Both A and B
   B. B only **D.** Neither A nor B

2. *Technician A* says repairs to the lighting system must meet all applicable standards. *Technician B* says repairs to the vehicle's lighting system must ensure vehicle safety. Who is correct?
   A. A only **C.** Both A and B
   B. B only **D.** Neither A nor B

3. *Technician A* says some dimmer switches are mounted on the floorboard. *Technician B* says that the dimmer switch is sometimes part of the multifunction switch. Who is correct?
   A. A only **C.** Both A and B
   B. B only **D.** Neither A nor B

4. A vehicle comes into the shop with dim headlights. *Technician A* says the cause could be a poor ground. *Technician B* says this can be caused by an excessive voltage drop in the circuit. Who is correct?
   A. A only **C.** Both A and B
   B. B only **D.** Neither A nor B

5. The turn signals of a vehicle operate only on one side. *Technician A* says the flasher is bad. *Technician B* says the fuse is blown. Who is correct?
   A. A only **C.** Both A and B
   B. B only **D.** Neither A nor B

6. Taillight assembly is being discussed: *Technician A* says that one bulb can be used for a tail, stop, and turn indicator. *Technician B* says that some systems use more than one bulb or light to indicate tail, stop, and turn. Who is correct?
   A. A only **C.** Both A and B
   B. B only **D.** Neither A nor B

7. None of the headlights come on with the headlight switch. *Technician A* says the problem could be the dimmer switch. *Technician B* says the most likely problem could be loss of power from the ignition switch. Who is correct?
   A. A only **C.** Both A and B
   B. B only **D.** Neither A nor B

8. *Technician A* says the dimmer switch is hooked in parallel to the headlights. *Technician B* says the dimmer switch is hooked in series to the headlights. Who is correct?
   A. A only **C.** Both A and B
   B. B only **D.** Neither A nor B

9. Composite headlight bulb replacement is being discussed: *Technician A* says care must be taken not to touch the glass with your fingers. *Technician B* says not to energize the bulb when it is not installed into the socket. Who is correct?
   A. A only **C.** Both A and B
   B. B only **D.** Neither A nor B

10. Only one lamp in the circuit is not operating. *Technician A* says the first check is to verify the bulb is functional. *Technician B* says to start at the switch because it is usually the source of power. Who is correct?
    A. A only **C.** Both A and B
    B. B only **D.** Neither A nor B

ASE Challenge

1. The front of a truck has been repaired following a collision. *Technician A* says the headlight alignment can be checked using a flat shop floor, a light-colored blank wall, and a tape measure. *Technician B* says that headlight alignment equipment only works on sealed beams. Who is correct?
   A. A only **C.** Both A and B
   B. B only **D.** Neither A nor B

2. When turned on, the headlights of a truck flash off and on. *Technician A* says the circuit breaker for the headlights may be weak. *Technician B* says there could be a short in the headlight circuit. Who is correct?
   A. A only **C.** Both A and B
   B. B only **D.** Neither A nor B

3. The signal lights on a trailer work in the opposite direction to the arm on the signal light switch. The tractor lights work okay. *Technician A* says the problem could be in the junction to the tractor plug. *Technician B* says the problem could be in the signal light switch itself. Who is correct?
   A. A only **C.** Both A and B
   B. B only **D.** Neither A nor B

4. *Technician A* says the grease on the terminals of a new sealed-beam taillight is there to shield moisture after installation. *Technician B* says sealed-beam taillights reduce the possibilities of poor ground connections. Who is correct?
   A. A only **C.** Both A and B
   B. B only **D.** Neither A nor B

5. *Technician A* says you should avoid joining marker and taillight connections to the same circuit on a semi-trailer. *Technician B* says that the more lights you connect in series, the more current (amperage) you will draw and the brighter the lights will glow. Who is correct?
   A. A only **C.** Both A and B
   B. B only **D.** Neither A nor B

Table 8-1 NATEF and ASE Task

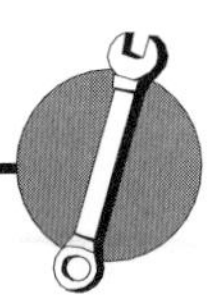

Diagnose the cause of brighter-than-normal, intermittent, dim, or no headlight operation

Problem Area	Symptoms	Possible Causes	Classroom Manual	Shop Manual
BRIGHTER-THAN-NORMAL OR INTERMITTENT HEADLIGHT OPERATION	Bright illumination, early bulb failure	1. Generator output too high 2. Defective dimmer switch	134	180
	Headlights flicker	1. Defective circuit breaker 2. Circuit overload 3. Bad connection 4. Defective headlight switch 5. Poor ground 6. High circuit resistance	186	234
NO HEADLIGHT OPERATION	Headlight inoperative	1. Burned out bulbs 2. Defective switch 3. Open circuit 4. Defective circuit breaker 5. Defective Maxi-fuse 6. Bad connection 7. Poor ground 8. Excessive resistance	177	230

Table 8-2 NATEF and ASE Task

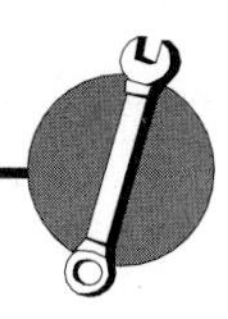

Test, aim, and replace headlights

Problem Area	Symptoms	Possible Causes	Classroom Manual	Shop Manual
HEADLIGHTS BURNED OUT OR OUT OF ADJUSTMENT	Improper road illumination	1. Burned out bulbs 2. Improper headlight aiming 3. Open circuit 4. Short in circuit 5. Bad connection 6. Poor ground 7. Wrong bulb	184	239

Table 8-3 NATEF and ASE Task

Test, repair, and replace headlight and dimmer switches, wires, connectors, terminals, sockets, relays, and control components

Problem Area	Symptoms	Possible Causes	Classroom Manual	Shop Manual
OPEN OR DEFECTIVE COMPONENTS	Improper or no headlight operation	1. Defective headlight switch 2. Defective relay 3. Defective dimmer switch 4. Open circuit 5. Poor connection	186	246

Table 8-4 NATEF and ASE Task

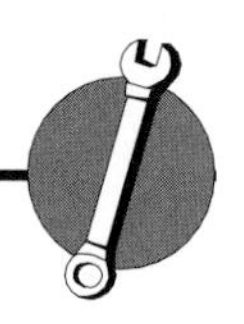

Inspect, test, repair, or replace switches, bulbs, sockets, connectors, terminals, relays, and wires of parking, clearance, and taillight circuits on trucks and trailers

Problem Area	Symptoms	Possible Causes	Classroom Manual	Shop Manual
OPEN OR DEFECTIVE COMPONENTS	Improper or no parking and/or taillight operation	1. Defective headlight switch 2. Defective relay 3. Defective bulb 4. Open circuit 5. Poor connection	191	253

Table 8-5 NATEF and ASE Task

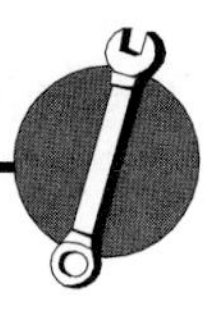

Inspect, test, repair, or replace interior cab light circuit switches, bulbs, sockets, connectors, terminals, and wires

Problem Area	Symptoms	Possible Causes	Classroom Manual	Shop Manual
OPEN OR DEFECTIVE COMPONENTS	Improper or no interior light operation	1. Wrong bulbs 2. Open circuit protection device (fuse) 3. Defective printed circuit board 4. Defective instrument panel computer 5. Defective relay 6. Defective bulb 7. Open circuit 8. Poor ground 9. Excessive resistance 10. Poor connection	203	230

Table 8-6 NATEF Task

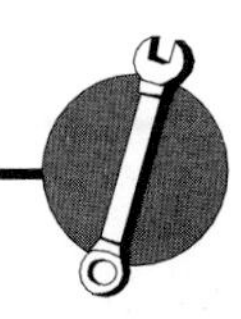

Inspect and test tractor-to-trailer multiwire connectors, repair or replace as needed

Problem Area	Symptoms	Possible Causes	Classroom Manual	Shop Manual
OPEN OR DEFECTIVE COMPONENTS	Improper or no trailer light operation	1. Wrong bulbs 2. Open circuit protection device (fuse) 3. Defective relay 4. Defective bulb 5. Open circuit 6. Poor ground 7. Excessive resistance 8. Poor connection	204	260

Table 8-7 NATEF and ASE Task

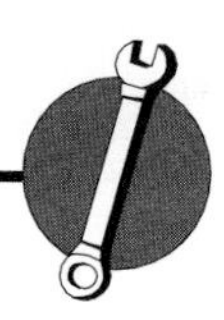

Inspect, test, adjust, repair, or replace stoplight circuit switches, bulbs, sockets, connectors, terminals, relays, and wires

Problem Area	Symptoms	Possible Causes	Classroom Manual	Shop Manual
OPEN OR DEFECTIVE STOP LIGHT COMPONENTS	Brake lights fail to operate some or most of the time	1. Misadjusted brake light switch 2. Faulty sockets 3. Faulty turn signal switch contacts 4. Defective bulb 5. Open circuit 6. Poor ground 7. Excessive resistance 8. Poor connection	198	255

Table 8-8 NATEF and ASE Task

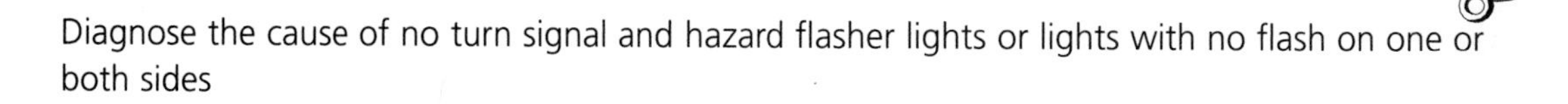

Diagnose the cause of no turn signal and hazard flasher lights or lights with no flash on one or both sides

Problem Area	Symptoms	Possible Causes	Classroom Manual	Shop Manual
NO TURN SIGNAL OPERATION	Turn signals do not operate in either direction or fail to flash	1. Blown fuse 2. Defective or worn flasher unit 3. Faulty turn signal switch 4. Improper bulb 5. Burned out bulb 6. Open circuit	199	196

Table 8-9 NATEF and ASE Task

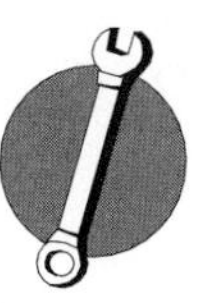

Inspect, test, repair, or replace turn signal and hazard circuit flasher, switches, bulbs, sockets, connectors, terminals, relays, and wires

Problem Area	Symptoms	Possible Causes	Classroom Manual	Shop Manual
NO HAZARD LIGHT OPERATION	Hazard lights do not operate in either direction	1. Blown fuse 2. Defective or worn flasher unit 3. Faulty turn signal switch 4. Improper bulb 5. Burned out bulb 6. Open circuit	199	196

Job Sheet 12

12

Name: ______________________ Date: ______________________

Checking a Headlight Switch

Upon completion of this job sheet, you should be able to check the operation of a headlight switch.

NATEF and ASE Correlation

This job sheet is related to ASE and NATEF Medium/Heavy Duty Truck Electrical/Electronic Systems List content area:

E. Lighting Systems Diagnosis and Repair, 1. Headlights, Daytime Running Lights, Parking, Clearance, Tail, Cab, and Dash Lights, ASE and NATEF Task 3: Test, repair, and replace headlight and dimmer switches, wires, connectors, terminals, sockets, relays, and control components.

Tools and Materials

A medium/heavy duty truck

A wiring diagram

A DMM

A number of colored highlighters (for outlining circuits)

A truck that has a master headlight switch with multiple lighting systems is preferred for this. If the truck uses multiple switches, incorporate this procedure using all the light switches. Usually multiple switches used for lighting have a relationship with each other anyway.

Procedure

1. Describe the truck being worked on: Year: ________ Make: ____________
   VIN: __________________ Model: __________________
2. Work the headlight switch or light switches through all the positions and observe which lights are controlled by each position. List each position and the controlled lights. ____
   ____
3. Locate the headlight switch in the wiring diagram and draw the switch with each possible connection and possible position. Label the lights controlled by each position of the switch. (If several copies of the diagram are available, the highlighters are very useful in outlining the various circuits in relation to the switch positions.)
4. Remove the fuse to the headlights and any other fuse or circuit protection device used for the lighting system. For increased safety of the circuit, disconnect the battery's negative cable. Remove the headlight switches according to the procedures outlined in the service manual. Describe the procedure to your instructor before removing it.

Instructor's okay to move to the next step ________

5. Identify the various terminals of the switch and list the different terminals that should have continuity in the various switch positions. ______________________
   ____

6. Connect the ohmmeter across these terminals, one switch position at a time, and record your reading: ______________________________

7. Based on the above test, what are your conclusions about the switches? __________
____

8. Ask your instructor if you should continue to the next procedure. It requires the connector and/or switch to have power and also to be accessed with a voltmeter. Access to the switch might require the switch to be out of the dashboard.

Instructor's okay to move to the next step __________

9. Reinstall the switch if it can be easily accessed or leave switch out if this is the only way it can be accessed. Reconnect power to the lighting system.

10. Using the voltmeter, check for available voltage to the switch or switches. List the terminal or terminals with available voltage and list the voltage? ______________
____

11. Operate the switch or switches through the various positions and list the terminals with available voltage and list the voltages. ______________________________
____

12. Do the paths of voltages parallel the ohmmeter test for continuity? Yes ____ No ____

13. Based on all the above tests, what are your conclusions about the switch? ________
____

14. If the switch was not reinstalled properly, do so now following proper installation procedures. Then check the operation of all the lights controlled by the headlight switch or any switch that was tested.

☑ **Instructor Check** ______________________________

Job Sheet 13

13

Name: ______________________ Date: ______________________

Checking a Truck's Lighting System

Upon completion of this job sheet, you should be able to visually check the integrity of a truck's lighting system.

NATEF and ASE Correlation

This job sheet is related to ASE and NATEF Medium/Heavy Duty Truck Electrical/Electronic Systems List content area:

E. Lighting Systems Diagnosis and Repair, 1. Headlights, Daytime Running Lights, Parking, Clearance, Tail, Cab, and Dash Lights, ASE and NATEF Task 3: Test, repair, and replace headlight and dimmer switches, wires, connectors, terminals, sockets, relays, and control components. 2. Stoplights, Turn Signals, Hazard Lights, and Backup Lights. ASE and NATEF Task 1: Inspect, test, adjust, repair, or replace stoplight circuit switches, bulbs, sockets, connectors, terminals, relays, and wires; ASE and NATEF Task 3: Inspect, test, repair, or replace turn signal and hazard circuit flashers, switches, bulbs, sockets, connectors, terminals, relays, and wires.

Tools and Materials

A medium/heavy duty truck

A wiring diagram for the truck

Procedure

1. Describe the truck being worked on: Year: ________ Make: ____________
   VIN: ________________ Model: ________________
2. Starting inside the cab of the truck, identify and list the number of switches needed to operate the truck's entire lighting system. ______________________
3. List and describe the condition of all of the switches.
   ____
4. Turn the headlights on and identify the number of lights that are turned on with this switch. ____
5. Inspect all of the lights noted above and describe their condition, including any wiring or connectors that are easily seen without disassembling any panels or harnesses. ____
   ____
   ____
6. Check and describe the operation of the dimmer switch. ______________________
   ____
   ____
7. Turn on any other switch that is related to parking, marker, and clearance lights. List all of these additional lights, describe their condition, and identify any wiring and connectors related to them.

   Note: Do not disassemble any panels or harnesses for this.
   ____
   ____
   ____

8. Operate the turn signals, list the number of lights turned on with this switch, and describe their condition. ____

____

____

9. Describe the operation of the flasher (slow, fast, normal):

____

10. Check the operation of the stop lights, list the number of lights that came on, and describe their condition. ____

____

For the following questions, be sure to explain your answer.

11. Did all the lights function properly? ____________________________

____

12. Did the truck have all of the required lights? ____________________________

____

13. Are all of the lenses correctly located in relationship to color? ____________

____

14. Does the truck's lighting system meet all safety and regulatory requirements? ______

____

☑ **Instructor Check** ____

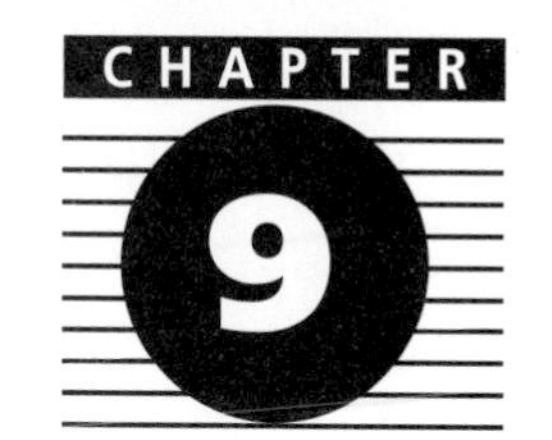

Instrument, Gauges, and Warning System Testing

Upon completion and review of this chapter, you should be able to:

- ❑ Remove instrument panels to access gauges.
- ❑ Remove and replace printed circuits.
- ❑ Diagnose and correct causes of noisy, erratic, and inaccurate speedometer readings.
- ❑ Diagnose problems with and calibrate electronic speedometer and tachometer systems.
- ❑ Diagnose and repair faulty gauge circuits.
- ❑ Diagnose and repair the cause of multiple gauge failures.
- ❑ Diagnose sender units, including thermistors, piezoresistive, and mechanical variable.
- ❑ Diagnose and repair warning systems.
- ❑ Diagnose engine shutdown systems.

Introduction

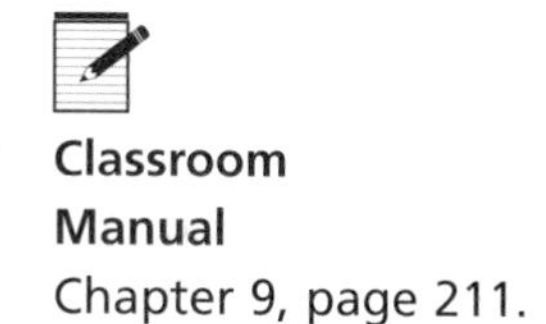
Classroom Manual
Chapter 9, page 211.

Instrument panels are used to house the gauges and warning lamps that provide information to the driver on the various operation conditions of the vehicle. The number of gauges and functions monitored varies between manufacturers and vehicle models. Most problems in the gauge and warning systems tend to be electrical. Examples of some problems are an open circuit in the wiring or printed circuit, improper gauge calibration, excessive resistance, connector problems, defective sending units, defective bulbs, or bad gauges. Some instrument clusters use simple electrical wiring. However, with the extensive use of onboard computers, the control and operation of today's trucks depend more and more on electronics. This requires the use of specific manufacturers' information, schematics, and wiring diagrams to diagnose instrumentation problems. A connector locator and pin chart can provide circuit information (Figure 9-1 and Figure 9-2). Looking at these figures it is easy to see how hard it would be to pinpoint problems without this

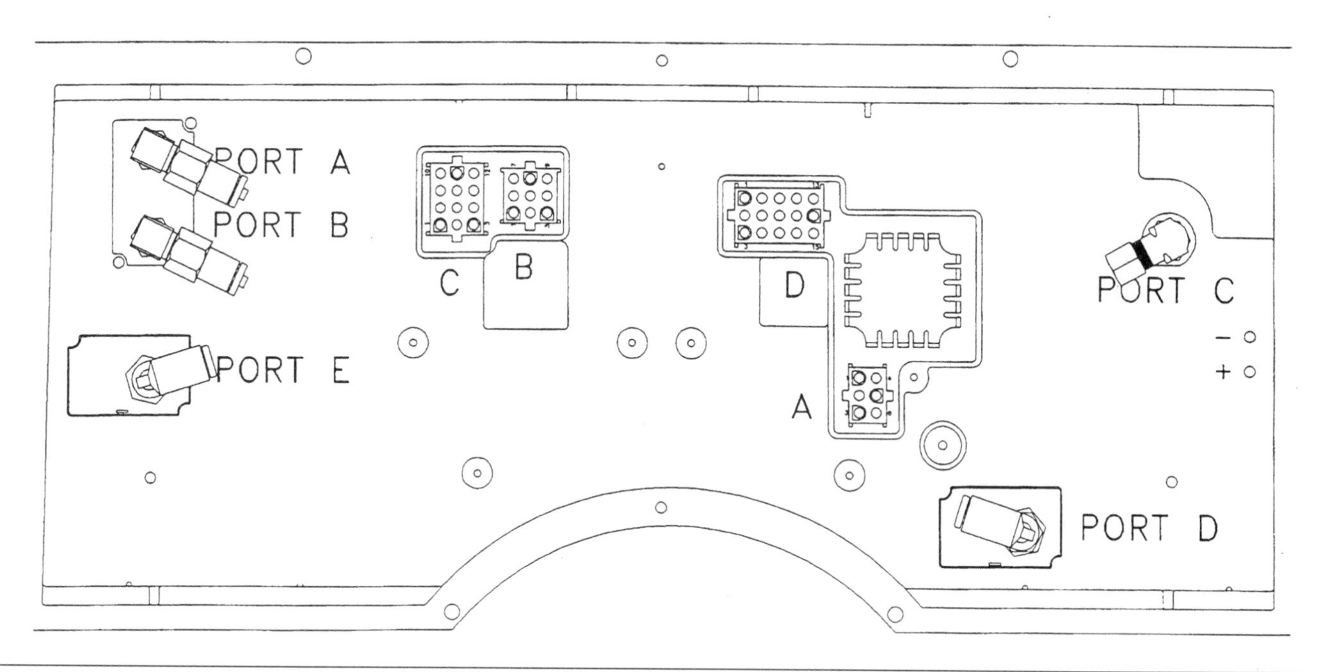

Figure 9-1 Connector locator guide for an instrument cluster. (Reprinted with permission. Courtesy of Volvo Trucks North America, Inc.)

Facelift pinout

Plug A

E1 = 6 pins

1. Power
2. Ground (power)
3. LCD OdO power
4. Lighting (+V)
5. Blank
6. Engine shutdown

Plug B

E2 = 9 pins

1. Speedo (+)
2. Speedo ground
3. J1708 (+)
4. J1708 (-)
5. Coolant warning switch
6. Two-speed axle
7. Tach (+)
8. Tach ground
9. Coolant shutdown switch

Plug C

E3 = 12 pins

1. Low coolant level
2. Coolant temperature gauge
3. Engine oil pressure warning switch
4. Battery charging
5. Engine oil pressure shutdown switch
6. Engine oil temperature
7. Fuel level
8. External air PSI switch
9. Trans temperature
10. R. rear axle temperature
11. F. rear axle temperature
12. Blank

Plug D

E4 = 15 pins

1. Right turn
2. Power take-off
3. Heated mirror
4. Air filter rest
5. Engine fluid
6. Check engine
7. Chassis ground
8. Stop engine
9. Light dimmer output
10. High beam
11. High beam
12. Do not shift
13. ABS
14. Engine preheat
15. Left turn

1 4
2 5
3 6

A

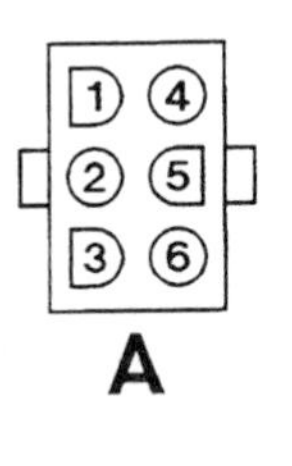

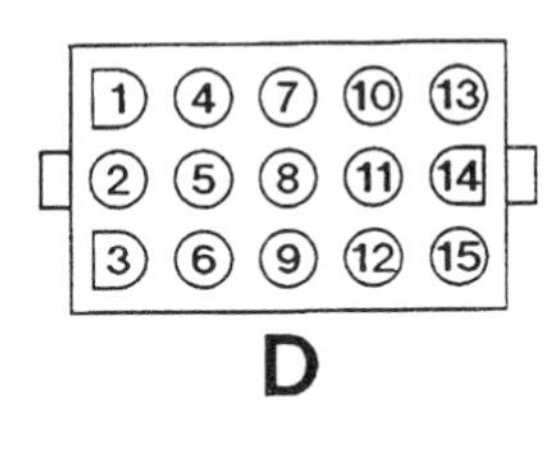

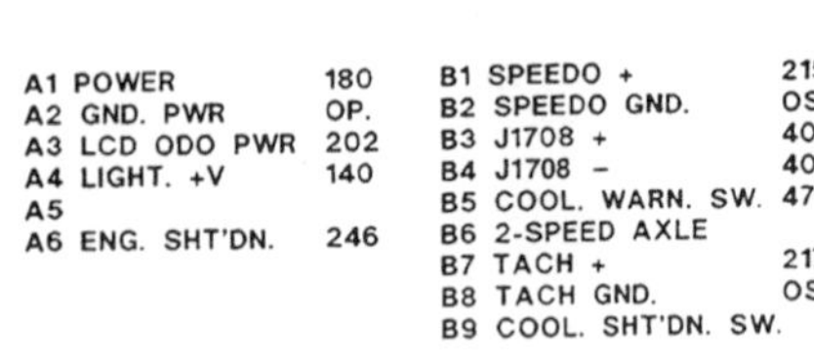

C1 LOW COOL. LEVEL 610A
C2 COOL. TEMP. GA. 181
C3 ENG. OIL PSI WARN. SW. 206
C4 BATT. CHARGING 14
C5 ENG. OIL PSI SHT'DN. SW. 608
C6 ENG. OIL TEMP. 204
C7 FUEL LEVEL 183
C8 EXT. AIR PSI SW. 198
C9 TRANS. TEMP. 314
C10 R. R. AXLE TEMP. 469
C11 F. R. AXLE TEMP. 468
C12

D1 RIGHT TURN 113
D2 PWR. TAKE-OFF 997
D3 HEATED MIRROR 702
D4 AIR FILT. REST. 199
D5 SPARE TELLTALE 207
D6 CHECK ENGINE
D7 CHASSIS GND. OP.
D8 STOP ENGINE
D9 LIGHT DIM. OPT. 141
D10 HIGH BEAM 31
D11 HIGH BEAM 338
D12 DO NOT SHIFT 325
D13 ABS 653
D14 ENG. PREHEAT 361
D15 LEFT TURN 112

Figure 9-2 Pinout legend. (Reprinted with permission. Courtesy of Volvo Trucks North America, Inc.)

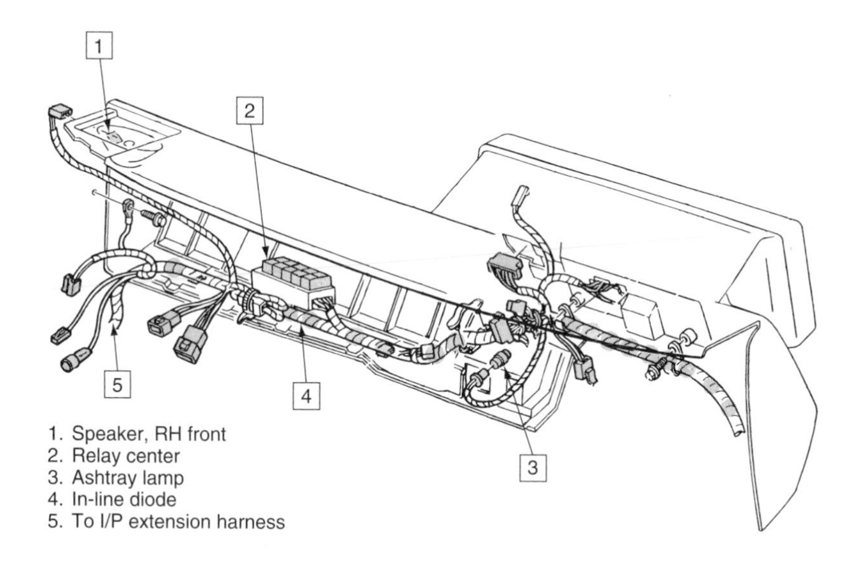

Figure 9-3 The number of connectors often used in dashboard areas. (Courtesy of General Motors Corporation, Service Operations)

information, especially when the instrument panel is removed and the technician is looking at the number of connectors as shown in Figure 9-3.

It is beyond the scope of this chapter to show every manufacturer's instrumentation system. We will, however, look at common individual gauges and circuits. The figures and procedures shown are from various manufacturers. These illustrate *general* procedures for testing and troubleshooting, so always refer to specific manufacturers' procedures and diagrams when performing any tests.

Instrument Panel and Printed Circuit Removal

Basic Tools

Basic mechanic's tool set

Service manual

Many times it may be necessary to remove the instrument panel to perform various functions such as circuit testing, or replacing gauges or lamps. Some systems are simple, requiring loosening a few panel screws to allow the panel assembly to come down and away from the dashboard for easy access of bulbs and the back sides of the gauges. Other vehicles have multiple panels and trim to contend with. Before removing panels, always disconnect the battery negative cable. Also, consult the service manual for the procedure for the vehicle on which you are working.

WARNING: Work performed on instrument clusters could subject the electronics on circuit boards to damage from electrostatic discharge. You should be grounded when working with these circuits. A grounding strap is one method used to guard against damage.

Special Tools

Battery terminal pliers

Safety glasses

Grounding strap

WARNING: Dashboards and fastening areas can be easily damaged if excessive force is used for prying or pulling. Use proper procedures and sequences.

The following is a common procedure for removing an instrument cluster and printed circuit with a number of covers and components to contend with:

1. Disconnect the battery negative cable.
2. Remove the retaining screws to the steering column opening cover. Then remove the cover (Figure 9-4).
3. Remove the finish panel retaining screws. Some vehicles use a cover like this to connect to air ducts. On some models, components such as a radio control are incorporated through the panel. In those cases it may be necessary to remove the radio knobs.
4. Remove the finish panel.
5. Remove the retaining bolts that hold the cluster to the dash.

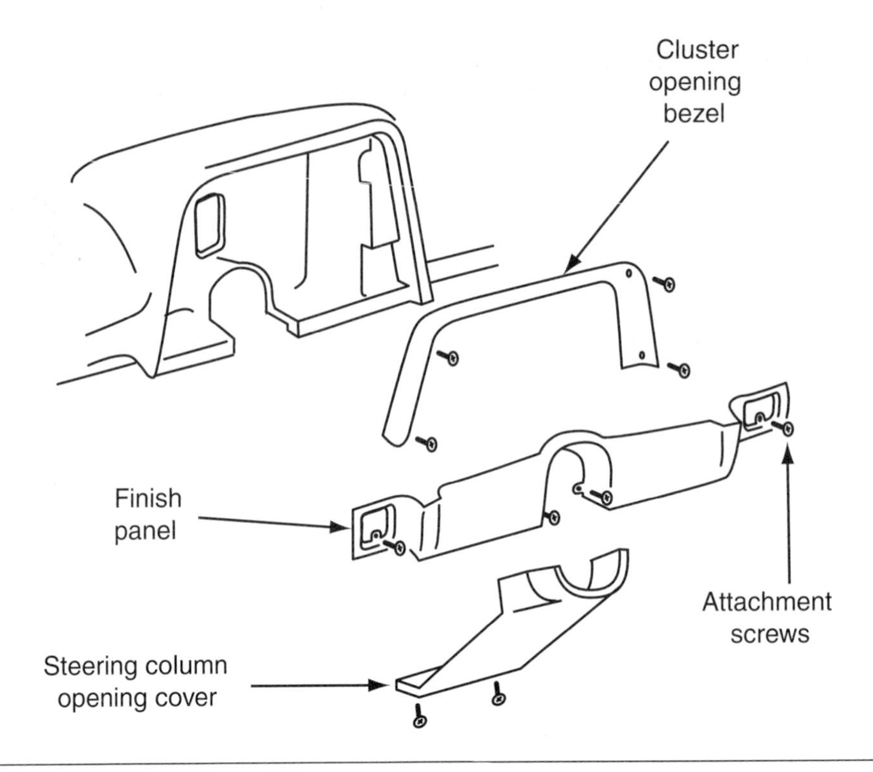

Figure 9-4 Instrument panel removal.

6. If the vehicle uses a mechanical speedometer and/or tachometer, reach behind and disconnect the cables.
7. Gently pull the cluster away from the dashboard.
8. Disconnect the cluster feed plug or plugs from the panel receptacles. Be careful not to damage the printed circuit. Note: The next steps are performed if required for further replacement of components.
9. Remove the instrument voltage regulator (IVR) (if applicable) and all illumination and indicator lamp sockets (Figure 9-5).
10. Remove the charging system warning lamp resistor if applicable.
11. Remove all printed circuit attaching nuts and remove the printed circuits.

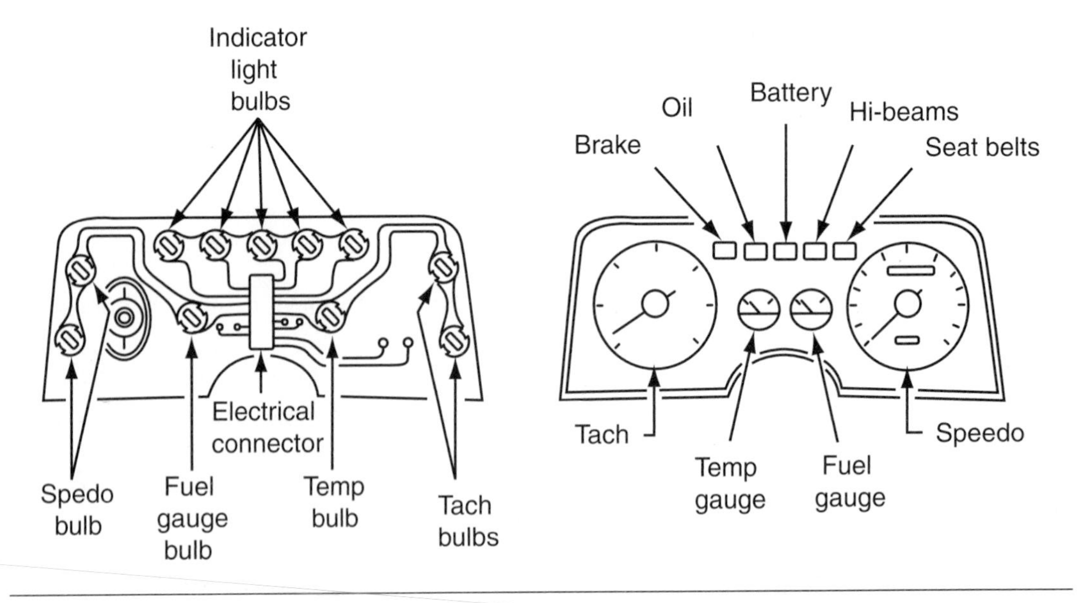

Figure 9-5 Instrument panel printed circuit board.

Speedometers and Tachometers

Classroom Manual
Chapter 9, page 222.

The **speedometer** indicates the speed of the vehicle.

Speedometers and tachometers used in trucks are generally of two types: mechanical or electrical. Complaints relating to speedometers depend on the type of speedometer being used. Complaints such as inaccurate reading or no readings are common to both types, while a complaint of chattering or noise is more common in mechanical speedometers and tachometers.

WARNING: Federal and state laws prohibit tampering with the correct mileage indicated on the odometer.

Electronic Speedometer and Tachometer

Classroom Manual
Chapter 9, page 223.

A **tachometer** is a measuring device used to measure the speed of the engine in revolutions per minute (RPM).

The **odometer** is a counter usually in the speedometer unit that indicates total miles driven by the vehicle.

In this section we are going to combine the diagnostics and testing for both the speedometer and tachometer. Problems that occur with both of these units involve:

- ❑ Checking for ignition voltage
- ❑ Checking for good ground
- ❑ Checking for signal input voltage and speed sensor operation

The relationship of the two units, power feed, ground, and sensors used is shown in the wiring diagram in Figure 9-6.

Most of today's computerized vehicles have the capability of performing initial power up or self-test of the speedometer/tachometer unit.

Usually the absence of a signal is indicated by the gauge pointer moving to its minimum reading.

The following is an example of an initial power-up test:

1. Turn the ignition key to the ON position (engine is off). Observe the speedometer and tachometer pointer movement. The pointer should perform as follows:
   - **A.** Both pointers will go to the nine o'clock position of the dial face for about two seconds.
   - **B.** Both pointers will then go to the zero position on the dial face, then advance through one complete gauge sweep.
   - **C.** Both pointers will then proceed to zero on the dial face in a smooth sweep before completing the test.

   Some vehicles require the cluster to be placed in a cluster diagnostic mode via a malfunction display.
2. If neither the speedometer or tachometer pointer moves during this test it usually indicates a lack of B+ voltage or a defective ground circuit. Looking at the diagram in Figure 9-6, you would suspect Circuit 47 or Circuit 11AG.
3. If cluster response is erratic, incomplete, or operates correctly you would perform the more detailed speedometer/tachometer unit self-test diagnostics.

Let's follow some test procedures:

- ❑ **Test 1 is the Initial Power-up Self-Test (Figure 9-7).**
- ❑ **Test 2 is the B+ Voltage and Ground Test (Figure 9-8).**

Special Tools

DVOM
Wiring diagram
Manufacturer's service manual

These tests use the diagram in Figure 9-6 as a guide. Test 2 is performed when neither the speedometer nor tachometer pointer moves during Test 1. Looking at the diagram you will notice that both the speedometer and tachometer rely on a single B+ voltage source and a single ground circuit. If there is either no feed or ground, the speedometer/tachometer will not function.

The following procedure is used to perform Test 2:

WARNING: When disconnecting connectors be sure to unlatch them properly. Any excess tugging or prying can result in a bad connection when it is reconnected.

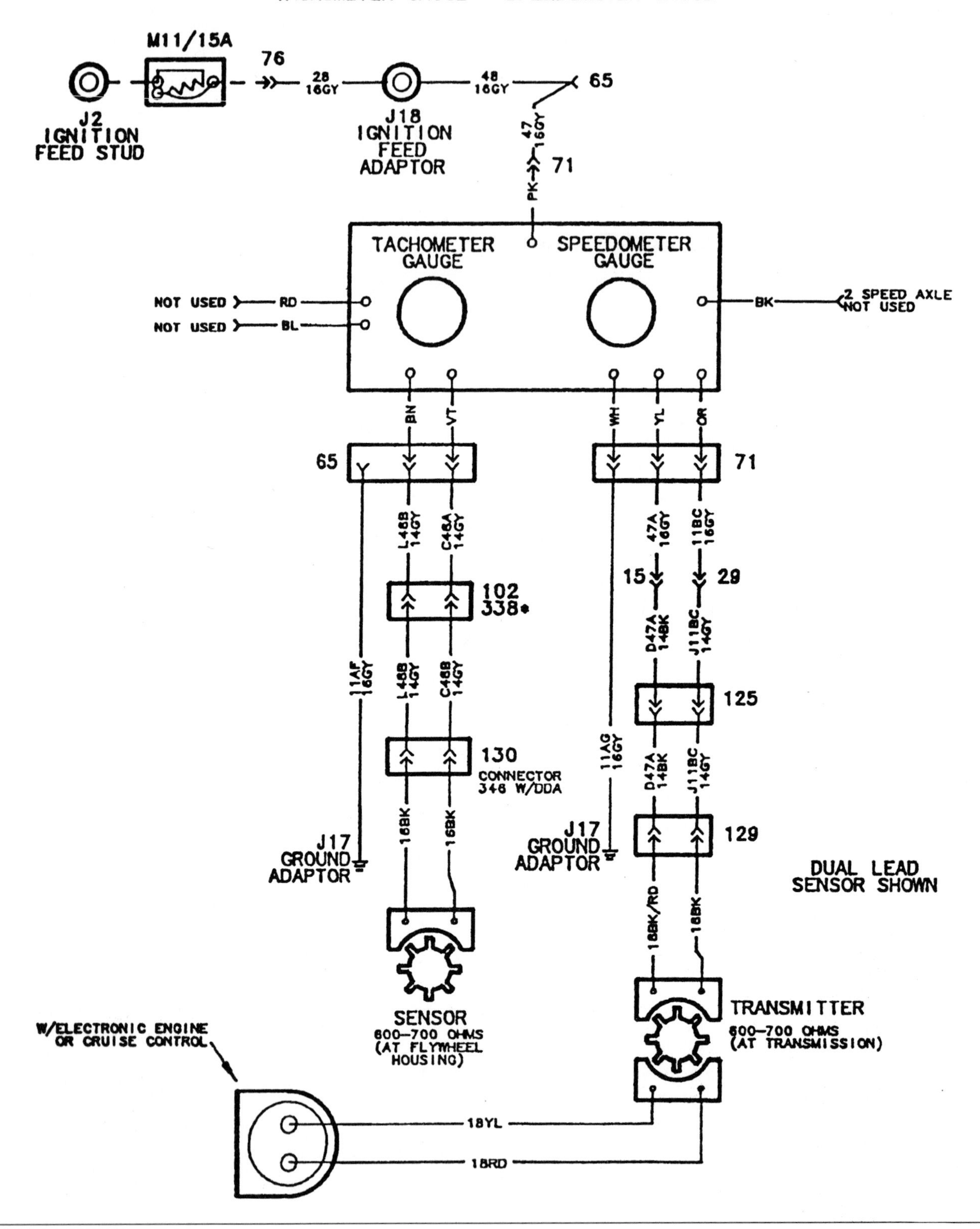

Figure 9-6 A speedometer and tachometer circuit diagram. (Courtesy of Navistar International Engine Group)

TEST 1 - INITIAL POWER-UP SELF TEST

NOTE: Turning the switch key ON causes the speedometer/tachometer unit to perform a brief self–test. The steps are as follows:

1. Turn switch key ON and observe gauge pointers.
2. Both pointers will go to nine o'clock position on dial face for about 2 seconds.
3. Both pointers will go to the zero position of the dial face, then advance through one complete gauge sweep.
4. Both pointers will then proceed to zero on the dial face in a smooth sweep before completing the test.

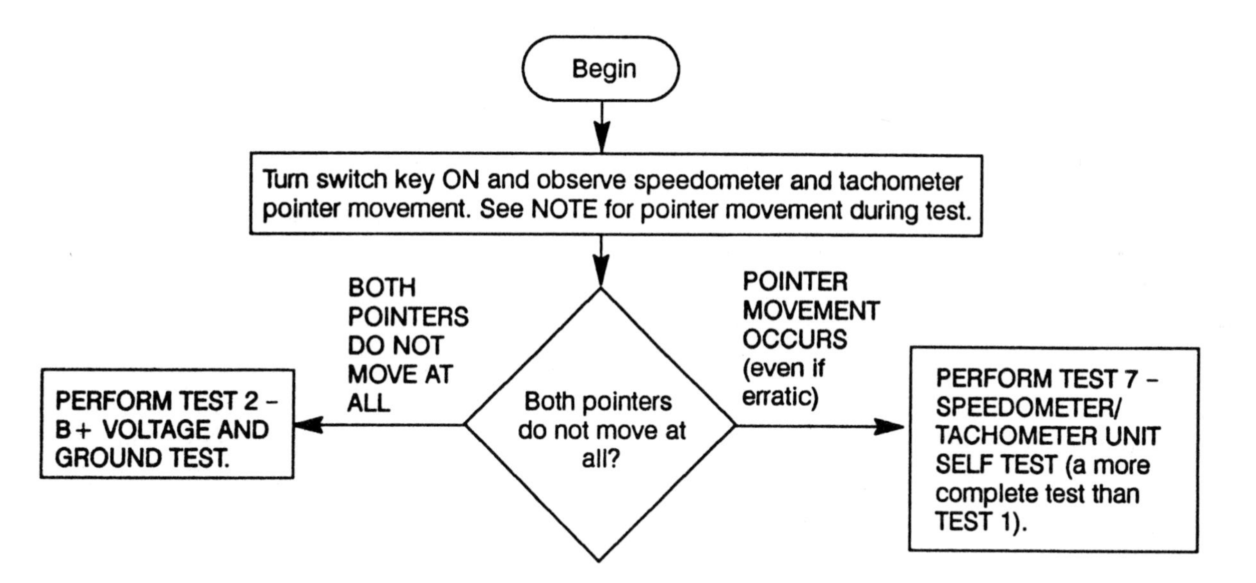

Figure 9-7 This initial power-up self-test is used to find a direction towards other possible tests. (Courtesy of Navistar International Engine Group)

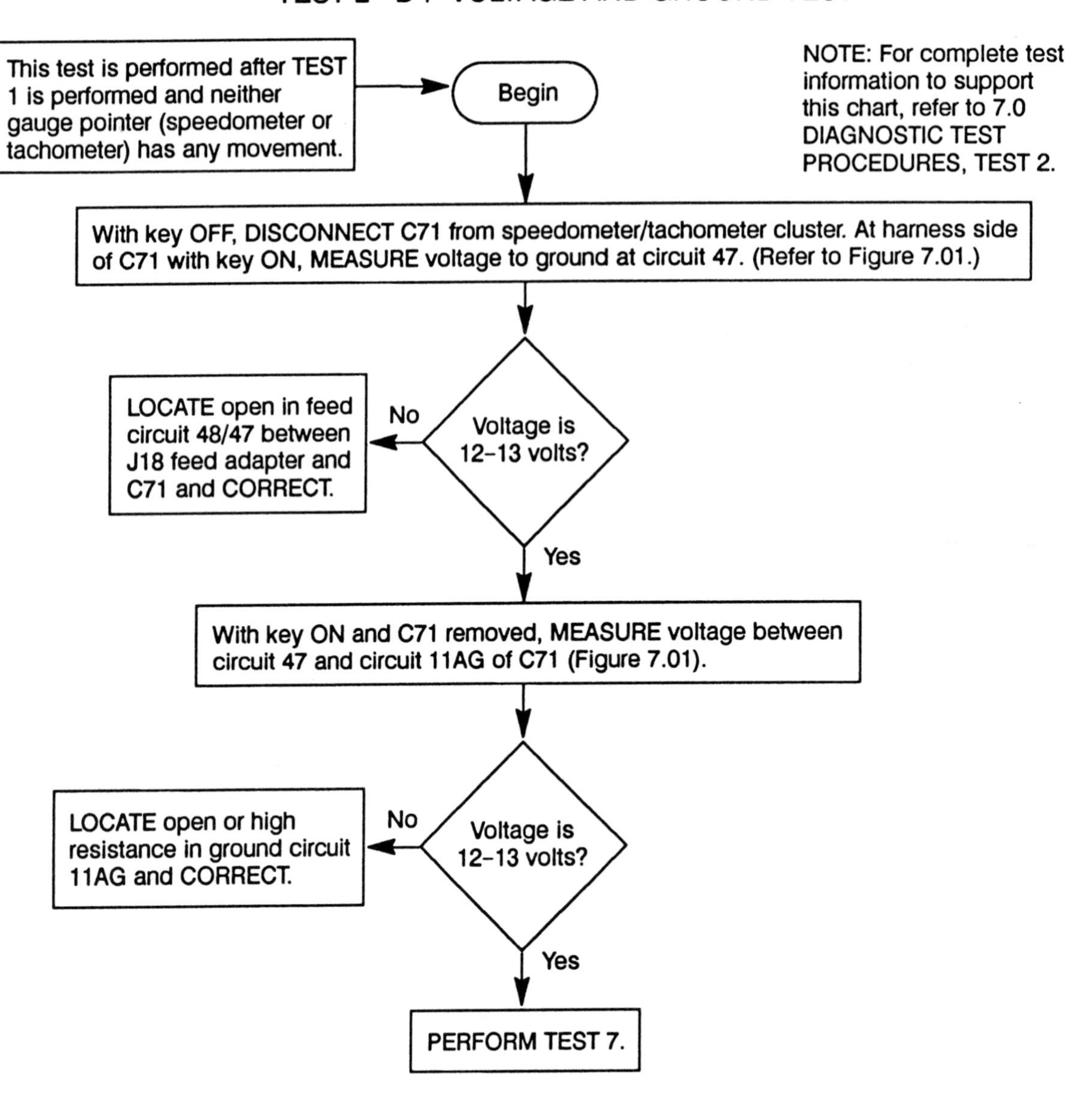

Figure 9-8 A flowchart to perform a power and ground test for the speedometer/tachometer circuit. (Courtesy of Navistar International Engine Group)

1. With the key OFF, disconnect cab harness connector C71, Circuit 47 (Figure 9-9).
   - **A.** If battery voltage is not present, locate an open or high-resistance condition in feed Circuit (47/48) and correct it.
   - **B.** If battery voltage is present go to Step 2.
2. With the key ON and C71 disconnected, measure voltage between Circuits 47 and 11AG at C71.
   - **A.** If battery voltage is not present, locate a high or open resistance condition in Circuit 11AG and correct it. Note: This can be a loose or damaged wire, connector, or terminal.
   - **B.** If battery voltage is present, perform an advanced Speedometer/Tachometer Unit Self-Diagnostic Test. Note that this test includes more self-test diagnostics than are performed in Test 1.

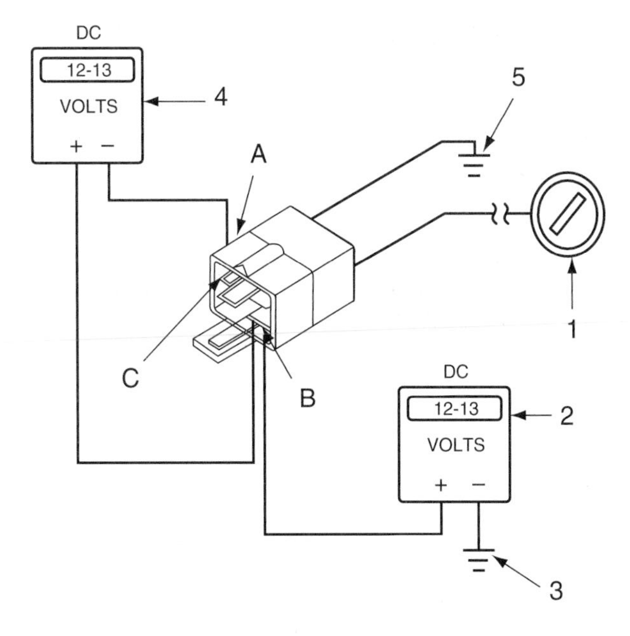

1. Key switch on
2. Checking B+ battery voltage
3. Good vehicle ground
4. Ground circuit test
5. Instrument panel ground
A. Vehicle harness speedometer connector C71
B. Circuit 47 (B+ voltage)
C. Circuit 11AG

Figure 9-9 Using a voltmeter to check for battery voltage and ground circuit.

❑ **Test 3 is the Sensor Circuit Ground and Resistance Check (at instrument panel) (Figure 9-10).**

Test 3 is performed after the advanced self-diagnostic test indicates that the speedometer/tachometer unit is operating properly. This test is used to check the sensor circuits for isolation from ground and for correct resistance. In this case, the test is performed from the instrument panel connector (C71 or C65) through the vehicle harness to the sensor. This test requires the use of an ohmmeter. The switch key has to be in the OFF position. Refer to Figures 9-11 and 9-12 for speedometer sensor testing.

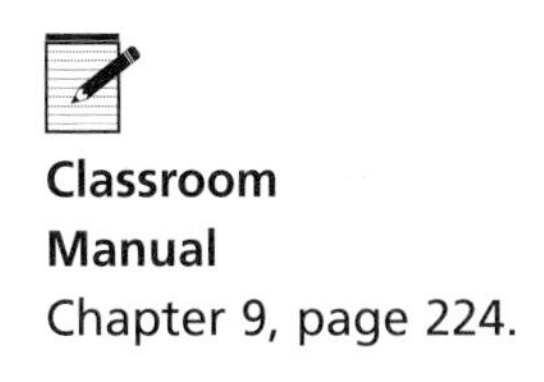

Classroom Manual
Chapter 9, page 224.

1. With the key OFF, disconnect C71 from the speedometer/tachometer unit. Measure resistance to ground at C71, Circuits 47a and 11BC.
   - **A.** If resistance in either circuit is less than 100,000 ohms, perform Test 5-Sensor Ground and Continuity Test (at sensor).
   - **B.** If resistance is greater than 100,000 ohms in both circuits, proceed to Step 2.
2. With the key OFF and C71 disconnected, measure resistance between Circuits 47A and 11BC and C71. Refer to the sensor resistance chart in Figure 9-13.
   - **A.** If sensor resistance is not as specified, Perform Test 5-Sensor Ground and Continuity Test (at sensor).
   - **B.** If sensor resistance is within specified limits, proceed to test 4-Sensor Output Voltage Test (at instrument panel). The tachometer portion of Test 3 is performed the same as the speedometer sensor, using the appropriate connectors of course.

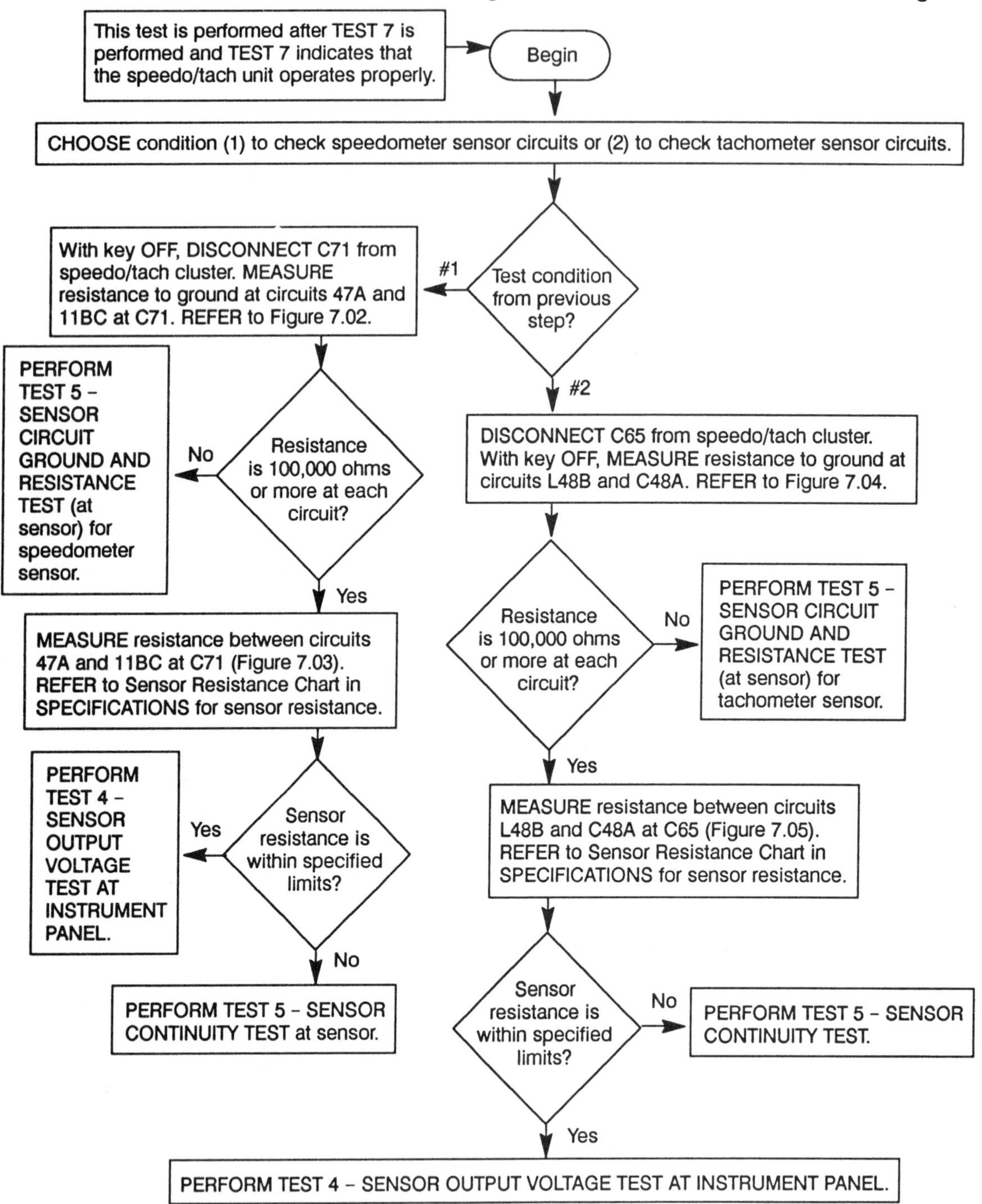

Figure 9-10 Flowchart used to check sensor ground and resistance. (Courtesy of Navistar International Engine Group)

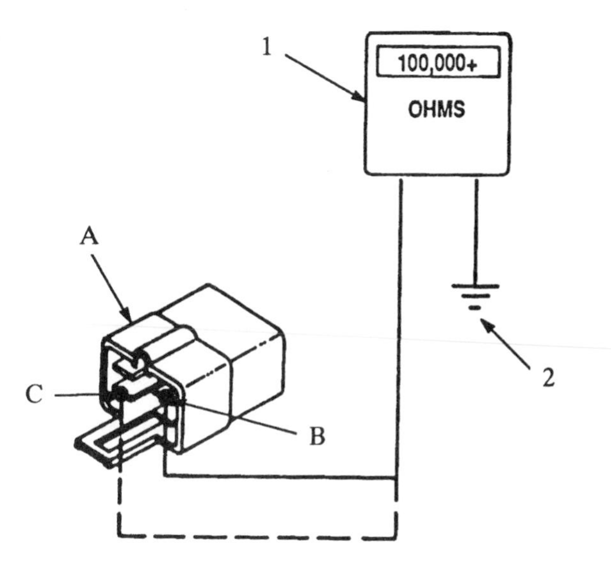

1. OHMMETER SET ON HIGHEST SCALE
2. GOOD VEHICLE GROUND

A. VEHICLE HARNESS SPEEDOMETER CONNECTOR C71
B. CIRCUIT 47A
C. CIRCUIT 11BC

Figure 9-11 Speedometer sensor circuit ground test. (Courtesy of Navistar International Engine Group)

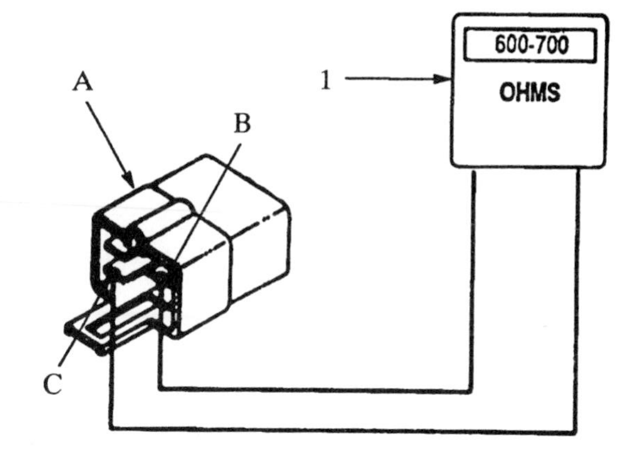

1. OHMMETER SET ON RX1 SCALE

A. VEHICLE HARNESS SPEEDOMETER CONNECTOR
B. CIRCUIT 47A
C. CIRCUIT 11BC

Figure 9-12 Speedometer sensor and circuit resistance check. (Courtesy of Navistar International Engine Group)

ENGINE	SENSOR TYPE	RESISTANCE
	Speedometer Sensor	
All Electronic Engines	Dual Coil - To Speedometer Dual Coil - To ECM	900 - 1100 OHMS 1200 - 1450 OHMS
Non-Electronic Engines	Single Coil - To Speedometer	600 - 700 OHMS
	Tachometer Sensor	
All Applications	Single Coil - To Tachometer	600 - 700 OHMS

Figure 9-13 A sensor resistance chart. (Courtesy of Navistar International Engine Group)

Special Tools

DVOM

Wiring diagram

Manufacturer's service manual

The following are further tests used by this vehicle manufacturer:

- **Test 4- Sensor Output Voltage (at instrument panel) (Figure 9-14)**
- **Test 5- Sensor Ground and Continuity Test (at sensor) (Figure 9-15).**

This test requires the use of an ohmmeter and the sensor unit must be disconnected from the vehicle wiring harness. The speedometer sensor is usually located at the rear of the transmission, and the engine **speed sensor** is located in the engine flywheel housing.

Classroom Manual
Chapter 9, page 225.

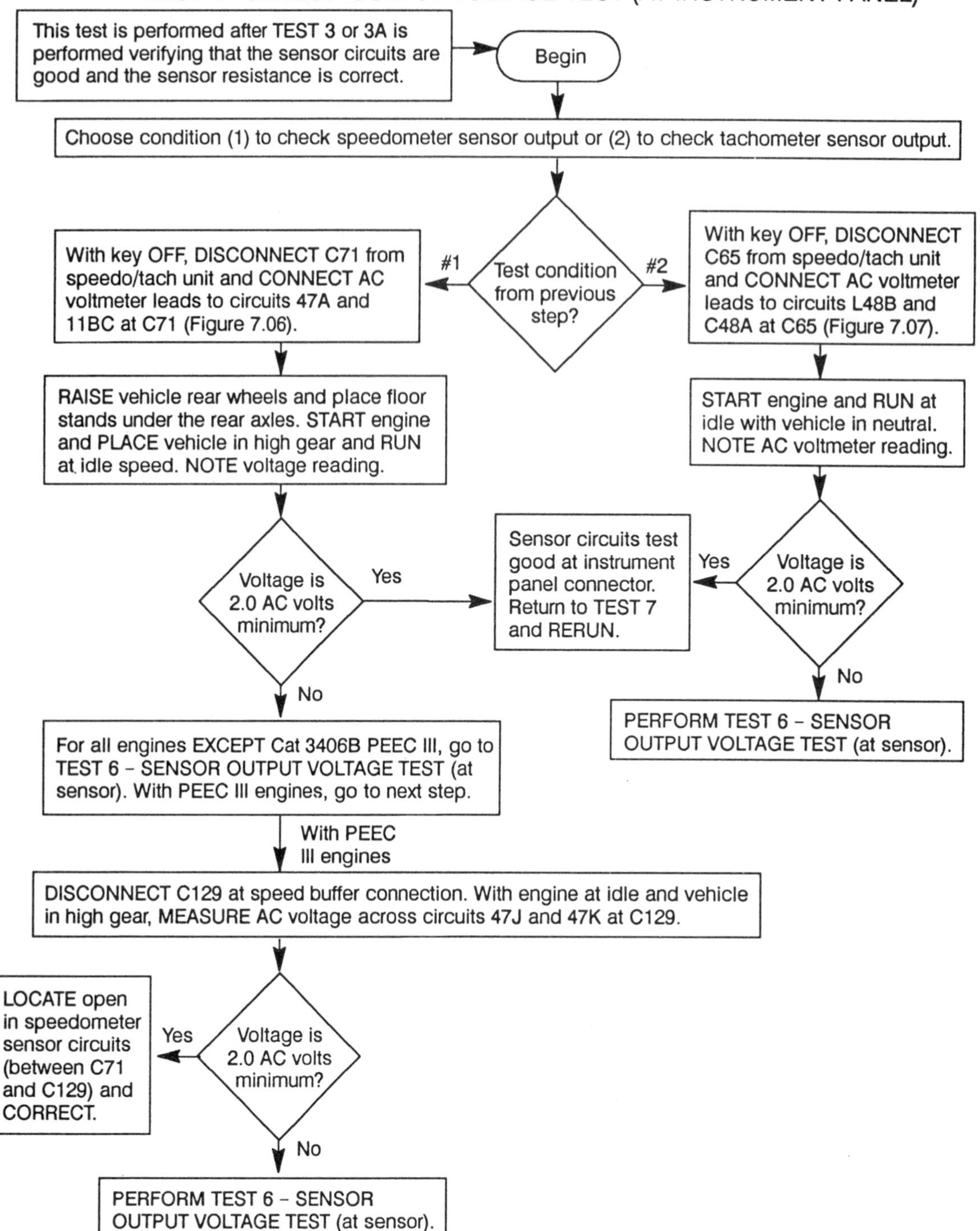

Figure 9-14 Sensor output voltage test flowchart. (Courtesy of Navistar International Engine Group)

Special Tools

Jack stands

Long meter leads

DVOM

Wiring diagram

Manufacturer's service manual

Safety glasses

Figure 9-16 shows the resistance to ground test being performed. We are looking for a resistance that is greater than 100,000 ohms in each terminal. If that value is not there, replace and adjust the sensor unit.

Figure 9-17 shows the sensor resistance test being performed. The sensor is disconnected from the harness and an ohmmeter is used. This test is used to ensure specified sensor resistance. Refer to the sensor resistance chart in Figure 9-13. If the sensor resistance is not within specified limits, replace and adjust the sensor.

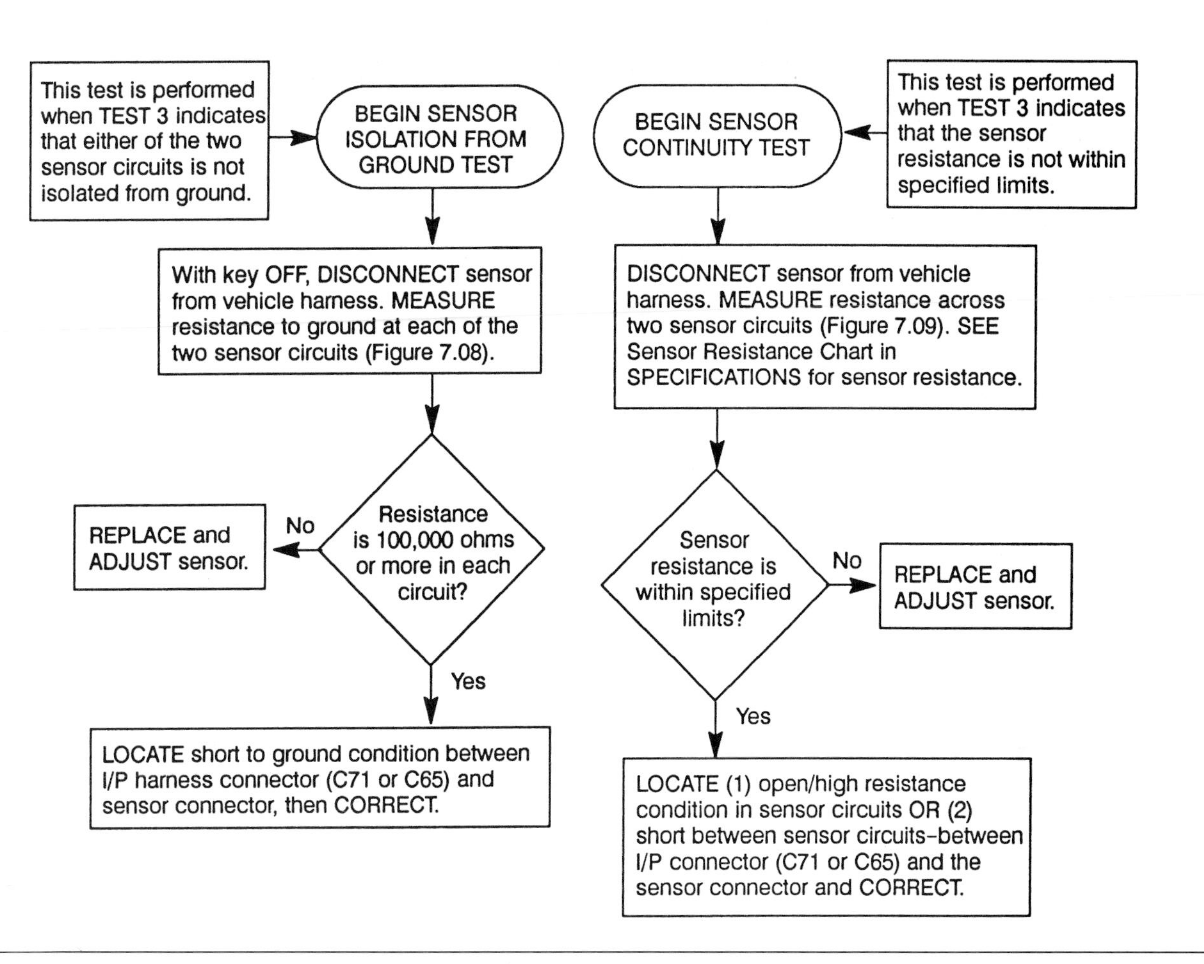

Figure 9-15 Sensor ground and continuity test flowchart. (Courtesy of Navistar International Engine Group)

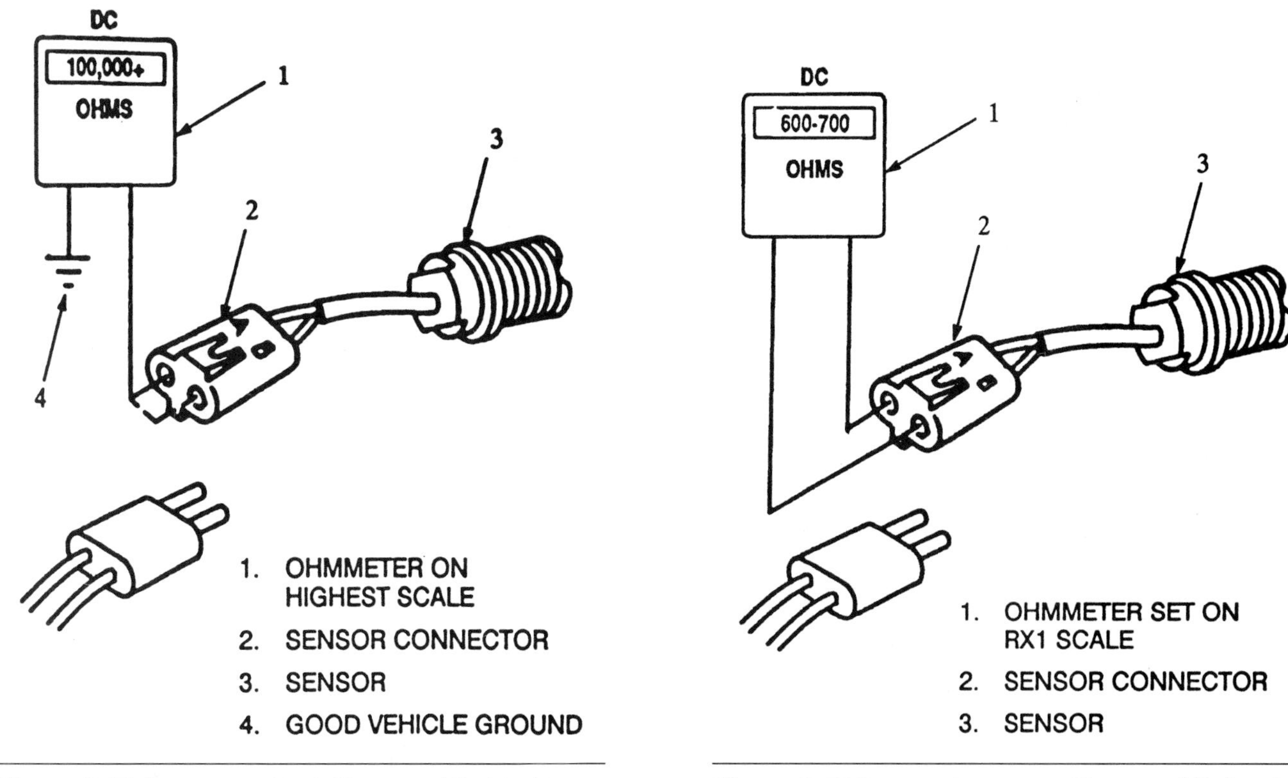

Figure 9-16 Sensor ground test. (Courtesy of Navistar International Engine Group)

Figure 9-17 Sensor resistance test. (Courtesy of Navistar International Engine Group)

Test 6—Sensor Output Voltage Test (at sensor) (Figure 9-18).

This test checks for proper AC voltage output at the sensor.

SERVICE TIP: An improper sensor adjustment causes a low or no voltage condition.

WARNING: To prevent damage to your test leads, use ones that are long enough to avoid contact with rotating shaft or vehicle wheels. You can make up your own various lengths of jumpers with alligator clips and probes for these conditions.

CAUTION: To avoid personal injury or property damage during this test requiring that the rear wheels are off the ground, you must be fully alert and use proper safety procedures. Do not use a jack when working under a vehicle. Always use jack stands to support the vehicle. Parking brakes should be applied when the transmission is in neutral, whether or not the engine is running.

For this test refer to Figure 9-19:

1. Disconnect the sensor lead from the vehicle harness. Connect the meter leads across the sensor unit connector terminals.
2. Checking speedometer sensor—With rear axles on floor stands (wheels off ground) and transmission in neutral, start the engine and run it at idle.

TEST 6 – SENSOR OUTPUT VOLTAGE TEST (AT SENSOR)

NOTE: This test checks for proper voltage output at the sensor (speedometer or tachometer). Improper sensor adjustment will cause low or no voltage output condition.

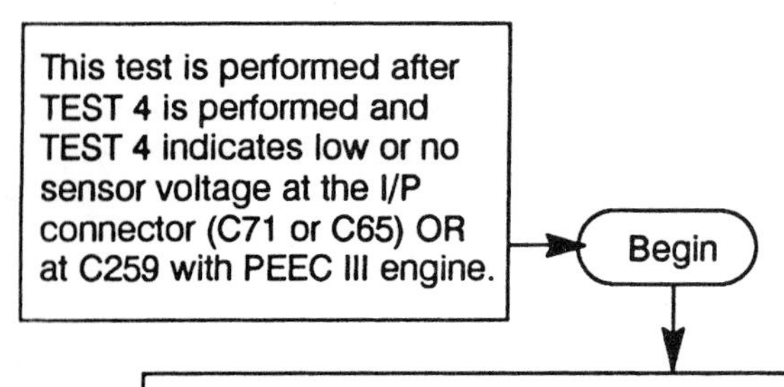

NOTE: To test speedometer sensor, rear axles are to be on floor stands and transmission in high gear as described in TEST 4.

NOTE: To test the tachometer, wheels will be blocked, vehicle park brake set, and the transmission in neutral as described in TEST 4.

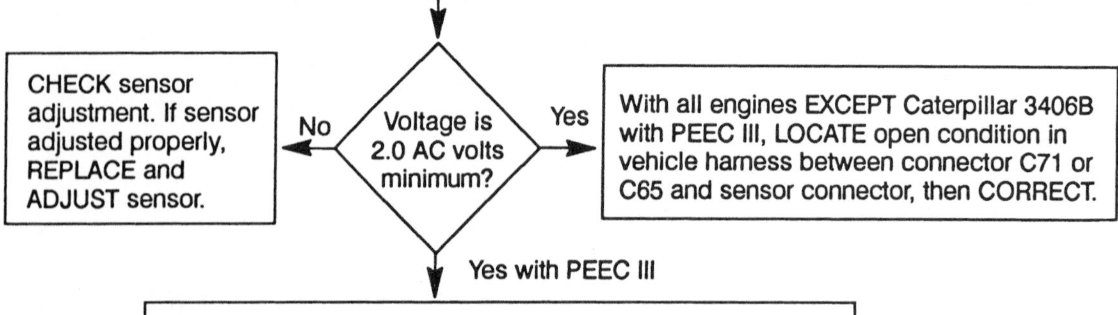

Yes with PEEC III

With Caterpillar 3406B PEEC III engines, have a Caterpillar Dealer check for proper operation of the vehicle speed buffer.

Figure 9-18 Sensor output voltage test (at sensor). (Courtesy of Navistar International Engine Group)

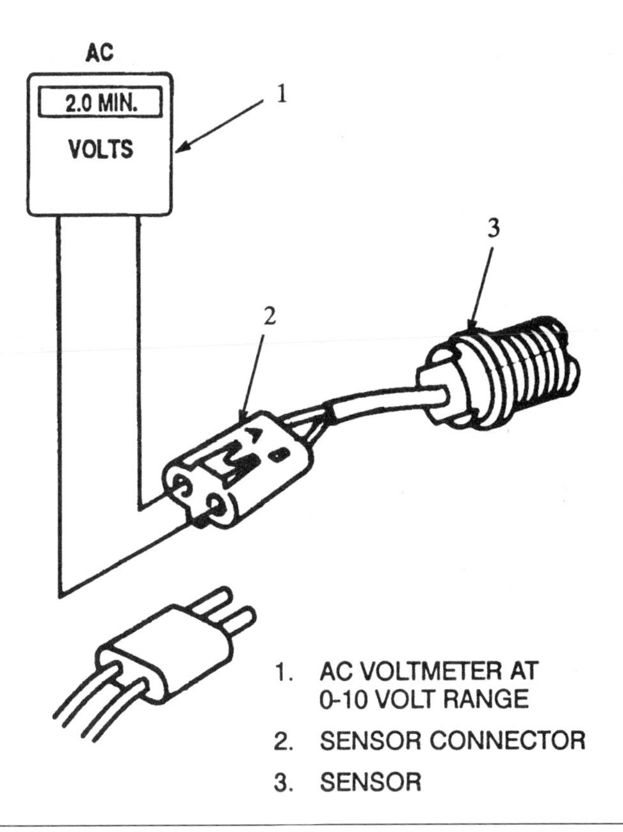

Figure 9-19 Sensor output voltage test (AC) at sensor. (Courtesy of Navistar International Engine Group)

   A. Checking tachometer sensor—With wheels blocked, vehicle park brake set, and transmission in neutral, start the engine and run it at idle.

3. Read AC voltage on voltmeter. Voltage should be 2.0 volts minimum.

   A. If no voltage is present, replace and adjust sensor.

   B. If low voltage is present, adjust sensor and retest to verify. If the adjustment fails to correct a low voltage condition, replace the sensor.

Speedometer and Tachometer Troubleshooting Tests

Classroom Manual
Chapter 9, page 223.

The following is a summary of speedometer and tachometer troubleshooting tests. We will consider performing the tests at the connectors at the back of the speedometer and tachometer heads. A wiring diagram is always required (Figure 9-20). We will refer to this diagram for the following tests:

Testing for Ignition Voltage (Figure 9-21)

1. Set the voltmeter to the VDC function.
2. Turn the key to the ACC or RUN position.
3. Connect the black (−) meter lead to a good ground and the red (+) meter lead to the 102 wire in the harness connector at the back of the gauge.

Observation: Ignition voltage should be present. If 0 volts or less than ignition voltage is indicated, check for an open or a source of high resistance in the circuit.

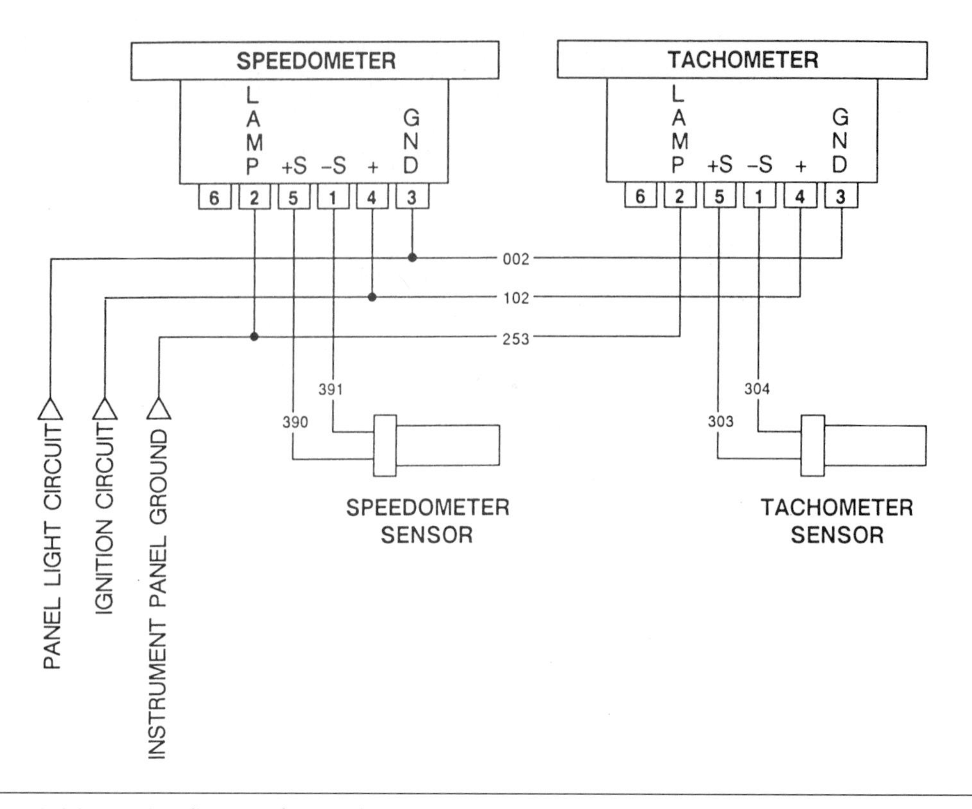

Figure 9-20 A wiring diagram of a speedometer/tachometer circuit. (Courtesy of Mack Trucks, Inc.)

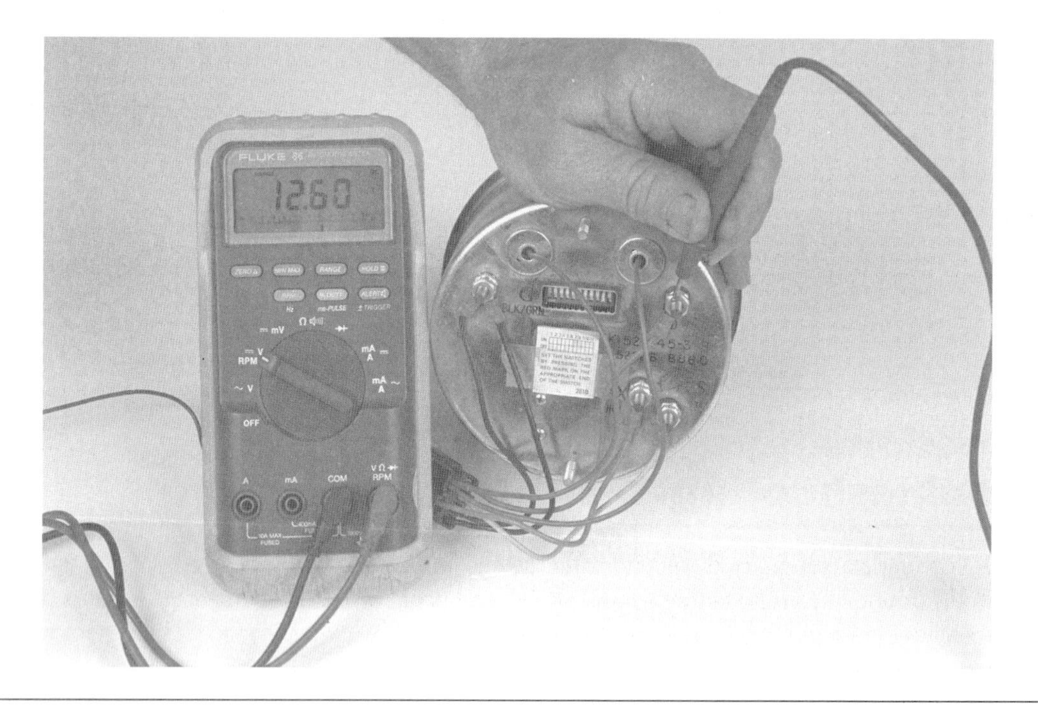

Figure 9-21 Testing for ignition voltage.

Testing the Ground (Figure 9-22)

1. Make sure the key switch is turned OFF.
2. Set the meter to the resistance function.
3. Connect one lead to 002 wire in the harness connector at the back of the gauge and the other lead to a common ground connection in the cab.

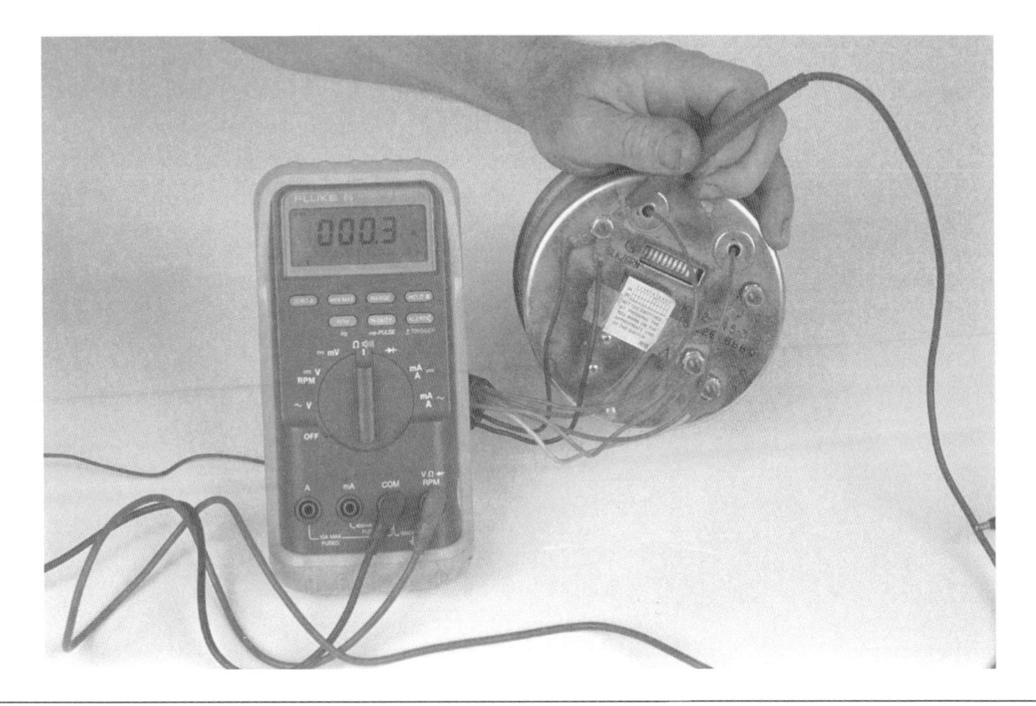

Figure 9-22 Testing the ground circuit.

Observation: The meter should show 0 or fractions of ohms resistance. Higher resistance readings indicate a poor connection, possibly resulting from a loose or damaged connection and broken or damaged wires.

Testing Speed Sensors (Figure 9-23)

Special Tools

DVOM with volts AC function

Metallic object (wrench)

SERVICE TIP: Make sure speed sensor is properly adjusted before condemning a sensor.

The following test measures sensor resistance:

1. Disconnect the wire from the sensor.
2. Set the multimeter to the resistance function.
3. Connect the meter leads to the sensor terminals and note the resistance reading indicated on the meter.

Observations: Sensor resistance should fall between 500 and 4000 ohms. If the resistance is not within this range, replace and adjust the sensor.

Testing Speed Sensors (Figure 9-24)

This test is a little bit different—it allows us to see the effectiveness of the sensor. The sensor has to be disconnected and removed from the vehicle.

1. Make sure the key switch is OFF and disconnect the wires from the sensor.
2. Loosen the jam nut, then unscrew the sensor from the housing or bracket.
3. Connect the multimeter leads to both terminals of the sensor.
4. Set the meter to the VAC function.
5. Pass a metallic object, such as a wrench, in front of the sensor approximately .05 inches away from the surface.
6. Observe the voltage reading as the object passes in front of the sensor.

Observation: When the metal object passes in front of the sensor, a pulse of AC voltage is indicated on the meter. If there is no voltage indication, replace and adjust the sensor.

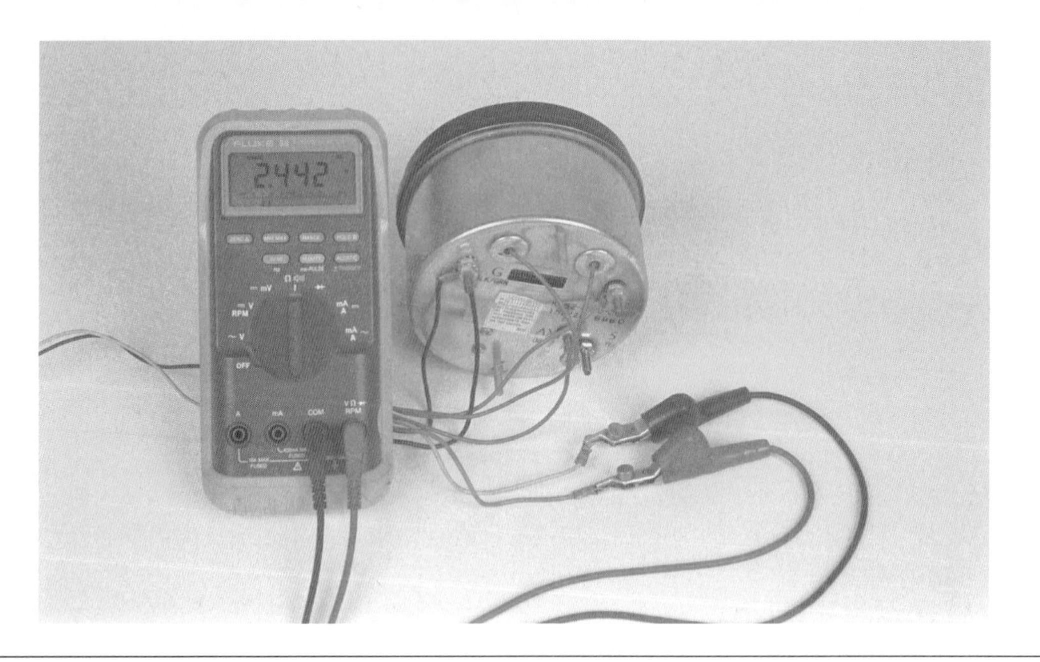

Figure 9-23 Measuring sensor resistance.

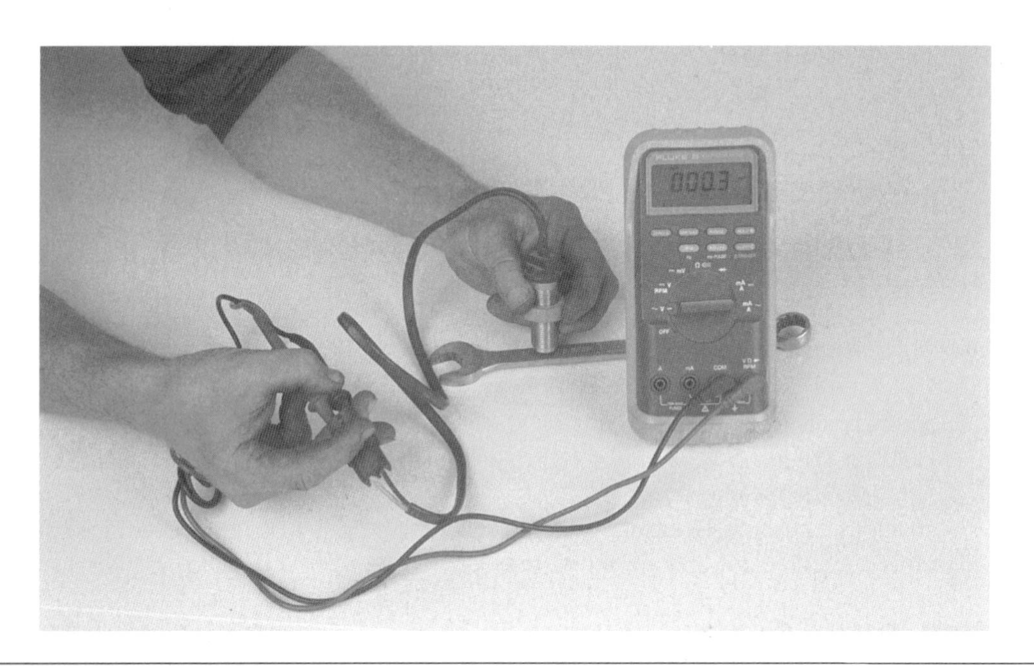

Figure 9-24 Testing sensor.

Sensor Adjustment (Figure 9-25)

A sensor will not output properly if the adjustment is not correct. The following is a typical sensor adjustment:

1. Install the sensor and turn by hand until it bottoms (contacts the tone wheel).
2. Back the sensor out 3/4 turn.
3. Tighten the jam nut.

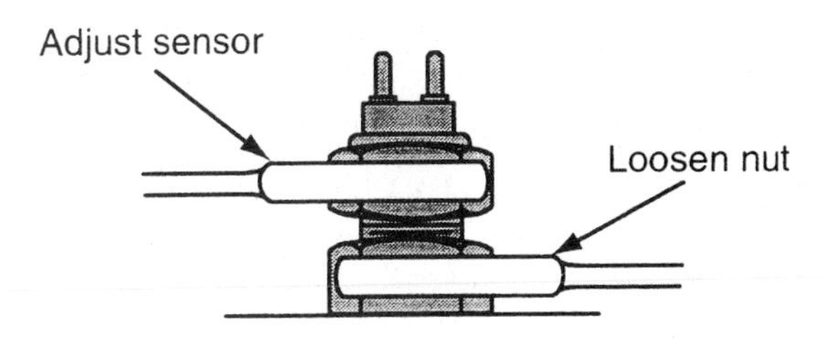

Figure 9-25 Adjusting sensor.

Testing Signal Voltage (Figure 9-26)

This test requires the vehicle's drive wheels to be raised off the ground.

Special Tools

DVOM

Jack stands

Safety glasses

CAUTION: Always use caution around a vehicle that is off the ground and running to prevent personal injury.

1. Block the front wheels.
2. Raise the rear wheels off the ground and use jack stands under the axles.
3. With the key switch OFF, disconnect the harness connector from the back of the speedometer.
4. Set the multimeter to the VAC function.
5. Connect the meter leads to the S+ and S− pins in the harness connector.
6. Start the engine.
7. Shift the transmission into the highest gear, release the park brake, and allow the vehicle to run at an idle.
8. With the engine running at an idle, note the AC voltage indicated on the meter.

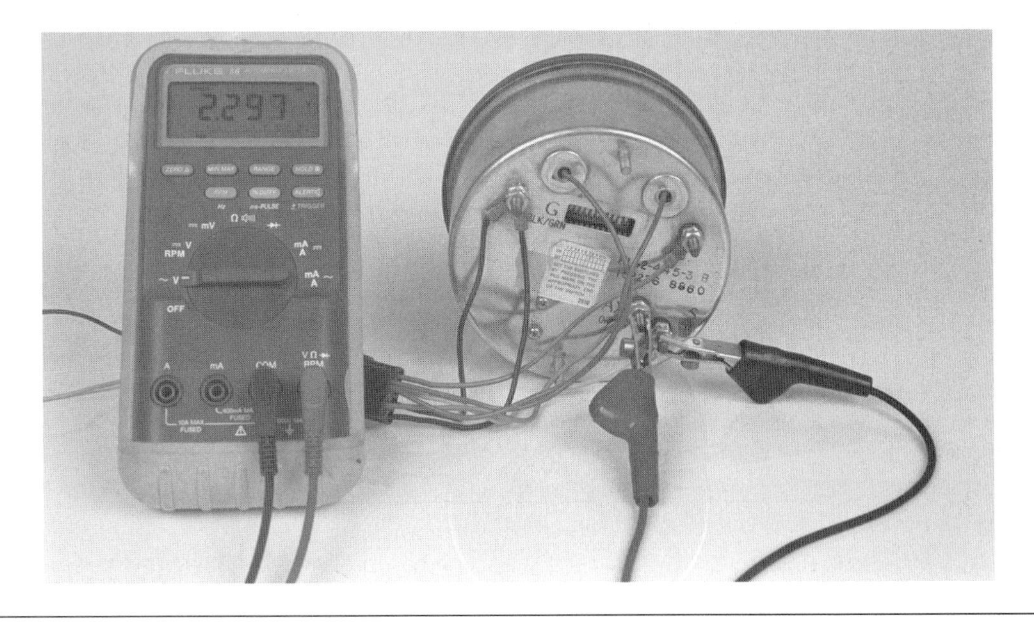

Figure 9-26 Testing signal voltage by connecting meter at S+ and S− connections.

Observations: Signal voltages should be approximately 1 to 3 volts AC. If there is no or low input voltage, adjust the sensor and recheck. Also check for an open in the wires' connection the speed sensor with the S+ and S− harness pins at the speedometer connector if no input signal was indicated. If readjustment does not bring the voltage within range and the circuit checks out good (including the sensor itself), replace the speed sensor.

Special Tools

Manufacturer's service manual

Small screwdriver or awl

Calibrating Speedometers and Tachometers

A set of **dip switches** is used for calibration. This function is usually performed when the speedometer or tachometer heads are replaced. The calibrations can often be found on the backs of the units.

SERVICE TIP: When replacing the speedometer or tachometer with a new unit, note the dip switch settings on the unit being replaced and set the dip switches on the replacement unit to the same setting.

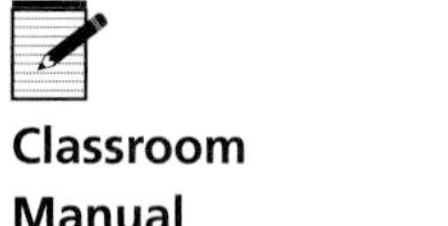

Classroom Manual
Chapter 9, page 226.

If tire sizes and/or rear axle ratios are changed, the speedometer calibration must be calculated as follows:

1. Calculate the pulses per mile (PPM). PPM is the product of the number of tire revolutions per mile times the rear axle ratio times sixteen (the number of teeth on the speedometer gear). Note: See each vehicle manufacturer's specifications.
   Axle Ratio × Tire Rev per Mile × 16 = PPM
   5.43 × 497 × 16 = 43,179 PPM
2. Calculate the calibration number. The speedometer calibration number is the quotient of the pulses per mile divided by 120.
   PPM/120 = calibration number
   43,179/120 = 360 The calibration number is used to determine the dip switch settings (Figure 9-27).

Calibration Number	DIPSWITCH SETTING									
	1	2	3	4	5	6	7	8	9	10
327	0	0	0	1	1	1	0	1	0	1
343	0	0	0	1	0	1	0	1	0	1
360	1	1	1	0	1	0	0	1	0	1
371	0	0	1	1	0	0	0	1	0	1
390	1	0	0	1	1	1	1	0	0	1
409	0	1	1	0	0	1	1	0	0	1

0 = Off position
1 = On position

Figure 9-27 Calibration numbers determine the dip switch setting. (Courtesy of Mack Trucks, Inc.)

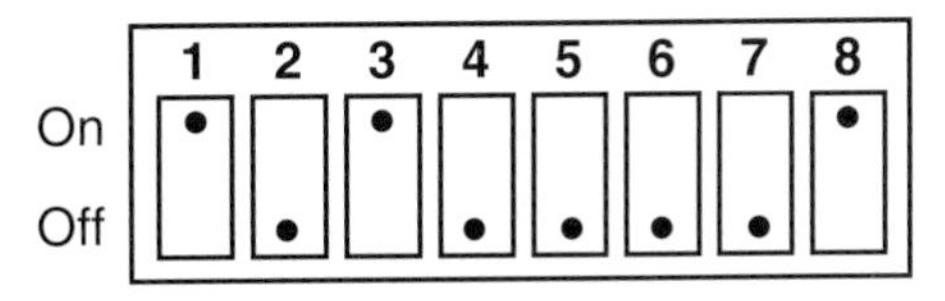

Figure 9-28 Standard tachometer dip switch settings. (Courtesy of Mack Trucks, Inc.)

Once the settings are determined, a set of switches is moved between the 0 and 1 positions, with 0 being OFF and 1 being ON. The speedometer has 10 switches and a tachometer has 8 switches (Figure 9-28).

A small screwdriver or fine-pointed awl can be used to either move or depress (depending on the type) the switches to their appropriate positions. Note: Some vehicles use speedometers and tachometers with the calibration unit in the heads while others have the calibration units separate from the heads.

A gauge is a device that displays the measurement of a monitored system. The typical gauge uses a pointer or needle that moves along a calibrated scale.

Testing Gauges and Gauge Circuits

Gauges react to inputs from senders, which determine a particular condition in the vehicle's or engine's operation. The sender is an input to the gauge, which indicates the specific operating condition.

Some instrument clusters use digital and linear displays to notify the driver of a monitored system condition.

WARNING: Many gauges are of the analog type because they use needle movement to indicate current levels. However, many of today's truck instrument panels use computer-driven analog gauges that operate under different principles. It is important that a technician follow specific manufacturer's procedures for testing the gauge and its circuits to prevent any damage to sensitive electronic components.

Some vehicles use an **instrument voltage regulator (IVR)** to provide a constant voltage to the gauges regardless of the voltage output of the charging system. A typical gauge is called an **electromechanical** device because it is operated electrically, but its movement is mechanical. Most gauges, with the exception of the voltmeter and ammeter, use some sort of variable resistance sending unit. The following broad tests performed will depend on the nature of the problem and if the system uses an IVR.

Bimetallic gauges (or thermoelectric gauges) are simple dial and needle indicators that transform the heating effect of electricity into a mechanical movement.

Single Gauge Failure

If the gauge system does not use an IVR, check the gauge for proper operation as follows:

1. Check the fuse panel for any blown fuses. A single gauge that is not operating may share a fuse with some other circuit that is separate from the other gauges.
2. Disconnect the wire connector from the sending unit of the malfunctioning gauge.
3. Check the terminal connectors for any physical problems such as damage or corrosion.
4. Use a test light or voltmeter to confirm that voltage is present to the connector with the ignition switch in the RUN position. If no voltage is indicated (if required), check the circuit back to the gauge and battery.
5. Connect a 10-ohm resistor to the lead. Connect the lead to ground (Figure 9-29).
6. With the ignition switch in the RUN position, observe the gauge. Depending on the gauge design the needle should indicate either high or low on the scale. Check the specific vehicle's service manual for the correct results.

Electromagnetic gauges use magnetic forces instead of heat to produce a needle movement.

Classroom Manual
Chapter 9, page 213.

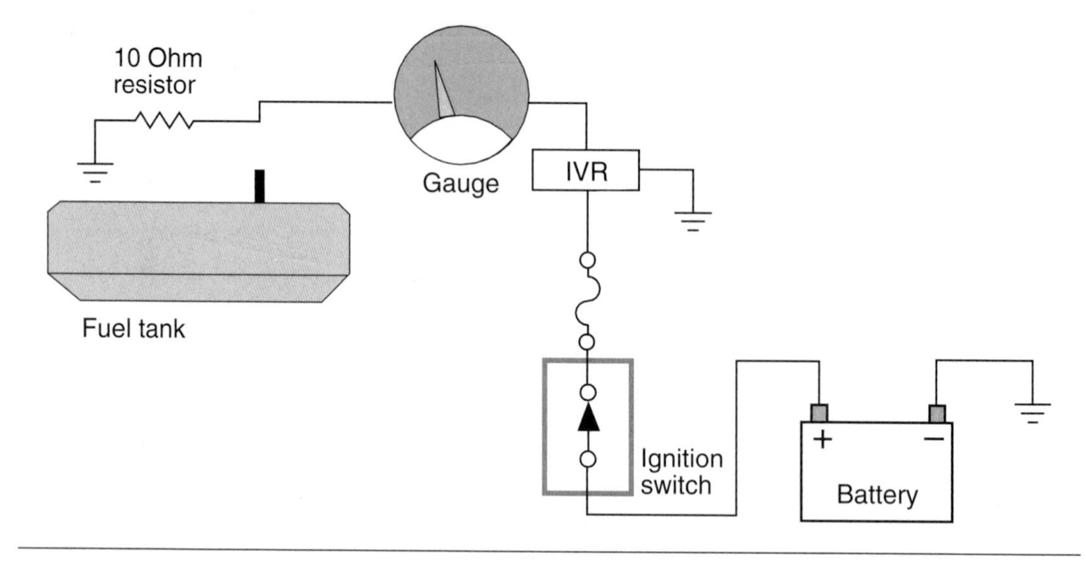

Figure 9-29 Testing gauge operation. The resistor shown in the circuit is there to protect the circuit.

Special Tools

12-volt test light

10-ohm resistor

70-ohm resistor or variable resistor

Wiring diagram

Safety glasses

7. Connect a 70-ohm resistor between the sensor lead and ground. Repeat Step 6. Note: There are commercially available tools that are used to inject various signals and resistance into circuits. These signals and resistances can be varied through a multiple of ranges.
8. If the test results are in the acceptable range, the sending unit is faulty.
9. If the gauge did not operate properly in Step 5, check the wiring to the gauge. If the wiring is good, replace the gauge.

It is important to know the operations of different gauges and their sending units. In most bimetallic gauges, the gauge reads high if resistance is low. With many electromagnetic gauges, the gauge reads low if sending unit resistance is low.

Classroom Manual
Chapter 9, page 214.

WARNING: Resistors are used instead of grounding the sender terminal directly to prevent gauge damage.

Classroom Manual
Chapter 9, page 215.

If the gauge circuits use IVR, follow Steps 1 through 4 as previously described. The test light should flicker on and off. If it did not illuminate, reconnect the sending unit lead and check for voltage at the **sender unit** side of the gauge (Figure 9-30). If there is voltage indicated at this point, repair the circuit between the gauge and the sending unit. If there is no voltage indicated, test for voltage at the battery side of the gauge. If voltage is indicated at this point, the gauge is defective and must be replaced. If no voltage is indicated at this terminal, continue to check the circuit between the battery and the gauge.

If the IVR was working properly and voltage was present to the sender unit, follow Steps 5 through 9.

Classroom Manual
Chapter 9, page 220.

SERVICE TIP: Most common problems occur at connections. Before replacing any sending unit or gauge, clean the connections, especially the ground sides.

Special Tools

12-volt test light

10-ohm resistor

70-ohm resistor or variable resistor

Wiring diagram

Safety glasses

Multiple Gauge Failures

A multiple gauge failure indicates a circuit failure affecting all the gauges. Begin by checking the circuit fuse. Next, test for voltage at the last point that is common to the entire circuit (Figure 9-31). If voltage is not present at this point, work toward the battery to find the fault.

If the system uses an IVR, use a voltmeter to test for regulated voltage at a common point to the gauges (Figure 9-32).

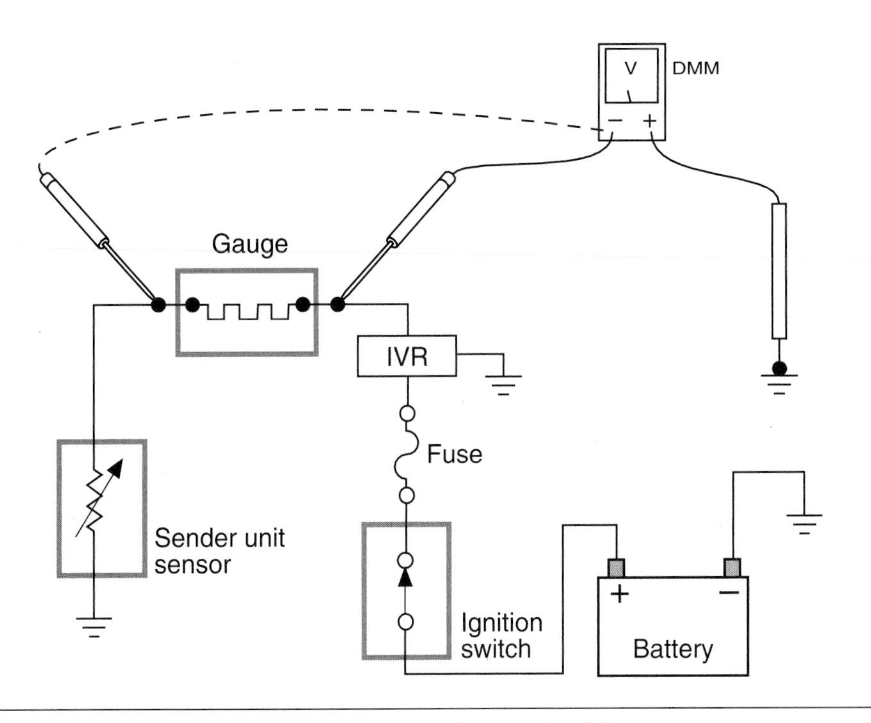

Figure 9-30 Checking for regulated voltage on the sender unit side of the gauge.

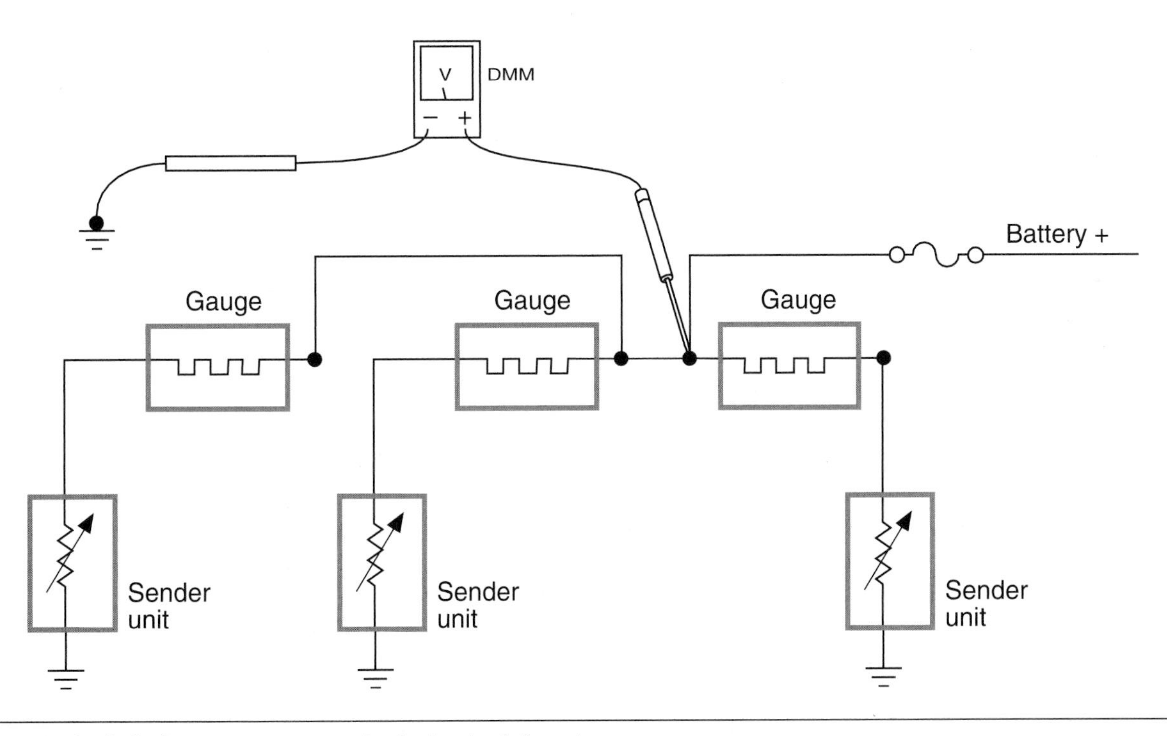

Figure 9-31 Check the last common connection in the circuit for voltage.

Any out of specification voltage will require the next step to be a check of the ground circuit of the IVR. If the ground circuit checks out good, replace the IVR. If there is no voltage present at the common point tested, check for voltage on the battery side of the IVR. If voltage is present at this point, replace the IVR.

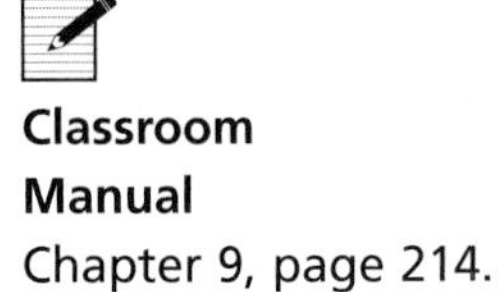

Classroom Manual
Chapter 9, page 214.

If regulated voltage is within specifications, test the printed circuit from the IVR to the gauges. If there is an open in the printed circuit, replace the board.

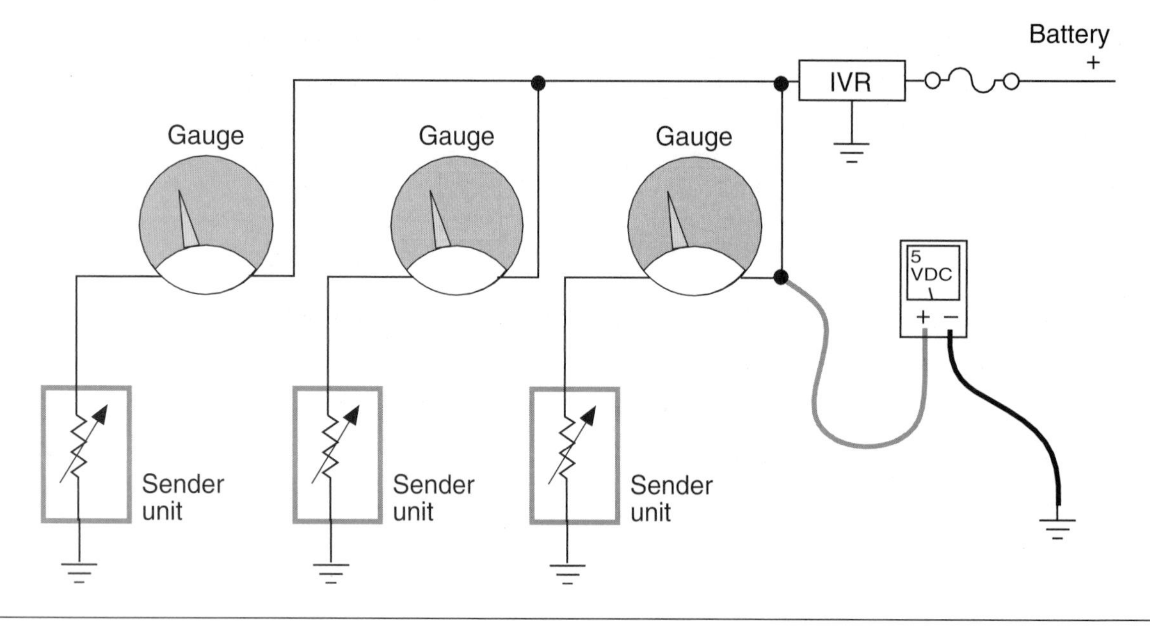

Figure 9-32 Testing for correct IVR operation.

Classroom Manual
Chapter 9, page 220.

SERVICE TIP: It is unlikely that all gauges will fail at the same time.

Special Tools

Digital multimeter (DMM)

12-volt test light

Safety glasses

Gauge Sending Units

The three most common types of sending units associated with electromechanical gauges are: a mechanical variable resistor, a thermistor, and a piezoresistive sensor. Most of these senders can be pinpoint tested to determine if they are at fault rather than replacing them to confirm a fault.

Classroom Manual
Chapter 9, page 221.

A mechanical variable resistor uses mechanical movement to change the resistance value. A good example is the fuel level sender unit.

Fuel Tank Sending Unit

The fuel tank sending unit is a mechanical variable resistor that can be tested in or out of the tank. Fuel level change in the fuel tank causes the resistance through the fuel level sending unit to change.

The following procedure can be used to check this sending unit:

1. Disconnect the wire from the center terminal of the fuel level sending unit at the fuel tank (Figure 9-33).
2. Set the multimeter to the resistance function.
3. Connect one meter lead to the center terminal of the sending unit and the other lead to ground.

Observation: The resistance reading indicated on the meter depends on the level of fuel in the tank. The most common reading will show a low resistance if the fuel level is low and increase with more fuel in the tank. Note: There is always going to be an exception to the rule. For example, you might run across a fuel sender whose resistances indicate the opposite in fuel levels.

If the sending unit is suspected of not functioning properly, it can be removed from the tank for further testing (Figure 9-34).

CAUTION: When working around fuel, use all necessary precautions such as wearing proper safety equipment and avoiding sparking.

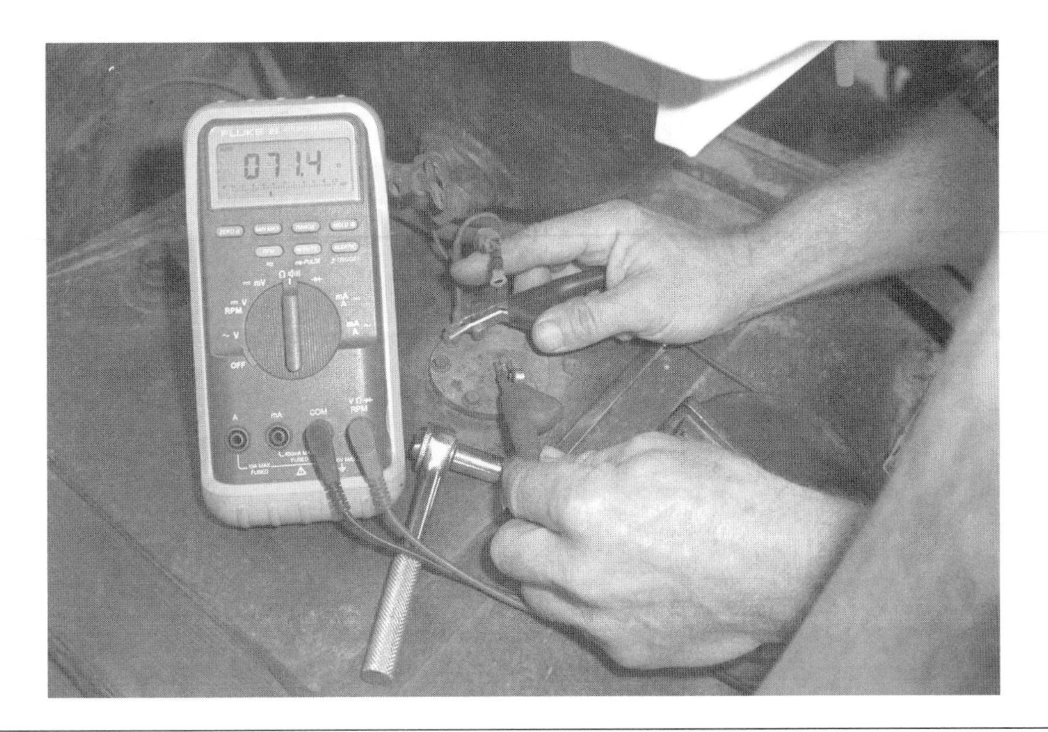

Figure 9-33 Checking resistance of a fuel tank sending unit.

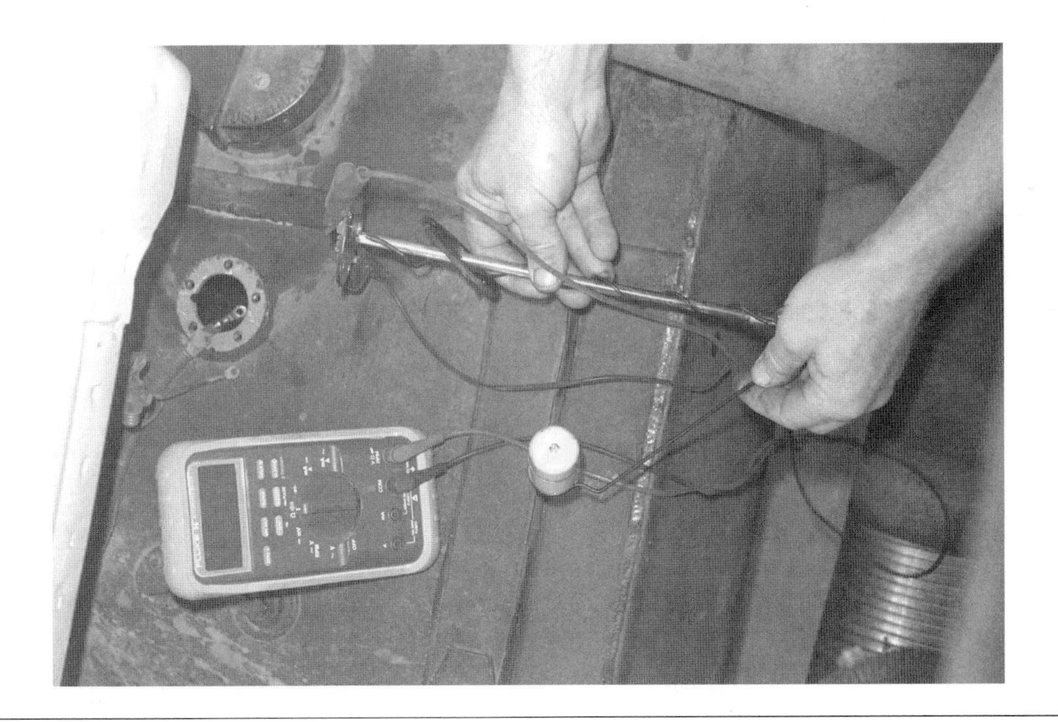

Figure 9-34 Testing fuel level sending unit out of the fuel tank.

With the sending unit out of the fuel tank, connect one meter lead to the center terminal and the other lead to the mounting flange. Move the float arm through a full swing. Resistance through the sending unit should change according to the position of the float as it is being moved from the lowest position to the highest position

A **thermistor** is a temperature-sensitive sensor that is often used as a coolant temperature sensor.

Classroom Manual
Chapter 9, page 220.

Temperature Sending Units

Temperature sending units react to changes in temperature, which cause the resistance through the sender to change accordingly. For this reason, an ohmmeter is the best tool to use to check this sensor (Figure 9-35). By measuring the resistance of the sender through various temperature changes we can determine if this sender is operating properly. The following procedure is used:

1. Disconnect the lead from the sending unit (Figure 9-36).
2. Set the multimeter to the resistance function.
3. Connect one lead to a good ground and the other lead to the sending unit terminal.
4. Measure and note the resistance through the sending unit while it is still cold.
5. Start the engine and allow the sending unit to heat up while observing the reading indicated on the meter.

Observation: The resistance values need to be compared to specifications. However, as a rule the resistance of a thermistor that is used to measure temperature should decrease as the sending unit gets hotter.

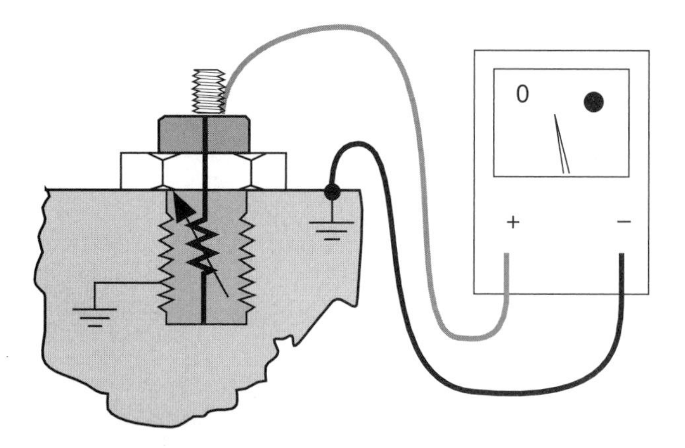

Figure 9-35 Testing a thermistor with an ohmmeter.

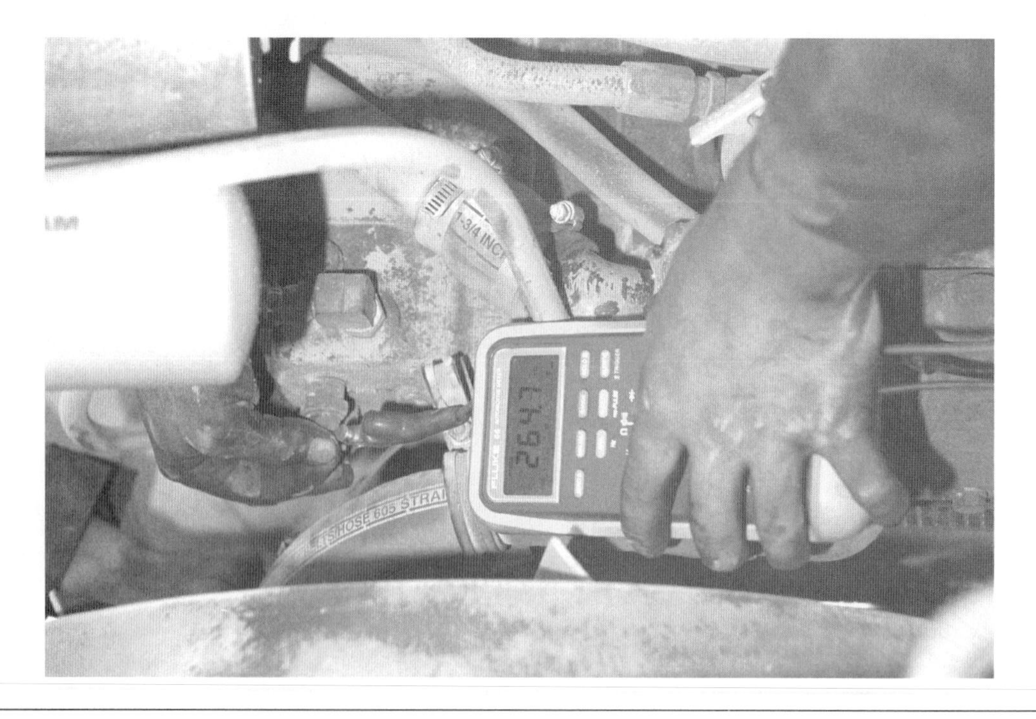

Figure 9-36 Testing temperature sensor.

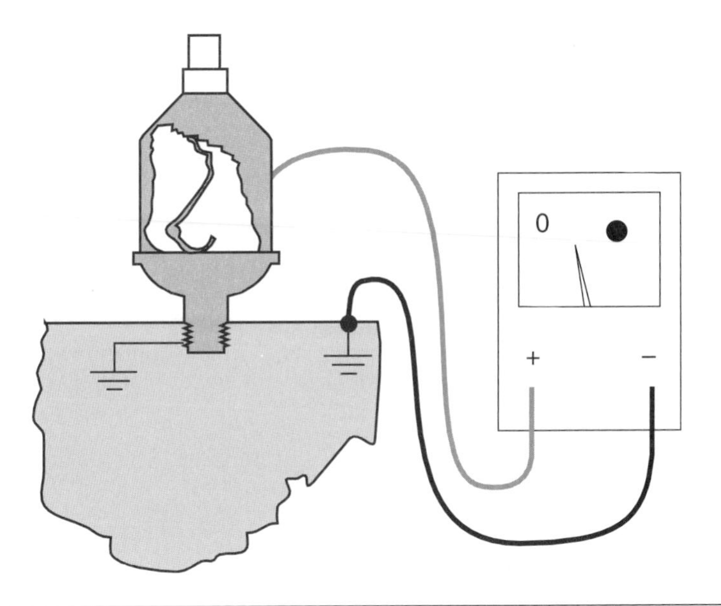

Figure 9-37 Using an ohmmeter to test a piezoresistive sensor.

Piezoresistive Sensor Sending Unit

A piezoresistive sensor is the type of sender often used for oil pressure gauges. An ohmmeter is used to check this type of sender by connecting the leads to the sending unit terminal and ground (Figure 9-37). First check the resistance with the engine off and compare to specifications. Next, start the engine and allow it to idle. Recheck the resistance value and compare it to specifications. Do not condemn this electrical sending unit immediately if the readings are not within specifications. First, connect a mechanical pressure gauge to the engine to confirm that it is producing adequate oil pressure.

Troubleshooting Warning Assemblies

Warning indicators, such as lights or **buzzers**, alert the driver of a problem that might occur so actions can be taken to prevent any further damage.

Faults that can occur are systems that are activated at all times or will not activate at all. A diagram of a simple warning system is shown in Figure 9-38 and will be used as a reference for the following tests:

Warning Assembly

This test checks for the presence of voltage:

1. Set a multimeter to the VDC function (Figure 9-39).
2. Turn the key switch to ON.
3. Connect the (+) lead of the meter to the 102 wire connection at the back of the suspect warning assembly and the (−) lead to a good ground.

Observation: A battery voltage should be indicated at the 102 wire connection when the key is turned ON. If no voltage is indicated, look for an open in Circuit 102 that is feeding the suspect warning assembly. If battery voltage is indicated to be present at the assembly, proceed to testing the respective sensor.

Low Oil Pressure Warning

The low oil pressure warning assembly should activate when the key is turned ON and either the engine is not running or oil pressure is below a set value of usually 10 PSI. If the warning

Classroom Manual
Chapter 9, page 220.

A **piezoresistive sensor** is sensitive to pressure changes. One use of this sensor is to measure engine oil pressure.

When it comes to an oil pressure warning system to indicate problems, a mechanical oil pressure gauge should always be used to confirm adequate oil pressure.

Special Tools

DMM

Jumper wire

Wiring diagram

Classroom Manual
Chapter 9, page 227.

A **buzzer** is a device similar to a relay that emits an audible warning to alert the vehicle operator of an impending system problem.

Warning systems usually have a warning light and/or buzzer that are activated to warn the driver of a possible problem or hazardous condition.

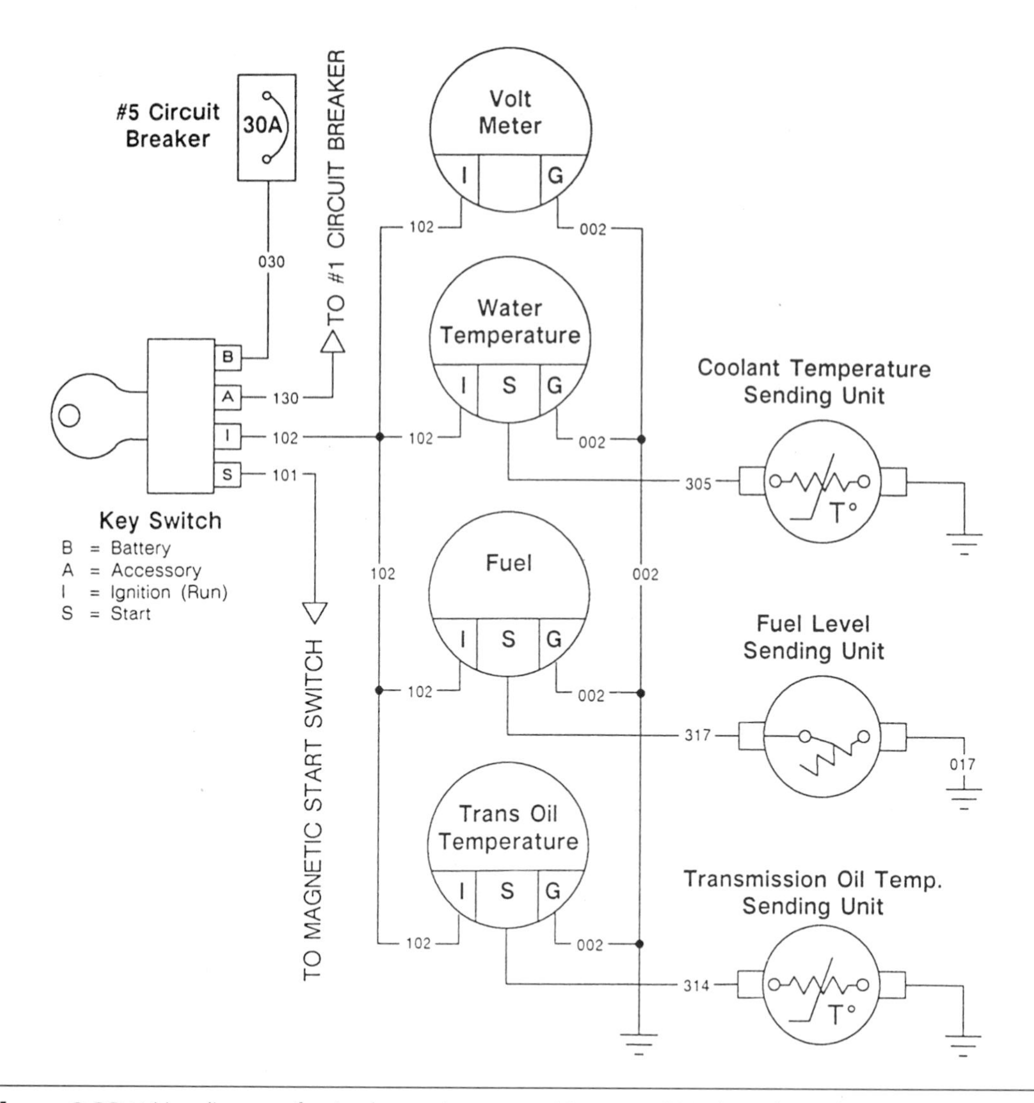

Figure 9-38 Wiring diagram of a simple warning system. (Courtesy of Mack Trucks, Inc.)

Shutdown systems are often used to actually shut an engine down if a problem exists or continues that can cause engine damage.

assembly fails to operate under these conditions and voltage was available at the warning assembly, the following quick test can be performed. Note: This is a nonelectronic system.

Disconnect Wire 807 from the oil pressure switch and touch it to ground. If the warning assembly activates, the problem lies with the oil pressure switch. If the warning assembly fails to activate, check for an open in Circuit 807.

Classroom Manual
Chapter 9, page 230.

Checking the Oil Pressure Switch

The oil pressure sensor is a normally closed pressure switch that opens when oil pressure is available at the switch port. The function of the switch can be checked by using a multimeter as follows:

1. Set the multimeter to the resistance function (Figure 9-40).
2. Disconnect the wires from the switch terminals.
3. Connect the meter leads across the disconnected switch terminals.
4. Start the engine.
5. Observe the resistance indicated on the meter when the oil pressure rises above 10 PSI.

Observations: With no oil pressure present at the switch, the contacts should be open and the meter should indicate infinite resistance. If the switch failed to operate, it must be replaced.

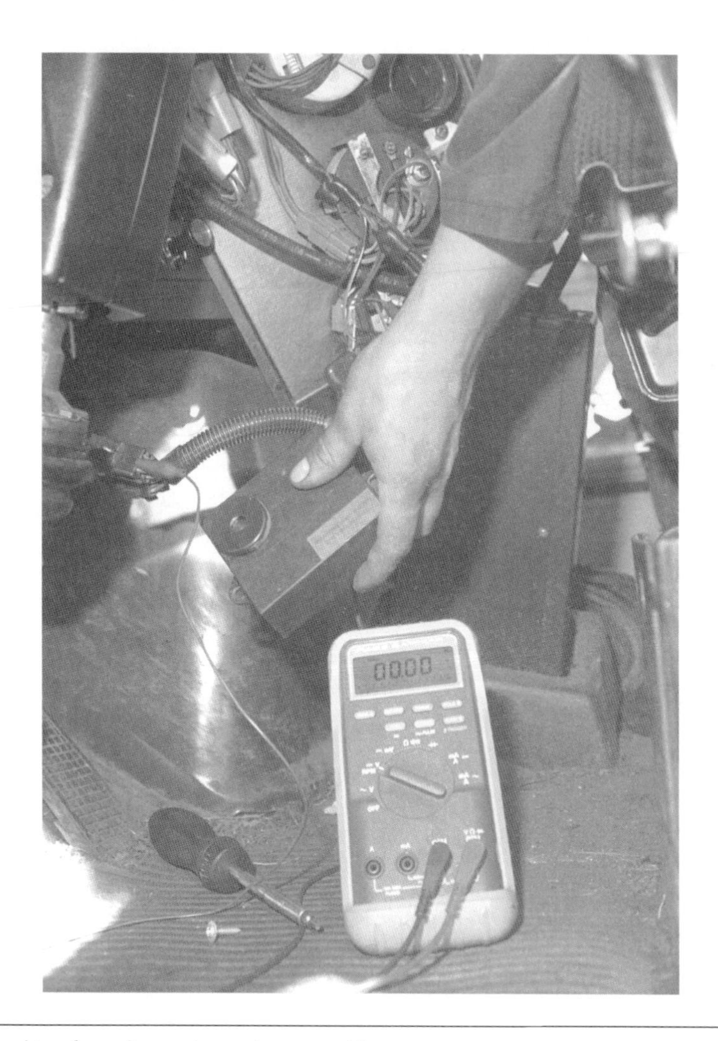

Figure 9-39 Checking for voltage at warning assembly.

Figure 9-40 Checking oil pressure switch.

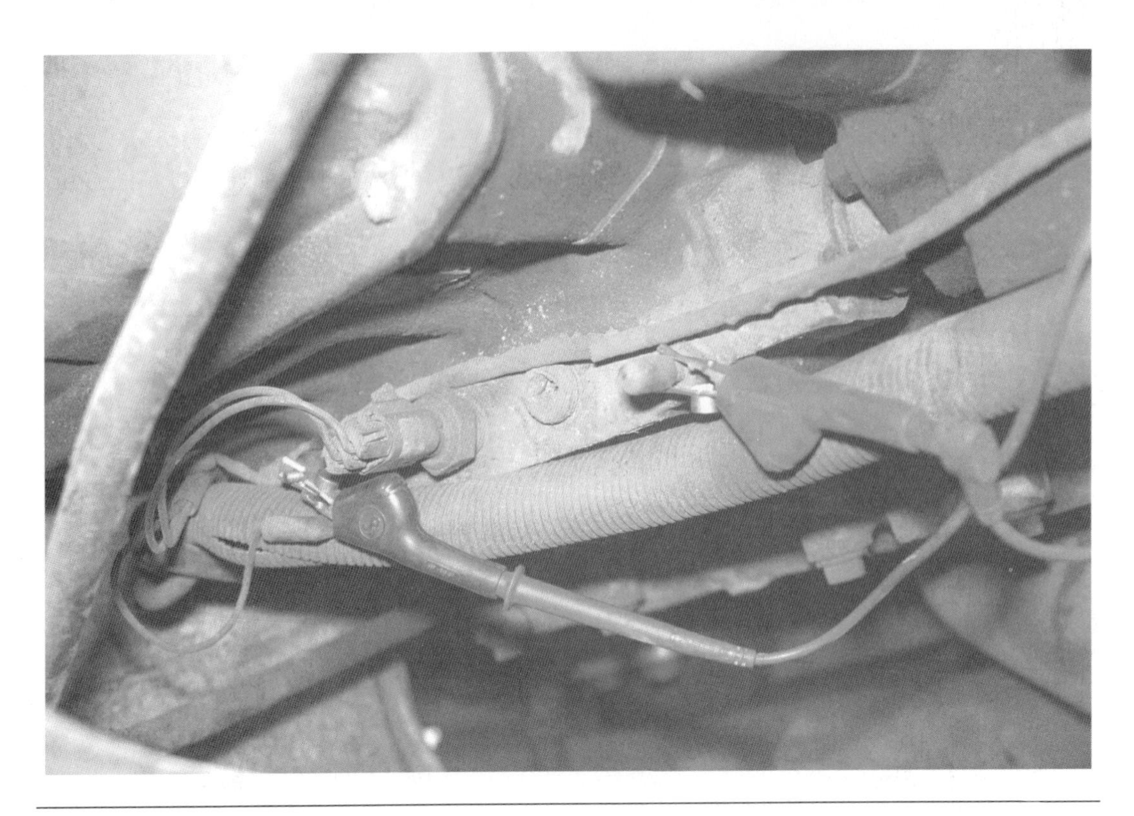

Figure 9-41 Quick check of coolant temperature circuit can be performed by grounding the high coolant temperature switch.

Classroom Manual Chapter 9, page 220.

Classroom Manual Chapter 9, page 233.

Classroom Manual Chapter 9, page 234.

High Coolant Temperature Warning

The high coolant temperature cannot be effectively tested unless a preset temperature of approximately 225°F can be reached. It is usually at this temperature that the high coolant temperature warning assembly activates. However, the circuit can be easily checked for integrity. This check is performed after ensuring voltage is present at the 102 wire connection.

Simply ground out the sensor (Figure 9-41). If the buzzer activates, the circuit is functioning properly and the problem is most likely caused by a faulty sensor. If the buzzer fails to activate, check for an open on Circuit 877.

Troubleshooting Kysor Shutdown Systems

Using the wiring diagram in Figure 9-42 as a reference, Test 1 (Figure 9-43) and Test 2 (Figure 9-44) can be used to troubleshoot a **Kysor Shutdown System**. *Always refer to the specific vehicle's and engine manufacturer's procedures.*

Special Tools

Jumper wires

Jumper wire with 20-amp in-line fuse

Test light

Classroom Manual Chapter 9, page 228.

Audible Warning Systems

Many warning systems use a buzzer to audibly alert a driver of an impending problem. The buzzer can be used in various circuits such as low air pressure, low oil pressure, key in ignition, and headlights on.

A buzzer circuit can also include a timer to control how long it is activated. If the buzzer does not shut off and it is known to have a timer, replace the timer with a known good unit. If the buzzer does not have a timer circuit and it remains on even after the parameters of the involved system are met, such as proper air pressure, check for a short to ground on the switch side of the buzzer and for a faulty switch.

If the buzzer does not sound, remove it and apply 12 volts across the terminals (Figure 9-45). If the buzzer sounds, it is good and the problem is most likely in the circuit.

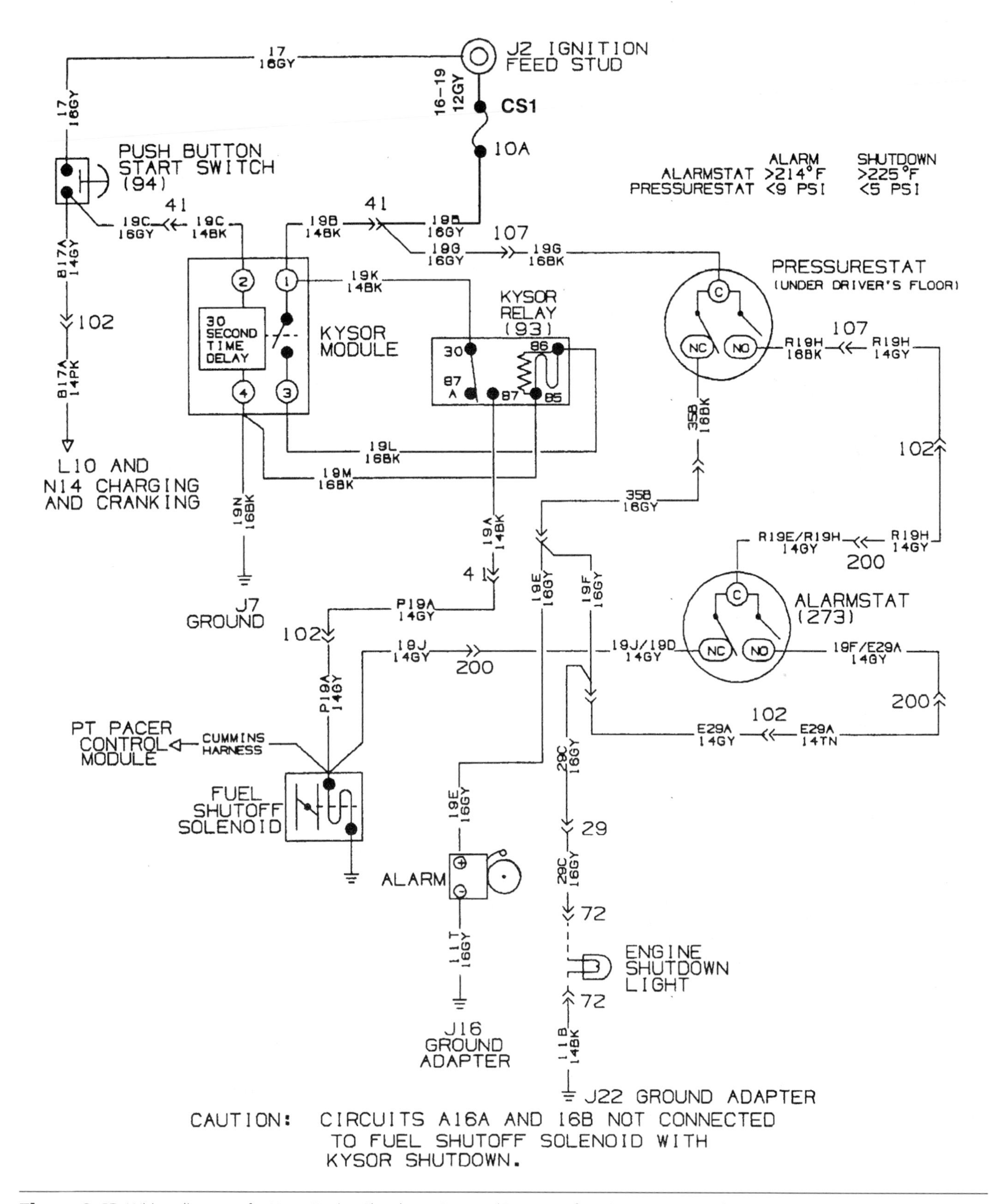

Figure 9-42 Wiring diagram of a Kysor Engine Shutdown System. (Courtesy of Navistar International Engine Group)

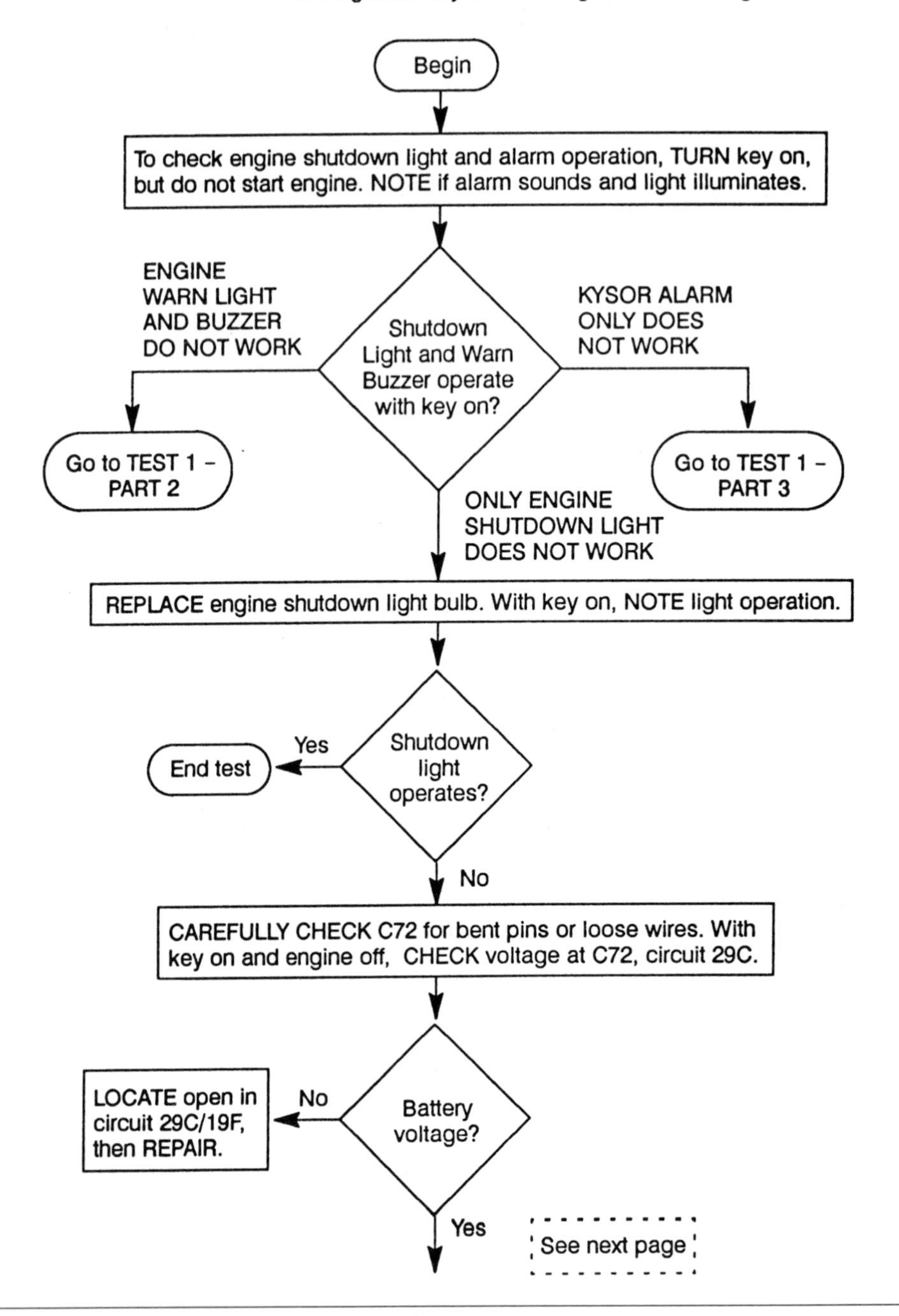

Figure 9-43 Kysor Shutdown flowchart. (Courtesy of Navistar International Engine Group)

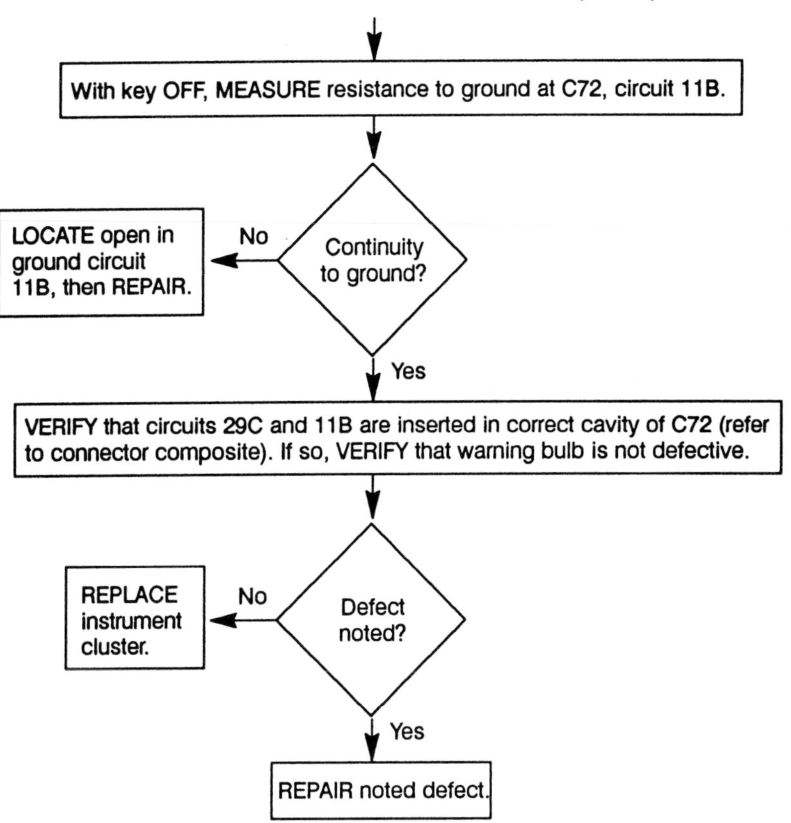

TEST 1 – PART 2 (continued from Part 1)

SYMPTOM: Engine Starts, but Kysor Engine Warning Light and Alarm DO NOT WORK.

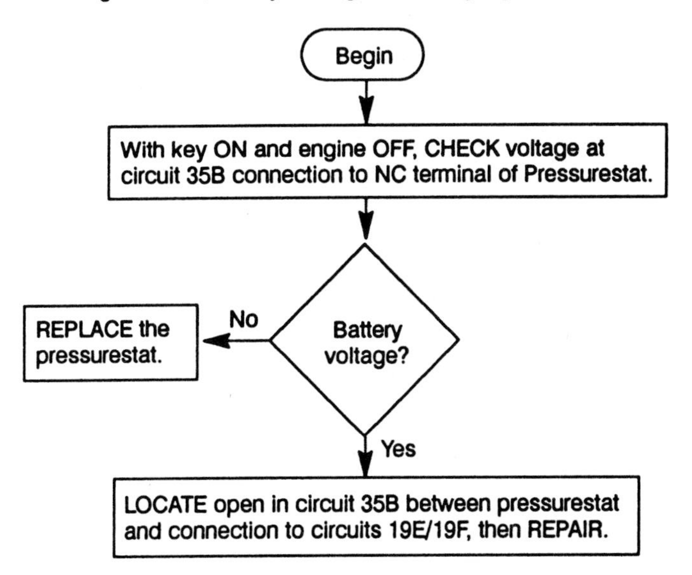

Figure 9-43 Continued.

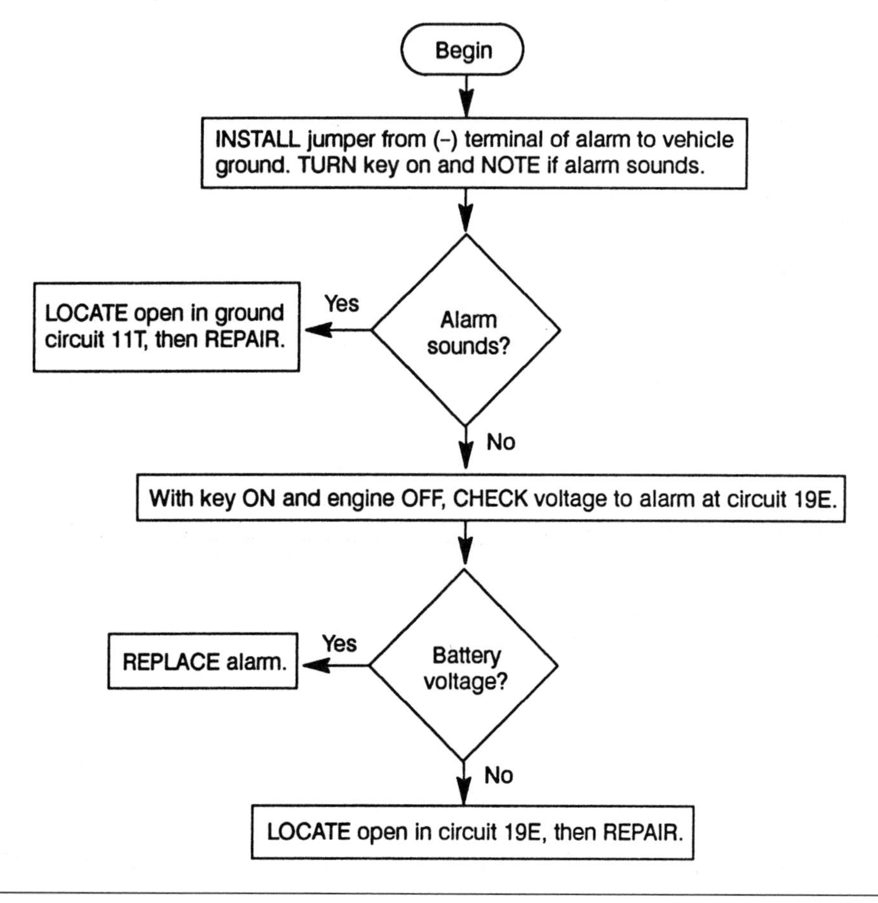

Figure 9-43 Continued.

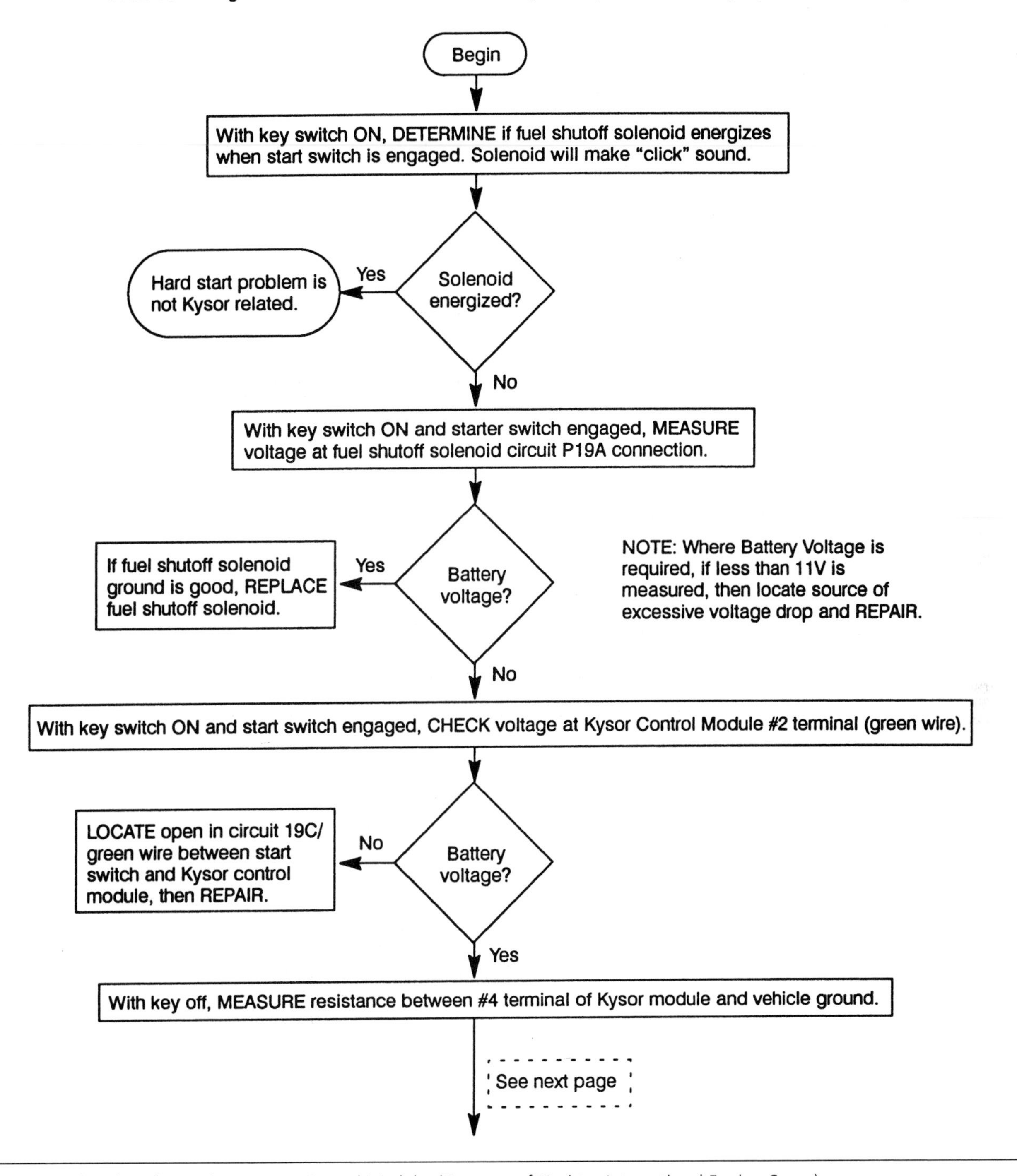

Figure 9-44 Flowchart for testing a Kysor Control Module. (Courtesy of Navistar International Engine Group)

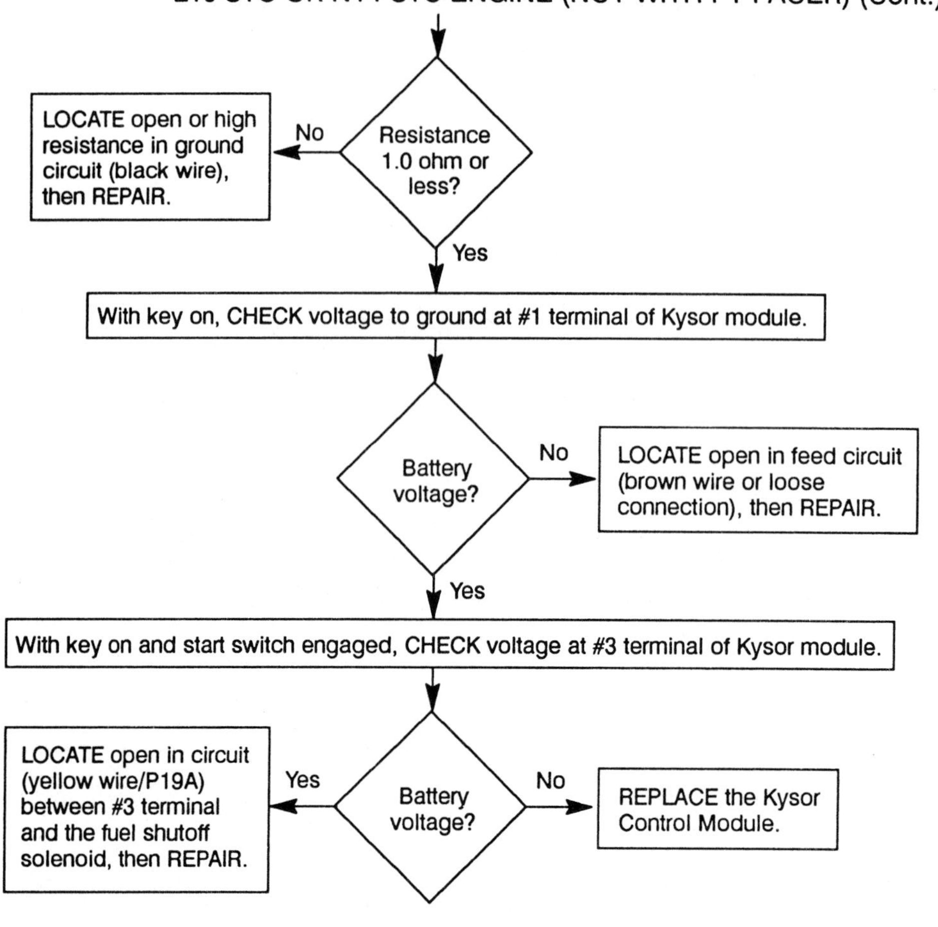

Figure 9-44 Continued.

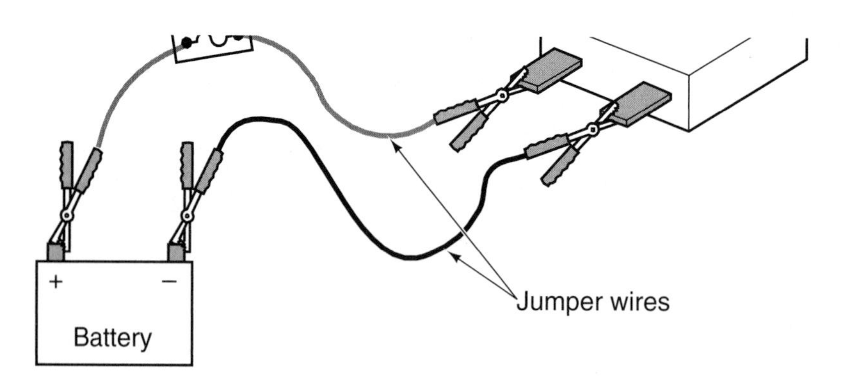

Figure 9-45 Testing the buzzer.

Voltmeter Testing

The voltmeter does not receive any signal voltage from a sending unit. A voltmeter uses ignition voltage as the input at the gauge terminal usually marked "I." If a voltmeter is suspected of being faulty, the troubleshooting procedure would simply require checking for power at the "I" terminal and making sure there is a good ground connection. Accuracy of the meter is tested by measuring the voltage across the "I" and "G" terminals with a multimeter (Figure 9-46).

Simply set the multimeter at the VDC function and compare the gauge readings with the readings indicated on the multimeter (Figure 9-47).

Classroom Manual
Chapter 9, page 216.

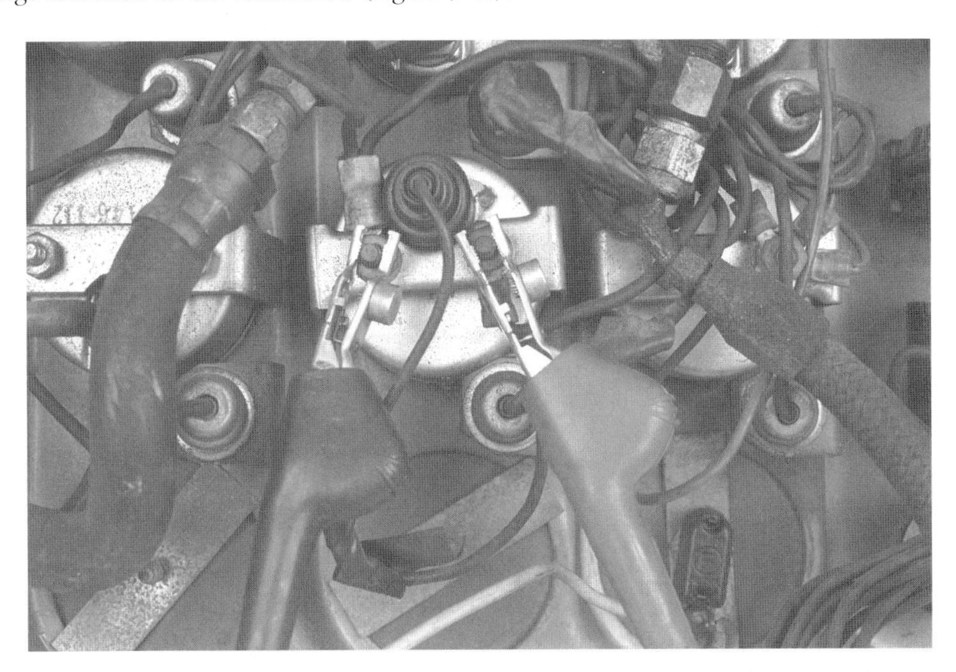

Figure 9-46 Measuring a vehicle's voltmeter accuracy with a multimeter.

Figure 9-47 Testing the voltmeter with the multimeter set at VDC function.

CASE STUDY

A customer brings a vehicle into the shop because the oil pressure warning light comes on intermittently. The technician could not get to the truck immediately, so when he finally had a chance to work on it, he ran the truck in the shop for a little while. He observed that the warning light did not get activated and decided that there must be an electrical problem in the circuit causing the light to come on intermittently. He knew that an electrical problem requires a good visual check, so at the engine he located various sensors. The technician checked all the connections and tightened them to prevent problems when the vehicle hit a bump. He also loosened and retightened many of the threaded sensors because they are often grounded through the threads. After trying unsuccessfully to activate the warning system, he assumed that the problem was fixed.

Unfortunately, the owner of the truck returned the next day and was very upset because he had the same problem. The original technician was busy on a service call so another technician was asked to look at the truck. First, she found a good wiring diagram and component locator. To verify that the circuit from the sender to the light was working properly, she disconnected the wire at the proper oil pressure sender and jumped it to ground. The light came on, indicating the potential was there for the circuit to operate. However, with everything hooked up and the vehicle running in the shop, the warning system did not activate. She then realized that the system was activated due to low oil pressure and that pressures change. Next, she installed a mechanical pressure gauge and took the vehicle down the road to heat it up. The longer the vehicle was driven and the warmer it got, the lower the oil pressure became. Bringing the vehicle back to the shop, the technician concluded that the warning system was working properly because it was being activated at certain times when the oil pressure became dangerously low. The owner of the truck was informed of an oil pressure loss problem that needed to be addressed before a complete engine failure occurred.

Terms to Know

Buzzer
Dip switch
Eddy currents
Electromagnetic gauges
Electromechanical
Instrument voltage regulator (IVR)
Kysor Shutdown System
Odometer
Piezoresistive sensor
Sender unit
Speed sensor
Speedometer
Tachometer
Thermistor
Warning system

ASE-Style Review Questions

1. Speedometer noises are being discussed: *Technician A* says that a noise can be heard that is due to a problem in the speedometer head. *Technician B* says that a noise related to the speedometer can be due to a dry cable. Who is correct?

A. A only
B. B only
C. Both A and B
D. Neither A nor B

2. Accuracy of a speedometer system is being discussed: *Technician A* says that an improper ratio adapter on an electronic speedometer system can cause an inaccurate reading. *Technician B* says that an improper ratio adapter on a mechanical speedometer system can cause an inaccurate reading. Who is correct?

A. A only
B. B only
C. Both A and B
D. Neither A nor B

3. *Technician A* says whenever a new electronic speedometer is installed, it needs to be calibrated. *Technician B* says that all new speedometers are calibrated right at the factory. Who is correct?
   A. A only **C.** Both A and B
   B. B only **D.** Neither A nor B

4. *Technician A* says that an electronic tachometer uses a piezoresistive concept for a pulse signal. *Technician B* says that both the electronic tachometer and electronic speedometer systems use the same basic concept and type of component to receive a speed signal. Who is correct?
   A. A only **C.** Both A and B
   B. B only **D.** Neither A nor B

5. *Technician A* says that the float arm on a fuel sender is connected to a potentiometer. *Technician B* says that the float arm is connected to a variable resistor unit. Who is correct?
   A. A only **C.** Both A and B
   B. B only **D.** Neither A nor B

6. *Technician A* says that a warning circuit can use a switch as a sensor. *Technician B* says variable resistors can also be used as sensors. Who is correct?
   A. A only **C.** Both A and B
   B. B only **D.** Neither A nor B

7. *Technician A* says that a negative resistance thermistor will show a high resistance as the temperature of the medium being measured increases. *Technician B* says that the resistance will decrease as the temperature of the medium being measured decreases. Who is correct?
   A. A only **C.** Both A and B
   B. B only **D.** Neither A nor B

8. Gauge malfunctions are being discussed: *Technician A* says a defective IVR can affect more than one gauge. *Technician B* says a fuse or fuses should be checked if one or more gauges are defective. Who is correct?
   A. A only **C.** Both A and B
   B. B only **D.** Neither A nor B

9. A warning buzzer does not seem to activate. *Technician A* says that jumper wires can be used to apply direct power to the buzzer to verify a possible circuit problem. *Technician B* says using jumper wires to power the buzzer verifies if the buzzer is defective or not. Who is correct?
   A. A only **C.** Both A and B
   B. B only **D.** Neither A nor B

10. *Technician A* says that a thermistor is used to measure pressures. *Technician B* says that an ohmmeter is used to test a thermistor sensor. Who is correct?
    A. A only **C.** Both A and B
    B. B only **D.** Neither A nor B

ASE Challenge

1. *Technician A* says that all trucks use constant voltage regulators (CVR) to stabilize voltage for the gauges. *Technician B* says that all heavy duty trucks use shunt resistors on the back of the gauge to adjust for CVR voltage. Who is correct?
   A. A only **C.** Both A and B
   B. B only **D.** Neither A nor B

2. A truck driver complains that his truck's fuel gauge reads empty for the last half of the tank. *Technician A* says that the likely cause could be a bad sending unit variable resistor. *Technician B* says it could be a faulty shunt resistor on the gauge. Who is correct?
   A. A only **C.** Both A and B
   B. B only **D.** Neither A nor B

3. A faulty fuel gauge is being diagnosed. *Technician A* says that with the signal wire removed from the tank sender and the key on, the gauge should move full circle one way. *Technician B* says you should always check for proper sender and tank ground. Who is correct?
   A. A only **C.** Both A and B
   B. B only **D.** Neither A nor B

4. A truck continually displays a low oil pressure warning, oil pressure fault code, and runs at reduced power. *Technician A* says the engine control module (ECM) should be changed and the engine fuel system checked out. *Technician B* says the oil pressure should be checked with a reliable mechanical gauge. Who is correct?
   A. A only **C.** Both A and B
   B. B only **D.** Neither A nor B

5. *Technician A* says that it is normal for the charge indicator light to be off when the engine is off as long as the generator is okay. *Technician B* says that if the generator malfunctions, a special relay locks the charge indicator light on. Who is correct?
   A. A only **C.** Both A and B
   B. B only **D.** Neither A nor B

Table 9-1 NATEF and ASE Task

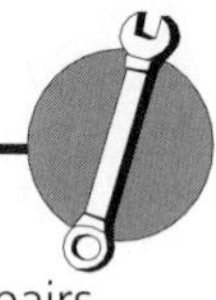

Diagnose the cause of intermittent, high, low, or no gauge readings; determine needed repairs (does not include charge indicators)

Problem Area	Symptoms	Possible Causes	Classroom Manual	Shop Manual
INTERMITTENT GAUGE READINGS	One or all gauges fluctuate from low or high to normal readings	1. Faulty sending unit 2. Open circuit 3. Poor ground 4. Excessive resistance 5. Defective printed circuit 6. Body computer fault	213	295
HIGH GAUGE READINGS	One or all gauges read high	1. Faulty IP regulator or computer 2. Shorted printed circuit 3. Faulty sending unit 4. Faulty gauge 5. Short to ground in sending unit 6. Poor sending unit ground	214	296

Table 9-2 NATEF and ASE Task

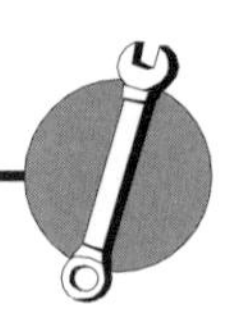

Diagnose the cause of data bus driven gauge malfunctions; determine needed repairs

Problem Area	Symptoms	Possible Causes	Classroom Manual	Shop Manual
LOW GAUGE READINGS	One or all gauges read low	1. Faulty IP regulator or computer 2. Improper connections 3. Improper bulb application 4. Poor sending unit ground	214	295

Table 9-3 NATEF and ASE Task

Inspect, test, adjust, repair, or replace gauge circuit sending units, gauges, connectors, terminals, and wires

Problem Area	Symptoms	Possible Causes	Classroom Manual	Shop Manual
NO GAUGE READINGS	One or all gauges fail to read	1. Blown fuse 2. Open in the printed circuit 3. Faulty gauge 4. Open in sending unit 5. Improper connections 6. Improper bulb application 7. Poor common ground	216	296

Table 9-4 NATEF and ASE Task

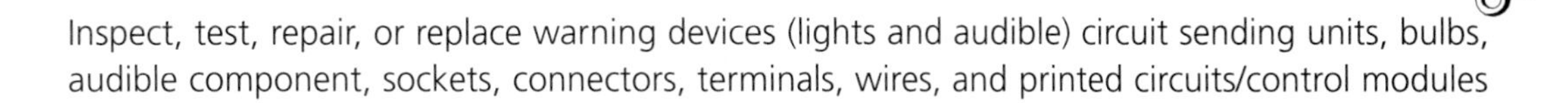

Inspect, test, repair, or replace warning devices (lights and audible) circuit sending units, bulbs, audible component, sockets, connectors, terminals, wires, and printed circuits/control modules

Problem Area	Symptoms	Possible Causes	Classroom Manual	Shop Manual
WARNING DEVICE OPERATES CONSTANTLY	Warning device operates at all times	1. Faulty device circuit 2. Faulty body system computer 3. Grounded switch circuit	227	301
WARNING DEVICE INOPERATIVE	Warning device does not operate at all	1. Faulty device circuit 2. Loose connection 3. Bad ground 4. Burned out bulb 5. Faulty audible device	230	304

Table 9-5 NATEF and ASE Task

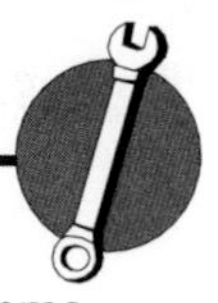

Inspect, test, replace, and calibrate electronic speedometer, odometer, and tachometer systems

Problem Area	Symptoms	Possible Causes	Classroom Manual	Shop Manual
SPEEDOMETER OR TACHOMETER INOPERATIVE OR INDICATES AN INCORRECT SPEED	Speedometer or tachometer does not indicate vehicle speed	1. Failed cable 2. Engine computer malfunction 3. Engine speed sensor failure 4. Vehicle speed sensor failure 5. Bad ground 6. Open circuit 7. IP Computer fault 8. Defective speedometer unit	226	294

Job Sheet 14

14

Name: ______________________ Date: ______________________

Checking a Fuel Gauge

Upon completion of this job sheet, you should be able to diagnose an inaccurate or inoperative fuel gauge.

NATEF and ASE Correlation

This job sheet is related to ASE and NATEF Medium/Heavy Duty Truck Electrical/Electronic Systems List content area:

F. Gauges and Warning Devices Diagnosis and Repair, ASE Task 3 and NATEF Task 4: Inspect, test, adjust, repair, or replace gauge circuit sending units, gauges, connectors, terminals, and wires.

Tools and Materials

A medium/heavy duty truck
Service manual for the truck
Wiring diagram for the truck
A DMM
A basic mechanic's tool set
Miscellaneous resistors
Variable resistor (purchased from an electrical store) or a signal injector with built-in variable resistance function (purchased from tool manufacturer)

Procedure

1. Describe the truck being worked on: Year: ________ Make: ____________
VIN: ______________ Model: ________________

2. Locate the procedures listed in the service manual and describe in your own words the procedures for testing the fuel gauge and gauge sending unit on the truck. Be sure to describe why the particular procedures are being used. ________________

____

____

3. Draw the circuit diagram below.

4. Describe the test point, procedure, and result if you were to check the gauge circuit.

____

____

____

5. Describe the test point, procedure, and result if you were to check the sender circuit with the tank full. ____

____

____

6. Describe the test point, procedure, and result if you were checking the sender circuit with the fuel tank empty. ____

____

____

7. Duplicate the test on the truck and compare both results. ______________________

____

____

8. Summarize the conclusion of the actual test. ______________________________

____

____

☑ **Instructor Check** ____

Job Sheet 15

15

Name: ______________________________ Date: ______________________________

Calibrating the Electronic Speedometer System

Upon completion of this job sheet, you should be able to calibrate and diagnose an electronic speedometer system.

NATEF and ASE Correlation

This job sheet is related to ASE and NATEF Medium/Heavy Duty Truck Electrical/Electronic Systems List content area:

F. Gauges and Warning Devices Diagnosis and Repair, ASE Task 5 and NATEF Task 7: Inspect, test, replace, and calibrate electronic speedometer, odometer, and tachometer systems..

Tools and Materials

A medium/heavy duty truck with an electronic speedometer system
A service manual for the truck
A wiring diagram for the truck
Jack stands
A tire marking device (chalk, soapstone)

Procedure

1. Describe the truck being worked on: Year: ________ Make: ______________
   VIN: ____________ Model: __________________

   The following procedure might require the removal of the speed sensor to access the speedometer gear for a count of the number of teeth on the speedometer gear.

2. Derive and note the following information: Tire size: ____________ (not required for calculation), Axle ratio: ________________, Number of teeth on the speedometer gear: ________________, Tire revolutions per mile: ___________

SERVICE TIP: To derive the tire revolutions per mile, measure the distance the truck moves in one tire revolution. Next, divide 5280 feet by the measured distance of one tire revolution (in feet).

SERVICE TIP: The axle ratio is usually stamped on the differential carrier or listed on the truck's line sheet. It can also be calculated by rotating the driveshaft one revolution and observing how many revolutions (full and partial) the tire makes.

3. Calculate the pulses per mile (PPM):
   Axle ratio ____________ × Tire revs. _____________ × No. of teeth _________ = PPM

4. Following the procedure in the truck's service manual, calculate the calibration number. ________

5. Determine the dip switch settings from the chart in the truck's service manual and write them out: ____

6. Access the dip switches on the truck following the manufacturer's procedures.

7. Do your determined dip switch settings and the actual dip switch settings match? Yes _______ No _______

☑ **Instructor Check** ____

Job Sheet 16

16

Name: ______________________ Date: ______________________

Diagnosing the Electronic Speedometer System

Upon completion of this job sheet, you should be able to diagnose an inoperative electronic speedometer system.

NATEF and ASE Correlation

This job sheet is related to ASE and NATEF Medium/Heavy Duty Truck Electrical/Electronic Systems List content area:

F. Gauges and Warning Devices Diagnosis and Repair, ASE Task 5 and NATEF Task 7: Inspect, test, replace, and calibrate electronic speedometer, odometer, and tachometer systems.

Tools and Materials

A medium/heavy duty truck with electronic speedometer
A service manual for the truck
A wiring diagram for the truck
A DMM
Jack stands
Wheel jocks

Procedure

1. Describe the truck being worked on: Year: __________ Make: ______________
   VIN: ____________ Model: __________________
2. Following the service manual procedures, access the electronic speedometer connections at the speedometer.
3. Using the wiring diagram and the DMM, check and record for the following:
   Ignition voltage ________________ Ground (verify if good or bad) ________________
4. Raise the truck and secure it on jack stands following proper safety procedures.
5. Describe the procedure for checking for signal input voltage to the speedometer.
   ____
   ____
   ____
6. Perform the procedure and record the result. ______________________________
   ____
   ____
7. Describe the procedure for checking the speedometer sensor's resistance. _________
   ____
   ____
8. Perform a resistance check of the speedometer speed sensor and record here.
   ____
   ____
   ____

9. Write down the conclusion of the entire test. ______________________________
____
____
____

☑ **Instructor Check** ____

CHAPTER 10

Truck Electrical Accessories Diagnosis and Repairs

Upon completion and review of this chapter, you should be able to:

- ❑ Identify the causes of constant or intermittent horn operation or no horn operation.
- ❑ Diagnose the cause of poor sound quality from the horn system.
- ❑ Diagnose and repair a windshield wiper that fails to operate in one speed only or in all speeds.
- ❑ Identify causes for slower-than-normal wiper operation.
- ❑ Remove and install a wiper motor, arms, and linkages.
- ❑ Determine the cause of windshield washer system improper operation and replace the motor/pump if required.
- ❑ Diagnose problems with blower motor systems.
- ❑ Diagnose problems with engine retarder systems, and adjust control switches.
- ❑ Diagnose typical thermostatic engine fan electro-pneumatic circuit problems.

Basic Tools

Basic mechanic's tool set

Service manual

Introduction

Classroom Manual Chapter 10, page 241.

A vehicle's electrical accessory system often consists of a major component, such as a blower motor, that is non-rebuildable. However, to access such a component, the technician needs to access areas of the vehicle that are sometimes difficult to reach. This could involve removing panels, ducts, and so on. Good diagnostic procedures and accurate information regarding correct system operation can help you avoid needlessly accessing hard-to-reach components. Remember that the fault is always easier to locate by first understanding how the system is supposed to operate. This chapter will cover general diagnostic tests and procedures relating to truck electrical system problems. However, you should always refer to the service manual to obtain correct procedures when it comes to working on the numerous accessory systems used in trucks.

Horn Diagnosis

Special Tools

Jumper wires

Voltmeter

Test light

Ohmmeter

Horn complaints generally fall into four areas: no operation, continuous operation, intermittent operation, or poor sound quality. Many of the tests for them overlap. Testing of the horn system also varies depending on whether or not it uses a relay.

No Horn Operation

Classroom Manual Chapter 10, page 241.

The common complaint of no horn operation can lead you to a simple cause such as a loose or broken connection. The connection can be at a location that can be easily accessed, such as at the horn itself, or it can involve a more difficult procedure, such as removing the steering wheel to access the contact to the slip ring.

A good indication that the problem is in the steering wheel area is a horn that sounds intermittently while the steering wheel is turned. The cause could be the sliding contact ring or a broken or worn tension spring. For a complete no-sound test, perform the following procedures:

1. Start by checking the fuse or circuit protection device. If it is defective, replace as needed. Operate the horn and any other circuit protected by it. Another circuit that is faulty can cause the fuse to blow, with the only initial symptom being the horn not activating.

2. Connect a jumper wire from the battery positive terminal to the horn terminal. If the horn sounds, continue testing; if the horn does not sound, check the ground connection. Replace the horn if the ground is good.

SERVICE TIP: Most trucks have the battery(s) quite a distance away from the horn. Try taking a spare battery near the electric horn and bypassing the truck's circuit by using jumper wires directly to the horn for both power and ground from the spare battery.

3. In a multiple horn system, test for voltage at the last common connection between the horn relay and the horns (Figure 10-1). On a single-horn system, test for voltage at the horn terminal. When the horn button is depressed, voltage should be present at that connection. If there is no voltage present, continue testing. If voltage was present, check the individual circuits from the connections to the horns and repair as needed.
4. Check for voltage at the power feed terminal of the relay (Figure 10-2). If voltage is present at the power feed terminal, continue testing. If no voltage is present, trace the wiring from the relay to the source of power to locate the problem.
5. Check for voltage at the switch control terminal of the relay. If voltage is present, continue testing. If voltage is not present, the relay is defective and must be replaced.
6. Check the ground to the relay. Use a jumper wire from the control terminal of the relay to ground. If the horn does not sound, it also indicates a bad relay that must be replaced. If the horn sounds, continue to test.
7. Check for voltage on the battery side of the horn switch. No voltage at this location indicates a fault between the relay and the switch. If voltage is present, continue testing.
8. Check for continuity through the switch. If there is continuity, check the ground connection for the switch and recheck operation. Replace the horn switch or required components(s) if there is no continuity when the button is depressed.

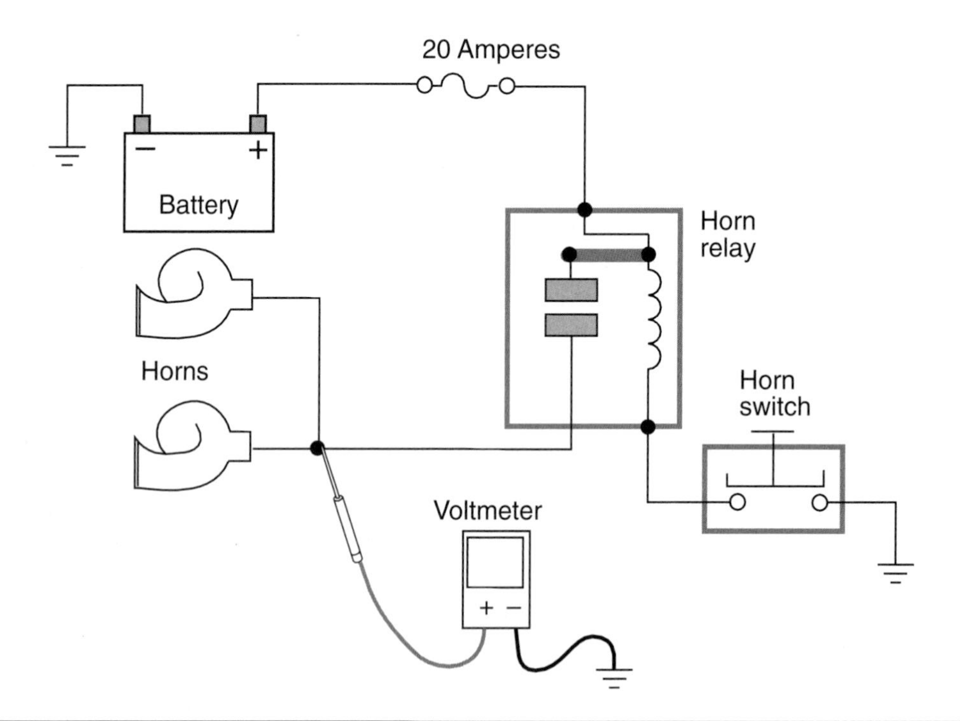

Figure 10-1 Testing for voltage at the last common connection.

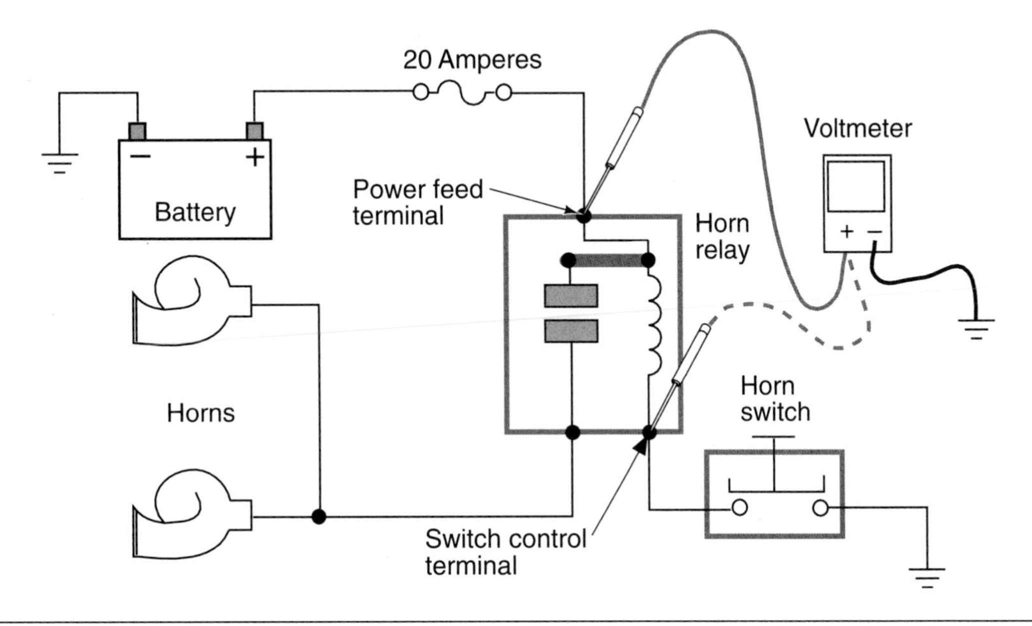

Figure 10-2 Testing for power into the relay and the relay coil.

Systems Without Relay

Before testing, confirm that the no-horn complaint is related to the horn circuit. If it is, perform the following tests to locate the fault.

1. Check the fuse or circuit protection device. If defective, replace as needed. After replacing the fuse, operate all other circuits protected by it.
2. Power the horn terminal with a jumper wire from the battery positive terminal. If the horn sounds, continue testing. If the horn does not sound, check the ground connection. Replace the horn if the ground is good.
3. In a dual-horn system, test for voltage at the last common connection between the horn switch and horns (Figure 10-3). On a single-horn system, test for voltage at the horn terminal. Voltage should be present at this connection only when the horn button is depressed. Continue testing if there is no voltage at this connection. If voltage is present, check the individual circuits from the connection to the horns. Repair as needed.
4. Check for voltage at the horn side of the switch when the button is depressed. If voltage is present, the problem is in the circuit from the switch to the horn(s). Continue testing if there is no voltage at this connection.
5. Check for voltage at the battery side of the switch. If voltage is present, the switch is faulty and must be replaced. If there is no voltage present at this terminal, the problem is in the circuit from the battery to the switch.

Special Tools

Voltmeter

Jumper wires

Poor Sound Quality

Several factors that can affect the sound quality of a horn or multiple horn system are: damaged **diaphragms,** poor ground connections, improperly adjusted horns, or excessive circuit resistance.

Sometimes a multiple horn system has only one horn that activates, resulting in poor sound quality. If this is the case, use a jumper wire from the battery positive terminal to the faulty horn's terminal to determine whether the fault is in the horn or in the circuit. If the horn sounds, the problem is in the circuit between the last common connection and the affected horn. If the circuit checks out and the sound quality cannot be improved by adjustment, replace the horn.

Classroom Manual Chapter 10, page 242.

The **diaphragm** is a thin, flexible, circular plate that is held around its outer edge by the horn housing, allowing the middle to flex.

Air horns used in trucks also use a flexible disc whose middle flexes in response to applied air pressure.

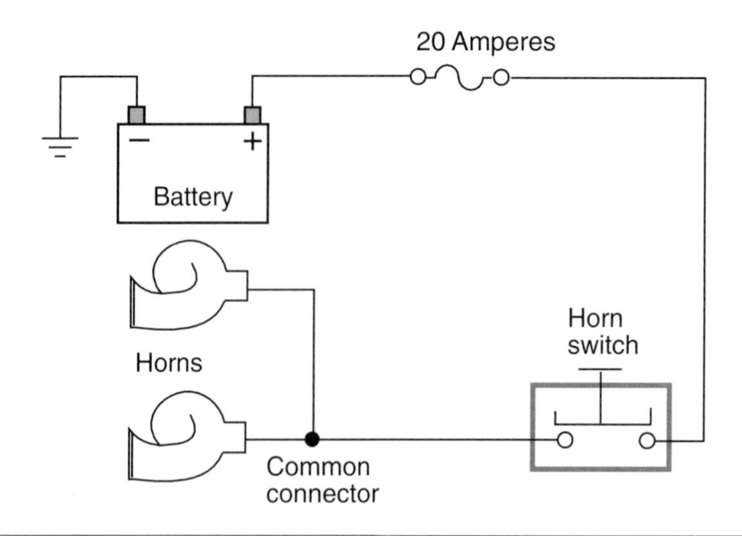

Figure 10-3 Simplified horn circuit without a relay.

CAUTION: There is no need to have the engine running when working with the horn circuit. Keep in mind that horns might be mounted at an area that might put you close to moving parts when the engine is running, increasing the risk of personal injury.

Special Tool

Ohmmeter

Horn Sounds Continuously

A horn that sounds continuously is usually caused by either a set of sticking contact points in the relay or a sticking horn switch. To find the fault, disconnect the horn relay from the circuit. Use an ohmmeter to check for continuity from the battery feed terminal of the relay to the horn circuit terminal (Figure 10-4). If there is continuity, the relay is defective. If there is no continuity through the relay, test the switch circuit for continuity. The switch circuit should have no continuity when the horn button is in the nondepressed position.

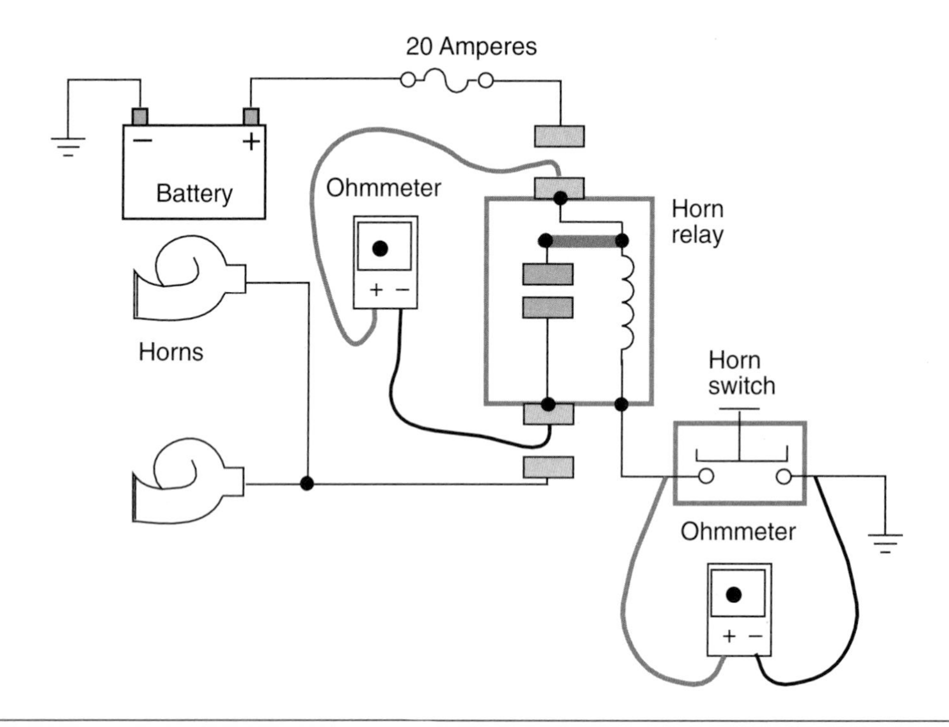

Figure 10-4 Ohmmeter testing to find the cause of continuous horn operation.

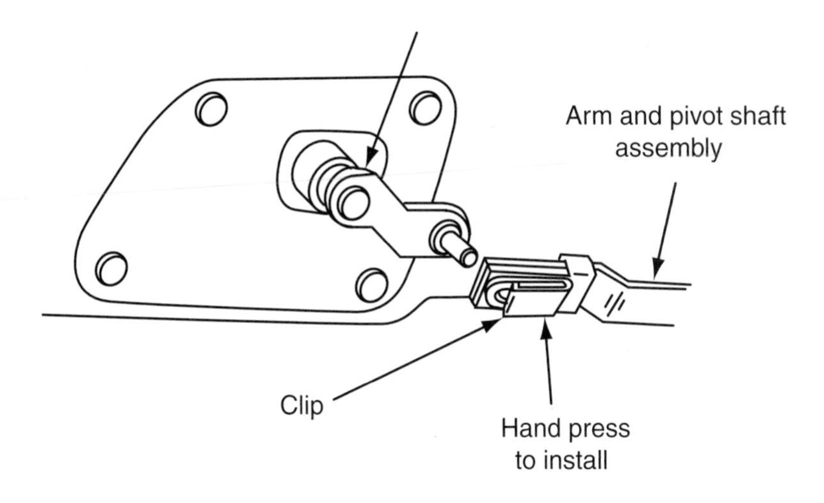

Figure 10-5 Disconnecting the mechanical arms from the motor. (Courtesy of Freightliner Corporation)

Wiper System Service

Classroom Manual
Chapter 10, page 243.

Classroom Manual
Chapter 10, page 244.

Wiper system complaints are numerous and include no operation, continuous operation, intermittent operation, partial operation, wipers that will not park, or improper wiper speed. Another type of complaint is the wiper blade hitting the molding, or the sweep of the blade not covering the required wipe area of the windshield. The latter type of complaint is generally an adjustment problem.

Any wiper problem brought to the shop will require the technician to determine if the problem is mechanical or electrical. To determine if a wiper problem complaint is due to an electrical or mechanical problem, disconnect the transmission linkage from the wiper motor (Figure 10-5). Turn on the wiper system and observe the operation of the motor. If the motor is operating properly, the problem is mechanical.

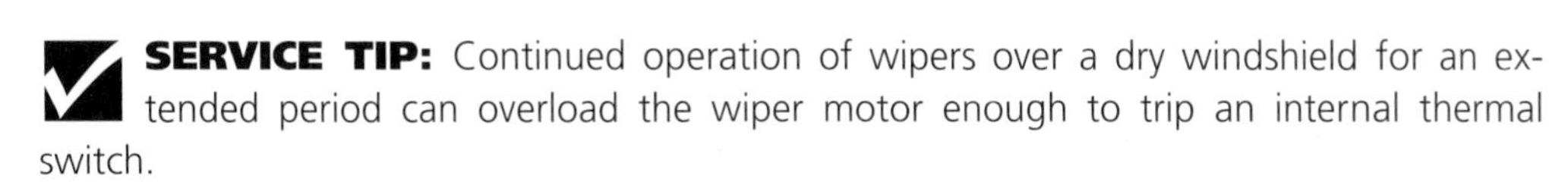

SERVICE TIP: Continued operation of wipers over a dry windshield for an extended period can overload the wiper motor enough to trip an internal thermal switch.

Preventive maintenance is a periodic maintenance procedure to ensure proper functions of various systems and components in trucks. Minor problems can turn into major problems if ignored.

As a truck technician, you will perform routine preventive maintenance (PM) procedures. The wiper system should be included as part of a preventive maintenance inspection. The following tests and checks can be used not only as a PM inspection but also to determine the causes of unusual noises, binding, and other operating problems.

Low Speed Test

1. Start the vehicle. Use the windshield washer to keep the windshield wet. Set the wiper motor to low speed.
2. Count the number of sweeps the wipers make in 15 seconds. A good wiper system should sweep a minimum of seven times (Bosch II wiper motor).
3. A speed that is too slow can indicate a problem in the pivots and linkages. Check, perform necessary repairs, and retest.

Classroom Manual
Chapter 10, page 245.

High Speed Test

1. With the vehicle running and the windshield washers on, set the wiper motor to high speed.

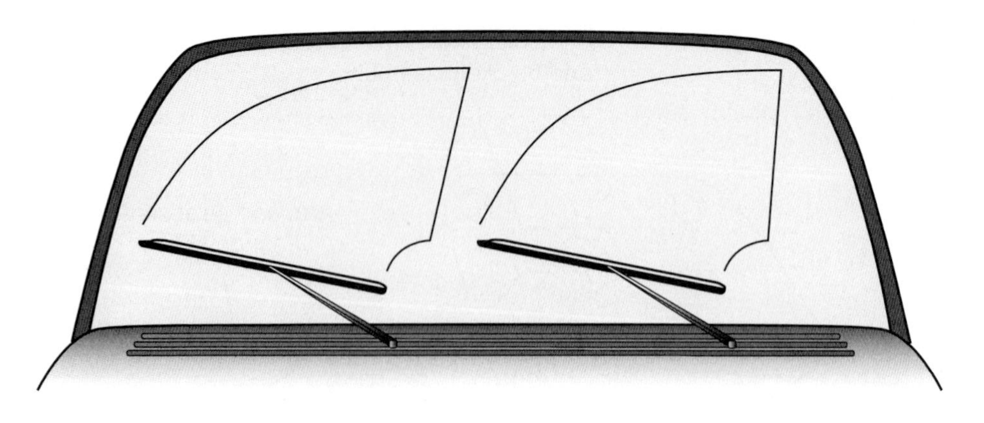

Figure 10-6 Windshield wipers in a parked position.

2. Count the number of sweeps the wipers make in 15 seconds. This time the wipers should sweep a minimum of eleven times.
3. If the speed is too slow, check the pivots and linkages for binding. Note: If the wipers oversweep (hitting the windshield moldings), check the linkage for excessive looseness or misadjustment.

Classroom Manual
Chapter 10, page 248.

Park Test

1. Turn the windshield wipers off while the blades are near the middle of the windshield. The arms should park as shown in Figure 10-6.
   - **A.** If the arms park at the position that the switch was turned off, repeat Step 1 but this time turn the switch off when the blades are in the up (outside) position on the windshield. If the wipers still do not go to the parked position but stop in a new position, there is an electrical problem in the switch or the motor.
   - **B.** If the blades return to the same park position each time (regardless of where they are when the switch is turned off) but are not in the correct position, the arms need to be removed and reinstalled to a correct position. Note: Some systems, such as the Bosch wiper system, need the "Wiper Motor Crank" indexed.
   - **C.** If the blades travel in the up direction to park by more than 0.5″, there is a problem with the position of the motor crank on the motor shaft. Note: Adjustment and service procedures are covered after the "Electrical Troubleshooting" section.

The following troubleshooting steps are explained thoroughly using various schematics. At the end there will be a specific vehicle's flowchart.

Special Tool
Voltmeter

Classroom Manual
Chapter 10, page 245.

No Operation in One Speed Only

Problems that cause a system not to operate in one of the switch positions or speeds are usually electrical in nature. To troubleshoot this type of problem requires the use of a service manual wiring schematic to determine proper operation. For this example we will use the three-speed wiper schematic in Figure 10-7. The problem with this vehicle is that the wipers do not operate in the MEDIUM speed position only. Looking at the figure, you will see that the problem could not be the shunt field in the motor because LOW and HIGH speeds do operate. The wiring to the motor is not the problem either, because this is shared by all speeds. However, the problem is a 7-ohm resistor that is open.

To verify an open resistor, use a voltmeter to measure voltage at the terminal leaking to the shunt field. If the voltage drops to 0 volts in the MEDIUM position, the switch has to be replaced. This was an example of using a schematic to locate a problem once the correct operation of the system is understood. The test equipment was used to verify the conclusion.

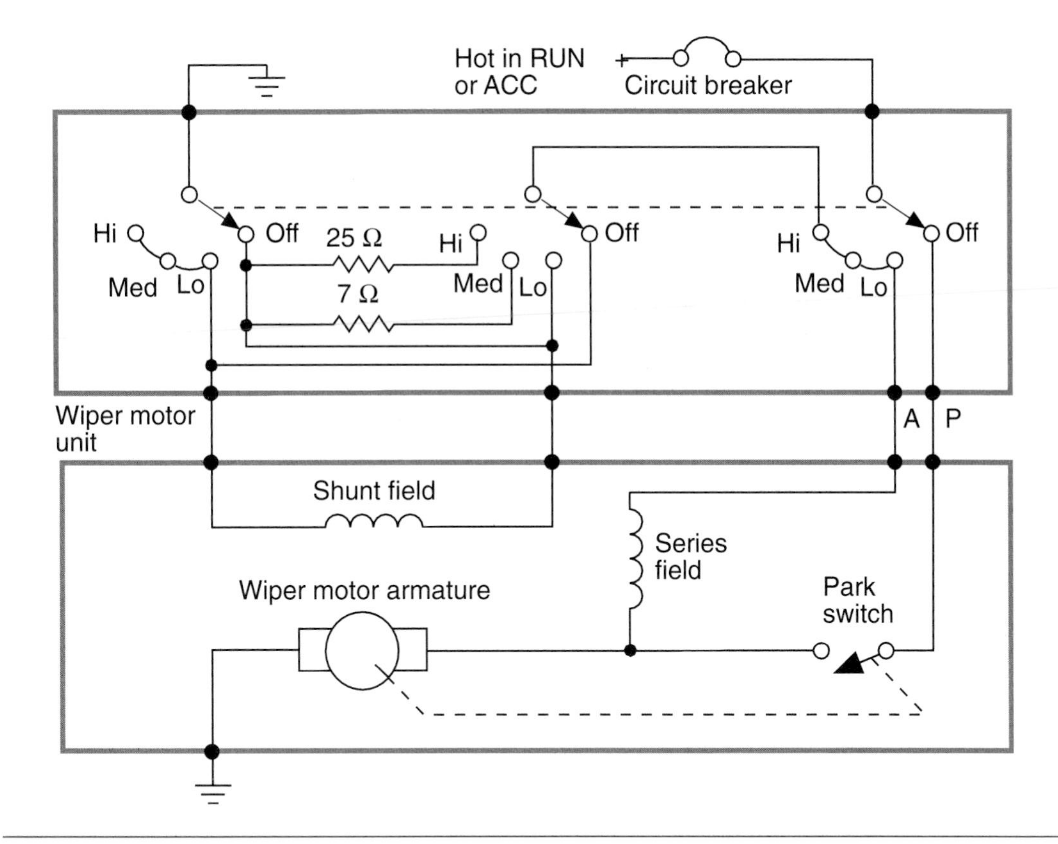

Figure 10-7 Three-speed windshield wiper system schematic.

Special Tool

DVOM

The following example involves a problem in a two-speed system with the motor operating in only a one-speed position. This problem can be caused by several different faults. It will require the use of schematics and test equipment to diagnose. The schematic in Figure 10-8 will be used as a reference to walk through a common test sequence to locate the fault for the motor not operating in the HIGH position.

Use the ACC position of the ignition switch if the wipers will operate in this position. If not, place the ignition switch in the ON or RUN position. Place the wiper switch in the HIGH position. Using a DVOM in the voltmeter function, check for voltage at circuit connector 56 of the motor. If voltage is present at this point, suspect a worn high-speed brush or an open wire from the terminal to the brush. This type of internal motor fault usually requires the replacement of the motor. Most shops do not rebuild wiper motors.

If no voltage is indicated at the Circuit connector 56, check for voltage at the connector for the switch at Circuit 56. If voltage is present at this point, the circuit from the switch to the motor is faulty. If no voltage is indicated at this point, replace the switch.

A no LOW speed operation complaint requires the same test procedure. However, the circuit to test is Circuit 58.

No Wiper Operation

The first thing to check in a no wiper operation in any speed is the fuse. Keep in mind that a binding in the mechanical portion of the system can cause an overload and blow the fuse. If the fuse is bad, replace it and recheck operation.

If the fuse is good, check the motor ground by using a jumper wire from the motor body to a good chassis ground. Turn the ignition ON and check for proper operation in all speed positions. If the motor operates, repair the ground circuit. If the motor does not operate, continue testing.

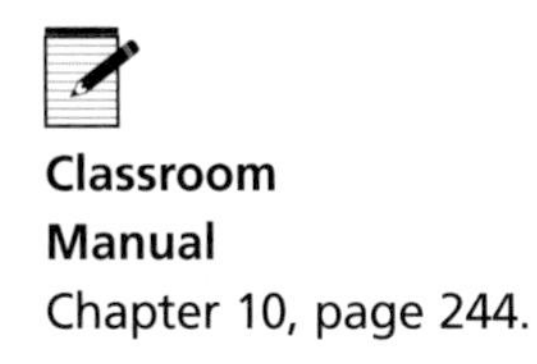

Classroom Manual
Chapter 10, page 244.

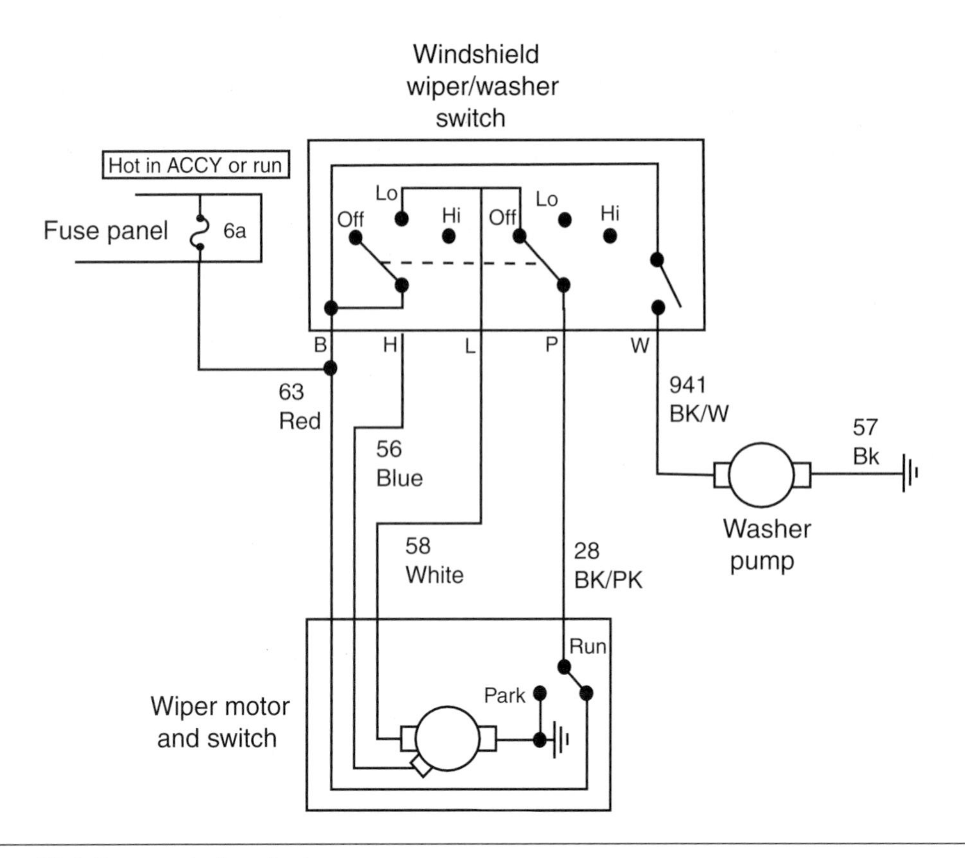

Figure 10-8 Two-speed wiper circuit.

Using a voltmeter, check for voltage at the LOW speed terminal of the motor with the ignition switch in the ON or ACC position and the wiper switch in the LOW position. If no voltage is indicated at this point, check for voltage on the LOW speed terminal of the wiper switch. If battery voltage is indicated at this terminal, the fault is in the circuit between the switch and the motor. Look for damaged or burned insulation of wiring or connectors that can affect both the HIGH and LOW speed circuits.

No voltage at the LOW speed terminal of the wiper switch indicates a fault in either the switch itself or the power feed circuit. Test for battery voltage at the battery supply terminal of the switch. If there is voltage indicated at this point, the switch is faulty and needs to be replaced. If no voltage is indicated at the supply terminal, trace the supply circuit back towards the battery to locate the fault.

If battery voltage was present at the LOW speed terminal of the motor, check for voltage at the HIGH speed terminal. Voltage indicated at both the terminals indicates a faulty motor that needs to be replaced. No voltage at the HIGH speed terminal require the same testing procedure as described for no voltage at the LOW speed circuit.

Special Tools

Ammeter

Voltmeter

Low Speed Circuit

Slower than Normal Wiper Speeds

In our initial low speed check we discussed the importance of looking for mechanical problems. If an electrical problem is suspected, an ammeter can be used to verify if the problem is mechanical or electrical. The test is performed with the mechanical system connected and not connected. A substantial change in current draw when the mechanical portion is disconnected indicates a problem in the arms linkages and/or wiper blades.

A problem that is electrical in nature is usually attributed to excessive resistance. If the complaint is that all speeds are slow, perform voltage drop tests of both the power and ground circuits. Also pinpoint check the switch for excessive resistance. For example, to check for excessive resistance on the ground circuit, connect the voltmeter positive lead to the ground terminal of the motor (or motor body) and the negative lead to a good chassis ground. The voltage drop should be no more than 0.1 volts. If excessive, repair the ground circuit connections Also check voltage drop through the motor by connecting the positive voltmeter lead to the LOW speed terminal and the negative lead to the vehicle chassis. With the ignition key ON, operate the motor on LOW speed and observe the voltmeter. Check for no or very low voltage reading (usually about 0.4-volt maximum). An excessive voltage drop indicates a faulty motor that should be replaced.

Wipers Will Not Park

A complaint of wipers that will not park is usually attributed to a faulty park switch. However, it will not always be the direct fault of the park switch. Operation of the park switch can often be observed if there is access to it by removing the motor cover (Figure 10-9). If that is possible, operate the wipers through a few cycles while observing the latch arm. When the wiper switch is turned off, the park switch latch must be in position to catch the drive pawl. If the drive pawl is good (not bent), replace the park switch.

A faulty wiper switch can also cause the park function not to operate properly. Looking at Figure 10-10 as a reference, if Wiper 2 is bent or broken, preventing an electrical connection with the contacts, the wipers will not park even with the park switch in the PARK position as shown.

A quick check is to see if there is voltage at Circuit L when the switch position is moved from the LOW to the OFF position. A properly operating switch has voltage present for a few seconds after the switch is in the OFF position, while a suspect switch has no voltage at this circuit when the wiper switch is turned off.

Another method of checking the park switch is to use a test light and probe for voltage at Circuit P when the wiper switch is turned off. Probing for voltage at this circuit should produce a pulsating light when the motor is running.

At times, the complaint is that the wiper blades continue to operate even though the wiper switch is in the OFF position. The most probable cause of this complaint is a set of welded contacts in the park switch (Figure 10-11). As long as the park switch does not open, current will

Special Tools

Test light

Voltmeter

Classroom Manual
Chapter 10, page 248.

Park contacts are located inside the motor assembly. They supply current to the motor after the wiper control switch has been turned to the PARK position. This allows the motor to continue operating until the wipers have reached their park position.

Most shops replace the wiper motor assembly rather than try to repair or rebuild a motor if there is an internal problem.

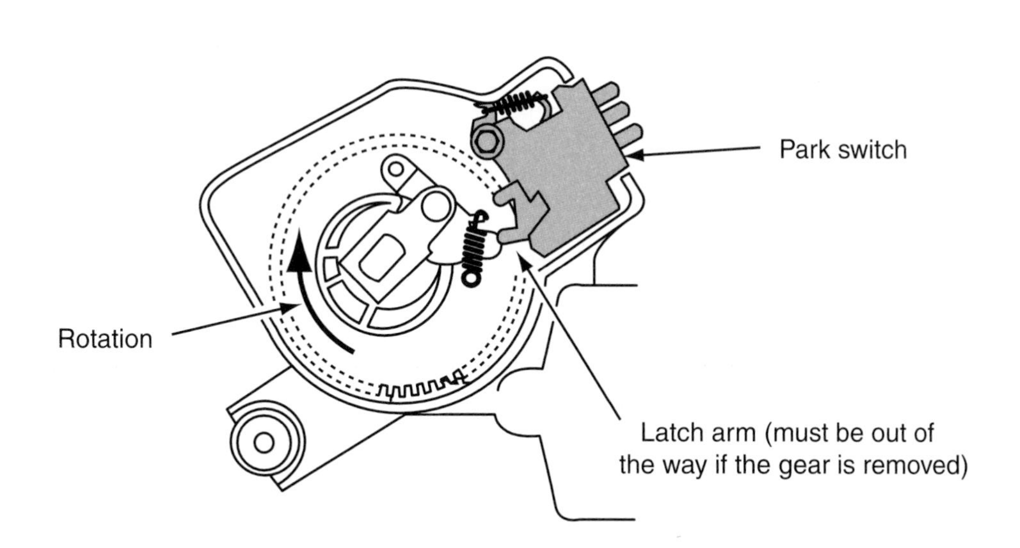

Figure 10-9 Checking the operation of the park switch while the motor is operating.

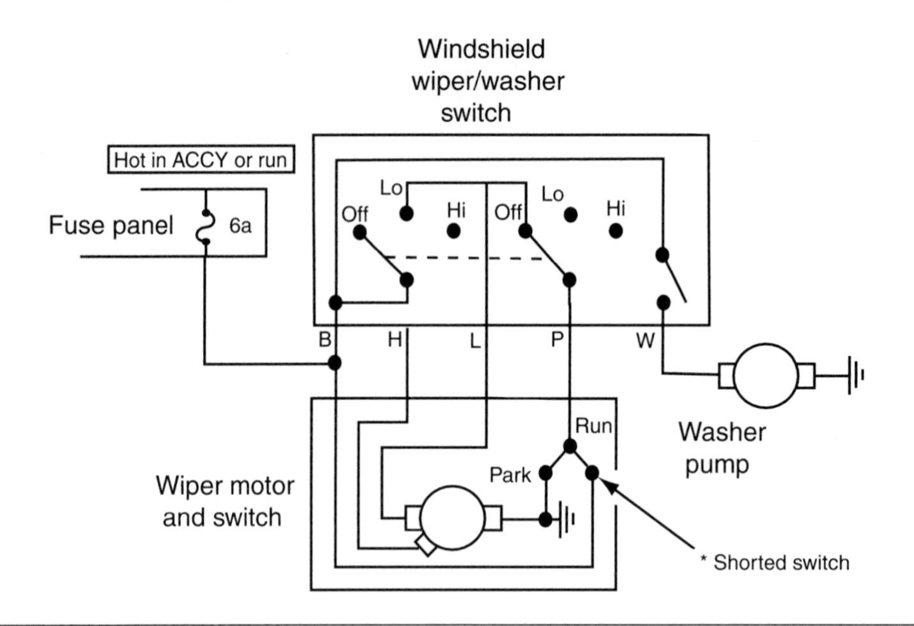

Figure 10-10 A faulty wiper in the switch can prevent the park feature from operating.

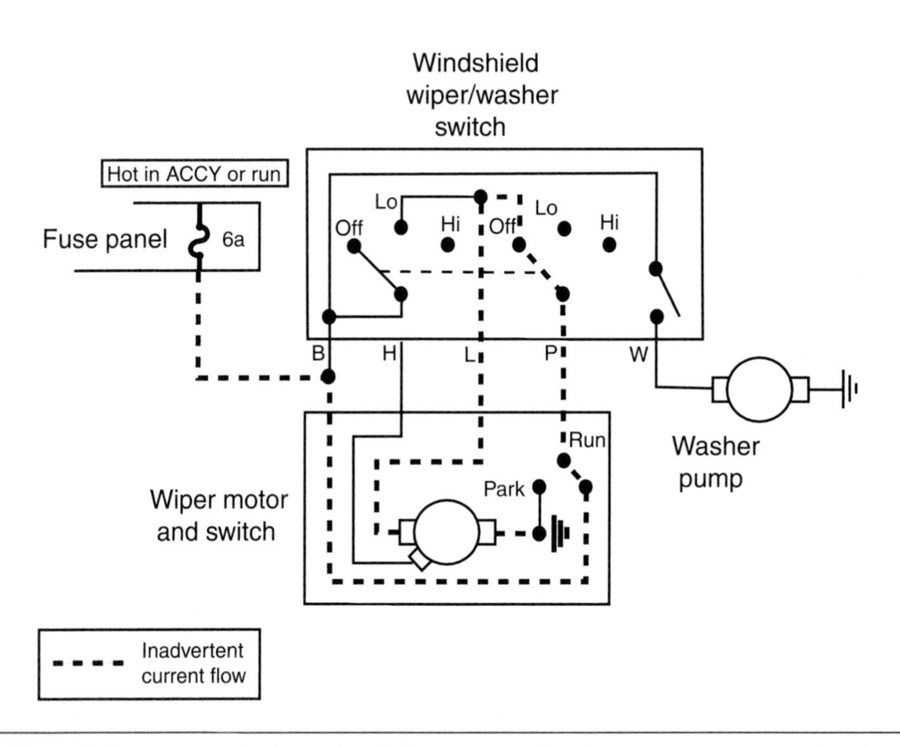

Figure 10-11 Sticking contacts in the park switch can cause the wipers to operate even after the switch is turned off.

flow to the wiper motor. The only way to turn the wipers off is to turn the ignition off or remove the wires to the motor.

The reason we have spent time walking through troubleshooting sequences is that many times the only help you will have from service manuals is a wiring diagram (Figure 10-12). This figure will be used as a reference for a set of flowcharts (Figures 10-13, 10-14, 10-15, 10-16).

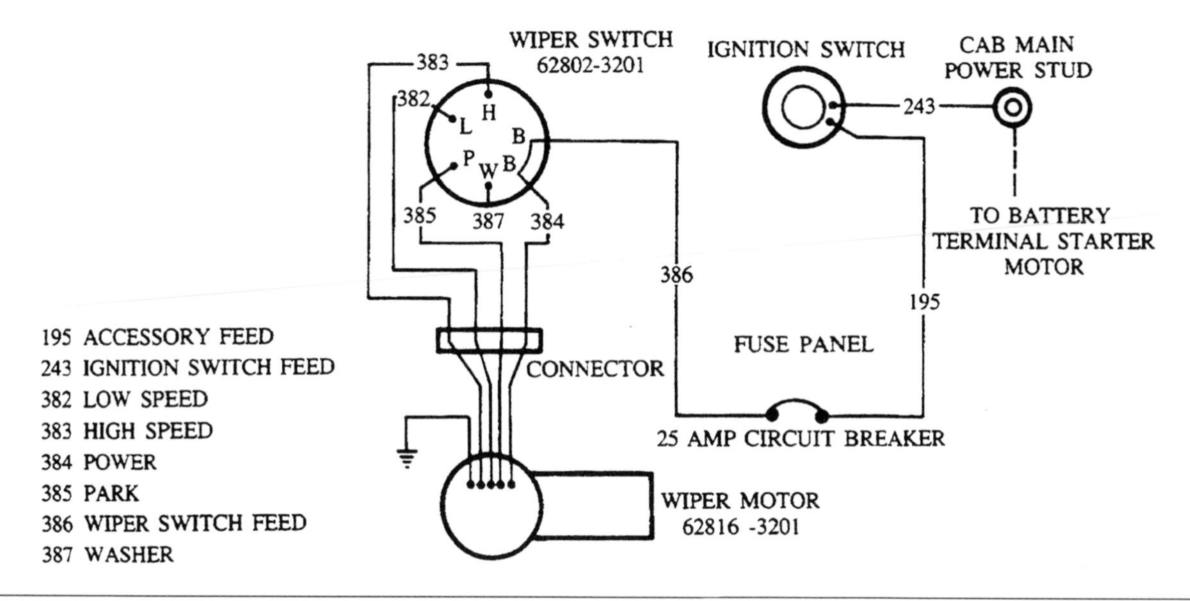

Figure 10-12 Electrical diagram of a windshield wiper system. (Reprinted with permission. Courtesy of Volvo Trucks North America, Inc.)

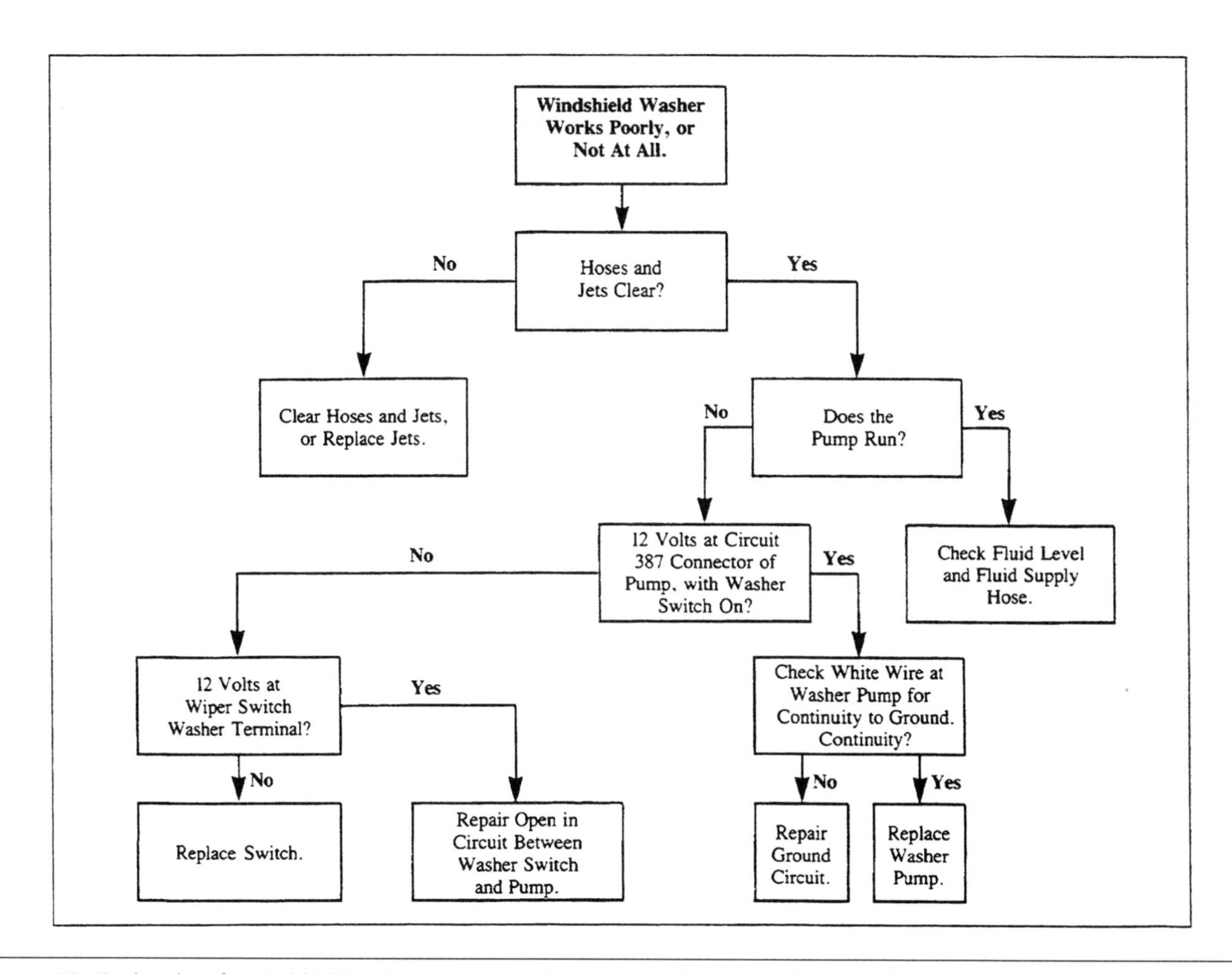

Figure 10-13 Flowchart for windshield washer working poorly or not at all. (Reprinted with permission. Courtesy of Volvo Trucks North America, Inc.)

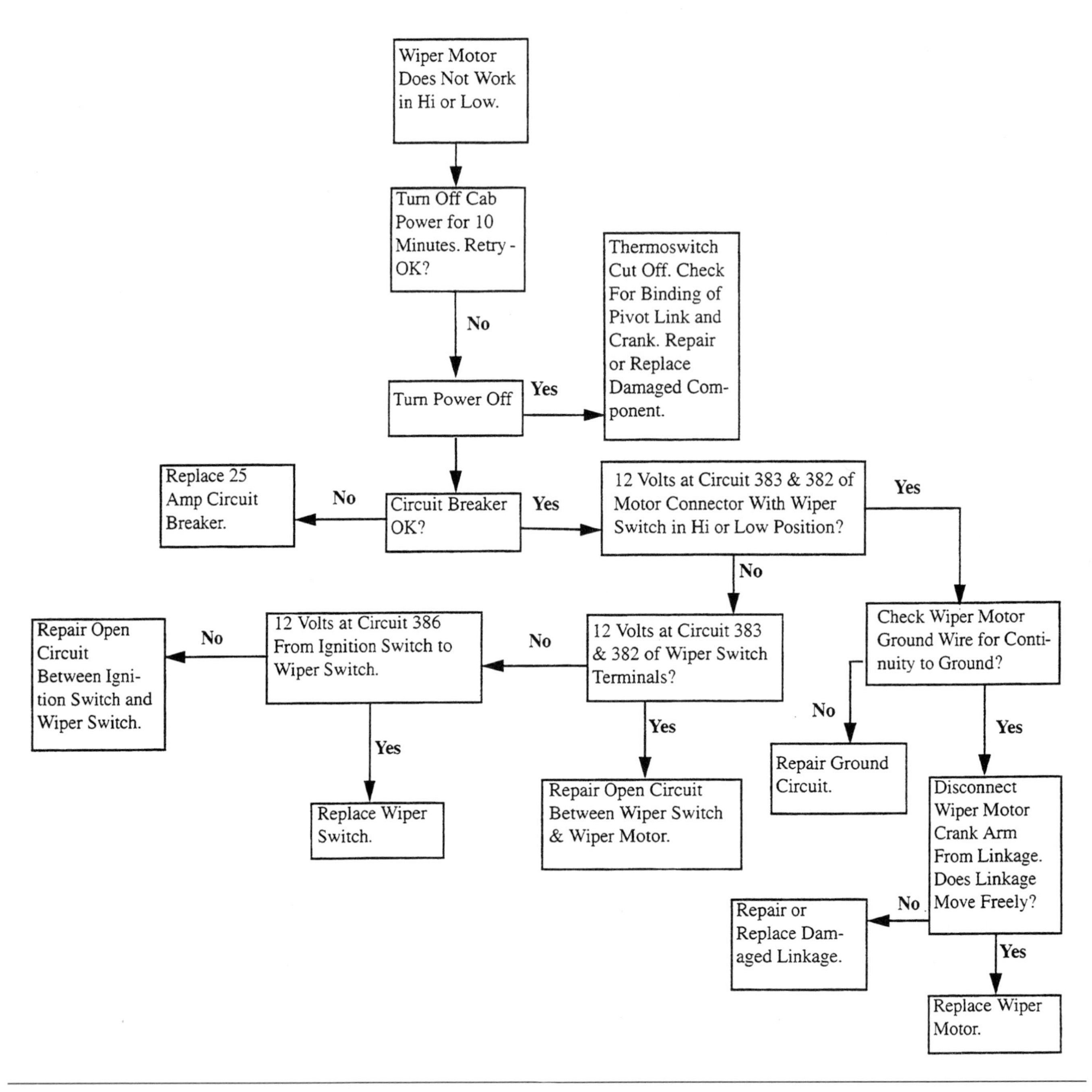

Figure 10-14 Flowchart for wiper motor not working in HI or LOW. (Reprinted with permission. Courtesy of Volvo Trucks North America, Inc.)

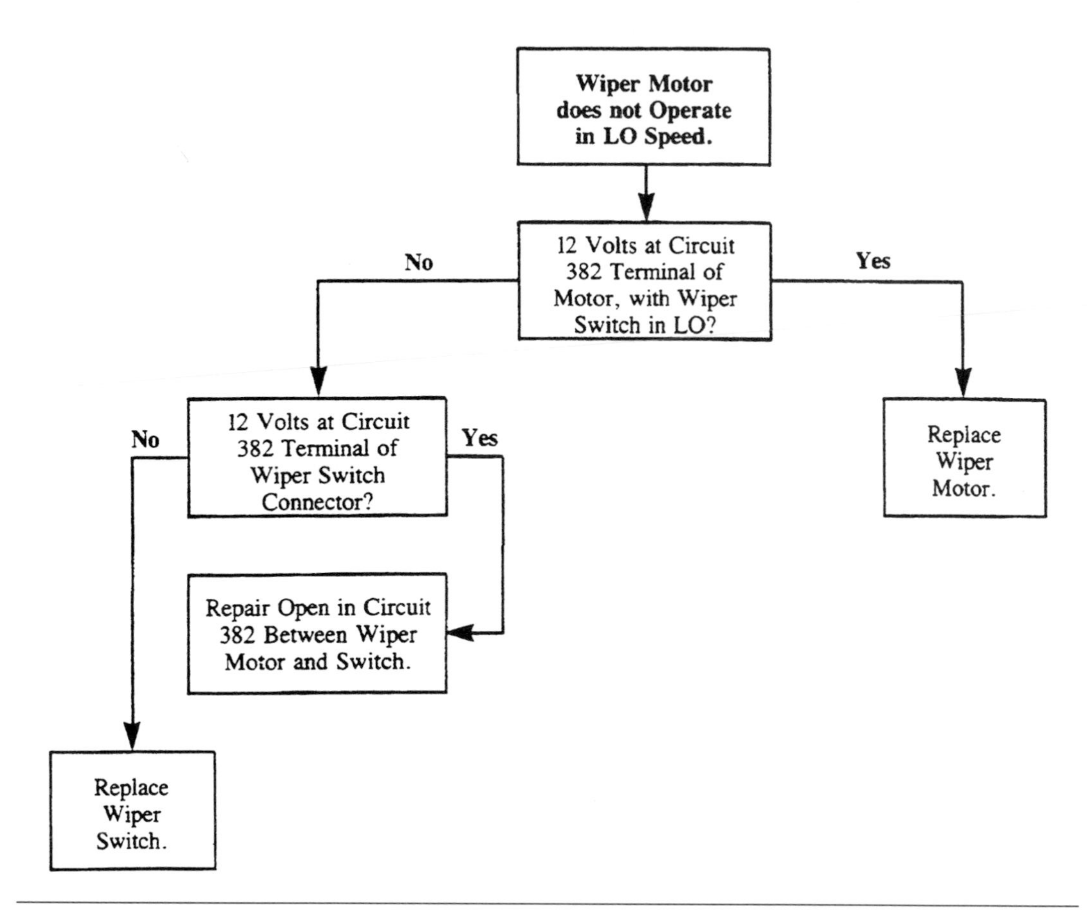

Figure 10-15 Flowchart for wiper motor not operating in LO speed. (Reprinted with permission. Courtesy of Volvo Trucks North America, Inc.)

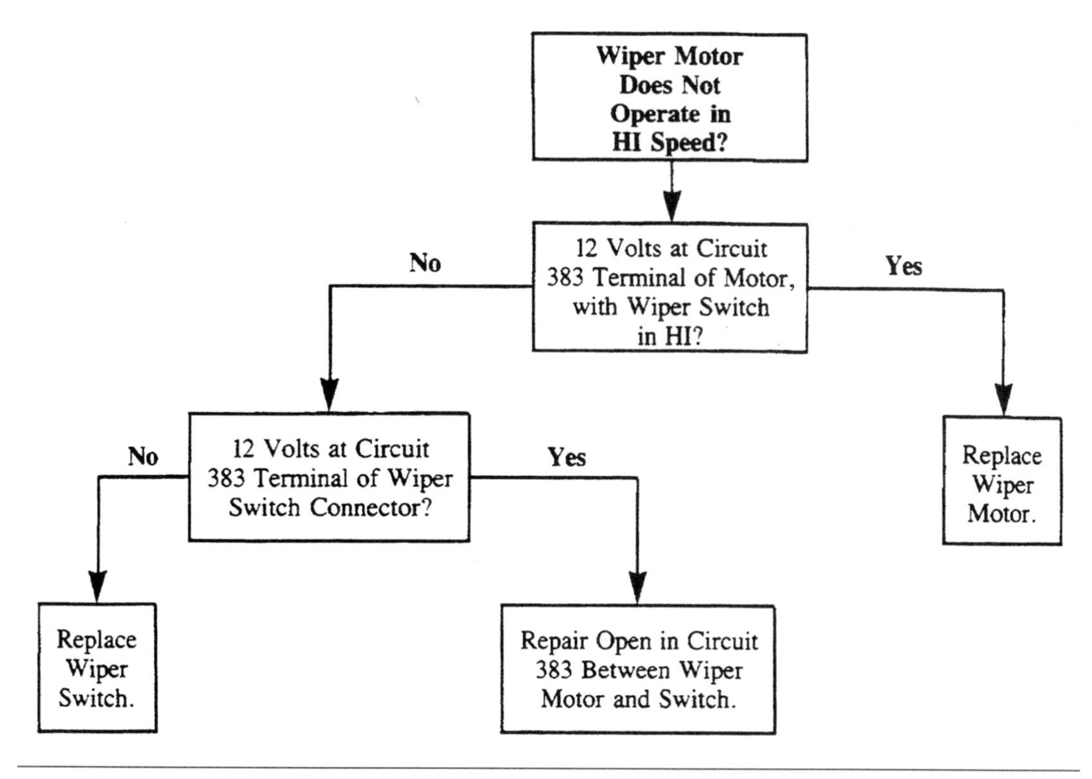

Figure 10-16 Flowchart for wiper motor not operating in HI speed. (Reprinted with permission. Courtesy of Volvo Trucks North America, Inc.)

Special Tool

Torque wrench

Wiper Motor Removal and Installation

The removal and installation process varies among manufacturers. We will look at one manufacturer's procedure so that you will know how to index a wiper motor crank if it is required. These particular procedures are performed on Volvo trucks with Bosch Wiper Systems.

Removal

CAUTION: When working on wiper motors or linkages, disconnect the battery negative cable to prevent the possibility of personal injury.

1. Disconnect the wiper motor wiring connectors.
2. Remove the four bolts that secure the wiper motor to the bulkhead.
3. Move the wiper motor far enough away from the bulkhead to access the crank retaining bolt.
4. Loosen the crank retaining bolt and slide the crank off the motor shaft (Figure 10-17).

WARNING: Do not drop or jar the motor. The internal permanent magnets of the motor are constructed of ceramic material and can be easily damaged.

Installation

1. Install the mounting plate on the motor assembly, if previously removed. Torque the motor to plate bolts to 4 ft. lb. (5 Nm).
2. Temporarily connect the wiper motor wiring and "park" the wiper motor by turning the wiper switch on and off. Disconnect the wiper motor wiring.
3. Position the wiper motor crank to the motor shaft so the crank is at a four-degree angle as shown in Figure 10-18.
4. Torque the crank pinch bolt to 6 ft. lb. (8 Nm).
5. Secure the wiper motor mounting plate to the vehicle bulkhead. Tighten bolts to 17 ft. lb. (23 Nm).
6. Connect the wiper motor wiring connectors.
7. Check for correct wiper operation. Note: If wiper motor park position is determined to be incorrect, index the wiper motor crank using the following procedure.

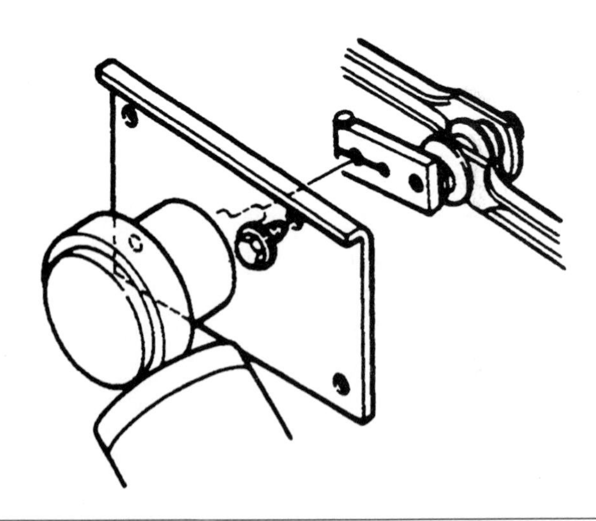

Figure 10-17 Removing crank motor. (Reprinted with permission. Courtesy of Volvo Trucks North America, Inc.)

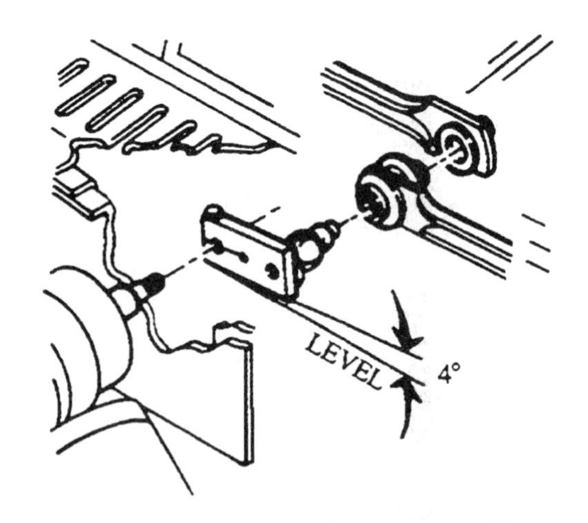

Figure 10-18 Positioning motor crank. (Reprinted with permission. Courtesy of Volvo Trucks North America, Inc.)

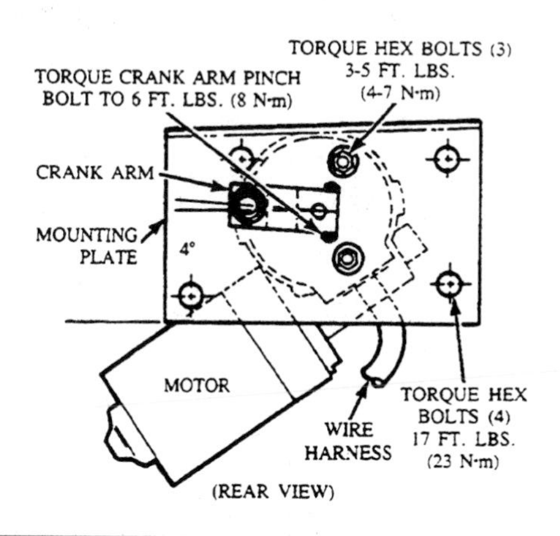

Figure 10-19 Crank mounting position. (Reprinted with permission. Courtesy of Volvo Trucks North America, Inc.)

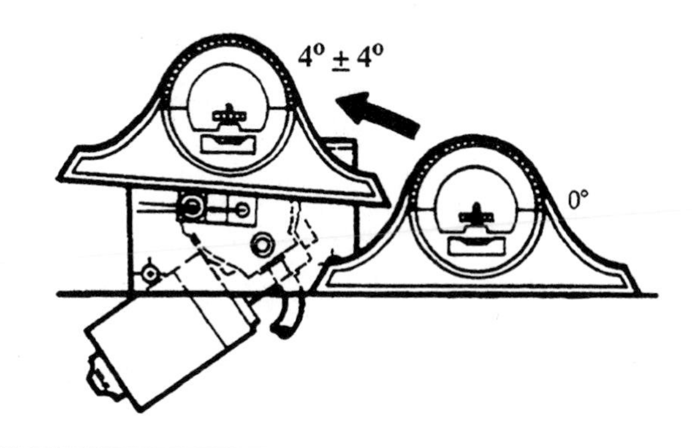

Figure 10-20 Checking crank angle with a bubble protractor. (Reprinted with permission. Courtesy of Volvo Trucks North America, Inc.)

Wiper Motor Crank Indexing

Disconnect the battery ground cable.

1. Remove the four wiper motor mounting plate bolts and pull the motor away from the bulkhead to gain access to the linkage.
2. Mark the end of the wiper motor shaft and crank for alignment.
3. Loosen the crank pinch bolt and slide the crank arm off the motor shaft.
4. Disconnect the wiper motor wiring connectors and remove the motor assembly from the vehicle.
5. Remove the crank arm from the wiper linkage.
6. Install the crank arm back to the wiper motor, aligning the marks made during removal.
7. Check the crank arm position to see if it is at a 4-degree (+ or − 4-degrees) angle to the bottom edge of the mounting plate as shown in Figure 10-19.
8. Using a **bubble protractor,** check the crank angle as follows:
   - **A.** Position the bottom edge of the mounting plate on a flat surface.
   - **B.** Place the bubble protractor on the flat surface and zero the protractor bubble.
   - **C.** Adjust the protractor to four degrees and place it on the arm of the wiper motor crank (Figure 10-20).
   - **D.** Adjust the crank arm to level the bubble at four degrees and torque the pinch bolt to 6 ft. lb. (8 Nm).
   - **E.** Repeat the procedure if necessary.

Special Tools

Bubble protractor

Torque wrench

A **bubble protractor** is a combination protractor/bubble leveling device used to accurately check any range of angle.

Wiper Arm Removal

With some systems, the wiper arms can be awkward to remove because of the tension of the arm against the window. Also, many arms use a serrated coupler that is very tight. The following sequence is used for most windshield wiper arm removal:

WARNING: Do not allow tools to slip when working around the windshield area. It can be expensive to replace a windshield.

1. Lift the arm off the window glass and snap open the retaining nut cover (if applicable).
2. While holding the wiper arm, remove the retaining nut and washer.
3. Disconnect the windshield washer hose from the cowl plate (if applicable).
4. Pull the wiper arm off the pivot shaft. If the wiper arm is stuck on the pivot shaft, use pliers (Figure 10-21).

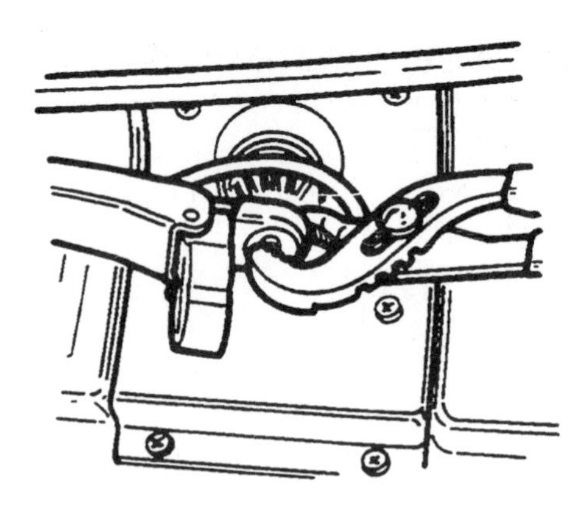

Figure 10-21 Using pliers to pull off the pivot shaft. (Reprinted with permission. Courtesy of Volvo Trucks North America, Inc.)

Linkage Removal and Installation

Some linkages have a clip with a locking tab (Figure 10-22). Simply lift the locking tab and pull the clip away from the pin. A flat-bladed screwdriver can be used to pry the linkage away from the pivot. Be careful not to damage the rubber seal and/or bushing that is used to keep the linkages together.

A pair of pliers can be used to snap the linkages together when reinstalling (Figure 10-23).

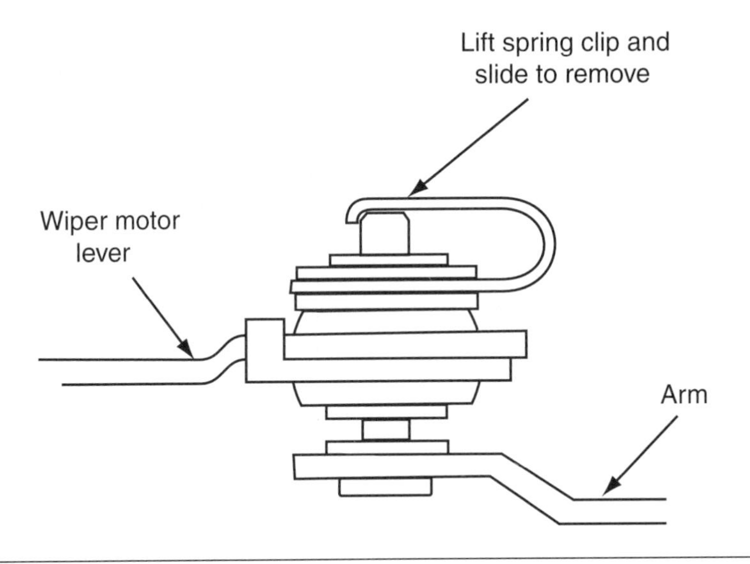

Figure 10-22 To remove the clip, lift up the locking tab and pull the clip.

Figure 10-23 Installing linkage to arm. (Reprinted with permission. Courtesy of Volvo Trucks North America, Inc.)

Windshield Washer System Service

Special Tools

Voltmeter

Test light

Ohmmeter

Safety glasses

The most common windshield washer problem is a restricted delivery system. A quick check for this problem is to disconnect the delivery hose from the pump and operate the system. If the pump ejects a stream of fluid, the fault is in the delivery system. If the pump does not eject any fluid, continue testing using the following procedure:

1. Perform a visual check for fluid level, contaminants in the reservoir that can restrict the pump intake, blown fuses, or disconnected wires.
2. Have someone activate the washer switch while you are observing the motor. If the motor operates but does not squirt fluid, check for blockage at the pump. Remove any foreign material. If there is no blockage, then replace the motor. Note: A working washer motor usually emits a whining noise that can be easily heard.
3. If the motor does not operate, use a test light or voltmeter to check for voltage at the washer motor with the switch activated. If there is no voltage, check the ground circuit. If the ground circuit is okay, replace the pump motor.
4. If no voltage was indicated in Step 3, trace the circuit back to the switch. Test the switch for proper operation. Check for power into the switch. If there is power going in but not out of it, replace the switch.

If the circuits and switch check out and the motor needs to be replaced, follow this procedure for the style of pump shown in Figure 10-24.

Disconnect the wire connector and hoses from the pump. If the pump can be easily accessed, remove the reservoir from the vehicle. Use a small screwdriver to pry out the retainer ring.

CAUTION: Take care not to allow the screwdriver to slip and injure you. Also, wear safety glasses to prevent the ring from striking your eyes.

Use a pair of pliers to grip one of the walls so the pump can be pulled out (Figure 10-25).

Before installing the new motor, lubricate the seal to prevent it from drying and sticking to the wall. Align the tab with the slot in the reservoir and install. Use a 12-point, 1″ socket or equivalent to the size of the retainer ring to press the ring into place. Reconnect the hoses and wire connector. Refill the reservoir and check the system operation while checking for leaks.

WARNING: Do not operate the washer pump without fluid. Doing so may prematurely damage the new pump motor.

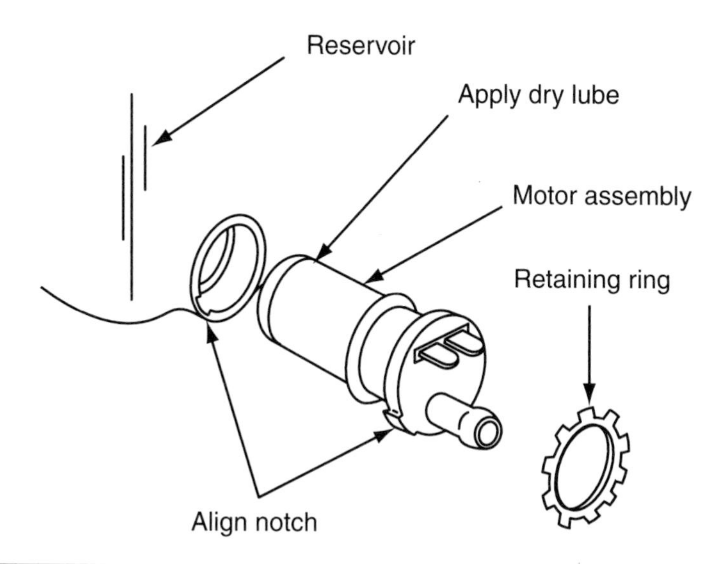

Figure 10-24 Reservoir-mounted washer pump and motor assembly.

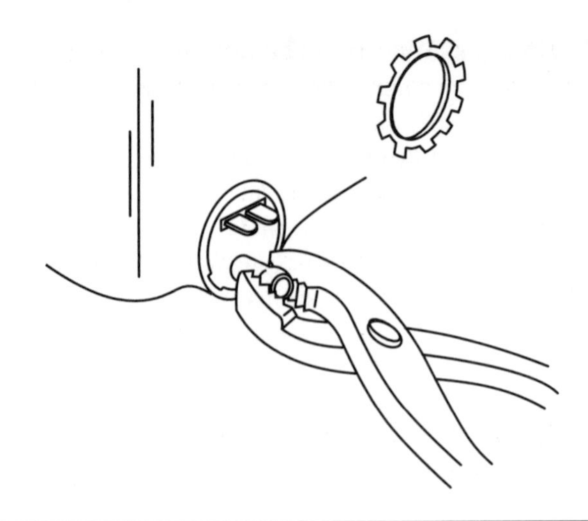

Figure 10-25 Use a pair of pliers to pull the motor out of the reservoir.

Special Tools

Test light

Voltmeter

Jumper wires

Classroom Manual

Chapter 10, page 269.

Power Window Troubleshooting

A good beginning sequence for diagnosing a power window problem is shown in Figure 10-26.

One major complaint is that a window does not operate. Begin the diagnosis by testing the circuit breaker, using a test light or voltmeter to test for voltage on both sides of the circuit breaker. If voltage is present on both sides, the circuit breaker is good. Voltage in with no voltage out indicates a bad circuit breaker. Of course, if there is no voltage into the circuit breaker, an open in the feed from the battery needs to be located.

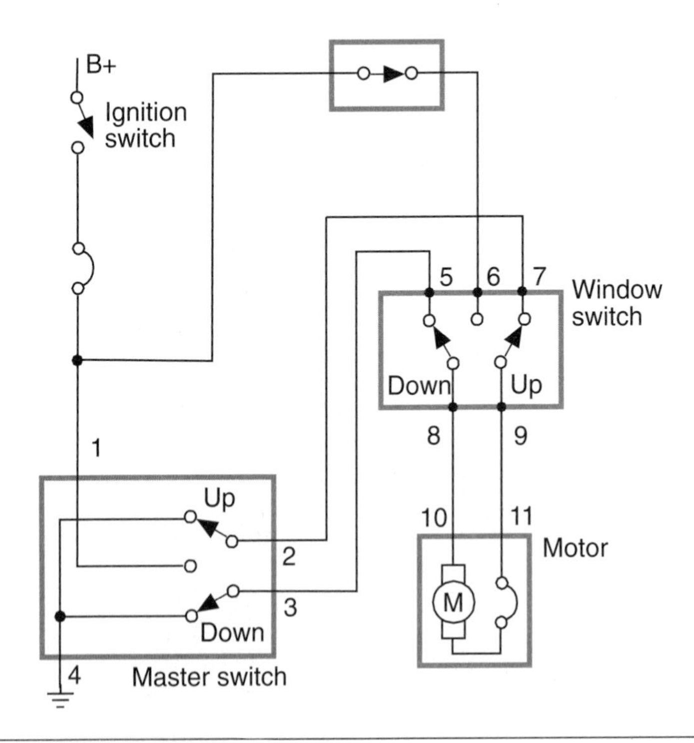

Figure 10-26 Simplified power window circuit.

SERVICE TIP: The following steps can be performed either on the motor or switches, depending on which component is easiest to access.

If the circuit breaker is good, use jumper wires to test the motor. The motor is a reversible type of motor, therefore connections to the motor terminals are not polarity sensitive. Disconnect the wire connectors to the motor. Connect battery positive to one of the terminals and ground the other. If the motor does not operate, reverse the jumper wire connections. The motor should operate in the reverse direction when the polarity is reversed. If the motor does not operate in one or both directions, it is defective and needs to be replaced.

CAUTION: When testing power window circuits and components, keep your hands out of the window operating area to prevent personal injury.

Circuits used in this manual tend to be typical of a system. However, for accuracy always use the service manual for the vehicle you are working on to get the correct wiring schematic.

If the motor operates when the switches are bypassed, the problem is in the control circuit. There are numerous methods for testing switches and control circuits. One method is to use a test light or voltmeter as shown in Figure 10-27.

In this example, the test light is connected across Terminals 5 and 6. The light should remain on until the switch is placed into the UP position. The DOWN position is tested by connecting the test light across Terminals 6 and 7. If you follow the circuit, you can easily see the reason why the test light is placed the way it is. The alligator clip is on the circuits' ground, which leads to a good point. Prior to diagnosing any system, determine which circuits are power, which circuits are ground, and what their relationship is to the operation of the component or system.

With these beginning sequences in mind, you can look at a manufacturer's diagram (Figure 10-28) as a reference. Now use their flowcharts in Figures 10-29 and 10-30 for diagnosing.

Any component used in trucks that involves movement or rotation usually has to overcome some sort of resistance. Any excess mechanical resistance has to be corrected prior to tackling electrical problems.

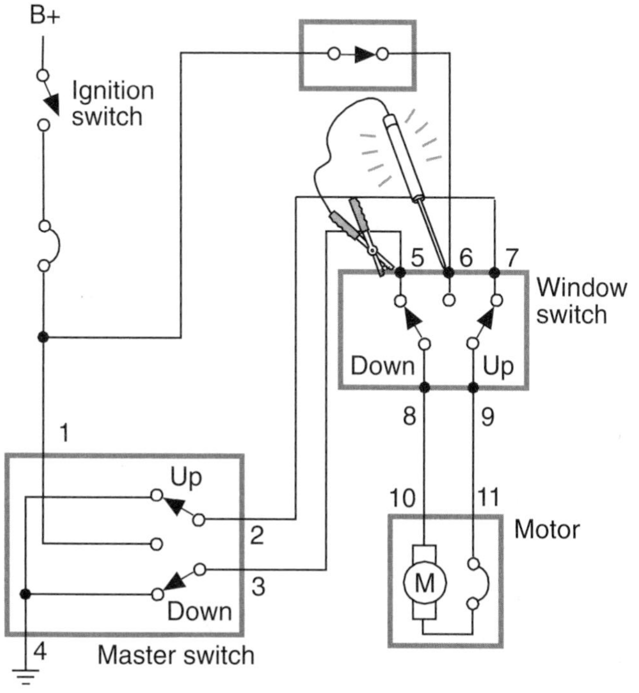

Figure 10-27 Test light connections for testing a window switch.

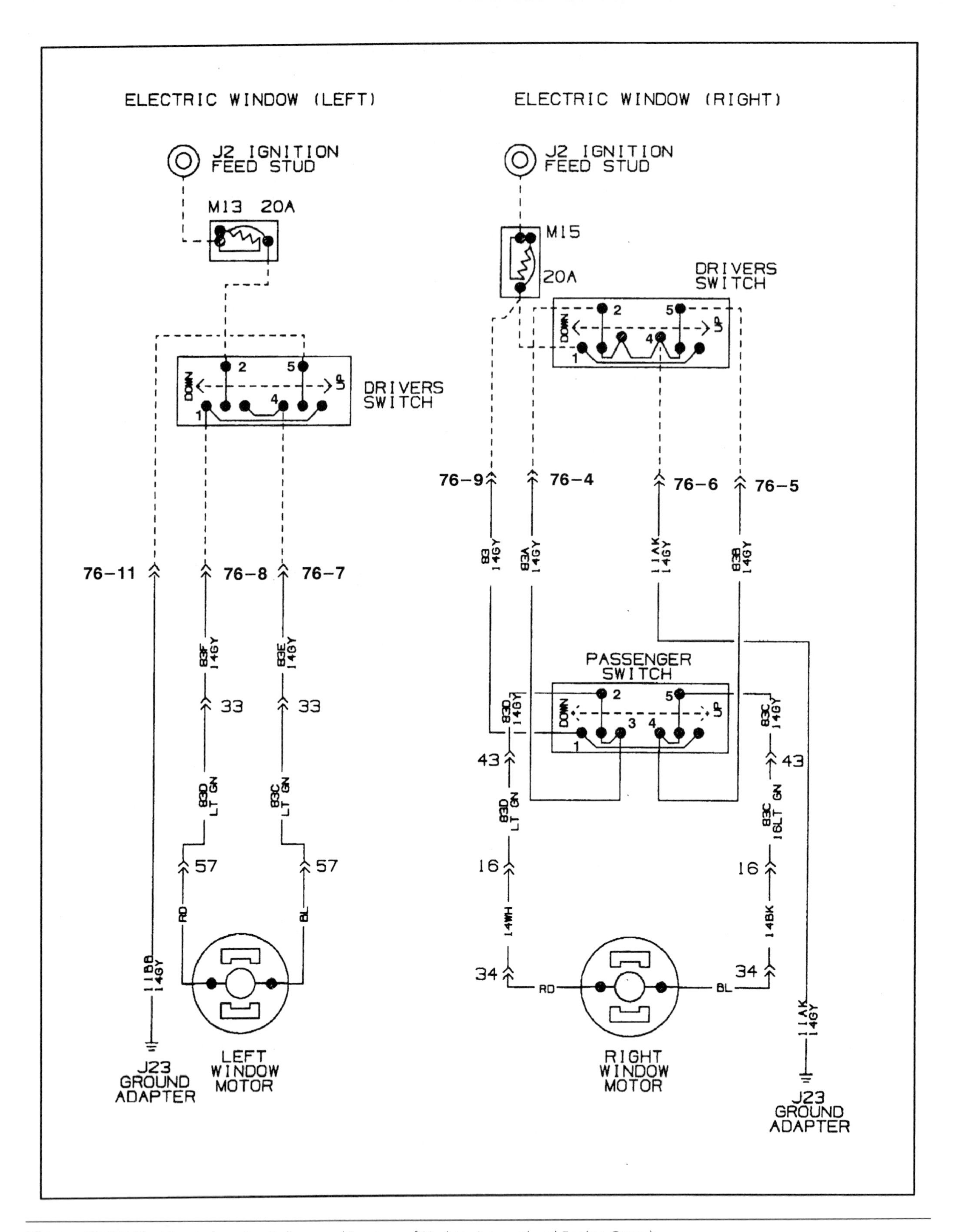

Figure 10-28 Electric window circuit diagram. (Courtesy of Navistar International Engine Group)

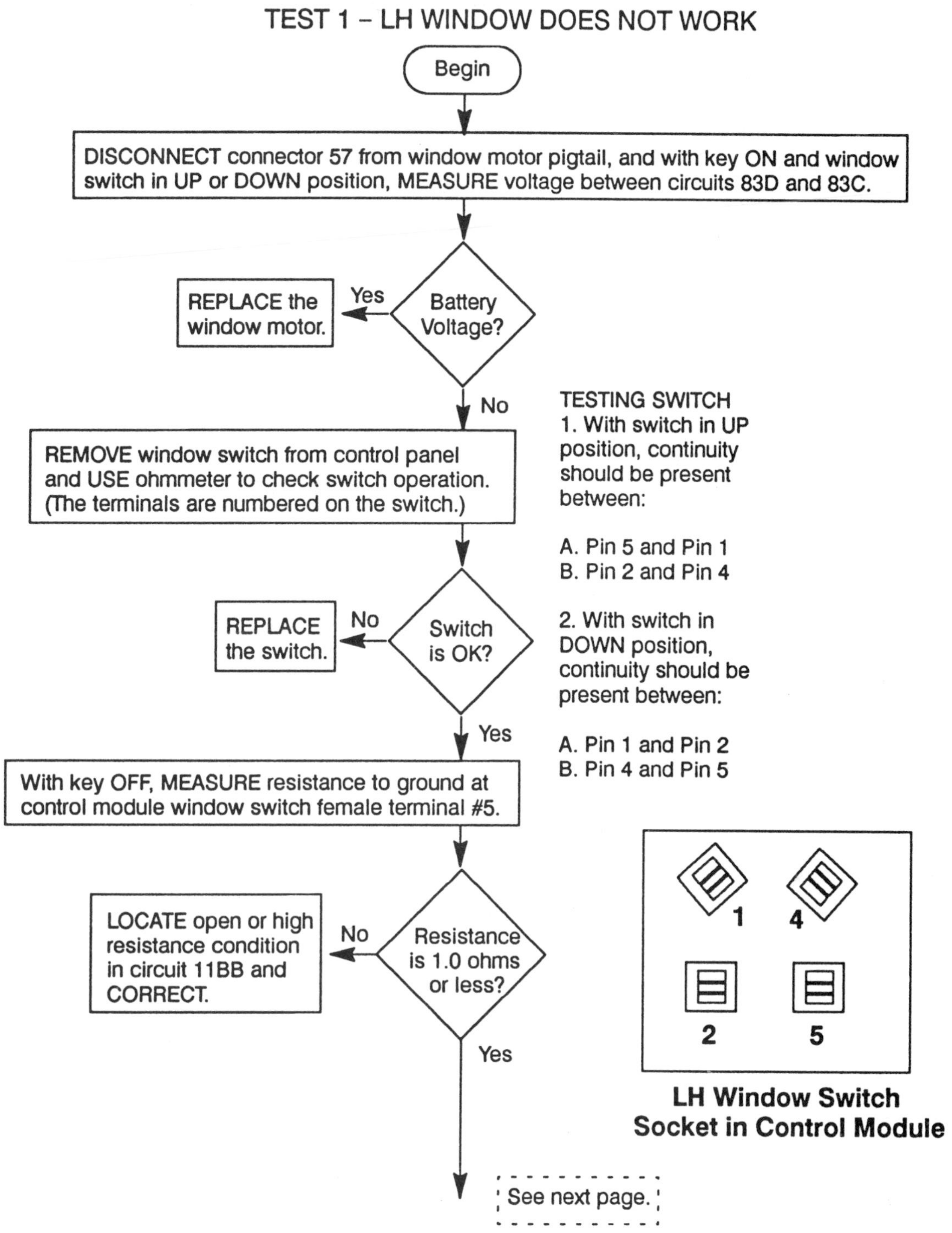

TEST 1 – LH WINDOW DOES NOT WORK

Begin

DISCONNECT connector 57 from window motor pigtail, and with key ON and window switch in UP or DOWN position, MEASURE voltage between circuits 83D and 83C.

Battery Voltage?

Yes → REPLACE the window motor.

No ↓

REMOVE window switch from control panel and USE ohmmeter to check switch operation. (The terminals are numbered on the switch.)

Switch is OK?

No → REPLACE the switch.

Yes ↓

With key OFF, MEASURE resistance to ground at control module window switch female terminal #5.

Resistance is 1.0 ohms or less?

No → LOCATE open or high resistance condition in circuit 11BB and CORRECT.

Yes ↓

See next page.

TESTING SWITCH

1. With switch in UP position, continuity should be present between:

A. Pin 5 and Pin 1
B. Pin 2 and Pin 4

2. With switch in DOWN position, continuity should be present between:

A. Pin 1 and Pin 2
B. Pin 4 and Pin 5

1 4

2 5

LH Window Switch Socket in Control Module

Figure 10-29 Flow chart for LH electric window not operating. (Courtesy of Navistar International Engine Group)

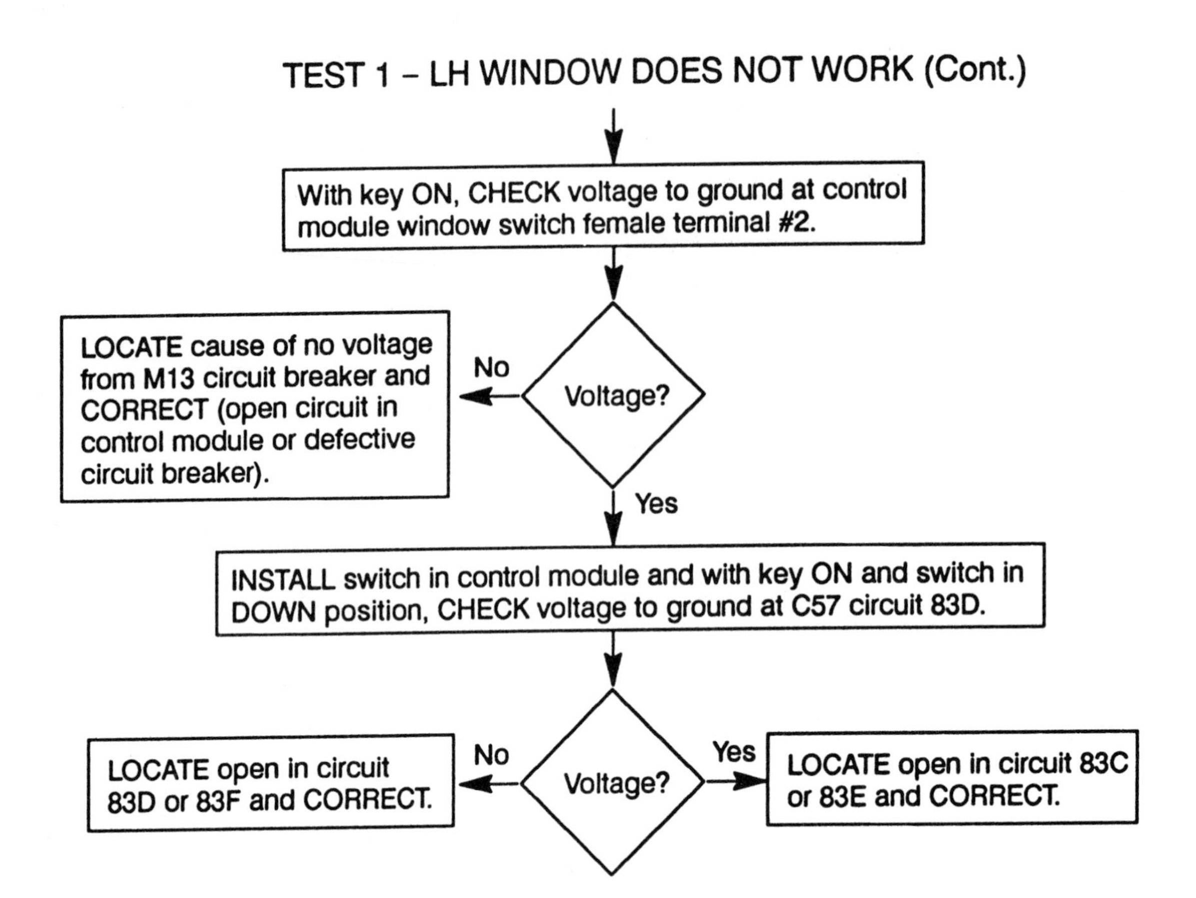

Figure 10-29 Continued.

Notice that in the flowchart in Figure 10-30 there is a referral to another area for testing switches. The referral is to Figure 10-31. This chart shows a method of checking the switches themselves by performing continuity tests. Another useful chart is one that can show expected voltages with various window switch positions (Figure 10-32).

One complaint is that the power windows are slower than normal. Any binding problems should be eliminated first. Electrically, a slow speed is usually an indication of excessive resistance. The cause of it can be determined using the voltage drop test. Excessive resistance can be in the switch circuit, the ground circuit, or in the motor. A good practice is to take the wiring diagram and see if you can properly place the leads on the diagram at the various locations and determine expected readings.

CAUTION: Follow manufacturer's recommended procedures when removing and installing power window motors. The springs used in window regulators can cause serious injury if improperly removed.

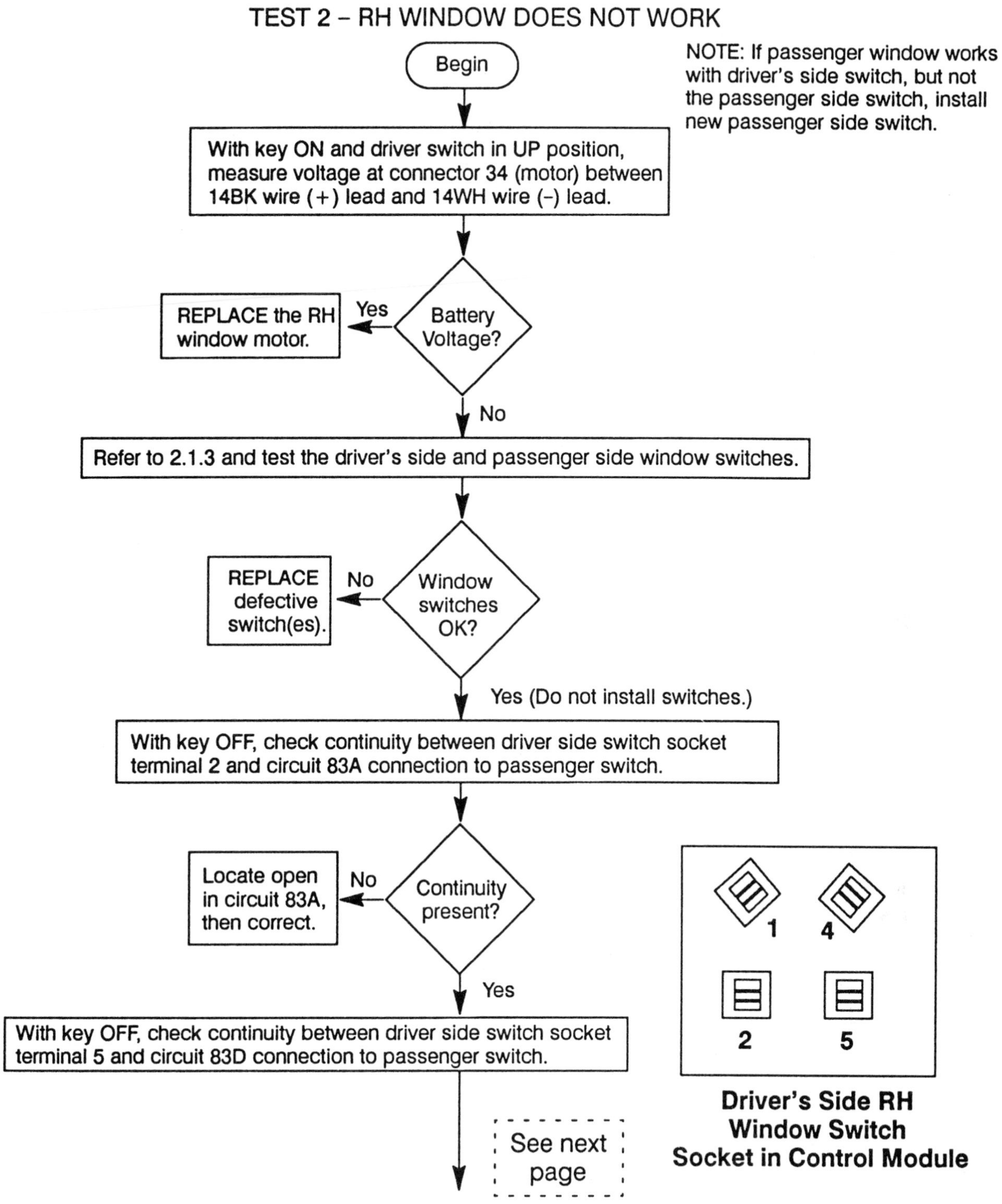

TEST 2 – RH WINDOW DOES NOT WORK

NOTE: If passenger window works with driver's side switch, but not the passenger side switch, install new passenger side switch.

Begin

With key ON and driver switch in UP position, measure voltage at connector 34 (motor) between 14BK wire (+) lead and 14WH wire (-) lead.

Battery Voltage?

Yes → REPLACE the RH window motor.

No ↓

Refer to 2.1.3 and test the driver's side and passenger side window switches.

Window switches OK?

No → REPLACE defective switch(es).

Yes (Do not install switches.) ↓

With key OFF, check continuity between driver side switch socket terminal 2 and circuit 83A connection to passenger switch.

Continuity present?

No → Locate open in circuit 83A, then correct.

Yes ↓

With key OFF, check continuity between driver side switch socket terminal 5 and circuit 83D connection to passenger switch.

See next page

1 4

2 5

Driver's Side RH Window Switch Socket in Control Module

Figure 10-30 Flow chart for RH electric window not operating. (Courtesy of Navistar International Engine Group)

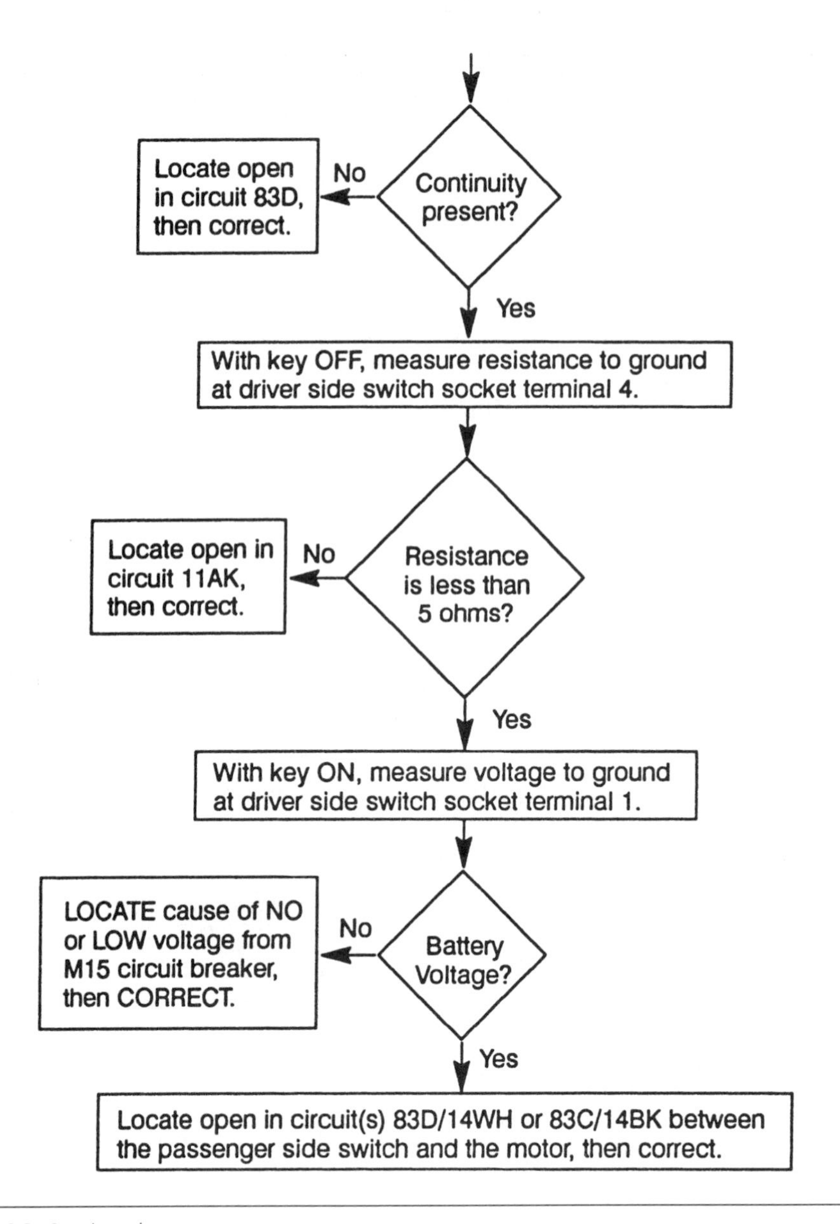

Figure 10-30 Continued.

Testing Passenger Door Window Switch	Testing Driver RH Window Switch
1. Neutral position A. Continuity between terminals 3, 4, 5 B. Continuity between terminals 1 and 6	1. Neutral Position A. Continuity between terminals 2, 3, 4, 5 B. Continuity between terminals 1 and 6
2. Up Position A. Continuity between terminals 6 and 5 B. Continuity between terminals 3 and 4	2. Up Position A. Continuity between terminals 1, 6, 5 B. Continuity between terminals 2, 3, 4
3. Down Position A. Continuity between terminals 1 and 6 B. Continuity between terminals 4 and 5	3. Down Position A. Continuity between terminals 2, 1, 6 B. Continuity between terminals 3, 4, 5

Figure 10-31 Continuity chart for various window switch positions. (Courtesy of Navistar International Engine Group)

Circuit	Both SW Neutral	Driver Side Up	Driver Side Down	Passenger Side Up	Passenger Side Down
83	12V	12V	12V	12V	12V
83A	0	0	12V	0	0
83B	0	12V	0	0	0
83C	0	12V	0	12V	0
83D	0	0	12V	0	12V
RD	0	0	12V	0	12V
BLUE	0	12V	0	12V	0

Circuit	Switch Neutral	Up	Down
11BB	0	0	0
83F/83D/RD	—	0	12V
83E/83C/BL	—	12V	0

Figure 10-32 Voltage chart for various window positions. (Courtesy of Navistar International Engine Group)

Blower Motor Circuit Diagnosis

Classroom Manual
Chapter 10, page 252.

Typical complaints relating to blower motors are: no blower motor operation, blower motor not operating in one or more of the speeds, motor speed seems to be slow, or a noisy blower motor.

The **resistor block** is a series of resistors with different values. There is usually one less resistor than there are fan speed positions because the high-speed circuit bypasses the resistors.

CUSTOMER CARE: You need to really communicate with the customer when it comes to heater problems. The complaint is often "no heat," so you need to ask questions such as: You have no heat but the blower is working? Is there heat available but the blower does not seem to be working? The answers to these questions will keep you from heading in the wrong direction with your diagnosis and will save you time and the customer money.

CUSTOMER CARE: Before attributing a blower problem to an electrical problem, make sure the integrity of the whole heater system is up to par. Often a lack of volume or speed of air movement is attributed to a slow-turning blower motor when in reality the heater core could be plugged externally by dirt and other contaminants. Check to see if the duct is not sealed properly.

A blower motor complaint can be easily diagnosed by carefully studying the vehicle's wiring diagram as shown in Figure 10-33. In this diagram you can see that blower motor speed is controlled by sending current through a resistor and thermo fuse assembly. The higher the resistance value, the slower the fan speed. The control switch position determines the number of resistors that are added to the circuit.

In this circuit, we also see a relay used for high speed control. When the control switch is in the high speed position, power from the "H" terminal energizes the high speed relay control coil. Energizing the relay closes the relay contacts and full voltage from the circuit breaker is applied to the blower motor.

Classroom Manual
Chapter 10, page 261.

If the customer complaint is that the blower only operates in certain speed positions, the most likely cause is an open resistor in the resistor assembly. Also, the circuit from the nonfunctioning speed terminal of the switch to the resistor assembly might be faulty. If the complaint is that all the speeds except HIGH work, the suspect circuit is the high speed relay, the circuits to it, and the circuit out of it. Either way, if the motor operates in at least one of the speeds, the fault is not with the motor. If the motor fails to operate at all, start with the fuse or circuit protection device.

Special Tools

Test light

Voltmeter

Jumper wires

More trucks than ever are now equipped with air conditioning. These vehicles extend their wiring diagrams to include the A/C circuit with the blower circuit (Figure 10-34). Using the

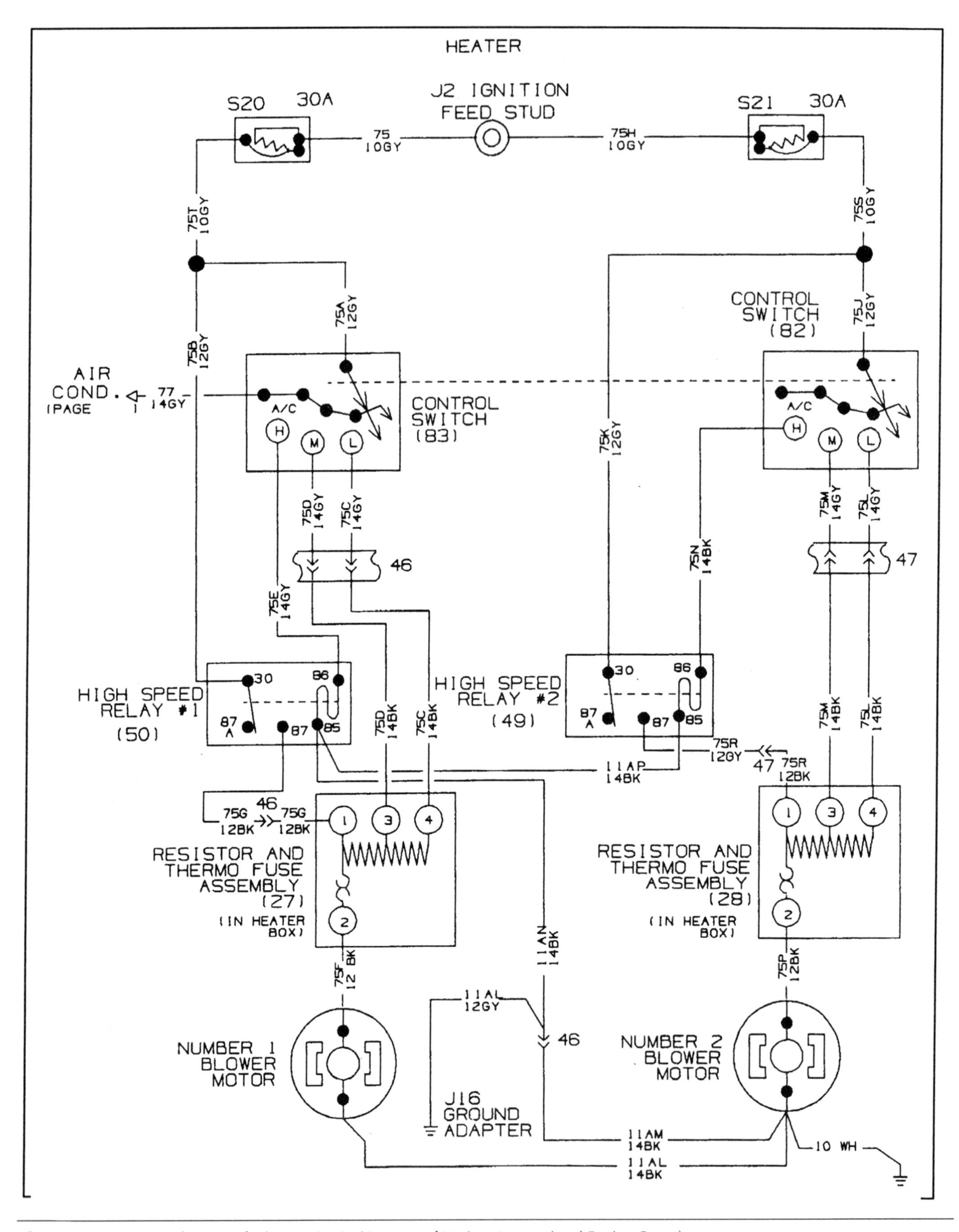

Figure 10-33 Wiring diagram of a heater circuit. (Courtesy of Navistar International Engine Group)

diagrams in Figures 10-33 and 10-34, you can easily diagnose the blower motor circuit system using the troubleshooting flowcharts shown in Figures 10-35 through 10-40.

To help with the diagnosis, look at the chart in Figure 10-41 for expected voltages and amperage draw readings.

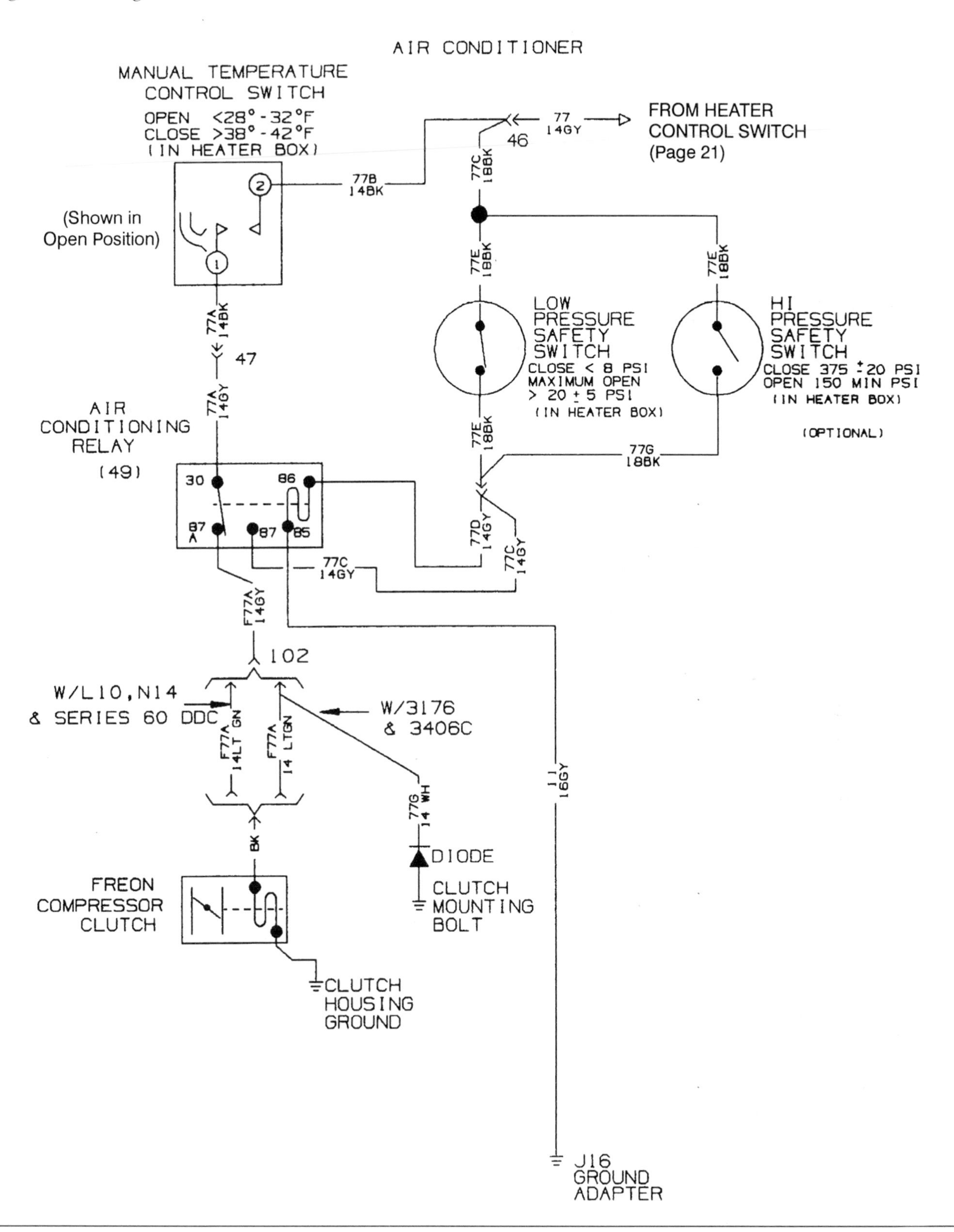

Figure 10-34 Wiring diagram extended from the heater circuit to include the A/C circuit. (Courtesy of Navistar International Engine Group)

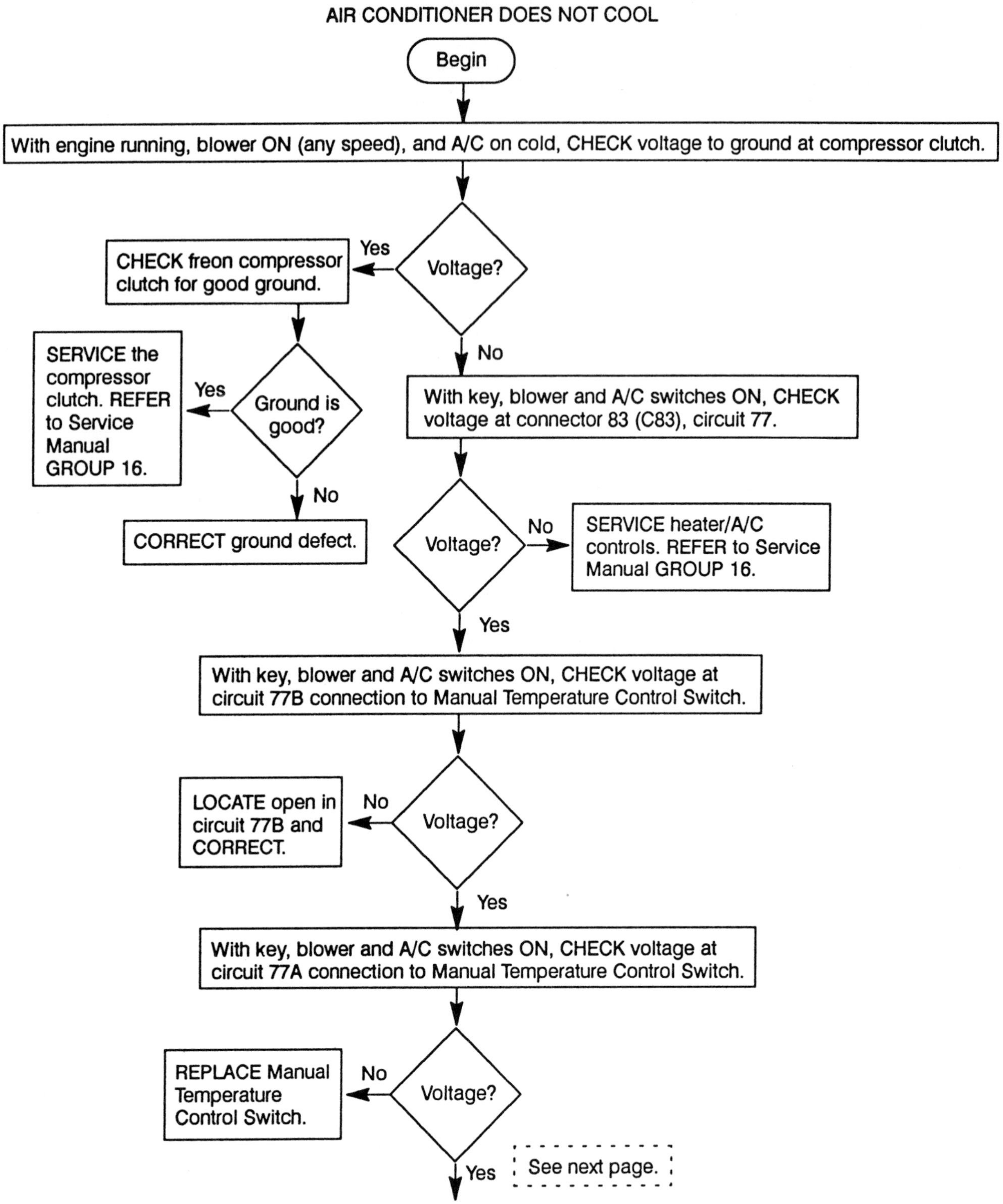

AIR CONDITIONER DOES NOT COOL

Begin

With engine running, blower ON (any speed), and A/C on cold, CHECK voltage to ground at compressor clutch.

Voltage?

Yes → CHECK freon compressor clutch for good ground.

Ground is good?

Yes → SERVICE the compressor clutch. REFER to Service Manual GROUP 16.

No → CORRECT ground defect.

No → With key, blower and A/C switches ON, CHECK voltage at connector 83 (C83), circuit 77.

Voltage?

No → SERVICE heater/A/C controls. REFER to Service Manual GROUP 16.

Yes → With key, blower and A/C switches ON, CHECK voltage at circuit 77B connection to Manual Temperature Control Switch.

Voltage?

No → LOCATE open in circuit 77B and CORRECT.

Yes → With key, blower and A/C switches ON, CHECK voltage at circuit 77A connection to Manual Temperature Control Switch.

Voltage?

No → REPLACE Manual Temperature Control Switch.

Yes ↓

See next page.

Figure 10-35 Flowchart for heater and air conditioner problem. This chart is used for an air conditioner that does not cool. (Courtesy of Navistar International Engine Group)

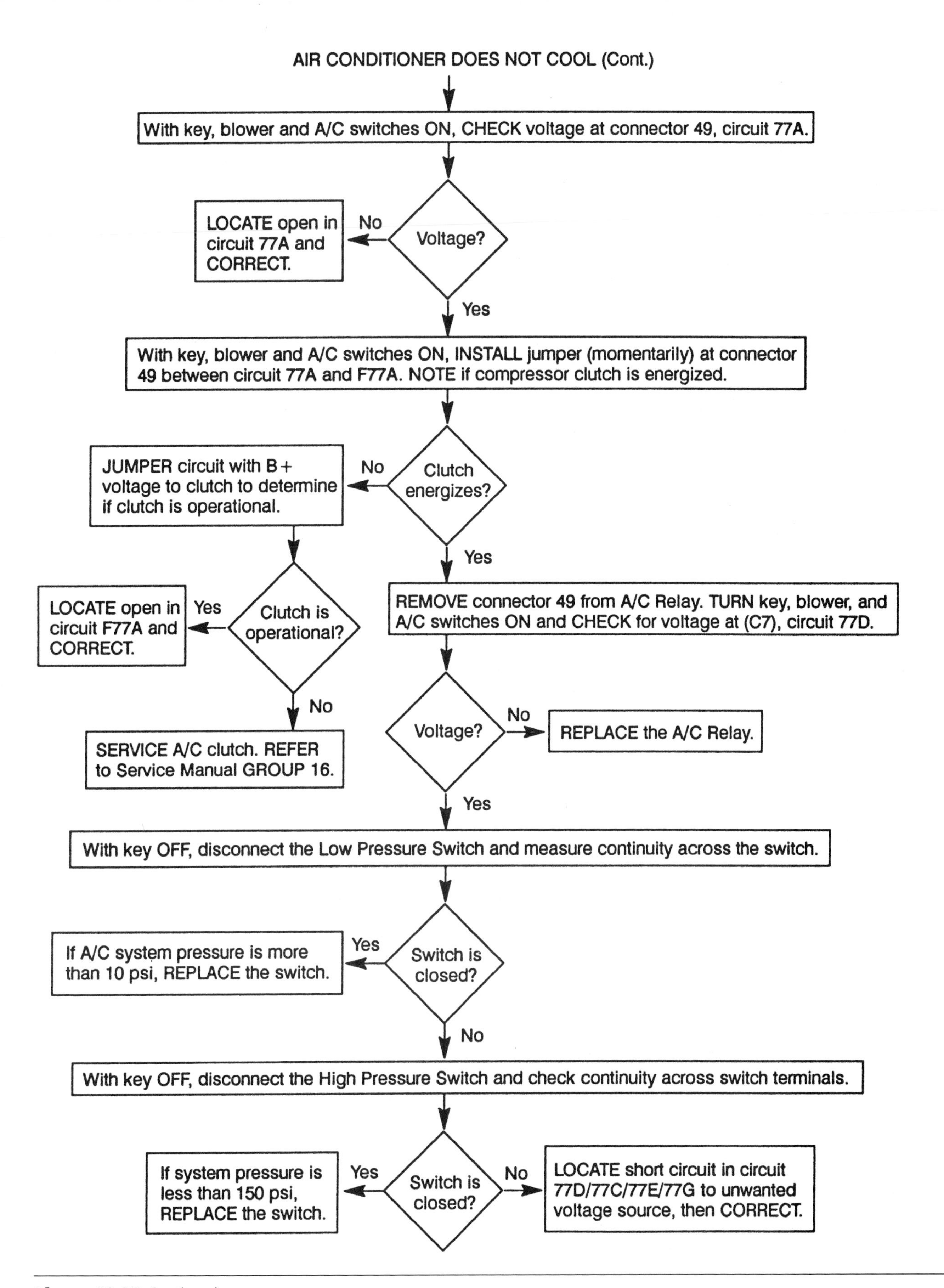

Figure 10-35 Continued.

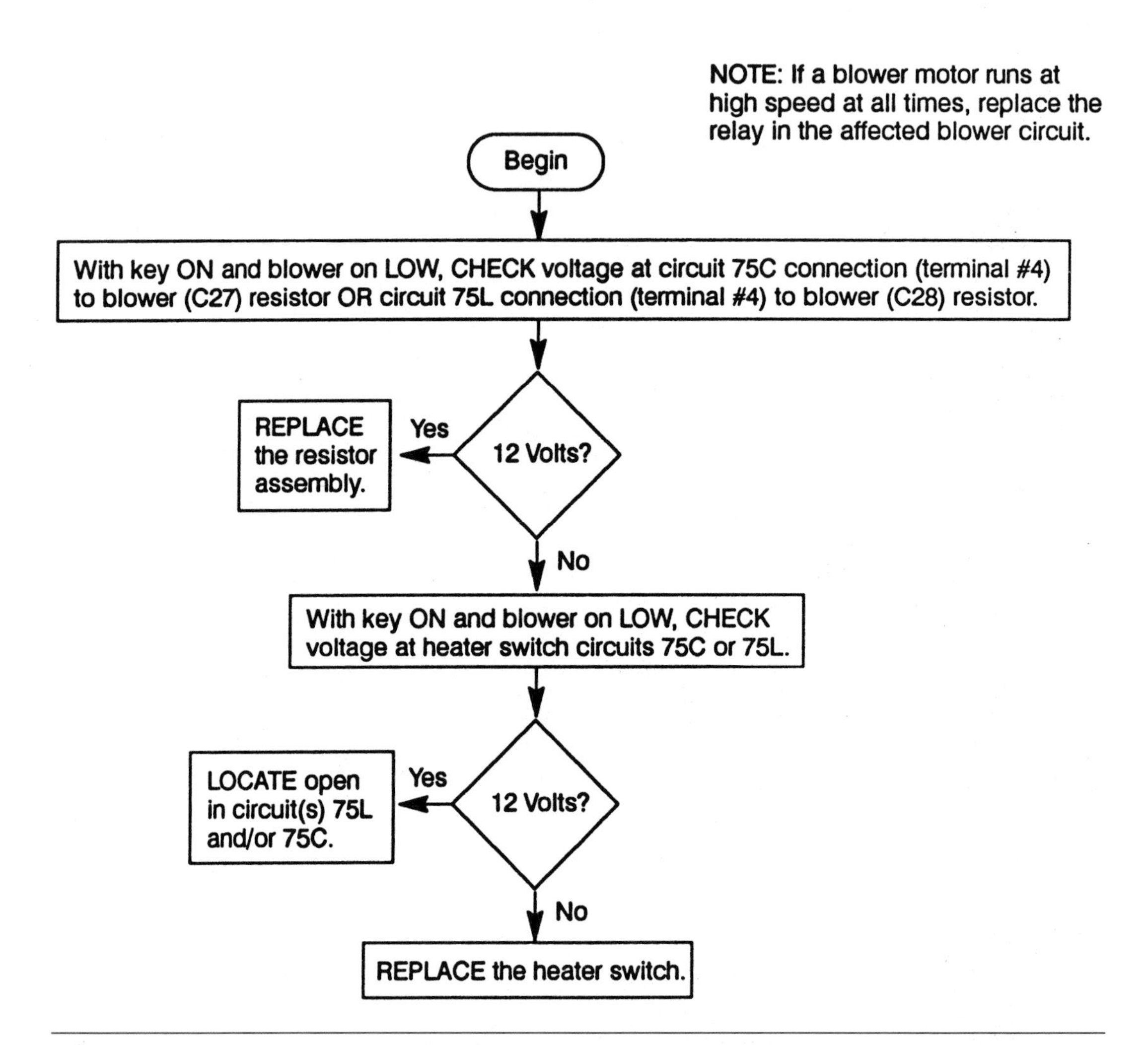

Figure 10-36 Flowchart for blower motor(s) low speed does not work. (Courtesy of Navistar International Engine Group)

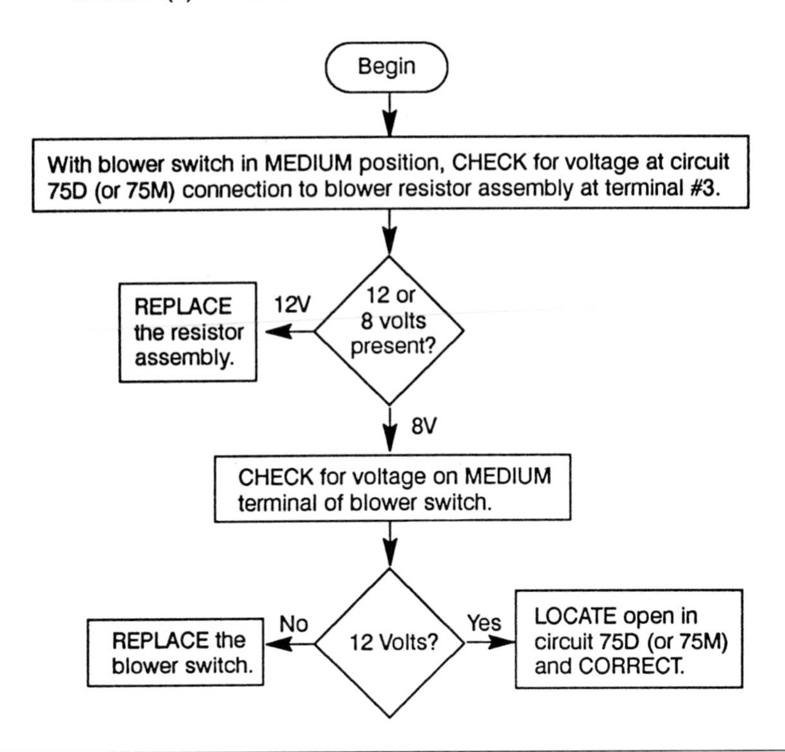

Figure 10-37 Flowchart for improper lower speed operation. (Courtesy of Navistar International Engine Group)

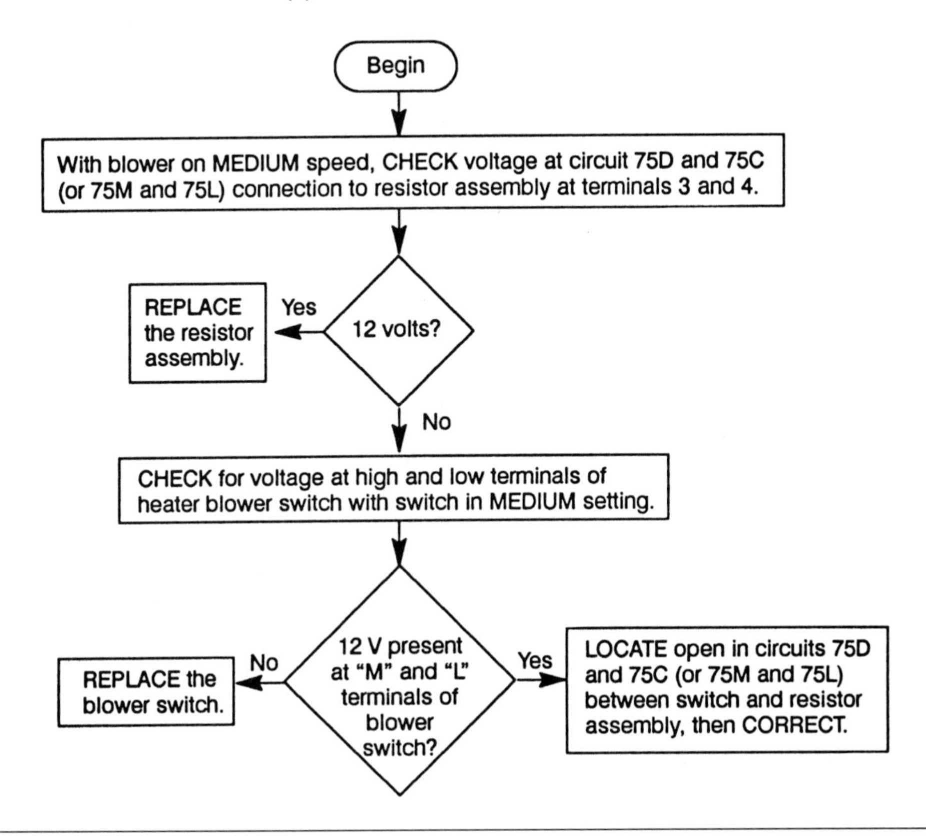

Figure 10-38 Flowchart for blower motor not working in low and medium setting. (Courtesy of Navistar International Engine Group)

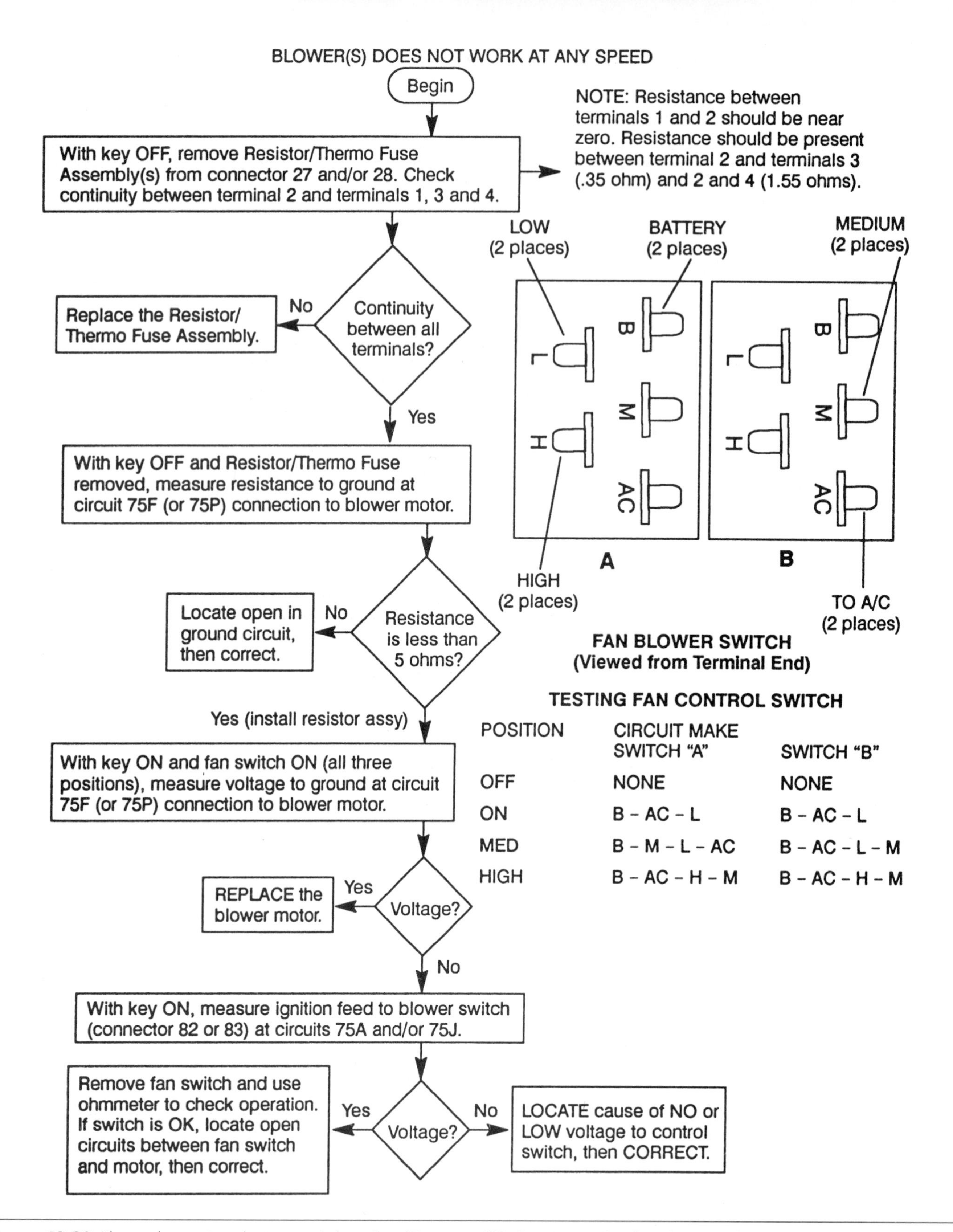

POSITION	CIRCUIT MAKE SWITCH "A"	SWITCH "B"
OFF	NONE	NONE
ON	B – AC – L	B – AC – L
MED	B – M – L – AC	B – AC – L – M
HIGH	B – AC – H – M	B – AC – H – M

Figure 10-39 Blower does not work at any switch setting. (Courtesy of Navistar International Engine Group)

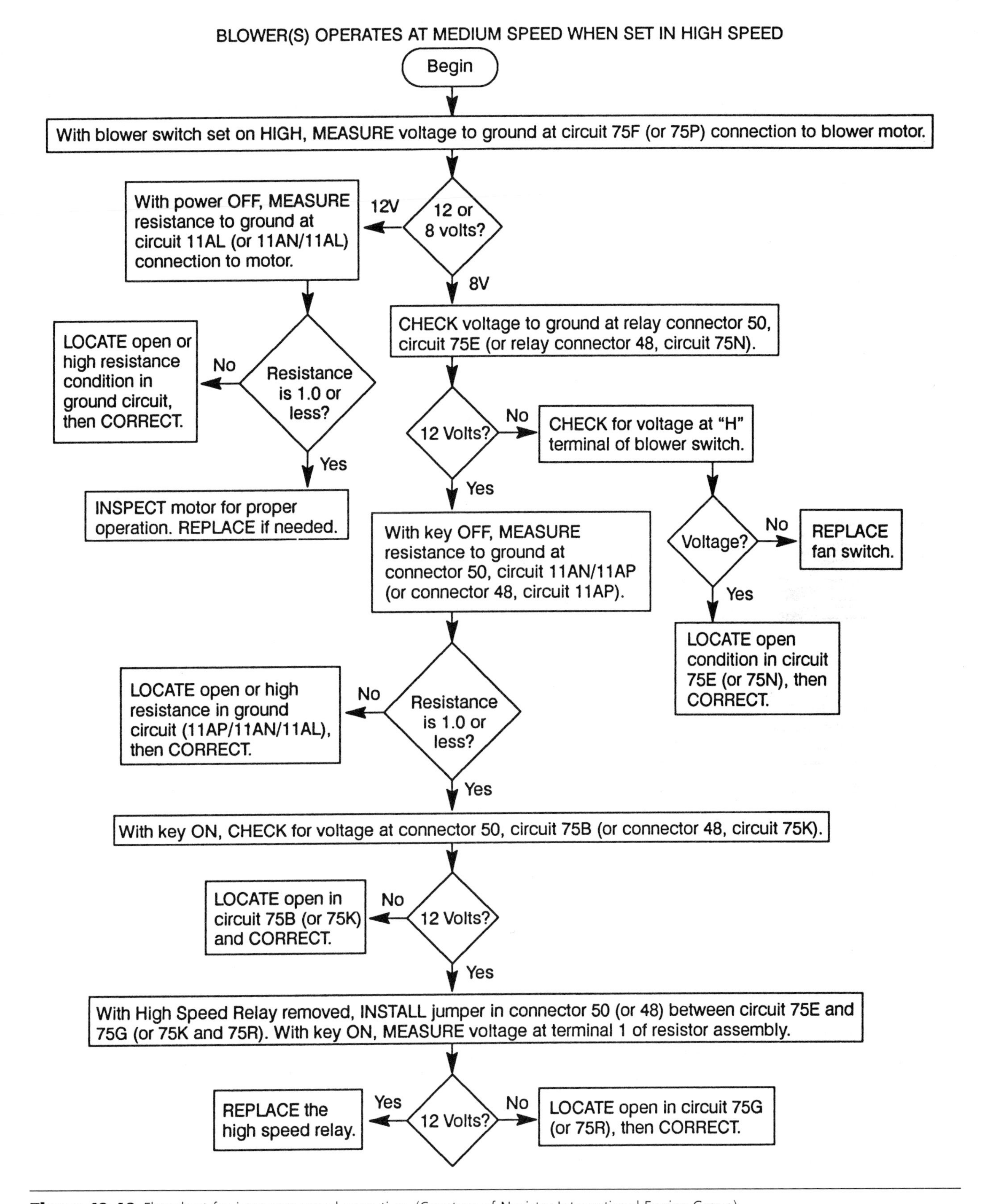

Figure 10-40 Flowchart for improper speed operation. (Courtesy of Navistar International Engine Group)

Fan Speed	Volts	Amperes
Low	3 to 5	3 to 5
Medium	6 to 7	8 to 11
High	11 to 13	12 to 16

Figure 10-41 Blower motor voltage/speed chart. (Courtesy of Navistar International Engine Group)

Special Tools

Jumper wires

Voltmeter

Ohmmeter

Inoperable Motor

Another quick method to check motor operation is to use a jumper wire to bypass a switch and/or resistor block. Connect the jumper wire from a battery positive supply to the motor terminal. If the motor does not operate, connect a second wire from the motor body or ground terminal (whichever is applicable) to a good ground. Replace the motor if it still does not operate.

If the motor operated in the bypass mode, trace the circuit up to the switch. A test light or voltmeter can be used to check for voltage in and out of the blower speed control switch. The switch is faulty if there is voltage at the input terminal but not at any of the output terminals. No voltage at the input terminal indicates an open in the circuit between the battery and the switch.

SERVICE TIP: It is very unlikely that open resistors would cause a no motor operation because the high speed circuit bypasses the resistor block. An open wire from the block to the motor could, however, cause a no motor operation problem. A switch that is suspected can be checked out by performing a continuity test.

Constantly Operating Blower Fans

Ground side switches control a circuit by completing the circuit to ground. Voltage is applied to the power terminal of the switch through the motor. In order for the motor to operate, the switch has to close the ground circuit.

Although rare, a complaint of a constantly operating blower fan does occur. This type of fault is more common with ground controlled switches (Figure 10-42).

It is easy to see in Figure 10-42 that a short to ground at any point on the ground side of the circuit will cause the motor to run. Other possible causes are the switch itself or problems in the circuit between the switch and the resistor block.

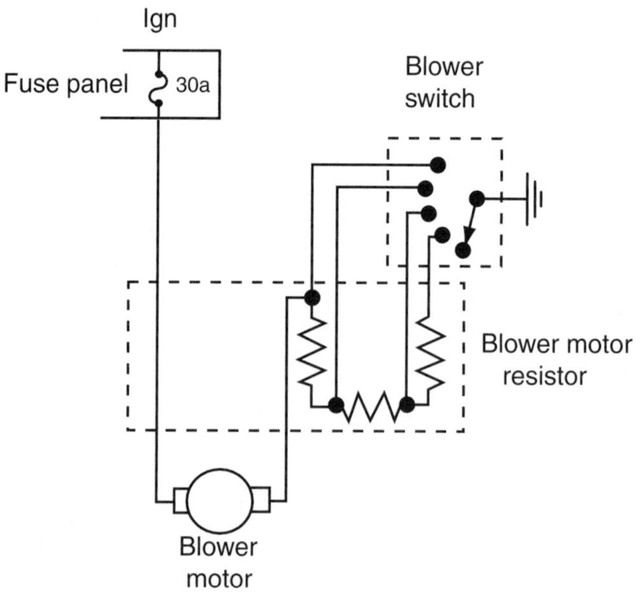

Figure 10-42 Blower motor circuit using negative side switch.

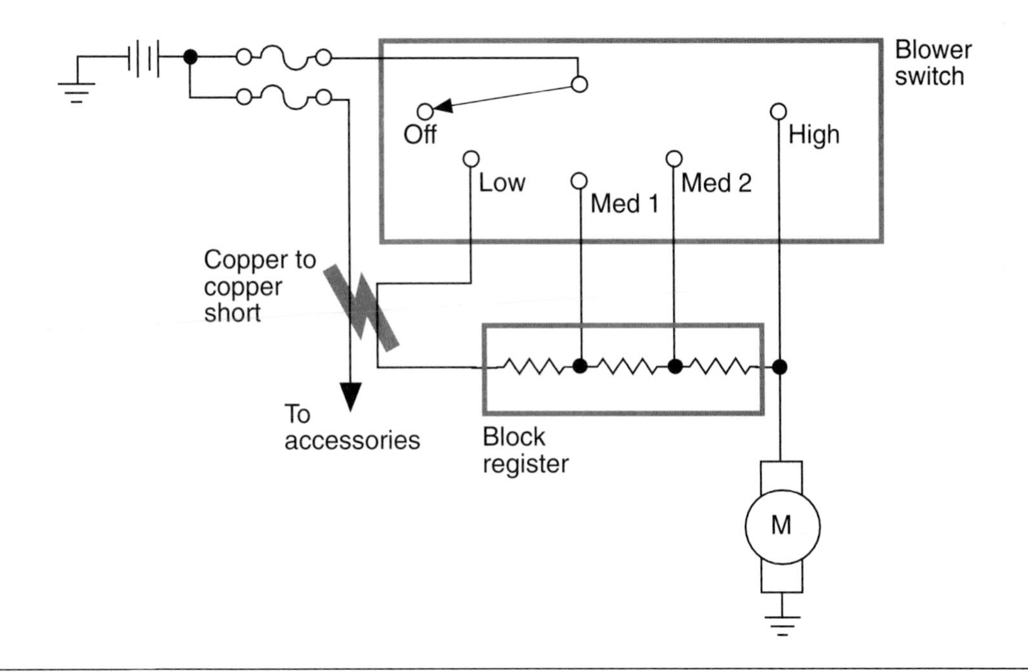

Figure 10-43 A copper-to-copper short can cause the fan motor to run when the switch is turned off.

A short to power can occur when multiple wires are running too close to a heat source and they tend melt the insulation, allowing the conductors to touch. The same thing can occur with wires rubbing together, wearing the insulation away enough for the conductors to touch each other.

Circuits with power-controlled switches need to be checked for copper-to-copper shorts (power shorts) in the power side of the system (Figure 10-43). In this diagram, it is easy to see that if the motor receives power from another circuit due to a power short, it will continue to run whenever current is flowing through that circuit. If another circuit is suspected, you might want to pull the fuses from various circuits one at a time. When the motor stops operating, you have probably found the suspect circuit.

Special Tools

OEM manuals

Wiring diagrams

Some systems incorporate a relay and if the contacts within them fuse together, the motor will also continue to operate.

Engine Brake Troubleshooting

Engine brake is a term used to describe an internal compression engine utilized as a retarder or supplemental brake.

It is beyond the scope of this manual to troubleshoot every manufacturer's **engine brake,** not to mention that engine brake applications vary between the numerous engine manufacturers. However, we will look at a few applications to see that with proper information and diagrams and system can be easily diagnosed.

Classroom Manual
Chapter 10, page 263.

Most of the components and switches in Figure 10-44 are common to many mechanically controlled engines.

For the engine retarder to become energized requires the interrelationship of the various control switches. Following this graphic circuit, you can see the number of switches that need to be electrically closed in order for the solenoid valve (Figure 10-45) to be energized.

Classroom Manual
Chapter 10, page 266.

One common problem of an engine brake that would cause it to be inoperative is switches that are not in adjustment. They can interrupt the flow of current to the solenoid valve.

Clutch Switch

To ensure that the clutch switch is capable of functioning, verify the following:

1. Verify that the clutch switch actuator arm is in contact with the clutch pedal arm or other clutch member (Figure 10-46).
2. Be sure that the switch adjustment is within specifications. The actuator arm should be deflected 1.0–1.5 inches (25–38mm) measured at the tip of the actuator when the

Special Tool

Tape measure

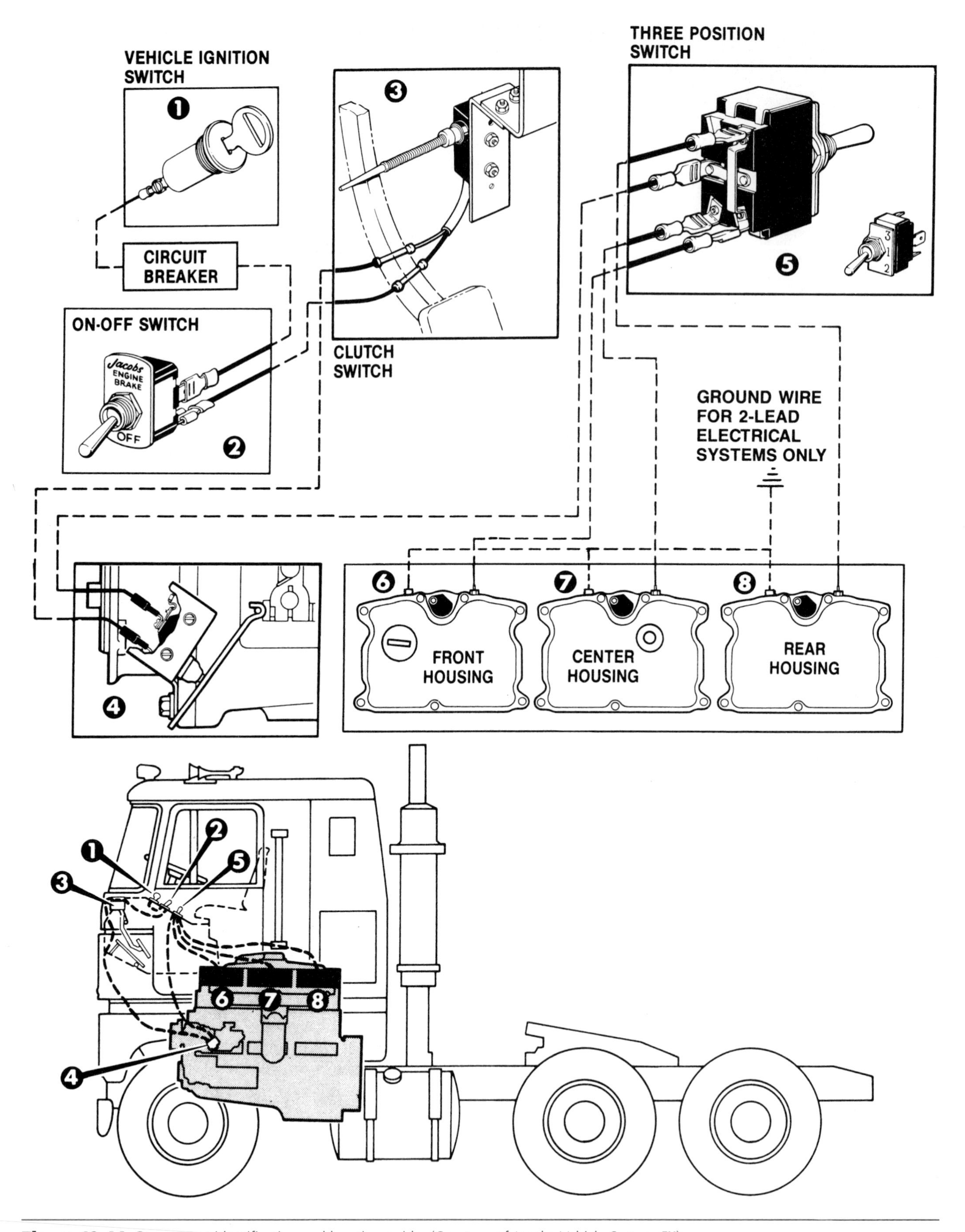

Figure 10-44 Component identification and location guide. (Courtesy of Jacobs Vehicle Systems™)

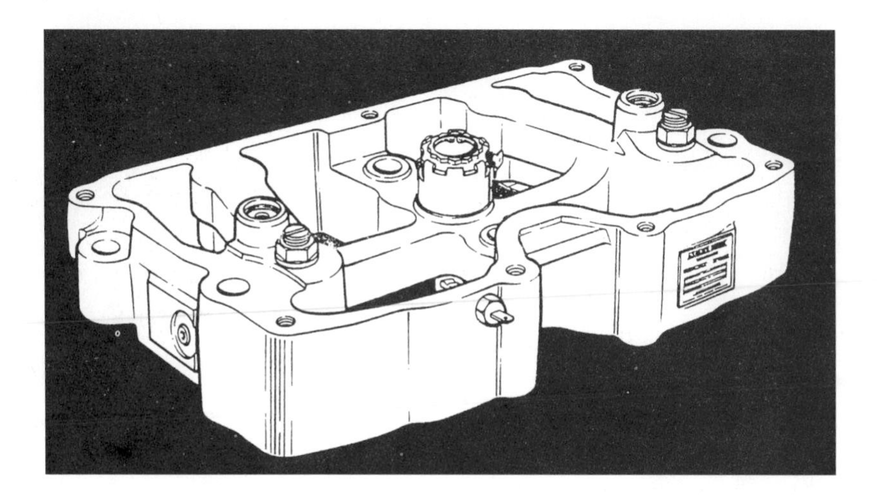

Figure 10-45 Solenoid mounted in housing. (Courtesy of Jacobs Vehicle Systems™)

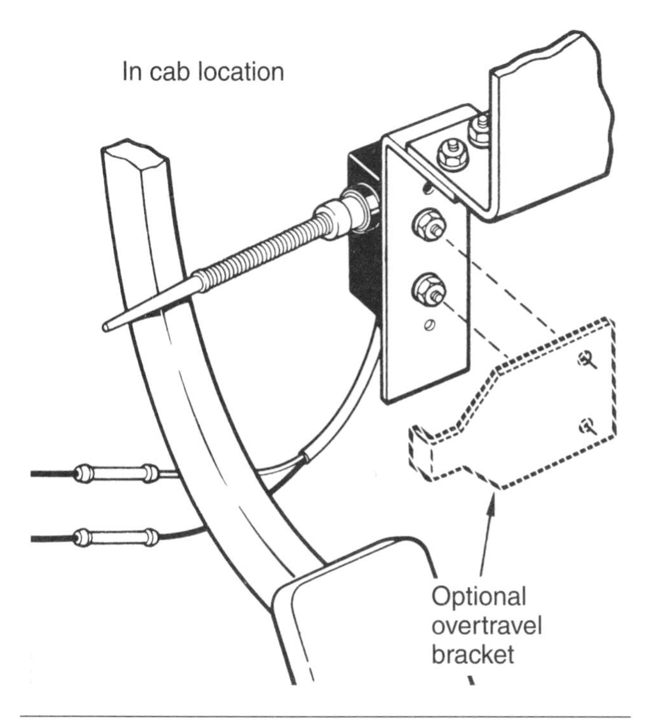

Figure 10-46 This is the proper position of the clutch switch. (Courtesy of Jacobs Vehicle Systems™)

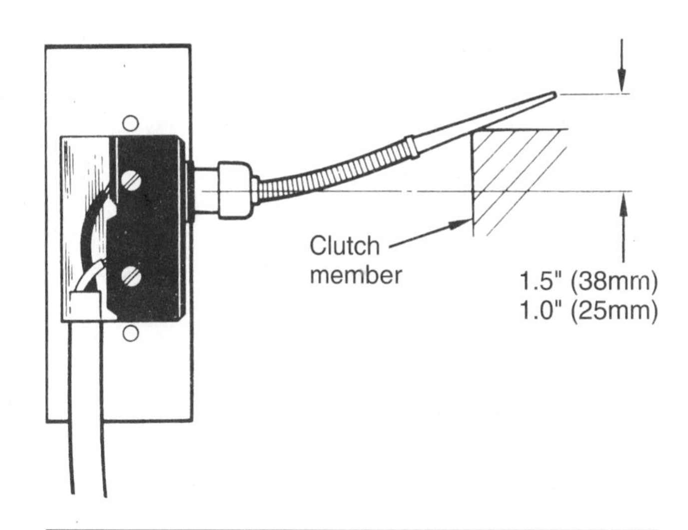

Figure 10-47 Checking clutch switch adjustment. (Courtesy of Jacobs Vehicle Systems™)

clutch pedal is in the up position (clutch engaged) (Figure 10-47). The switch is adjusted by moving the switch along the mounting bracket.

3. Verify proper installation and adjustment by moving the clutch pedal. The switch should click in the freeplay motion of the clutch pedal before actual clutch disengagement takes place.

Fuel Pump Switch

A failure of the fuel pump switch requires replacement and proper adjustment. This procedure is shown in Figure 10-48.

Special Tools

Voltmeter

Ohmmeter

Loosen the lower capscrew on the fuel pump housing and remove the center capscrew as shown.

Install the Jacobs fuel pump switch. The slotted area in the switch bracket is for easier installation and for minor bracket adjustments. Tighten the cap screws.

Remove Cummins nut, washers and capscrew and install the Jacobs tee bolt, guide, actuating lever and nut. Install parts in correct order as shown.

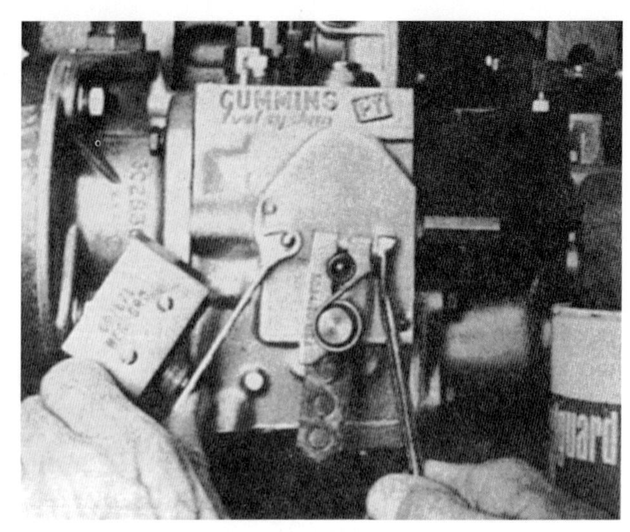

With the throttle shaft in the idle position, move actuating arm until switch actuates. Hold in this position and tighten nut to 10 lbft (14 N•m).

⚠ CAUTION

DO NOT BEND ACTUATING ARM TO ADJUST SWITCH. BENDING THE ARM WILL RESULT IN PREMATURE FAILURE OF THE ARM AND LOSS OF ENGINE BRAKING.

ALTERNATE SWITCH AND THROTTLE LEVER POSITIONS.

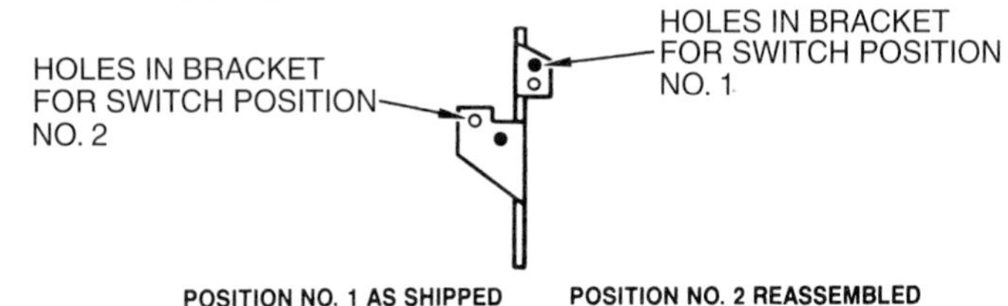

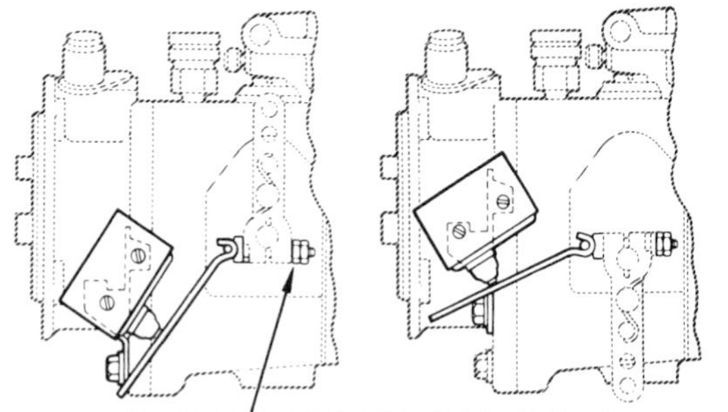

⚠ CAUTION

CHECK THE FUEL PUMP THROTTLE SHAFT TO ENSURE THAT THE THROTTLE PEDAL WILL MOVE THE SHAFT TO THE FULL FUEL POSITION AFTER INSTALLING THE ACTUATING ARM.

Figure 10-48 Fuel pump switch replacement procedure. (Courtesy of Jacobs Vehicle Systems™)

The engine brake system for a particular engine has to interface with the vehicle's wiring harness as shown in Figure 10-49. Using that figure as a reference, the flowcharts in Figures 10-50 and 10-51 can be used to troubleshoot this system.

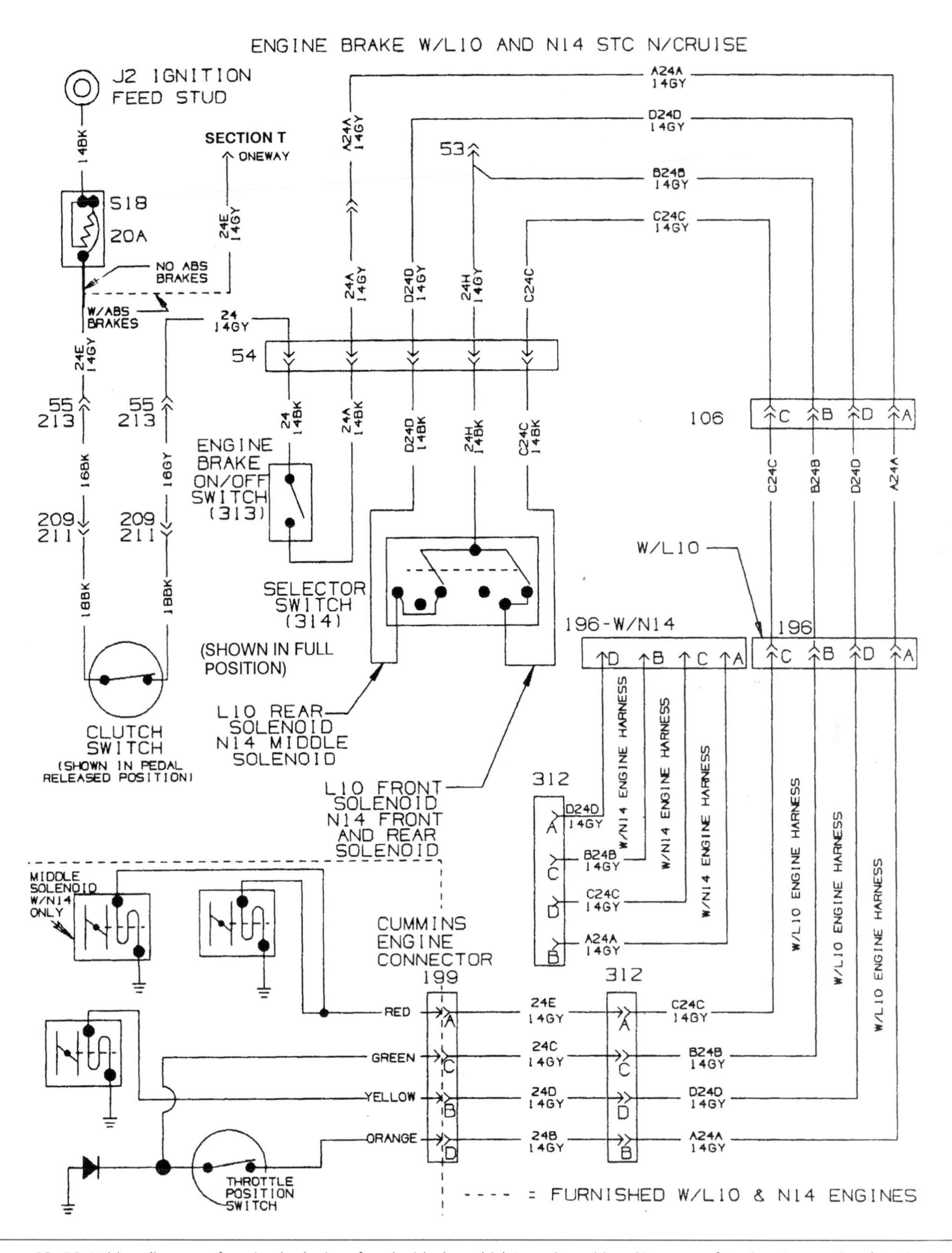

Figure 10-49 Wiring diagram of engine brake interfaced with the vehicle's engine wiring. (Courtesy of Navistar International Engine Group)

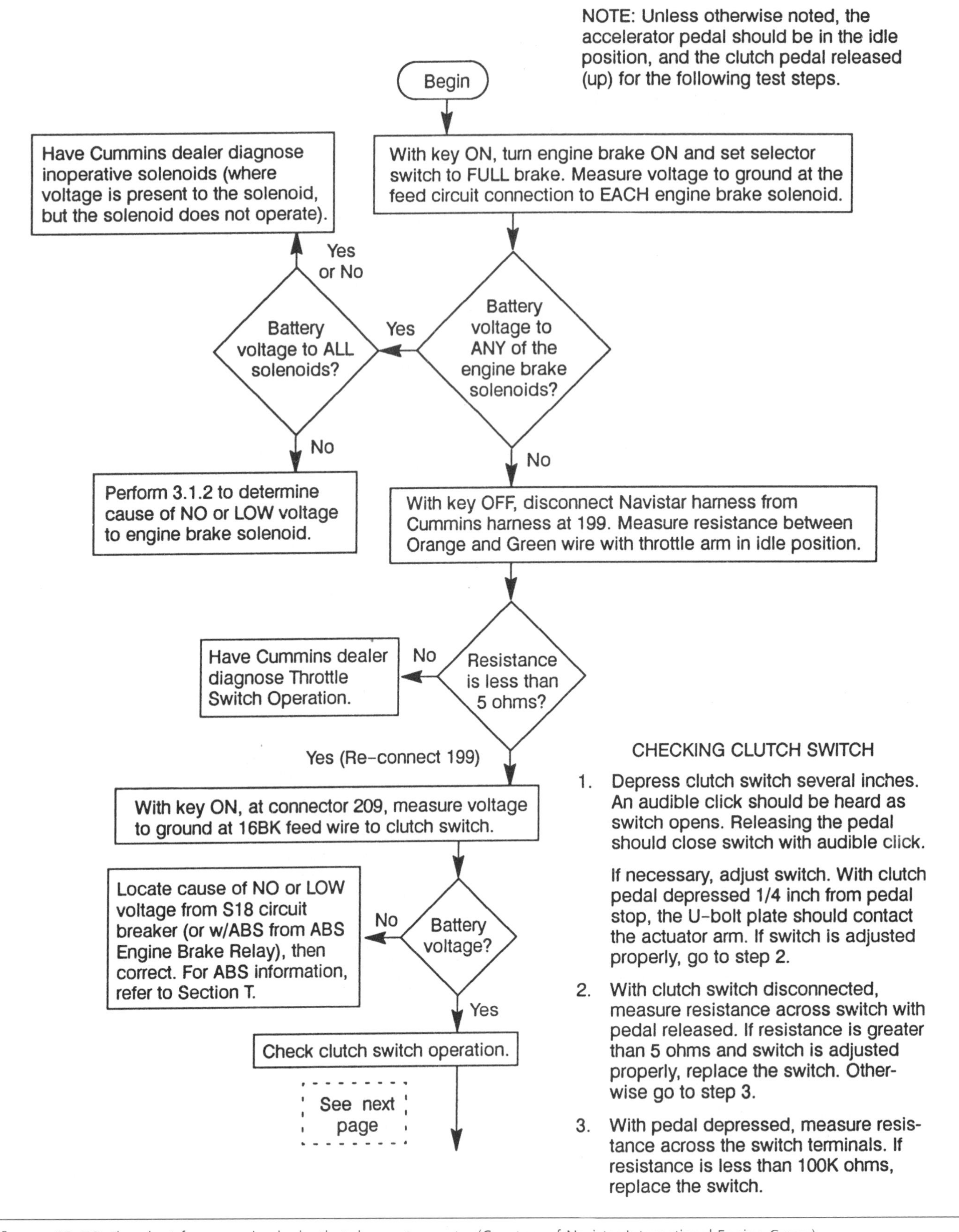

Figure 10-50 Flowchart for an engine brake that does not operate. (Courtesy of Navistar International Engine Group)

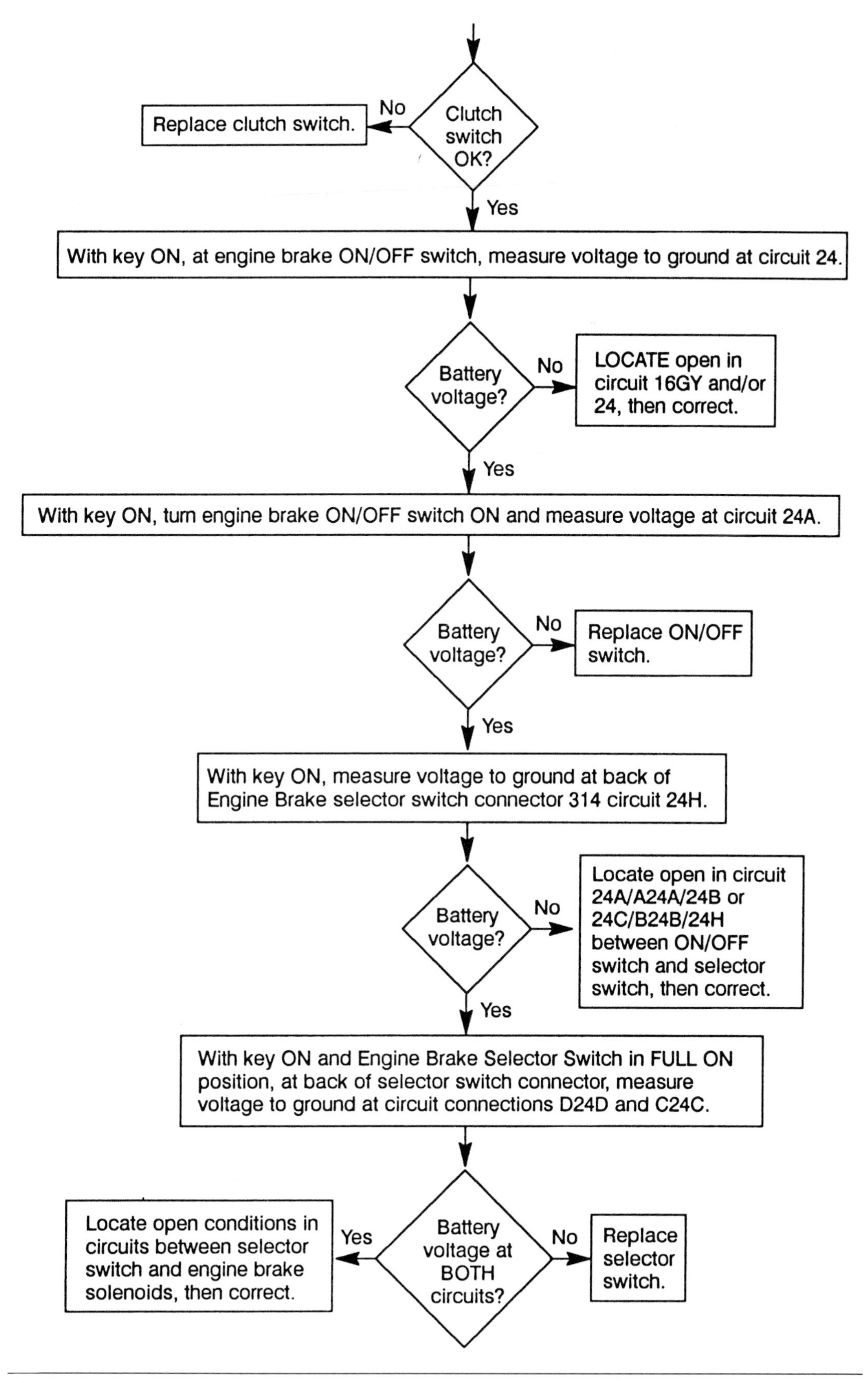

Figure 10-50 Continued.

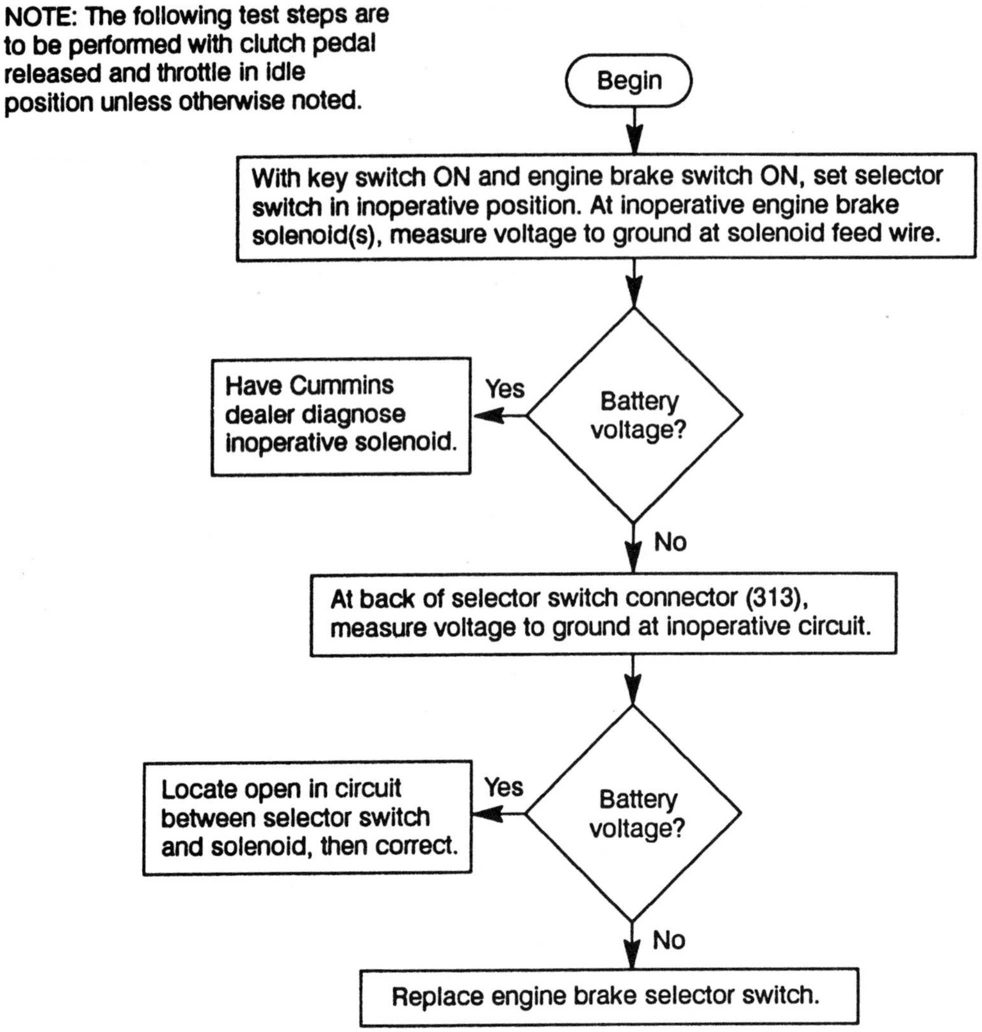

NOTE: The following test steps are to be performed with clutch pedal released and throttle in idle position unless otherwise noted.

Begin

With key switch ON and engine brake switch ON, set selector switch in inoperative position. At inoperative engine brake solenoid(s), measure voltage to ground at solenoid feed wire.

Battery voltage?

Yes → Have Cummins dealer diagnose inoperative solenoid.

No ↓

At back of selector switch connector (313), measure voltage to ground at inoperative circuit.

Battery voltage?

Yes → Locate open in circuit between selector switch and solenoid, then correct.

No ↓

Replace engine brake selector switch.

Figure 10-51 Flowchart for engine brake not operating in all selector switch positions. (Courtesy of Navistar International Engine Group)

Classroom Manual
Chapter 10, page 269.

Electronically Controlled Engines Using Engine Retarders

Most vehicles today use engines that are computer controlled, eliminating some of the control switches such as the clutch switch and fuel pump switch. However, the dash or in-cab switches are still used as shown in Figure 10-52. Notice that this system uses a relay.

Figure 10-53 shows a Detroit DDEC II electronic engine wiring diagram using an engine brake control module.

Looking at the diagram, notice there is an optional in-line switch module. If it is used, it is connected to the power input side of the control module.

- The ON/OFF power switch connects the red wire to the +12 volt vehicle electrical system, providing power to the control module.
- The purple and gray inputs select which one of the dark blue or yellow outputs will be active. The (AUX LO) green/yellow, (AUX HI) orange, (#508) brown, and (CLUTCH SWITCH) black and white inputs control when the dark blue and yellow will be active. To allow outputs to be active, the following must be true:
  1. The clutch switch closed connecting the BLACK and WHITE wires together.
  2. The (AUX LO) GRN/YEL wire switched to ground (0 VDC).

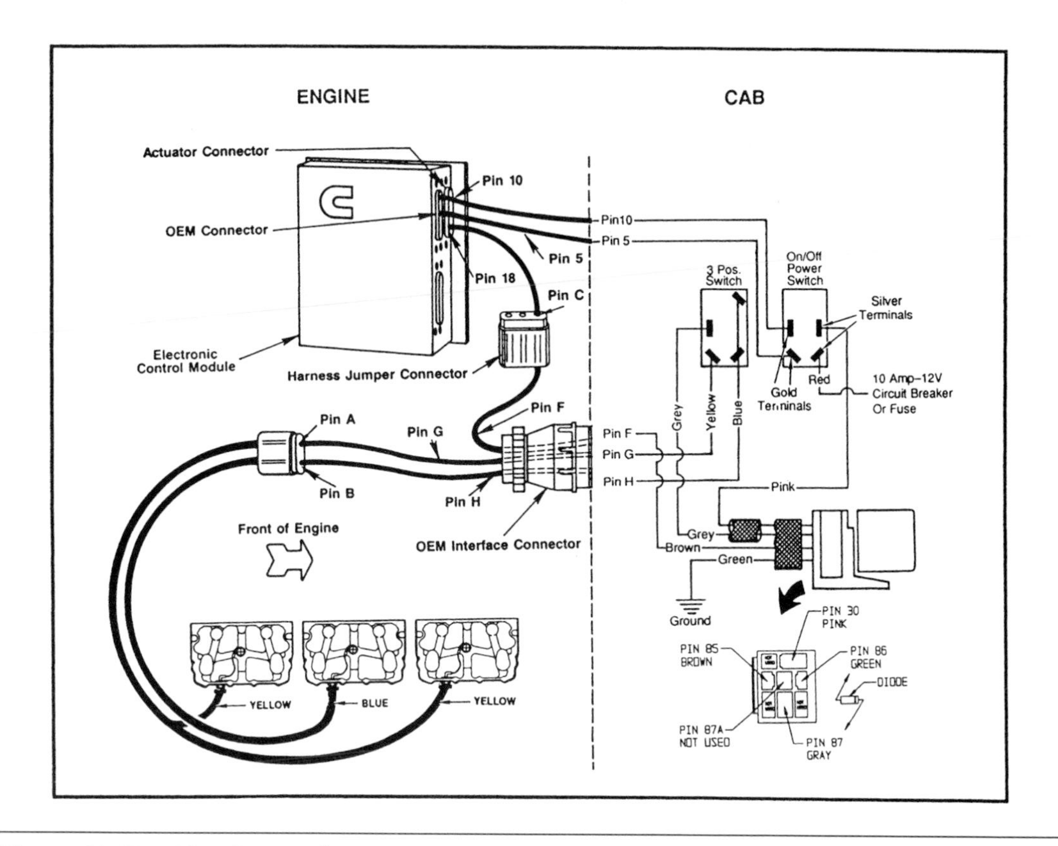

Figure 10-52 Wiring diagram of engine brake system interfaced with an electronically controlled engine. (Courtesy of Jacobs Vehicle Systems™)

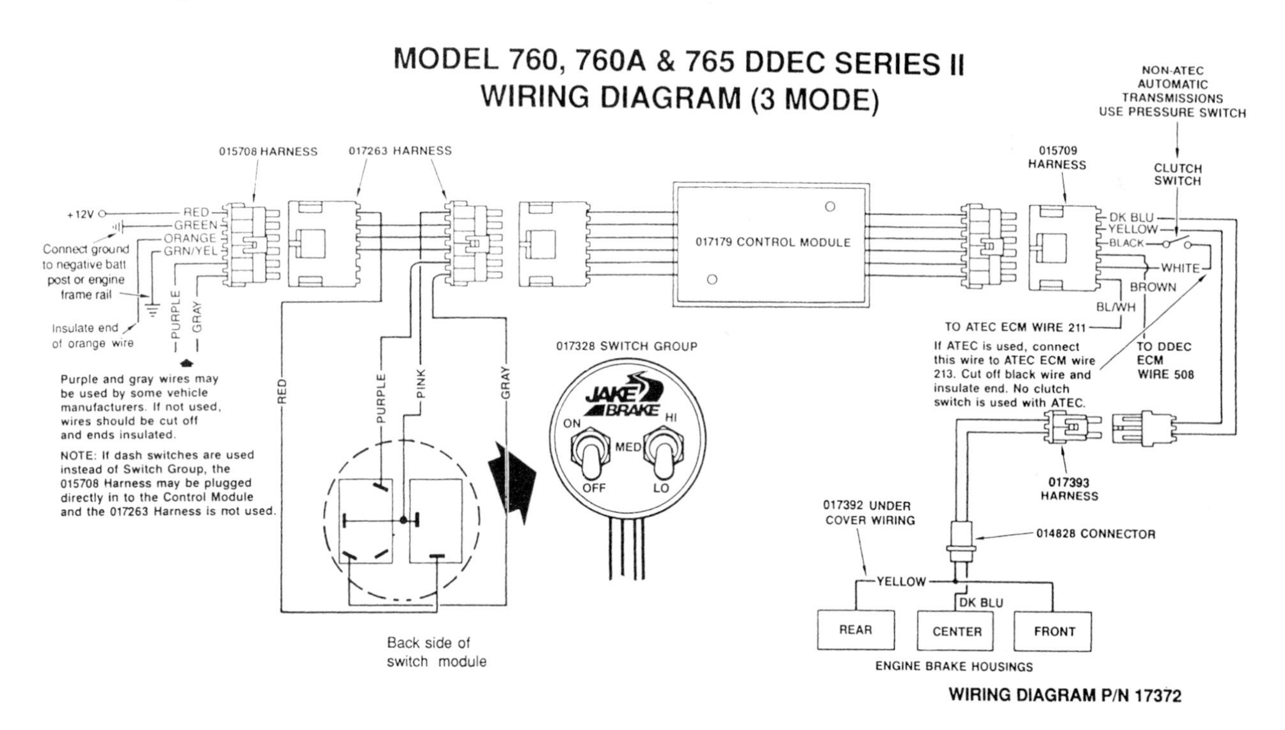

- Voltage on PURPLE wire controls voltage on DK BLUE wire for LO operation.
- Voltage on GRAY wire controls voltage on YELLOW wire for MED operation.
- Voltage must be on both YELLOW and DK BLUE for HI operation.

Figure 10-53 Wiring diagram of Jacobs brake used on a DDEC Series II engine. (Courtesy of Jacobs Vehicle Systems™)

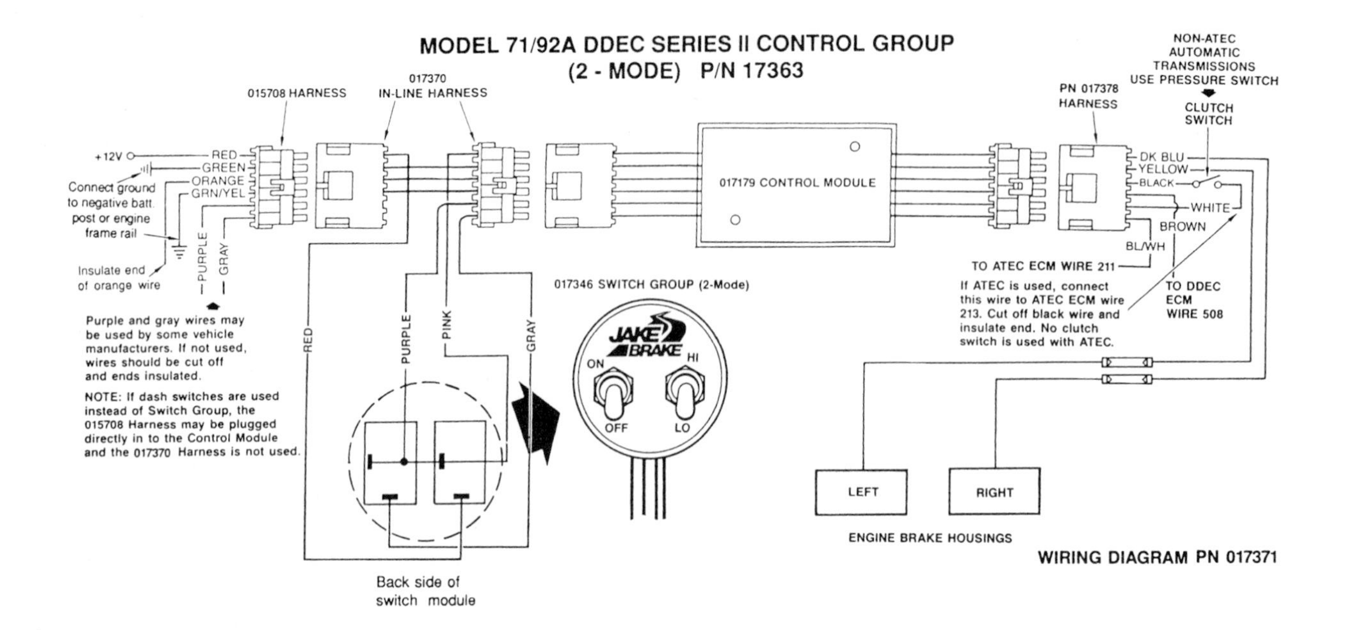

- Voltage on PURPLE wire controls voltage on DK BLUE wire for LO operation (right bank).
- Voltage on GRAY wire controls voltage on YELLOW wire for HI operation.

Figure 10-53 Continued.

Special Tool

DVOM with 20,000 ohm/volt input impedance, minimum

3. The (#508) BROWN wire switched to ground (0 VDC).
4. The (AUX HI) ORANGE wire not connected to ground (insulated).

Armed with this information the following graphic troubleshooting sequences shown in Figure 10-54 can be used for various engine brake problems with this particular system.

Testing systems that operate with delicate electronic circuitry require the use of high-impedance meters to prevent damage to the circuitry and components.

Thermostatic Engine Fan Troubleshooting

A number of **thermostatic engine fan** manufacturers' systems are used by the major vehicle manufacturers. However, they all have one objective: to maintain engine temperature by engaging or disengaging the cooling fan. The result is greater engine efficiency, better fuel economy, and a quieter engine.

Features common to many of the thermostatic fan manufacturers are: they use temperature as a control medium, which requires some type of sensor, and air is used as the energy source to engage or disengage the clutch fan, which requires some type of electrically controlled solenoid. Figure 10-55 shows a Horton fan clutch system and Figure 10-56 shows a Bendix fan clutch system. Note that Figure 10-55 shows the system interfaced with an air-conditioning system. This configuration requires an electro-pneumatic override for the air conditioning.

Classroom Manual
Chapter 10, page 271.

CAUTION: The engine fan can engage at any time when the engine is running, causing personal injury.

"Pneumatic" is another word for compressed air that is used as a source of energy to perform work.

Before trying to diagnose a clutch fan problem, you need to know the type of control used. Some systems use air to disengage the fan clutch and air off to engage, while some systems use the reverse of air application for clutch disengage and engage.

Before active troubleshooting is begun, check the integrity of all wiring and harness connections to verify that connections are tight and that wires are not pinched or have scraped insulation.

Problem	Probable Cause	Correction
A. Engine Brake will not activate	1. Check supply voltage.	With the ignition switch on, disconnect the P/N 15708 harness from the control module connector. Measure the voltage at the RED wire. Place the positive probe (+) of the voltmeter on the terminal of the RED wire and the negative probe (-) to ground. The voltmeter should read +12 VDC (Fig. 3). If this condition is not present, check that system is energized and check power supply.

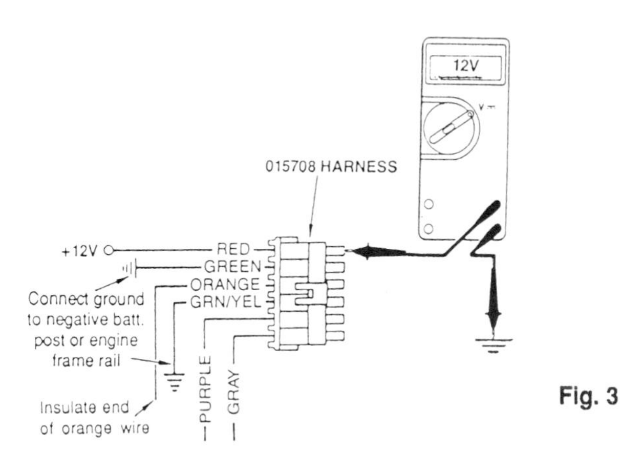

Fig. 3

2. Check switches and connections.

Optional Selector Switch

Disconnect P/N 15708 harness from control module. Measure voltage at both PURPLE and GRAY wires. With selector switch in HI position, both wires should read +12 VDC (Fig. 4). If this condition is not present, check power supply, connections and switches. Repair or replace as required.

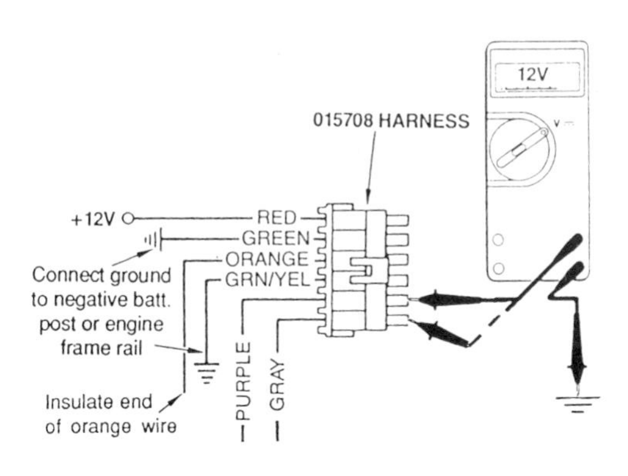

Fig. 4

***Jacobs* Switch Group**

Disconnect P/N 17263 (17370) harness from *Jacobs* control module. Measure the voltage at the RED wire. The voltmeter should read +12 VDC when the main power supply is ON and 0 VDC with main power supply OFF (Fig. 5). If these conditions are not present, check power supply and connections.

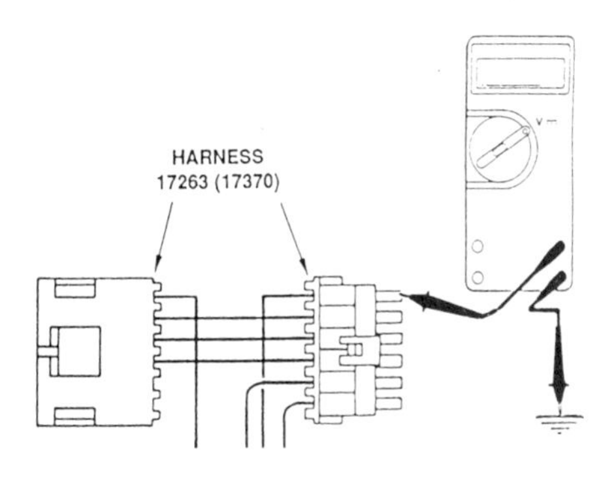

Figure 10-54 Troubleshooting sequence for engine brake problems used with DDEC II engine control systems. (Courtesy of Jacobs Vehicle Systems™)

Problem	Probable Cause	Correction
A. Engine Brake will not activate (Cont'd.)	2. Check switches and connections.	(Fig. 6) With main power supply ON and selector switch in LO, the PURPLE wire should read +12 VDC and GRAY wire 0 VDC. With selector switch in MED position, GRAY wire should measure +12 VDC; PURPLE wire 0 VDC. With selector in HI position, both PURPLE and GRAY wires should measure +12 VDC. If these conditions are not present, check connections, check wiring schematic for proper position of wires to switch and/or replace switch.
	3. Check clutch switch.	With the P/N 15709 (17378) harness connected to the control module, measure the voltage at the terminal of the WHITE wire. With the clutch engaged, (pedal not depressed) a reading of 0 VDC should be measured. With the clutch disengaged, (pedal depressed) a reading of +5 VDC +/ - 0.5 VDC should be measured (Fig. 7).If this condition does not exist, check continuity of clutch switch and black and white wires.

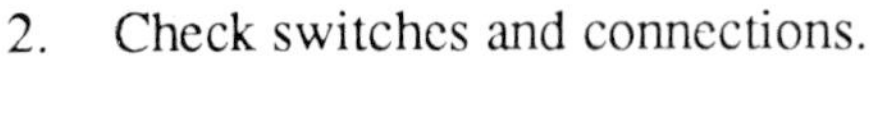

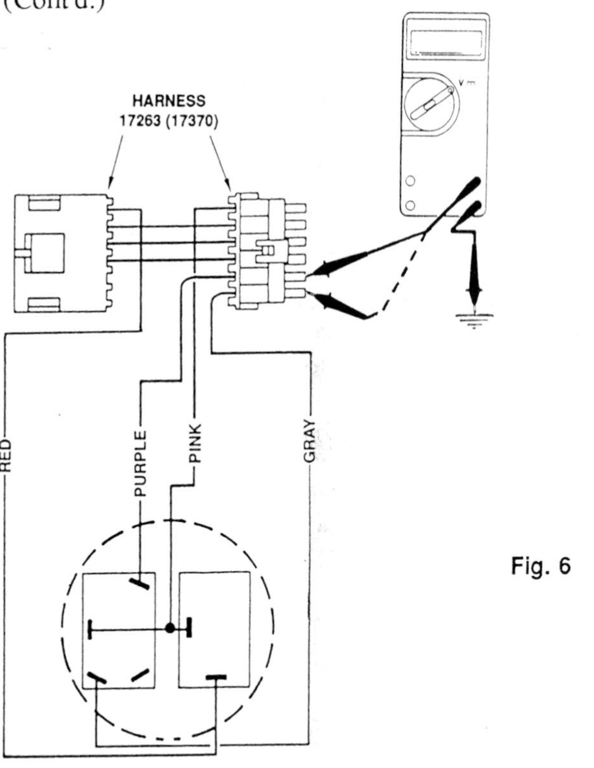

Fig. 6

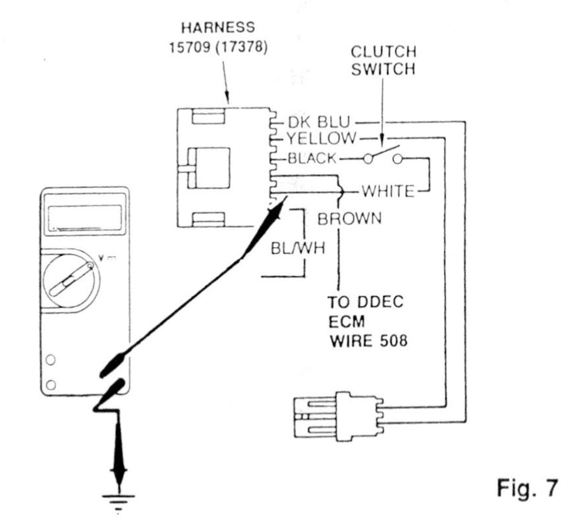

Fig. 7

Figure 10-54 Continued.

Problem	Probable Cause	Correction
A. Engine Brake will not activate (Cont'd.)	4. Check engine brake enable signal.	Disconnect 15709 (17378) harness from 17179 module. Start the engine. Turn the engine brake switch OFF. Place the positive probe of the voltmeter at the terminal of the BROWN wire and the negative probe on ground (Fig. 8). Increase engine RPM to rated engine speed. The voltmeter should measure +12 VDC. Release throttle; voltage should drop to 0 VDC. When the engine reaches idle, the voltage should again read +12 VDC. If the voltage does not change, check connections and wiring. If problem continues, have the engine ECM checked.

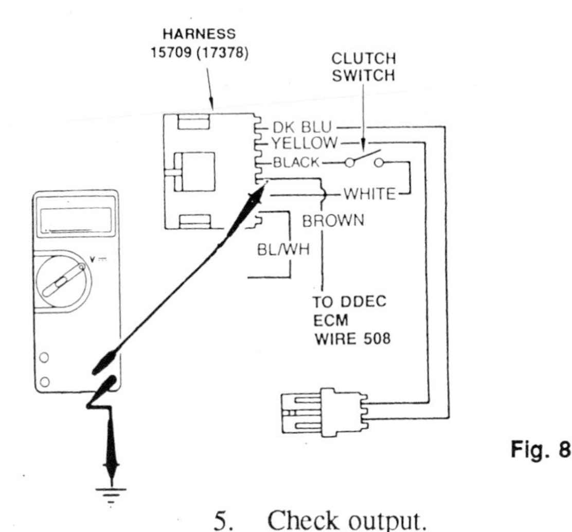

Fig. 8

	5. Check output.	Inspect DARK BLUE and YELLOW wires leading to solenoid valve connectors. Check for loose contacts, pinched wires or scraped insulation. Start the engine, turn the engine brake switch ON and select HI. Advance the throttle to rated speed and then release the throttle. Voltage at both YELLOW and DARK BLUE wires should measure +12 VDC (Fig 9).

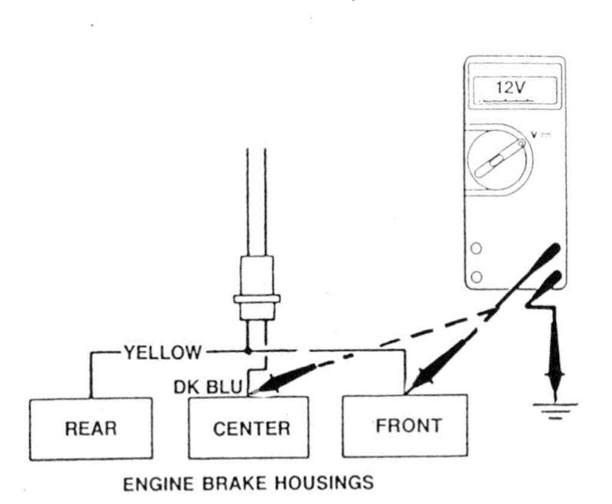

Fig. 9

NOTE:
When measuring voltage, check that all harness connections are tight. If the voltage is measured with the harness from the solenoid loose or disconnected, both the DARK BLUE and YELLOW wires will measure approx. +1 VDC. This is an internal voltage established by the control module for reference.

	6. Check *Jacobs* control module.	Measure the voltage at the ORANGE wire of the control module. With system power ON, the voltage should measure +5 VDC +/- 0.5 VDC (Fig. 10). If this condition is not present, replace module.

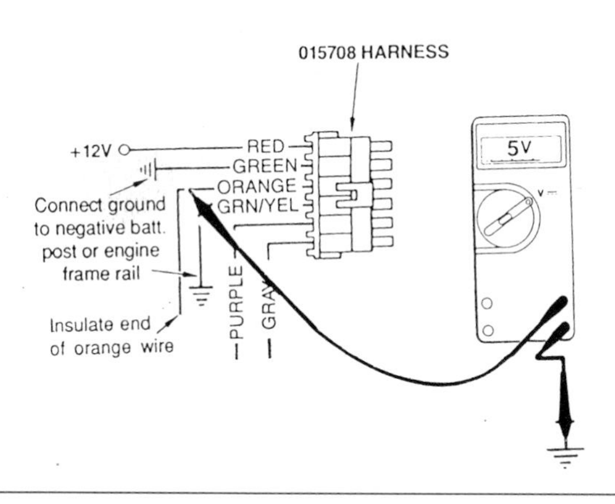

Figure 10-54 Continued.

Problem	Probable Cause	Correction
B. Engine Brake performance erratic/intermittent	1. Check ground connection.	The resistance between the engine block and the negative terminal of the battery must be less than 1 OHM (Fig. 11). The resistance between the GREEN wire of the engine brake control module and the negative terminal of the battery must be less than 5 OHMs for proper module operation (Fig. 12). **If vehicle is equipped with ABS system:** The GREEN/YELLOW wire must be grounded when not in use, preferably to the same point as the GREEN wire. These wires should be isolated from other system ground wires. The orange wire must not be grounded and must be insulatedwhen not in use.

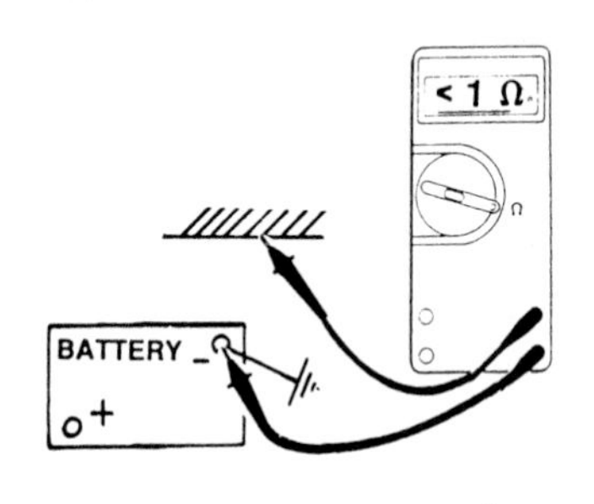

Fig. 11

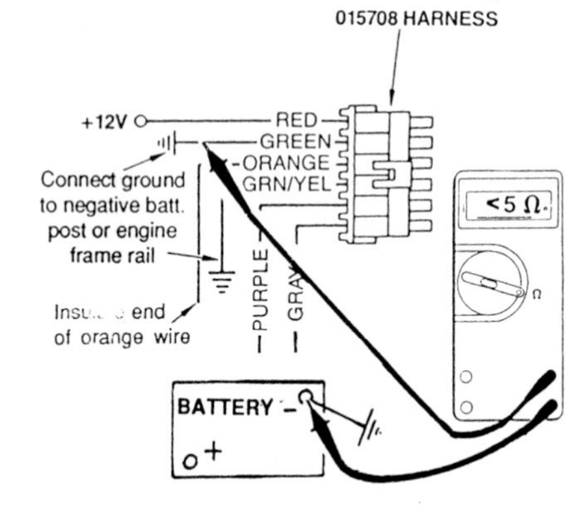

Fig. 12

Problem	Probable Cause	Correction
	2. Check undercover wiring.	Make sure solenoid wires are securely attached to the solenoid valves.
	3. Check for solenoid failure.	Measure resistance of each solenoid valve. The current solenoid, P/N 16440, for 12 VDC applications should have a cold measurement of 9.75 to 10.75 OHMs (Fig. 13). Solenoid valves not within these values must be replaced. NOTE: Resistance may increase significantly when solenoid valves are above 100 deg. F.

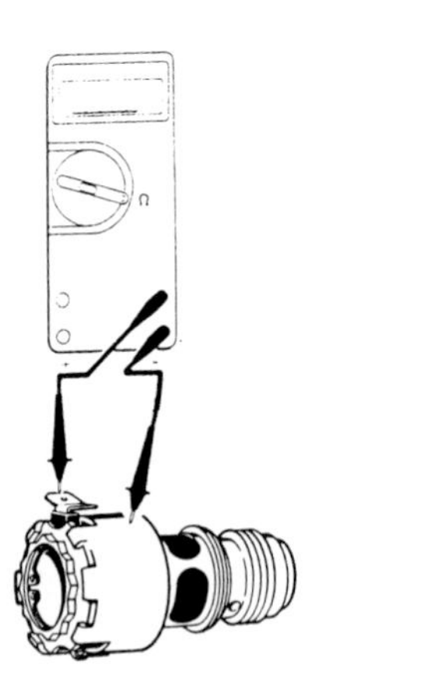

Fig. 13

Problem	Probable Cause	Correction
	3. Check Allison ATEC automatic transmissions.	Check that the BLUE/WHITE wire from the control module is connected to the ATEC ECM wire #211. The WHITE wire from the control module is connected to ATEC ECM wire #213. The BLACK wire from the control module must be insulated.

Figure 10-54 Continued.

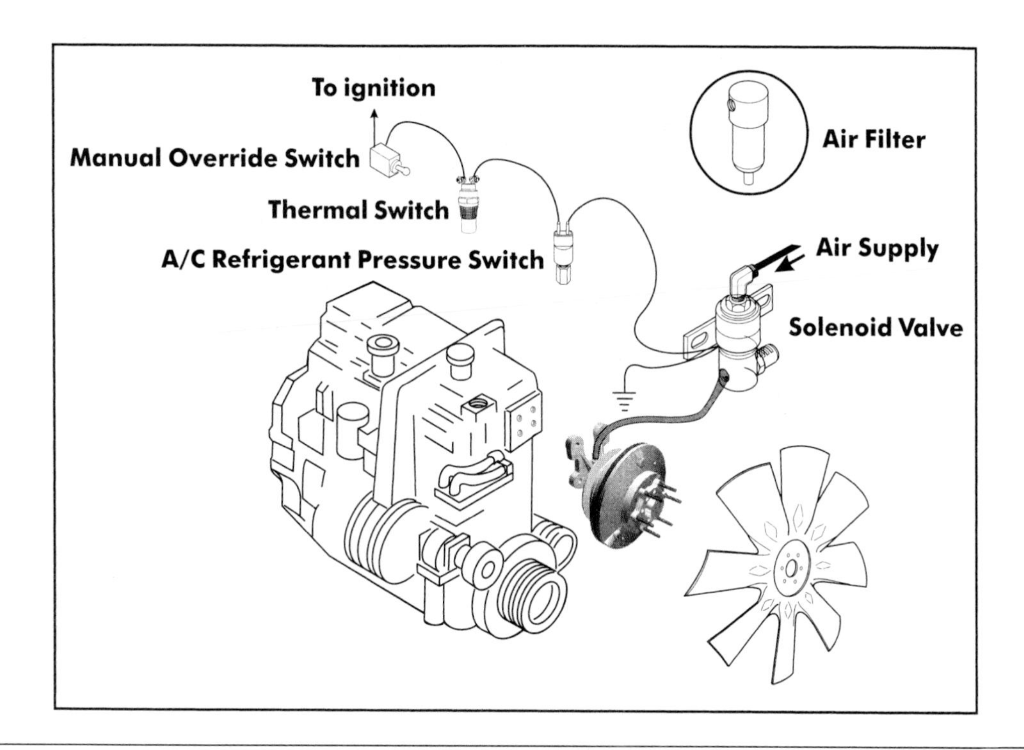

Figure 10-55 Typical Horton fan clutch system. (Courtesy of Horton Industries, Inc.)

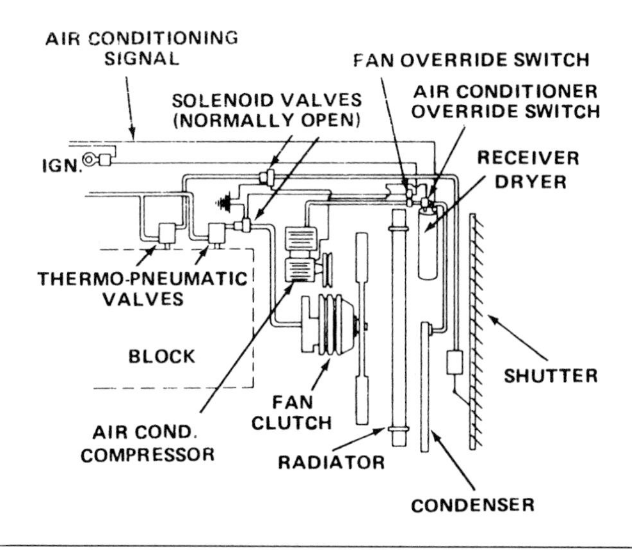

Figure 10-56 Typical Bendix fan clutch system. (Courtesy of Bendix Brakes by Allied Signal)

Classroom Manual
Chapter 10, page 272.

Special Tools

Manufacturer's manual

DVOM

Figure 10-57 shows a sampling of the types of systems various manufacturers tend to use. For a vehicle using a Horton clutch fan, it is a simple task to determine if the system is a normally open or normally closed electrical control system. Look at the solenoid valve. If the air inlet enters the side of the solenoid and the exhaust air releases out the end, the electrical control system is normally open. On the other end, if the air inlet enters the solenoid from the top and the exhaust air releases out the side, the electrical control system is normally closed.

We will look at troubleshooting a Horton fan clutch used in a Navistar truck. In this case, the Horton fan clutch is spring loaded in the FAN OFF position, requiring air pressure to engage the fan clutch. A normally open (NO) solenoid is used. Normally open means when no voltage is applied to the fan solenoid, the air valve is open, applying air to the fan clutch and causing

Normally Open Electrical System

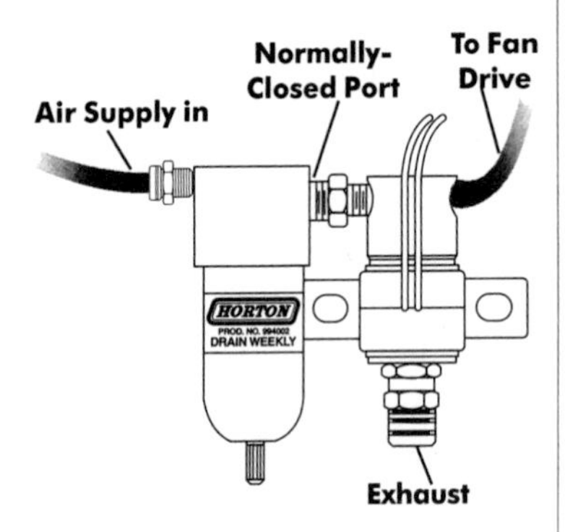

- Electrical switches are wired in parallel.
- Air supply plumbed to solenoid's normally-closed port (the side port).
- Test fan Drive by installing a jumper wire across a sensor.

Normally Closed Electrical System

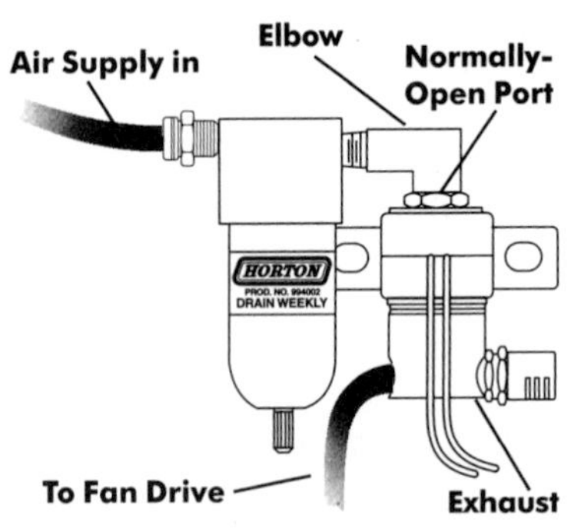

- Usually found in Blue Bird, Ford, GMC, Mack, Marmon, Navistar, Peterbilt, Kenworth, Freightliner,Volvo/WHITEGMC, and Western Star vehicles.
- Electrical switches are wired in series.
- Air supply plumbed to solenoid's normally-open port (the end port).
- Test fan Drive by disconnect ing a wire from a sensor.

Normally-Open Electrical System with Normally-Closed Pneumatics	
All switches open Solenoid de-energized Drive disengaged	Any switch closed Solenoid energized Drive engaged

Figure 10-57 Types of control systems used by various manufacturers. (Courtesy of Horton Industries, Inc.)

the fan to be engaged. Use the diagram in Figure 10-58 as a reference. This diagram incorporates the fan clutch system with A/C.

When the key is ON, the Fan Temperature Switch and the refrigerant pressure are both closed (engine does not require fan cooling), and ignition voltage is applied to the fan solenoid, causing it to close. In this condition the spring-loaded off fan does not operate. When certain conditions are met, such as engine coolant exceeding a specific temperature or the head pressure in the A/C compressor exceeding the limits needed to open the fan pressure switch, the open switch(s) causes voltage not to be applied to the fan solenoid. This causes the solenoid to go from a closed position to the open position, applying air pressure to engage the fan clutch.

The following tests are used for the more common symptoms:

❑ **Horton fan stays on all of the time**

Perform Test 1 (Figure 10-59).

❑ **Horton fan does not come on**

Perform Test 2 (Figure 10-60).

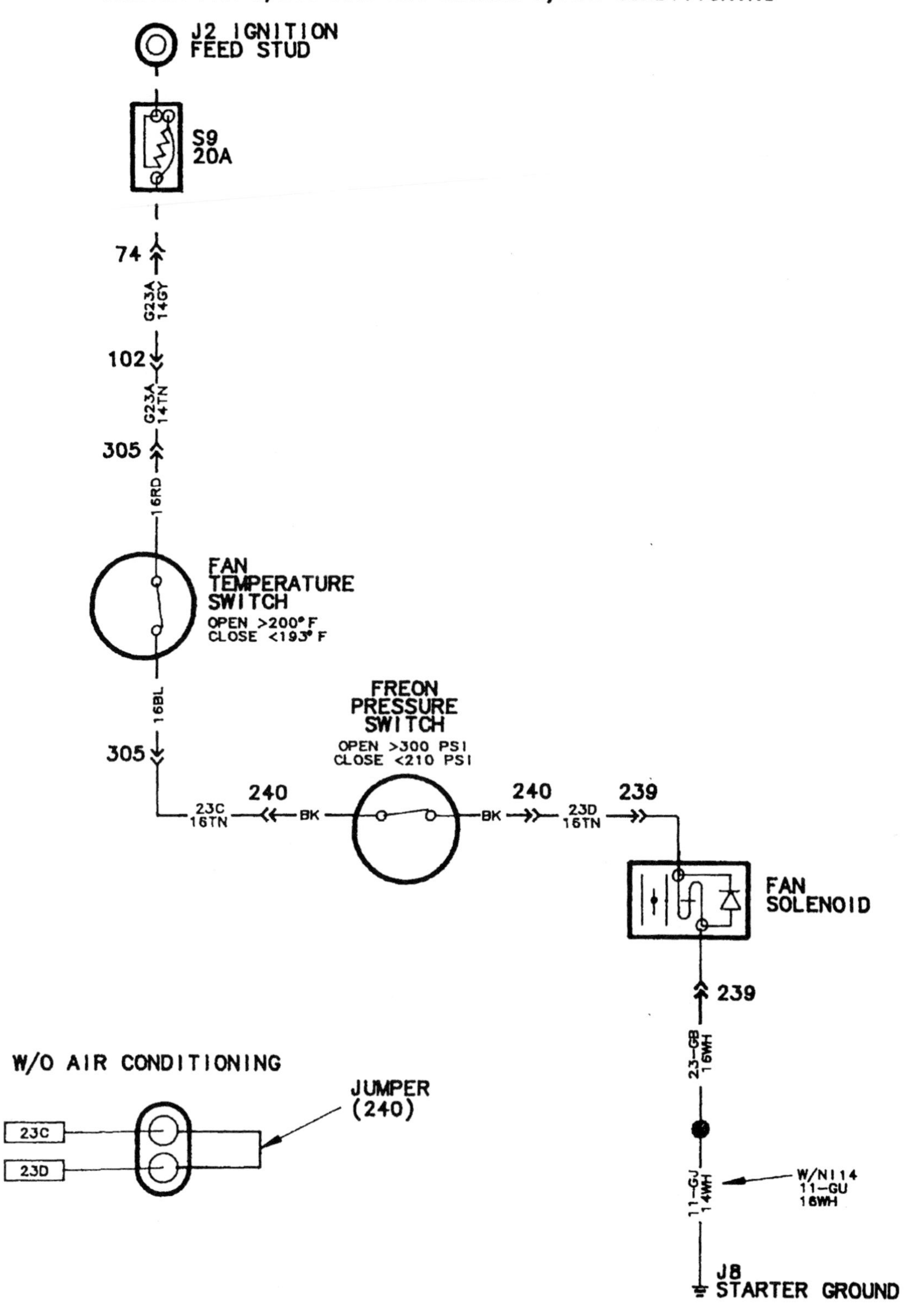

Figure 10-58 Wiring diagram of a Horton fan clutch system interfaced with the A/C system. (Courtesy of Navistar International Engine Group)

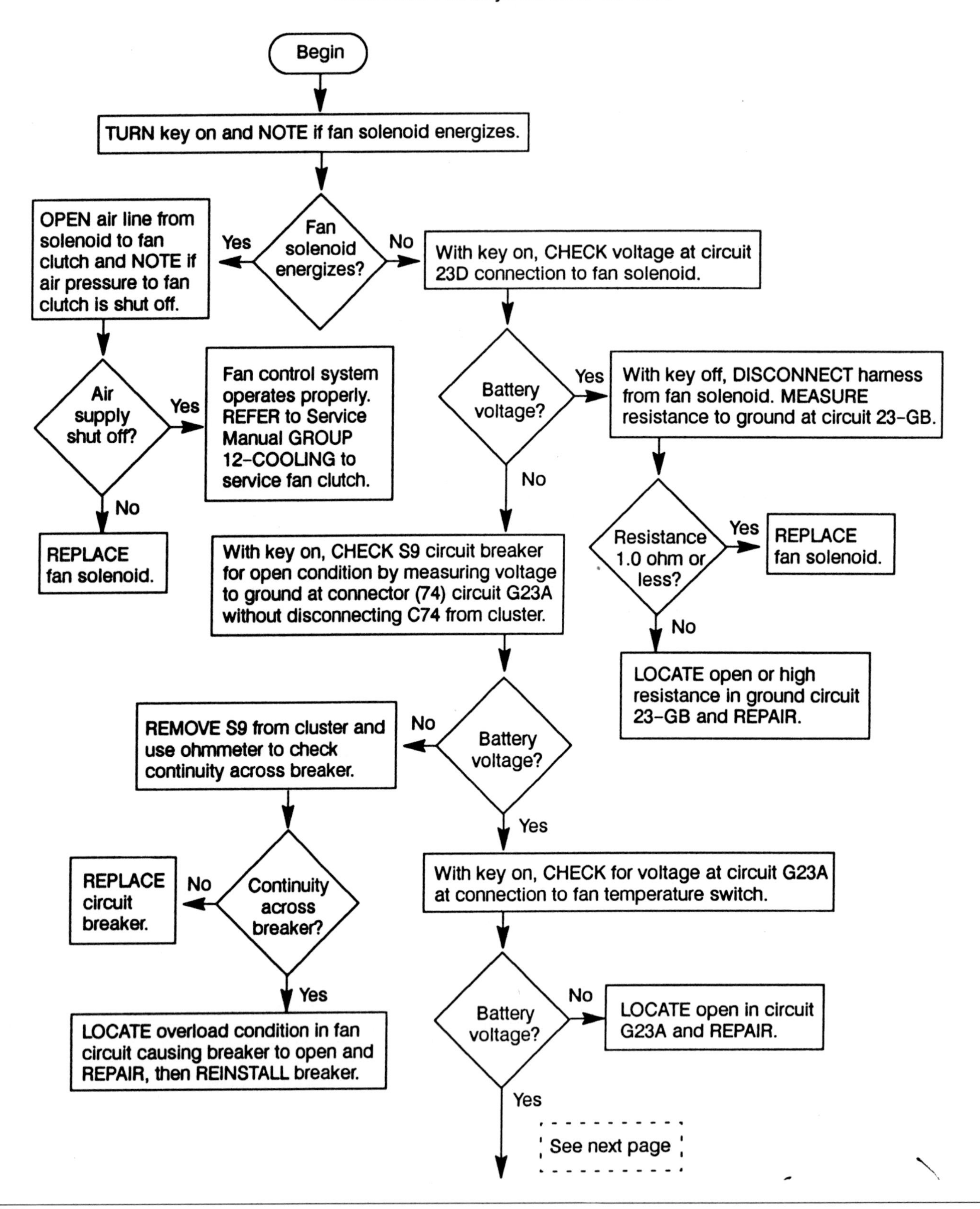

Figure 10-59 Flowchart for fan staying on all the time. (Courtesy of Navistar International Engine Group)

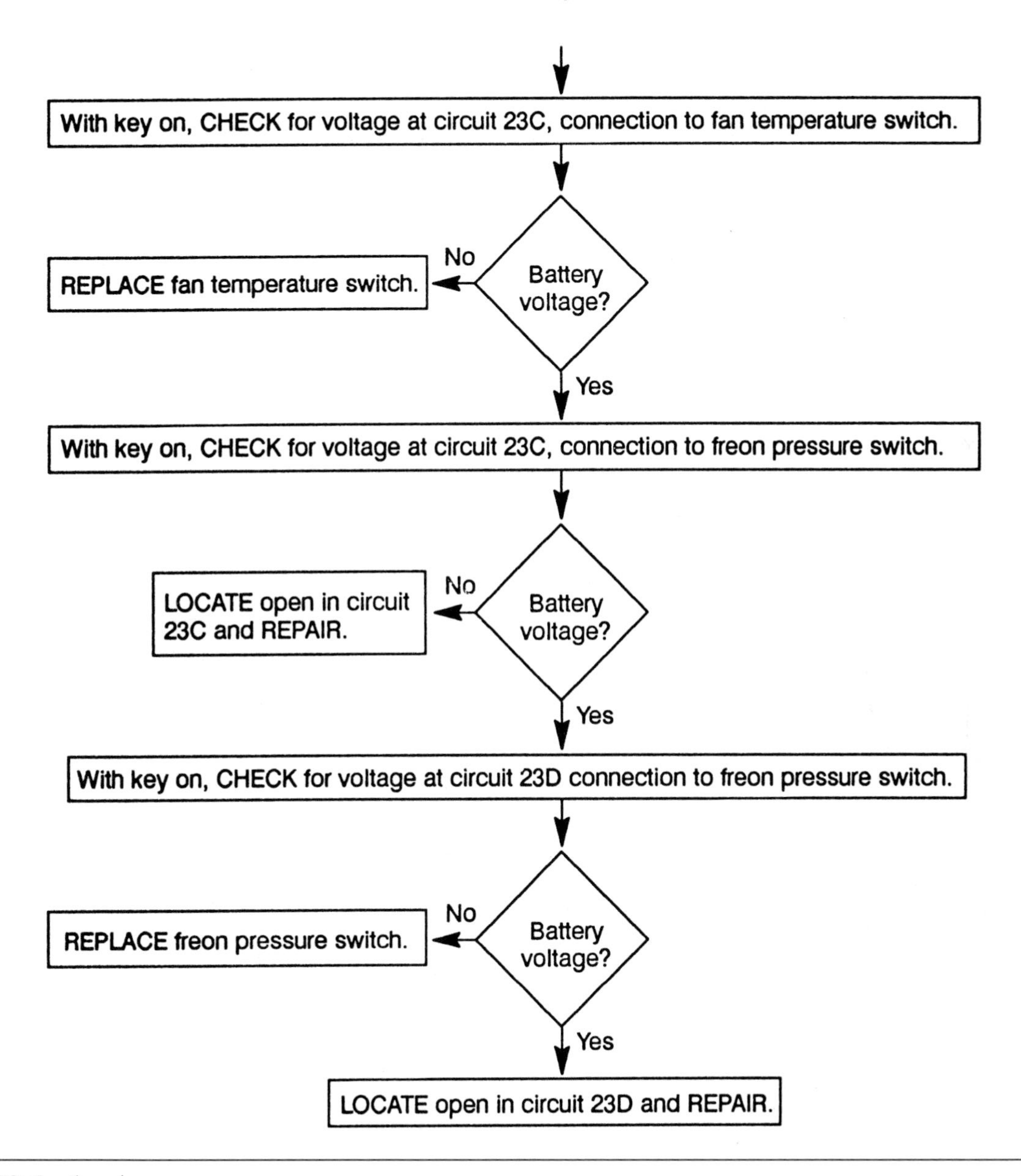

Figure 10-59 Continued.

The likely electrical causes for the Horton fan not coming on are:

1. Defective fan solenoid or
2. The fan control circuit is shorted to power (another hot circuit), keeping the fan solenoid energized and preventing the fan clutch from engaging.

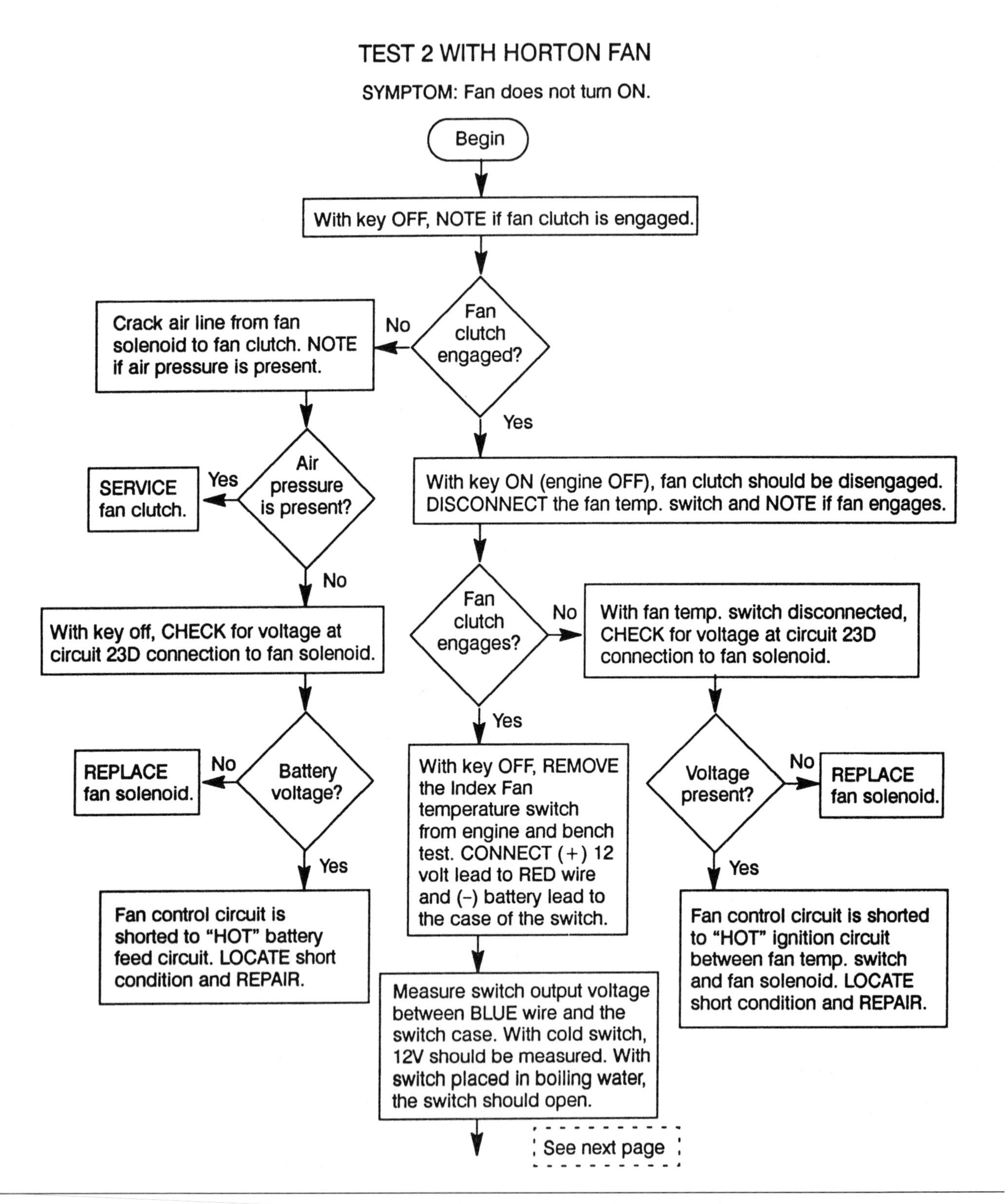

Figure 10-60 Flowchart for fan does not turn on. (Courtesy of Navistar International Engine Group)

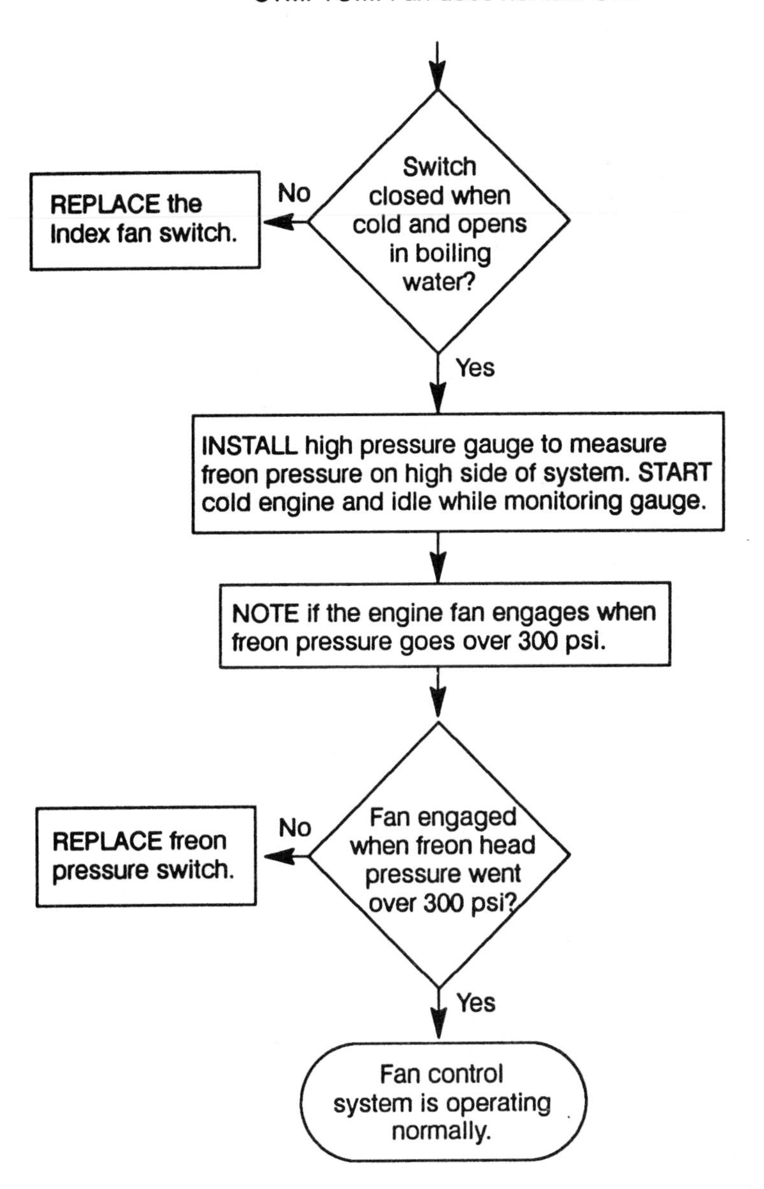

Figure 10-60 Continued.

Working on fan clutches requires you to handle components that function with compressed air, so the following precautions must be observed:

1. Never connect or disconnect a hose on line that contains air pressure. It may whip as air escapes. Never remove a plug or component with air pressure behind it. Make certain all system pressure has been depleted.
2. Always wear safety glasses when working with air pressure. Never direct air from a component at anyone or look into air jets such as from a line.
3. Never disassemble a component until you have read and understood all the recommended procedures. Some components contain powerful springs that can cause personal injury.
4. When working in the engine compartment, the engine should be shut off. If the engine is required to be running for certain tests and procedures, use **EXTREME CAUTION** to prevent personal injury resulting from contact with rotating, moving, or heated components.

CASE STUDY

Two technicians have worked in the same shop for years, retaining many loyal customers. One such customer has been coming in several times a year with a no blower complaint. His regular technician had always been able to make the blower system operable by replacing the control switch. On occasion, he has also had to replace the feed wire because of excessive heat damaging it. However, one very cold day the same vehicle was back with the same problem, but this time the frozen customer wasn't very happy. His regular technician was on the road that day, so his co-worker took a look. She determined that the fault to the blower switch was caused by a burned contact where the power wire was connected. However, she noticed that the wire had been sectioned in quite a few places and the wire gauge size was too small for the load that needed to be carried to the switch. Replacing the switch and installing the proper size wire in the proper manner got the blower system up and running again and this time the cause of the problem was also remedied permanently.

Terms to Know

Bubble protractor
Diaphragms
Electro-pneumatic
Engine brake
Ground side switches
Park contacts
Resistor block
Thermostatic engine fan

ASE-Style Review Questions

1. A vehicle's horn does not turn off. *Technician A* says the cause could be a faulty relay. *Technician B* says the cause could be welded diaphragm contacts inside the horn. Who is correct?
A. A only **C.** Both A and B
B. B only **D.** Neither A nor B

2. The two-speed windshield operates in HIGH position only. *Technician A* says the LOW speed brush may be worn. *Technician B* says the park contact is bent, preventing grounding of the circuit. Who is correct?
A. A only **C.** Both A and B
B. B only **D.** Neither A nor B

3. A slower than normal wiper speed is being discussed. *Technician A* says the problem may be in the mechanical linkage. *Technician B* says the problem may be excessive electrical resistance. Who is correct?
A. A only **C.** Both A and B
B. B only **D.** Neither A nor B

4. A customer complains of having no windshield washers. *Technician A* say the problem could be a restricted hose. *Technician B* says the problem could be an open in the ground circuit. Who is correct?
A. A only **C.** Both A and B
B. B only **D.** Neither A nor B

5. The customer complains that the windshield wipers do not park properly. *Technician A* says the wiper switch is faulty. *Technician B* says the park switch is faulty. Who is correct?
A. A only **C.** Both A and B
B. B only **D.** Neither A nor B

6. The heater blower motor does not operate in HIGH speed position. *Technician A* says that it is very unlikely that a faulty resistor block is the cause. *Technician B* says it is unlikely that the high speed relay is bad because the other speeds do operate. Who is correct?
A. A only **C.** Both A and B
B. B only **D.** Neither A nor B

7. A fan motor that does not shut off is being discussed. *Technician A* says that a grounded resistor in the resistor block can prevent the motor from shutting off. *Technician B* says that a ground side switch that has the ground side of the switch shorting to ground will prevent the motor from shutting off. Who is correct?
A. A only **C.** Both A and B
B. B only **D.** Neither A nor B

8. Jake brakes are being discussed. *Technician A* says that one condition of the system is for the throttle to be at idle in order to complete the electrical circuit through the microswitch. *Technician B* says the clutch pedal needs to be up (not disengaged) for that circuit to be complete. Who is correct?
   A. A only **C.** Both A and B
   B. B only **D.** Neither A nor B

9. Jake brakes are being discussed. *Technician A* says that if one of the solenoids in a multiple solenoid (multiple cylinder heads) system is not functioning, a likely cause is an open between the engine brake ON/OFF switch and the selector switch. *Technician B* says the likely cause is an open between the selector switch and the affected solenoid. Who is correct?
   A. A only **C.** Both A and B
   B. B only **D.** Neither A nor B

10. Thermostatic engine fans are being discussed. *Technician A* says a normally open solenoid means when no voltage is applied to the air valve, the valve is open, applying air to the fan clutch. *Technician B* says a normally open solenoid means when voltage is applied to the air valve, the valve is open, applying air to the fan clutch. Who is correct?
   A. A only **C.** Both A and B
   B. B only **D.** Neither A nor B

ASE Challenge

1. The electric horn of a truck occasionally sounds when a rough spot on the road is crossed. *Technician A* says that an intermittent short between the horn relay and horn switch could be the cause. *Technician B* says that low resistance in the horn relay coil could be the cause. Who is correct?
   A. A only **C.** Both A and B
   B. B only **D.** Neither A nor B

2. A very weak low tone emits from an electric horn when sounded. *Technician A* says the cause could be high resistance in the horn button. *Technician B* says the cause could be low resistance in the horn relay contacts. Who is correct?
   A. A only **C.** Both A and B
   B. B only **D.** Neither A nor B

3. A passenger-side power window in a truck moves only on command from the passenger side switch. *Technician A* says that the problem could be the master switch power feed. *Technician B* says it could be that the master to passenger side wiring is open. Who is correct?
   A. A only **C.** Both A and B
   B. B only **D.** Neither A nor B

4. Jake brake (compression brake) controls on a late model truck are being discussed. *Technician A* says the Jake solenoids are activated by the electronic/engine control unit (ECU). *Technician B* says the system requires a clutch switch. Who is correct?
   A. A only **C.** Both A and B
   B. B only **D.** Neither A nor B

5. An electro-pneumatic engine cooling fan circuit is being discussed. *Technician A* states that most systems are designed to give "fan on" in case of electrical failure. *Technician B* states that most systems will default to "fan on" in case of compressed air loss. Who is correct?
   A. A only **C.** Both A and B
   B. B only **D.** Neither A nor B

Table 10-1 NATEF and ASE Task

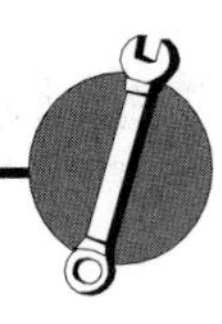

Diagnose the cause of constant, intermittent, or no horn operation.

Problem Area	Symptoms	Possible Causes	Classroom Manual	Shop Manual
CONSTANT HORN OPERATION	Horn sounds even when switch is not activated	1. Faulty horn switch 2. Horn control circuit shorted to ground 3. Faulty horn relay	241	326
INTERMITTENT HORN OPERATION	Horn only operates some of the time	1. Faulty horn switch 2. Poor horn contacts 3. Poor ground connection at horn 4. Poor ground connection at switch 5. Faulty horn relay	242	324
NO HORN OPERATION	Horn fails to sound when the horn switch is activated	1. Blown fuse 2. Faulty horn switch 3. Poor horn contacts 4. Poor ground connection at horn 5. Poor ground connection at switch 6. Faulty horn relay	242	323

Table 10-2 NATEF and ASE Task

Diagnose the cause of constant, intermittent, or no wiper operation; diagnose the cause of wiper speed control and/or park problems.

Problem Area	Symptoms	Possible Causes	Classroom Manual	Shop Manual
CONSTANT WIPER OPERATION	Wipers operate anytime the ignition switch is in the RUN position	1. Faulty wiper switch 2. Defective park switch or activation arm 3. Shorted control circuit	245	331
INTERMITTENT WIPER OPERATION	Wipers operate some of the time	1. Poor ground 2. Poor control circuit connection 3. Faulty switch 4. Worn motor contacts or brushes	250	330
NO WIPER OPERATION	Wipers fail to function	1. Blown fuse 2. Open in the control circuit 3. Short in the control circuit 4. Bad wiper motor 5. Poor ground 6. Mechanical linkage binding 7. Faulty wiper switch	244	329

Table 10-3 NATEF and ASE Task

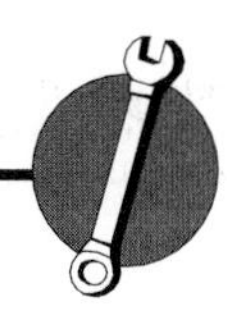

Inspect and replace wiper motor transmission linkage, arms, and blades.

Problem Area	Symptoms	Possible Causes	Classroom Manual	Shop Manual
WIPERS DO NOT OPERATE	Wiper arm will not move or hits the other wiper arm	1. Linkage disconnected 2. Linkage arm broken 3. Wiper transmission faulty	245	327
WIPERS DO NOT CLEAN	Wiper blades fail to wipe clean	1. Defective wiper blades		

Table 10-4 NATEF and ASE Task

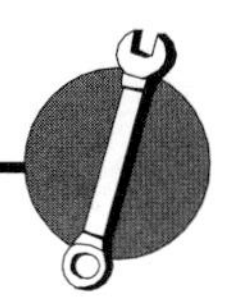

Inspect, test, repair, or replace windshield washer motor or pump/relay assembly, switches, connectors, terminals, and wires.

Problem Area	Symptoms	Possible Causes	Classroom Manual	Shop Manual
CONSTANT WASHER OPERATION	Washers operate without switch activation	1. Faulty switch 2. Shorted control circuit	250	339
INTERMITTENT WASHER OPERATION	Washers operate some of the time	1. Poor ground 2. Poor control circuit connection 3. Faulty switch 4. Faulty pump	250	339
NO WASHER OPERATION	Washers fail to function	1. Blown fuse 2. Open in the control circuit 3. Short in the control circuit 4. Poor ground 5. Faulty motor 6. Faulty switch	250	333

Table 10-5 NATEF and ASE Task

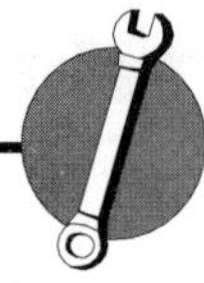

Inspect, test, repair, or replace heater and A/C electrical components including: A/C clutches, motors, resistors, relays, switches, controls, connectors, terminals, and wires.

Problem Area	Symptoms	Possible Causes	Classroom Manual	Shop Manual
AC SYSTEM OPERATION	No AC because clutch will not engage	1. Defective clutch 2. Defective control circuit 3. Open in circuit 4. ECM/PCM/ECU control fault 5. Bad connector	258	347

Table 10-6 NATEF and ASE Task

Diagnose the cause of slow, intermittent, or no power side window operation.

Problem Area	Symptoms	Possible Causes	Classroom Manual	Shop Manual
SLOW POWER SIDE WINDOW OPERATION	Power windows operate slower than normal	1. Control circuit problem 2. Faulty motor 3. Poor ground 4. Defective or misadjusted regulators 5. Binding linkage	269	340
INTERMITTENT POWER SIDE WINDOW OPERATION	Power windows fail to operate sometimes	1. Faulty switch or control circuit 2. Faulty motor 3. Poor ground 4. Poor control circuit connection 5. Binding linkage		341
NO POWER SIDE WINDOW OPERATION	Power windows fail to operate	1. Faulty switch or control circuit 2. Faulty motor 3. Poor ground 4. Poor control circuit connection 5. Faulty circuit protection device		343

Table 10-7 NATEF and ASE Task

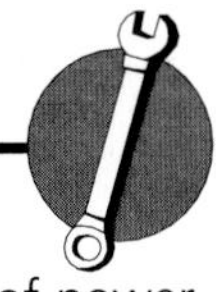

Inspect, test, repair, and replace motors, switches, relays, connectors, terminals, and wires of power side window circuits.

Problem Area	Symptoms	Possible Causes	Classroom Manual	Shop Manual
POWER SIDE WINDOW OPERATION	Power windows fail to operate	1. Faulty switch or control circuit 2. Faulty motor 3. Poor ground 4. Poor control circuit connection 5. Faulty circuit protection device	269	343

Table 10-8 ASE Task

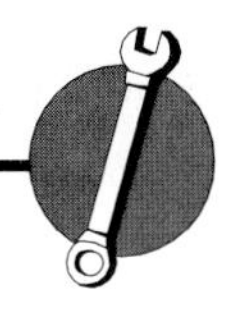

Inspect block heaters; determine needed repairs.

Problem Area	Symptoms	Possible Causes	Classroom Manual	Shop Manual
ENGINE BLOCK HEATER	Engine cranks slowly	1. Faulty block heater 2. Faulty control circuit 3. Faulty switch		

Table 10-9 ASE Task

Inspect, test, repair and replace engine cooling fan electrical control components.

Problem Area	Symptoms	Possible Causes	Classroom Manual	Shop Manual
LOSS OF ENGINE COOLANT	Overheating	1. Fan does not engage 2. Incorrect Fan-on temperature setting	271	366

Job Sheet 17

17

Name: ______________________ Date: ______________________

Testing Blower Motor Circuits

Upon completion of this job sheet you should be able to diagnose problems in the blower motor circuit.

NATEF and ASE Correlation

This job sheet is related to ASE and NATEF Medium/Heavy Duty Truck Electrical/Electronic Systems List content area:

G. Related Systems, ASE and NATEF Task 8: Inspect, test, repair, or replace heater and A/C electrical components including: A/C clutches, motors, resistors, relays, switches, controls, connectors, terminals, and wires.

Tools and Materials

A medium/heavy duty truck
Wiring diagram for the truck
Service manual for the truck
A DMM

Procedure

1. Describe the truck being worked on: Year: ________ Make: ____________
   VIN: __________________ Model: __________________
2. With the service manual as a reference, locate the switch, resistor block, and blower fan. Describe the procedure for accessing these components for testing or replacement.
   ____
   ____
3. Draw the blower circuit below.

4. Describe how the blower motor is controlled to operate at different speeds. ________
   ____
   ____
   ____
5. Locate the resistor block or the closest harness connector to perform a voltage drop test across each added resistor (for example, the positive voltmeter lead to the low speed input terminal of the resistor assembly, and the negative meter lead to the output terminal of the resistor assembly or a feed connection to the motor). Record the readings and describe what they indicate. ______________________
   ____
   ____

6. Connect the voltmeter across the ground circuit and energize the blower motor. What was your reading on the meter, and what does this indicate? ________________
____
____
____

7. If this circuit is part of the A/C circuit, describe how the blower circuit works in relation to the A/C circuit. ________________
____
____
____

8. If this truck does not have A/C, draw a simple diagram to show where the blower switch ties in with the air conditioner.

9. Describe the general operation of the blower motor. Include in your description whether air is also moving through the proper outlets during all the selected functions such as heat, defrost, etc. ________________
____
____
____

☑ **Instructor Check** ________________

Job Sheet 18

18

Name: ______________________ Date: ______________________

Testing a Thermactor Engine Fan System

Upon completion of this job sheet you should be able to diagnose an inoperative fan clutch and check it for proper operation.

NATEF and ASE Correlation

This job sheet is related to ASE and NATEF Medium/Heavy Duty Truck Electrical/Electronic Systems List content area:

G. Related Systems, ASE Task 13: Inspect, test, and repair or replace engine cooling fan electrical control components.

Tools and Materials

A medium/heavy duty truck with an electro-pneumatically controlled fan clutch
A serivce manual for the truck
A wiring diagram for the truck
A DMM

Procedure

1. Describe the truck being worked on: Year: ________ Make: ____________ VIN: __________________ Model: __________________
2. Without referring to the wiring diagram or service manual, identify the manufacturer's system you are working on. ______________________________
3. Without referring to the service manual, are you able to identify if this system's fan clutch is engaged with air to it or when air is removed from the fan clutch? Yes _________ No _________
4. Without referring to the service manual, are you able to identify if the fan solenoid is normally open or normally closed? Yes _________ No _________
5. Use the truck's service manual to verify the type of system used and how it operates.

WARNING: Before conducting any tests, be sure to re-read the *Warning* at the end of the last chapter pertaining to fan clutch troubleshooting.

6. Find a troubleshooting procedure in the service manual or see your instructor if the one in this manual does not apply to your truck. Proceed with a test for a complaint of "The fan does not turn on."
7. Briefly describe each step of the test and what the result was. ______________________________
8. Describe the operating condition of the fan clutch system. Include a brief description of the integrity of the components, including wires, air lines, and the fan itself.

Instructor Check ______________________________

CHAPTER 11

Ignition System Diagnosis and Service

Upon completion and review of this chapter, you should be able to:

- ❏ Explain a logical ignition systems diagnostics procedure.
- ❏ Identify probable problem areas by relating symptoms.
- ❏ Identify and describe the major ignition components, both primary and secondary.
- ❏ Perform a no start/no spark test procedure.
- ❏ Test individual ignition components using test equipment such as a voltmeter, ohmmeter, and test light.
- ❏ Remove, read, and reinstall spark plugs.
- ❏ Test and set (when possible) ignition timing.
- ❏ Remove and reinstall a distributor.

Basic Tools

Hand tools

High impedance multimeter

Classroom Manual

Chapter 11, page 284.

Introduction

This chapter deals with testing ignition systems and their components. Manufacturers' system variations and the constant technological changes in this field make it impossible to cover each particular system and change. Therefore, the tests covered in this chapter are those *generally* used as basic troubleshooting procedures. For exact test procedures, refer to the manufacturer's service manual for the particular vehicle and engine being tested or serviced. However, there are *two important precautions that should be taken during all ignition system tests:*

1. Turn the ignition switch off before disconnecting any system wiring.
2. Do not touch any exposed connections while the engine is cranking or running.

Logical Troubleshooting

The complexity of today's engine control systems to meet emissions standards and at the same time provide maximum power with minimum fuel consumption requires a logical troubleshooting approach in order to diagnose ignition and other related engine problems.

With the high cost of electronic parts and other components, replacing parts until the problem is solved is not very cost effective. Keep in mind that the owner of the truck or fleet relies on his overhead to be cost effective in order to provide a service and provide jobs.

Before we even begin to think about troubleshooting, it should be understood that you as a technician should meet the following minimum requirements:

1. Have a basic understanding of how to use basic hand tools and basic test tools such as meters.
2. Have a basic understanding of electricity; how a circuit operates; the relationship of volts, amps, and resistance to each other; and be able to read and understand electrical diagrams.
3. Have a basic knowledge of the system on which you are working.

Lack of any of the above criteria could result in misdiagnosis or damage to electronic circuitry and components used in today's trucks.

CUSTOMER CARE: Unfortunately, competition has led to quite a spread when it comes to ignition components costs. Most often the old adage of "you get what you pay for" still applies. To retain your customer, it is worth investing in quality parts, rather than losing him or her due to a premature failure of the ignition system. Even if it is your own fleet vehicle, the time lost can be costly.

Diagnostics is an art. It separates the technician from the mechanic. It is more than just following certain steps in order to find a fault or solve the problem of a symptom. Diagnosis is a way of looking at nonfunctioning systems in order to find out why they are not functioning properly. This requires the technician to know how the system is supposed to work correctly. This concept applies to all vehicle systems. The following are some basic rules that will enable you to find the cause of any condition or conditions, hopefully on the first try. These rules were put together by instructors, trainers, and industry people.

Diagnostic Theory

1. Know the System

This means knowing how the system operates and what its limits are and what happens when something goes wrong. It also means knowing how the parts go together to make the system function properly. At times, it could mean getting your hands on a system that is working properly and checking it against the one you are working on that is not functioning properly.

2. Know the History of the System

How new or old is the system? Has it been previously serviced in a manner to cause a certain condition? Remember, you might not be the only one working on the vehicle. What kind of treatment has it had? All drivers don't necessarily treat a vehicle the same. What is the service history? Good communication is necessary with the driver, the one most apt to notice a change from what he or she is used to in the operation of the vehicle.

3. Know the History of the Condition

Again, good communication is needed. Questions to ask include: Did it start all of a sudden? Did it appear gradually? Was it related to some other occurrence such as a panic stop, hitting an object, etc., or another part that was replaced? Find out how the condition made itself known. Remember, you are trying to find clues.

4. Know the Possibilities of Certain Conditions Happening

Most conditions are caused by simple things rather than by complex ones, and they occur in a predictable manner or pattern. As an example, electrical problems usually occur at connections and not components. An engine no-start is more likely to be caused by a loose wire or by a minor component or adjustment rather than a broken crankshaft or camshaft.

Keep in mind the difference between "improbable" and "impossible." Much time can be wasted because you assumed a certain failure was impossible only to find out it was improbable, and actually happened. A good example is a new part that has been installed. Even though it is a new part, it could be the cause of the problem.

5. Follow a Fixed Sequence of Diagnosis

The very first thing to do is eliminate the obvious and then move systematically until you pinpoint the source of the condition.

Following a process of elimination can lead you to a factual conclusion. A hit-and-miss or random approach might work sometimes but a fixed sequence will always lead you to a proper conclusion.

6. Don't Cure the System and Leave the Cause

Just as adding air to a low tire does not repair the leak, installing a new battery does not cure a no-charge situation. Make sure you don't simply fix the problem temporarily without addressing the underlying cause.

7. Be Positive on Your Findings

Double check your findings. If you find a worn or broken component or something out of adjustment, stop and ask yourself what could have caused this condition. If one component is bad could a mating component also be worn? Build up a picture in your mind of relationships of parts.

8. Double Check Your Work

Always road test the vehicle after completion of your work, to make sure the problem you have is corrected. Remember, it is very difficult to simulate road conditions while the truck is in the shop.

As a technician, when you are diagnosing you are expressing a relationship between logic and a physical system of components. This means an organized approach.

Classroom Manual
Chapter 11, page 279.

Today's engines are computer controlled. This requires the technician to consider how a computer processes information and what is the expected result of this. If you think about it, the computer and the technician function very similarly when it comes to diagnosing problems: they both gather information and make decisions based on it. An example of this is when a vehicle is first started. The computer determines the engine mode. Is it cranking, idling, cruising, or accelerating? Because air/fuel mixture is always of the highest priority, the computer can choose the best strategy for that particular mode. A computer would determine that the vehicle is under a heavy acceleration when the sensor's inputs indicate high RPM, warm engine, manifold absolute pressure is high, and the throttle plate is wide open. With this information, the computer determines that a rich air/fuel mixture is required.

A perfect air/fuel ratio is 14.7 to 1 and is referred to as stochiometric.

When operation in open loop fuel control, the computer uses a look-up table similar to that shown in Figure 11-1. With the throttle wide open, manifold pressure high, and coolant temperature at 170°, the table indicates that the air/fuel mixture should be 13:1. This ratio is by weight 13 pounds of air for every 1 pound of fuel. Air does have weight. An air/fuel ratio of 13:1 creates the rich air/fuel mixture needed for acceleration.

Less air in weight as compared to fuel in weight creates a rich condition. An example of this is an air/fuel ratio of 13:1.

The computer will act upon its decision to achieve the desired goal by increasing fuel injector pulse width. This results in the injector nozzle remaining open longer so more fuel can be drawn into the cylinder. More fuel provides the additional power needed for hard acceleration.

A good technician understands the system and with proper readings or inputs through the use of meters would likely anticipate the same results as the computer. If any of the parameters are off, it should be looked at as a clue or cause of the improperly operating system.

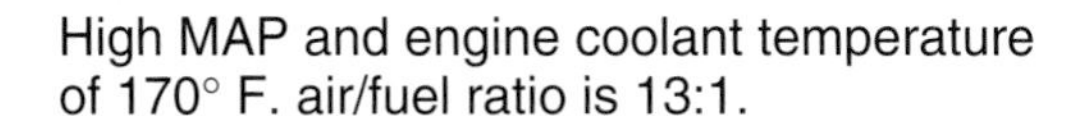

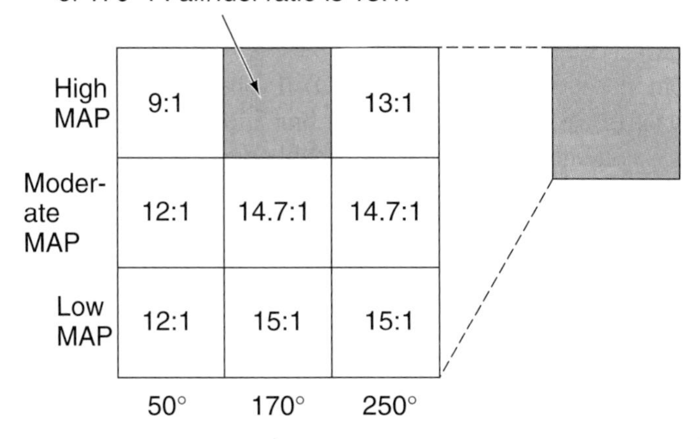

Figure 11-1 A typical look-up table for open loop fuel control.

One word of caution: Never lose sight of the basics. For example, a low battery voltage might result in faulty sensor readings, and engine control systems and components cannot make an unhealthy (no compression) engine healthy again

Visual Inspection

There is no substitute for a good visual inspection as a first step in diagnosis. This first step can often detect the problem without any other test steps. Perform the visual/physical inspection carefully and thoroughly. If an ignition problem is suspected, check for obvious problems such as disconnected, loose, or damaged secondary cables; loose, disconnected, or damaged primary wires, and connections; damaged distributor cap or rotor; a damaged or worn primary system switching device; or any other component or system with a problem that can lead to an ignition-related complaint.

Inspection of Secondary and Primary Wiring

Classroom Manual
Chapter 11, page 283.

Spark plug and ignition coil wires are a more common source of driveability problem. They are not meant to last the life of the vehicle. First of all, the high voltages carried by the wires and the extreme environment they are in limits their usefulness in time. Most often, they are run at a close proximity to the hot exhaust manifolds or directly on the manifold if improperly installed. That alone is enough to damage the wires. An indication of that is usually a whitish/grayish color due to excessive heat and arcing (voltage leak) where the wire is too close or has made contact with the manifold. The secondary wires should also be firmly inserted into the distributor cap and coil and onto the spark plugs. Any signs of burning, arcing, or brittleness will require replacing the secondary wires.

CAUTION: Handle secondary wires with great care. They have enough high voltage to shock you. That shock can force you to move a portion of your body, such as your hand, into a moving component.

The coil also should be checked for cracks or any evidence of arcing or burning. Evidence of arcing and burning will be more evident starting at the coil tower. It is a good idea to pull the coil wire from the tower and look inside for corrosion or chemical buildup, which can be the cause of high resistance. Coils also contain oil that at times can leak out. If the oil leaks out, air space is present in the coil, allowing condensation to form inside the coil container. Condensation in an ignition coil will cause high voltage leaks and misfiring. **E-core** type of coils with the windings exposed should be carefully inspected for signs of burning.

Secondary wires that leak can often be heard when the engine is running; in the proper lighting condition (dim to dark) the arcing can also be seen. If in doubt, you can take a spray bottle with water and mist the secondary wires and secondary components to induce voltage leaks to verify a fault with the suspect circuit.

Another problem with secondary wires is their arrangement according to the firing order of the engine. Actually, there are two problems that occur. The first one is when the wires are replaced and a technician does not reinstall them in the proper order.

SERVICE TIP: It is much easier in the long run to replace the wires one at a time than to pull them all off on the assumption that you are good enough to remember where each one belongs. All it takes is one wire to be misplaced for the driver to notice it while driving down the road.

Crossfiring occurs when one wire induces a firing voltage in another wire, causing it to fire at the wrong time.

The second problem is **crossfiring,** which can occur because of the way spark plug wires are arranged. Spark plug wires from consecutively firing cylinders should cross, rather than run parallel to, one another (Figure 11-2).

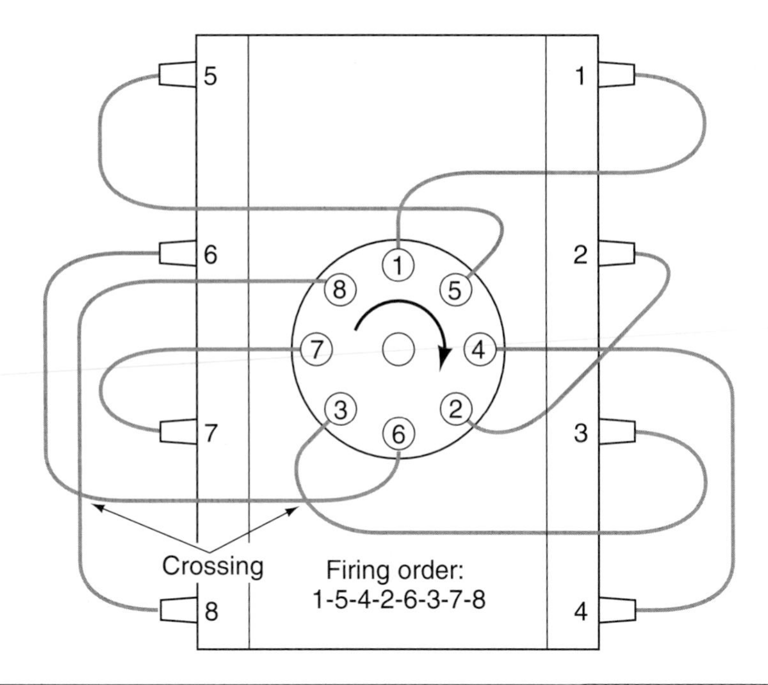

Figure 11-2 Spark plug wires from consecutively firing cylinders should cross each other rather than run parallel to one another.

Primary ignition wiring should also be checked. Remember that you cannot have secondary voltage unless the primary circuit is working properly. With the low voltages used by electronic-controlled ignitions, it is even more critical that the integrity of the primary wires is there.

Any resistance due to corrosion or loose connection can cause a voltage drop, creating a running or starting problem. Make sure you take that one extra step to wiggle, tug, or tap on connections while the engine is running. Any problem that might occur from vehicle vibration or engine movement is duplicated this way. If the engine hesitates, misses, or runs rough while performing this physical test, it will require you to check the suspect connector or component. Repair and/or replace as needed and reverify it again by the physical test.

Do not forget to go back to the basics. That means to check the source of power and the circuitry to the coil of distributor. Examples of circuits and components to check are the battery, starter solenoid (voltage source to some coils), and ignition switch. A bad connection in some of that circuitry can result in ignition interruption.

Ground Circuits

These tend to be the forgotten circuits. For every circuit that is fed with power, there has to be a ground back to the battery. The ground side of the vehicle can incorporate body panels, frame members, and the engine block as the ground path. Many circuits are connected to these major parts as a source for the ground path. One way of connecting the ground paths is the use of ground straps. These ground straps are often neglected, left disconnected after routine service due to ignorance. This can leave quite a few subcircuits with a bad path or no ground path at all. Current might try to seek a ground, however, the path might be through another circuit. This could cause the circuits to operate erratically or fail altogether.

SERVICE TIP: At times you might run across metal components that are prematurely worn or feel warm to the touch when there is no reason for it. Suspect a bad ground system. Wheel bearing troubleshooting charts often show wheel bearings that are bad due to electrical current problems.

With bad or no grounds, current may also be forced through other components that are not meant to carry current. Some examples of bad ground types of ignition failures are: burnt ignition modules, erratic performance due to a poor distributor to engine block ground, and intermittent ignition operation resulting from a poor ground at the control module

Classroom Manual
Chapter 11, page 285.

Many vehicles require the use of **dielectric grease** on connector terminals to seal the connections from contaminants.

Carbon tracking is the formation of a line of carbonized dust between the distributor cap terminals or between a terminal and the distributor housing.

Distributor Cap

Distributors are fastened in a number of ways. The initial check is to ensure that the cap is properly secured.

Next, the distributor cap and rotor should be removed for a good visual inspection. Any physical or electrical damage is easily recognized. Cracks in the cap are easily seen (Figure 11-3). Other damages that you need to look for are corroded or burned metal terminals and **carbon tracking** inside the distributor cap. Carbon tracking is an indication that the high-voltage secondary electricity has found a low resistance path over or through the plastic. This can cause a cylinder to fire at the wrong time or misfire. Any damaged cap has to be replaced.

The rotor should also be inspected for discoloration and other physical damage (Figure 11-4). The end of the rotor never touches the distributor cap terminals. This gap between the rotor and the terminals is called an air gap. It is small, usually a few thousands of an inch, and it is necessary because if the rotating rotor comes into contact with the distributor cap terminals, damage will occur. However, the arcing that occurs across this gap will eventually cause corrosion, requiring the cap and rotor to be replaced. Some manufacturers recommend coating the end of the rotor with a silicon dielectric compound to reduce the arcing. This helps in the prevention of radio frequency and extends the life of the cap and rotor.

Some distributor caps, such as the one used by GM, have a high-energy ignition coil housed inside the top of the distributor cap (Figure 11-5).

Replacement of this distributor cap is a little more involved. The coil assembly, unless being replaced, is removed from the old distributor cap and reinstalled onto the new one. Make sure the carbon center cap button under the coil is properly installed with the tension spring and rubber washer type seal.

Timing Advance Mechanisms

A good time to check the timing advance mechanism is when the distributor cap is off.

Centrifugal advance mechanisms should be checked for free motion. Move the rotor on the distributor shaft clockwise and counterclockwise. The rotor should rotate in one direction approximately 1/4 inch and then spring back. If not, the centrifugal advance mechanism might

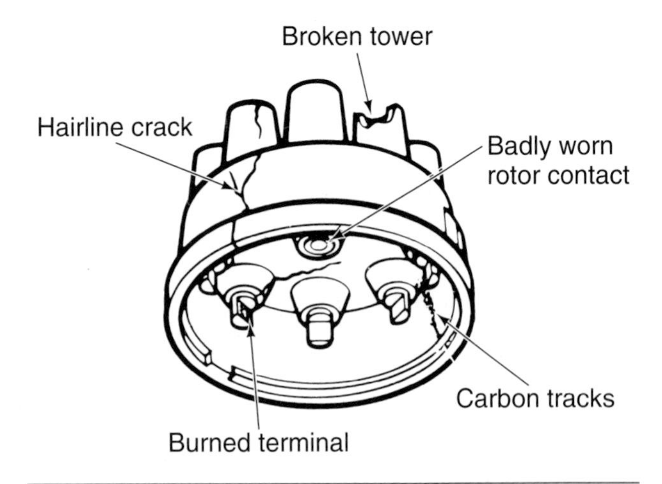

Figure 11-3 Types of distributor cap problems.

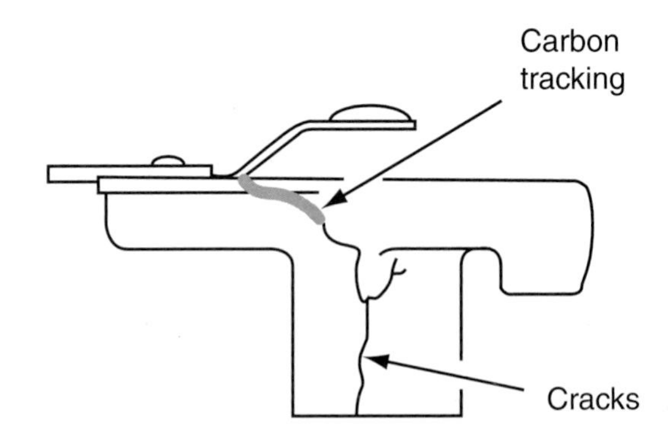

Figure 11-4 Inspect the rotor for cracks and evidence of high voltage leaks.

Figure 11-5 HEI coil mounted in the distributor cap.

be binding or rusty. A quality penetrating oil or lubricant can be used to free up the pivot points.

Don't forget to check the vacuum hoses that are part of the advance system. Check for damage, leaks, or disconnection. Disconnected hoses can cause poor fuel economy, poor idle, stalling, and performance.

Electronically controlled timing systems eliminated the use of centrifugal and vacuum advance/retard mechanisms within distributors. The wiring to and from the distributor and ignition module should also be checked for faults.

Primary Circuit Switches and Sensors

Electronic and computer-controlled ignitions use transistors as switches. These transistors and supporting electronic circuitry are housed inside a **control module** that is mounted to or inside the distributor or is remotely mounted, depending on the vehicle manufacturer. Regardless of where it is mounted, the control module should be mounted tightly on a clean surface. A loose mounting can cause a heat buildup that can damage the electronic components contained within the housing. Some manufacturers recommend the use of a special heat-conductive silicone grease between the module unit and its mounting. This helps to dissipate the heat away from the module, reducing the chance of heat-related failures.

The transistors in the control module rely on voltage pulses from a crankshaft position sensor in order to cause primary triggering. The sensor can be either a magnetic pulse generator or Hall-effect sensor. These sensors are usually mounted in the distributor for truck engines. In light duty vehicle application they are often mounted near the crankshaft.

The reluctor is often called a pole piece.

Most sensors are trouble free. The reluctor in the pulse generator type is replaced only if it is broken or cracked. A problem area could be the pick-up coil wire leads due to breaker plate movement with the vacuum advance unit. Inspect these leads carefully for broken wires or worn insulation causing the leads to be grounded (Figure 11-6).

On some systems, the gap between the pickup and the reflector must be checked and adjusted to manufacturer's specifications. To do this, use a nonmagnetic feeler gauge of the proper size to check the air gap between the coil and reflector (Figure 11-7). Adjust the gap if it is out of specifications.

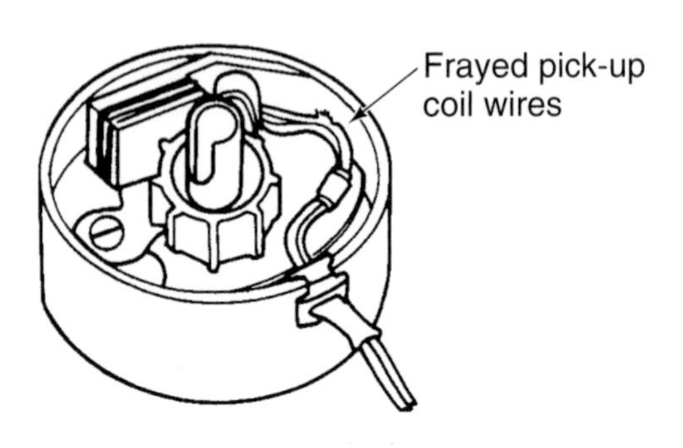

Figure 11-6 Inspect the pickup coil's wiring for damage and worn insulation.

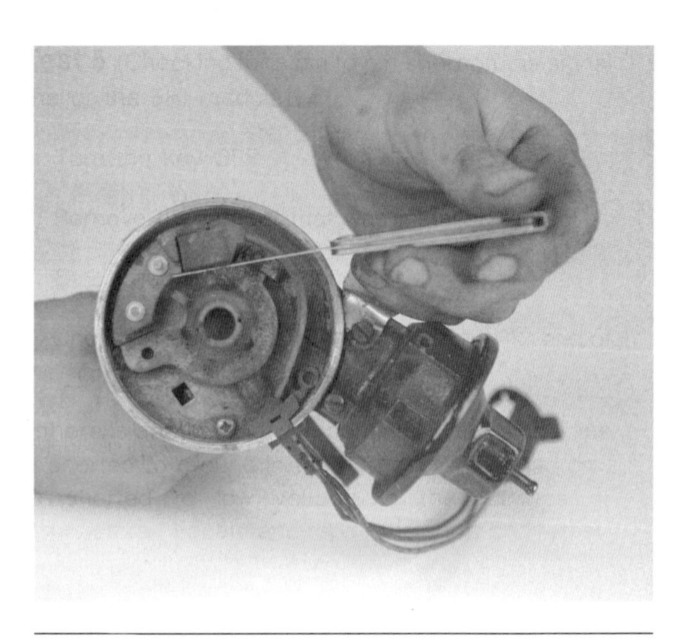

Figure 11-7 A brass feeler gauge is used to set electronic ignition pick-up air gap.

SERVICE TIP: A weak magnet in the pick-up coil can affect ignition timing. A quick check to detect a weak magnet can be performed if the distributor is removed from the engine for any reason. With the distributor removed, rotate the shaft by hand. You should feel the magnetic poles align on a good distributor. If you cannot feel the poles align, the magnet is probably weak.

Hall-effect sensors are also prone to similar problems as magnetic pulse generators. These include worn or broken wires and distributor shaft bushings and drive coupling problems.

Spark Plug Service

Classroom Manual
Chapter 11, page 287.

Special Tools

Spark plug sockets

Torque wrench

Spark plug gapper

Spark plugs are replaced for several reasons. One reason is to follow the manufacturer's recommendation of replacing them between 20,000 and 100,000 miles. The service interval depends on a number of factors such as type of ignition system, type of spark plug, engine design, operating condition, type of fuel used, and the types of emission-control devices used.

The other reason for removing spark plugs is if they are suspected of causing a driveability problem.

To remove and reinstall spark plugs requires a good spark plug socket. Spark plug sockets are available in two sizes: 13/16 inch (for 14-mm gasketed and 18-mm taper seat plugs) and 5/8 inch (for 14 mm tapered seat plugs). Many of the sockets feature an external hex so that they can be turned using an open end or box wrench. Some sockets also have a rubber insert within the socket to grip the spark plug, preventing it from dropping out of the socket and being damaged.

SERVICE TIP: Frozen spark plugs can break off, requiring extra work to extract them. In the long run, it will be more beneficial to soak a spark plug with penetrating oil if you have any doubts about it coming out.

To remove spark plugs:

1. Remove the cables from each spark plug, being careful not to pull on the cables. Instead, grasp the boot and twist it off gently. Note: Spark plug boot pullers are commercially available.
2. Using a spark plug socket and ratchet, loosen each plug a couple of turns.
3. Use compressed air to blow any dirt away from the base of the plugs.
4. Remove the plugs.

SERVICE TIP: To save time and ensure proper plug wire arrangement, use masking tape to mark each of the cables with the number of the plug to which it attaches. Many electronics stores have sheets of numbered tape strips that can be used for automotive and truck application.

SERVICE TIP: Using baby powder in plug wire boots can make future spark plug wire removal an easier task. Silicone dielectric grease is also used at the contact points.

SERVICE TIP: Take a piece of cardboard, and punch holes in it to reflect the number of cylinders and distinguish each side of the engine. Place the spark plugs in order in the holes, inserting the terminal end into each hole. That way, if a problem is indicated by the spark plug, you will know which cylinder it came from for further diagnosis.

Once the spark plugs have been removed, you must be able to look at them and determine their condition, which can relate to the condition under which the plug has been operating. Spark plug manufacturers provide charts that show the more common looking plugs and the engine condition that caused the plug to look the way it does (Figure 11-8).

A plug in good condition is usually light tan or gray in color with minimal gap wear. A plug that shows excessive wear (no more than 0.0001 inch for every 10,000 miles) should be replaced and the cause of the excessive wear repaired.

Many engine problems can be diagnosed by reading the plugs properly. Usually, if there is an engine problem, it will be in only one or a few cylinders. For that reason, you should first be looking for differences between cylinders. An example of this is a plug that is oil fouled, indicating oil is entering the combustion chamber past the rings or down past the valve guides. This is the type of engine condition that will not be taken care of by a tune-up. More than likely, some major engine work will be required.

Another example is a complete set of spark plugs that are carbon fouled. This is indicated by a layer of black, dry fluffy carbon deposits on the insulator tips. Looking at the chart, you will see some engine conditions that can be the cause of the plug condition. A common cause of this condition is a rich running system, meaning that the amount of gas in the combustion chamber is greater than it should be. Replacing the spark plugs might be only a temporary fix. What might be required is an air/fuel system repair or maintenance in addition to new plugs.

Preignition damage is characterized by melting of the electrodes and possible blistering of the insulator. It is caused by excessive combustion temperatures. When this occurs, look for anything that can cause temperatures to rise in the engine. Some of the causes could be overadvanced ignition timing, carbon buildup on pistons and valves, or the exhaust gas recirculation (EGR) valve not working when it is supposed to.

Regapping Spark Plugs

The gap of a spark plug is the distance in thousandths of an inch between the ground (side) electrode and the center electrode. The gap has a direct effect on the operation of the ignition system and the firing voltage. A new spark plug will have clean, square electrode surfaces

Figure 11-8 Normal and abnormal spark plug conditions.

(Figure 11-9). It is hard to maintain the flatness of the surface once the plugs accumulate miles due to wear. However, when plugs (new or old) are gapped, the squareness must always be maintained. A plug feeler gauge is used for an accurate measurement (Figure 11-10). The ground (side) electrode is bent down until the proper size feeler gauge or wire gauge passes through the gap, touching both electrode surfaces. Every engine has its own gap setting. Even new spark plugs should be checked for proper gap. Once the gap is properly set, it should look like the one in Figure 11-11.

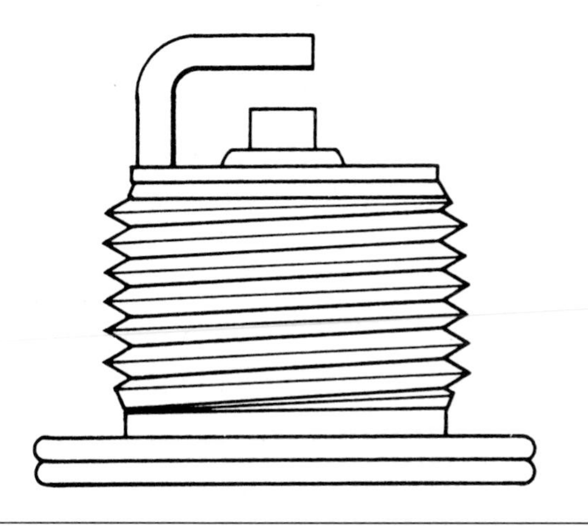

Figure 11-9 New plugs will have a flat, straight surface.

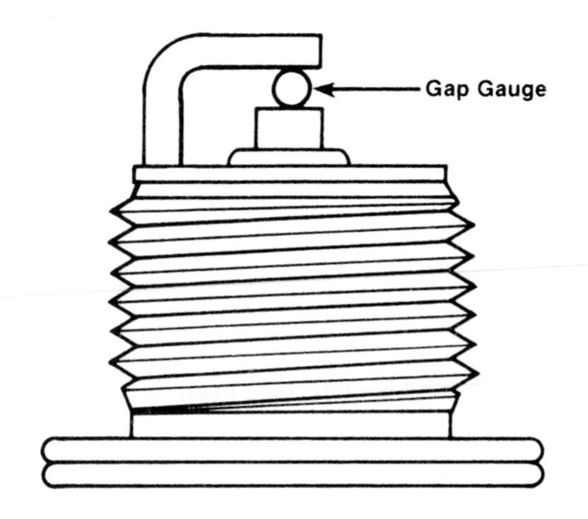

Figure 11-11 A squared off gap is correctly set.

Figure 11-10 (Top) plug feeler gauges; (bottom) flat feeler gauges.

Plug Type	Cast-Iron Head	Aluminum Head
14-mm Gasketed	25 to 30	15 to 22
14-mm Tapered Seat	7 to 15	7 to 15
18-mm Tapered Seat	15 to 20	15 to 20

Figure 11-12 Spark plug installation torque values.

To install spark plugs:

1. Make sure the plugs are clean. (Dirt could have been picked up during the service process.) Thread contact areas between plug and cylinder also have to be rust-free (proper grounding).
2. Adjust the air gap to specifications.
3. Install the plugs by hand first. By finger tightening, it assures that the threads are okay and you are not crossthreading the spark plugs. If they will not go in by hand, the threads should be checked and might require cleaning and rethreading with a thread-chasing tap.
4. Tighten the plugs with a torque wrench, following the vehicle manufacturer's specifications.

Examples of values are shown in Figure 11-12.

Special Tools

Heli coil kit

On rare occasions, you might run across a situation where the spark plug is literally welded into the cylinder head threads. In the process of removing the plug, the threads might be destroyed and a tap will not even work. Damage can also occur trying to extract a broken spark plug. Ideally, the cylinder head should be pulled off so any broken or drilled pieces can be removed to prevent further engine damage. However, sometimes the owner will make a decision to do whatever it takes short of pulling the cylinder head realizing that there is a gamble involved with this. When this situation occurs, a process called installing an insert thread can be performed. **Heli-coiling** is such a process. This process is shown in Figure 11-13. The old threads of the head are drilled out (A) and the hole is rethreaded to a larger size (B). A special steel insert (C) with the new larger size outside threads is installed into the newly threaded hole with a special tool called a mandrel (D). When the mandrel is removed, the steel insert remains in the cylinder head (E). Again, ideally this process should be performed with the cylinder head out of the vehicle. Maintaining proper spark plug replacement intervals could have prevented this process.

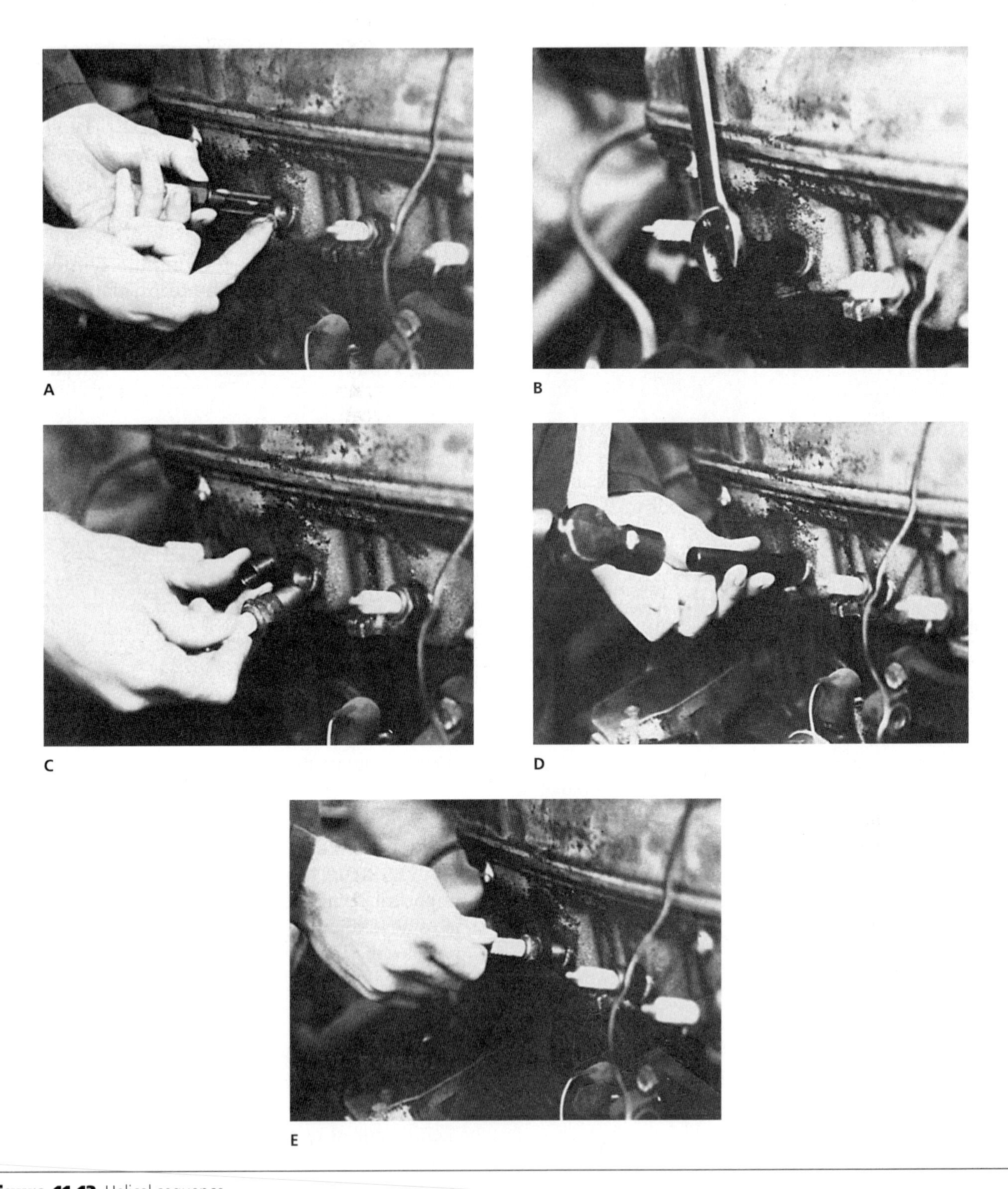

Figure 11-13 Helical sequence.

Setting Ignition Timing

Classroom Manual
Chapter 11, page 280.

The next common procedure that you might be required to perform is to ensure that ignition timing is correct. Procedures for checking and adjusting timing depend on the make and year of the vehicle and the type of ignition system. Most vehicles have an underhood decal with proper procedures and settings. For more detailed instructions, you should use the vehicle's service manual. Note: Some vehicles with computer-controlled ignition timing are non-adjustable and an ignition timing problem usually indicates a problem in the computer circuit.

CAUTION: Setting ignition timing often involves working in the proximity of moving components, so be sure to review all safety precautions before starting work.

Timing Light

Special Tools
- Timing light
- Service manual

The **timing light** is the most commonly used tool to check timing. By being able to check timing, this tool needs to be looked at as a diagnostic tool. It can verify and pinpoint many driveability complaints. The two common types of timing lights used by technicians are the standard timing light and the **advance timing light.**

We will discuss the standard timing light first. It has three connections: two for the battery and one for the spark plug. The spark plug connection is usually inductive and is placed around the number one spark plug wire. When the number one cylinder fires, the timing light's strobe light is also fired. The effect of the strobe light stops or freezes the action of the vibration dampner or flywheel where the timing marks are located. A number of different timing marks are shown in Figure 11-14.

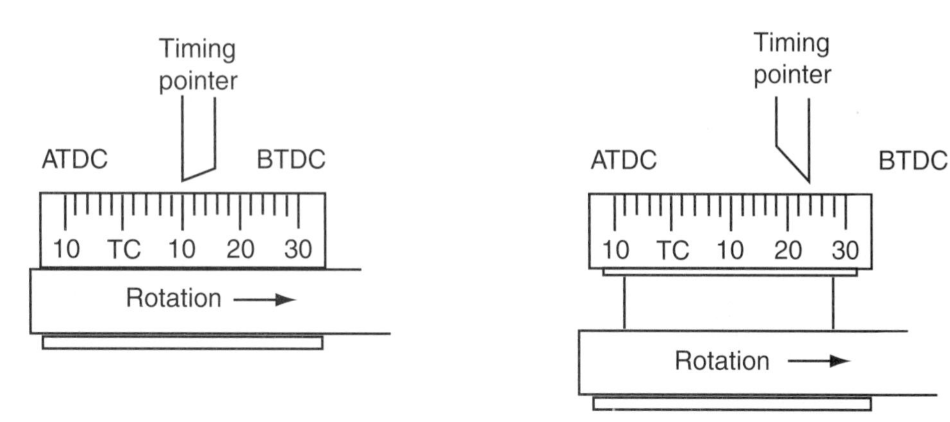

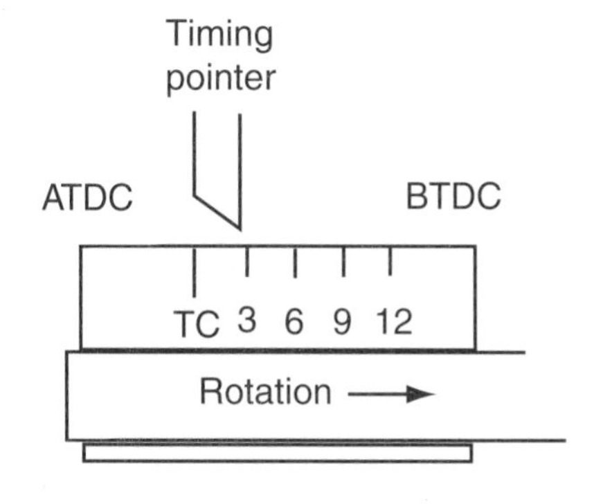

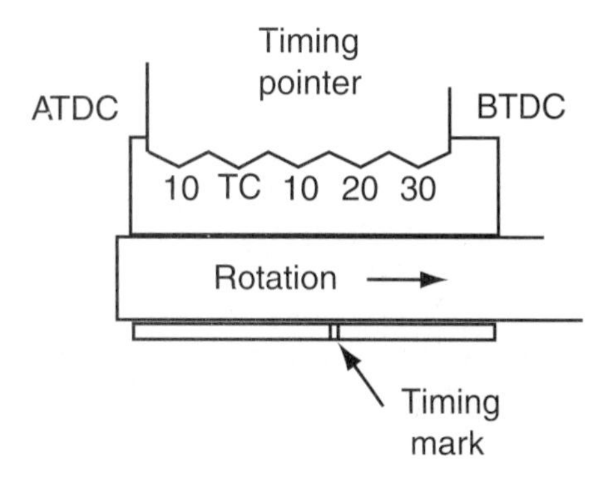

Figure 11-14 Typical timing marks.

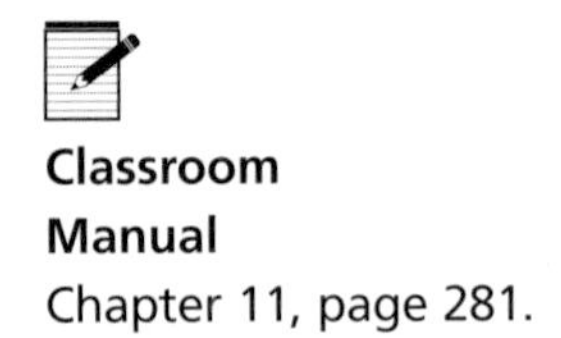

Classroom Manual
Chapter 11, page 281.

As you can see, timing marks vary with different engine sizes. Notice also that some engines use a marker or pointer with the settings notched on the vibration dampner while others have a notch on the dampner and the settings are on a stationary plate. Usually the top dead center (TDC) will be marked or a "0" will be used. When the notch and the 0 or TDC line up, cylinder number one will be in top dead center. The strobe of the timing light will freeze the position of the vibration dampner and allow us to see what position cylinder number one is in relative to top dead center. This concept is used to check engine timing and/or it allows us to properly time the engine. Timing must be correct. When checking timing, check to ensure that base timing is correct. Before checking or setting base timing, the engine must be in the proper mode. The following are some general guidelines:

- Base timing is normally checked and adjusted at idle speed. Verify the engine speed first. The speed required is usually low enough so any mechanical advance mechanism is not activated and is not affecting the base timing reading.
- Some vehicles require the vehicle to be in neutral, park, or even in drive. Check the decal or manual for the vehicle being tested.
- The vacuum advance unit must also be disabled so it does not affect base timing. Remove the vacuum hose to the unit and plug it.
- Computer-controlled systems might have other requirements. Because the computer controls timing advance, the computer must be eliminated in most cases from the timing control circuit. Usually this can be accomplished by disconnecting the appropriate connector at the distributor. Ford calls this connector the **SPOUT** (spark out) connector (Figure 11-15). General Motors carbureted vehicles are timed with the four-wire connector disconnected (Figure 11-16). Always keep in mind that if **base timing** is incorrect, all the automatic adjustment of timing during all modes of engine operation will be incorrect. This is especially true with computer-controlled systems, where the computer will make adjustments based on incorrect data.

Once the timing light is hooked up and the engine is running and all the conditions have been met, shine the light down on the set of timing marks and observe where they align. For example, if the specifications call for a 10 degrees before top dead center reading (Figure 11-17a) and the reading found is 3 degrees before top dead center (Figure 11-17b), the timing is retarded or off by 7 degrees.

Figure 11-15 Opening the SPOUT connector to check timing.

A

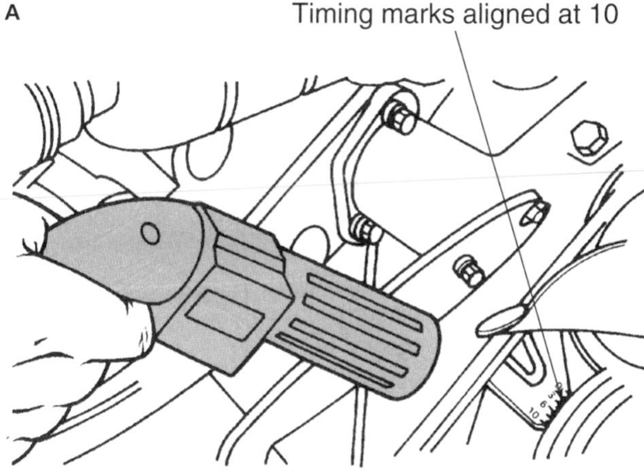

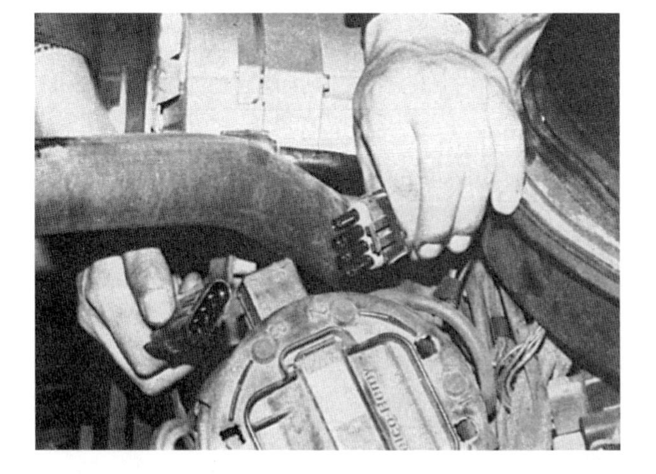

Figure 11-16 Opening the four-wire connector to check timing.

B

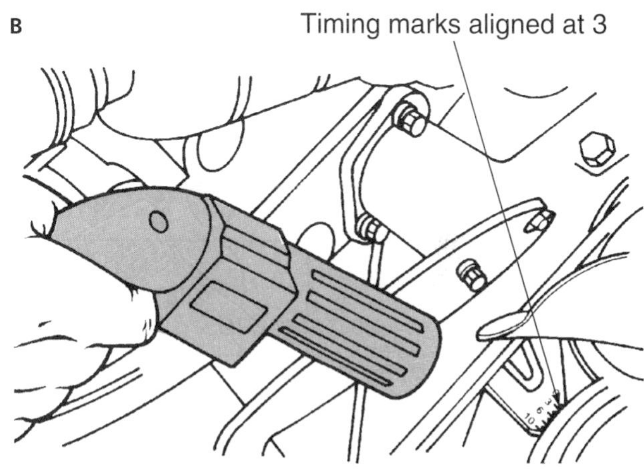

Figure 11-17 (A) Timing marks illuminated by a timing light and showing 10° BTDC; (B) Timing marks at 3° BTDC.

To correct this, the timing must be advanced by 7 degrees. To do this, rotate the distributor while observing the timing marks with the timing light until the timing marks align at 10 degrees (Figure 11-18).

The Advance Timing Light

Special Tools

Advance timing light

Service manual

An advance timing light is very useful for checking advancing systems. Figure 11-19 shows a typical advance timing light. You will notice that the back of the light has a meter calibrated from 0 degrees to 45 degrees. A knob controls this meter. It is part of the handle and controls the amount of delay between the time the spark plug fires and the time that the light flashes. By turning the knob, you can delay the flash of the light by any number of degrees shown on the back of the meter. The delay of the light allows the use of the engine's timing mark as a reference and enables you to note a change in timing. It is an accurate way of reading how many degrees advance the engine is running at.

Let us walk through a typical procedure:

The initial timing has to be checked and reset if needed. Using Figure 11-20 as a reference, the engine is timed to 9 degrees before top dead center (BTDC) at 650 RPM. We wish to determine if the mechanical advance will advance the timing by an additional 10 degrees by the time the engine has reached 1500 RPM. With the vacuum advance disconnected and plugged, we raise the engine speed to 1500 RPM and turn the knob on the timing light handle until the timing marks line up at 9 degrees. The meter on the back will then indicate how many degrees

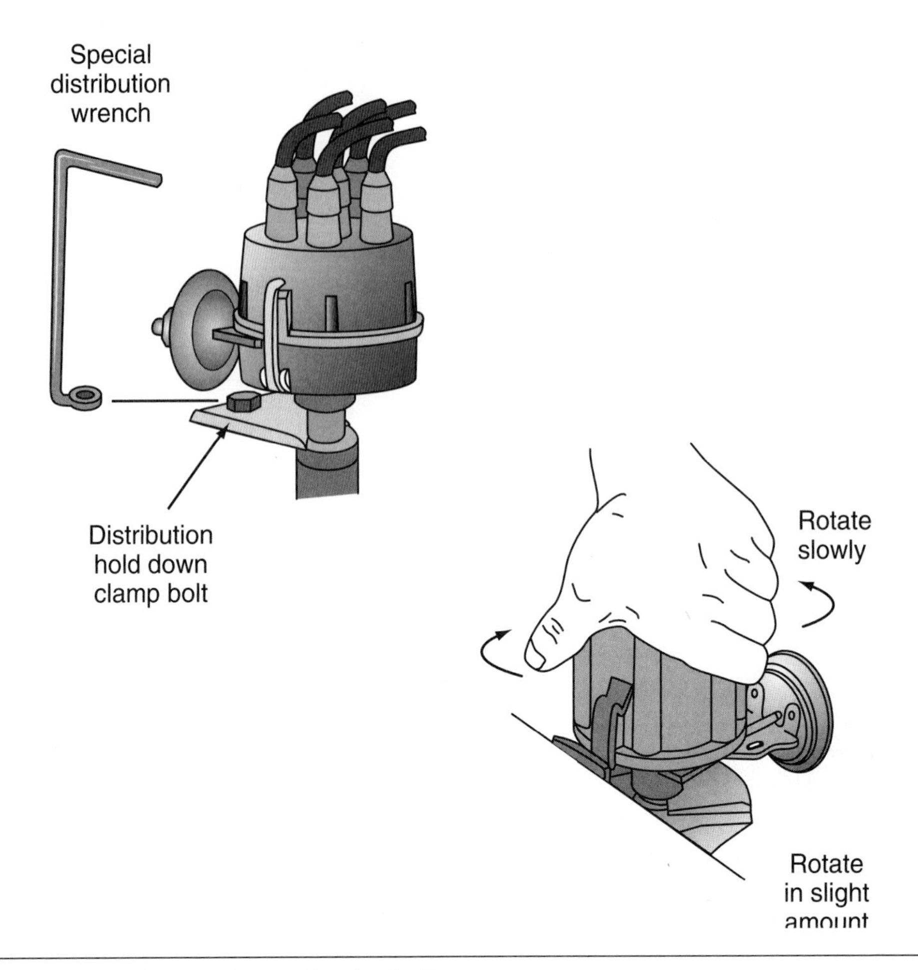

Figure 11-18 Set the timing by rotating the distributor.

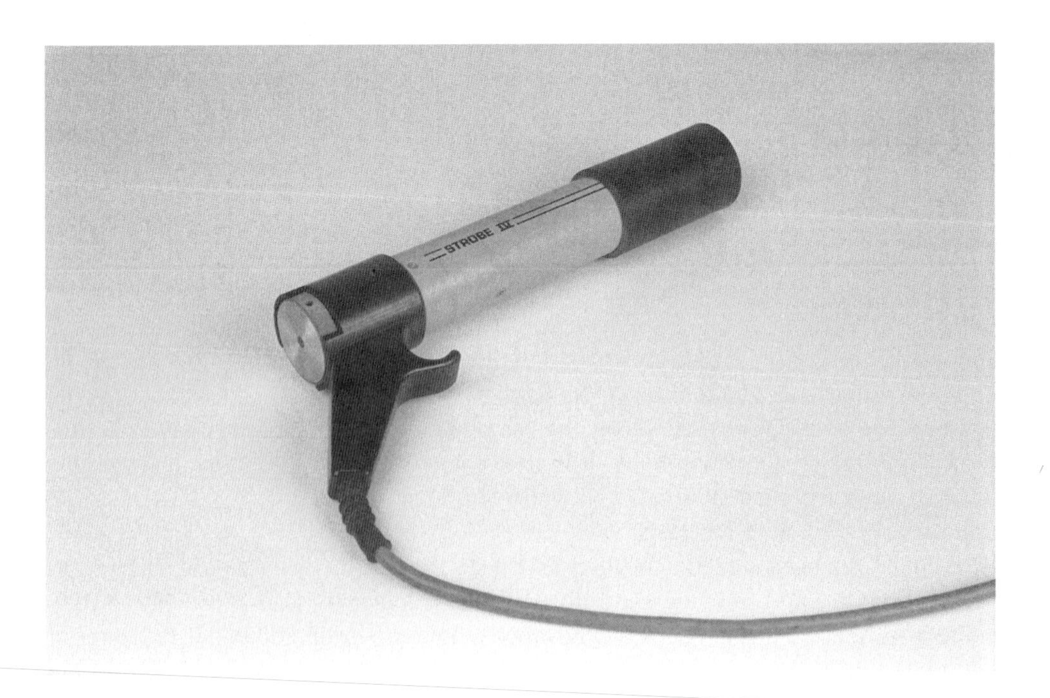

Figure 11-19 Advance timing light.

Speed	Actual	Specification
650	9 BTDC	9 BTDC
1000	16 BTDC	12 BTDC
1500	25 BTDC	15 BTDC
2000	30 BTDC	20 BTDC
2500	30 BTDC	28 BTDC

Figure 11-20 Comparison of specifications and speed.

we have delayed the flash of the light. Those degrees represent the amount of mechanical advance at 1500 RPM. Note: The timing light does not change the actual ignition timing. It only delays the flash.

With an advance timing light, we can advance curve the distributor against manufacturer's specifications. Advance curving is another way of saying we checked mechanical advance at various speeds and vacuum advance at various vacuum levels or load conditions. By comparing our readings and the manufacturer's reading, we can determine if a driveability problem is timing advance related.

Symptoms of overly advanced timing include pinging or knocking. Insufficient advance or retarded timing at higher RPMs could cause hesitation and poor fuel economy. Many computer-controlled systems use knock sensors to retard timing when the engine is experiencing pinging or knocking. A faulty knock sensor can produce problems that are ignition related. A quick check is to watch the ignition timing while the engine is running and the engine block is tapped (near the sensor if possible). The noise should cause a change in timing.

Individual Component Testing (No Spark)

Medium and heavy duty truck shops don't usually use oscilloscopes for diagnostics. However, there are checks that can be easily performed with normal tools in case you are confronted with an ignition problem in a truck with a gasoline (spark-ignited) engine. The following tests are to be used as guidelines. Always refer to manufacturer's procedures. Most tests and checks involve the knowledge you have already gained in the use of typical digital volt ohmmeter (DVOM) meters and basic understanding of electricity and electronic.

Figures 11-21 and 11-22 outline a procedure for isolating ignition-related problems without an oscilloscope.

Ignition Switch

Special Tools

Digital multimeter (DMM)

Test light

Service manual

The ignition switch is usually the source of power to the ignition control module or ignition coil. This can be seen in Figure 11-23, which illustrates an ignition system with two wires connected to the run terminal of the ignition switch. One is connected to the module and the other is connected to the primary resistor and coil. The start terminal of the switch is also wired to the module.

The main concern here is that voltage from the switch reaches the control module and the ignition coil. A test light or a digital voltmeter can be used to check for voltage. When using a test light, turn the ignition off and disconnect the wire connector at the module. Making sure that the engine will not crank, turn the key to the run position and probe the red wire connection to check for voltage.

SERVICE TIP: Disconnect the "S" terminal of the starter solenoid to prevent the engine from cranking.

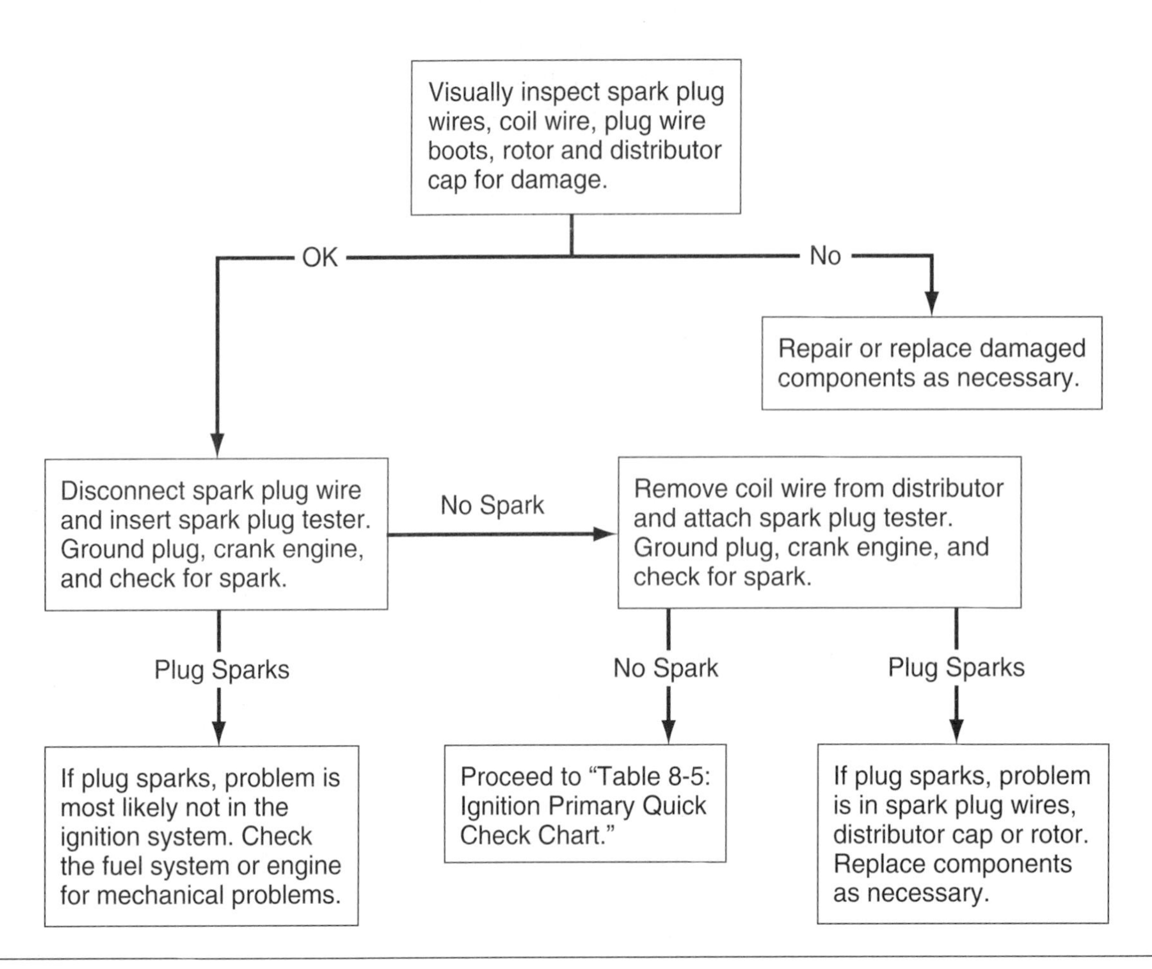

Figure 11-21 Ignition secondary quick check chart.

While the key is in the run position, also check for voltage at the ignition coils BAT or + terminal.

This next step requires the engine not to crank. Turn the key to the start position and check for voltage at the white wire connector at the module and the BAT terminal of the ignition coil.

The same test can be performed using a digital voltmeter. This time, a small pin can be inserted into the appropriate module wire. Another method is to back probe the wire at the connector rather than piercing the wire. Connect the voltmeter's positive lead to the pin and ground the negative lead to a good ground, preferably the distributor base. Turn the ignition to the run or start position as needed and observe the indicated voltage. The reading should be within 90% of battery voltage.

These tests ensure that power is reaching the control module and ignition coil. If no voltage was present, find the open and repair as needed.

Primary Resistor

If the ignition system uses a primary resistor, it should also be tested. An ohmmeter is used to check the resistor. Make sure there is no power to the resistor or your meter will get damaged. Using Figure 11-24 as a reference, the ohmmeter leads should be connected at the BAT terminal of the coil and the wiring harness connector wire that joins the red wire in the ignition module connector.

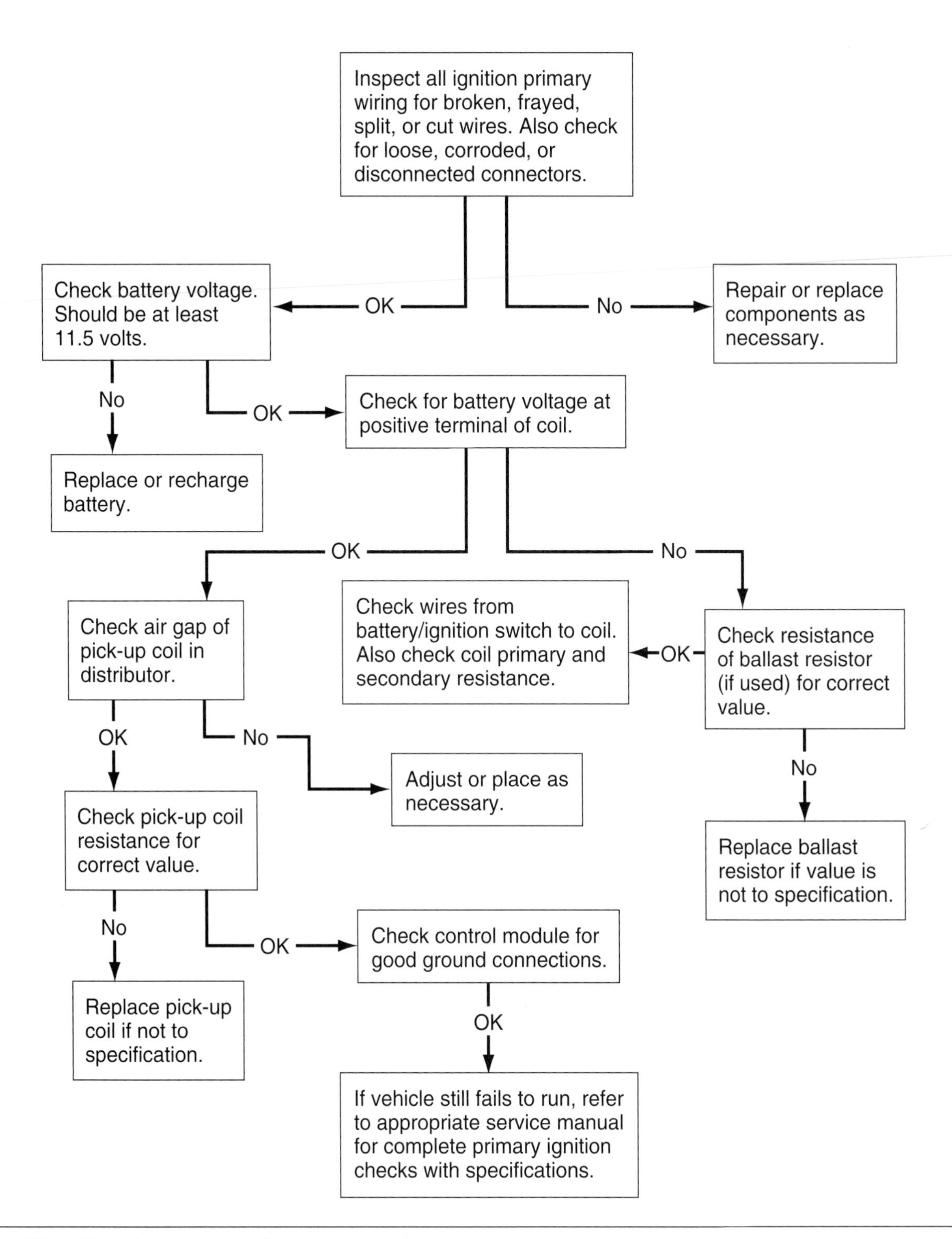

Figure 11-22 Ignition primary quick check chart.

Ignition Coil Resistance

By now, you should realize that all these tests are more than likely performed to diagnose a no start, no spark symptom. Once available voltage has been verified, the next test would be to measure the primary resistance of the ignition coil (Figure 11-24). Look for the correct specifications in the vehicle's service manual. For this example, the listed specification is 3 to 4 ohms. If the

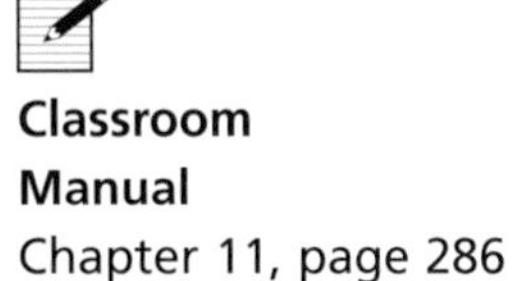

Classroom Manual
Chapter 11, page 286

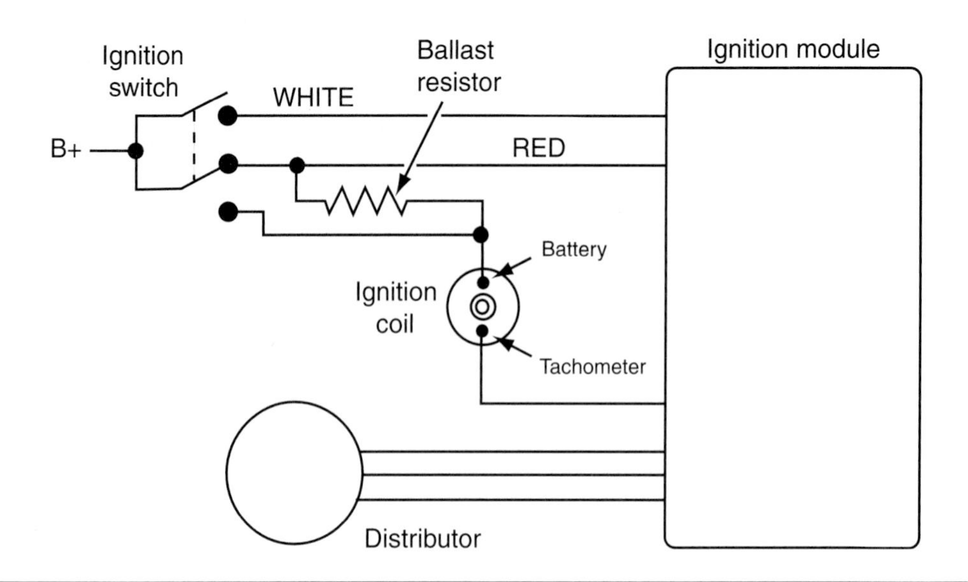

Figure 11-23 Simple ignition switch circuit.

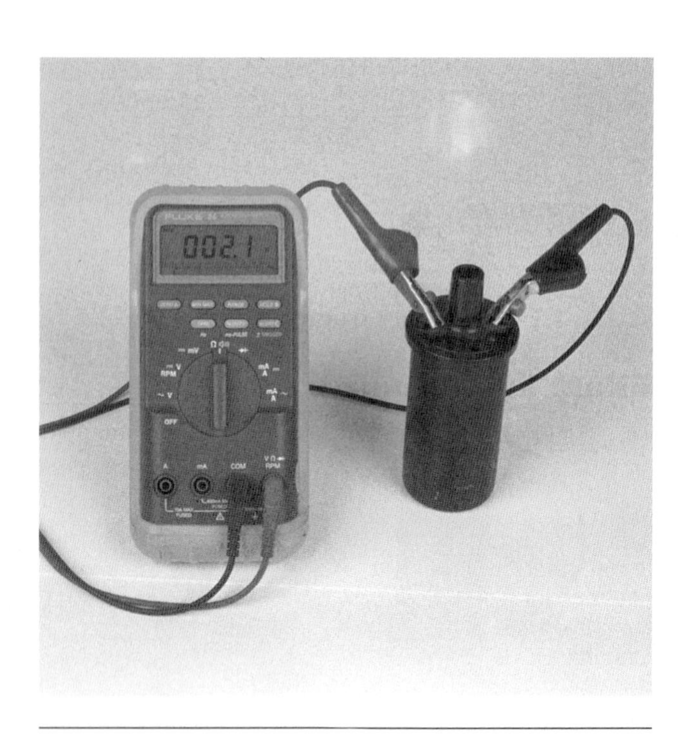

Figure 11-24 Measuring coil primary resistance.

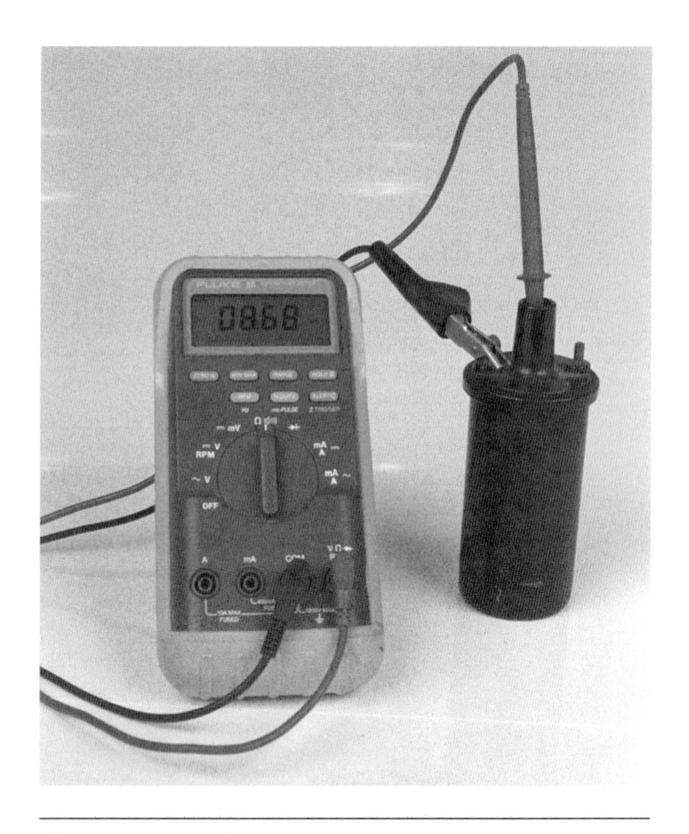

Figure 11-25 Coil secondary resistance.

Special Tools

Ohmmeter

Service manual

specification is verified to be correct, move one of the leads over to the coil tower (Figure 11-25). The reason that we expect continuity and resistance is because the coil secondary winding is connected to the primary winding inside the coil case. Secondary resistance will be measured regardless of which primary terminal is used for one of the ohmmeter lead connections (Figure 11-26). The specification for this system is 7,000 to 13,000 ohms. These tests ensure that the coil is capable of producing the required kilovolts of output as long as no cracks or insulation breakdowns (in E-coils) are present.

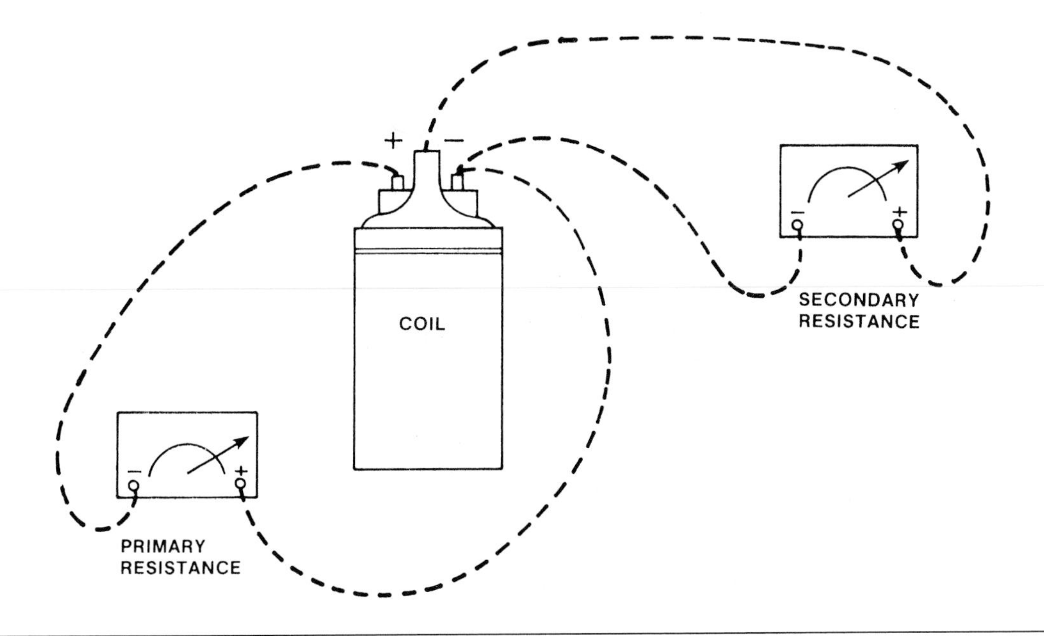

Figure 11-26 Ohmmeter connections for primary and secondary resistance.

Pickup Coil

The pickup coil of a magnetic pulse generator can also be checked for proper resistance. The AC pickup can also be easily checked for AC voltage output with a scope or voltmeter. For any of the tests, use Figure 11-27 as a wiring reference.

For a pickup coil resistance check, you will use a DVOM with the meter function set to read resistance. With the connector disconnected (Figure 11-28), hook the meter leads to the purple and orange/yellow wires and measure the resistance. The specification is 400 to 800 ohms. If the resistance is correct, verify that the distributor does turn when the engine cranks.

Classroom Manual
Chapter 11, page 291.

Special Tools

DVOM

Labscope

SERVICE TIP: When the symptom is a no start/no spark, one of the first visual checks is to see if the distributor shaft turns. It can save a lot of unnecessary diagnostic time.

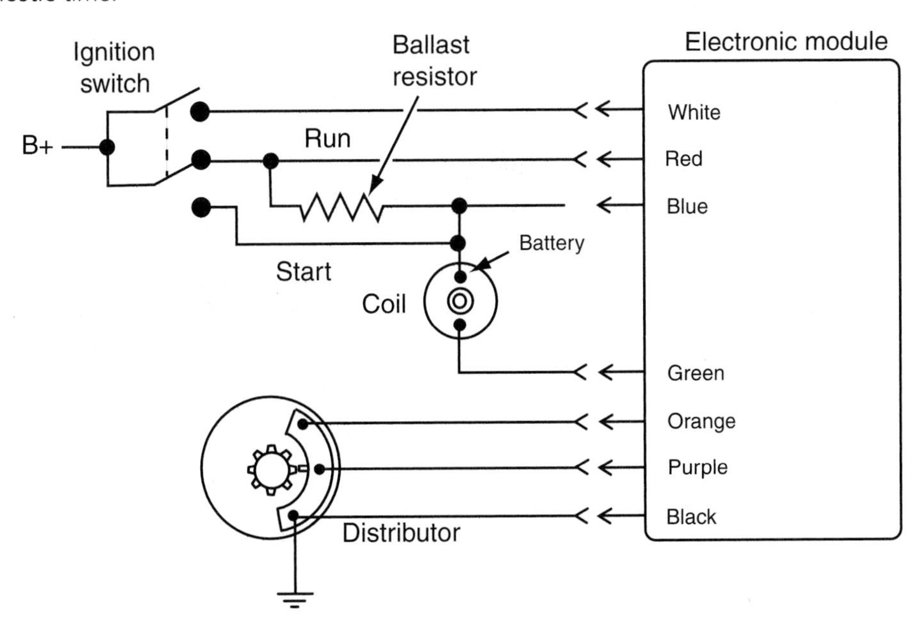

Figure 11-27 Dura-Spark.

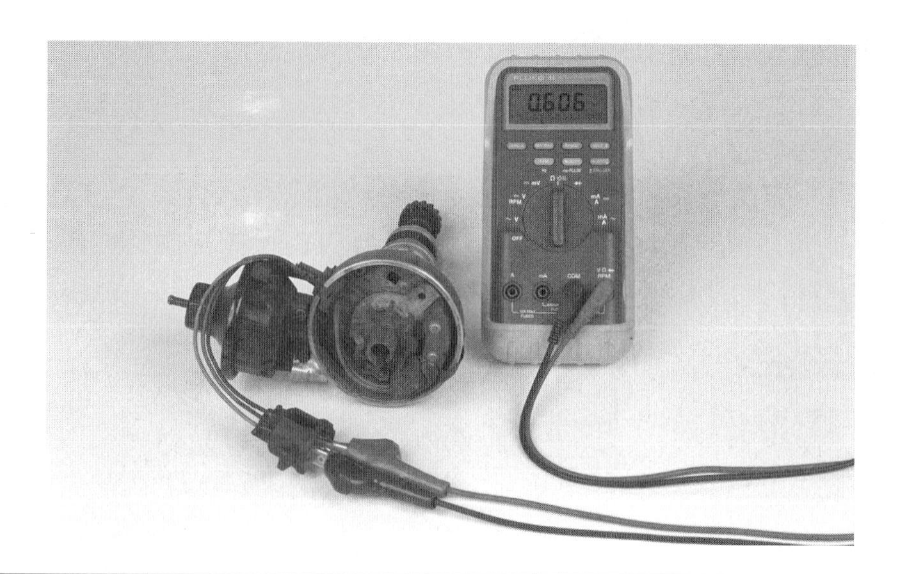

Figure 11-28 Ohmmeter checking stator.

Special Tools

Distributor wrench

Spark plug socket

With the leads of the meter connected to the pickup coil, set the DVOM function to read AC voltage. Crank the engine while observing the meter. The indicated voltage should read 1 to 3 volts.

A labscope can also be used to check pickup coil operation. With a scope, you can check for voltage and integrity of the signal. The scope can indicate problems such as an out of adjustment pickup coil assembly, a bent distributor shaft, or worn shaft bushings. Connect the lab scope's leads across the same wires as the previous AC voltage test leads were connected and set the labscope to a low AC voltage scale. When the distributor shaft spins, an AC waveform should appear on the screen (Figure 11-29). The waveform should have both a positive and negative pulse. The tops and bottoms of each cycle should be even with each other. If not, it could be an indication of a worn distributor shaft and/or bushings or a bent shaft.

Removal and Replacement of a Pickup Unit

If a pickup unit is determined to be faulty, it will have to be replaced. Photo Sequences 8 and 9 show two different distributors and how the pickup units are replaced.

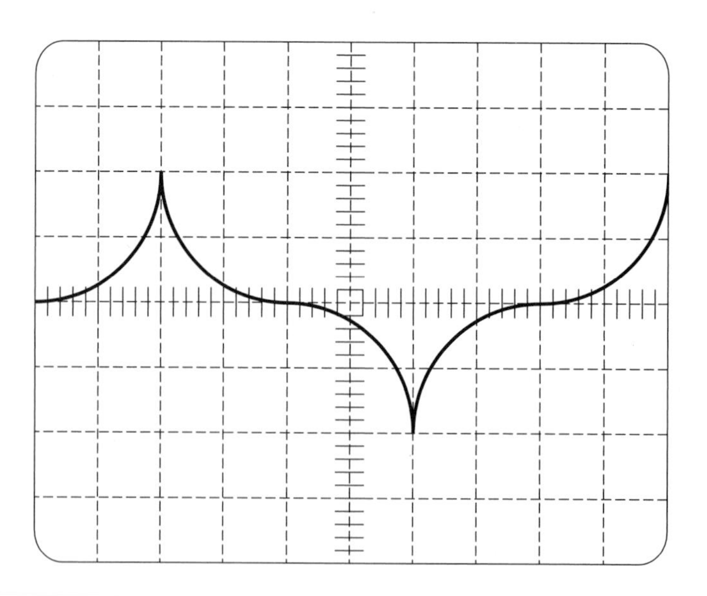

Figure 11-29 An AC waveform from a PM generator pickup unit.

Photo Sequence 8
Replacing Pick-Up Coil

P8-1 Release the hold-down clamps.

P8-2 Remove the distributor cap and rotor.

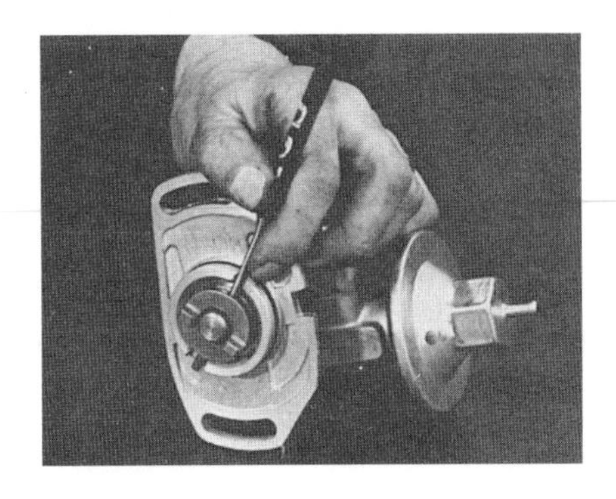

P8-3 Drive pin out of gear.

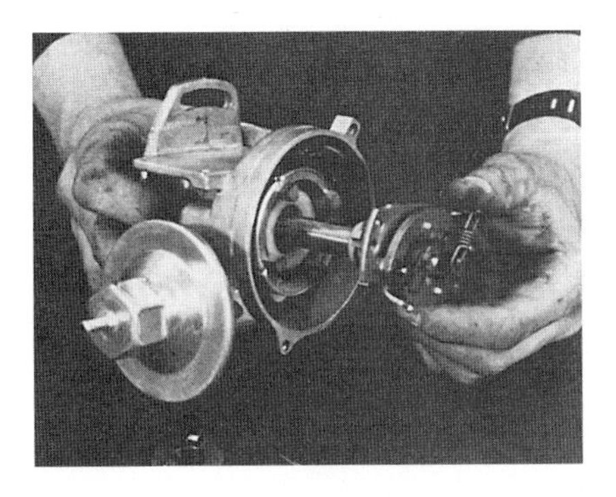

P8-4 Remove the distributor shaft.

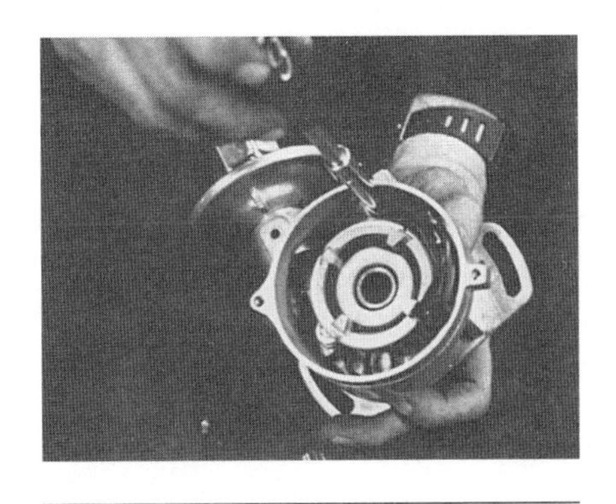

P8-5 Remove the pick-up coil hold-down screws.

P8-6 Remove the pick-up coil.

P8-7 Install a new pick-up coil.

P8-8 Replace the distributor shaft and drive gear.

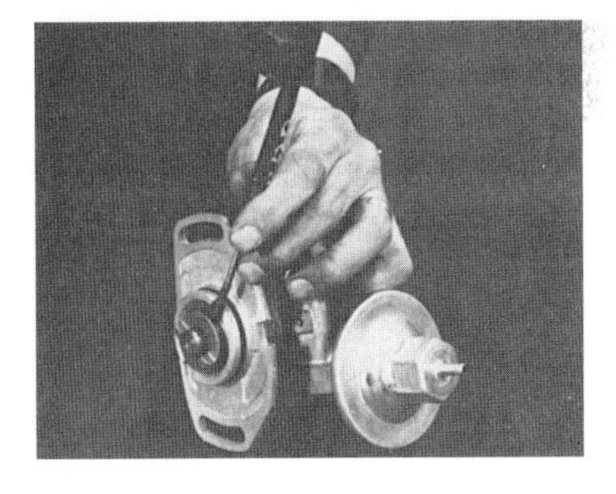

P8-9 Drive the pin in place.

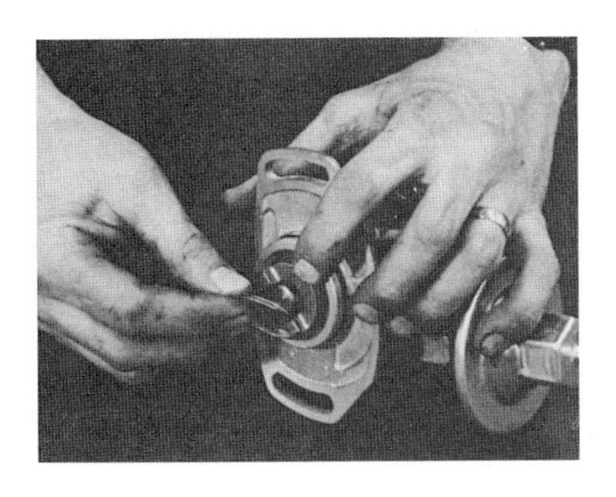

P8-10 Wind the spring around the pin.

Photo Sequence 9
Procedure for Removing and Replacing GM Pick-Up Coil

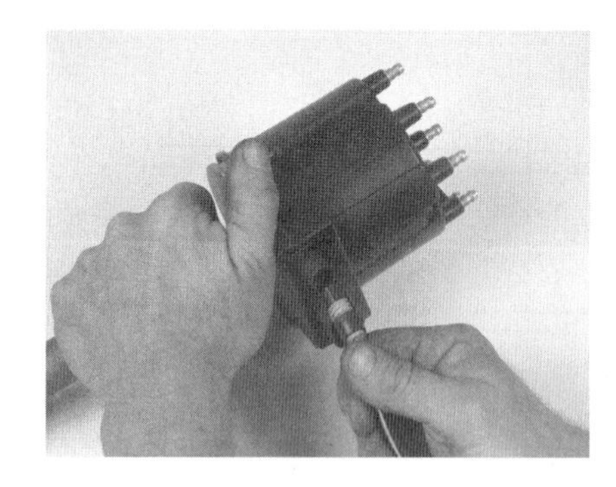

P9-1 Remove both connectors prior to pulling distributor.

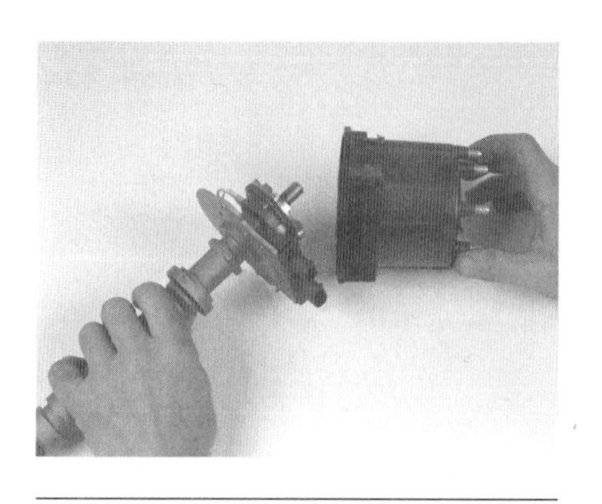

P9-2 Remove distributor and rotor.

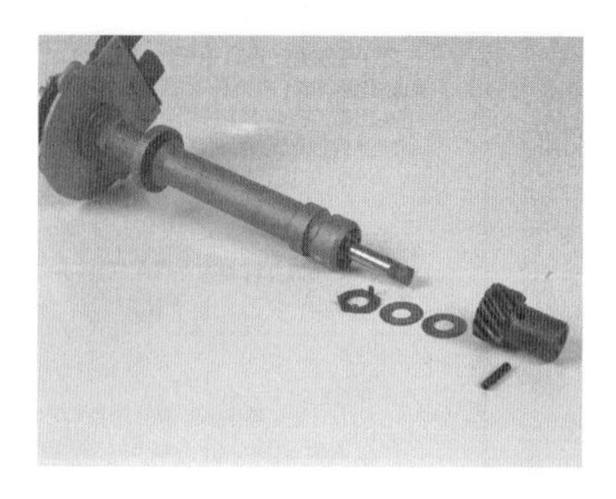

P9-3 Remove distributor dry pin and gear.

P9-4 Pull shaft out.

P9-5 Remove locking washer. Disconnect pick-up coil from module and pull off.

Classroom Manual
Chapter 11, page 292.

Hall-Effect Sensors

Hall-effect switches are used on spark-ignited engines and they can be also found on some diesel engines. One such example is in a Navistar DT 466E where it is used as a camshaft position sensor (CMP).

The most important concept to remember about a Hall-effect switch is that it needs power and ground to operate. The third wire is the signal wire back to the control module or computer. When a metallic shunt is inserted, the signal goes from low to high.

Therefore, most Hall-effect sensors can be tested by connecting a 12-volt battery across the plus (+) and minus (−) voltage terminals of the sensor and a voltmeter across the (−) and signal voltage terminals (Figure 11-30).

The shunting of the switch is simulated by inserting a steel feeler gauge or knife blade between the Hall layer and magnet (Figure 11-31).

If the sensor is good, the voltmeter should read within 0.5 volts of battery voltage when the metal blade is inserted. When the metal blade is removed, the voltage should read less than 0.5 volts.

This sensor can also be measured with a labscope. With the engine running, it will appear to vary (varies around 1/2 battery voltage). Any increase of RPM will not change the voltage. However, the frequency of the signal changes. If the ranges are out of specifications, replace the sensor.

Note: It is necessary to remove the distributor from the vehicle if the coil is surrounding the distributor shaft.

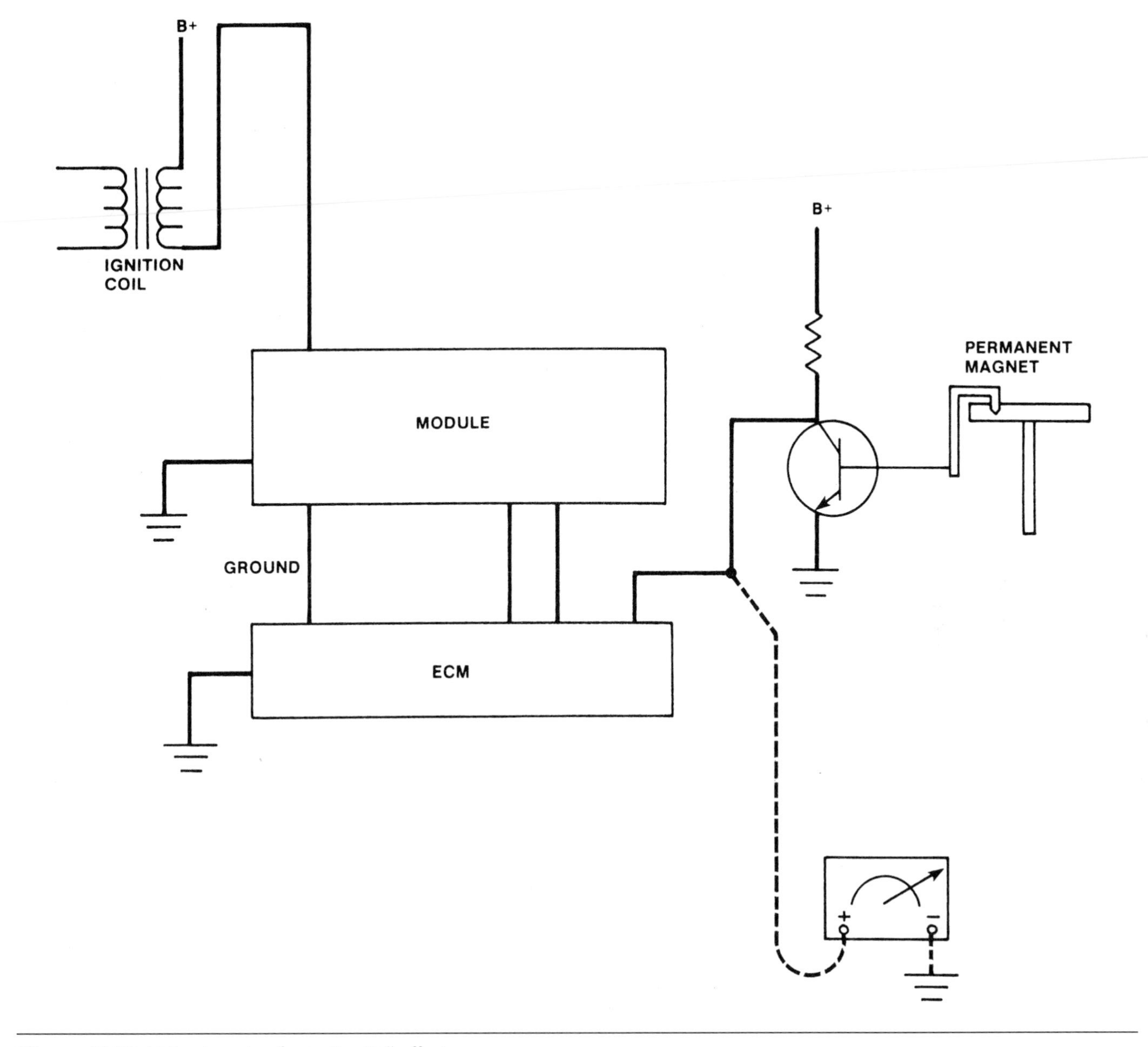

Figure 11-30 Voltmeter setup for testing Hall-effect sensors.

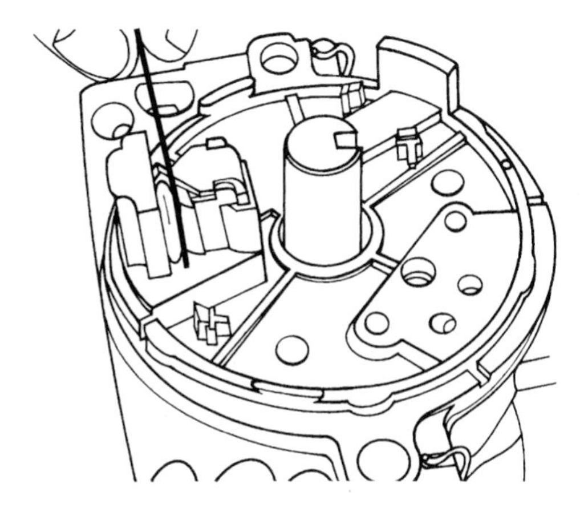

Figure 11-31 Test a Hall-effect switch with a steel feeler gauge.

Procedure for Removing and Replacing Distributor Assemblies

Any time a distributor is removed from the engine, a few common steps need to be taken in order to ensure a quick and timely reinstallation. To remove a typical distributor, begin by disconnecting the distributor wiring connector and vacuum hose(s) if applicable. Next, remove the distributor cap and note the position of the rotor. Someplace on the distributor you can make a chalk mark that is in relation to the rotor as a future reference for reinstalling.

Another chalk mark or similar mark should be made on the engine block or other component to reference the position of the distributor itself. The vacuum housing unit is usually a good distributor housing reference. Remove the hold down bolt and clamp if used. Remove the distributor. If the engine is not cranked before reinstallation, you can reverse the procedure by aligning your reference marks and then re-time the engine. This procedure is shown in Photo Sequence 10.

The service often involves the engine having to be cranked over once the distributor is removed. This involves a procedure that requires timing the distributor to the engine. This procedure is shown in Photo Sequence 11.

Stress Testing Components

Stress testing is a method of diagnosing intermittent problems that might be due to temperature, moisture (rainy or humid weather), and vibration. By stressing various ignition components, you might be able to reproduce certain driving conditions. Proper questions asked from the customer can also lead to the proper stress test. The following are sample questions: Does the problem occur on cold mornings? Does it occur when the engine is hot? Is the problem more evident on a rainy day? The rainy day problem was addressed in the visual inspection segment where a mist of water was sprayed to induce a voltage leak in the plug wires, distributor cap, and rotor.

Special Tools

Spray bottle (water)

Heat gun

Component cooler

Ideally, a good stress test involves monitoring an oscilloscope secondary pattern. A scope can visually show a problem as certain tests are performed. However, without a scope you can often feel or hear a change in the engine.

To stress ignition components requires minimum pieces of equipment. You are already aware of the use of a water spray or misting bottle. For the other two tests you will need a **heat gun** or hair dryer of about 500 watts and a means of cooling components. Cooling components can be performed with a component cooler spray that can be purchased from an electronics store. These component coolers can cool to below −60° F.

WARNING: Do not use liquid freon (R-12) to cool components. It is illegal.

Let's say you decided through good communication that an ignition-related problem seems to occur only during cold periods, you can duplicate the cold condition by using the Component Cooler. Begin to cool the major electronic ignition components, such as the ignition module or the pickup unit. By directing the cold air onto the components to a freezing level, you might be able to feel or hear a change in the engine if an affected component reacts to the cold. Allow the affected component to warm up again to see if the symptom disappears. If the component is found to react to the changes, it is an indication that it is bad and should be replaced.

If the complaint leads you to believe it is heat related, the heat condition can be duplicated by using a heat gun (Figure 11-32). Again, hit all the major ignition components with heat while observing for a change in the running of the engine. If a component reacts to the heat, perform a pinpoint test of the component (if applicable). Replace or repair as needed.

Photo Sequence 10
Procedure for Removing and Replacing Distributor Assemblies

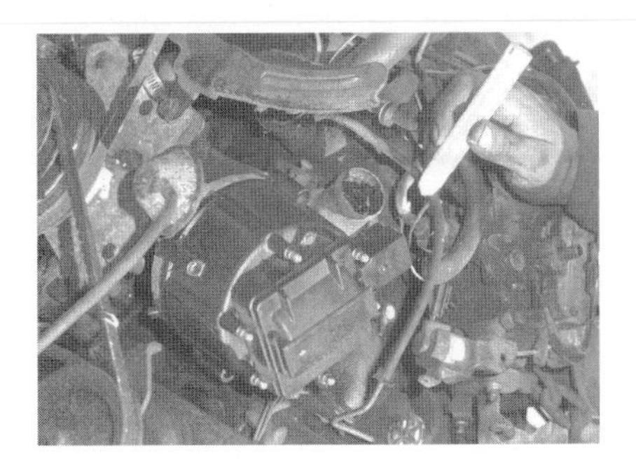

P10-1 Chalk distributor vacuum advance, base or wires for position of distributor base in engine.

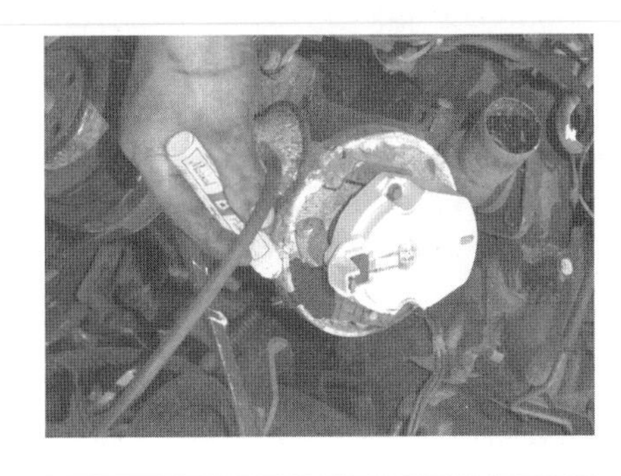

P10-2 Chalk rotor position.

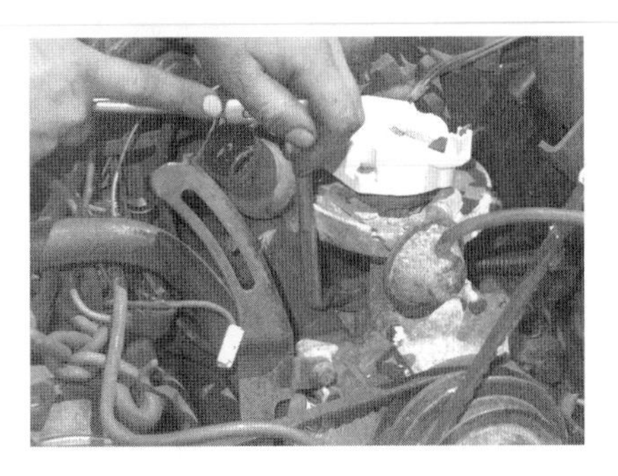

P10-3 Remove hold-down bolt.

P10-4 Pull distributor assembly out.

P10-5 Time engine after reversing removal for installation.

Photo Sequence 11
Typical Procedure for Timing the Distributor to the Engine

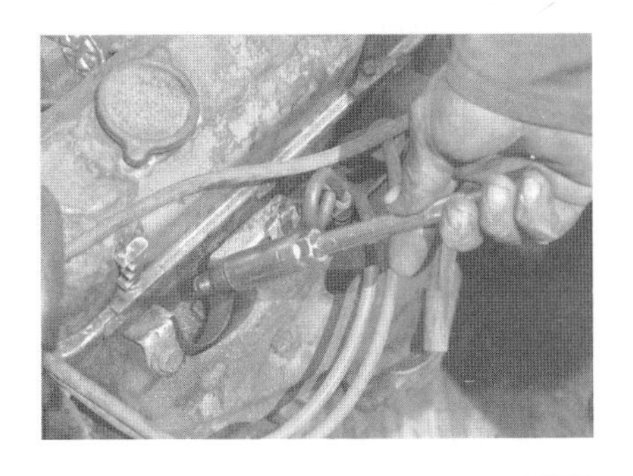

P11-1 Remove the number one spark plug.

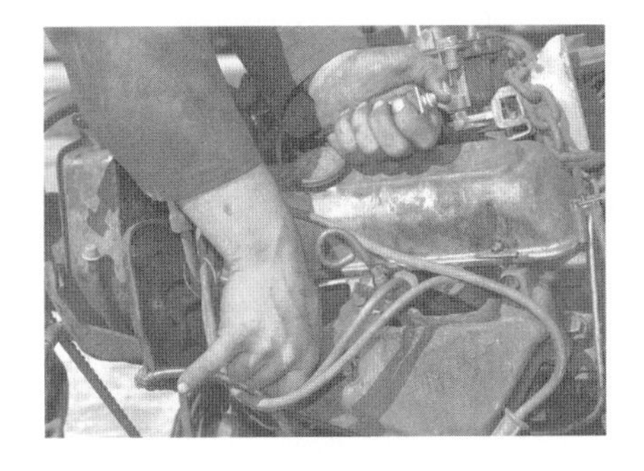

P11-2 Place your thumb over the number one spark plug opening and crank the engine until compression is felt.

P11-3 Crank the engine a very small amount at a time until the timing marks indicate that the number one piston is at TDC on the compression stroke.

Photo Sequence 11 (Continued)
Typical Procedure for Timing the Distributor to the Engine

P11-4 Determine the number one spark plug wire position in the distributor cap.

P11-5 Install the distributor with the rotor under the number one spark plug wire terminal in the distributor cap and one of the reluctor high points aligned with the pickup coil.

P11-6 After the distributor is installed in the block, turn the distributor housing slightly so the pickup coil is aligned with the reluctor.

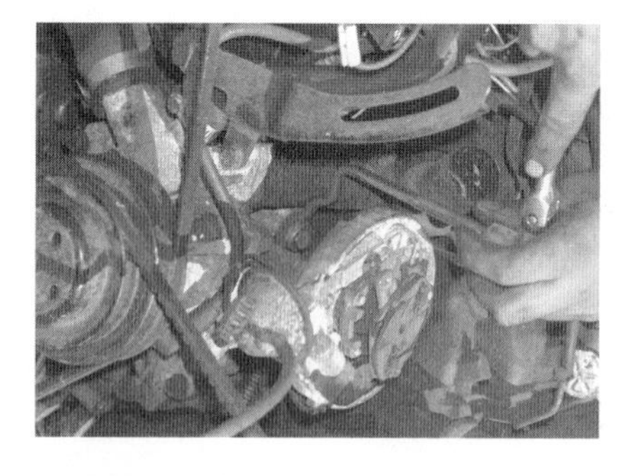

P11-7 Install the distributor clamp bolt but leave it slightly loose.

P11-8 Install distributor cap. Connect all leads to the wiring harness.

P11-9 Install the spark plug wires in the proper cylinder firing order and in the direction of distributor shaft rotation.

Figure 11-32 Heating the system to simulate hot weather.

Open-Circuit Precautions

Before electronics became a part of the ignition system, it was a common procedure to remove a spark plug or coil wire while the engine was running to check for available secondary voltage. Another procedure was to stick a screwdriver or similar object in the spark plug wire boot and bring it close to the engine block (ground) and look for spark if a no start/no spark symptom was suspected. This was performed while the engine was being cranked over and also worked for checking each individual wire for an open circuit.

All of these tests forced the system to produce a maximum voltage. Performing these types of tests on today's vehicles can damage the electronic circuits and components. To prevent damage, similar tests can be performed but a spark tester has to be used (Figure 11-33).

Special Tools

Test spark plug

Grounding probe

Jumper wire

This spark tester is very useful as a quick test tool for a NO START problem. Two spark testers are used: one for standard systems (15kv) and another one for high energy ignition (HEI) or high output systems (20kv).

In a NO START problem, the tester is used to check for spark distribution at the spark plugs. Remove a spark plug wire from the vehicle's spark plug and install it on the spark tester. Ground the spark tester to a good engine ground with the plug's alligator clip. Crank the engine over while looking for a spark.

SERVICE TIP: Check spark at two wires. This eliminates the chance that you might have the spark tester on an open or shorted wire.

If there is no spark at the plugs, it will indicate one of two possible problems:

1. Spark distribution
2. Spark production

To close in on the problem area, perform the same test on the coil wire (distributor end). If there is no spark, it indicates a primary circuit problem. If there is spark, the likely problem is spark distribution.

SERVICE TIP: If there was spark at the plugs, check the condition of the vehicle's plugs. Extremely soaked or worn plugs will not fire. If the problem is wet plugs and it is recurring, you might want to consider a higher heat range or try regapping the plugs 5 thousandth of an inch smaller.

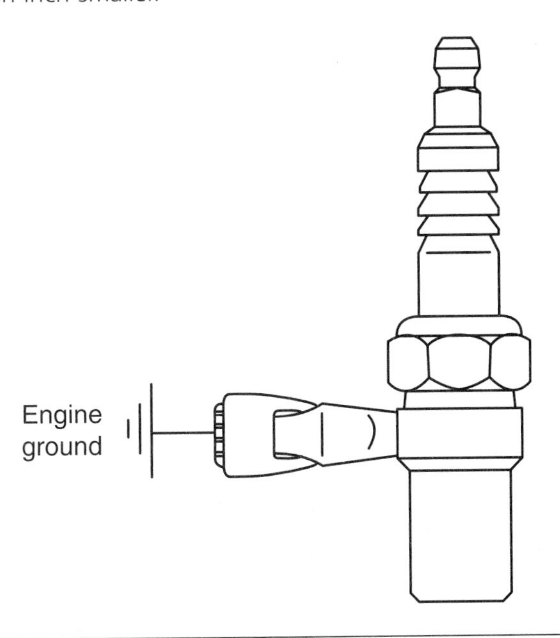

Figure 11-33 Test spark plug.

SERVICE TIP: Check the distributor shaft. If the shaft is not turning, there can be no triggering for spark production.

Special Tools

Scan tool

DMM

Jumper wire

Classroom Manual

Chapter 11, page 290.

Computerized Engine Control Diagnostics

When it comes to diagnostics, we need to look at computerized engine control systems in two ways:

1. To determine a problem requires a thorough understanding of how the system works.
2. Computerized systems can be placed into a self-test mode and provide certain information through the use of fault codes.

Any troubleshooting process requires a logical flow and specific information from an appropriate factory service manual. The following guidelines can help you with diagnosis.

Most electronic engine control problems are usually caused by sensors and their circuits, but output devices can also be the cause of a problem. Realizing this, the logical procedure in most cases is to check the input sensors and wiring first, then the output devices and their wiring, and lastly the computer itself. Do not lose sight of the fact that noncomputer components and systems can affect the sensor's outputs. For this reason, all mechanical and noncomputer-controlled system integrity should be checked before any computer system checks are performed.

If your diagnosis does lead to a malfunctioning component, it can usually be checked, depending upon the type of component, with a voltmeter, ohmmeter, labscope, or scan tool. A good visual check can often find an obvious problem. Any part that is burned, broken, or corroded, or has any other visible problem must be replaced or repaired as needed before any further diagnosis is performed.

Most sensors and output devices can be checked with an ohmmeter or voltmeter. A thermistor is a perfect example of this. First, determine the relationship of resistance to temperature (Figure 11-34). Then use an ohmmeter to verify the resistance as the vehicle engine is started when cold and allowed to warm up. Look at the table in the figure for typical values for a negative temperature coefficient (NTC)-type of thermistor. Output devices such as solenoids can also be checked with an ohmmeter.

Service manuals often list voltage values for their sensors and output devices. This allows the use of the voltmeter as a test tool. Any device that works with voltages can also be checked with a labscope. A scan tool allows us to see sensor and output device activity.

Self-Diagnostic Systems

Computerized vehicles will typically have a means of entering into self-test mode. This enables the computer to evaluate the condition of various electronically controlled engine systems. If problems are found, they are identified as either hard faults (on demand) or intermittent failures. Each fault is assigned a number that is referred to as a trouble code and is stored in the computer.

Figure 11-35 shows examples of trouble codes and the devices to which they pertain. Note that these are pre OBDII codes.

Resistance	Temperature
100,000	−40°F
25,000	32°F
1000	100°F
500	180°F
150	212°F

Figure 11-34 Typical thermistor values.

Inputs	Codes
Coolant Temperature Sensor	14, 15
Vacuum Sensor	34
Barometric Pressure Sensor (if used)	32
Throttle Position Sensor	21
Distributor Reference (crank position—engine speed)	12, 41
Oxygen Sensor	13, 44, 45
Vehicle Speed Sensor	24
Ignition On	
Air Conditioner On/Off	
Park/Neutral Switch	
System Voltage	
Transmission Gear Position Switch(es)	
Spark Knock Sensor (if used)	43
EGR Vacuum Indicator Switch (only on 1984 and later models)	53
Idle Speed Control Switch	35

ECM
ROM and PROM process information (inputs) and issue commands (outputs).
RAM monitors indicated circuits, sets codes, and reads codes out when put in diagnostics.

Codes	Outputs
23	Fuel Mixture
42	Electronic Spark Timing
	Electronic Spark Retard (if used)
	Idle Speed
	AIR Management
	EGR
	Canister Purge
	Torque Converter Clutch (or shift light on manual transmission)
	Air Conditioner
	Early Fuel Evaporation
	Diagnosis (check engine light) (test terminal) (serial data)

Figure 11-35 Overview of computer command control. Inputs and outputs with code numbers are monitored for faults. Those without are not.

If a problem is found in a system at the time of the self-test, it is referred to as a "hard fault." An intermittent or history fault refers to a malfunction that has previously occurred but is not present at the time of the self-test. An example of this is an intermittent open or short caused by a poor connection. Nonvolatile random access memory (RAM) allows intermittent faults to be stored for up to a specific number of ignition on/off cycles. Many times, if the trouble does not reappear during that period, the code is erased from the computer's memory.

There are various methods of retrieving and/or accessing the trouble codes generated by the computer. Vehicle manufacturers have diagnostic equipment designed to monitor and test the electronic components of their vehicle. Aftermarket companies also manufacture scan tools that monitor inputs and outputs and access trouble codes. Some vehicles also flash trouble codes on dash lights.

For each individual make and model, you should check the appropriate service manual on how to access and interpret trouble codes. In the next section, we will look at the two most common vehicles that use spark-ignited engines. Each vehicle uses a different method of retrieving trouble codes.

General Motors Systems

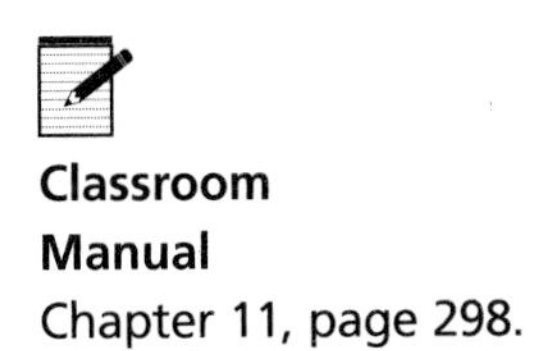

Classroom Manual
Chapter 11, page 298.

The ECM monitors major inputs, outputs, and their respective circuits for proper operation. If the ECM sees a fault such as an open, short, or a voltage value that stays too high or too low beyond a specified time in any of the monitored circuits, it will turn on a "check engine" light located in the instrument panel. A code number will also be recorded in its memory. Some models have a "service engine now" light and a "service engine soon" light, either of which can be selected depending upon the severity of the problem. The light warns the driver that a fault exists. Note: The "check engine" light should come on anytime the ignition is on without the engine running. This is known as a bulb check. A warning light system is shown in Figure 11-36.

A technician can retrieve these stored trouble codes using either of two methods. Both methods require the use of an assembly line communication link (ALCL), which is also referred to as the assembly line diagnostic link (ALDL) (Figure 11-37).

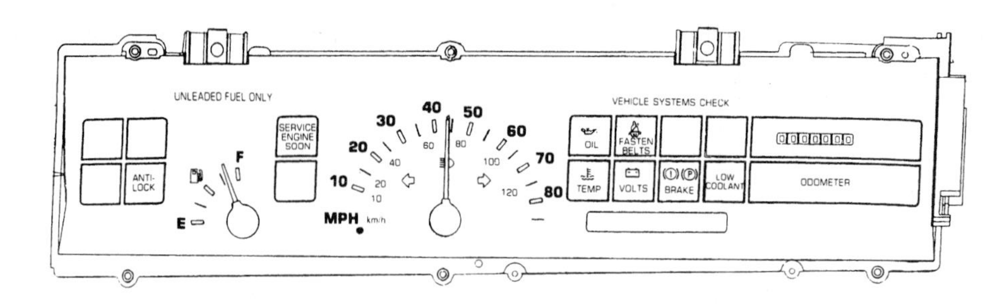

Figure 11-36 General Motors' CCC warning light system.

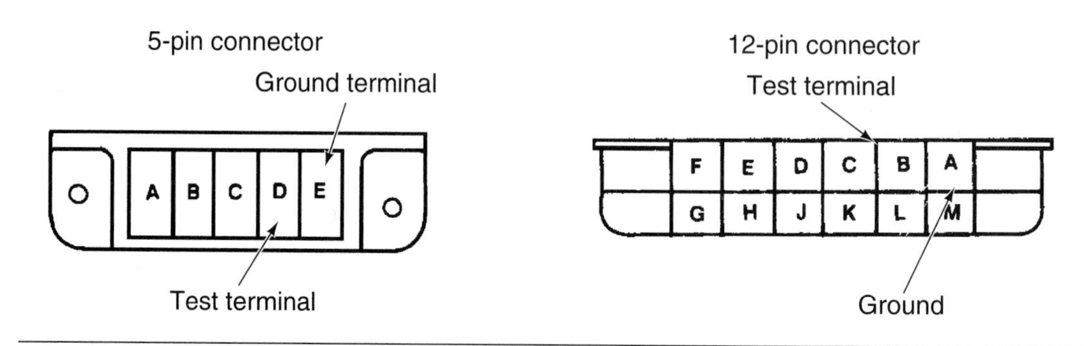

Figure 11-37 To initiate self-diagnostics, connect the ground and test terminals on the DLC connector.

In order to start standardizing acronyms, SAE recommended standard J1930 recommends that an electronic control module (ECM) be called a power train control module (PCM). The check engine light (CEL) and the service engine soon light should be called the malfunction indicator lamp (MIL).

The ALCL is also called the ADL; however, according to J1930 standards, it should be called the data link connector (DLC).

The first method of retrieving a trouble code is to ground the test terminal in the ALDL. Using the 12-pin connector as an example, you would connect pin A (a ground) to pin B (test terminal). This is accomplished using a paper clip, small jumper wire, or an aftermarket code key. The ignition key needs to be on for this check. Watch the "check engine" light to see if it displays a code 12. Code 12 is one flash, a short pause, and two flashers (Figure 11-38). A code 12 indicates the computer's diagnostic program is working properly. If any trouble codes are stored in the ECM memory, they will be displayed in the same manner. To exit diagnostic mode, turn the ignition key off and remove the jumper. Several codes can often be in the ECM memory. The different codes will appear with different sets of two numbers, such as 12, 21, 32, etc. The flashes are read as a set. Once a two-digit number completes, there will be a slightly longer pause and then the next two-digit code will appear. When a code displays, it repeats twice before the next new code is displayed.

Once the codes have been retrieved and noted, the next step is to check the service manual. Each code that is identified begins a flow of troubleshooting that is specific to that vehicle and code. The troubleshooting tree has to be followed precisely. If you skip a step, you might miss the problem.

The second method of obtaining stored codes is to plug a scan a type of test equipment into the ALCL. Test equipment of this nature can be hand held or a large engine analyzer. With the ignition, these test units display the stored codes on a digital screen. Most scan tools will also

Figure 11-38 The "check engine" light signals the codes. A code 12 signals that the computer's diagnostic program is working properly.

define the code and are able to see and display what the computer sees. Note: A scan tool will only receive and display what the computer allows it to.

After all the faults and problems are corrected, clear the ECM (PCM) memory of any current codes by pulling the PCM fuse(s) for 10 seconds. Verify that the codes are cleared by running the engine for 2 to 5 minutes. A road test is more conclusive.

SERVICE TIP: The ignition should be on before grounding the test terminal, otherwise the "check engine" light does not function properly.

Ford's System

Classroom Manual
Chapter 11, page 302.

Special Tools

Scan tool

Analog voltmeter

Service manual

Ford's main computer is referred to as an electronic control assembly (ECA) or PCM with the advent of onboard diagnostics II (OBD II). As with most engine control systems, the ECA monitors most of the input and output circuits. It expects to see certain voltage values within a specified range for any set of driving conditions. If it fails to see the expected voltages and conditions, a service code is recorded in its memory. If the fault does not reoccur within a specific set of ignition on/off cycles, the ECA will erase the recorded service code.

To retrieve these codes, Ford's EEC-IV systems provide a self-test connector somewhere in the engine compartment (Figure 11-39) or under the dash. This test connector can be connected to an analog voltmeter, the "Star Tester" (a Ford specific tester), or generic scan tools. The ECA displays service codes on the test equipment as a needle sweep on the analog voltmeter or as a digital readout on the scan tools. Figure 11-40 shows a hook-up using an analog voltmeter. Consult the service manual or the test equipment manufacturer's instructions for specific directions on how to conduct the self-test.

The following lists the different types of service codes stored in the ECA:

- On demand codes (also referred to as hard codes): These refer to existing faults when the self-test is ongoing.
- Separator codes: This code is represented by the number 10 and it only occurs during the key on/engine off (KOEO) segment of the self-test. It indicates to the technician that the on-demand codes are completed, and the memory codes are coming up next.
- Memory codes: They are sometimes referred to as intermittent codes or continuous codes. They refer to faults the ECA detected during normal vehicle operation but are not present during the self-test.
- Dynamic response code: This code occurs during the key on/engine running (KOERR) segment of the self-test and is represented also by the number 10 (the meaning of code 10 depends on the self-test segment in which it appears). With a scan tool, this segment is prompted step by step with instructions displayed on the screen. Either

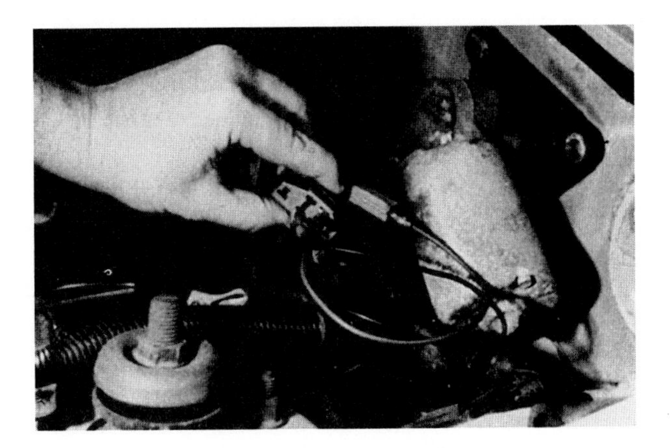

Figure 11-39 Ford's self-test connectors are used to connect test equipment to the system's computer.

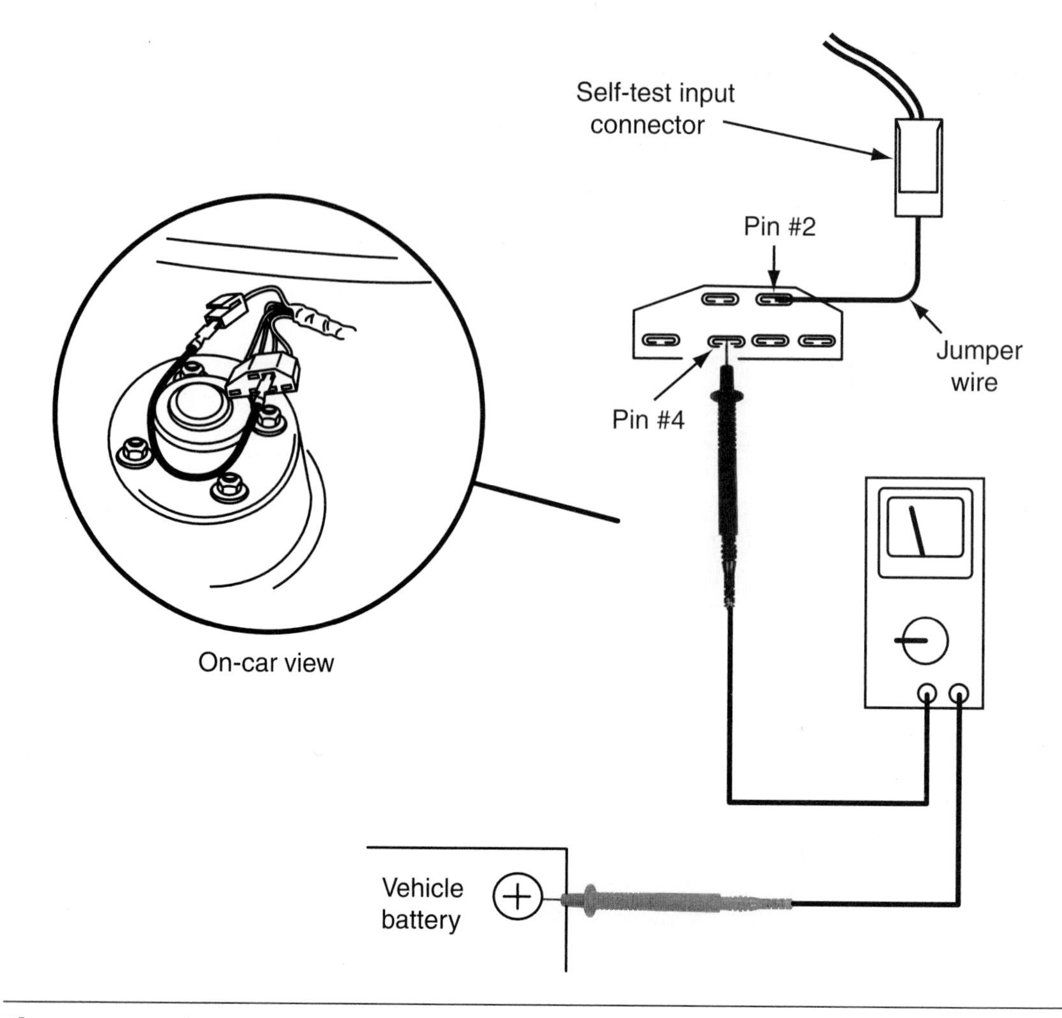

Figure 11-40 Voltmeter connection to Self-Test connector.

way, the technician has 15 seconds to quickly press the accelerator pedal to the floor and release it. This allow the ECA to check the throttle position sensor (TPS), the manifold absolute pressure (MAP), and the vane air flow (VAF) for a proper response. This test is often referred to as the "goose test."

- Fast codes and engine identification codes: These codes are intended for assembly plant tests.

Reading Service Codes with a Voltmeter

When reading codes with a voltmeter, you have to pay attention because it is easy to miscount the readings. You will have to observe and count the needle sweeps and estimate the time between sweeps. The following is an example of a typical sequence of a code 21 (hard code) followed by a separator code 10 and an 11, which represent no memory codes stored.

With the voltmeter properly hooked up and the key turned on/engine off, the needle sweeps upscale and back (0 to approximately 12 volts). Immediately, it will repeat this movement again. A two-second pause indicates the first digit (2) is complete. After another two-second pause, one sweep occurs followed by a four-second pause. The single sweep indicated that the second digit is 1. The four-second pause indicated the code is complete. After the four-second pause, the 21 is repeated. A six- to nine-second pause precedes the memory codes. A single sweep followed by a two-second pause and another sweep indicates a code 11. A four-second pause occurs and the code 11 is repeated.

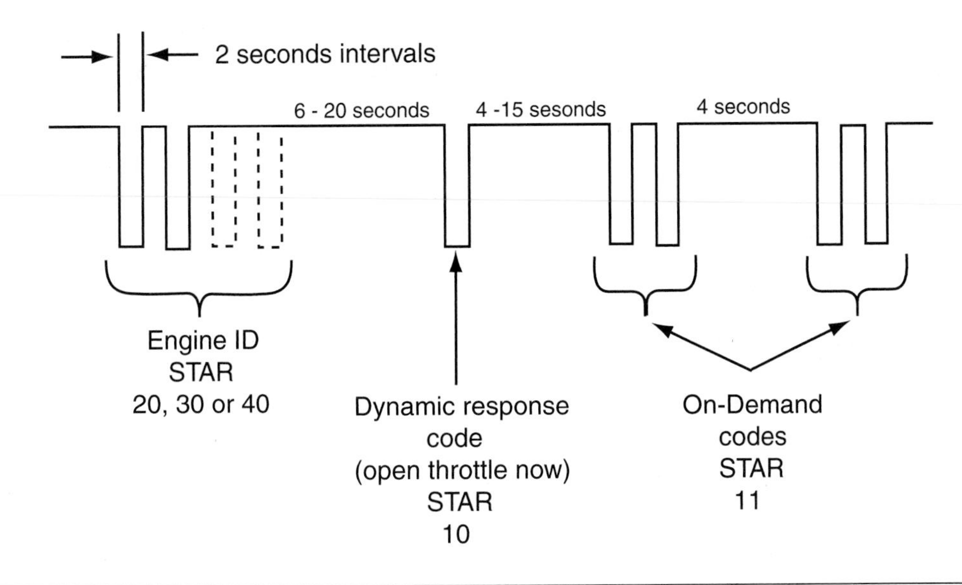

Figure 11-41 Typical service code display during Engine Running segment of the Self-Test.

An engine running segment of the self-test is presented in the same format, except a 6- to 20-second pause separates the engine identification code from the dynamic response code. A 4- to 15-second pause separates the dynamic response code from the hard codes (Figure 11-41).

Points to remember:

- ❑ A code 11 indicates all clear, system pass, or no memory codes.
- ❑ When performing a dynamic response code procedure, you have 15 seconds to perform functions such as goose the throttle, step on the brake pedal, and turn the steering wheel 180 degrees. If you fail to perform this within the allotted time, a code 77 will be set. This means operator did not respond or you blew it.
- ❑ Always check the vehicle's service manual for code interpretation. Code numbers may change meaning from year to rear.
- ❑ Do not measure voltage or connect a test light at the ECA harness connector unless specifically directed to do so by a particular test procedure. Damage to the ECA or to sensors can occur.
- ❑ The self-test also provides for a "wiggle test" to help you locate intermittent faults. This test can be performed in both the engine off or engine running mode.

OBD II Standards

The purpose of the onboard diagnostics II (OBD II) system is to closely monitor emissions related systems for faults, activate the MIL, and set a code. This second generation OBD goes one step farther because it also contains imbedded procedures that define how emission systems and components are tested. Included in the OBD II system are how codes are cleared and how a drive cycle is performed.

OBD II also addresses many of the industry concerns. With cooperation between SAE and various regulator agencies, the following standards were achieved:

- ❑ All vehicles would have a standard diagnostic connector in a standard location.
- ❑ All vehicles implemented a set of industry standard diagnostic test modes that can be executed over a standard communication data link.
- ❑ An OBD II scan tool is used to utilize a minimum set of standardized diagnostic test modes and information.

Environmental Protection Agency 40 CFR Part 86 Control of Air Pollution from New Motor Vehicles and New Motor Vehicle Engines, regulations required On-Board Diagnostic Systems on 1994 and later model year light duty vehicles and light duty trucks.

California Code of Regulations Section 1968. 1, Title 13 required a Malfunction and Diagnostic System for 1994 and subsequent model year passenger cars, light duty trucks, and medium duty vehicles with Feedback Control Systems.

Since many federal regulations tend to have their beginnings with California regulations, you might consider getting comfortable with OBD II. The following is an example showing General Motors coverage of OBD II relating to trucks.

- ❑ Vehicle Coverage for Truck, Vans, C/K Body, Commercial Van, and Motor Home (1998 only), 1996 C/K Series, Pickup, Sierra.

Engine: 4.3L V6 MFI	Vin W
Engine: 5.0L V8 MFI	Vin M
Engine: 5.7L V8MFI	Vin R
Engine: 6.5L V8 Diesel (1996-97)	Vin F
Engine: 6.5L V8 Turbo Diesel (1996-97)	Vin P
Engine: 6.5L V8 Turbo Diesel (1997-98)	Vin S
Engine: 7.4L V8 MFI	Vin J

The following are guidelines of OBD II that all vehicles will have:

1. A universal diagnostic test connector known as the Data Link Connector (DLC) with dedicated pin assignments.
2. A standard location for the DLC. It must be under the dash on the driver's side of the vehicle and must be visible.
3. A standard list of diagnostic trouble codes (DTCs).
4. Vehicle identification that is automatically transmitted to the scan tool.
5. Stored trouble codes that can be cleared from the computers memory with the scan tool.
6. The ability to record and store in memory a snapshot of the operating conditions that existed when a fault occurred.
7. The ability to store a code whenever something goes wrong and affects emissions.
8. A standard glossary of terms, acronyms, and definitions used for all components in the electronic control system.

The standard DLC, 16 pin-connector is shown in Figure 11-42. It is a D-shaped connector with a guide key that allows the scanner to be connected in only one way. Standardizing the connector and designating the pins allow data retrieval with any scan tool designed for OBD II.

Seven of the sixteen pins have been assigned OBD II standards. The remaining nine pins can be used by the individual manufacturers for their own needs. The DLC is designed only for scan tool use. It cannot be jumped across any of the terminals to display codes by flashes as pre-

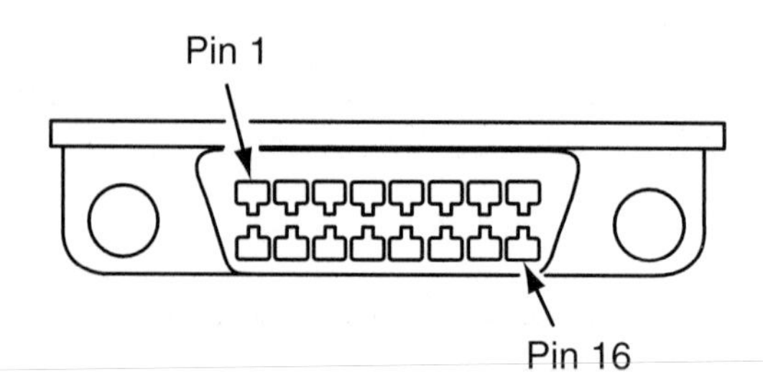

Figure 11-42 Standard DLC for OBD II.

The SAE J2012 standards specify that all DTCs will have a five-digit alphanumeric numbering and lettering system. The following prefixes indicate the general area to which the DTC belongs:

1. P — power train
2. B — body
3. C — chassis

The first number in the DTC indicates who is responsible for the DTC definition.

1. 0 — SAE
2. 1 — manufacturer

The third digit in the DTC indicates the subgroup to which the DTC belongs. The possible subgroups are:

0 — Total system
1 — Fuel-air control
2 — Fuel-air control
3 — Ignition system misfire
4 — Auxiliary emission controls
5 — Idle speed control
6 — PCM and I/O
7 — Transmission
8 — Non-EEC power train

The fourth and fifth digits indicate the specific area where the trouble exists. Code P1711 has this interpretation:

P — Power train DTC
1 — Manufacturer-defined code
7 — Transmission subgroup
11 — Transmission oil temperature (TOT) sensor and related circuit

Figure 11-43 OBD II trouble codes.

OBD II systems did. The MIL is only used to inform the driver that the vehicle needs to be serviced soon. For a technician, it means that the computer has set a trouble code.

The trouble codes have also been standardized to mean the same thing regardless of the manufacturer. Some of the codes pertain only to a particular system or have a different meaning with each system. The DTC is a five-character code combining letters and numbers (Figure 11-43).

The first letter defines the system to which the code was set. The four current possibilities are: "B" for body, "C" for chassis, "P" for powertrain, and a "U" for future use.

The second character is a number and defines the code as a mandated code or a special manufacturer code. A "0" code means the fault is a mandated OBD II type of fault. A "1" code means that the code is manufacturer specific.

The third through fifth characters (numbers) describe the fault. The third character of a power train code identifies where the fault occurred. The remaining two characters describe the condition that set the code. All identified codes will require you to still follow step-by-step procedures as prescribed in the vehicle's service manual.

Although a lot of time has been spent on the ignition portion of the spark-ignited engines, you will find that the diagnostic procedures are similar to those used with diesel engine-related electrical problems. Many of the sensors and output devices can be tested similarly in both engine type applications.

CASE STUDY

In one shop, the majority of jobs were related to diesel-powered vehicles. However, some of the customers did have trucks with spark-ignited engines. One such customer needed a routine maintenance requiring the replacement of plugs, wires, distributor cap, rotor, and any necessary components to achieve a tune up. A technician was assigned this job. He rarely worked on these types of engines, but he wondered, "How complicated can it be to change a bunch of parts?" He completed the job, and the customer picked up the truck. The customer did not get very far before the truck lost power and began misfiring. He returned to the shop but found that the original technician was on another call.

To please the customer, another technician got to work on his truck right way. A visual inspection showed that the secondary wires were tied together tightly on the assumption that this would prevent them from rubbing against each other. Experience had taught the technician to get the service manual and verify the firing order arrangement of the wires. Rechecking the spark plug wiring arrangement, he discovered that two wires were out of order. The original technician must have walked away from the job temporarily and assumed he had a good memory.

The new technician arranged the wires properly and rerouted the wires that were tightly tied together back into their proper separators. He took the truck out with the customer to verify that the truck now had full power. As the customer was leaving, he heard the customer say that the original technician would never touch his truck again.

Terms to Know

Advance timing light
Base timing
Carbon tracking
Control module
Crossfiring
Dielectric grease
E-core
Heat gun
Heli-coiling
SPOUT (spark out)
TDC
Timing light

ASE-Style Review Questions

1. Two technicians are looking at a set of plugs just pulled out of an engine. *Technician A* says that light tan or gray deposits on the tip indicate a plug that is in normal condition. *Technician B* says this indicates the plug is carbon fouled. Who is correct?
A. A only **C.** Both A and B
B. B only **D.** Neither A nor B

2. To test a magnetic pickup coil unit, *technician A* says to apply power to one lead, with the test light on the other lead, a pulsing light should occur when the shaft is spun. *Technician B* says an ohmmeter can be used to test a pickup coil for an open or shorted condition. Who is correct?
A. A only **C.** Both A and B
B. B only **D.** Neither A nor B

3. A mechanical advance unit is being checked. *Technician A* says that a labscope is the most common tool used. *Technician B* says that an advance timing light is the most useful tool for this check. Who is correct?
A. A only **C.** Both A and B
B. B only **D.** Neither A nor B

4. *Technician A* says that dielectric grease is used to seal contaminants out of electrical terminals. *Technician B* says it is used to dissipate heat away from sensitive electrical components. Who is correct?
A. A only **C.** Both A and B
B. B only **D.** Neither A nor B

5. Checking engine base timing is being discussed. *Technician A* says that a vacuum advance unit must have its vacuum hose on in order to ensure proper advance. *Technician B* says the RPM does not affect base timing. Who is correct?
A. A only **C.** Both A and B
B. B only **D.** Neither A nor B

6. *Technician A* says that a temperature sensor cannot be checked with a voltmeter. *Technician B* says that a temperature sensor can be checked with a voltmeter. Who is correct?
A. A only **C.** Both A and B
B. B only **D.** Neither A nor B

7. Checking an NTC temperature sensor with an ohmmeter is being discussed. *Technician A* says that the resistance reading will be high on a cold engine. *Technician B* says that the resistance reading will be low on a cold engine. Who is correct?
   A. A only **C.** Both A and B
   B. B only **D.** Neither A nor B

8. Observing a voltmeter hooked to the output of a properly working crankshaft position sensor while the engine is cranking: *Technician A* says the meter will fluctuate between low and high readings. *Technician B* says the meter will show a constant low voltage. Who is correct?
   A. A only **C.** Both A and B
   B. B only **D.** Neither A nor B

9. Looking at probable causes in electronic engine control systems: *Technician A* says that defective sensors are usually at fault. *Technician B* says that sensors can usually be checked with ohmmeters. Who is correct?
   A. A only **C.** Both A and B
   B. B only **D.** Neither A nor B

10. A no start symptom is being discussed. *Technician A* says a very quick check is to pull any plug wire and use a screwdriver in the boot and bring it close to the engine ground to see if there is spark while the engine is being cranked. *Technician B* says to use a special spark tester to check for spark at the spark plug. Who is correct?
   A. A only **C.** Both A and B
   B. B only **D.** Neither A nor B

ASE Challenge

1. A couple of technicians are diagnosing a hard start (slow starting) condition on a gasoline-powered delivery van. Cranking speed seems normal; there is pressure at the fuel rail and the engine will eventually start. *Technician A* says it would be a good idea to check spark condition with a dummy plug. *Technician B* says you can check spark condition with an amp meter. Who is correct?
   A. A only **C.** Both A and B
   B. B only **D.** Neither A nor B

2. A General Motors HEI system distributor cap is being inspected. *Technician A* says cracks radiating from any of the secondary terminals in or on the cap are unacceptable. *Technician B* says to inspect the cap for trace marks, similar to pencil marks between terminals since these are unacceptable. Who is correct?
   A. A only **C.** Both A and B
   B. B only **D.** Neither A nor B

3. *Technician A* says that on a standard ignition coil the primary and secondary coil windings are common at the (dist.) or (–) terminal. *Technician B* says all terminals of an ignition coil will show some continuity to each other. Who is correct?
   A. A only **C.** Both A and B
   B. B only **D.** Neither A nor B

4. A set of ignition wires is being tested with an ohmmeter. *Technician A* says that if the wire shows any continuity (end to end) it is okay to reuse. *Technician B* says wires should show 6K to 12K ohms per foot length. Who is correct?
   A. A only **C.** Both A and B
   B. B only **D.** Neither A nor B

5. *Technician A* claims that it demonstrates good workmanship to install secondary ignition wires in a neat parallel manner. *Technician B* claims the new high-performance silicon ignition wires are designed to withstand close contact with exhaust manifolds. Who is correct?
   A. A only **C.** Both A and B
   B. B only **D.** Neither A nor B

Job Sheet 19

19

Name: ______________________ Date: ______________________

Visual Ignition System Inspection

Upon completion of this job sheet, you should be able to perform a visual check of a typical ignition system and to verify proper wiring order.

Tools and Materials

A medium/heavy duty truck
Service manual for the truck
Basic hand tools
Spray bottle (water)

Procedure

1. Describe the truck being worked on:
   Year______________ Make ______________________ Vin ______________
   Model______________ Engine Type ______________ Size ______________
2. Describe the general appearance of the engine. ______________________
3. Inspect ignition secondary wires for cracks, rubbing through, oil soaking, and signs of flashing over. Record your findings. ______________________
4. Inspect the distributor cap for cracks, carbon tracking, moisture, loose terminals, and bad contacts. Record your findings. ______________________
5. Check the rotor condition and record your findings. ______________________
6. Inspect the pulse generator or Hall effect (physical condition only). Record your findings. ______________________
7. Inspect the coil for any indication of problems such as cracks, tracking, or leaking oil, if applicable. Record your findings. ______________________

8. Look up the correct firing order in the service manual, and verify the truck's wire arrangement. Draw the proper arrangement below.

9. Using the spray bottle, spray a mist of water on the wires and secondary components to check for voltage leaks. Record your findings. ______________________
____
____
____

10. Based upon your findings, what are your recommendations and conclusions? ______
____
____
____

☑ **Instructor Check** ____

Engine and Vehicle Computer Troubleshooting and Service

CHAPTER 12

Upon completion and review of this chapter, you should be able to:

- ❑ Describe the service procedures associated with servicing the ECM and any electronic device.
- ❑ Describe the more common ways of retrieving fault codes.
- ❑ Describe the different electronic service tools (ESTs) used when troubleshooting or servicing engine control systems.
- ❑ Perform diagnostics with various ESTs.
- ❑ Perform basic sensor tests.
- ❑ Properly replace computer PROM chips.

Basic Tools

Basic mechanic's tool set

Appropriate electronic service tools

Service manuals

Safety glasses

Introduction

The object of this chapter is to familiarize you with diagnosing and troubleshooting electronically controlled engines and vehicles. It is beyond this chapter to cover all the various manufacturer's systems completely. However, we will look at portions of some of the systems used in trucks in order for you to see that there are some commonalities between systems.

As with any electronically controlled system, you cannot always assume that the nature of the problem is electrical. Any sound diagnostic procedure requires a good understanding of both the mechanical aspect and electrical aspect of the engine and/or vehicle. Without this understanding, it is hard to find a direction in your diagnostics. Most original equipment manufacturers (OEMs) produce excellent sequential troubleshooting manuals. The use of the manual still requires that you have a basic understanding of electrical and electronic circuits, with the capability of interpreting schematics and wiring diagrams.

Classroom Manual
Chapter 12, page 307.

You also need to be comfortable using various tools, which can range from a digital multimeter (DMM) to PC-based diagnostic equipment. Many OEMs have gone to PC-based sequential troubleshooting programs for their systems. These types of programs are user friendly and can save a lot of time, versus leafing through pages in manuals.

As with any electronic system, you need to keep in mind that certain tools such as test lights or analog type meters should not be used unless recommended by the manufacturer for a specific test procedure.

Before you begin any troubleshooting procedure, verify that the problem does exist. Not all faults are registered by the computer. If the problem does exist, it is important that OEM procedures are followed. OEMs use a sequential step-by-step procedure. Do not skip any steps—you might inadvertently invalidate the next step or stage. Many procedures also depend on which diagnostic tools are available to you.

Electronic Service Precautions

There can never be enough reminders to technicians about the precautions that must be taken before and during servicing computer and electronic systems in trucks and buses. Any electronic circuit and component is designed to withstand only a normal current draw required with normal operation. Any current above normal can result in damage to the ECM or related electronic devices and circuits. The following service precautions need to be adhered to in order to prevent ECM and circuit damage:

1. Consult a service manual first before performing any tests or checks for specific warnings and precautions. Go through the vehicle's diagrams and schematics so you

have a good idea of how the circuit works. This can prevent you from probing or loading down circuits that can be damaged.

2. Do not ground or apply voltage to any controlled circuits unless the service manual instructs you to do so.
3. Use only high-impedance multimeter (10 megaohm or greater) to test the circuits. Test lights should be used only if instructed to do so by the service manual.
4. Unless instructed otherwise in the service manual, the ignition switch has to be in the OFF position before disconnecting a circuit or connector to sensors or actuators. The same applies for connecting the sensors, actuators, and circuits.
5. Make sure the ignition switch is turned off before making or breaking electrical connections to the ECM. In some cases, the battery needs to be disconnected. Consult the service manual for the vehicle on which you are working.
6. Do not connect any other electrical accessories of the computer-controlled systems.
7. Use only OEM specific tests and service procedures for the vehicle.

Static electricity can be 25,000 volts or higher.

Electrostatic Discharge

Electrostatic discharge is a major concern with electronics components manufacturers. Many manufacturers mark certain components and circuits with a symbol to warn technicians that the component is sensitive to electrostatic discharge (Figure 12-1). Static electricity can destroy electronic devices.

Use the following guidelines whenever you are handling electronic devices:

1. Always touch a good known ground before handling the component. This should be repeated during the course of the service, especially after sliding across a seat, sitting down from a standing position, or walking a distance. Note: Aftermarket grounding wrist straps can be purchased from a tool vendor.
2. Do not touch electrical terminals unless you are instructed to do so in the service manual procedures. Grease and oil from your skin can cause corrosion.
3. When you are using a DMM, especially in the voltmeter function, always connect the negative meter lead first.
4. Do not remove a part from its protective package until it is time to install it.
5. Before removing the part from its package, ground yourself and the package to a known good ground on the vehicle.

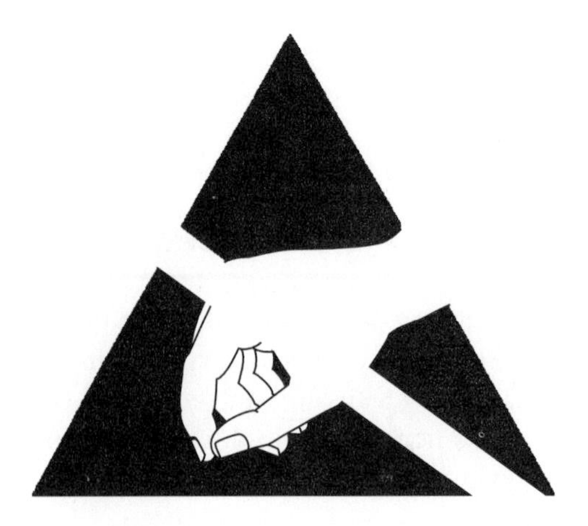

Figure 12-1 General Motors' Electrostatic Discharge (ESD) symbol that warns the technician that the component or circuit is sensitive to static. (Courtesy of General Motors Corporation, Service Operations)

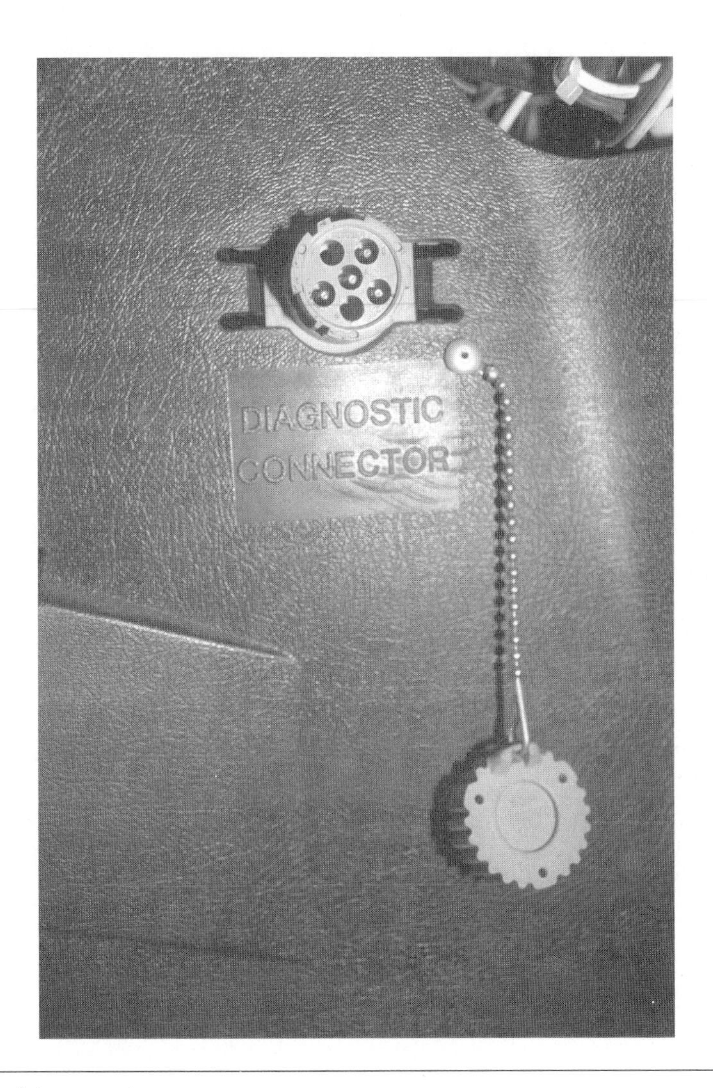

Figure 12-2 ATA data connector.

Electronic Service Tools (EST)

Classroom Manual
Chapter 12, page 308.

Electronic service tools (EST) can range from onboard diagnostic/malfunction lights to sophisticated computer-based communications equipment. These tools can also be further categorized as proprietary ESTs that are designed to work with an OEM's specific electronics or as generic ESTs that can be cost effectively upgraded.

ESTs that are designed to read ECM data are connected to the onboard electronics by means of a Society of Automotive Engineers (SAE)/American trucking Association (ATA) J1584/ J1708/ 1939 6-pin Deutsch connector in all current systems (Figure 12-2).

A **scan tool** is a microprocessor designed to communicate with the ECM. It accesses trouble codes and performs various other functions, short of programming the vehicle's ECM system.

This common connector and the adherence by the many electronics OEMs to SAE software protocols enables proprietary software by any of the manufacturers to read their competitors' parameters. The following are categories of electronic service tools that are used:

- Onboard Diagnostic Lights: These lights are also often referred to as Blink, or Flash, codes. They are a means of onboard troubleshooting using a dash or ECM-mounted electronic malfunction light or **check engine light (CEL).** As a rule, only **active codes** (indicated malfunction at the time of the reading) can be read from the truck or engine ECM. However, there is always an exception to the rule.

Generic **reader/ programmers** are also microprocessor designed to read and reprogram all proprietary systems.

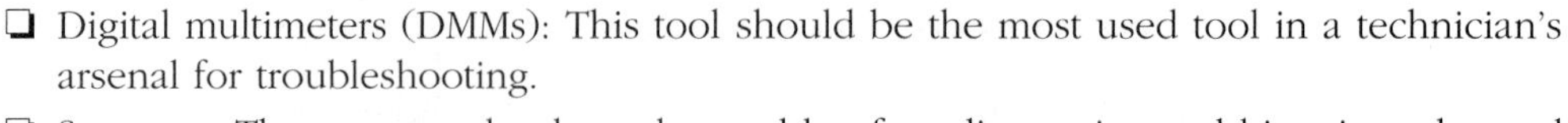

- ❑ Digital multimeters (DMMs): This tool should be the most used tool in a technician's arsenal for troubleshooting.
- ❑ Scanners: These are read-only tools capable of reading active and historic codes and sometimes system parameters. They are obsolete as a diagnostic tool for today's truck and bus diesel engines.
- ❑ Generic reader/programmers: These are microcomputer based ESTs designed to read and reprogram all proprietary systems in conjunction with the appropriate software cartridge.
- ❑ Proprietary reader/programmers: These are usually PC-based test instruments designed and programmed to be used exclusively on a specific OEM system. Even though they are user friendly, they are system specific. Some examples are: Cummins-Compulink, Cat-ECAP, and DDC-DDR Programming Station.
- ❑ Personal computer (PC): Most OEMs have made the generic PC and its operating systems such as MS Windows 3.1/ Windows 95/ Windows 98 or DOS as their diagnostic and programming tool of choice. PCs are relatively low in cost, have a large computing power, and are easily upgradable. They are connected to a vehicle ECM through the ATA 6-pin connector serial link or interface module.

Classroom Manual
Chapter 12, page 307.

Classroom Manual
Chapter 12, page 322.

Codes and Protocols

In order to standardize the use of various tools, SAE/ATA developed a set of standard codes that has been adopted by all the North American truck engine/electronics OEMs. SAE J 1587 covers common software protocols in electronic systems, SAE J 1708 covers common hardware protocols and the more recent SAE J 1939 covers both hardware and software protocols. Using these protocols enables interfacing of electronic systems manufactured by different OEMs on truck and bus chassis and allows reading of OEMs electronic systems by any manufacturer's software. These standard codes are useful for technicians who work on multiple OEM systems, rather than using the proprietary codes. Samples of the number of standards are shown in Figure 12-3.

These tables represent message types that are most likely available on the serial line, if generic tools are use.

MESSAGE IDENTIFIER (MID)	DESCRIPTION
128	Engine Controller
130	Transmission
136	Brakes—Antilock/Traction Control
137–139	Brakes—Antilock, Trailer #1, #2, #3
140	Instrument Cluster
141	Trip Recorder
142	Vehicle Management System
143	Fuel System
162	Vehicle Navigation
163	Vehicle Security
165	Communication Unit—Ground
171	Driver Information System
178	Vehicle Sensors to Data Converter
181	Communication Unit—Satellite

The acronym **PID (parameter identifier)** is used to code components within an electronic subsystem.

PARAMETER IDENTIFIER (PID)	DESCRIPTION
65	Service Brake Switch
68	Torque Limiting Factor
70	Parking Brake Switch
71	Idle Shutdown Timer Status
74	Maximum Road Speed Limit
81	Particulate Trap Inlet Pressure
83	Road Speed Limit Status
84	Road Speed

Figure 12-3 Partial list of SAE/ATA codes.

PARAMETER IDENTIFIER (PID)	DESCRIPTION
85	Cruise Control Status
89	PTO Status
91	Percent Accelerator Pedal Position
92	Percent Engine Load
93	Output Torque
94	Fuel Delivery Pressure
98	Engine Oil Level
100	Engine Oil Pressure
101	Crankcase Pressure
102	Manifold Boost Pressure
105	Intake Manifold Temperature
108	Barometric Pressure
110	Engine Coolant Temperature
111	Coolant Level
113	Governor Droop
121	Engine Retarder Status
156	Injector Timing Rail Pressure
157	Injector Metering Rail Pressure
164	Injection Control Pressure
167	Charging Voltage
168	Battery Voltage
171	Ambient Air Temperature
173	Exhaust Gas Temperature
174	Fuel Temperature
175	Engine Oil Temperature
182	Trip Fuel
183	Fuel Rate
184	Instantaneous MPG
185	Average MPG
190	Engine Speed

The acronym **SID (subsystem identifier)** is used to identify the major subsystems of an electronic circuit.

SUBSYSTEM IDENTIFIERS (SID) COMMON TO ALL MIDS	DESCRIPTION
242	Cruise Control Resume Switch
243	Cruise Control Set Switch
244	Cruise Control Enable Switch
245	Clutch Pedal Switch
248	Proprietary Data Link
250	SAE J1708 (J1587) 1939 Data Link

SUBSYSTEM IDENTIFIERS (SID) FOR MID 128, 143	DESCRIPTION
01–16	Injector Cylinder #1 through #16
17	Fuel Shutoff Valve
18	Fuel Control Valve
19	Throttle Bypass Valve
20	Timing Actuator
21	Engine Position Sensor
22	Timing Sensor
23	Rack Actuator
24	Rack Position Sensor
29	External Fuel Command Input

SUBSYSTEM IDENTIFIERS (SID) FOR MID 130	DESCRIPTION
1–6	C1–C6 Solenoid Valves
7	Lockup Solenoid Valve
8	Forward Solenoid Valve
9	Low Signal Solenoid Valve
10	Retarder Enable Solenoid Valve
11	Retarder Modulation Solenoid Valve
12	Retarder Response Solenoid Valve
13	Differential Lockout Solenoid Valve
14	Engine/Transmission Match
15	Retarder Modulation Request Sensor
16	Neutral Start Output
17	Turbine Speed Sensor
18	Primary Shift Selector
19	Secondary Shift Selector
20	Special Function Inputs
21–26	C1–C6 Clutch Pressure Indicators
27	Lockup Clutch Pressure Indicator
28	Forward Range Pressure Indicator
29	Neutral Range Pressure Indicator
30	Reverse Range Pressure Indicator
31	Retarder Response System Pressure Indicator
32	Differential Lock Clutch Pressure Indicator
33	Multiple Pressure Indicators

Figure 12-3 Continued.

SUBSYSTEM IDENTIFIERS (SID) FOR MID 136-139	DESCRIPTION
1	Wheel Sensor ABS Axle 1 Left
2	Wheel Sensor ABS Axle 1 Right
3	Wheel Sensor ABS Axle 2 Left
4	Wheel Sensor ABS Axle 2 Right
5	Wheel Sensor ABS Axle 3 Left
6	Wheel Sensor ABS Axle 3 Right
7	Pressure Modulation Valve ABS Axle 1 Left
8	Pressure Modulation Valve ABS Axle 1 Right
9	Pressure Modulation Valve ABS Axle 2 Left
10	Pressure Modulation Valve ABS Axle 2 Right
11	Pressure Modulation Valve ABS Axle 3 Left
12	Pressure Modulation Valve ABS Axle 3 Right
13	Retarder Control Relay
14	Relay Diagonal 1
15	Relay Diagonal 2
16	Mode Switch—ABS
17	Mode Switch—Traction Control
18	DIF 1—Traction Control Valve
19	DIF 2—Traction Control Valve
22	Speed Signal Input
23	Warning Light Bulb
24	Traction Control Light Bulb

SUBSYSTEM IDENTIFIERS (SID) FOR MID 136-139	DESCRIPTION
25	Wheel Sensor, ABS Axle 1 Average
26	Wheel Sensor, ABS Axle 2 Average
27	Wheel Sensor, ABS Axle 3 Average
28	Pressure Modulator, Drive Axle Relay Valve
29	Pressure Transducer, Drive Axle Relay Valve
30	Master Control Relay

SUBSYSTEM IDENTIFIERS (SID) FOR MID 162	DESCRIPTION
1	Dead Reckoning Unit
2	Loran Receiver
3	Global Positioning System (GPS)
4	Integrated Navigation Unit

FMIs (failure mode identifiers) are indicated whenever an active or historic code is read using ProLink or a PC.

FAILURE MODE IDENTIFIERS (FMI)	DESCRIPTION
0	Data valid but above normal operating range
1	Data valid but below normal operating range
2	Data erratic, intermittent, or incorrect
3	Voltage above normal or shorted high
4	Voltage below normal or shorted low
5	Current below normal or open circuit
6	Current above normal or grounded circuit
7	Mechanical system not responding properly
8	Abnormal frequency, pulse width, or period
9	Abnormal update rate
10	Abnormal rate of change
11	Failure mode not identifiable
12	Bad intelligent device or component
13	Out of calibration
14	Special instructions

Figure 12-3 Continued.

Diagnosing with Electronic Service Tools

This section deals with the common usage of various service tools as applied to truck/bus engine and vehicle electronics.

Onboard Diagnostic Lights

We will use a Mack truck's V-MAC system as an example. V-MAC stands for Vehicle Management and Control system. The following information and procedures are derived from the V-MAC II Service Manual.

The V-MAC module is capable of blinking a two-digit blink code for each of the detectable active faults in the V-MAC system. These codes are displayed on the electronic malfunction lamp, which is located on the dashboard. The primary reason for the blink code is to allow for a quick diagnosis of an active fault in the system without requiring an expensive troubleshooting tool. The blink codes can be used for isolating and troubleshooting any active faults in the V-MAC system.

Trouble codes are a method of using one to three digital characters to indicate a fault in a system as detected by the computer.

To properly activate and use the blink codes, follow these steps:

1. Turn the key ON and wait until the electronic lamp's two-second power-up test is finished.
2. There must be an active fault that keeps the light ON after the two-second power-up test.
3. With the speed control ON/OFF switch in the OFF position, press and hold the SET/DECEL or the RESUME/ACCEL SWITCH until the fault lamp goes OFF.
4. The fault lamp will remain OFF for approximately one second.
5. Immediately after the wait time, the V-MAC module will begin to flash a two-digit blink code. The two digits of the code will be separated by a one-second idle time (fault lamp OFF).
6. Each digit of the blink code may consist of up to eight ON/OFF flashes. The ON and OFF time for each flash is one-fourth of a second.
7. The ON flashes of the fault lamp must be counted in order to determine the two-digit blink code.
8. Only one active fault is blinked per request. Where there are multiple active faults in the system, there must be a separate request for each active fault. To request that another fault be displayed, hold in the SET/DECEL or the RESUME/ACCEL switch until the fault lamp goes OFF. The blinking sequence will begin again after a one-second delay.
9. If the fault blinking request is repeated while V-MAC is in the process of blinking an active fault, that sequence will stop and the next active fault will be blinked.
10. If an active fault is cleared while V-MAC is blinking that fault, the procedure will not stop.
11. After every complete blinking sequence, the fault lamp will return to normal functions. It will remain ON for active faults and OFF for inactive faults.

Hard codes (active faults) are failures that are indicated at the time of the reading or the last time the ECM tested the circuit.

Intermittent (inactive) codes have occurred in the past but were not present during the last ECM test of the circuit.

Note: If more than one active fault is present, continue the blink code sequence until the first active fault is redeployed to be sure all faults have been recovered.

The V-MAC blink code software will not provide codes for inactive faults. To access this data requires the use of a hand-held reader or diagnostic computer. A list of possible blink codes are shown in Figure 12-4.

Detroit Diesel Electronic Controls (DDEC) systems are also set up to use blink codes. The operator is alerted to system problems by illuminating dash lights, the check engine light (CEL), and the **stop engine light (SEL).** The CEL is illuminated when a system fault is logged to alert the driver that DDEC fault codes have been generated. The SEL is illuminated when the ECM detects a problem that can result in serious damage.

MACK BLINK CODE IDENTIFICATION TABLE

BLINK CODE	PROTOCOL PID	SID	FMI	ASSIGNMENT LISTING	FAILURE	MID
1–1	100		4	Engine Oil Pressure	Voltage below normal or shorted low	142
1–2	100		3	Engine Oil Pressure	Voltage above normal or shorted high	142
1–3	108		4	Barometric Pressure	Voltage below normal or shorted low	142
1–4	108		3	Barometric Pressure	Voltage above normal or shorted high	142
*1–7	111		3	Engine Coolant Level	Voltage above normal or shorted high	142
2–1	110		4	Engine Coolant Temperature	Voltage below normal or shorted low	142
2–2	110		3	Engine Coolant Temperature	Voltage above normal or shorted high	142
2–3	105		4	Intake Manifold Air Temperature	Voltage below normal or shorted low	142
2–4	105		3	Intake Manifold Air Temperature	Voltage above normal or shorted high	142
2–5	102		4	Boost (Air Inlet) Pressure	Voltage below normal or shorted low	142
2–5	106		4	Boost (Air Inlet) Pressure	Voltage below normal or shorted low	142
2–6	102		3	Boost (Air Inlet) Pressure	Voltage above normal or shorted high	142
2–6	106		3	Boost (Air Inlet) Pressure	Voltage above normal or shorted high	142
3–2	190		2	RPM/TDC (Engine Position)	Data erratic, intermittent or incorrect	142
3–4		1	2	Timing (TEM) Sensor	Data erratic, intermittent or incorrect	142
3–5		2	7	Timing Actuator	Mechanical system not responding properly, or out of adjustment	142
4–1	84		4	MPH (Road Speed) Sensor	Voltage below normal or shorted low	142
4–2	84		3	MPH (Road Speed) Sensor	Voltage above normal or shorted high	142
4–3	84		8	MPH (Road Speed) Sensor	Abnormal frequency, pulse width or period	142

Figure 12-4 List of fault codes. (Courtesy of Mack Trucks, Inc.)

The CEL and SEL will flash codes when the diagnostic request switch is toggled. The CEL will flash inactive codes and the SEL will flash active codes.

One more example of blink code usage is the Volvo engine with an Electronic Diesel Control (EDC) system. Vehicles using this engine control system have a diagnostic request button (Figure 12-5). The current diagnostic light is also provided on the dashboard.

BLINK CODE	PROTOCOL			ASSIGNMENT LISTING	FAILURE	MID
	PID	SID	FMI			
5–1	91		4	Throttle Position Sensor (TPS)	Voltage below normal or shorted low	142
5–2	91		3	Throttle Position Sensor (TPS)	Voltage above normal or shorted high	142
5–3		4	5	Rack Actuator	Current below normal or open circuit	142
5–4		4	7	Rack Actuator	Mechanical system not responding properly, or out of adjustment	
5–5		4	6	Rack Actuator	Current above normal or grounded circuit	142
5–6		4	11	Rack Actuator	Failure mode not identifiable	142
6–6		254	12	Internal Module Communications	Bad intelligent device or component	142
7–1		246	4	Brake Pedal Switch #1 (Service Brake)	Voltage below normal or shorted low	142
7–5		3	5	Rack Position Sensor	Current below normal or open circuit	142
7–6		3	6	Rack Position Sensor	Current above normal or grounded circuit	142
7–7		3	11	Rack Position Sensor	Failure mode not identifiable	142
Red Light & Alarm	110		0	Engine Coolant Temperature	Data valid but above normal operating range	142
Red Light & Alarm	100		1	Engine Oil Pressure	Data valid but below normal operating range	142

* Red Light and Amber Light and Alarm

Figure 12-4 Continued.

Figure 12-5 Diagnostic request button location on a Volvo truck.

Flash codes in this system make it possible to locate a faulty functional path. In the event of a fault, the diagnosis light either flashes continuously, lights up continuously, or remains off, depending on the severity of the fault. If there are several faults present, flashing has priority over continuously lit, which has priority over off.

The flashing and continuous lights indicate a current or intermittent faults. Both faults are stored in the electronic control unit (ECU) and can be called up by the flashing code even after the engine has been switched off. The following steps are used for retrieving codes:

1. Press the diagnosis prompt button for at least 2–4 seconds with the engine stopped and ignition switched ON. The diagnosis lamp lights up in the process.
2. The flashing process starts, following a brief OFF phase. Each detected fault is assigned a number between 1 and 23. For example:

   Fault 1 = 1 flashing pulse of the diagnostic light
   Fault 8 = 8 flashing pulses of the diagnostic light

3. Repeat the process and determine if another fault is stored. If the first indicated fault appears again, all faults have been covered.
4. Clear the fault memory after the fault(s) have been eliminated. Press the request button for at least 2 seconds with the ignition switched ON and at the same time press down on the service brake or clutch pedal.

This manufacturer's service manual will direct you to a troubleshooting chart with codes as shown in Figure 12-6. Also, a wiring diagram is included as a reference (Figure 12-7). Take note that in the flowchart there is mention of a 15-pin breakout box. In the wiring diagram, you will also notice an engine diagnostic connector. This socket is used to check the performance of the vehicle. Voltage measurements are made on the 9-pin test socket. For example, if you are doing a performance check on a stationary vehicle, the ignition key is turned ON and the voltage readings taken should fall within a given range as shown in the chart in Figure 12-8.

As you can see, many manufacturers use onboard diagnostics systems with blinking lights to guide you to a fault. Check each manufacturer's service manual for the proper procedure.

WARNING: Before using a DMM, make sure you understand the circuit you are measuring. Also use the correct meter function for the desired test and all appropriate precautions. If you do not, damage to your meter or the circuit being tested can occur.

Using Digital Multimeters (DMM)

High **impedance** is the combined opposition to current created by the resistance, capacitance, and inductance of the meter.

The use of the DMM has been shown throughout various sections of this shop manual. However, repetition is often needed to familiarize yourself with tools used for electronics system and component testing.

The DMM is a portable, simply operated tool for taking electrical measurements. For delicate electronic devices and circuits, a high impedance digital meter should replace the analog multimeter and test light in the truck/bus technician's collection of tools (Figure 12-9). The cost of the multimeter is proportional to its features, resolution, and display quality. The following figures show many of the functions and features that can be obtained, depending on your needs (Figures 12-10, 12-11, 12-12, and 12-13).

The DMM uses the principles of Ohm's law to measure and display values in an electrical circuit. The following measurements can be arrived at and displayed by a DMM:

Special Tool

DMM

- Measuring voltage—Most meters are capable of measuring both DC and AC current. This function is useful because most of the components and circuits are powered by DC voltage. However, some sensors do send out an AC signal. Therefore, we can say that a meter can determine the following:
  1. Source voltage
  2. Voltage drop
  3. Voltage imbalance
  4. Ripple voltage
  5. Sensor voltages

Step	Testing For	Instructions	Set Values
1.	Fault in sensor or wiring	Turn ignition key off. Disconnect harness from the EDC relay. Connect harness to 15-pin test box. Do not attach 15-pin test box to relay. Check relay harness terminals 7 and 8. If resistance is OK, go to step 2. If not OK, disconnect terminal at speed sensor and check the sensor itself for resistance. If resistance is OK, check wires (V307 and V308).	950-1150 ohms 950-1150 ohms
2.	Fault in wiring between ECU and EDC relay	Ignition key still off. Disconnect the harness from the ECU and connect the 35-pin test box to the harness. Check terminal 15 of relay test box and terminal 32 of the ECU test box. If resistance is OK, go to step 3. If not OK, connect the 9-pin test box to the test socket. Check terminal 15 of relay harness and terminal 2 of test socket. If not OK, wire V315 is broken. If resistance is OK, check terminal 32 of the ECU harness and terminal 2 of the test socket. If not OK, wire V232 is broken.	0-1 ohm. 0-1 ohm 0-1 ohm
3.	Fault in EDC-relay	Ignition key still off. Connect 15-pin test box also to the relay. Check terminals 1 and 32 of the ECU harness. Check terminals 2 and 32 of the ECU harness. If resistance is OK, go to step 4. If not OK, check the resistance between the ECU terminal 1 and relay harness terminal 14 (= wire V201 and V314).	900-1100 ohms 900-1100 ohms 0-1 ohm
4.	Fault in EDC-relay	Check for faulty resistance in relay itself. Check terminals 6 and 15 of the relay harness. Check terminals 14 and 15 of relay harness. If resistance is OK, go to step 5. If resistance is not OK, replace the EDC relay.	900-1100 ohms 900-1100 ohms

Figure 12-6 Fault code diagnostic flowchart. (Reprinted with permission. Courtesy of Volvo Trucks North America, Inc.)

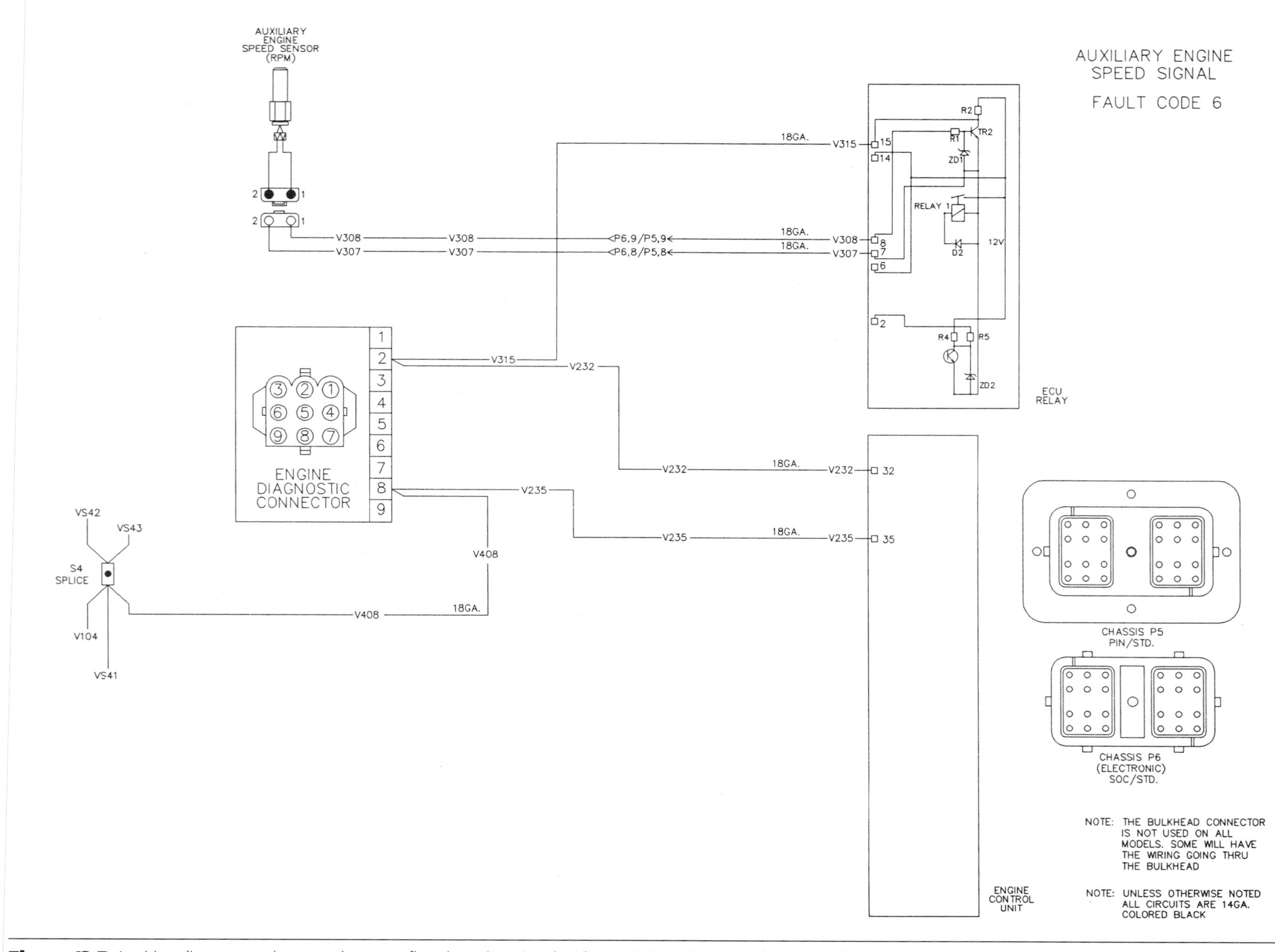

Figure 12-7 A wiring diagram used to complement a flowchart. (Reprinted with permission. Courtesy of Volvo Trucks North America, Inc.)

Plug Term.	Note	Range
NOTE: Engine off.		
6–8	Reference voltage	4.5–5.0V
Basic check of accelerator pedal and turbo boost sensor		
7–8	Control rack travel (idle)	0.3–0.6V
7–8	Control rack travel (full)	1.4–3.3V
3–8	Acclerator pedal (full)	3.7–4.6V
3–8	Accelerator pedal (idle)	0.3–0.6V
4–8	*Turbo boost sensor	1.4–1.6V

*Refer to Temperature compensating boost pressure reading.

Figure 12-8 Voltage range chart. (Reprinted with permission. Courtesy of Volvo Trucks North America, Inc.)

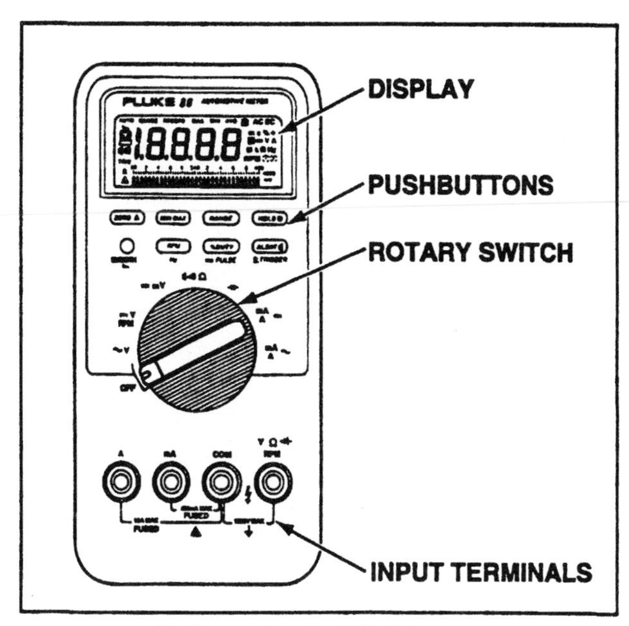

Figure 12-9 Fluke 88 DMM. (Courtesy of Navistar International Engine Group)

Input Terminals The digital multimeter has four input terminals as shown in the following illustration. The terminals are protected against overload to the limits shown in the User's Manual.

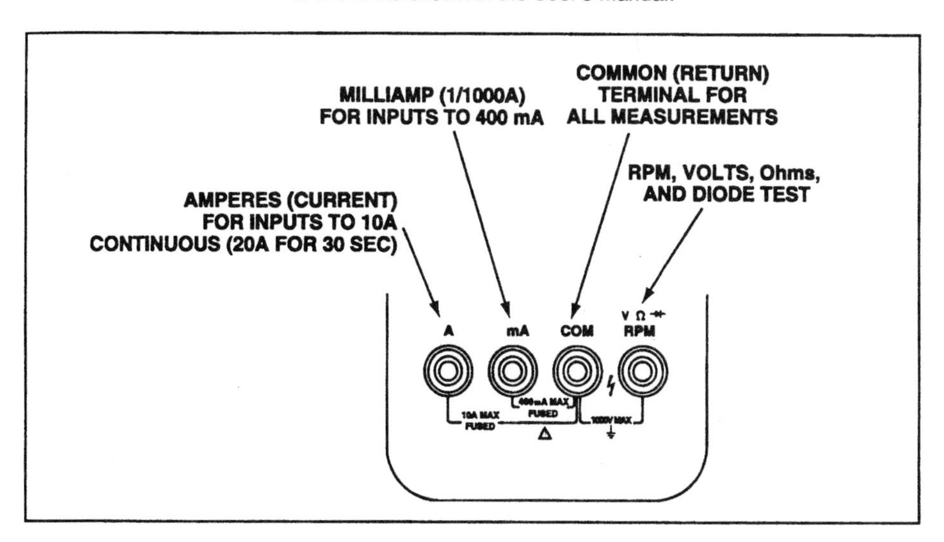

Figure 12-10 DMM input terminals. (Courtesy of Navistar International Engine Group)

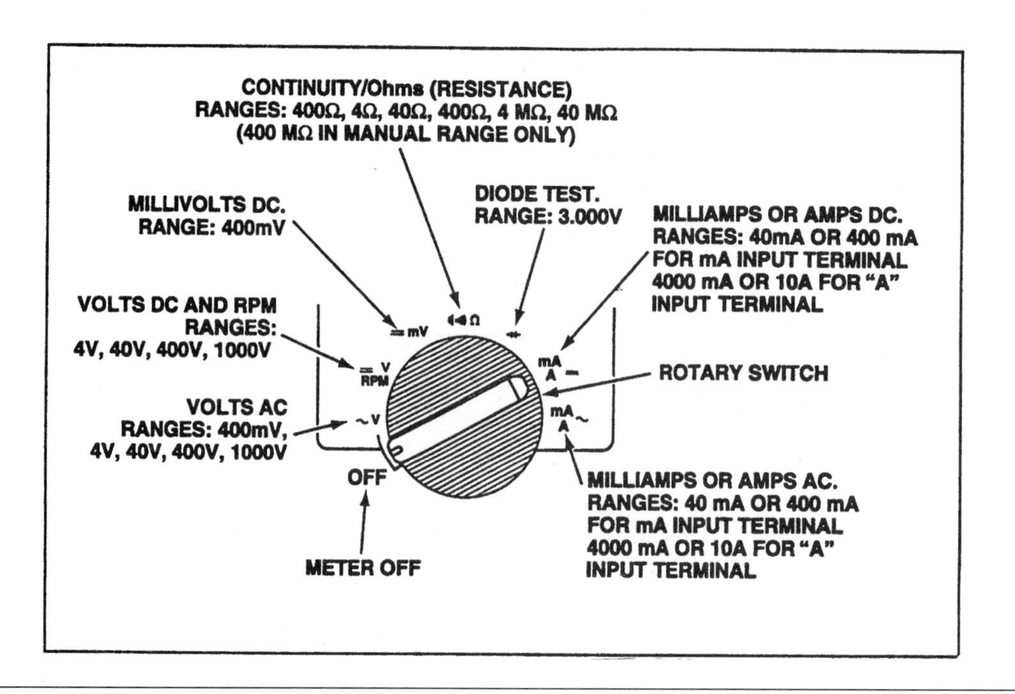

Figure 12-11 DMM rotary switch. (Courtesy of Navistar International Engine Group)

The pushbuttons are used to select meter operations. When a button is pushed, a display symbol will appear and the beeper will sound. Changing the rotary switch setting will reset all pushbuttons to their default settings.

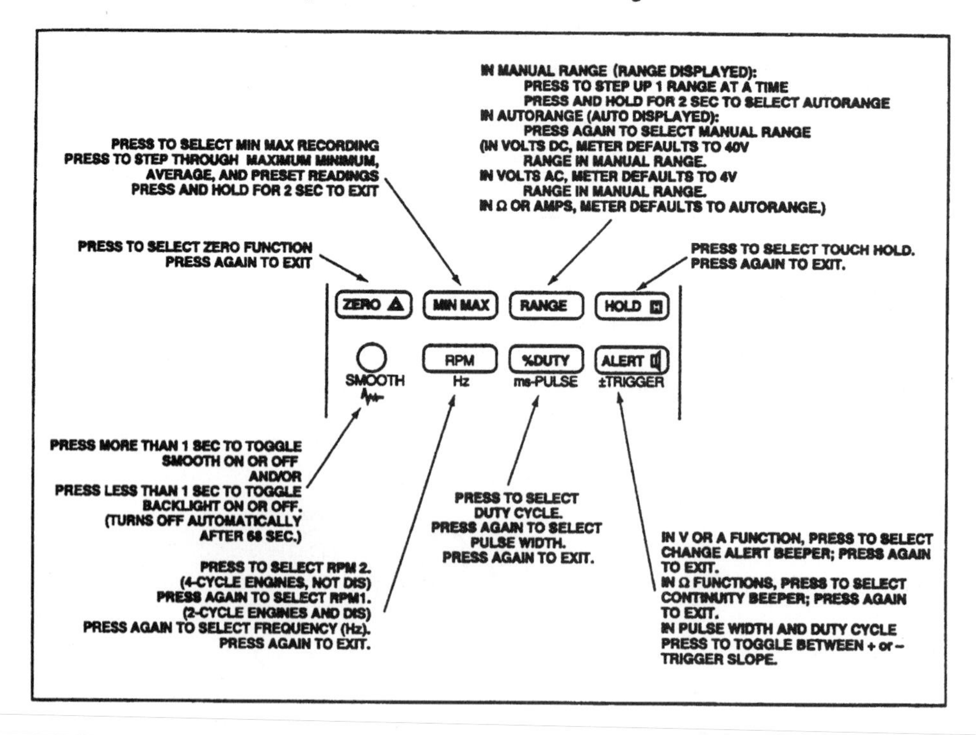

Figure 12-12 DMM pushbuttons. (Courtesy of Navistar International Engine Group)

Display The digital multimeter has a digital and analog display capability. The digital display should be used for stable inputs, while the analog display should be us ed for frequently changing inputs. If a measurement is to large to be displayed, "**OL**" is shown on the digital display.

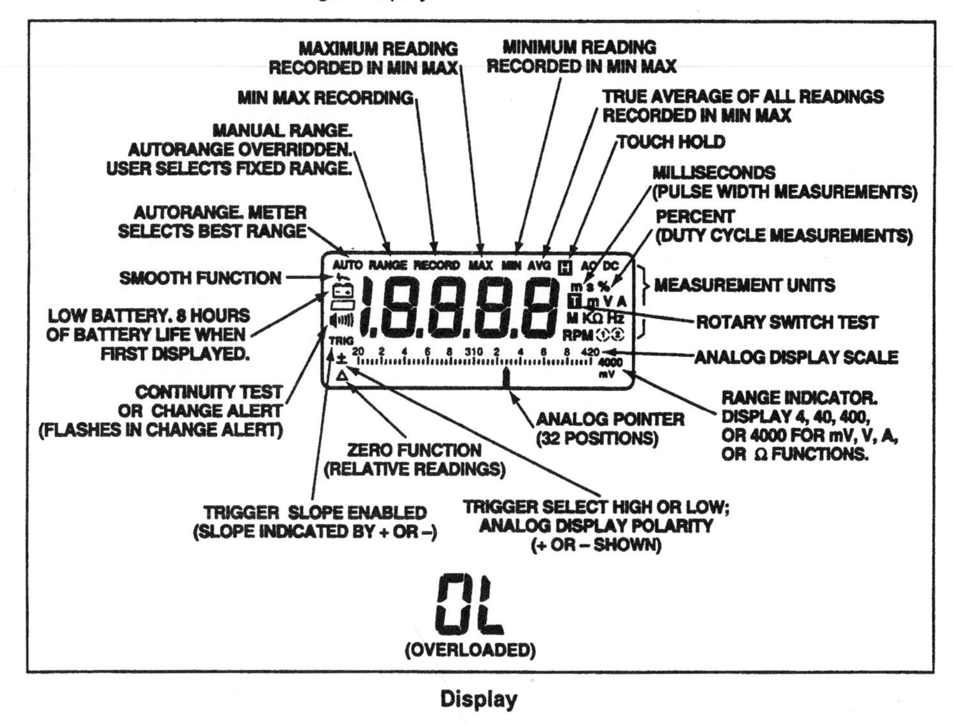

Figure 12-13 DMM data display. (Courtesy of Navistar International Engine Group)

- Measuring resistance—Most DMMs measure resistance values as low as 0.1 ohms and as high as up to 300 megaohms. Infinite resistance or greater than the instrument can measure is indicated as "OL" on the display. An example of this is an open circuit. Resistance and continuity tests should be performed on only nonpowered components and circuits. An energized component or circuit can damage your meter. Resistance measurements determine the following:
  1. Resistance of a load
  2. Resistance of conductors
  3. Value of a resistor
  4. Operation of variable resistors
- Measuring current—Current measurements are made in series (Figure 12-14) unless the meter has an inductive pickup. Voltage and resistance measurements are made in parallel to the load or circuit. Most meters have a separate fused input jack for the test leads. Current measurements determine the following:
  1. Circuit draw or operating current
  2. Circuit overloads
  3. Current in different branches of a circuit

WARNING: Before measuring current, anticipate the expected current draw in order not to exceed the meter's amperage measuring limit and cause meter damage.

Typical DMM Tests

The following examples of tests are usually performed when required to do so by a service manual procedure or in conjunction with a generic reader/programmer. Notice that these tests have been performed in previous sections of this manual.

Special Tools

DMM

Service manual with wiring diagrams

Measuring Current

Amperage is the measurement of current flow. To measure amperage, set the digital multimeter to the **mA/A** function. Plug the black (negative) lead into the **COM** input jack and the red (positive) lead into the **A** or **mA** input jack. Place the meter in series with the circuit so that current passes through the meter. Use the correct type of current probes for this purpose. Power the circuit and note the reading.

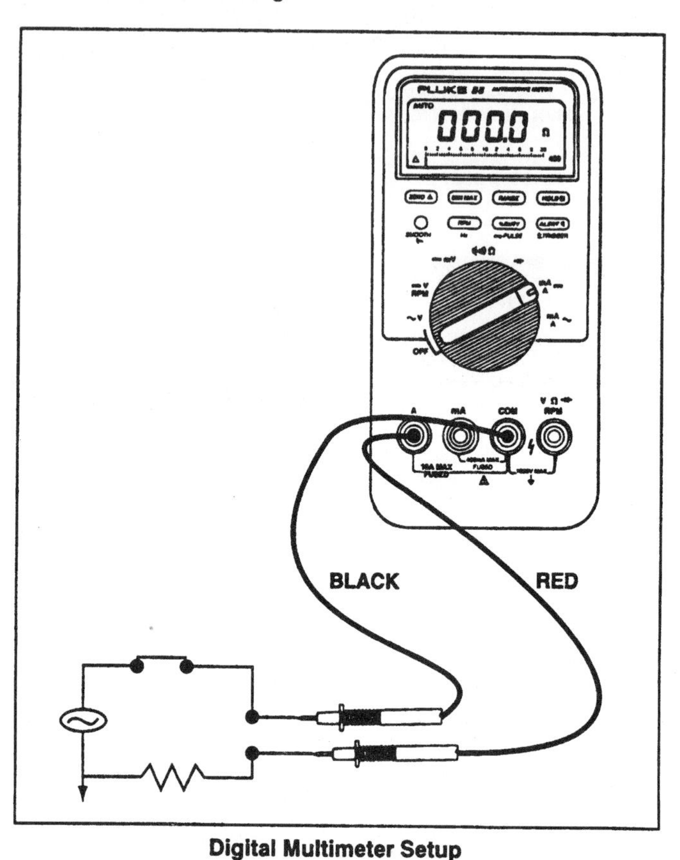

Figure 12-14 DMM setup for measuring current flow. (Courtesy of Navistar International Engine Group)

Sensors convert some measurement of vehicle operation into an electrical signal.

Hall-effect switches operate on the principal that if current is allowed to flow through thin conducting material that is exposed to a magnetic field, another voltage is produced.

Classroom Manual
Chapter 12, page 329.

Classroom Manual
Chapter 12, page 329.

- ❑ Engine position (fuel injection pump camshaft), cam, and crank position sensors: Hall effect sensors:
  - **A.** Switch meter to measure Volts DC/RPM (using a Fluke DMM).
  - **B.** Identify the ground and signal terminals at the Hall sensor. Connect the positive (+) test lead to the ground terminal.
  - **C.** Crank the engine. At cranking speeds, the analog bar graph should pulse. If the engine were to run at idle speeds or above, the pulses would be too fast to read.
  - **D.** Press the duty cycle button once. Duty cycle can indicate square wave quality, with poor quality signals having a low duty cycle. Functioning Hall sensors should have a duty cycle of around 50%. Check specifications.
- ❑ Potentiometer type TPS—Figure 12-15 shows a typical electronic throttle. It is a potentiometer that varies the output voltage in response to throttle position as shown in the graphs in Figures 12-16 and 12-17.

Two tests can be performed on this component:

Resistance test:

1. Key has to be off.
2. Disconnect the TPS.
3. Select ohms on the DMM. Connect the test leads to the signal and ground terminals.

A potentiometer is a variable resistor that provides accurate voltage drop readings to the computer.

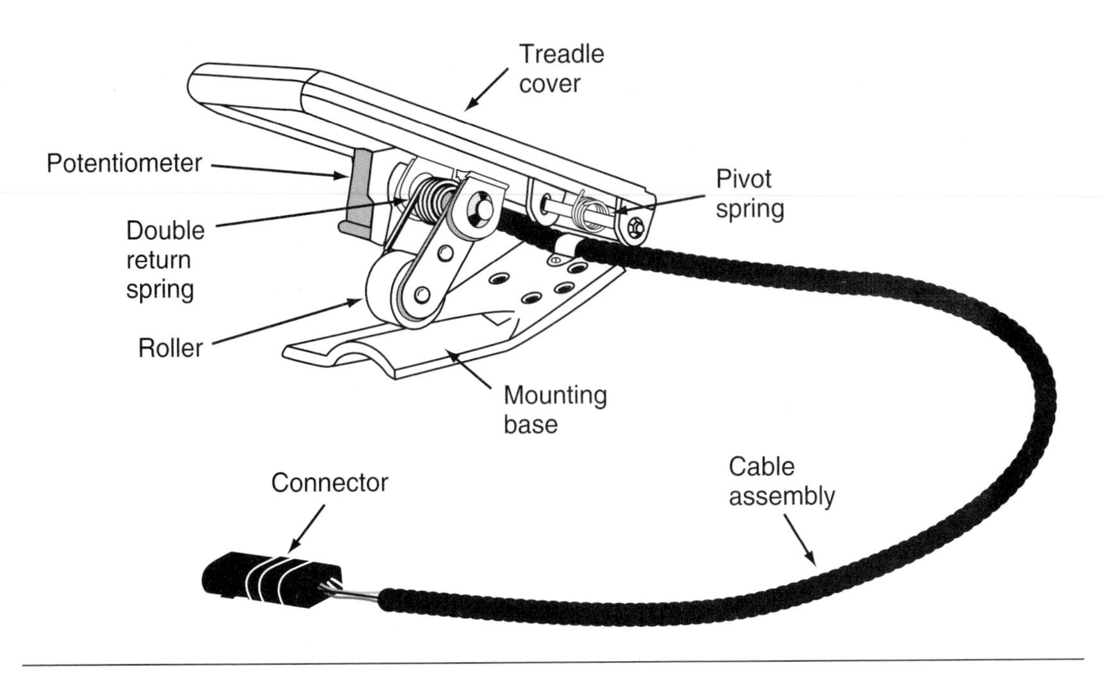

Figure 12-15 Electronic throttle.

	DETROIT DIESEL	MACK
CLOSED THROTTLE (IDLE)	6% - 14%	10% - 20%
OPEN THROTTLE (FULL)	86% - 94%	70% - 85%

Figure 12-16 Output voltage as a percentage of supply voltage. (Courtesy of Bendix Brakes by AlliedSignal)

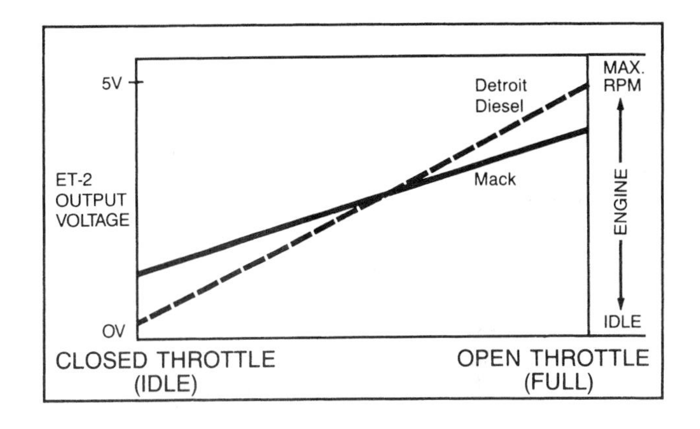

Figure 12-17 Output voltage, throttle position, and RPM relationship. (Courtesy of Bendix Brakes by AlliedSignal)

4. Move the accelerator through its stroke while observing the meter's bar graph.
5. The analog bar should move smoothly without jumps or steps. If a step or glitch is noticed, it could indicate a bad spot in the sensor.

Voltage Test

Classroom Manual
Chapter 12, page 326.

This is an operational test recommended by Bendix for its throttle position sensor:

1. Unplug the cable from the potentiometer's integral connector. Inspect the cable for any damage such as corrosion, wear, or loose terminals. Check the end-to-end electrical continuity at terminals. Note: Potentiometer pin location will remain constant (Figure 12-18). However, cable assembly connector pin out may vary from engine to engine.
2. Remove the assembly from the vehicle.
3. Secure it to a smooth, flat surface.
4. Connect the potentiometer to the voltmeter as shown in Figure 12-18. Connect a power supply also as shown. This can be a 12 VDC battery in good condition.

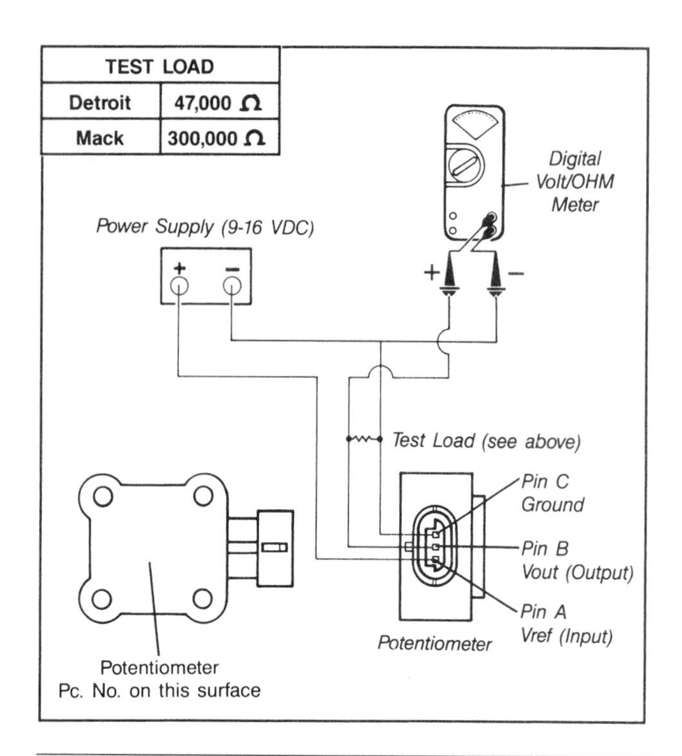

Figure 12-18 Electrical schematic of potentiometer. (Courtesy of Bendix Brakes by AlliedSignal)

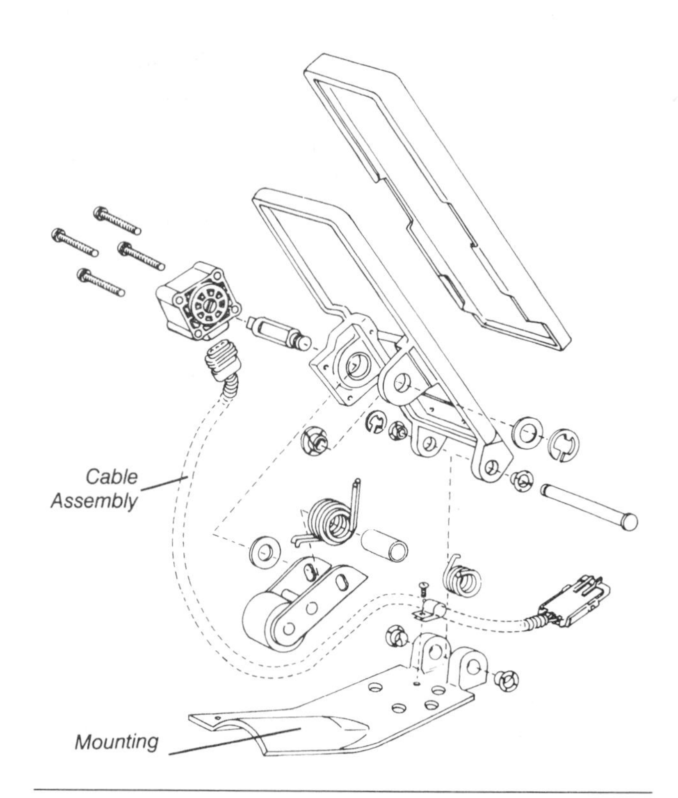

Figure 12-19 Exploded view of a complete acceleration pedal sensor assembly. (Courtesy of Bendix Brakes by AlliedSignal)

Special Tool

DMM

Magnetic pulse generators use the principle of magnetic induction to produce a voltage signal.

Classroom Manual
Chapter 12, page 327.

A **thermistor** is a solid variable resistor made from a semi-conductor material.

Classroom Manual
Chapter 12, page 328.

5. Verify that the closed throttle (idle) output voltage, as a percentage of supply voltage, is within limits as shown in Figure 12-16.
6. Depress the treadle to its full throttle position. Again, verify the output voltage as a percentage of supply voltage as listed in Figure 12-16. Note: A resistor can be installed as a test load to achieve proper results. The following is for testing a Detroit diesel TPS potentiometer: Battery = 10 VDC. Full throttle = 9 VDC. 9/10 × 100 = 90%.
7. Make several full applications and record idle position voltage each time. Verify that idle position voltage does not vary by more than .4% (.02 volts). If the unit fails to operate within the specified ranges, service the unit or replace it. With the above test, the integrity of the complete assembly was taken into consideration as can be seen in Figure 12-19.

- Magnetic sensors—The output of these sensors is an AC voltage pulse whose value rises proportionally with rotational speed increase. Voltage values can range from 0.1 V up to 5.0 V, depending on the rotational speed an type of sensor. Vehicle speed sensors (VSS), engine speed sensors (ESS), and ABS wheel speed sensors all use this method to determine rotational speed. Set the meter to the VAC function and connect the test leads across the appropriate terminals. Observe the readings.
- Thermistors—Most thermistors used in computerized engine systems have a negative temperature coefficient (NTC). This means that as the sensor temperature increases, its resistance decreases. They can be checked to specifications by using a DMM ohmmeter function and an accurate temperature measurement instrument. By knowing the temperature of the thermistor you can verify the expected resistance reading.

Breakout Boxes

Many OEMs require the use of a breakout in conjunction with the DMM. Breakout devices are designed to break into an electrical circuit (Teed) in order to enable circuit measurements on active or inactive (de-energized) circuits. Using this method of interfacing a testbox into a circuit, allow for mea-

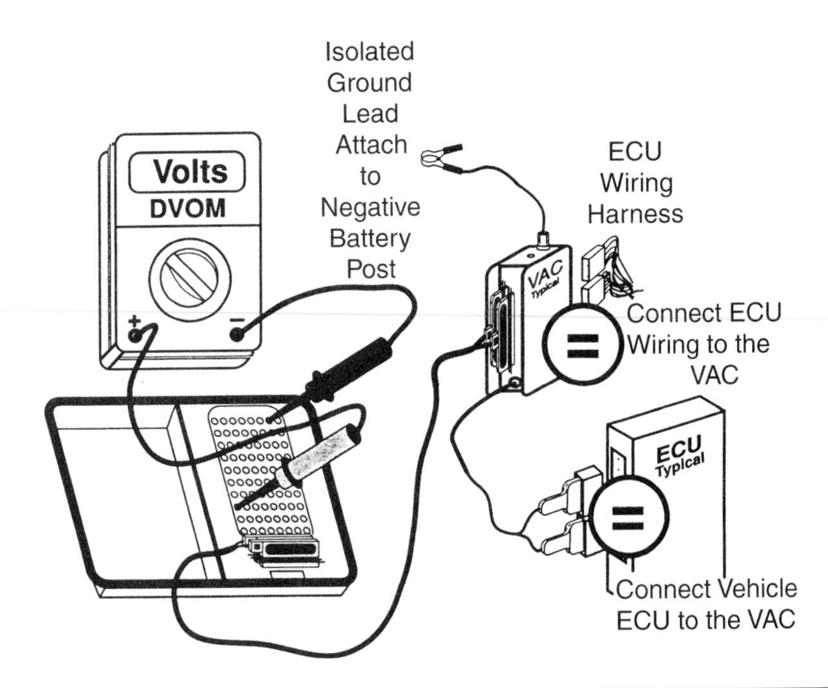

Figure 12-20 Typical breakout setup used in conjunction with a DVOM. (Courtesy of OTC Tool and Equipment, Division of SPX Corporation)

suring components and activities without interrupting the circuit. The advantage of using this device is that instant access is provided to multiple circuits. Time can be saved in tracing the circuit wiring.

The face of the breakout box displays a number of coded sockets into which the probes of a DMM can be safely inserted to read circuit conditions (Figure 12-20).

Breakout boxes allow the technician to test circuits, sensors, and actuators by providing test points.

WARNING: Always use the recommended tools for troubleshooting sequences. Never puncture wiring or electrical harnesses in order to obtain readings. The corrosion damage that can result will create problems later.

Diagnostic Connector Jumpers

In order to prevent damage to wires or connectors, some electrical/electronic connector manufacturers sell harness or wire jumpers (Figure 12-21). The jumper wires provide a means of accessing the circuits with a DMM without piercing the wires or damaging the connector sockets and pins. If these jumpers are available, make sure that you use the proper ones.

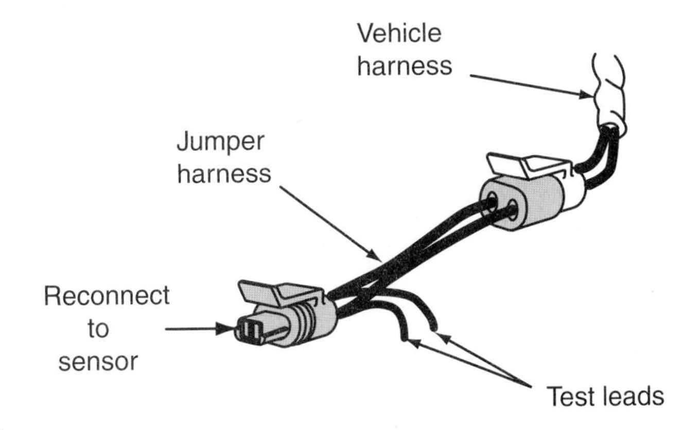

Figure 12-21 Jumper wires connected between the sensor and the harness allow the technician to probe for voltage or to test resistance without damaging the wiring.

SERVICE TIP: You can often accumulate enough of a variety of male and female connectors due to replacement services. Save all the connectors and start fabricating your own jumpers. After a while, you will be building up a collection of these jumpers to have a full coverage of the various connectors used in truck/bus circuits.

Classroom Manual
Chapter 12, page 323.

Generic Reader/Programmers

Most OEMs recommend the MPSI Pro Link 9000 generic reader/programmer as their electronic service tool of choice. This tool has become the industry standard, EST capable of reading most current systems with the correct software cartridge. The Pro Link 9000 is completely user friendly in that first time users can get comfortable with it by exploring through the menus while connected to an ECM. These are microprocessor-based test instruments that depend on the system to do the following:

1. Access active and inactive codes
2. Erase inactive (historic) codes
3. View all system identification data
4. View data on engine operation with the engine running
5. Perform diagnostics tests on system subcomponents such as EUI testing
6. Reprogram customer data parameters on engine and chassis systems
7. Act as a serial link to connect the vehicle ECM via modem to a centrally located mainframe for proprietary data programming (some systems only) (Figure 12-22)
8. Snapshot system data parameters to assist in finding solutions to intermittent problems

Snapshot Tests

Most systems will allow a snapshot test readout from the ECM if no codes are logged and there seems to be a problem and even if codes are logged with no clear reason. Snapshot mode troubleshooting can be triggered by a variety of methods (codes, conditions, manually). When triggered, this feature can record data frames before and after the trigger. This data can be recorded to a reader/programmer or a laptop PC while the vehicle is running stationary or down the road

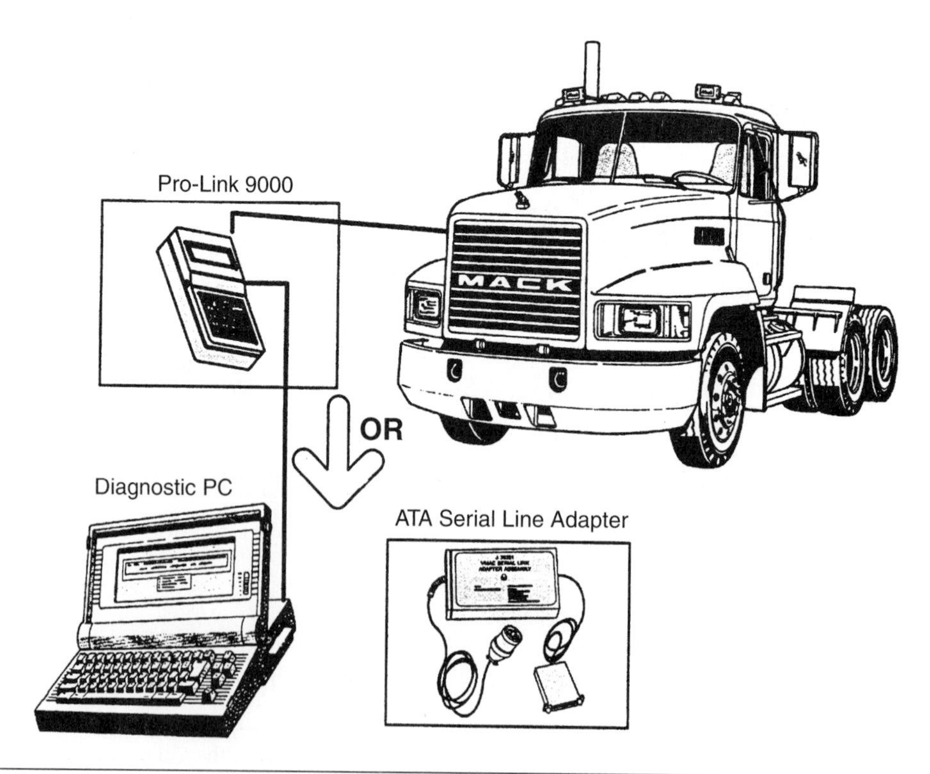

Figure 12-22 Pro Link used as a service tool. (Courtesy of Mack Trucks, Inc.)

to duplicate certain operating conditions. For example, if you are driving down the road and you notice or feel the intermittent problem, manually hit the trigger. Certain amounts of data frames will be recorded before and after you hit the trigger. This allows each data frame to be studied individually after the event and drive cycle to compare circuit or component activity at the moment of the intermittent problem.

The use of a snapshot test is sometimes referred to as flight recording.

R S-232 Serial Port

This port is located on the right side of the Pro Link head. You have three choices when this menu item is selected:

1. Printer output
2. Terminal output
3. Port setup

The four parameters that need to be set up for the instrument to communicate with a printer or PC terminal are:

1. Handshake—This refers to how the data will be transmitted between two electronic components. The two commonly used methods are: BUSY and XON/XOFF. Select the method specified on the printer specifications.
2. Baud rate—This refers to the speed at which data is transmitted. Both the output device and the receiving device must agree on the transmission speed.
3. Parity—This refers to the even or odd parity of the number of 0s and 1s. The menu options are: none, odd, or even.
4. Stop bits—This is the last element in the transmission of a character. The output and receiving devices must agree on how many stop bits to expect to determine the end of a transmission character.

Figures 12-23 and 12-24 show the Pro Link data connection hardware, cables, and the software cartridge. The Pro Link will transmit to a printer or PC terminal any data that it can read itself from a system ECM. Printouts are very useful for analyzing specific actions that can be the cause of a condition.

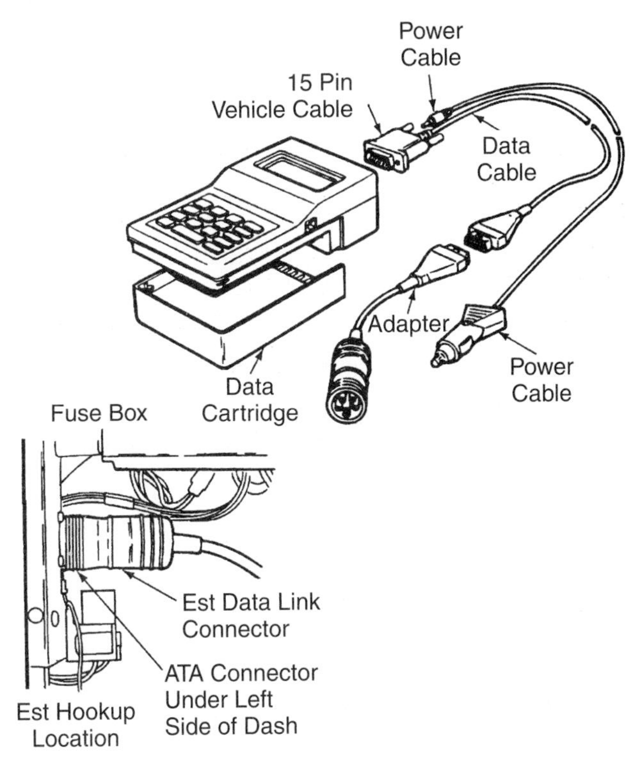

Figure 12-23 Pro Link data connection hardware. (Courtesy of Navistar International Engine Group)

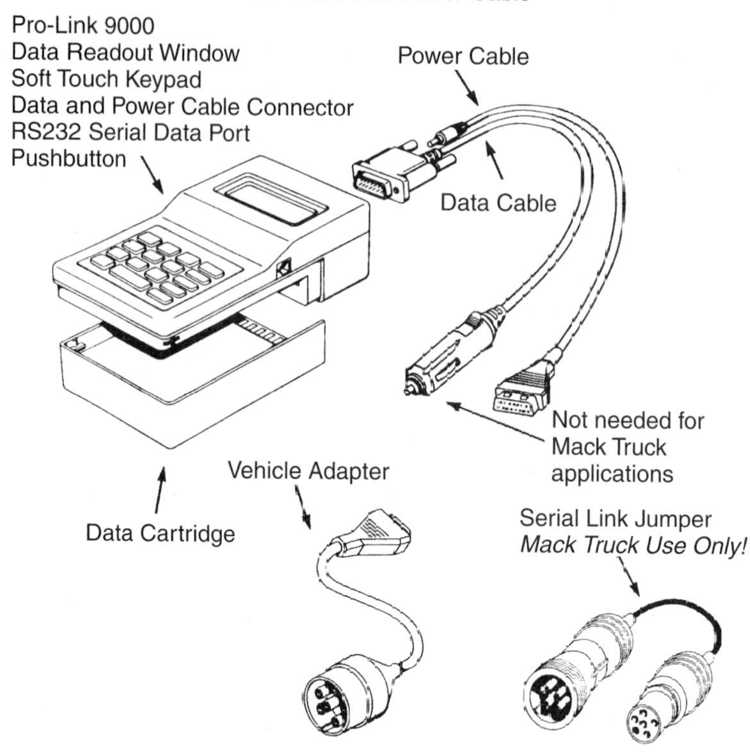

Figure 12-24 Pro Link head, software cartridge, and cables. (Courtesy of MPSI)

Special Tools

PROM removal tool

Grounding strap

Programmable read only memory (PROM) contain specific data that pertain to the exact vehicle or device in which the computer is installed.

Classroom Manual
Chapter 12, page 323.

PROM Replacement

On occasion, you will be required to replace a PROM in the computer. Usually this occurs due to a required update. Some computers are replaced, requiring the transfer of the previous computer's PROM to the new computer. It is critical that all service precautions are observed while servicing an ECM. Follow Photo Sequence 12 for a correct method of replacing the PROM.

WARNING: Installing a PROM chip in backward will immediately destroy it. **Electrostatic discharge (ESD)** will also destroy the chip. Wear static straps to prevent ESD while you are working on the unit.

Proprietary Engine Electronics Features and Troubleshooting

It is beyond this manual to cover every vehicle and engine electronics diagnostics. However, we will look at a few systems to have a better idea of what to expect when these vehicles are approached to solve a problem.

Mack Trucks V-Mac I and II

The V-MAC I system incorporates two modules electronically connected to each other in order to manage engine fueling, diagnostics, and chassis functions (Figure 12-25). As you can see, the fuel injection control (FIC) module in this V-MAC I system is a Bosch system component. This Bosch electronic diesel control (EDC) module is found in applications such as John Deere, Volvo, and other OEMs. In these earlier systems, the FIC module sensed and switched most of the func-

Photo Sequence 12
Typical Procedure for Replacing the PROM

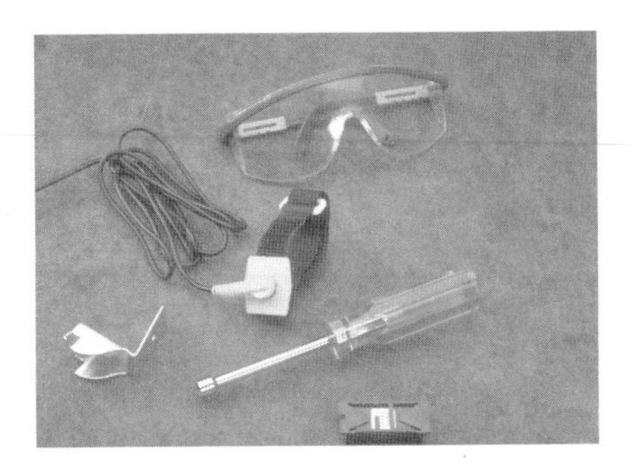

P12-1 Tools required to remove and replace the PROM: rocker type PROM removal tool, ESD strap, safety glasses and replacement PROM.

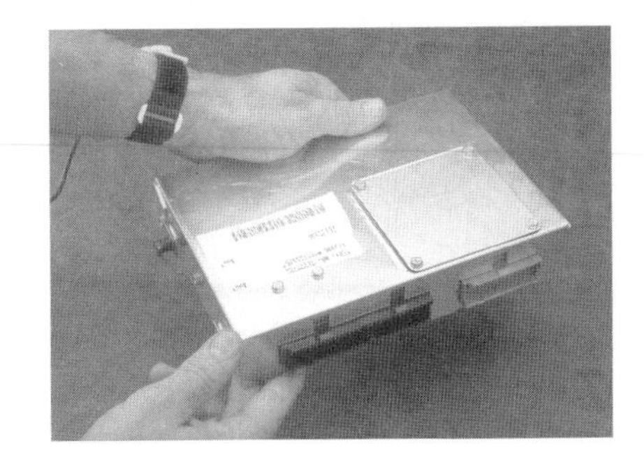

P12-2 Place the control module (unit, assembly) onto a clean work bench with the PROM access cover facing up. Be careful not to touch the electrical connectors with your fingers.

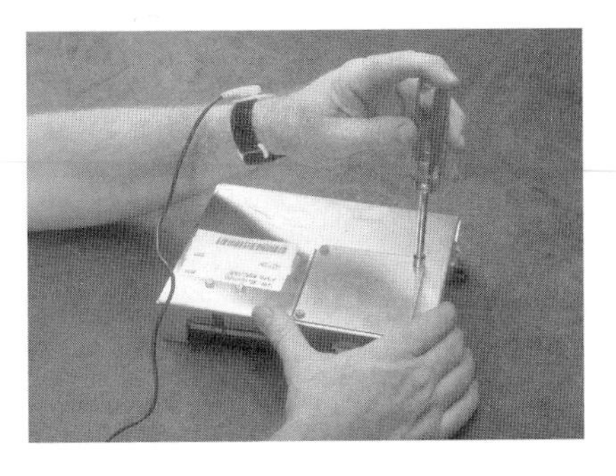

P12-3 Remove the PROM access cover.

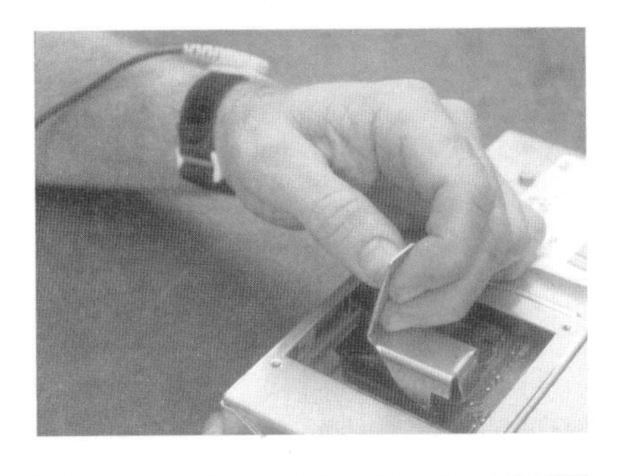

P12-4 Using the rocker–type PROM removal tool, engage one end of the PROM carrier with the hook end of the tool. Grasp the PROM carrier with the tool only at the narrow ends of the carrier.

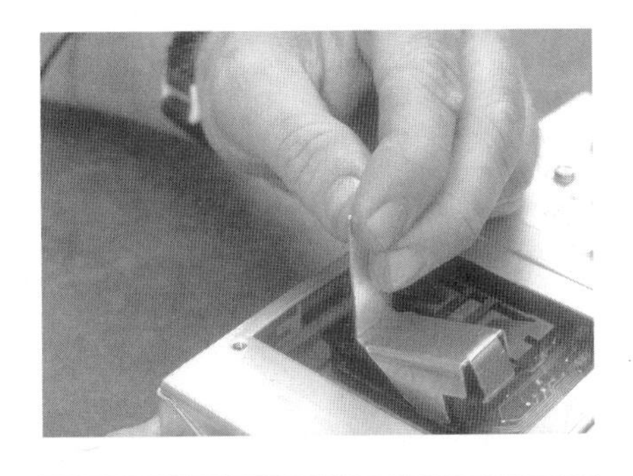

P12-5 Press on the vertical bar end of the tool. Rock the end of the PROM carrier up as far as possible.

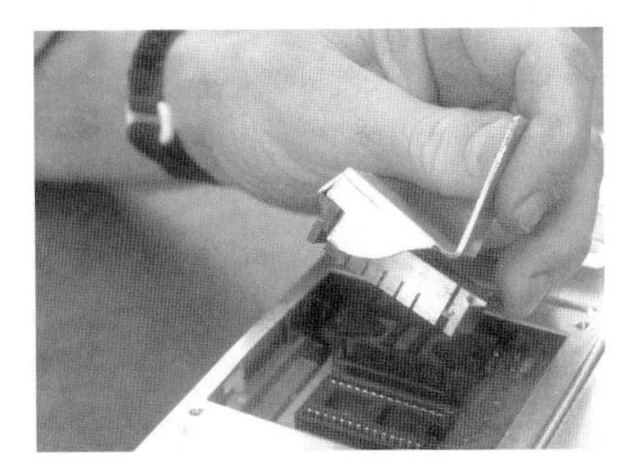

P12-6 Repeat the process on the other end of the carrier until the PROM carrier is removed from the socket.

P12-7 Inspect the replacement PROM part number for proper calibration.

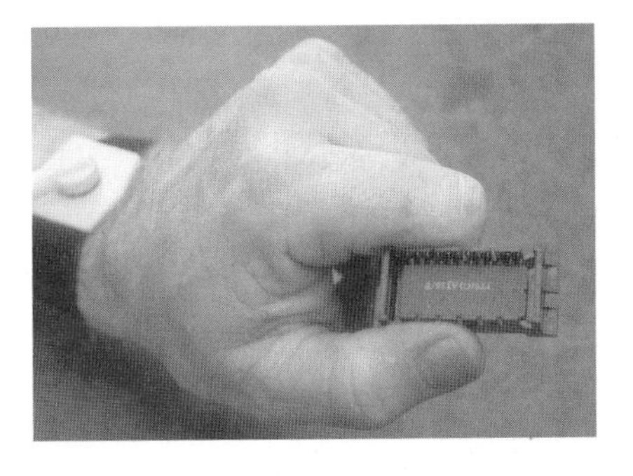

P12-8 Check for proper orientation of the PROM in the carrier. The notch in the PROM should be referenced to the smaller notch in the carrier. Be careful not to bend the pins.

P12-9 Align the PROM carrier with the socket. The carrier notch and socket notch must be aligned.

Photo Sequence 12 (Continued)
Typical Procedure for Replacing the PROM

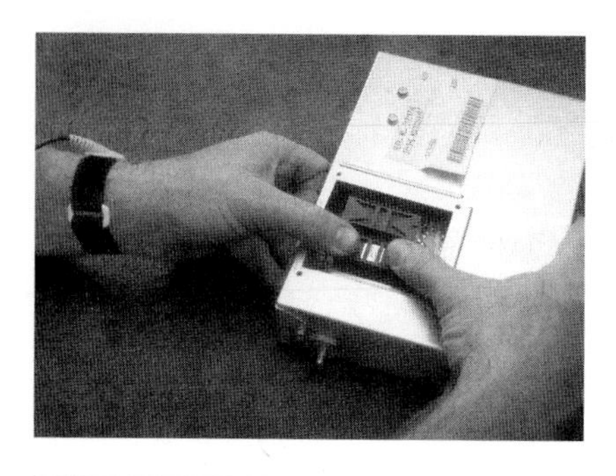

P12-10 Press the PROM carrier until it is firmly seated in the socket.

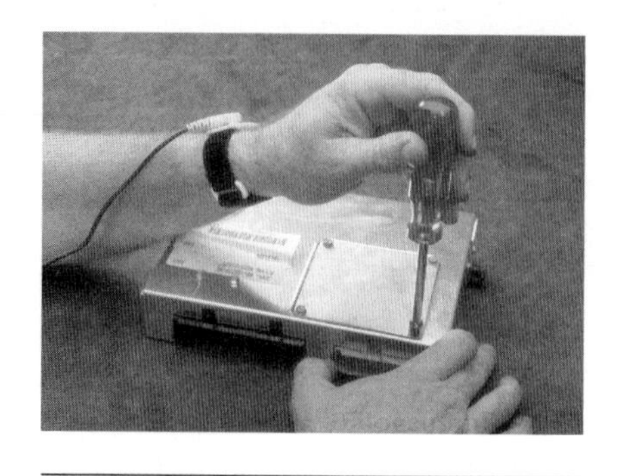

P12-11 Reinstall the access cover.

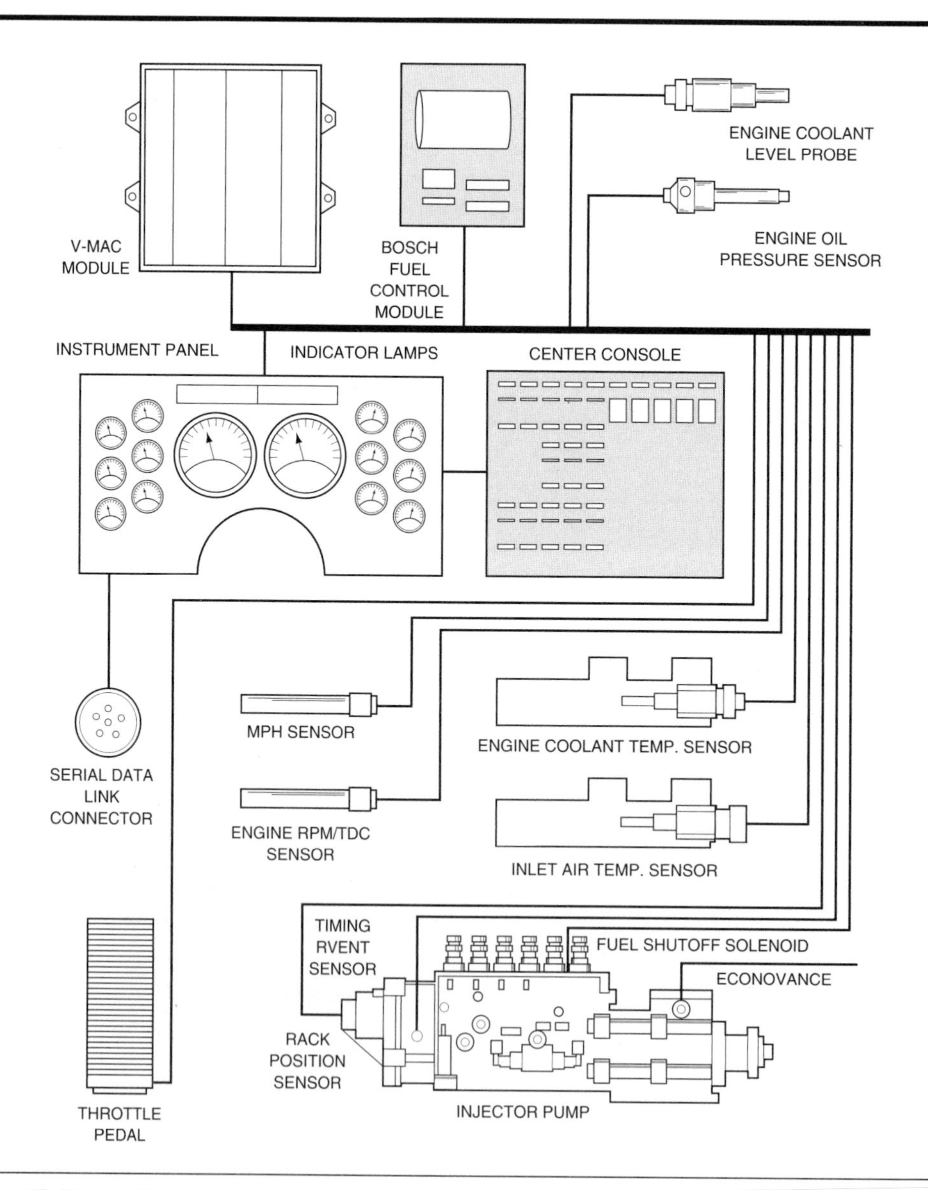

Figure 12-25 V-MAC I system overview. (Courtesy of Mack Trucks, Inc.)

tions directly associated with engine fueling. However, the V-MAC module manages the FIC module in order to control fuel management to vehicle application, performance, emissions management, and command inputs such as cruise control and PTO.

The more current V-MAC II eliminated the separate FIC module, incorporating all the functions in a single V-MAC II module currently manufactured by Motorola. Both versions of V-MAC had the V-MAC module(s) located in the cab dash compartment on the passenger side of the vehicle. In the V-MAC I system, the FIC module was located on top of the V-MAC module.

V-MAC modules can be accessed for fault code display, reading system parameters, and customer and proprietary data programming via the standard 6-pin ATA/SAE serial communications connector located under the dash on the driver's side of the vehicle.

The V-MAC system uses a Bosch injection pump as shown in Figure 12-26. The PE7100 was used with the first V-MAC I system and the more recent version is a PE8500. Both pumps are driven off the engine timing gear train at camshaft speed and use an electronically controlled, electric over hydraulic timing device, "Econovance," located between the pump and the engine accessory drive gear.

The rack actuator housing (Figure 12-27) incorporates sensors that input data to the ECM (V-MAC module) and a switched output from the ECM controlling rack position and thus, injected fuel quantity.

The rack assembly is spring loaded to the rack "no fuel" position (Figure 12-28). This no fuel rack position rotationally positions all the pump plungers so their vertical slots register with their barrel spill ports throughout their cam-actuated stroke. This plunger position allows the fuel to be displaced only as the plungers are actuated.

When the rack actuator is energized by the V-MAC module switching, the rack moves linearly, opposing the spring pressure. This alters the rotational position of the plungers, increasing plunger effective stroke.

Increasing current flow through the rack actuator coil forces the rack further in. In the full fuel position, the rack retraction spring would be maximally compressed. The rack actuator is an ECU module switched output that responds to command signals.

Within the assembly is a rack travel sensor that reports the rack position to the ECM. The sensor assembly consists of a measuring coil that is energized with a 5-volt reference voltage.

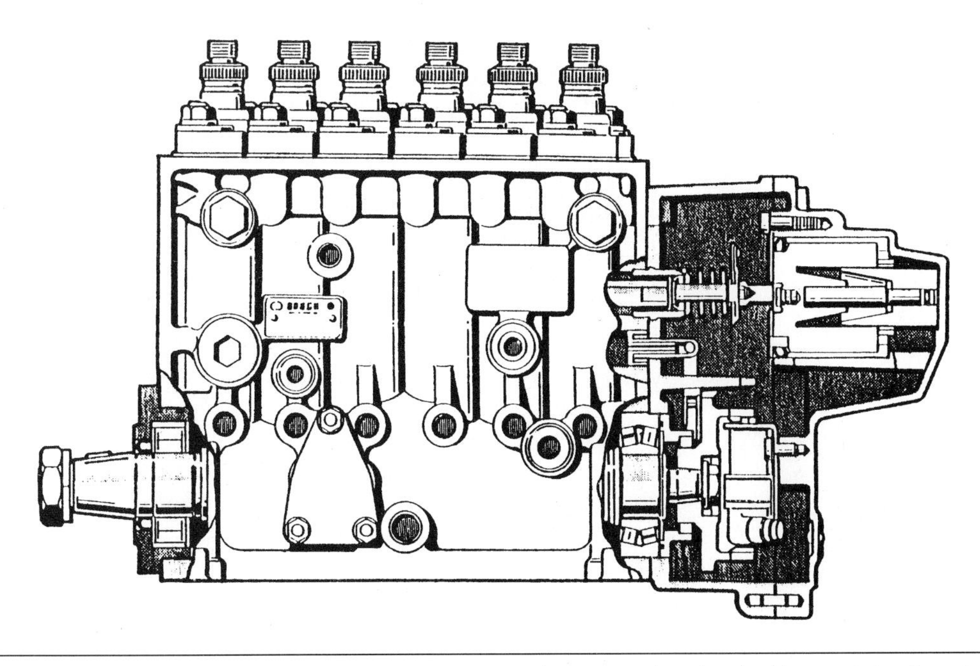

Figure 12-26 Bosch PE7100 injection pump with RE30 rack actuator. (Reprinted with permission from Robert Bosch Corporation)

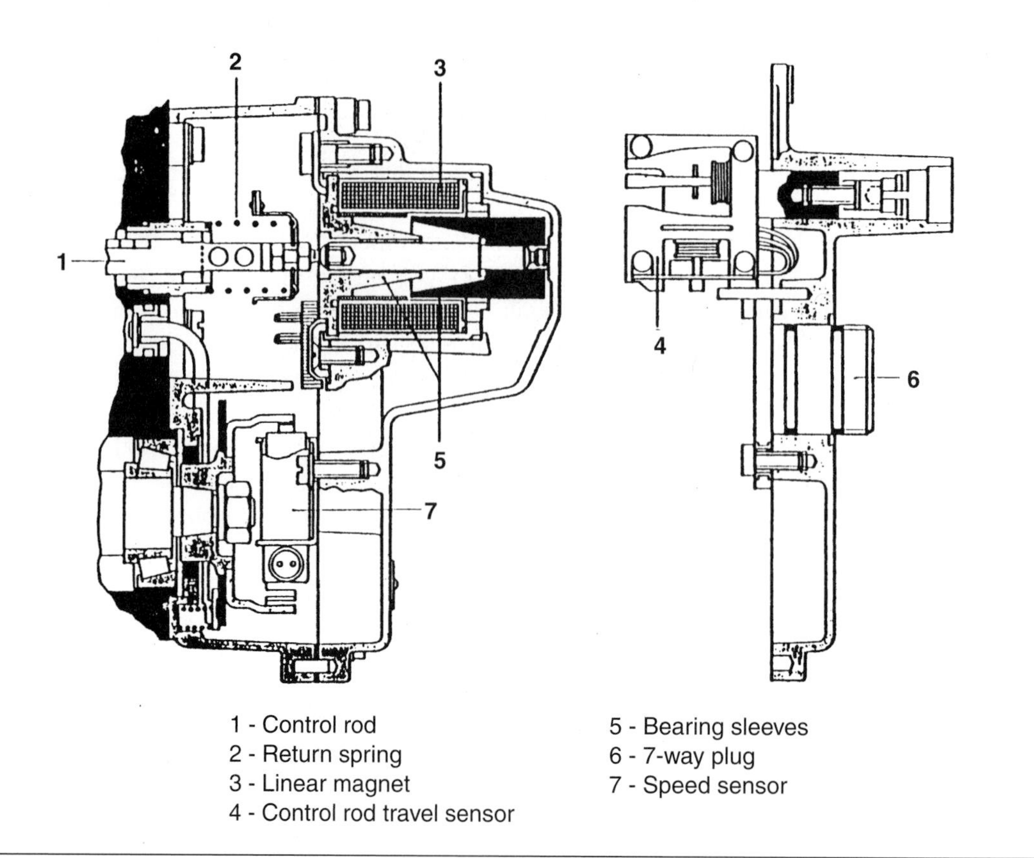

Figure 12-27 Sectional view of Bosch RE30 rack actuator housing. (Reprinted with permission from Robert Bosch Corporation)

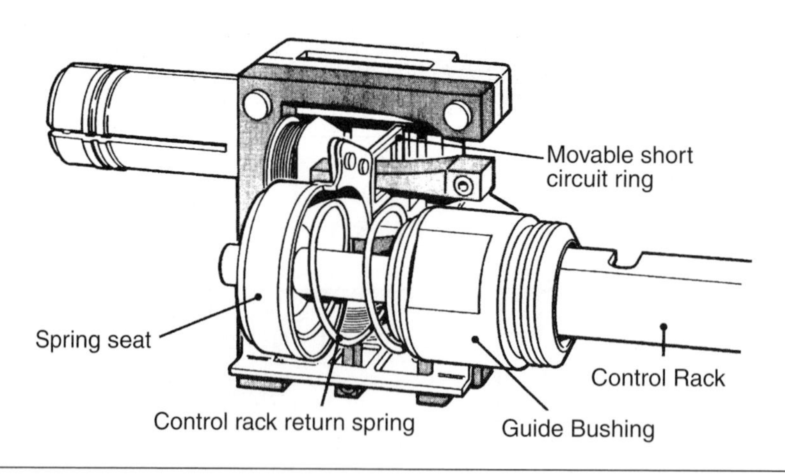

Figure 12-28 Rack actuator assembly components. (Reprinted with permission from Robert Bosch Corporation)

Also within the assembly is a short circuit ring; protruding from the fixed measuring coil is an iron bar (Figure 12-29).

The short circuit ring is designed to slide over the iron bar without physically contacting it. As the fuel control rack is moved by the rack actuator, the short circuit ring will either move closer or farther away from the measuring coil, thereby varying the electromagnetic field spread by the coil and, therefore, the voltage signal returned to the V-MAC module. The sensor system conveys very precise rack position data to the ECM (Figure 12-30 and 12-31).

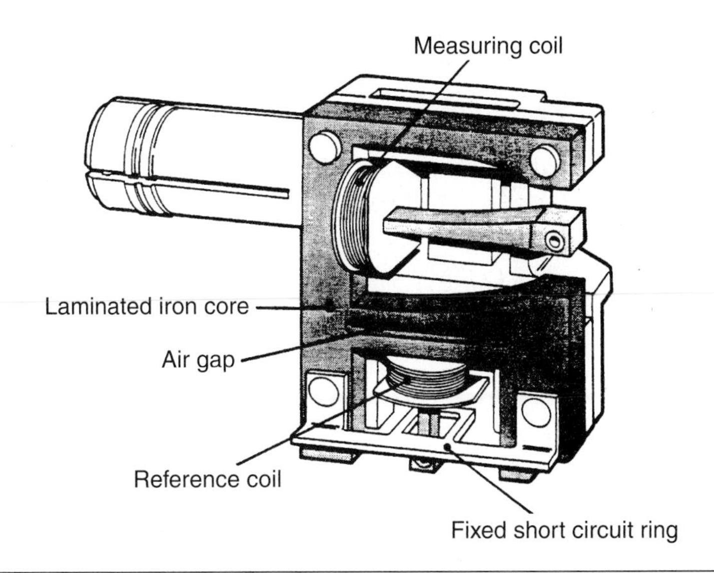

Figure 12-29 Rack travel sensor components. (Reprinted with permission from Robert Bosch Corporation)

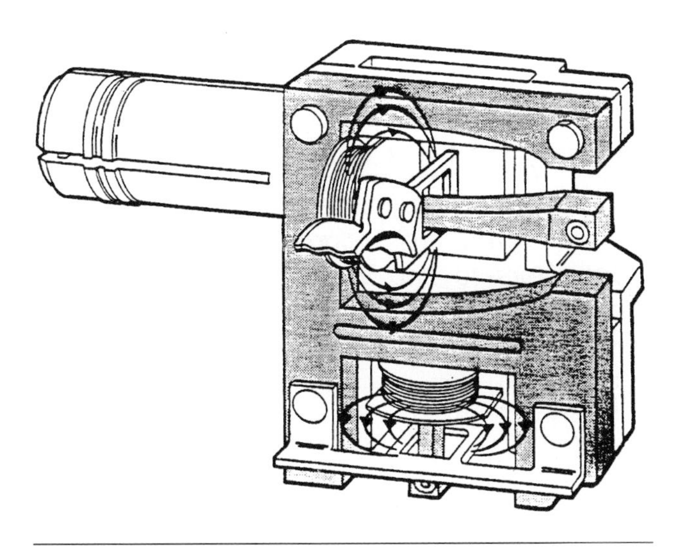

Figure 12-30 Rack travel sensor at idle fuel rack position. (Reprinted with permission from Robert Bosch Corporation)

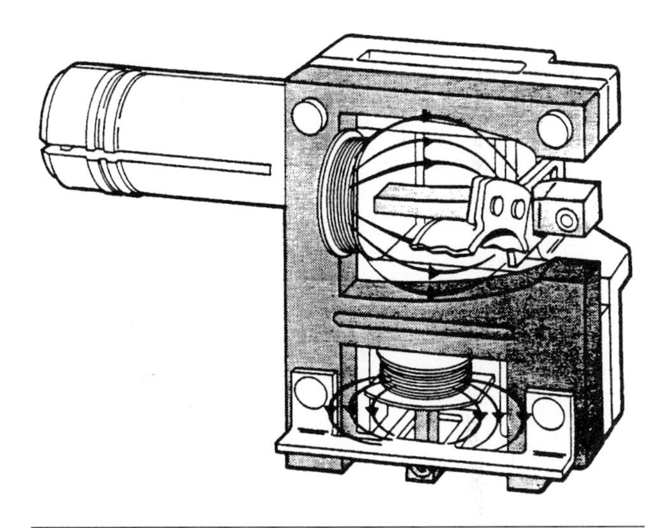

Figure 12-31 Rack travel sensor at maximum rack (peak fuel per cycle) position. (Reprinted with permission from Robert Bosch Corporation)

To troubleshoot the V-MAC electronic control system requires the use of the V-MAC service manual. The diagnostic flow is pretty straightforward, just remember not to skip any steps. Figure 12-32 is an example of a diagnostic flowchart used if a throttle position low voltage fault code was retrieved. The code is a Mack 5-1 or SAE P1091, FM14.

Navistar with Cummins CELECT Engine Controls

This section will be a sampling of the Cummins CELECT Electronic Engine Control system as applied in a Navistar vehicle. The CELECT system provides an electronic fuel control system, electronic engine protection, and other operational controls.

CELECT is used on L10 and N14 engines and controls the following functions:

- ❑ Road speed governing
- ❑ Gear down protection

Special Tools

DMM

Vehicle service manual

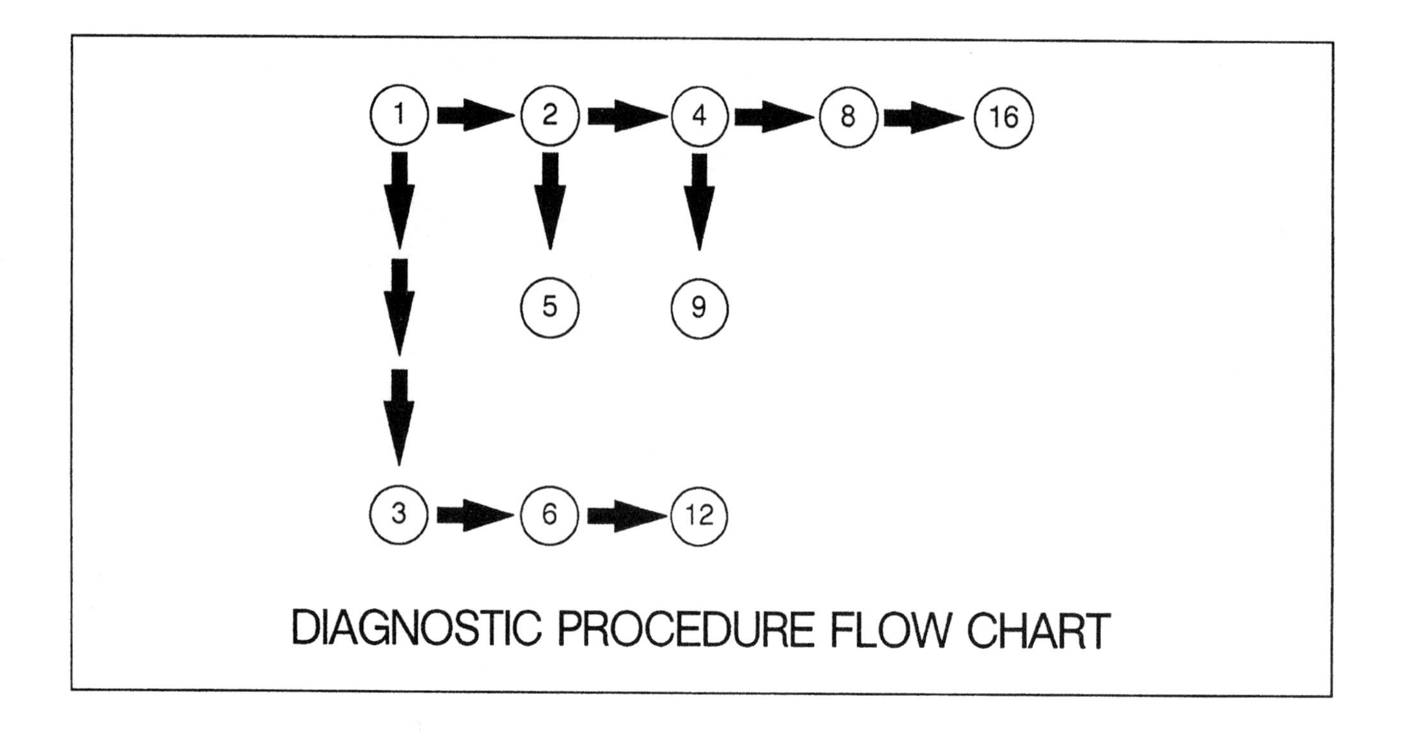

NOTE

When performing your electrical tests, it is important to wiggle wires and connectors to find intermittent connection problems.

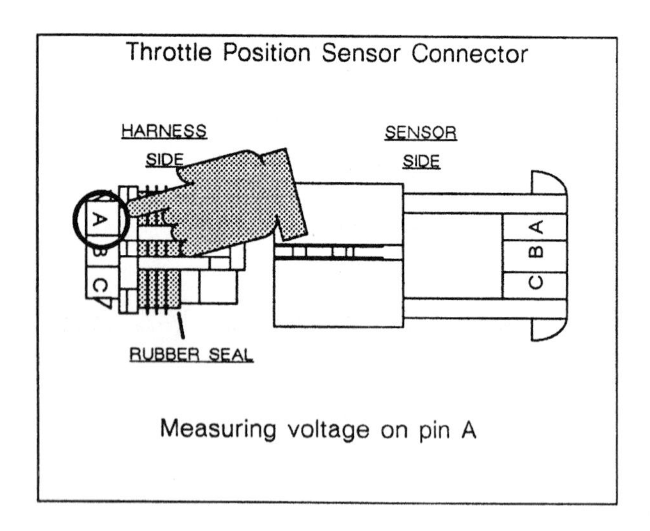

Test 1 (Checking for proper supply voltage)

1. Turn the key OFF.
2. Disconnect the TPS from the harness.
3. Turn the key ON.
4. Read the voltage from the pedal PLUS (+) (Pin A) at the sensor end of the harness to a good ground.

If the voltage is greater than 4.35 volts, proceed to test 2.

If the voltage is less than 4.35 volts, proceed to test 3.

Figure 12-32 V-MAC Diagnostic Flowchart: TPS code 5-1. (Courtesy of Mack Trucks, Inc.)

Test 2 (Checking for the signal line pulled low by a short)

1. Turn the key OFF.

2. Connect a jumper between Pin A and Pin B at the sensor end of the harness.

3. Turn the key ON.

4. Read the voltage from the jumper to a good ground.

If the voltage is between 4.35 and 5.25 volts, proceed to test 4.

If the voltage is NOT between 4.35 and 5.25 volts, proceed to test 5.

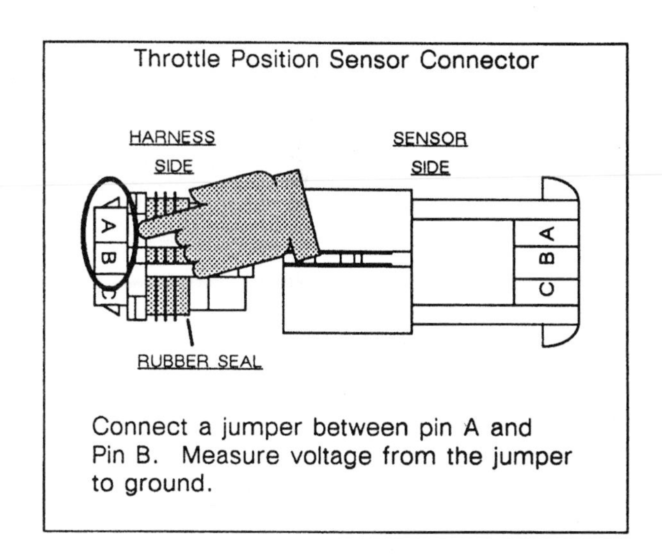

Connect a jumper between pin A and Pin B. Measure voltage from the jumper to ground.

Test 3 (Checking for a short to voltage in the harness)

1. Turn the key OFF.

2. Disconnect the FIC connector.

3. Turn the key ON and read the voltage from Pin 33 at the FIC end of the harness to a good ground.

If there is NO voltage, proceed to test 6.

If the voltage is greater than 0.5 volts, the pedal PLUS (+) line (pin 33) is shorted to low voltage in the harness or connector. Locate and repair the short.

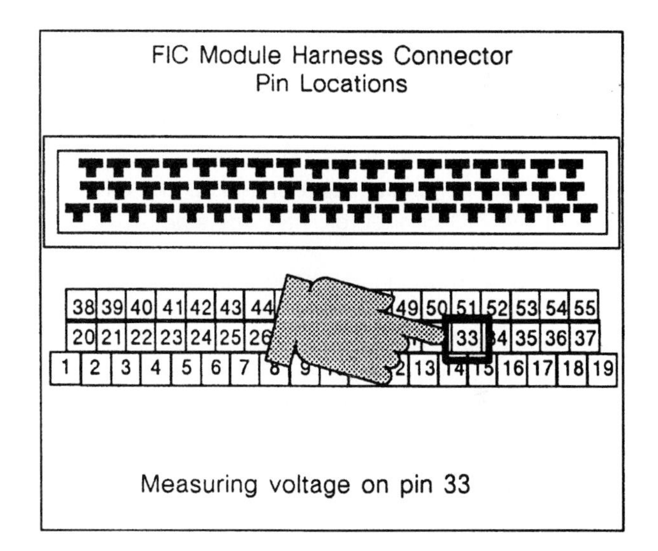

Measuring voltage on pin 33

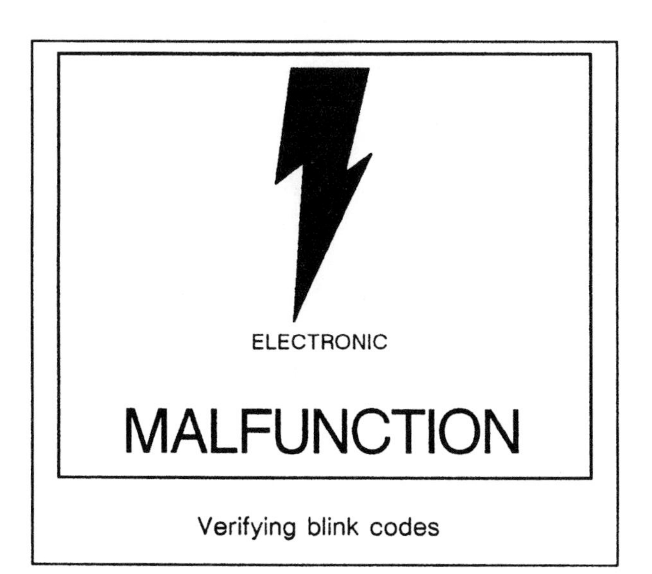

Verifying blink codes

Test 4 (Verifying blink codes)

If the blink code has changed to 5–2, proceed to test 8.

If the blink code has NOT changed to 5–2, proceed to test 9.

Figure 12-32 Continued.

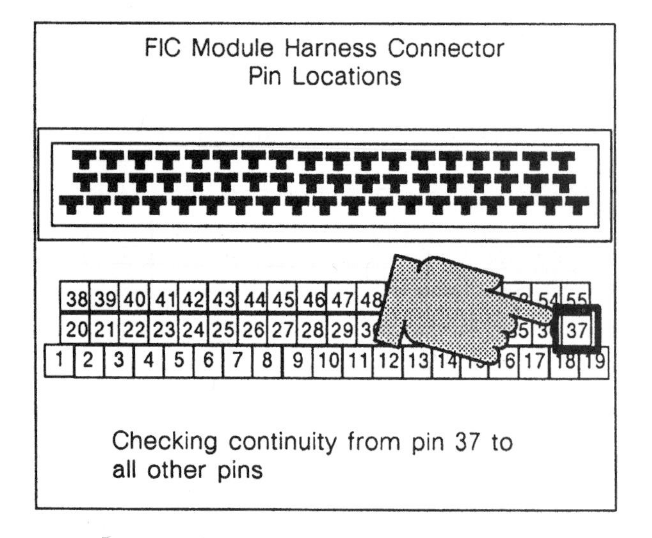

Test 5 (Checking for a pin to pin short on the signal line or the FIC module)

1. Turn the key OFF.

2. Remove the jumper that was installed in Test 2.

3. Disconnect the FIC connector.

4. Check for continuity from Pin 37 at the FIC end of the harness to all other pins.

If there is another pin with continuity, the signal line is shorted to ground in the harness or connectors, or shorted to low voltage in the harness or connectors. Locate and repair the short and retest.

If there is NOT continuity, replace the FIC module.

Test 6 (Checking for a pin to pin short or short to ground on the pedal PLUS (+) line)

1. Turn the key OFF.

2. Check for continuity between Pin 33 and all other pins at the FIC connector, harness side.

If there is NO continuity, proceed to test 12.

If there is continuity, the pedal PLUS (+) line is shorted to ground in the harness or connector, or there is a pin to pin short on the pedal PLUS (+) line at the connectors.

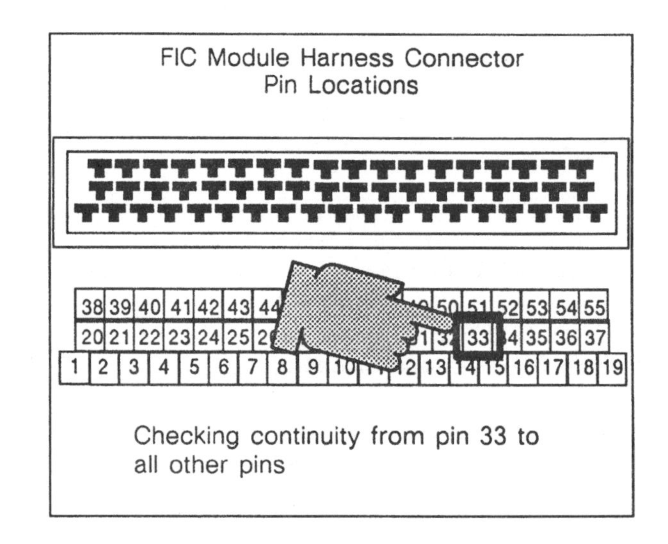

Test 8 (Checking for a pin to pin short on the signal line in the harness)

1. Turn the key OFF.

2. Remove the jumper.

3. Disconnect the FIC connector.

4. Check for continuity between Pin 37 and all other pins at the FIC end of the harness.

If there is NO continuity, proceed to test 16.

If there is continuity, the pedal signal line is shorted in the harness or connectors.

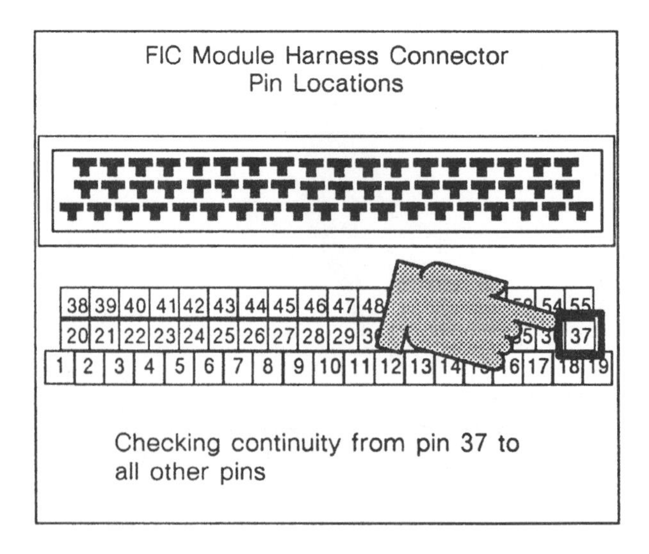

Figure 12-32 Continued.

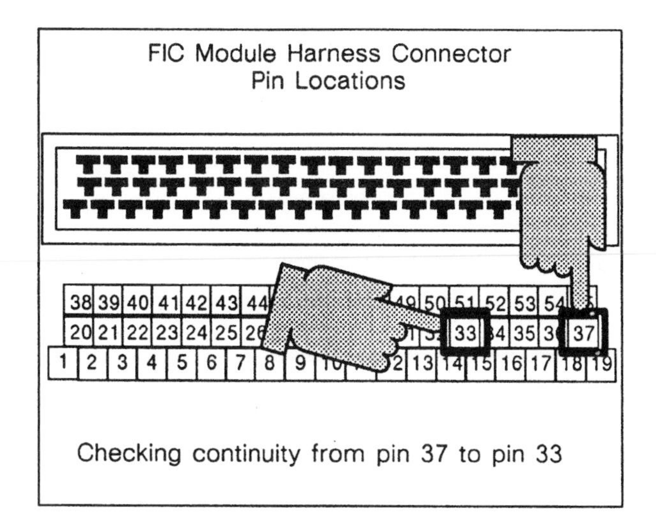

Checking continuity from pin 37 to pin 33

Test 9 (Checking for an open in the signal line, or a bad FIC module)

1. Turn the key OFF.

2. Disconnect the FIC connector.

3. Check for continuity between Pins 37 and 33 at the FIC end of the harness.

If there is continuity, replace the FIC module and retest.

If there is NOT continuity, the signal line is open in the harness or connectors.

FIC Module Harness Connector
Pin Locations

Checking continuity from pin 37 to pin 33

Test 12 (Checking for an open on the pedal PLUS (+) line in the harness or FIC module)

1. Connect a jumper between Pins A and B at the sensor end of the harness.

2. Check for continuity between Pins 33 and 37 at the FIC end of the harness.

If there is NO continuity, the pedal PLUS (+) line is open at the harness or connector.

If there is continuity, replace the FIC module and retest.

Test 16 (Checking for a bad sensor or FIC module)

1. Replace and reconnect the pedal sensor. Reconnect the FIC connector.

2. Turn the key ON.

3. Press the pedal to the floor.

If blink codes 5–1 or 5–2 are present, replace the FIC module and retest.

If blink codes 5–1 or 5–2 are NOT present, replacing the TPS has corrected the problem.

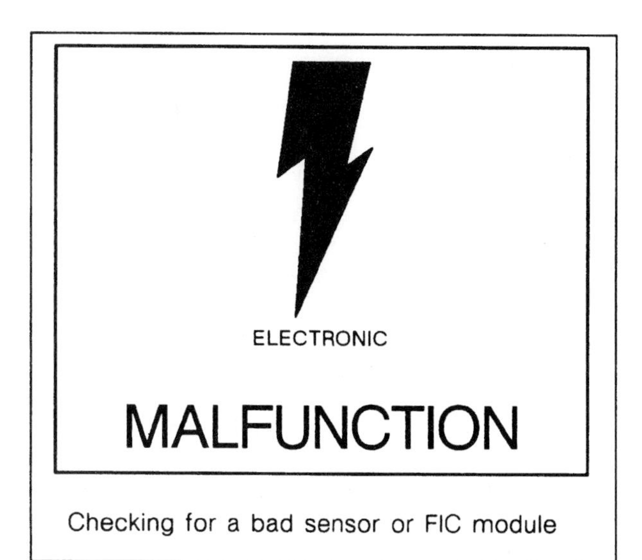

Checking for a bad sensor or FIC module

Figure 12-32 Continued.

- ❑ Cruise control
- ❑ PTO control
- ❑ Progressive shifting
- ❑ Idle shutdown
- ❑ Adjustable low speed (in cab)
- ❑ Choice of governor type (electronically selectable)
- ❑ Self diagnostics
- ❑ Engine protection system

Special Tools

Service manual

Cummins inspection tool (part #3823383)

The CELECT Plus electronics are used to manage current M11 Plus and N14 Plus engines. CELECT Plus is a full authority, computerized engine management system that uses cam-actuated, electronically controlled CELECT injectors.

The CELECT Plus system can be divided into the following four interconnected systems:

1. Fuel subsystem
2. Input circuit sensors
3. ECM
4. Output circuit (CELECT injectors)

Fuel Subsystem Fuel is drawn in from the fuel tank by a gear pump. It is pulled through the cooling plate (heat sink) on which the CELECT ECM is mounted (Figure 12-33). Circulating fuel through this heat exchanger helps to dissipate heat from the ECM. From the ECM cooling plate, fuel passes through the fuel filter. Both the ECM cooling plate and filter are under suction. From the filter head, the fuel is piped to the gear pump, which is pressurized and then discharges fuel at a constant charge pressure of 150 PSI to tubings that deliver the fuel to the cylinder head and its fuel galleries. This fuel is always available to the CELECT injectors. After circulating through the CELECT injectors, fuel exits by means of a fuel drain and enters the cylinder head return gallery. From there it is piped back to the fuel tank.

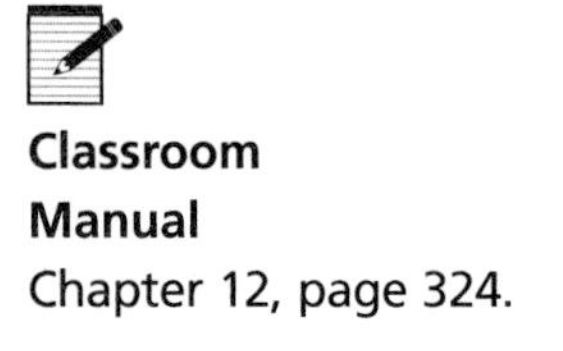

Classroom Manual

Chapter 12, page 324.

Input Circuit Components in this circuit are connected to the ECM by a single wiring harness integrating the separate sensor and actuator harnesses of the CELECT system (Figure 12-34). The following are the engine monitoring sensors:

1. Ambient pressure sensor—piezoelectric type
2. Oil pressure sensor—piezoelectric type
3. Engine position sensor—magnetic pulse generator type
4. Oil temperature sensor—thermistor
5. Intake manifold temperature sensor—thermistor

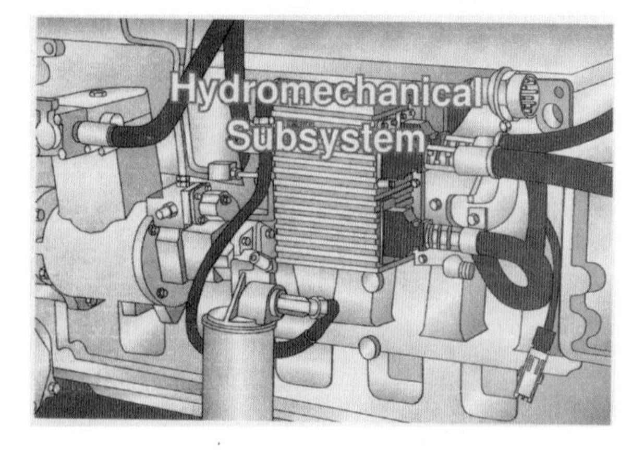

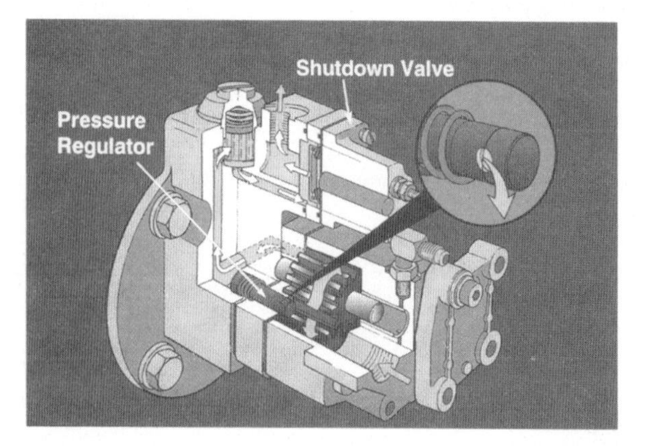

Figure 12-33 The hydromechanical fuel subsystem on a CELECT engine. (Courtesy of Cummins Engine Company)

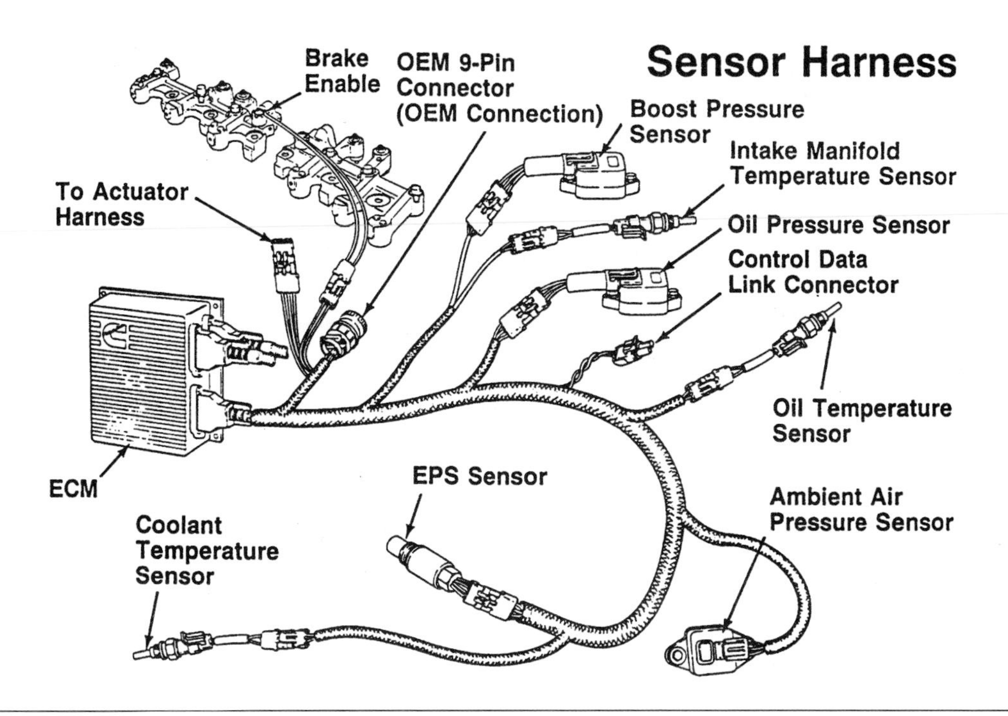

Figure 12-34 The CELECT sensor harness. (Courtesy of Cummins Engine Company)

6. Boost pressure sensor—piezoelectric type
7. Coolant temperature sensor—thermistor
8. Coolant level sensor—a switch type sensor that ground through the engine coolant with dual probes to indicate high or low coolant levels. This is an OEM-supplied sensor.

The following are command sensors and switches:

1. Throttle pedal assembly: OEM-supplied accelerator pedal assembly that incorporates a 5-volt reference potentiometer and an idle validation switch.
2. Cruise control
3. PTO control: This feature controls engine RPM at a constant speed selected by the operator.
4. Jacobs C-brake

Electronic Control Module The ECM receives and processes information in order to output appropriate signals to the injector drive circuitry. A CELECT Plus injector driver unit uses coils to spike the EUI solenoid voltage to 78 volts and is the primary output of the ECM. Figure 12-35 shows the ECM and Figure 12-36 shows a variety of modes graphically.

Output Circuit The primary ECM output function is to control the CELECT injectors. The CELECT ECM drives output circuit functions through the actuator harness (Figure 12-37). Figure 12-38 shows a sectional view of the CELECT injector. Figure 12-39 illustrates a CELECT injector operation.

Classroom Manual
Chapter 12, page 333.

To troubleshoot the CELECT and CELECT Plus system requires the use of Cummins Troubleshooting and Repair Manual. The required tools are a DMM and the appropriate EST, preferably a PC with INSITE software.

WARNING: Never attempt electronic engine management repairs or diagnostics without proper information. It can lead to misdiagnosis and possible improper repairs that can be very costly.

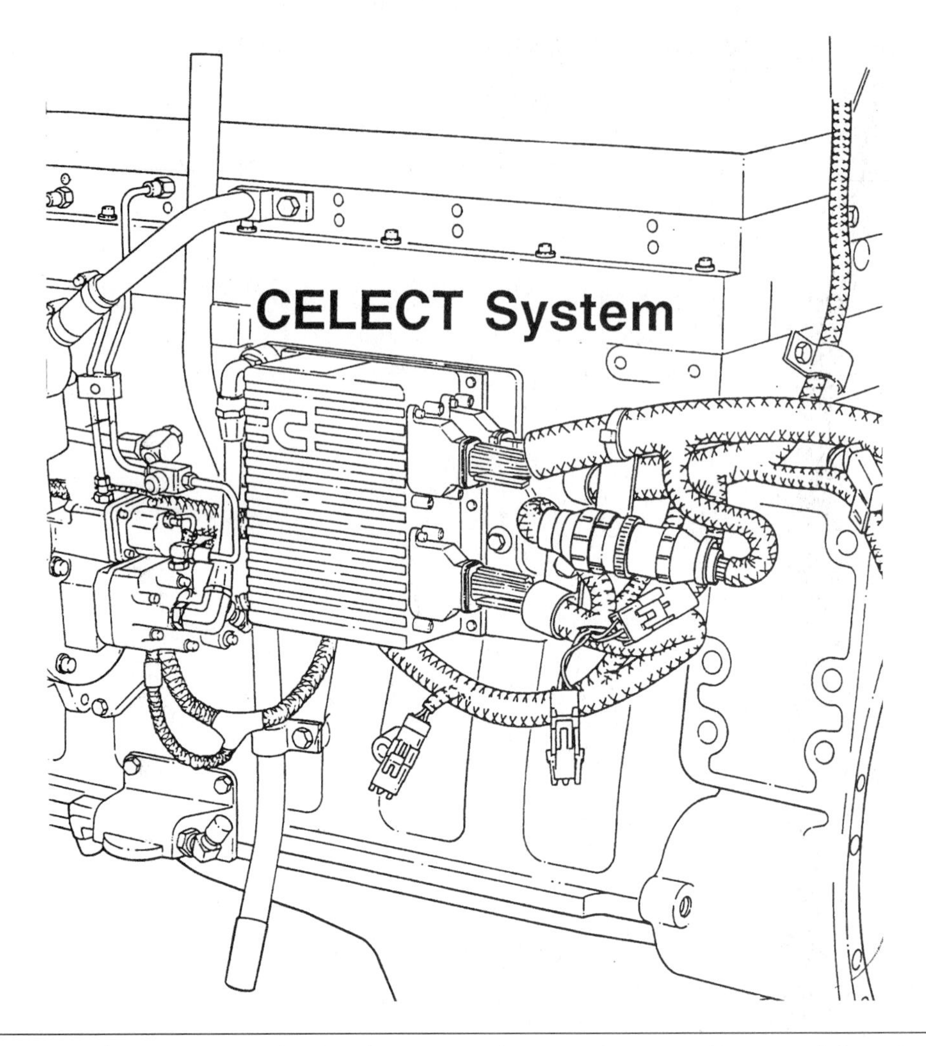

Figure 12-35 The CELECT ECM showing the sensor and actuator harnesses. (Courtesy of Cummins Engine Company)

The following is an example of troubleshooting a system where the OEM (Navistars) and CELECT circuits work together:

❑ Inspecting Navistar Engine Harness ECM connector

1. Remove connector from the ECM by first removing the two cap screws. Carefully pull the connector straight out of the ECM. The pins are numbered and correspond to the numbers on the circuit diagram. Figure 12-40 shows the connector. Figure 12-41 shows the wiring diagram that can be used as a reference.
2. Inspect the connector for pushed back or expanded terminals in the following manner:
   - **A.** Insert the tool (Cummins part number 3823383) into each Amp connector (Figure 12-42).

WARNING: DO NOT force the tool into the connector or damage to the connector can occur.

   - **B.** A terminal that is not pushed back or expanded will have a surface-to-tool gap (Figure 12-43) of approximately 0.050 inches (1.3 mm). If the tool touches the connector, replace the terminal.

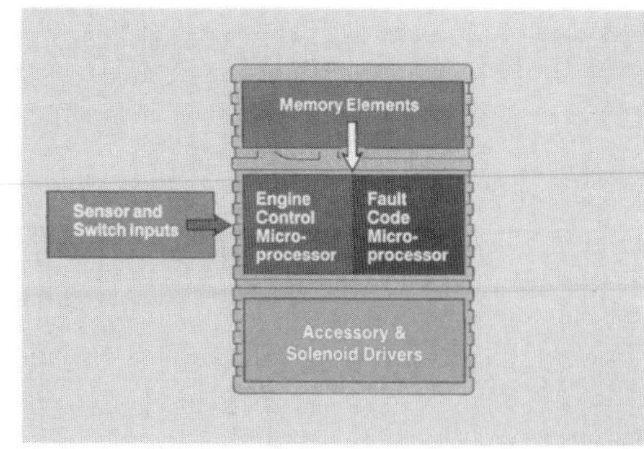

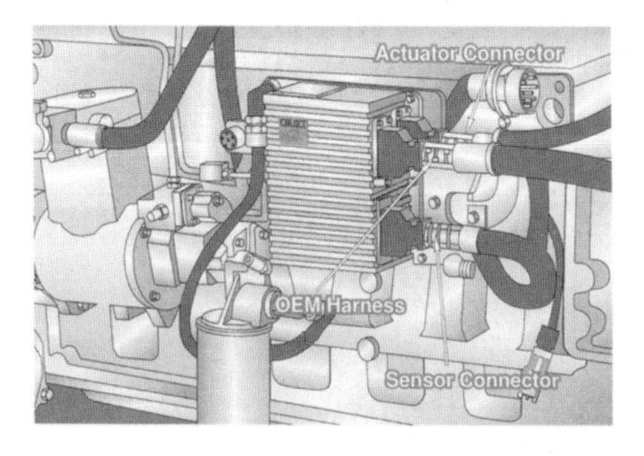

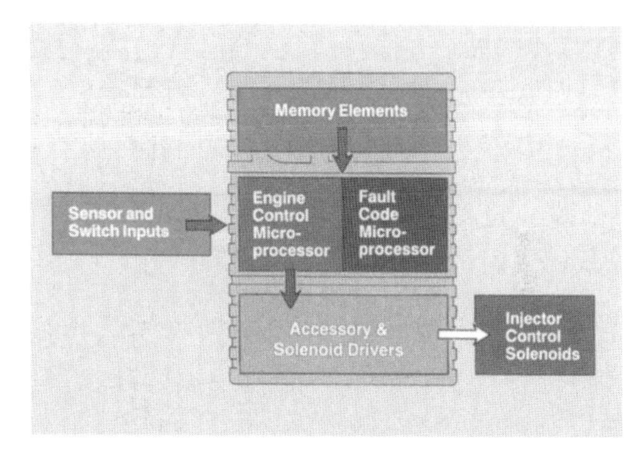

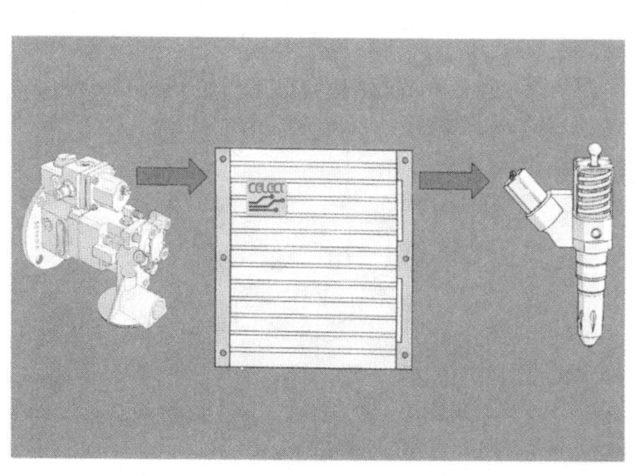

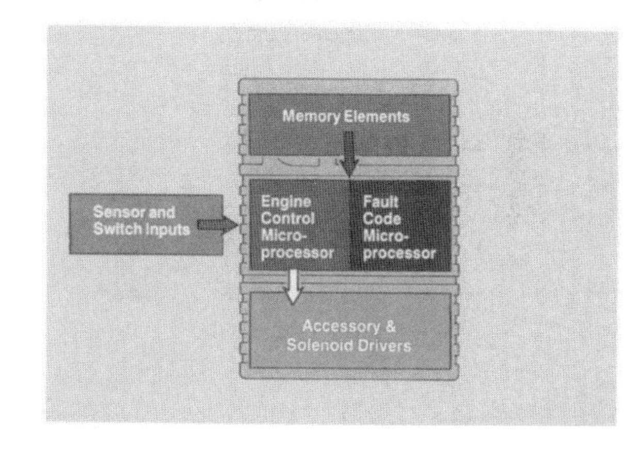

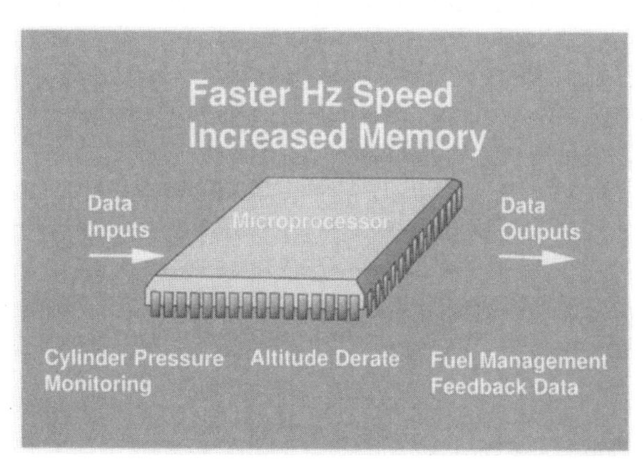

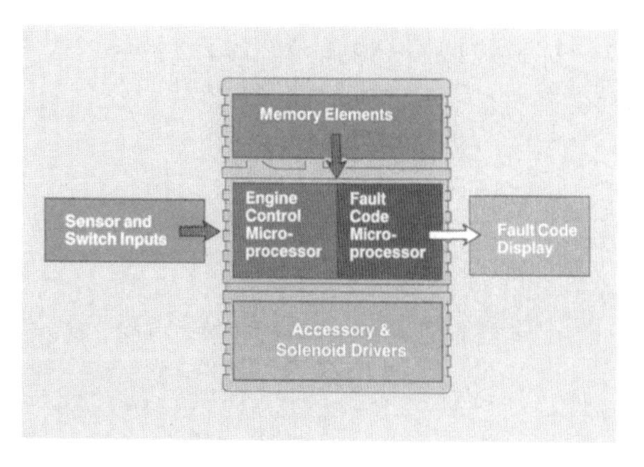

Figure 12-36 CELECT Plus processing cycle graphics. (Courtesy of Cummins Engine Company)

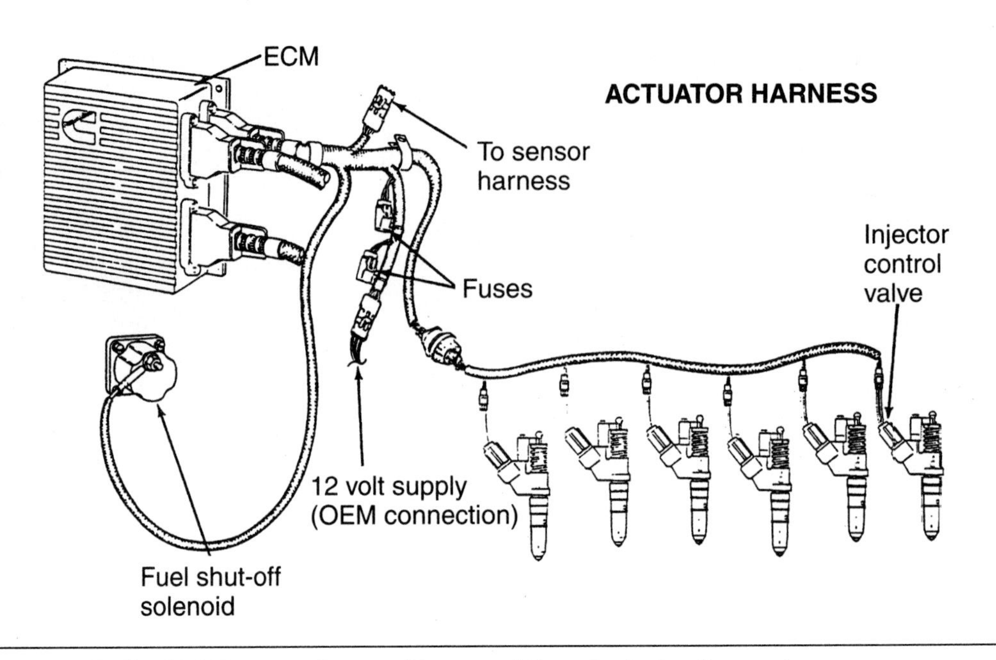

Figure 12-37 The CELECT actuator harness. (Courtesy of Cummins Engine Company)

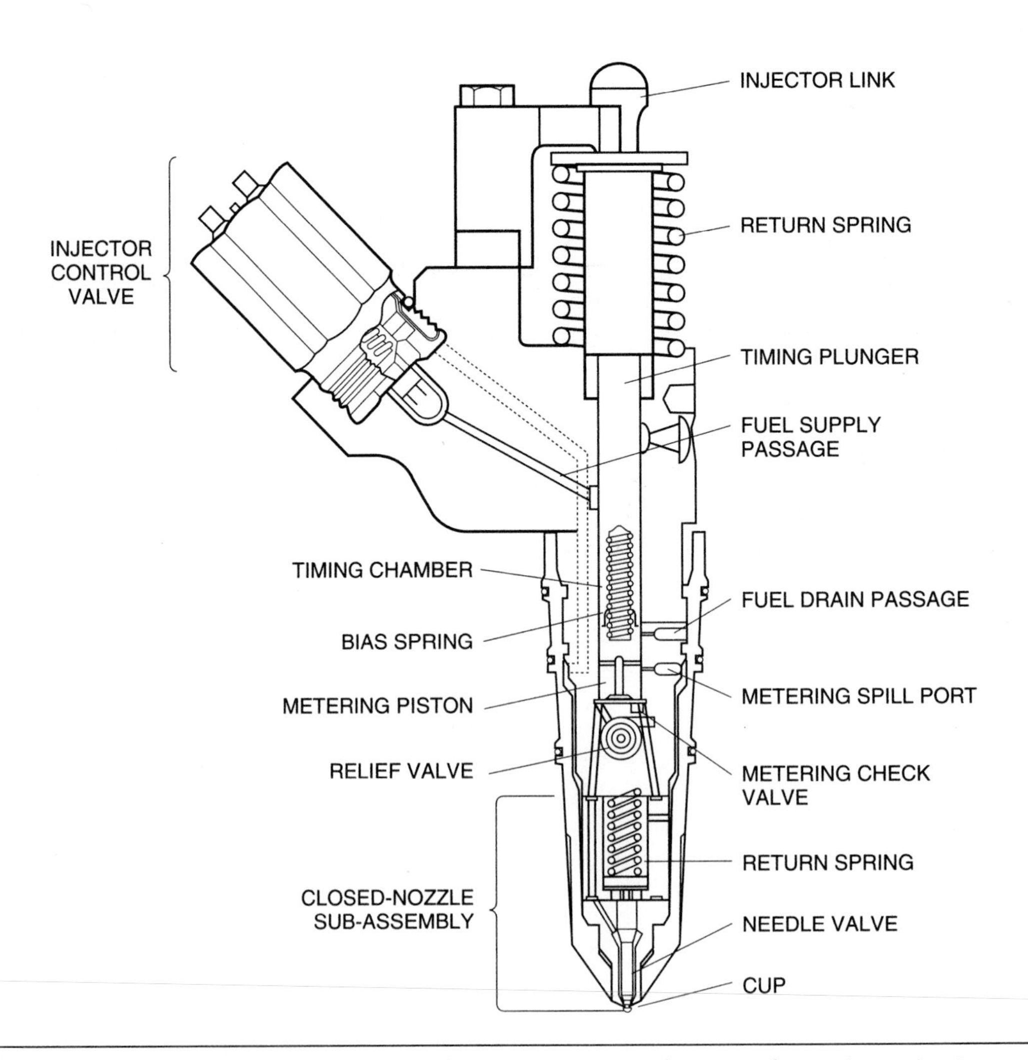

Figure 12-38 Sectional view of the key CELECT injector components. (Courtesy of Cummins Engine Company)

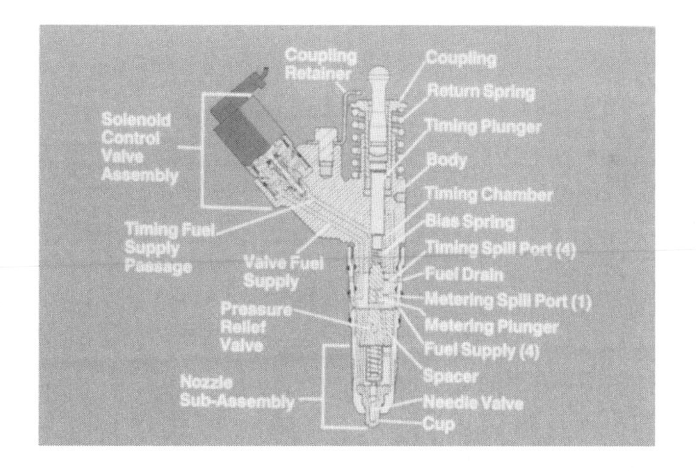

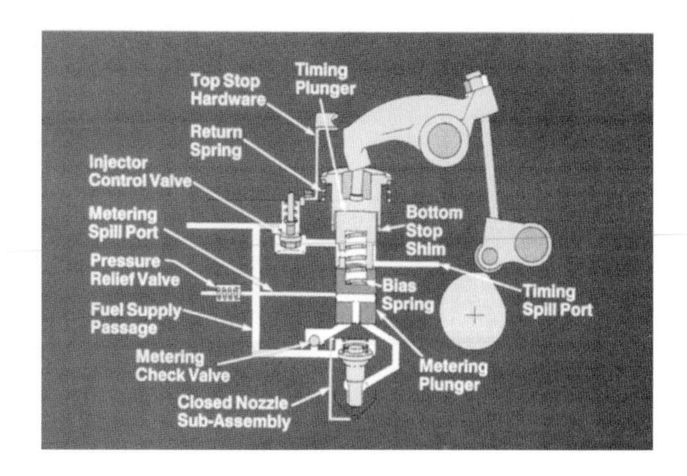

1. At the start of metering, both the metering plunger and the timing plunger are at their lower travel limit as their actuating cam is at peak lift. The injector control valve closes, actuated by a 78 V induction coil derived spike delivered from the CELECT injector driver unit.

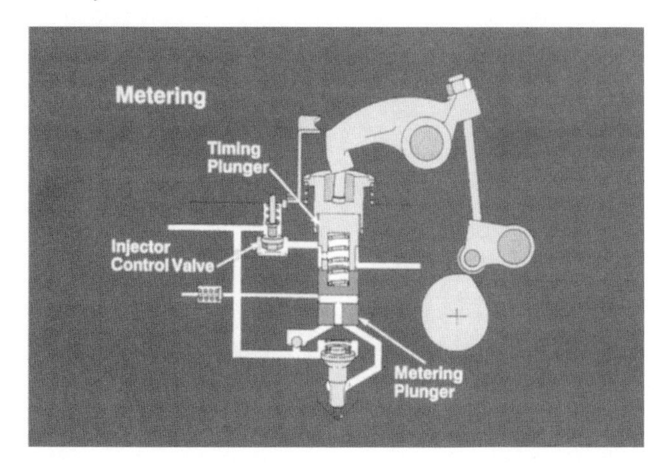

2. As the cycle continues, the cam ramps off the nose toward base circle, unloading the injector train and permitting the timing plunger return spring to lift the timing plunger. This enables fuel to flow past the metering checkball into the metering chamber. This flow continues as long as the timing plunger is moving upward and the injector control valve is closed; supply pressure acting on the bottom of the metering piston forces it to maintain contact with the timing plunger.

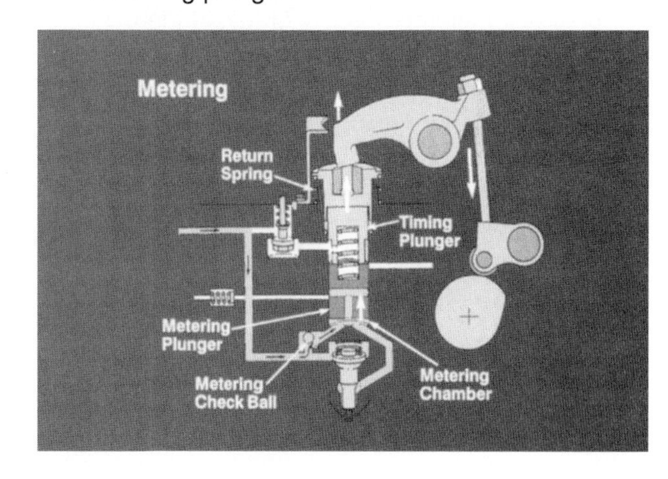

Figure 12-39 The CELECT Plus injector. (Courtesy of Cummins Engine Company)

3. The ECM will determine the end of metering by switching the injector control valve to its *open* position. This action causes the metering checkball to seat and permits fuel to pass around the injector control valve.

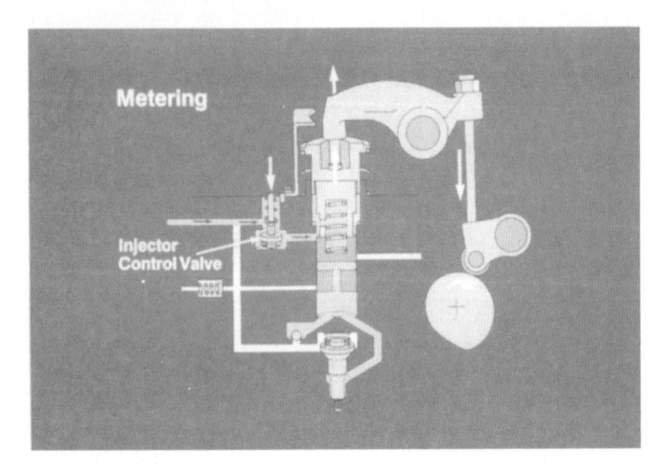

4. Fuel at supply pressure then flows into the timing chamber stopping metering piston travel. The bias spring ensures that the metering plunger remains stationary, preventing it from drifting upward as the timing plunger moves upward. This same force against the metering plunger results in sufficient fuel pressure below the metering piston to keep the metering checkball seated. The result is a precisely metered quantity of fuel in the metering chamber.

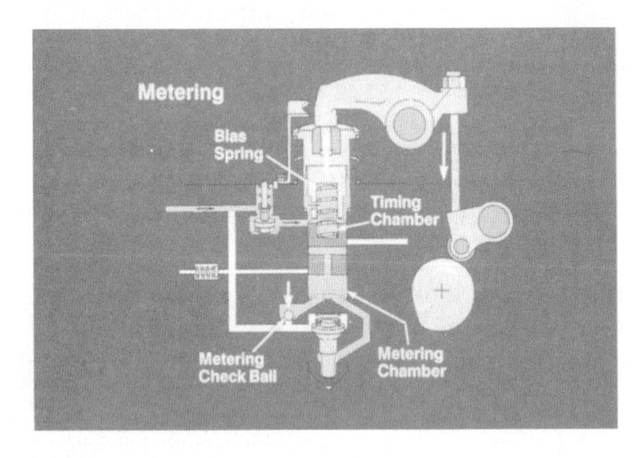

5. As the cycle continues, the injector train continues to ride toward cam inner base circle permitting the timing chamber to fill with fuel.

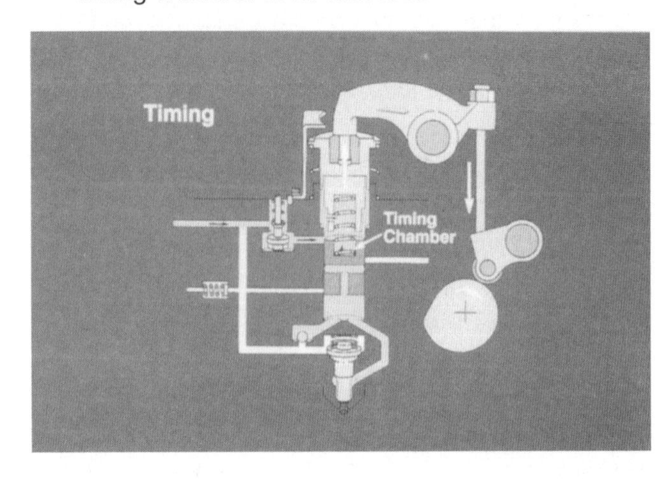

6. Next, the injector cam passes over the inner base circle location and begins ramping toward outer base circle. This action loads the injector train and consequently the timing plunger begins its downstroke. Initially the injector control valve remains open, allowing fuel to spill from the timing chamber reverse flowing it through the fuel supply passage.

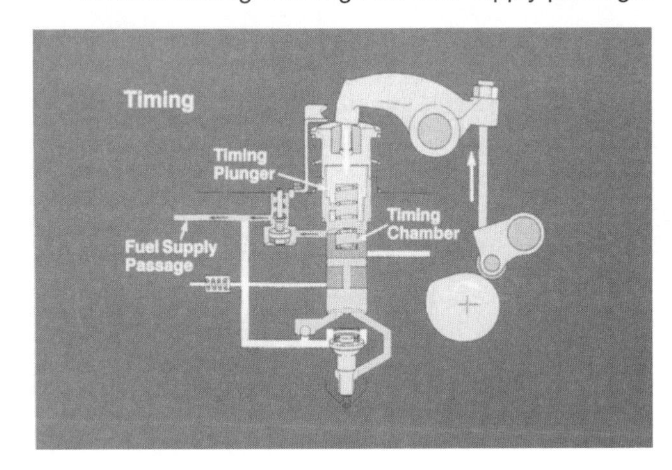

Figure 12-39 Continued.

7. The delivery sequence begins when the ECM switches the injector control valve to its closed position trapping fuel in the timing chamber. This trapped fuel acts as a hydraulic link between the timing plunger and the metering plunger; this forces the metering plunger downward with the timing plunger.

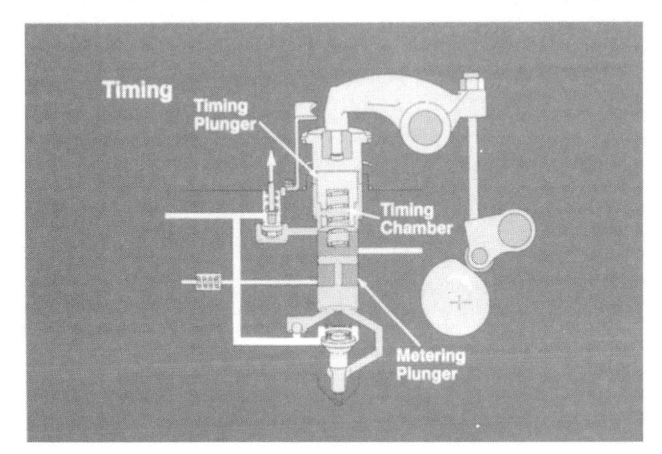

8. Because the metering plunger is being driven downward (hydraulically), rapid pressure rise begins in the metering chamber.

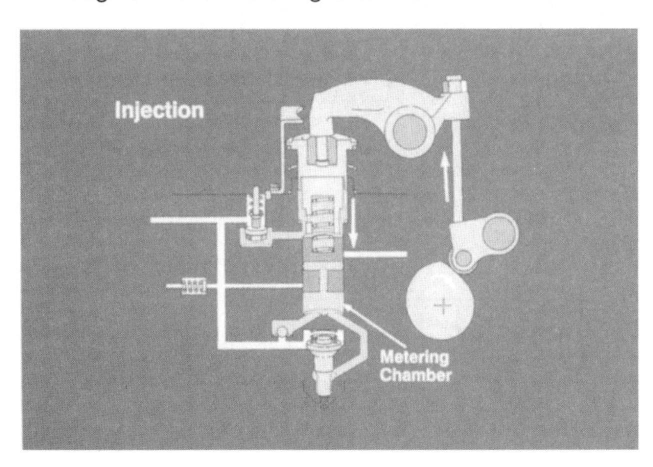

9. Ducting connects the metering chamber with the pressure chamber of the hydraulic, multi-orifii injector nozzle located at the base of the CELECT injector. When the pressure in the metering chamber (and therefore in the nozzle pressure chamber) reaches the NOP value, approximately 5,000 psi (340 atms/34.442 MPa), the nozzle valve opens and the injection begins. Fuel is forced through the nozzle orifii and atomized directly into the engine cylinder. The minute sizing of the nozzle orifii means that they are unable to relieve the pressure as fast as it is created and peak pressure is capable of rising well above the NOP value depending on the length of the effective stroke.

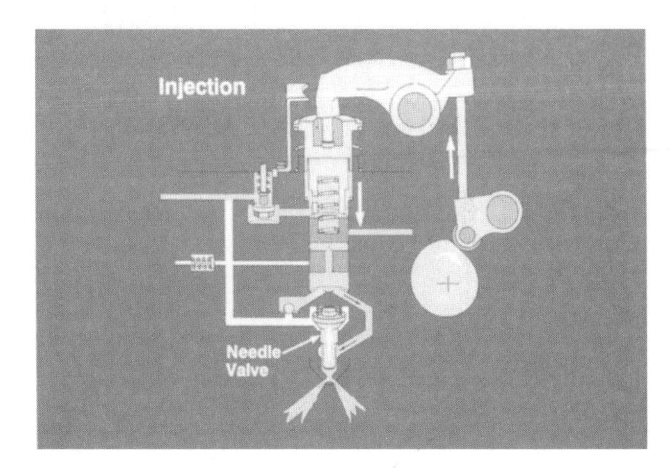

10. Injection continues until the metering plunger passes the spill passage. This action causes a collapse of metering chamber pressure permitting abrupt nozzle valve closure. At this moment, the pressure relief valve will relieve, minimizing the effect of the high pressure spike that occurs at metering spill—the relief valve passage connects to the fuel drain line.

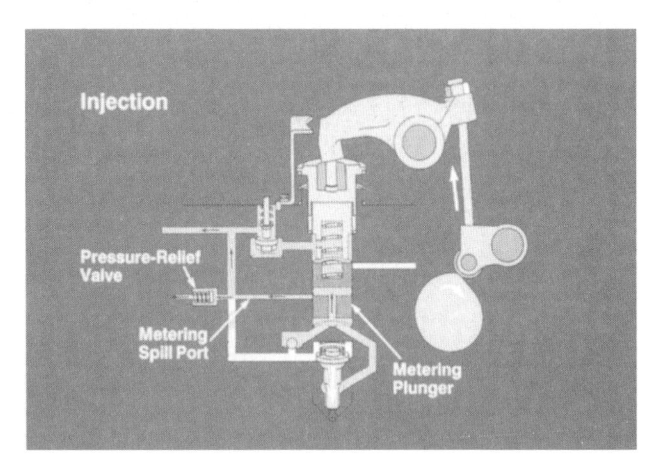

Figure 12-39 Continued.

11. Immediately after the metering spill port is exposed by the downward travel of the metering plunger, its upper edge exposes the timing spill port.

12. This action permits the fuel in the timing chamber that was used as hydraulic medium to spill to the fuel drain. This completes the cycle.

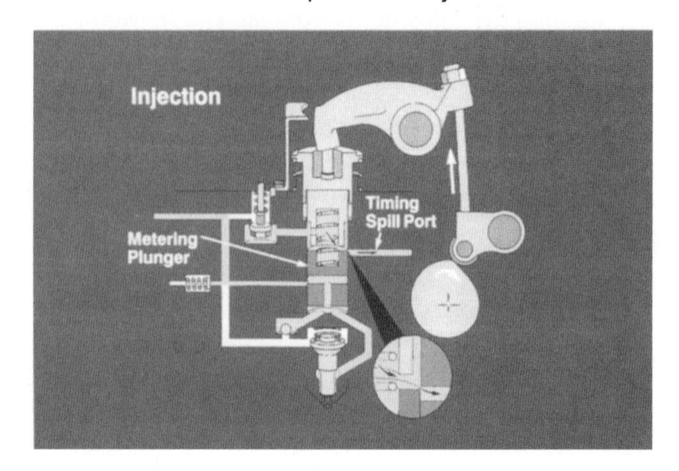

Figure 12-39 Continued.

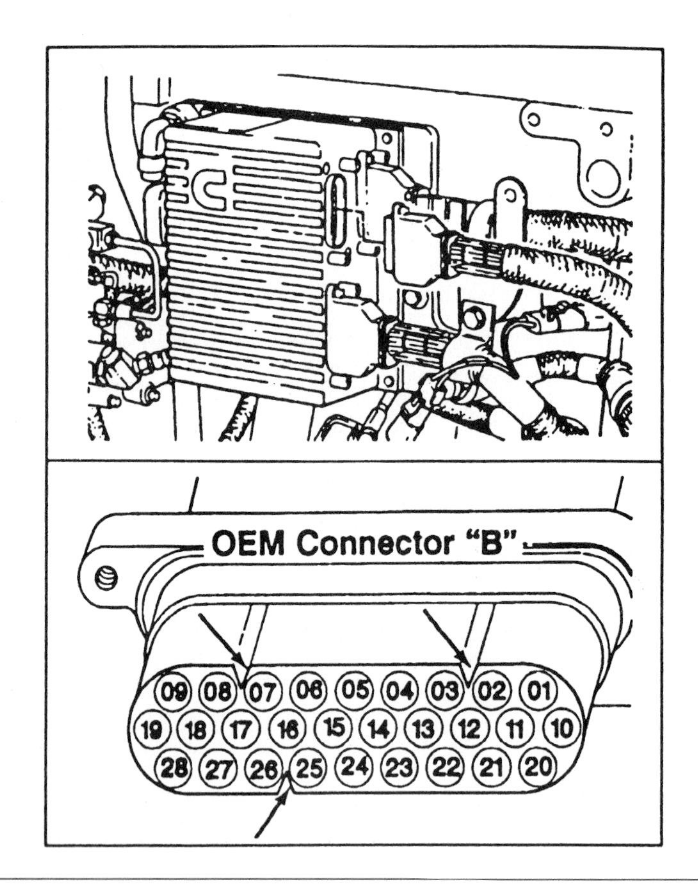

Figure 12-40 Navistar ECM connector. (Courtesy of Navistar International Engine Group)

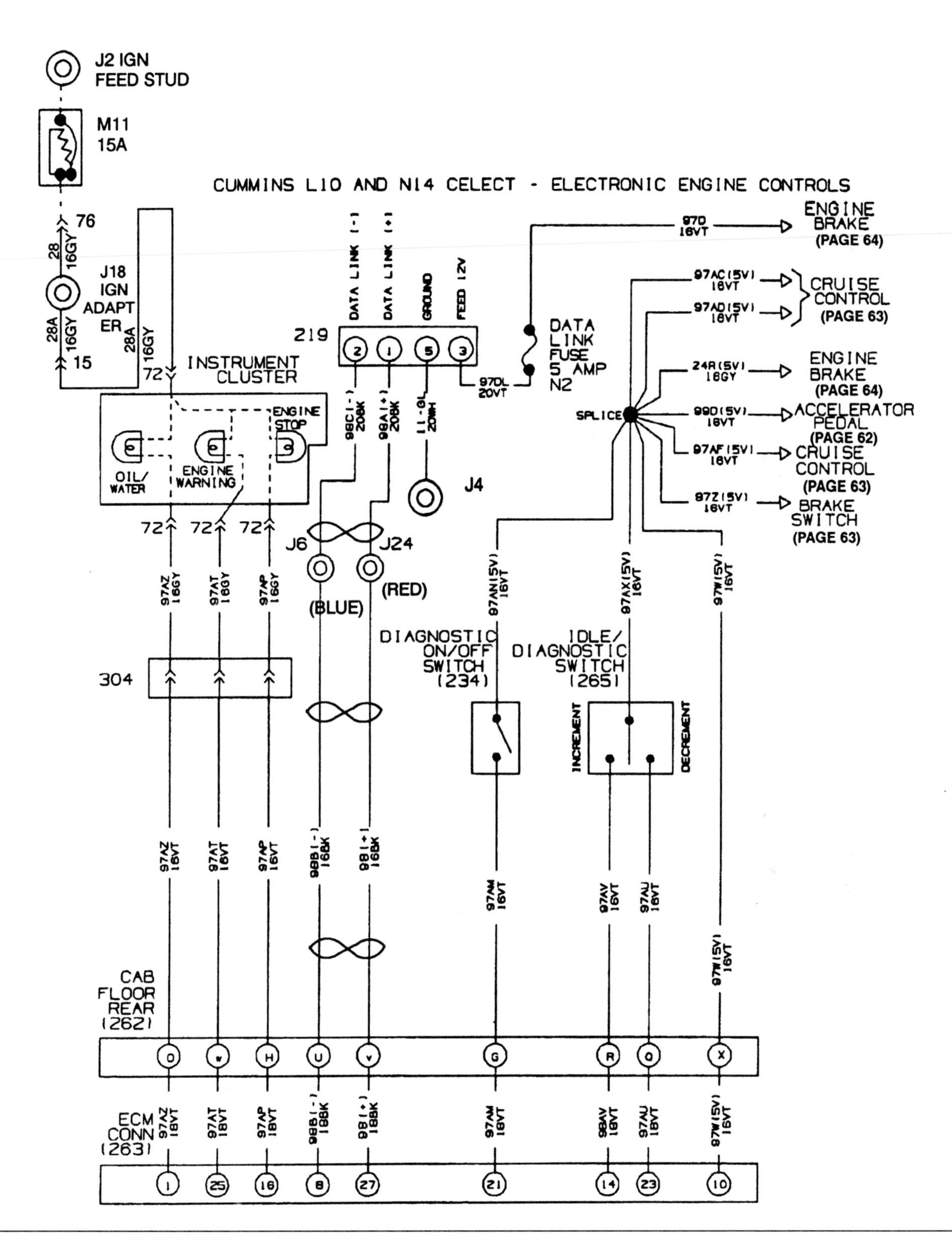

Figure 12-41 Circuit diagram of a Cummins Electronic Engine Control System used in a Navistar truck. (Courtesy of Navistar International Engine Group)

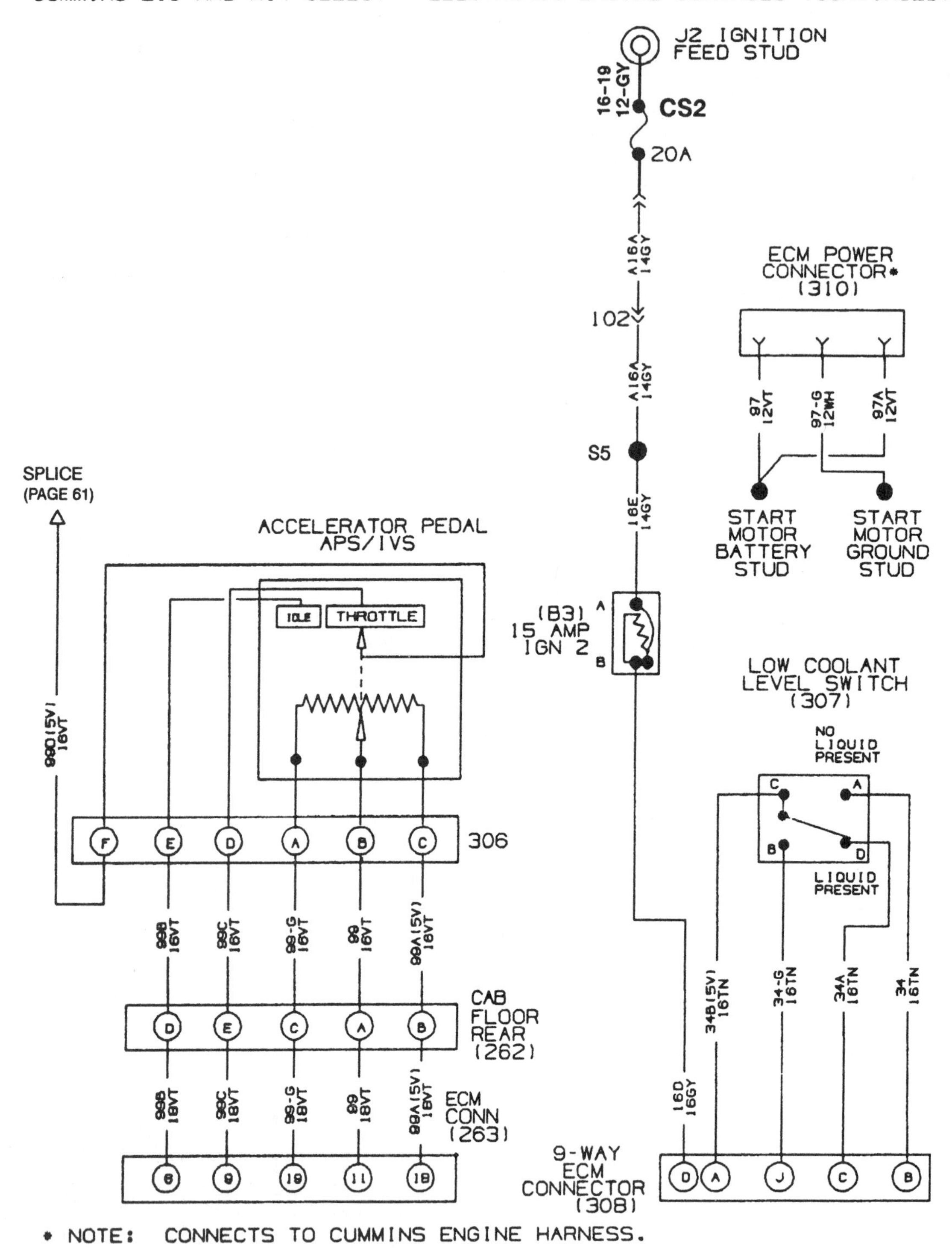

Figure 12-41 Continued.

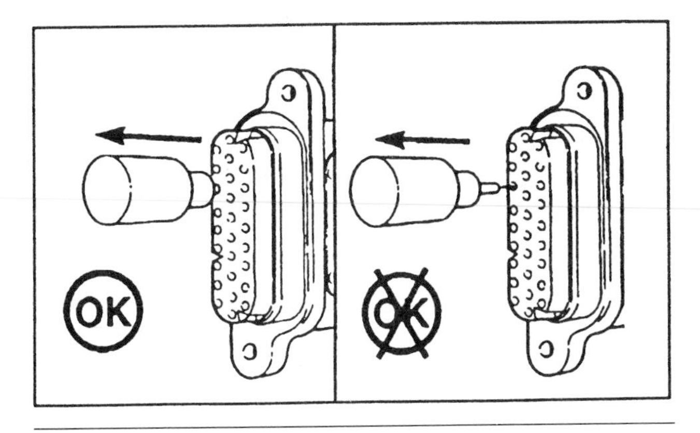

Figure 12-42 Inserting the inspection tool. (Courtesy of Navistar International Engine Group)

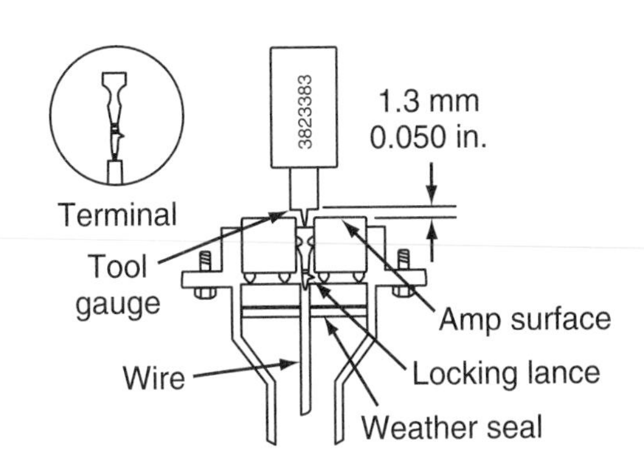

Figure 12-43 Inspecting for pushed back terminals. (Courtesy of Navistar International Engine Group)

C. Push the tool into the terminal until it touches the connector (Figure 12-44). Slowly remove the tool from the terminal. A small resistance (6 to 10 ounces) must be felt. If the resistance is not felt, the terminal is expanded and must be replaced.

D. Repeat steps A through C to check for pushed back or expanded terminals for all terminals in the ECM connector. Repair if required.

E. Inspect the connector pins in the ECM. If the pins are bent (Figure 12-45), the recommendation is to have a Cummins dealer replace the ECM.

The next example of troubleshooting is the Accelerator Pedal APS/IVS. APS refers to the Accelerator Position Sensor and IVS refers to the Idle Validation Switch. Previously, we looked at a TPS without the IVS. This particular sequence looks at the APS, IVS, and related circuitry that is important because of the relationship between the OEM and engine manufacturer's wiring.

WARNING: The following tests have to be performed in sequence from A through E. As with any test or procedure, a skipped or out of sequence procedure can invalidate the result.

- ❑ Figure 12-46: Test A
- ❑ Figure 12-47: Test B
- ❑ Figure 12-48: Test C
- ❑ Figure 12-49: Test D
- ❑ Figure 12-50: Test E

Special Tools

Service manual

DMM

Jumper leads

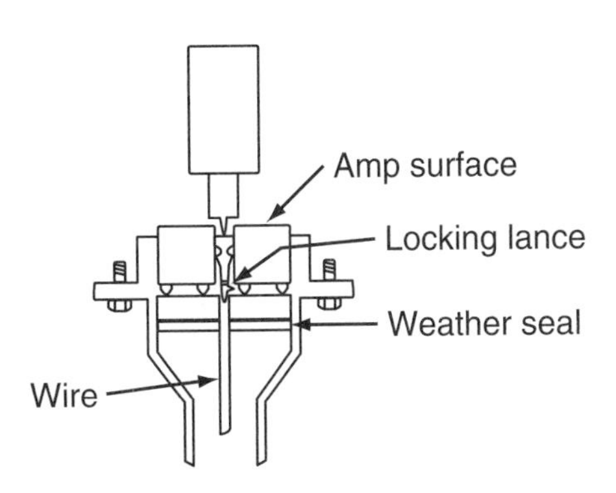

Figure 12-44 Inspecting for expanded terminals. (Courtesy of Navistar International Engine Group)

Figure 12-45 Inspecting for bent pins. (Courtesy of Navistar International Engine Group)

No matter how complicated a system might seem, the most important concepts are the following:

1. All electrical circuits have similar properties.
2. Because the circuits and components operate in similar fashion, you need to ensure for:
   - **A.** Power
   - **B.** Ground
   - **C.** Continuity
   - **D.** (or) Change in voltages.
3. To determine all the above requires basic diagnostic tools such as:
   - **A.** DMM
   - **B.** Test light
   - **C.** Jumper wire(s)
4. Understanding the operation
5. Being able to read wiring diagrams and schematics

CUSTOMER CARE: Electronics has entered into trucks' systems at a very rapid pace. Never assume that your customer understands the vehicle's electronic systems and how they are supposed to work, their functions, and their limitations. There are also do's and don'ts of which an operator should be aware. The most likely person to educate the operator is the knowledgeable technician.

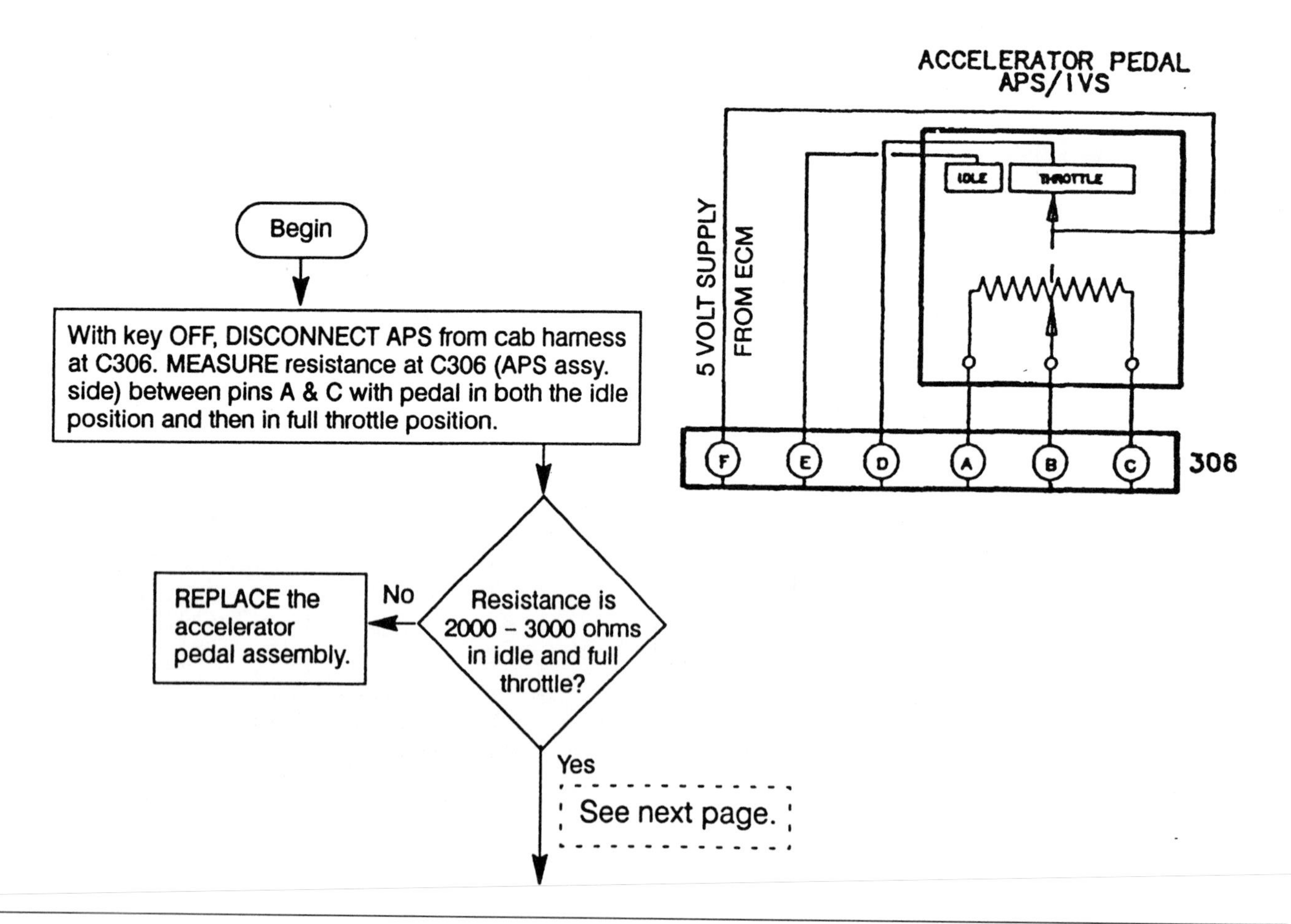

Figure 12-46 Test A—Accelerator Position Sensor (APS) flowchart. (Courtesy of Navistar International Engine Group)

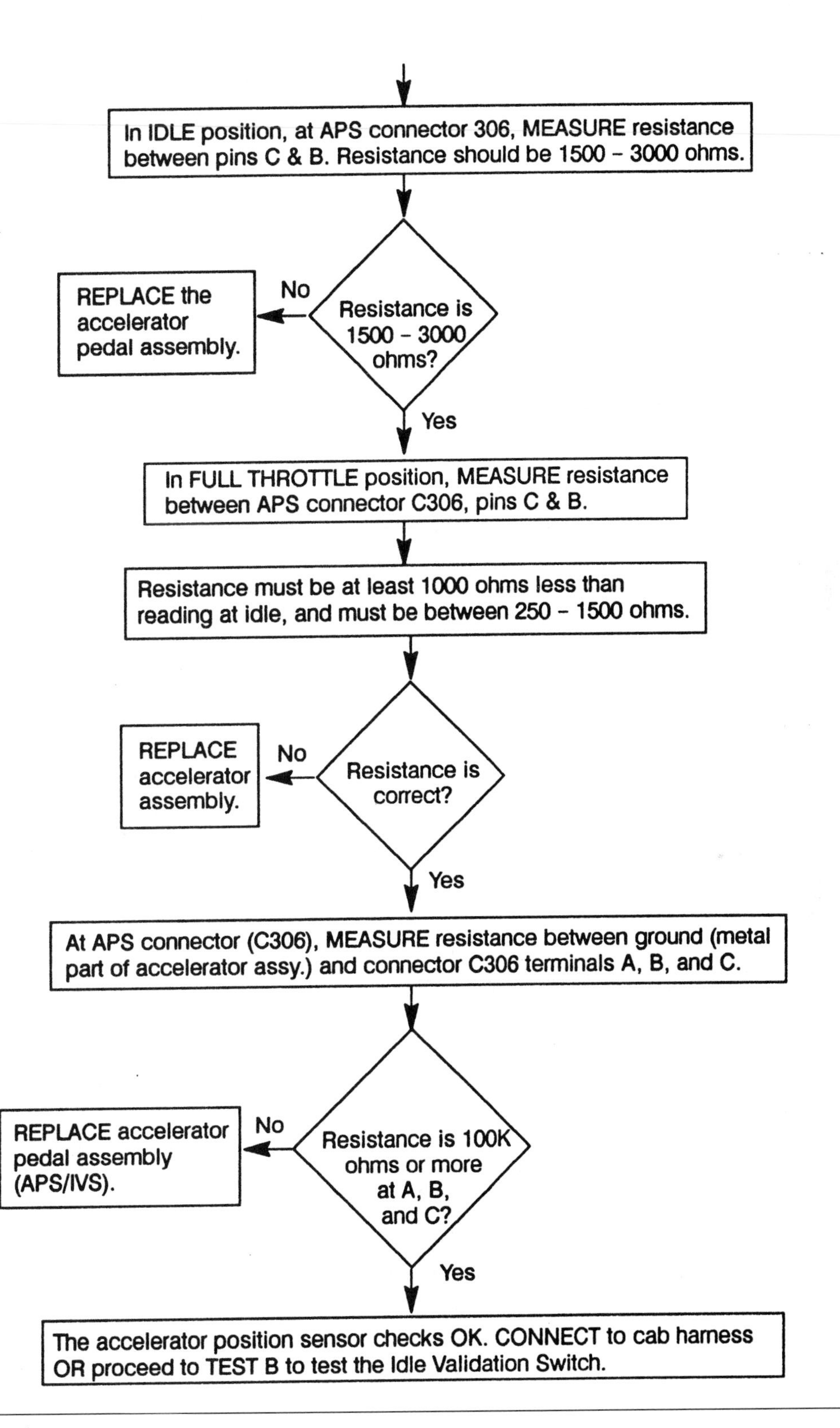

Figure 12-46 Continued.

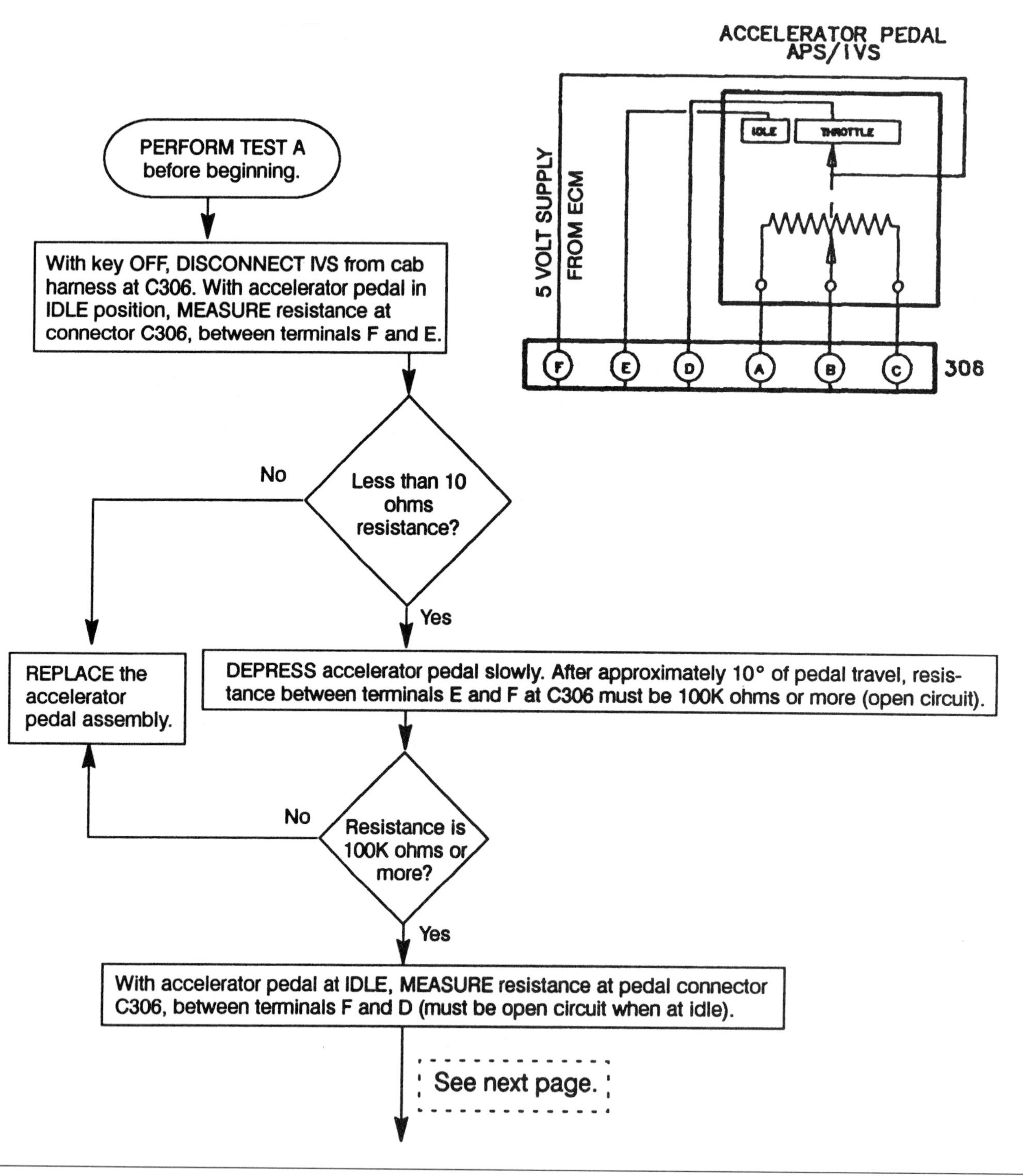

Figure 12-47 Test B—Idle Validation Switch (IVS) circuit test. (Courtesy of Navistar International Engine Group)

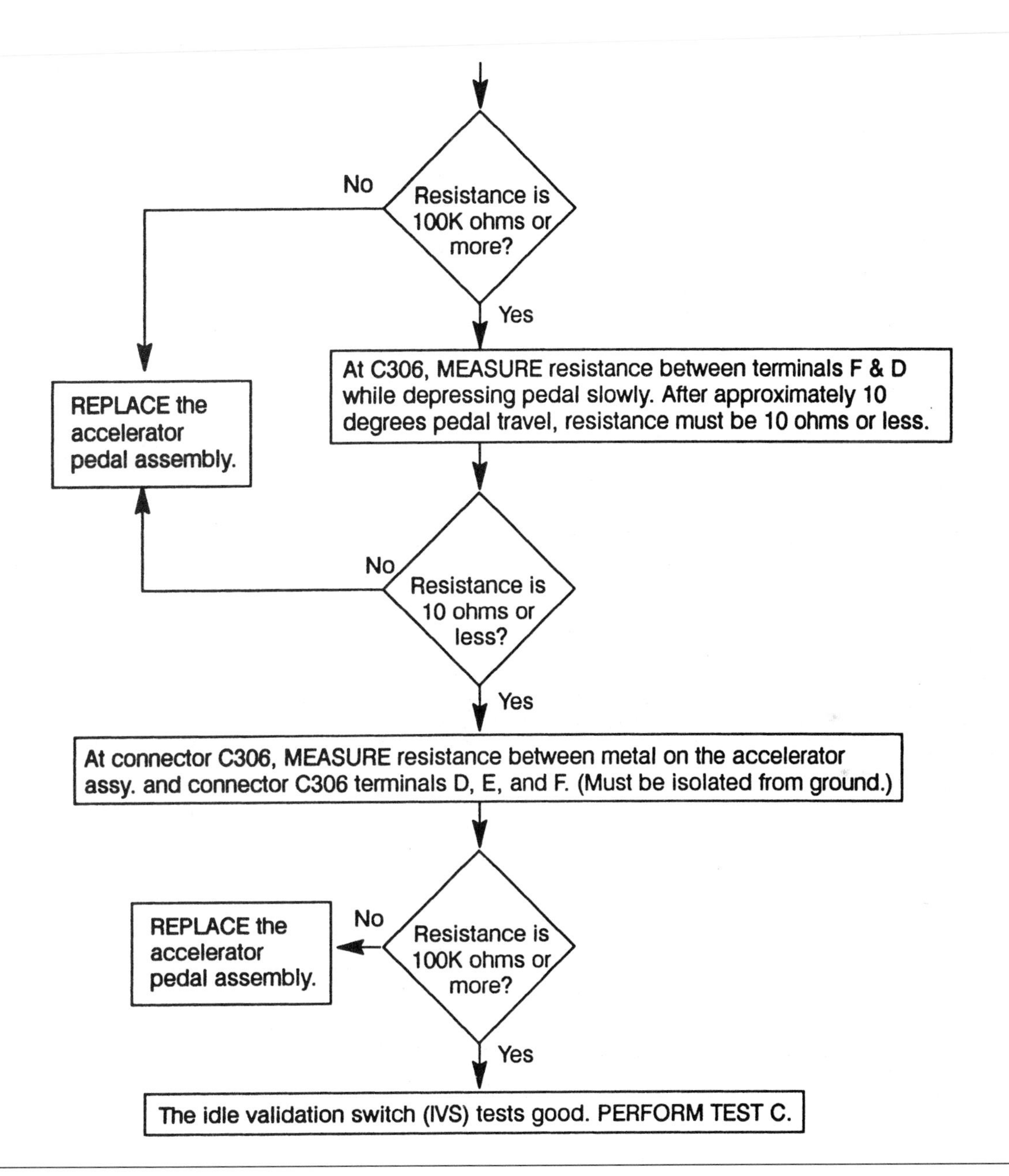

Figure 12-47 Continued.

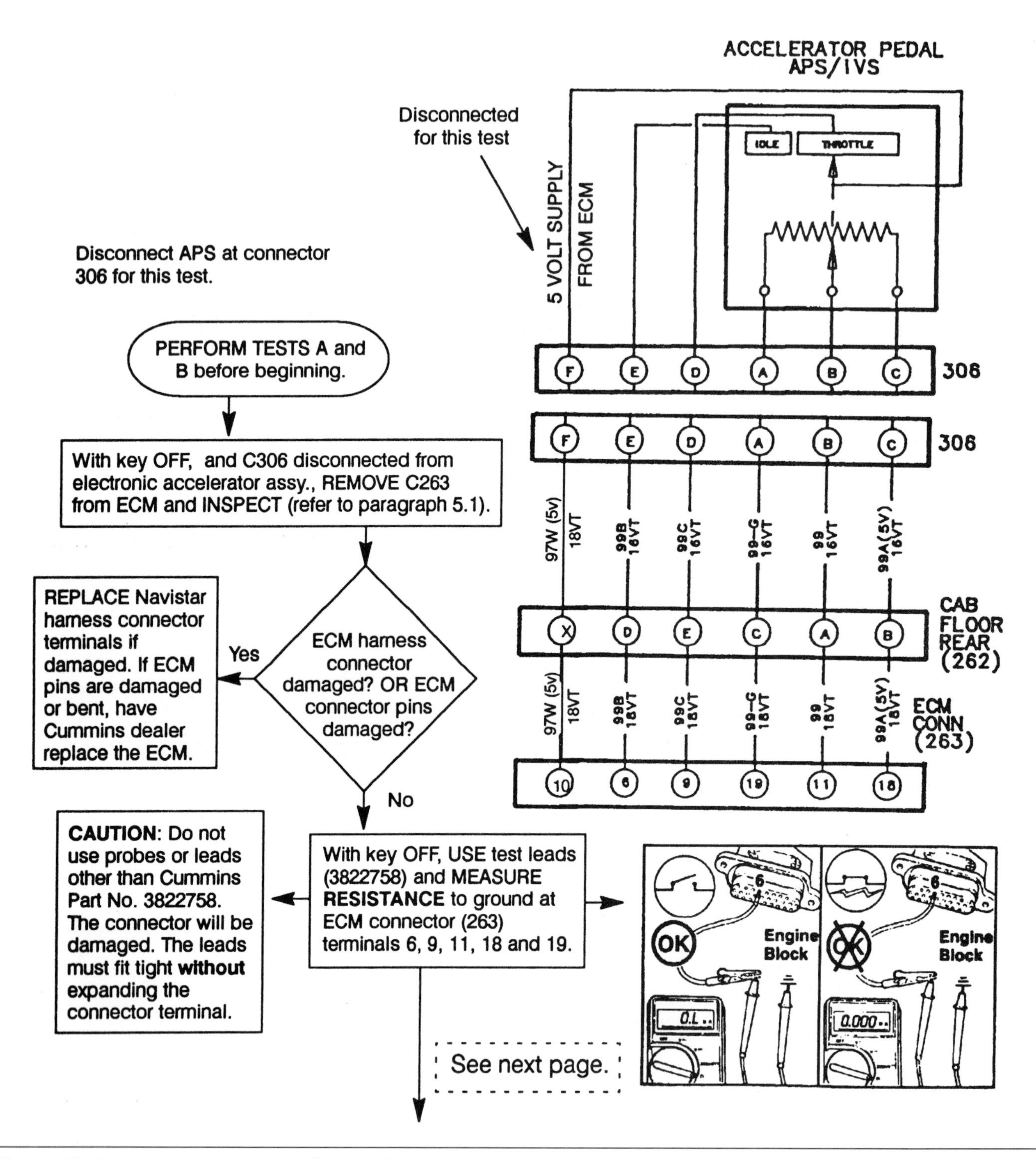

Figure 12-48 Test C—Accelerator Pedal (APS/IVS) short and open circuit tests. (Courtesy of Navistar International Engine Group)

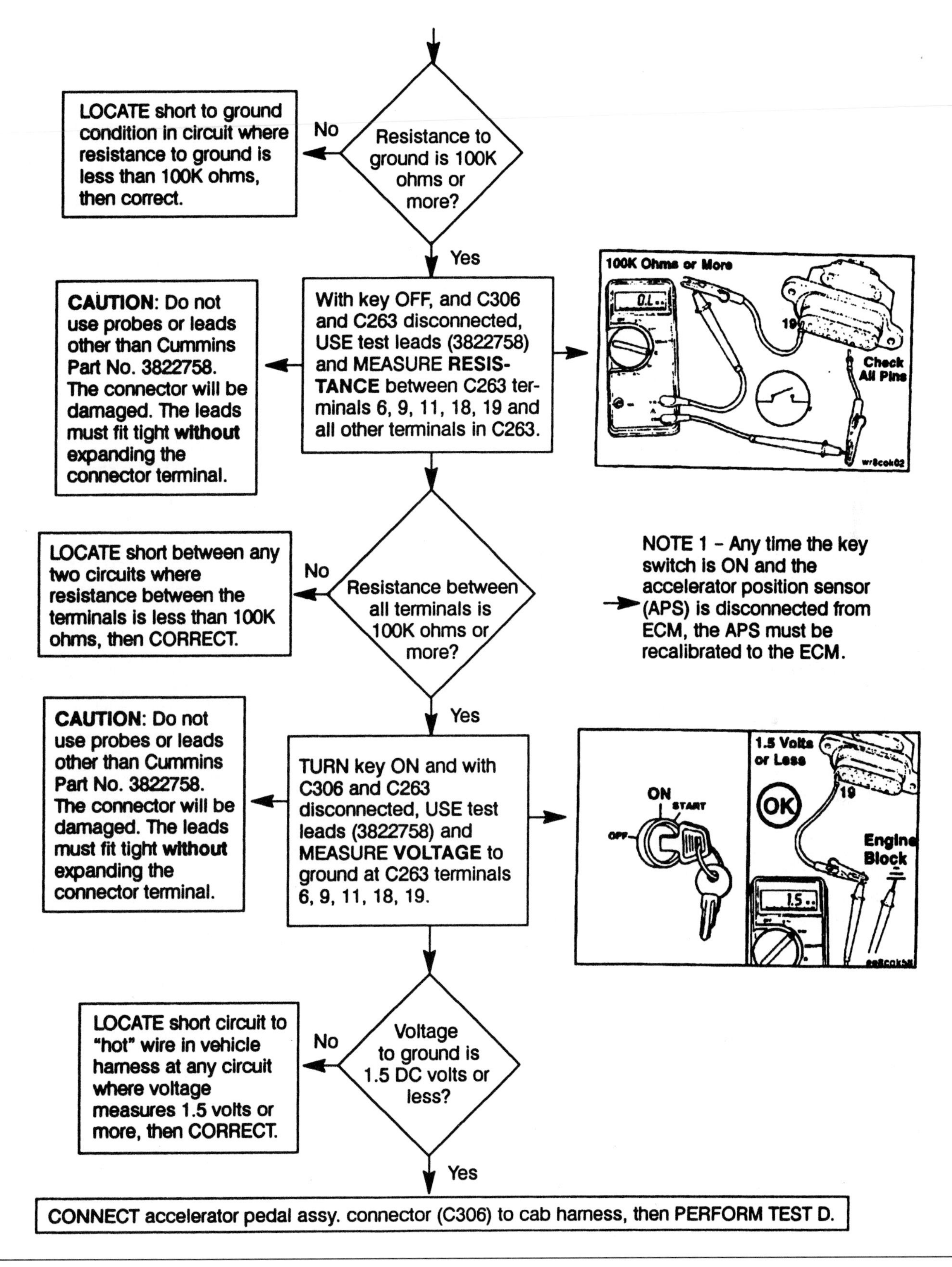

Figure 12-48 Continued.

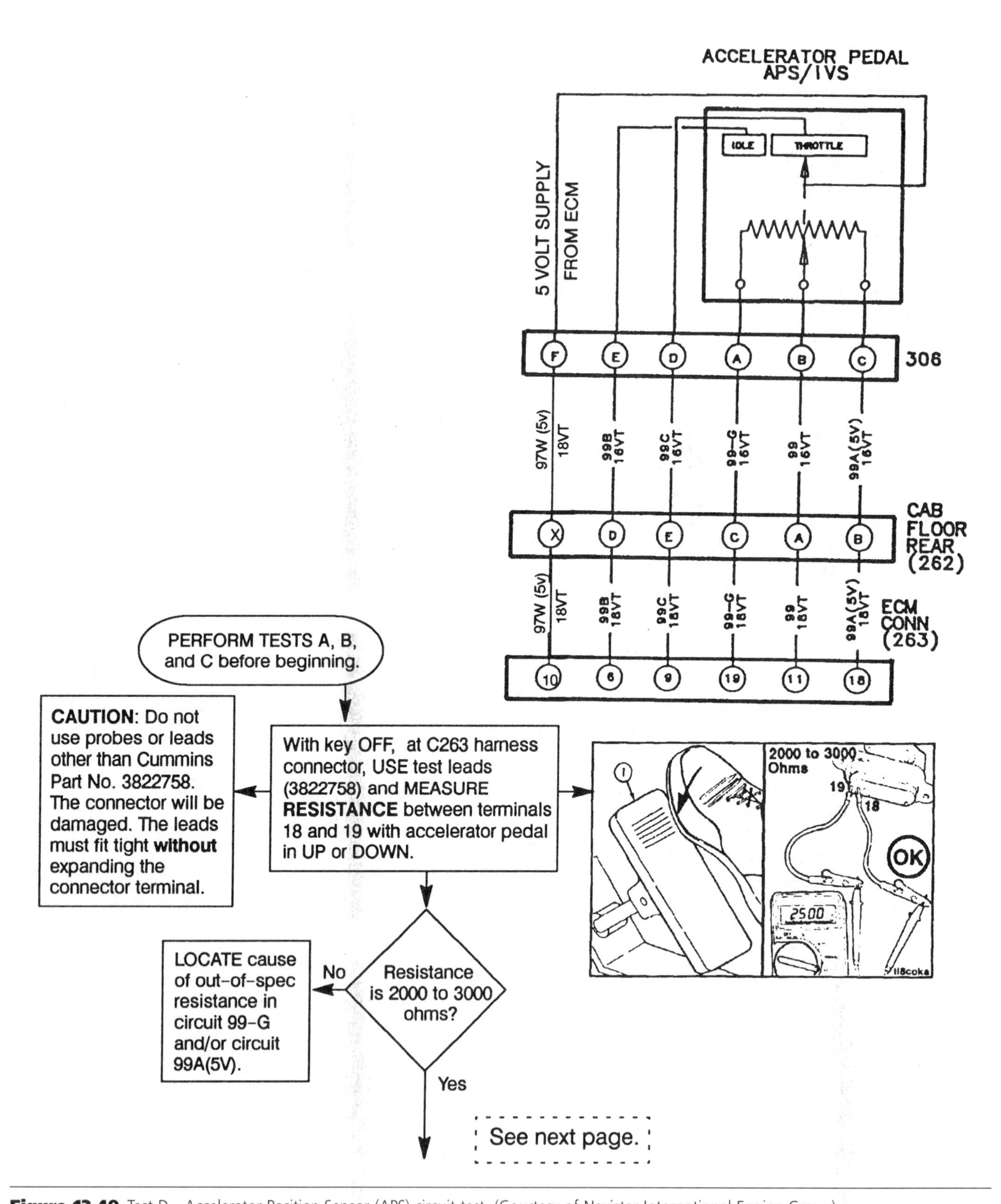

Figure 12-49 Test D—Accelerator Position Sensor (APS) circuit test. (Courtesy of Navistar International Engine Group)

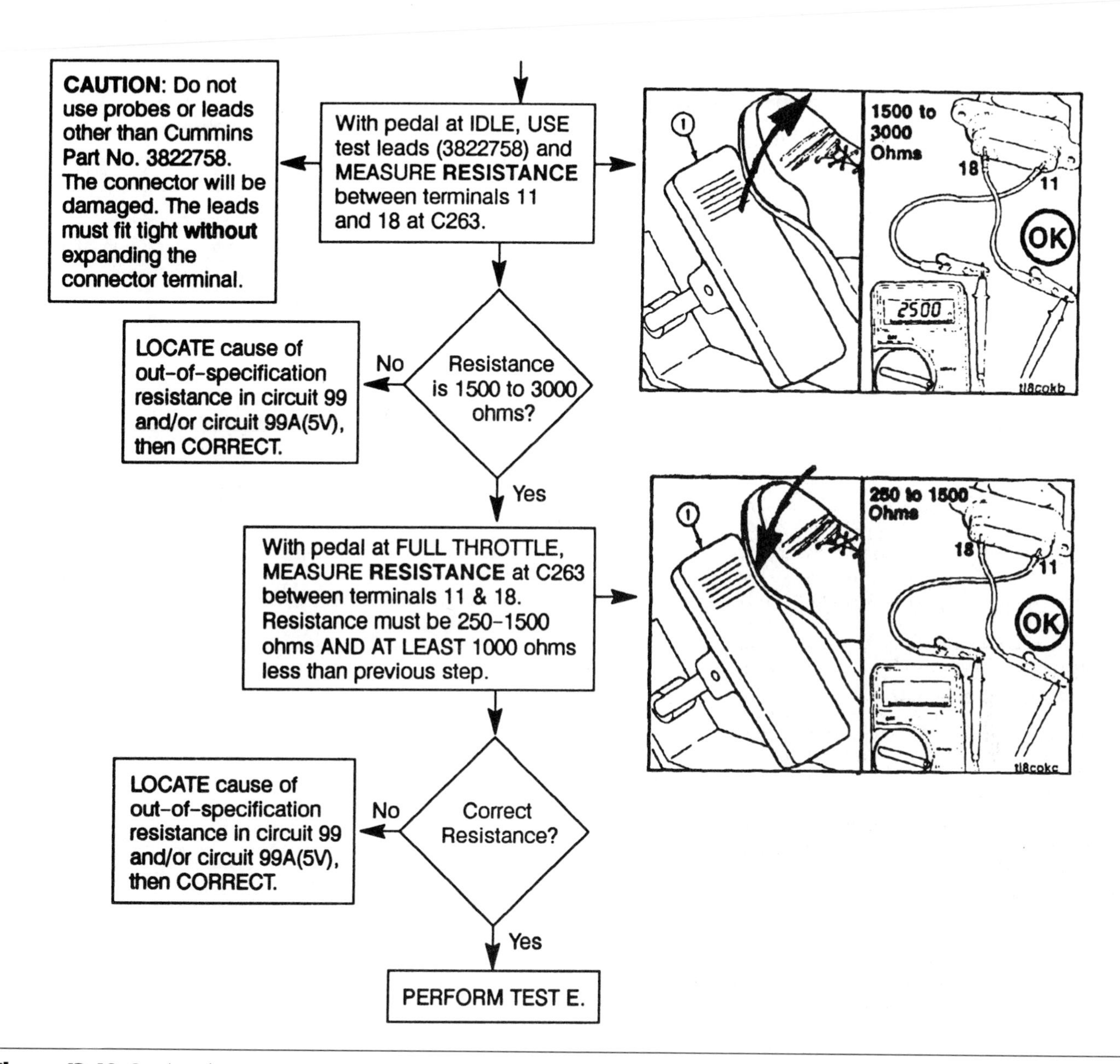

Figure 12-49 Continued.

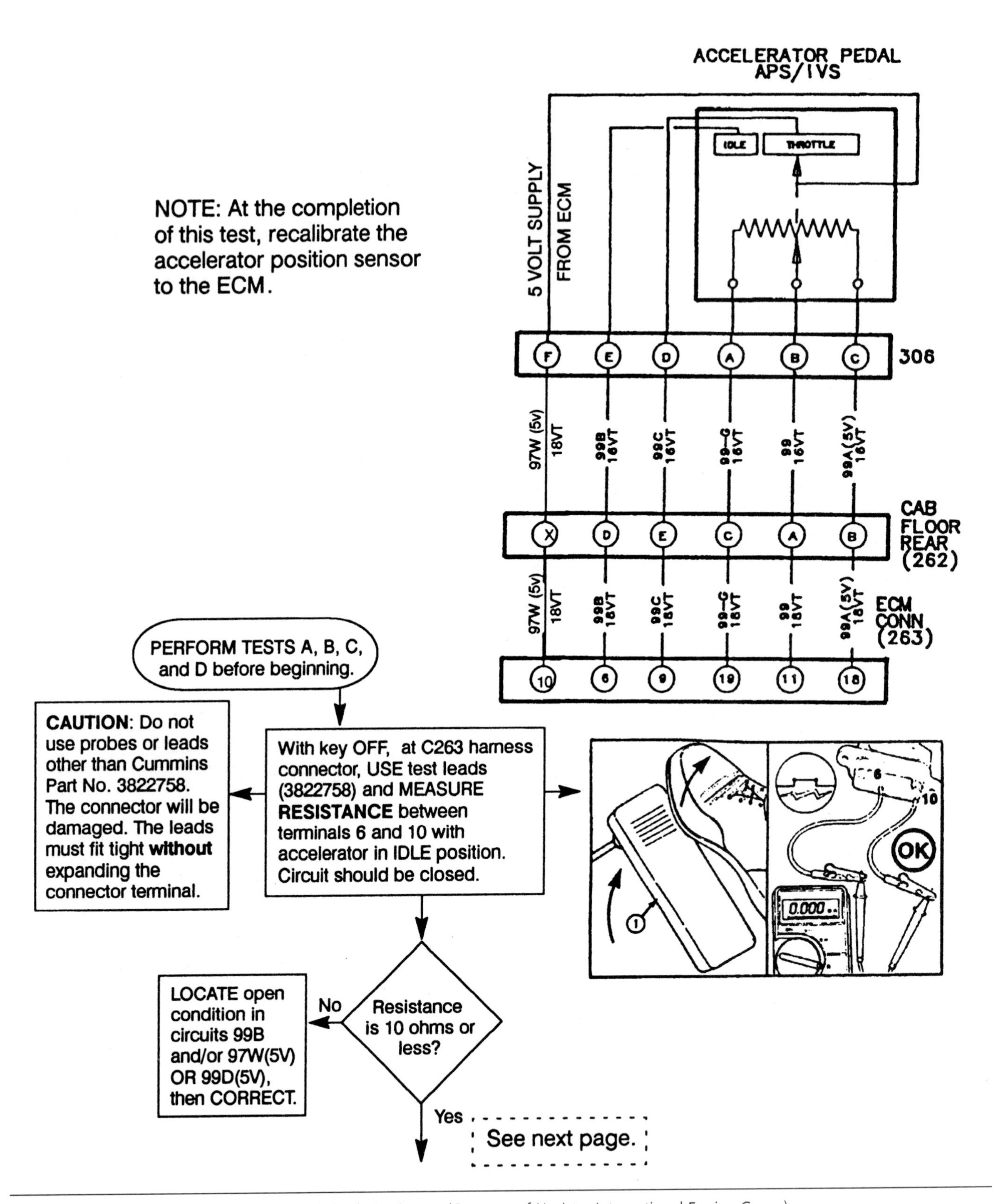

Figure 12-50 Test E—Idle Validation Switch (IVS) circuit test. (Courtesy of Navistar International Engine Group)

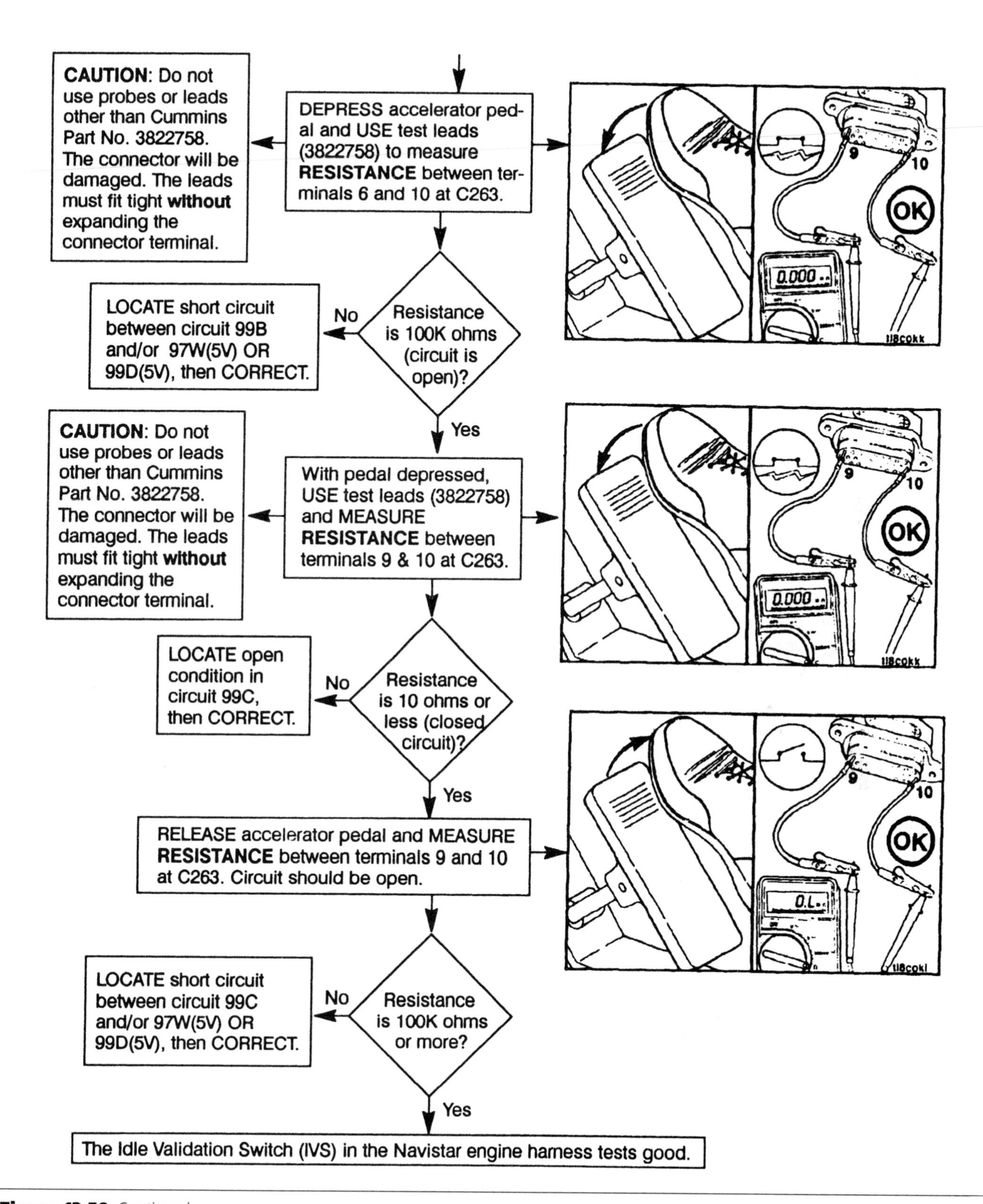

Figure 12-50 Continued.

CASE STUDY

A truck was brought into the shop with a drivability problem of no power/slow or no acceleration. The problem was intermittent. The technician pulled out the service manual and followed the proper procedures for retrieving fault codes. The fault codes indicated an accelerator position sensor problem. Feeling confident enough that at least he had found the area of the problems, he asked another technician to follow through with the pinpoint diagnosis. To save time, the second technician unplugged the harness, hooked up his ohmmeter to the APS, and observed the bar graph. She thought she saw a hesitation. She had been striving to become an electronics ACE. Proud of herself that she had discovered the problem to the APS, she then performed a quick replacement procedure. After reconnecting the connector at the APS, she even hit the pedal a few times to verify that the problem was solved.

The next day the same truck was back with the same intermittent problem. The technician who "corrected" the problem was at a scheduled electronics seminar, so her coworker had to diagnose and repair the problem. Being familiar with this vehicle, he verified for faults and found the APS circuit still had a problem. He got out the proper service manual and performed a very specific test of the complete circuit. He discovered the problem to be a corroded pin problem in a connector that caused the electrical connection to be temporarily lost whenever the truck flexed enough over certain road conditions. He performed the proper repairs, cleared the codes, road tested, and rechecked for any more fault codes.

Terms to Know

Active codes
Breakout boxes
Check engine light (CEL)
Electronic service tool (EST)
Electrostatic discharge (ESD)
Hall-effect switches
Hard codes
Impedance
Intermittent (inactive) code
Magnetic pulse generators
Programmable read only memory (PROM)
Reader-programmer
Scan tool
Sensors
Stop engine light (SEL)
Thermistors

ASE-Style Review Questions

1. Throttle position sensors on electronically controlled engines are being discussed: *Technician A* says that the APS on a Cummins CELECT system is on the electronic fuel pump. *Technician B* says that the APS is part of the accelerator pedal assembly. Who is correct?

A. A only
B. B only
C. Both A and B
D. Neither A nor B

2. Cooling of the ECM unit is being discussed. *Technician A* says that some systems circulate diesel fuel around the housing for cooling purposes. *Technician B* says that fluids cannot be used around the ECM area for cooling due to a possibility of short circuiting. Who is correct?

A. A only
B. B only
C. Both A and B
D. Neither A nor B

3. Sensors on a CELECT system are being discussed: *Technician A* says the coolant level sensor is an NTC thermistor type. *Technician B* says the coolant level sensor is a PTC thermistor type. Who is correct?

A. A only
B. B only
C. Both A and B
D. Neither A nor B

4. Testing a throttle position sensor is being discussed: *Technician A* says that a TPS can be tested with an ohmmeter. *Technician B* says that a TPS can be tested with a voltmeter. Who is correct?
   A. A only **C.** Both A and B
   B. B only **D.** Neither A nor B

5. Fault codes are being discussed: *Technician A* says in order to retrieve fault codes in today's trucks, you need an electronic reader. *Technician B* says that blink codes are available on today's trucks to quickly identify faults. Who is correct?
   A. A only **C.** Both A and B
   B. B only **D.** Neither A nor B

6. *Technician A* says installing a PROM chip in backward will immediately destroy the chip. *Technician B* says electrostatic discharge will destroy the chip. Who is correct?
   A. A only **C.** Both A and B
   B. B only **D.** Neither A nor B

7. *Technician A* says that today's electronic circuits require a reader-programmer to test the sensors. *Technician B* says the tool of choice to test sensors in trucks is the DMM. Who is correct?
   A. A only **C.** Both A and B
   B. B only **D.** Neither A nor B

8. *Technician A* says trouble codes indicate the exact failure in the circuit. *Technician B* says trouble codes will direct the technician to the circuit with a fault in it. Who is correct?
   A. A only **C.** Both A and B
   B. B only **D.** Neither A nor B

9. Testing a magnetic pulse generator is being discussed: *Technician A* says that an ohmmeter can be used to test the resistance value of the coil. *Technician B* says that the generated voltage can also be tested by using a voltmeter. Who is correct?
   A. A only **C.** Both A and B
   B. B only **D.** Neither A nor B

10. To test a Hall-effect sensor: *Technician A* says that hooking an ohmmeter across the signal terminal and ground terminal of the switch and then cranking the motor should produce a pulsating resistance reading (high to low). *Technician B* says that a Hall-effect sensor needs power in order to properly test it. Who is correct?
   A. A only **C.** Both A and B
   B. B only **D.** Neither A nor B

ASE Challenge

1. *Technician A* says that most heavy truck engine ECUs (also called ECMs) are mounted on the engine to eliminate and reduce wiring. *Technician B* says mounting an ECU on the engine makes it easier to adapt wiring for different truck manufacturers. Who is correct?
   A. A only **C.** Both A and B
   B. B only **D.** Neither A nor B

2. Fault code retrieval for heavy trucks is being discussed. *Technician A* says many OEMs provide fault code retrieval by pressing several dash buttons in a given sequence. *Technician B* says some OEMs require that fault codes be accessed by jumping particular pins in the data link connector. Who is correct?
   A. A only **C.** Both A and B
   B. B only **D.** Neither A nor B

3. *Technician A* says that to change or modify engine operating parameters, a specific type of electronic service tool and authorization code must be used. *Technician B* says that in some cases software is available to allow engine electronics fault code reading and service using a laptop computer. Who is correct?
   A. A only **C.** Both A and B
   B. B only **D.** Neither A nor B

4. An engine oil pressure sensor is being tested. *Technician A* says that such sensors are usually variable resistance units. *Technician B* says that some pressure sensors are Hall-effect sensors. Who is correct?
   A. A only **C.** Both A and B
   B. B only **D.** Neither A nor B

5. Updating or changing operating parameters on electronically controlled engines is being discussed. *Technician A* says that with proper access codes all operating parameters can be modified. *Technician B* says that on older electronically controlled systems it may be necessary to change the PROM chip in order to achieve desired modifications. Who is correct?
   A. A only **C.** Both A and B
   B. B only **D.** Neither A nor B

Table 12-1 ASE Task

Connect computer programming equipment to vehicle/engine; access and change customer parameters; determine needed repairs

Problem Area	Symptoms	Possible Causes	Classroom Manual	Shop Manual
DIAGNOSTIC PROCESS	Engine fault light on	1. System malfunction active/historic code set 2. On Board Diagnostics not working		
	System fault diagnosis	3. Wrong scan tool or DDR cartridge 4. Incorrectly loaded OEM software in PC		
	Customer parameter change request PTO change	5. Improper OEM authorization		

Table 12-2 ASE Task

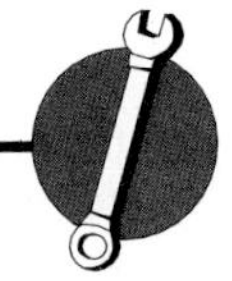

Check and record engine electronic diagnostic codes and trip/operational data; clear codes; determine needed repairs

Problem Area	Symptoms	Possible Causes	Classroom Manual	Shop Manual
DIAGNOSTIC PROCESS	Flash codes stores	1. System malfunction active/historic code set 2. On Board Diagnostics not working		
	Engine fault light on	1. System malfunction 2. Engine active or historic code set 3. Wrong scan tool or DDR cartridge 4. Incorrectly loaded OEM software in PC 5. Scan tool or diagnostic data reader (DDR) malfunction		

Job Sheet 20

20

Name: ______________________ Date: ______________________

Retrieving Blink Codes

Upon completion of this job sheet, you should be able to use the blink code function to retrieve existing fault codes to quickly determine if a problem circuit exists.

Tools and Materials

A medium/heavy duty truck
A service manual for the truck

Procedure

1. Describe the truck being worked on:
   Year: ____________ Make: ____________ VIN: ____________
   Model: ____________
2. Describe the procedure for retrieving blink codes: ______________________
   ____
   ____
3. Go through the procedure and describe your findings: ______________________
   ____
   ____
   ____

☑ **Instructor Check** ______________________

Job Sheet 21

21

Name: ______________________ Date: ______________________

Retrieving Fault Codes Using an Electronic Service Tool

Upon completion of this job sheet, you should be able to enter a computer with an EST and retrieve fault codes.

Tools and Materials

A medium/heavy duty truck
A service manual for the truck
A scanner or reader/programmer

Procedure

1. Describe the truck being worked on: Year: ________ Make: ____________
   VIN: __________________ Model: __________________
2. Describe the procedure for retrieving fault codes using an EST. ______________
   ____
   ____
   ____
3. Go through the procedure and describe your findings. ______________
   ____
   ____
   ____

Instructor Check ______________________________

Job Sheet 22

22

Name: ______________________ Date: ______________________

Performing a Complete Throttle Position Circuit Test

Upon completion of this job sheet, you should be able to test a complete throttle position circuit and components following an OEM diagnostics test procedure.

Tools and Materials

A medium/heavy duty truck with a throttle position circuit
A DMM
Service manual for the truck
Proper wiring diagrams for the truck

Procedure

1. Describe the truck being worked on: Year: ________ Make: ____________
   VIN: __________________ Model: __________________
2. Locate the possible harness connector(s) location(s) that would be used for a full TPS circuit test and note them here. ______________________________
   ____
   ____
3. Going through the service manual, locate and read all relevant information pertaining to the TPS circuit of that truck.
4. Draw a simple schematic of the circuit and show the various test points.
5. Using your schematic as a reference, record your anticipated resistance and voltage readings of the TPS unit itself during various operating conditions. _______________
   ____
   ____
6. Locate the TPS troubleshooting flow charts in the manual and perform all tests related to the TPS circuit. Note the readings of each test that would relate to your anticipated reading. ______________________________
   ____
   ____
7. When the tests are complete, operate the accelerator to verify proper operation. Check for any fault codes.
8. Record your conclusions from this test. ______________________________
   ____
   ____

☑ **Instructor Check** ______________________________

Appendix A

ASE Practice Examination

1. A nonfunctioning right front turn signal light is being diagnosed on a heavy duty truck. *Technician A* says it would be a good idea to check the bulb and socket first, then the signal light switch. *Technician B* says that it may be necessary to check for proper ground connections at the front of the truck. Who is correct?
   A. Technician A **C.** Both A and B
   B. Technician B **D.** Neither A nor B

2. A heater/AC blower circuit is being tested using a DVOM. *Technician A* says to check for a voltage drop across the blower motor; it should be less than 0.04 volts. *Technician B* says to check resistance through the fuse; it should be 50% in ohms of the fuse rating in amps. Who is correct?
   A. Technician A **C.** Both A and B
   B. Technician B **D.** Neither A nor B

3. *Technician A* says that when checking amperage, the meter must always be connected in the circuit being tested in parallel. *Technician B* says that checking for amperage with a DVOM only must be limited to circuits with less than 10 amps. Who is correct?
   A. Technician A **C.** Both A and B
   B. Technician B **D.** Neither A nor B

4. *Technician A* says that prior to checking a circuit using an ohmmeter you should insure that all power is off or disconnected. *Technician B* says that the reading "OL" on the ohm scale of a DVOM indicates extremely low resistance. Who is correct?
   A. Technician A **C.** Both A and B
   B. Technician B **D.** Neither A nor B

5. To determine the voltage drop between the battery and the starter switch, you would do which of the following while cranking the engine?
   A. Connect an ammeter between the battery ground post and the live side of the starter switch
   B. Connect a voltmeter between the positive battery post and the live side of the starter switch
   C. Connect a voltmeter between the start post of the solenoid and the solenoid ground post
   D. None of the above; it is impossible to determine the voltage drop of this section of the starter

6. The fuse on an interior light continues to blow when the light is turned on. *Technician A* says to remove the fuse and bulbs, then check for continuity from any connection to ground with an ohmmeter. Very low resistance will indicate a definite short to ground. *Technician B* says that as long as resistance is very low from the switch to the bulb connections, the problem is not in the wiring. Who is correct?
   A. Technician A **C.** Both A and B
   B. Technician B **D.** Neither A nor B

7. *Technician A* says that you can check a capacitor or condenser with an ohmmeter using the same techniques and procedures as used in checking a diode. *Technician B* says that when a capacitor is charged it will show "OL" on a DVOM ohm scale. Who is correct?
   A. Technician A **C.** Both A and B
   B. Technician B **D.** Neither A nor B

8. You suspect a broken wire on a gas tank gauge circuit between the gauge and sending unit; however, the suspect wire is very difficult to trace the entire distance. Which of the following scales on a DVOM would be most helpful in proving your suspicion?
   A. AC volts **C.** HTC
   B. Ohms **D.** Milliamp

9. *Technician A* says that the symbol for a common point on most electrical diagrams is the letter "C." *Technician B* says that the symbol for a positive connection or hot point on a diagram is the letter "P." Who is correct?
   A. Technician A **C.** Both A and B
   B. Technician B **D.** Neither A nor B

10. *Technician A* claims that electrical components that do a lot of work or give a lot of light require a significant amperage. *Technician B* claims that electrical components that draw significant amperage from a 12-volt system must have low electrical resistance. Who is correct?
    A. Technician A **C.** Both A and B
    B. Technician B **D.** Neither A nor B

11. *Technician A* says that when checking electronic components you should be sure to use a DVOM with high internal resistance. *Technician B* says you can check for current flow through electronic components using a 12-volt test light. Who is correct?
    A. Technician A **C.** Both A and B
    B. Technician B **D.** Neither A nor B

12. *Technician A* says that voltmeters are connected in parallel with components in order to measure voltages. *Technician B* says that ammeters should also be connected in parallel. Who is correct?

A. Technician A **C.** Both A and B
B. Technician B **D.** Neither A nor B

13. *Technician A* says that older analog meters should always be set to the highest voltage scale, then worked down until a good reading is obtained when measuring voltages. *Technician B* says that most digital meters are self-ranging and will automatically give proper readings. Who is correct?

A. Technician A **C.** Both A and B
B. Technician B **D.** Neither A nor B

14. When performing a load test on a battery with 440 CCAs, *Technician A* sets the test load at 440 amps. *Technician B* removes the surface charge by applying a 220-amp load for 5 minutes. Who is correct?

A. Technician A **C.** Both A and B
B. Technician B **D.** Neither A nor B

15. *Technician A* says that if a battery is to be disconnected from a negative ground truck, the positive post connection should be removed first. *Technician B* says that if you remove the negative connection first you could cause arcing and personal injury. Who is correct?

A. Technician A **C.** Both A and B
B. Technician B **D.** Neither A nor B

16. *Technician A* says that a battery stores energy as free electrons in acid trapped between lead plates. *Technician B* says that a battery stores energy by containing and controlling a chemical reaction. Who is correct?

A. Technician A **C.** Both A and B
B. Technician B **D.** Neither A nor B

17. *Technician A* says that a pressure wash with warm water is a good way to clean batteries and battery boxes. *Technician B* says that whenever cleaning batteries, care should be used to avoid getting contamination into the battery vents of the fill spout. Who is correct?

A. Technician A **C.** Both A and B
B. Technician B **D.** Neither A nor B

18. A truck battery is being load tested. After several attempts to charge, remove service charge, then load, the battery still cannot quite meet specifications. *Technician A* says to continue the charge/test cycle until the battery either fails by a large margin or finally passes. *Technician B* says to drop several aspirin tablets or other battery additives then try the test several more times. Who is correct?

A. Technician A **C.** Both A and B
B. Technician B **D.** Neither A nor B

19. An off-highway truck with a 24-volt negative ground system is being jump-started using a service truck with a 12-volt system. *Technician A* says that the positive of the service truck must be connected to the positive of the 24-volt system closest to the starter and the negative to a good ground location. *Technician B* says that only one of the 12-volt batteries used in a 24-volt system can be jumped by a single 12-volt battery at a time. Who is correct?

A. Technician A **C.** Both A and B
B. Technician B **D.** Neither A nor B

20. Which is the *least* likely cause of slow or erratic cranking by a starter motor?

A. Poor connection at the battery
B. Bad brushes
C. Poor ground connection
D. Faulty key switch

21. Which of the following is a typical symptom of a faulty overrunning clutch in a starter circuit?

A. Starter drags, engine turns slowly
B. Starter spins freely, engine fails to turn
C. Starter makes clicking noise, engine turns in jerks
D. Starter hums, smoke and heat are present

22. A starter on a heavy truck fails to crank the engine and makes no noises. The batteries and cable connections have been checked and found to be okay. *Technician A* says the problem is a faulty starter and it should be replaced. *Technician B* says the problem is a bad ignition/start key switch and that the switch should be replaced. Who is correct?

A. Technician A **C.** Both A and B
B. Technician B **D.** Neither A nor B

23. A voltage drop test for battery cables is being performed on a large truck. *Technician A* has attached a DVOM lead to the battery's most common positive point and the other to the battery post of the starter solenoid. She set the meter to DC volts and says that while cranking the reading should be less than 0.02 volts. *Technician B* says you can check the ground cables using a similar procedure but attaching leads to battery negative and starter ground post. Who is correct?

A. Technician A **C.** Both A and B
B. Technician B **D.** Neither A nor B

24. A large bus with a "no crank" problem will start when a remote starter button service tool is used. *Technician A* says the bus key/start switch and wiring should be checked. *Technician B* says that the start circuit should be examined to see if it has a magnetic start switch (relay), and if so, have it checked. Who is correct?

A. Technician A **C.** Both A and B
B. Technician B **D.** Neither A nor B

25. *Technician A* claims that it is acceptable practice to change an alternator without disconnecting the battery as long as caution is used not to allow the output terminal or wire to contact ground. *Technician B* says that if the battery is disconnected, electronic fault codes may be erased from computerized systems on the truck. Who is correct?

A. Technician A **C.** Both A and B
B. Technician B **D.** Neither A nor B

26. *Technician A* says that the stator leads should be disconnected from the rectifier bridge before testing the diodes. *Technician B* says it doesn't matter whether or not the diodes are separated if you use the right scale on the ohmmeter. Who is correct?

A. Technician A **C.** Both A and B
B. Technician B **D.** Neither A nor B

27. Which of the following would most likely cause an excessively high charging rate?

A. Slipping drive belt
B. Worn brushes
C. Loose wiring connections
D. Faulty regulator

28. Which of the following *would not* be a possible cause of a low or unsteady charging rate when troubleshooting an alternator?

A. Belt loose
B. Poor connections
C. Worn brushes
D. Open rotor coil
E. Bad regulator

29. *Technician A* says that in some charging circuits the regulator creates a charge by completing the path to ground after the field or rotor winding. *Technician B* says that regulators vary the charge rate by gradually increasing or decreasing the field voltage. Who is correct?

A. Technician A **C.** Both A and B
B. Technician B **D.** Neither A nor B

30. *Technician A* says that an electric horn circuit works by grounding the horn relay coil via the horn ring or button of the steering wheel. *Technician B* says that power from the hot side of the signal light flasher is usually used to operate the horn relay. Who is correct?

A. Technician A **C.** Both A and B
B. Technician B **D.** Neither A nor B

31. A taillight on a medium duty truck glows dimly when lit. The lens and bulb are removed and the leads of a voltmeter are placed between the pigtail contact and socket wall. A reading of 12.8 volts is obtained. *Technician A* says there is no problem with the wiring; a new bulb is needed. *Technician B* says probing the plug or connector nearest the light with a voltmeter and leaving the bulb in the socket is a better procedure. Who is correct?

A. Technician A **C.** Both A and B
B. Technician B **D.** Neither A nor B

32. A signal light switch is being replaced on an old heavy truck. An OEM switch is not available so an aftermarket bolt to the column switch is being substituted. *Technician A* says to double-check the operation of the switch using an ohmmeter to ensure proper connection to the truck harness. He also states that color matching may need to be ignored. *Technician B* says that a wiring diagram for the truck and diagram instructions for the new switch should provide sufficient information to complete the job. Who is correct?

A. Technician A **C.** Both A and B
B. Technician B **D.** Neither A nor B

33. The service brake stoplight switch is being inspected on a heavy truck. *Technician A* says that with air applied to the switch an ohmmeter test should show infinite resistance or "OL" on a DVOM across the switch terminal. *Technician B* says to swap the old service brake switch with the emergency or park brake switch and check for proper operation. If the service brake light works correctly, then the old switch is bad. Who is correct?

A. Technician A **C.** Both A and B
B. Technician B **D.** Neither A nor B

34. The back-up warning horn continually sounds in all gear positions on a large dump truck. *Technician A* says that the likely cause is stuck contacts in the warning horn assembly. *Technician B* says the problem could be a short to ground in the wiring to the warning horn transmission switch. Who is correct?

A. Technician A **C.** Both A and B
B. Technician B **D.** Neither A nor B

35. The turn signal lights on a medium duty truck flash at an extremely rapid rate in both left and right turn position. *Technician A* says the cause could be the addition of extra lamps in the circuit. *Technician B* says the cause could be high resistance in the circuit wiring. Who is correct?

A. Technician A **C.** Both A and B
B. Technician B **D.** Neither A nor B

36. A courtesy or floor light continues to glow after the doors are shut on a truck. *Technician A* says the cause could be a grounded wire to a door jam switch. *Technician B* says the problem could be a stuck-open door jam switch. Who is correct?

A. Technician A **C.** Both A and B
B. Technician B **D.** Neither A nor B

37. *Technician A* says that when using an electronic reader (scan tool) it may be necessary to enter engine model numbers and horsepower ratings prior to information retrieval. *Technician B* says that with the proper scan tool software most readers will automatically find engine numbers and ratings. Who is correct?

A. Technician A **C.** Both A and B
B. Technician B **D.** Neither A nor B

38. *Technician A* says that all electronic engine warnings systems are designed to shut down the engine in the event of catastrophic failure, such as no oil pressure or lost coolant. *Technician B* says that many electronic warning systems rely on a signal for the alternator to prevent key on, engine off (KOEO) false warnings. Who is correct?

A. Technician A **C.** Both A and B
B. Technician B **D.** Neither A nor B

39. Which of the following conditions is *least* likely to eventually initiate a fault code and appropriate warning on an electronically controlled diesel engine?

A. Partially plugged radiator
B. Partially restricted fuel filter
C. Partially restricted oil filter
D. Faulty crank angle or crank position sensor

40. Troubleshooting and repair of electronically controlled diesel engines with multiple fault codes is being discussed. *Technician A* says you should attempt to solve the first displayed code problem first. *Technician B* says you should always check for mechanical problems before troubleshooting electronic problems. Who is correct?

A. Technician A **C.** Both A and B
B. Technician B **D.** Neither A nor B

41. A fuel gauge on a large truck reads empty all the time. The instrument fuse is okay. *Technician A* says to jump 12 volts to the positive of the sending unit and watch for meter movement. *Technician B* says to ground the negative of the sending unit and watch for meter movement in the opposite direction. Who is correct?

A. Technician A **C.** Both A and B
B. Technician B **D.** Neither A nor B

42. A windshield wiper arm fails to return to the park position when the wipers are shut off. *Technician A* says the problem is most likely in the wiper switch. *Technician B* says the problem could be a loose park wire or connection to the motor/cam assembly. Who is correct?

A. Technician A **C.** Both A and B
B. Technician B **D.** Neither A nor B

43. A medium duty truck is equipped with a large electric engine cooling fan. *Technician A* says that when the cooling fan switch reaches its preset temperature the fan should turn on. *Technician B* says that whenever the air conditioner is turned on the engine fan should turn on. Who is correct?

A. Technician A **C.** Both A and B
B. Technician B **D.** Neither A nor B

44. *Technician A* says that when checking mirror heater circuits the grid should show a fairly low resistance from one end connection to the other. *Technician B* says the grid should show very low resistance from either end connection to ground. Who is correct?

A. Technician A **C.** Both A and B
B. Technician B **D.** Neither A nor B

45. The driver of a late model heavy truck complains of lack of power at high altitudes on the high passes of Colorado. A check with a reader/scan tool reveals no obvious problems and the fuel filter and other mechanical components of the engine seem okay. *Technician A* says the problem is likely a faulty turbo bypass switch. *Technician B* says the problem could be the MAP or Baro Sensor. Who is correct?

A. Technician A **C.** Both A and B
B. Technician B **D.** Neither A nor B

46. A driver is complaining that his gasoline-powered delivery truck just quits after it is warmed up. If the truck is allowed to cool down it will restart. *Technician A* says that since the engine is EFI he suspects a vapor lock condition is causing the problem. *Technician B* says she suspects the problem to be with the ignition module. Who is likely to be correct?

A. Technician A **C.** Both A and B
B. Technician B **D.** Neither A nor B

47. A driver is complaining that his large, late model diesel-powered truck has poor compression brake (Jake) hold back on hills. He further states that it makes very little, if any, extra noise coming down hills. *Technician A* says that clutch and dash selections switch operation should be checked. *Technician B* says that with late model trucks a properly closed throttle signal from the TPS (throttle position switch) is critical to Jake operation. Who is correct?

A. Technician A **C.** Both A and B
B. Technician B **D.** Neither A nor B

48. A medium duty truck with a printed circuit dash has erratic operating gauges; sometimes they work, sometimes they don't. *Technician A* says to check the harness connector and pins to the dash, particularly the ground. *Technician B* says to always replace the constant voltage regulator (CVR) or instrument voltage regulator (IVR) before replacing the printed circuit. Who is correct?

A. Technician A **C.** Both A and B
B. Technician B **D.** Neither A nor B

49. The driver side door window on a heavy truck moves down slowly and up extremely slowly. Occasionally it takes several minutes before it will move up at all. The track and lift mechanism have been checked and lubricated. *Technician A* says to check the switch and connectors for any resistance, replacing or repairing such items if necessary. *Technician B* says to jump the motor with a good 12-volt source and ground; if it is still slow, the motor should be replaced. Who is correct?

A. Technician A **C.** Both A and B
B. Technician B **D.** Neither A nor B

50. The driver of a diesel-powered motor home complains of hard starting in cold weather, particularly after standing overnight. He claims he plugs in the block heater several hours prior to starting. The glow plug system has been checked and no problems have been found. *Technician A* says to check the block heater with an ohmmeter. It should have moderately low resistance. *Technician B* says to check the cord and plug with an ohmmeter. It should have high resistance. Who is correct?

A. Technician A **C.** Both A and B
B. Technician B **D.** Neither A nor B

Appendix B

Metric Conversions

	to convert these	to these,	multiply by:
TEMPERATURE	Centigrade Degrees	Fahrenheit Degrees	1.8 then + 32
	Fahrenheit Degrees	Centigrade Degrees	0.556 after − 32
LENGTH	Millimeters	Inches	0.03937
	Inches	Millimeters	25.4
	Meters	Feet	3.28084
	Feet	Meters	0.3048
	Kilometers	Miles	0.62137
	Miles	Kilometers	1.60935
AREA	Square Centimeters	Square Inches	0.155
	Square Inches	Square Centimeters	6.45159
VOLUME	Cubic Centimeters	Cubic Inches	0.06103
	Cubic Inches	Cubic Centimeters	16.38703
	Cubic Centimeters	Liters	0.001
	Liters	Cubic Centimeters	1000
	Liters	Cubic Inches	61.025
	Cubic Inches	Liters	0.01639
	Liters	Quarts	1.05672
	Quarts	Liters	0.94633
	Liters	Pints	2.11344
	Pints	Liters	0.47317
	Liters	Ounces	33.81497
	Ounces	Liters	0.02957
WEIGHT	Grams	Ounces	0.03527
	Ounces	Grams	28.34953
	Kilograms	Pounds	2.20462
	Pounds	Kilograms	0.45359
WORK	Centimeter Kilograms	Inch Pounds	0.8676
	Inch Pounds	Centimeter Kilograms	1.15262
	Meter Kilograms	Foot Pounds	7.23301
	Foot Pounds	Newton Meters	1.3558
PRESSURE	Kilograms/Sq. Cm	Pounds/Sq. Inch	14.22334
	Pounds/Sq. Inch	Kilograms/Sq. Cm	0.07031
	Bar	Pounds/Sq. Inch	14.504
	Pounds/Sq. Inch	Bar	0.06895

Appendix C

Medium/Heavy Duty Truck Electricity and Electronics Tool and Equipment Suppliers List

Bend-Pak Incorporated
1645 Lemonwood Drive
Santa Paula, CA 93060
www.bend-pak.com

Carquest Corporation
12596 W. Bayaud Avenue
Suite 400
Lakewood, CO 80228
www.carquest.com

Fluke Corporation
PO Box 9090
Everett, WA 98206
www.fluke.com

Mac Tools
4635 Hilton Corporate Drive
Columbus, OH 43232
www.mactools.com

Matco Tools
4403 Allen Road
Stow, OH 44224
www.matcotools.com

Midtronics, Inc.
7000 Monroe Street
Willowbrook, IL 60521
www.midtronics.com

Micro Processor Systems, Inc.
6405 Nineteen Mile Rd.
Sterling Heights, MI 48314-2115
www.mpsilink.com

OTC
SPX Corporation
655 Eisenhower Drive
Owatonna, MN 55060
www.otctools.com

Prestolite (Leece-Neville)
400 Main Street
Arcade, NY 14009
www.prestolite.com

Snap-on Incorporated
PO Box 1410
Kenosha, WI 53141-1410
www.snapon.com

Vetronix Corporation
2030 Alameda Padre Serra
Santa Barbara, CA 93103
www.vetronix.com

GLOSSARY/GLOSARIO

Note: Terms are highlighted in bold followed by Spanish translation in bold italic.

Accelerator position sensor (APS) or throttle position sensor (TPS) A variable resistor (potentiometer) input sensor used with both gasoline and diesel powered engines. Its input affects most ECM functions such as fuel and timing control. Gasoline engines have the TPS mounted in a manner to work with the throttle plate. The APS used on diesels is normally mounted as part of the accelerator pedal assembly. That is why many technicians refer to todays electronic diesel powered vehicles as "Drive by wire."

Sensor de posición del acelerador (APS o TPS) Sensor de entrada de resistor de variable (potenciómetro) que se utiliza en motores de gasolina y de gasóleo. Su entrada afecta a la mayoría de las funciones ECM como el control de combustible y de sincronización. Los motores de gasolina tienen el TPS montado de forma que trabaje con la placa del acelerador. El APS utilizado en motores de gasóleo se suele montar como parte del conjunto de pedal de acelerador. Por eso, muchos técnicos utilizan la expresión "sustentación electrónica" para referirse a los vehículos de motor diesel electrónico.

Adjustable voltage regulator Refers to AC generator systems that have provisions to "fine tune" the charging system output.

Regulador de tensión ajustable Se refiere a sistemas generadores de CA con capacidad para "poner a punto" la salida del sistema de carga.

Advance timing light A type of timing light that enables a technician to note the timing change. By turning a knob, the flash of the light can be delayed by the number of engine degrees shown on the meter.

Luz de comprobación de avance de encendido Tipo de luz de prueba del encendido mediante la cual el técnico advierte el cambio de sincronización. Accionando un botón, se puede retrasar el destello de la luz en el número de grados de motor que muestre el indicador.

Ammeter A test meter used to measure current draw.

Amperímetro Medidor de comprobación que se utiliza para medir el paso de corriente.

Amplitude The usage of this is more prevalent when discussing wave forms and it refers to the height of a wave form.

Amplitud Su uso es más habitual en el estudio de las formas de onda y se refiere a la altura de éstas.

Analog meter Refers to a meter with a needle movement. This makes sense when you think of analog as a voltage signal that is infinitely variable.

Instrumento analógico Se refiere a un instrumento que indica la tensión con el movimiento de una aguja. Su mecanismo se basa en que una señal de tensión es infinitamente variable.

Analog scope A voltmeter that displays voltage changes over a period of time in the form of waves. An analog scope displays "love" voltages because of its ability to react extremely fast.

Pantalla analógica Voltímetro que muestra cambios de tensión en un período de tiempo en forma de ondas. Una pantalla analógica muestra tensiones "cercanas a cero" por su capacidad de reaccionar muy deprisa.

Antilock brake systems (ABS) A brake system that modulates braking effort automatically to prevent wheel lockup, thus maintaining steering control.

Sistemas de antibloqueo de frenos (ABS) Sistema de frenado que modula el esfuerzo de frenado automáticamente para evitar el bloqueo de las ruedas, con lo que se mantiene el control de la dirección.

Armature The rotating component of an electric motor, consisting of a conductor wound around a laminated iron core so a magnetic field can be created.

Inducido Componente giratorio de un motor eléctrico que consiste en un conductor bobinado alrededor de un núcleo de hierro laminado para crear un campo magnético.

ATA 6-pin connector A common method adopted by industry for a data transmission point or link.

Conector ATA de 6 contactos Método habitual adoptado por la industria como interfaz o punto de transmisión de datos.

Battery leakage test Used to determine if current is discharging across the top of the battery case.

Prueba de detección de fugas en la batería Se utiliza para determinar si la corriente se descarga por la parte superior de la caja de la batería.

Base timing Timing the engine with the computer or vacuum not effecting timing.

Sincronización base Sincronizar el motor con la computadora o vacío que no lleva a cabo la sincronización.

Belt tension gauge A tool used to determine proper belt tension when adjusting belts.

Indicador de tensión de la correa Instrumento que sirve para determinar si la tensión de las correas es adecuada al ajustarlas.

Bobtailing Refers to driving a tractor without the semi-trailer hooked to it.

Cabeza sola Se refiere a la conducción de una cabeza tractora sin llevar enganchado el remolque.

Bubble protractor A protractor with a leveling unit housed in it.

Transportador de ángulos de burbuja Transportador de ángulos que incluye una unidad niveladora.

Bus-bar A common electrical connection to which all of the fuses or circuit breakers are connected. The bus bar is connected to battery voltage.

Barra colectora Conexión eléctrica de uso habitual a la que se conectan todos los fusibles o cortacircuitos. La barra colectora está conectada a la tensión de la batería.

Camshaft position sensor (CPS) A sensor that is used to provide the ECU with specific piston position.

Sensor de posición de árbol de levas (CPS) Sensor que se utiliza para indicar a la ECU la posición del pistón en un momento determinado.

Capacitor An electric storage device sometimes referred to as a condenser.

Capacitor Dispositivo de almacenamiento eléctrico conocido también como condensador.

Capacity test A test that checks the battery's ability to perform when loaded down.

Prueba de capacidad Prueba que verifica la capacidad de la batería para funcionar cuando se descarga.

Carbon monoxide An odorless, colorless and toxic gas that is produced as a result of combustion. An indicator of a "rich" running engine.

Monóxido de carbono Gas inodoro, incoloro y tóxico que se produce durante la combustión. Indica que la mezcla de combustible y aire es "rica".

Carbon pile Refers to an adjustable carbon material device incorporated into load testers. By adjusting a knob, pressure is applied on the carbon pile. The more pressure that is applied to the carbon pile, the less resistance the carbon will offer to current flow.

Pila de carbono Se refiere a un dispositivo de material de carbono ajustable que se incorpora en los comprobadores de carga. Ajustando un botón, la presión se aplica a la pila de carbono. Cuanta más presión se ejerza sobre la pila de carbono, menos resistencia ofrecerá el carbono al paso de corriente.

Check engine light (CEL) A means of alerting the vehicles operator of a system problem when this dash light is illuminated.

Luz de comprobación del motor (CEL) Medio que avisa a los operadores del vehículo acerca de un problema del sistema cuando se enciende en el tablero.

Closed circuit A complete circuit with no breaks that allows current flow from and to its source.
Circuito cerrado Circuito completo y sin interrupciones que permite el paso de corriente desde y hacia su origen.

Commutator A series of conducting segments located around one end of an armature.
Conmutador Serie de segmentos conductores ubicada a un extremo de un inducido.

Component locator An aid many manufacturers use to help a technician find the actual component in the vehicle that was shown of a diagram.
Localizador de componentes Medio que utilizan muchos fabricantes para ayudar a los técnicos a localizar el componente de un vehículo que se muestra en un diagrama.

Connector composite Found as part of many diagrams to indicate exact wire location in connectors. This makes following wires and troubleshooting much easier.
Compuesto de conector Se utiliza con frecuencia en los diagramas para indicar la ubicación exacta de los cables en los conectores. Hace mucho más fácil la localización de los cables y la solución de problemas.

Control circuit Usually this refers to circuits that control other circuits. A relay is an example of a control circuit, whereas a low current circuit is used to control a higher current circuit.
Circuito de control Suele referirse a circuitos que controlan otros circuitos. Un relé sería un circuito de control, ya que utiliza un circuito de bajo voltaje para controlar un circuito de voltaje superior.

Current drain Used in reference to when a battery loses its charge with the ignition key off.
Fuga de corriente Se refiere a cuando una batería pierde su carga con la llave de encendido desconectada.

Current draw This is related to the amount of current a component draws when activated. It can be used as a base for test measurements.
Llamada de corriente Está relacionada con la cantidad de corriente que atrae un componente cuando se activa. Se puede utilizar como base de mediciones de prueba.

Current output test A diagnostic test used to determine the maximum output of a generator.
Prueba de salida de corriente Prueba de diagnóstico que determina la salida máxima de un generador.

Cycle Refers to a complete set of changes in a signal that repeats itself.
Ciclo Se refiere a un conjunto completo de cambios en una señal que se repite.

Data programming A means of making changes to computer parameters.
Programación de datos Medio para realizar cambios en los parámetros de una computadora.

Diaphragm A flexible member that has force applied on one side of the diaphragm in order to accomplish movement.
Diafragma Miembro flexible sobre el que se ejerce fuerza en un lado para obtener movimiento.

Dielectric grease A type of chemical material usually applied to electrical connections to provide a seal from adverse conditions or help in the dissipation of heat.
Grasa dieléctrica Tipo de material químico que se suele aplicar a las conexiones eléctricas para impermeabilizarlas o para ayudar a disipar el calor.

Digital meter Meters that take advantage of electronics to display actual digital numbers like a watch and with a very high input resistance.
Medidor digital Medidor electrónico dotado de una pantalla numérica digital como la de un reloj, y con una resistencia de entrada muy alta.

Dimmer switch A control switch for cycling headlights between low and high beams.
Regulador de intensidad de luz Conmutador de control para cambiar cíclicamente entre el alumbrado de cruce y de carretera.

DMM A digital multimeter that is capable of measuring voltage, resistance, low amounts of current, and in many cases, other measurements such as frequency and R.P.M.
DMM Multímetro digital capaz de medir la tensión, la resistencia, corrientes bajas y, en muchos casos, otras magnitudes como la frecuencia o las R.P.M.

Diode trio Used by some AC generator manufacturers to rectify stator AC so it can be used to create the magnetic field in the field coil of the rotor.
Trío de diodos Lo utilizan algunos fabricantes de generadores de CA para rectificar la CA del estator de forma que se pueda usar para crear el campo magnético en la bobina del rotor.

Dip switches Refers to the type of very small switches used in electronic speedometer and tachometer systems for setting specific parameters.
Conmutadores DIP Se refiere al tipo de conmutadores muy pequeños que se utilizan en los sistemas de velocímetros y cuentarrevoluciones electrónicos para definir parámetros específicos.

Direct injection diesel A diesel whose fuel is directly injected into the combustion chamber instead of a pre-combustion chamber.
Diesel de inyección directa Motor diesel cuyo combustible se inyecta directamente en la cámara de combustión en lugar de en una precámara de combustión.

Dry-floor absorbents Commercially sold products to absorb waste such as oil, making it easier to clean the floor of spilled liquids.
Absorbentes para el suelo Productos comerciales que absorben residuos como el aceite, lo que hace que sea más fácil limpiar líquidos derramados en el suelo.

DVOM Digital volt ohm meter.
DVOM Voltímetro y ohmiómetro digital.

E-core Refers to two E-shaped laminated components face to face with the ignition coil slipped on to the center leg.
Núcleo en E Se refiere a dos componentes laminados en forma de E dispuestos uno frente al otro con la bobina de encendido puesta en la pata central.

Electrolyte A solution of 64% of water and 36% sulfuric acid used in the truck's battery.
Electrolito Solución de 64% de agua y 36% de ácido sulfúrico que lleva la batería del camión.

Electromagnetic gauge A type of gauge that uses magnetic forces for needle movement.
Indicador electromagnético Tipo de indicador que utiliza fuerzas magnéticas para mover la aguja.

Electronic service tools (EST) Categorizes tolls that are used to diagnose and service electronic systems.
Herramientas de reparación electrónica (EST) Se refiere a las herramientas que se utilizan en el diagnóstico y reparación de los sistemas electrónicos.

Electrostatic discharge This is a reference to static electricity that can be harmful to sensitive electronic components.
Descarga electrostática Hace referencia a la electricidad estática que puede dañar los componentes electrónicos sensibles.

Engine brake Any type of engine retarder. Usually refers to an internal engine compression brake although an exhaust brake falls under this category as well.
Freno de motor Cualquier tipo de retardador del motor. Se suele referir a un freno de motor interno por compresión, aunque el freno por compresión de aire también pertenece a esta categoría.

Engine shutdown system A type of system used in trucks with diesel engines to prevent engine damage related to air or temperature problems.
Sistema de apagado del motor Tipo de sistema utilizado en camiones con motores diesel para daños en el motor relacionados con problemas de funcionamiento o de temperatura.

Exhaust brake An external type of engine compression brake. This type of brake cuts off the flow of exhaust gas.
Freno por compresión de aire Freno de motor por compresión de tipo externo. Este tipo de freno interrumpe el flujo de los gases de escape.

Fast charging The most often used method of recharging a battery. High amps are delivered in a short period of time to the battery.
Carga rápida El método más utilizado para recargar la batería. Se realiza suministrando un amperaje alto en un breve período de tiempo a la batería.

Field coil A component of a motor and generator that provides a magnetic field to accomplish the desired effect of the specific motor or generator.
Bobina de inducción Componente de un motor y generador que crea un campo magnético para alcanzar el efecto deseado en un motor o generador específico.

Filament A poor conductor inside a bulb usually made of tungsten. It will glow red or white hot giving off light when current flow through it.
Filamento Material poco conductor en el interior de una lámpara que suele estar hecho de tungsteno. Se pone al rojo o al blanco vivo y emite luz cuando la corriente pasa por él.

Fire extinguisher A device that contains chemicals, foam, or special gas that is discharged to extinguish a fire.
Extintor Dispositivo que contiene sustancia químicas, espuma o un gas especial que se descarga para apagar un fuego.

Flasher These units can be of the electromechanical type with a bimetallic strip, contacts, and a coil heating element or an electronic type. Either component's function is to cycle the turn signal lights.
Intermitente Estas unidades pueden ser de tipo electromecánico con una banda bimetálica, contactos y un elemento térmico de bobina, o de tipo electrónico. En cualquiera de los dos tipos su función es accionar cíclicamente las luces indicadoras de giro.

Floor jack A portable hydraulic tool used to raise and lower parts of the vehicle.
Gato de suelo Herramienta hidráulica portátil utilizada para elevar y bajar partes de un vehículo.

Forward bias When a positive voltage is applied to the P-type material and negative voltage to the N-type material of a semiconductor.
Polarización directa Cuando se aplica una tensión positiva al material del tipo positivo y una tensión negativa al material de tipo negativo de un semiconductor.

Fuel solenoid Usually an electromagnetic device used to shut off fuel to the diesel engine.
Solenoide de combustible Dispositivo electromagnético que se utiliza habitualmente para cortar el suministro de combustible al motor diesel.

Full field test A test whereas the field windings are energized to produce a maximum AC generator output.
Prueba de inducción completa Prueba en la que se excitan las bobinas de inducción para producir el máximo rendimiento del generador de CA.

Fusible link A special wire with a heat resistant insulation. If an overload occurs in the circuit, the wire melts and opens the circuit.
Cable fusible Cable especial con un aislamiento resistente al calor. Si se produce una sobrecarga en el circuito, el cable se funde y el circuito se abre.

Ground circuit test A test performed to check the integrity of the ground side of a circuit. A voltage drop test is the preferred method of testing.
Prueba del circuito de tierra Prueba que se realiza para comprobar la integridad de la parte conectada a tierra de un circuito. El método de prueba más utilizado es una caída de tensión.

Grounding strap A commercially sold device that a technician straps to his or her wrist and grounds to the vehicle to protect against electrostatic discharge.
Muñequera de puesta a tierra Dispositivo comercial que un técnico se ata a la muñeca y a la toma de tierra del vehículo para protegerse de descargas electrostáticas.

Growler Specific test equipment used to test starter armatures for shorts and grounds. The growler produces a strong magnetic field capable of inducing a current flow and magnetism in a conductor.
Probador de inducidos Equipo de prueba específico que se utiliza para comprobar los cortocircuitos y conexiones de tierra de los inducidos del motor de arranque. El probador de inducidos produce un fuerte campo magnético capaz de inducir un flujo de corriente y magnetismo en un conductor.

Hall-effect switch A magnetic switching device that operates on the principle that current, when allowed to flow through a thin conducting material is exposed to a magnetic field, another voltage is produced.
Conmutador de efecto Hall Dispositivo magnético de conmutación que funciona según el principio de que, cuando se expone a un campo magnético una corriente que fluye a través de un material conductor estrecho, se produce otra tensión.

Halogen bulb A bulb using any of a group of chemical elements such as chlorine, flouring, and iodine.
Lámpara halógena Lámpara que utiliza un elemento químico del grupo del cloro, flúor y yodo.

Hard codes Described as codes that indicate a problem that is existing at the present time.
Códigos duros Se definen como códigos que indican un problema que se está produciendo en el momento presente.

Headlight switch The main on-off control of the vehicles headlights. The same switch can also control other lights.
Conmutador de alumbrado Control principal de encendido y apagado de los faros de un vehículo. El mismo conmutador puede controlar otras luces.

Heat shrink tubing A hollow insulation like material fit over a connection that will shrink to an air tight fit when exposed.
Tubo termorretráctil Tubo hueco de material aislante que recubre una conexión y que se contrae hasta hacerse hermético cuando se expone al calor.

Heat gun A tool similar to a hair dryer that is used to induce heat to circuits or components for diagnostic purposes.
Pistola de calor Herramienta parecida a un secador de pelo que se utiliza para inducir calor en circuitos o componentes con el fin de realizar un diagnóstico.

HEUI Hydraulically actuated, electronic unit injector used presently by Navistar and Caterpillar.
HEUI Inyector electrónico de mecanismo hidráulico utilizado actualmente por Navistar y Caterpillar.

Hydrometer A test instrument for measuring specific gravity of a liquid. It is most commonly used to check specific gravity of electrolyte in a battery to determine the battery's state of charge.
Hidrómetro Instrumento de prueba para medir la densidad relativa de un líquido. Se suele utilizar sobre todo para comprobar la densidad relativa del electrolito de una batería para determinar su estado de carga.

Idle validation switch Used with some systems in combination with the TPS. Used to inform the ECU when the accelerator pedal is in the idle position.
Conmutador de validación de ralentí Se utiliza en algunos sistemas en combinación con el TPS. Se utiliza para indicar a la ECU que el pedal del acelerador está en posición inactiva (ralentí).

Impedance The combined opposition to current created by the capacitance, resistance, and inductance of a circuit. This is often applied to reference test meters.
Impedancia Oposición combinada a la corriente creada por la capacidad, resistencia e inductancia de un circuito. Se suele aplicar con relación a medidores de prueba.

Infinity Refers to resistance beyond the measurement capability of a meter.
Infinito Se refiere a la resistencia que supera la capacidad de medición de un instrumento.

Injection pressure regulator (IPR) An ECM controlled device used by certain Navistar engines to manage HEUI actuating oil pressure.
Regulador de presión de inyección (IPR) Dispositivo controlado por ECM que utilizan algunos motores Navistar para regular la presión del combustible en el HEUI.

Injector Driver Module (IDM) A separate injector driver unit used in versions of HEUI systems up to 1997.
Módulo controlador de inyección (IDM) Unidad de control de inyección independiente utilizada en versiones de sistemas HEUI anteriores a 1997.

Instrument voltage regulator (IVR) Provides a constant voltage to a gauge, regardless of the output voltage of the charging system.
Regulador de tensión de instrumentos (IVR) Proporciona una tensión constante a un indicador, independientemente de la tensión de salida del sistema de carga.

Insulated circuit test Refers to tests performed on the portion of the circuit from the positive side of the source to the load components. The preferred test would be a voltage drop test.
Prueba de circuito aislado Se refiere a las pruebas realizadas en la parte del circuito que va desde el lado positivo del origen a los componentes de carga. La prueba más aconsejable es la de caída de tensión.

Intermittent code Usually refers to a code that is attributed to a problem that had occurred and the problem was inadvertently fixed. Example of this is a lose connection that gets affected when a bump is hit by the vehicle. Many computers store these types of codes.
Código intermitente Se suele referir a un código que se atribuye a un problema que se soluciona de forma espontánea. Un ejemplo sería una conexión suelta que se arregla cuando se da un golpe al vehículo. Muchas computadoras almacenan estos tipos de códigos.

Jack stands These are support devices to hold the vehicle or parts of the vehicle off the floor after it has been raised by a hydraulic jack.
Soportes del gato Dispositivos de soporte que mantienen levantado el vehículo o partes del vehículo cuando se utiliza un gato hidráulico.

Jump starting Usually performed on service calls by using another battery or sets of batteries to start a vehicle with dead batteries. Commercial starting devices designed for this purpose are also used.
Arranque con pinzas Se suele realizar en reparaciones a domicilio utilizando otra batería u otro conjunto de baterías para arrancar un vehículo con la batería agotada. También se utilizan dispositivos de arranque comerciales para este fin.

Magnetic pulse generator A device that uses the principle of magnetic induction to produce a voltage signal.
Generador magnético de impulsos Dispositivo que utiliza el principio de inducción magnética para producir una señal de tensión.

Magnetic start switch Some manufacturers refer to this as the starter solenoid or starter relay.
Conmutador magnético de arranque Algunos fabricantes lo denominan solenoide del motor de arranque o relé del motor de arranque.

Molded connector Molded connectors are a one piece design that can not be separated for repairs.
Conector moldeado Los conectores moldeados son de una pieza y no se pueden desmontar para su reparación.

No-crank A term used to describe a no start condition.
Sin arranque Término que describe un problema de arranque.

No-load test Refers to a free spinning test performed on a starter that is out of the vehicle. It is used to determine the free rotational speed of the armature.
Ensayo en vacío Se refiere a una prueba de giro de un motor de arranque situado fuera del vehículo. Sirve para determinar la velocidad de giro libre del inducido.

Non-powered test light This type of test light requires the circuit's power being tested in order for the light to come on.
Luz de prueba no alimentada Este tipo de luz de prueba necesita que se compruebe la alimentación del circuito para que se encienda la luz.

Normally closed (NC) Designation of a switch to its contact position. A NC switch's contacts are closed until acted upon by an outside force.
Normalmente cerrado (NC) Designación de un conmutador en su posición de contacto. Los contactos de un conmutador NC permanecen cerrados hasta que los acciona una fuerza externa.

Normally open (NO) This designated switch's contacts are open until acted upon by an outside force.
Normalmente abierto (NO) Los contactos de este conmutador designado permanecen abiertos hasta que se accionan con una fuerza externa.

Odometer A mechanical counter that is usually part of the speedometer unit to indicate total miles accumulated on the vehicle.
Cuentakilómetros Contador mecánico que suele formar parte del velocímetro e indica el kilometraje total acumulado de un vehículo.

Ohm The unit of measurement for resistance.
Ohmio Unidad de medida de la resistencia.

Ohmmeter A test meter used to measure resistance and continuity in a circuit.
Ohmiómetro Medidor de prueba que se utiliza para evaluar la resistencia y continuidad de un circuito.

Ohm's law Defines the relationship between current, voltage, and resistance.
Ley de Ohm Define la relación entre corriente, tensión y resistencia.

Open circuit Used to define a circuit where current flow is stopped. The path for current flow is broken.
Circuito abierto Se utiliza para definir un circuito en el que se ha detenido el paso de corriente interrumpiendo la continuidad.

Open circuit voltage test A test used to determine the battery's state of charge. A voltmeter is used for this test.
Prueba de tensión de circuito abierto Prueba que determina el estado de carga de una batería. En esta prueba se utiliza un voltímetro.

Overload Excess current flow in a circuit.
Sobrecarga Paso de corriente excesiva por un circuito.

Overcharge A condition that allows the AC generator to output beyond a preset level. This is evident with a battery showing excess gassing.
Carga excesiva Condición que permite que la salida del generador de CA supere un nivel preestablecido. Esto es evidente en una batería que tenga exceso de gases.

Parallel circuit A circuit that provides two or more paths for electricity to flow.
Circuito paralelo Circuito que cuenta con dos o más rutas para el flujo de la electricidad.

Piezoresistive A term used to indicate the ability of certain crystals to produce voltage when subjected to stress.
Piezorresistente Término que indica la capacidad de ciertos cristales para producir tensión cuando se los somete a un esfuerzo.

Pinion gear Refers to a small drive gear of the starter drive assembly that is engaged to the engine's flywheel ring gear in order to start the engine.
Engranaje con piñones Se refiere al pequeño engranaje de mando de la transmisión del motor de arranque que actúa con la corona dentada para arrancar el motor.

Pinion clearance A measurement that is taken to assure proper pinion to ring clearance.
Holgura de piñón Medida que se toma para mantener una distancia adecuada entre piñón y corona.

Pneumatic tools Power tools that require compressed air to operate.
Herramientas neumáticas Herramientas mecánicas que necesitan aire comprimido para funcionar.

Power tools Designates tools that use forces such as compressed air, electricity, or hydraulic pressure to generate and multiply forces.
Herramientas mecánicas Designa las herramientas que utilizan fuerzas como el aire comprimido, la electricidad o la presión hidráulica para generar y multiplicar fuerzas.

Preventive maintenance A periodic inspection and service of the vehicle's systems to ensure no "on the road" breakdowns occur and also to prolong the life expectancy of the vehicle.
Mantenimiento preventivo Inspección y reparación periódicas de los sistemas del vehículo que evita las averías "en carretera" y prolonga la duración del vehículo.

Primary wiring Conductors that carry low voltage and current.
Cableado primario Conductores que transportan tensión y corriente bajas.

Pulley ratio Used to maintain a certain charging output at certain Rpm's by changing the pulley diameter of the AC generator.
Relación de polea Se utiliza para mantener una salida de carga determinada a unas rpm determinadas cambiando el diámetro de la polea del generador de CA.

Ratio adapter A gearing device that allows for an accurate speedometer reading by adopting different gear ratios to compensate for differential ratios and tire sizes. Used in older cable type speedometer and tachometer systems.
Adaptador de relación Dispositivo de transmisión que permite una lectura precisa del velocímetro adoptando diferentes relaciones de transmisión para compensar las relaciones diferenciales y los distintos tamaños de ruedas. Se utilizaba en los antiguos sistemas de velocímetros y cuentarrevoluciones de tipo cable.

Rectifier bridge The housing that hold the diodes that rectify AC to DC on an AC generator.
Puente rectificador Cuerpo que alberga los diodos que rectifican la CA a CD en un generador de CA.

Reverse-bias A positive voltage is applied to the N-type material and negative voltage is applied to the P-type material of a semiconductor.
Polarización inversa Una tensión positiva se aplica al material de tipo negativo y una tensión negativa al material de tipo positivo de un semiconductor.

Reversible DC motor A motor whose direction of rotation can be changed, usually by reversing the current flow. An example of this is a stepper motor often used for idle control.
Motor de CD reversible Motor cuya dirección de rotación se puede cambiar, normalmente invirtiendo el paso de corriente. Un ejemplo sería un motor de velocidad gradual de los que se utilizan para el control del ralentí.

Ripple pattern A waveform pattern that is used to diagnose bad diodes in AC generators. Often referred to as AC riding over DC.
Patrón de ondulaciones Forma de onda que se utiliza para diagnosticar diodos defectuosos en los generadores de CA. A menudo se denomina CA sobre DC.

Rotor test Refers to a visual check and resistance test with an ohmmeter of the AC generator rotor. This test is usually performed once the AC generator is disassembled.
Prueba de rotor Se refiere a una comprobación visual y de resistencia con un ohmiómetro del rotor del generador de CA. Se suele realizar una vez desmontado el generador de CA.

Scan tool A diagnostic test tool that is designed to communicate with the vehicle's on-board computer.
Instrumento de exploración Aparato de diagnóstico diseñado para comunicarse con la computadora de a bordo del vehículo.

Sealed beam taillight An environmentally sealed taillight assembly. If the assembly fails, it has to be replaced.
Faro trasero sellado Conjunto de faros traseros sellados al exterior. Si el conjunto falla, hay que reemplazarlo.

Self-powered test light Similar to a regular test light, except it has a small internal battery. When the ground clip is attached to the negative side of a component and the probe is touched to the positive side, the lamp will light if there is continuity in the circuit.
Luz de prueba autoalimentada Se parece a una luz de prueba normal, salvo en que tiene una pequeña batería interna. Cuando la abrazadera de tierra está conectada a la parte negativa del componente y el lado positivo toca la sonda, la lámpara se encenderá si hay continuidad en el circuito.

Sending unit A sensor for a gauge. The sensor is used to monitor various conditions.
Unidad de envío Sensor para un indicador. El sensor se utiliza para supervisar diferentes condiciones.

Sensor Any device that provides an input to the computer.
Sensor Cualquier dispositivo que proporciona una entrada a la computadora.

Series circuit A circuit that provides a single path for current flow from the power source through all the circuit's components and back to the source.
Circuito en serie Circuito que constituye una única ruta para el paso de corriente desde la fuente de alimentación hacia todos los componentes del circuito y de vuelta al origen.

Shorted circuit A circuit where the current flow has taken a different path than it was intended to take.
Circuito cortocircuitado Circuito en el que el paso de corriente siguió una ruta diferente de la que se pretendía.

Slow charging Refers to setting a battery charger at a very low amperage over a long period of time to charge a battery.
Carga lenta Se refiere a ajustar un cargador de batería a un amperaje muy bajo durante un largo período de tiempo para cargar una batería.

Soldering A process where heat and a mixture of lead and tin (solder) is used to make a splice or connection. This is always the preferred method for wiring repairs.
Soldadura Proceso en el que se utiliza calor y una mezcla de plomo y estaño (soldadura) para hacer un empalme o una conexión. Es el método preferido para reparar el cableado.

Solderless connectors Hollow metal tubes that are covered with insulating plastic. Used to splice wires or as part of a terminal end to be connected to the wire. A crimping tool is required.
Conectores sin soldadura Tubos huecos de metal cubiertos de plástico aislante. Se utilizan para empalmar cables o para preparar un terminal que se va a conectar al cable. Se necesita una herramienta engarzadora.

Solenoid circuit test Any test that is used to isolate a starter circuit test to the solenoid. The preferred test is a voltage drop test.
Prueba de circuito solenoide Cualquier prueba que se utilice para aislar una prueba de circuito de motor de arranque del solenoide. La prueba más utilizada es la de caída de tensión.

Speedometer An instrument gauge that indicates vehicle speed.
Velocímetro Instrumento que indica la velocidad del vehículo.

Speed sensor A pulse generator type of sensor whose output voltage is used to calculate vehicle, engine, or wheel speed.
Sensor de velocidad Tipo de sensor generador de impulsos cuya tensión de salida se utiliza para calcular la velocidad del vehículo, del motor o de las ruedas.

Spout The Ford spark output (spout) signal that comes from the ECA and indicates to the ignition module when to fire the next spark plug.
Spout Señal de salida de chispas de Ford que proviene del ECA e indica al módulo de encendido cuándo debe activar la siguiente bujía.

Splice The joining of single wire ends or the joining of two or more wire ends at a single point.
Empalme Unión de dos extremos de cables sencillos o de dos o más extremos de cables en un punto.

Splice clip A special connector when used with soldering to assure a good connection.
Pinza de empalme Conector especial que se utiliza en la soldadura para asegurar una correcta conexión.

Stabilize Used in reference when performing a battery test. After a charge the battery should be allowed to stabilize before performing any tests.
Estabilizar Se utiliza para referirse a una prueba de batería. Tras la carga, se debe dejar que la batería se estabilice antes de realizar una prueba.

Starter current draw test A test to determine the amount of current the starter draws when it is actuated. This test is performed to determine the electrical and mechanical condition of the starting system.
Prueba de corriente del motor de arranque Prueba para determinar la cantidad de corriente que el motor de arranque atrae cuando se acciona. Esta prueba se realiza para determinar el estado eléctrico y mecánico del sistema de arranque.

Starter solenoid The device that is responsible for engaging the starter drive and turning the armature.
Solenoide del motor de arranque Dispositivo que establece la transmisión del motor de arranque y gira el inducido.

State of charge The condition of a battery's electrolyte and plate materials at any given time.
Estado de carga Condición del electrolito y de los materiales de la placa de una batería en un momento dado.

Stator test A set of tests usually performed with an ohmmeter to determine the stator's integrity.
Prueba de estator Conjunto de pruebas que se suele realizar con un ohmiómetro para determinar la integridad de los estatores.

Stop engine light (SEL) This light is illuminated in some engine protection systems when the ECM detects a problem that could result in a serious failure.
Luz de paro del motor (SEL) Esta luz se enciende en algunos sistemas de protección del motor cuando el ECM detecta un problema que podría provocar una avería grave.

Sulfation This is an unwanted battery condition that reduces its output. The sulfate tends to harden on the plates, resulting in irreversible damage to the battery.
Sulfatación Condición no deseada de la batería que reduce su rendimiento. El sulfato tiende a endurecerse en las placas, lo que causa un daño irreversible a la batería.

Tachometer An instrument that measures engine speed in revolutions per minute (rpm).
Cuentarrevoluciones Instrumento que mide la velocidad del motor en revoluciones por minuto (rpm).

TDC Top dead center refers to the position of the piston at the very end of its power stroke.
TDC Punto muerto superior, o posición del pistón justo al terminar el tiempo de combustión.

Thermistor A solid state variable resistor made from a semiconductor material that changes resistance in proportion to temperature changes.
Termistor Resistor variable de estado sólido hecho de un material semiconductor que cambia la resistencia de forma proporcional a los cambios de temperatura.

Thermatic engine fan Refers to a type of engine fan used in many heavy duty trucks. The fan is engaged when the engine attains a preset operating temperature.
Ventilador térmico de motor Se refiere a un tipo de ventilador utilizado en muchos camiones pesados. El ventilador se activa cuando el motor alcanza una temperatura de funcionamiento predeterminada.

Timing light A type of strobe light that is configured to light when the number one cylinder fires. It freezes the action of the vibration dampener or flywheel where the timing marks are located.
Luz de sincronización Tipo de luz estroboscópica configurada para encenderse cuando se activa el cilindro número uno. Bloquea la acción del amortiguador de vibraciones o volante de inercia cuando se encuentran las marcas de sincronización.

Three minute charge test A test that is performed when a battery is suspected of being sulfated. (See sulfation).
Prueba de carga de tres minutos Prueba que se realiza cuando se sospecha que una batería está sulfatada (Véase Sulfatación).

Torque wrench Torque wrenches measure resistance to turning effort. Torque wrenches are used on fasteners to ensure specific clamping force is achieved between two components.
Llave dinamométrica Las llaves dinamométricas miden la resistencia al esfuerzo de giro. Se utilizan con las abrazaderas para asegurar que se alcance una fuerza de agarre específica entre dos componentes.

Trailer cord plug Refers to the connector on the end of the electrical cord from the tractor that gets connected to the trailer.
Enchufe del cable del remolque Se refiere al conector del extremo del cable eléctrico que va de la cabeza tractora al remolque.

Trailer cord receptacle This is the connector that the trailer cord gets mated to. It is mounted to the nose of the trailer.
Receptáculo del cable del remolque Es el conector al que se une el cable del remolque. Está instalado en el morro del remolque.

Undercharge A condition that is indicated when a battery's state of charge is low.
Carga insuficiente Condición que se observa cuando el nivel de carga de una batería es bajo.

Variable resistor A resistor capable of providing an infinite number of resistance values within a range.
Resistor variable Resistor capaz de proporcionar un número infinito de valores de resistencia dentro de un rango.

Vehicle personality module (VPM) The function of this module (Navistar) is to trim engine management to a specific vehicle application and customer requirements.
Módulo de comportamiento del vehículo (VPM) La función de este módulo (Navistar) es ajustar la gestión del motor a la aplicación específica del vehículo y a los requisitos del cliente.

Voltage drop Used to define the concept that a resistance in the circuit reduces the electrical pressure available after the resistance. The resistance can be the designed resistance such as a load component or an unwanted resistance due to a bad connection.
Caída de tensión Se utiliza para definir el concepto de que una resistencia en el circuito reduce la presión disponible tras la resistencia. La resistencia puede ser la diseñada, como un componente de carga, o una resistencia no deseada debida a una conexión defectuosa.

Voltmeter A meter used to indicate volts.
Voltímetro Instrumento que sirve para medir el voltaje.

Warning system A subsystem used to monitor system operations and alert the operator of an impending problem.
Sistema de aviso Subsistema que supervisa el funcionamiento de un sistema y alerta al operador de un problema inminente.

Waveform The electronic trace that appears on a scope, representing voltage over time.
Forma de onda Figura electrónica que aparece en una pantalla, y que representa la tensión en un tiempo.

Weather pack connector An electrical connector with rubber seals on the terminal ends and on the covers of the connector's half to protect the complete circuit connection from corrosion.
Conector impermeabilizado Conector eléctrico con protectores de goma en los extremos y en las tapas de los conectores que protege la conexión del circuito de la corrosión.

Wiring diagram A representation of electrical and/or electronic wiring and components of the vehicle's electrical system or systems.
Diagrama de cables Representación de los cables y componentes eléctricos y/o electrónicos del sistema o sistemas eléctricos de los vehículos.

INDEX